THE ATLAS OF MOUSE DEVELOPMENT

THE ATLAS OF MOUSE DEVELOPMENT

Matthew H. Kaufman

Department of Anatomy,
University of Edinburgh, UK

ACADEMIC PRESS
Harcourt Brace & Company, Publishers

LONDON SAN DIEGO NEW YORK BOSTON
SYDNEY TOKYO TORONTO

ACADEMIC PRESS LIMITED
24–28 Oval Road
London NW1 7DX

US edition published by
ACADEMIC PRESS, INC.
San Diego, CA 92101

Second printing with index 1994

This book is printed on acid-free paper.

A catalogue record for this book is available from
the British Library.

ISBN 0–12–402035–6

Typeset by Columns Design and Production Limited, Reading
Printed and bound in Great Britain by the University Press, Cambridge

To

Claire, Simon and David

Contents

B POSTIMPLANTATION PERIOD 17

C SCANNING ELECTRON MICROGRAPHS, AND SPECIAL SYSTEMS 371

Note:
Plates are located within the text to which they refer. When this has not been possible, the pages where the relevant plates can be found is indicated.

Tables and figures

Preface to the second printing of the First Edition

The rapid sales and favourable reports of the Reviewers have led to the preparation of a revised version of the First Edition of *The Atlas of Mouse Development*. This has enabled a number of typographical errors to be corrected and references to be updated. More importantly, it has allowed the inclusion of a comprehensive index. The final form of the index has gradually evolved, and is along the lines suggested by Dr Jonathan Bard. While the preparation of the index was a joint venture between the author and Dr Bard, the former, of course, accepts full responsibility for any errors present.

Preface

The aim of this book is to provide an histological account of the development of the mouse spanning the period from fertilization to term, though the main emphasis relates to the postimplantation period. In order to facilitate this exercise, selected scanning electron micrographs have been provided which clearly illustrate the three-dimensional form of intact and dissected embryos, their isolated external components and a variety of their internal organs and organ systems. The majority of the book, however, consists of plates which display intermittent serial histological sections through a selection of embryos specifically chosen to display the sequential but nevertheless continuous changes that take place shortly after implantation, during the period of embryogenesis and organogenesis, and the relatively short period of consolidation which occurs during the latter part of gestation in this species. An additional selection of plates displays, using both scanning electron micrographs and representative histological sections, the sequential changes that take place in relation to specific organs and organ systems, such as the eyes, the gonads, the kidneys, the lungs and the heart, where the detailed changes that occur during their development and differentiation are not easily observed in the relatively low magnification intermittent serial sections.

It is hoped that this book will serve as a laboratory "workshop manual", which provides readers with the means to recognize, with the minimum of difficulty, all of the major structures that are likely to be encountered in their own serially sectioned material. In order to facilitate this exercise, every effort has been made to select micrographs which illustrate the normal features that may be observed in symmetrically sectioned embryos.

While a very wide range of standard and special fixatives and histological stains is available to the embryologist, there are distinct advantages to be gained by using a simple fixative (e.g. Bouin's fixative), and a single, simple staining regimen (e.g. haematoxylin and eosin). The latter gives a minimum, but acceptable, level of shrinkage artefact, associated with an excellent degree of cellular clarity. In those few instances where other fixatives or histological or histochemical stains have been used, generally for a specific purpose, e.g. to demonstrate the presence of primordial germ cells, attention is drawn to the fixative used and the particular staining regimen employed. Similarly, in nearly all instances, the material has been embedded in paraffin wax (Paraplast) and the wax blocks cut at a nominal thickness of either 6 μm, 7 μm or 8 μm. While it might have been useful for those that are new to the field to have access to colour plates, for most purposes when the principal aim of the exercise is to recognize a specific structure, this was considered to be an unnecessary extravagance, and would, without question, have made the book far too expensive and, in all probability, not a commercial proposition.

While some individuals may wish to browse through the plates in order to gain a general impression of the interrelationship which exists between the different components of a specific region of the embryo, backing this up with a detailed account of the same events, most individuals are likely to want to use the Atlas to recognize specific structures or features present in their own material. For this reason, in addition to a few observations on methodology, the majority of the introductory text consists of an introduction to the plates, complemented by a brief descriptive account of the events that occur during mouse embryogenesis, organogenesis etc. While some readers may consider that the introductory text is too brief, I believe that by its brevity it serves the useful purpose of directing the reader's attention to the plates (with their associated explanatory notes) which are the raison d'etre of this monograph. It is, however, only by frequent reference to the reader's own material, supplemented by reference to the plates presented here, that familiarity with the three-dimensional arrangements within the embryo, and the sequential changes that occur during development and differentiation, can be most readily appreciated. As with any technical activity, it is possible to gain a certain limited degree of expertise by carefully following an instruction manual. However, to gain mastery of the subject, it is essential to try to understand, in this instance, how individual organs and systems relate to each other, both at the tissue and cellular level, during their development and differentiation.

Clearly, simply looking at conventionally stained individual histological sections, or even at serially-sectioned material, has its limitations; it is nevertheless the first and possibly the most critical approach that should be used in post implantation embryological studies, as it provides essential baseline morphological information which can be gained by no other means. Additional more specialized information can be gained, in certain circumstances, by utilizing a variety of other conventional stains, or by using more specific histochemical stains. If applicable, immunohistochemical means may also be employed to visualize certain cells or tissues. However, whatever the information gained, this must still be viewed in its proper context, and for this to be the case, it is essential that the reader should understand that the two-dimensional sections that are being viewed are only useful if they provide the reader with a three-dimensional image, and confer an essentially dynamic view of the events being studied. Most individuals can learn to "read" serially-sectioned histological material, but it is only after this elementary degree of competence is gained, that one can truly appreciate the value of this particular analytical approach.

It has been argued, erroneously in my view, that the analysis of serially-sectioned material is both time-consuming and tedious. This can undoubtedly be the case, but such comments often come from individuals who have never made any concerted effort to try to reconstruct in their minds the three-dimensional architecture of the material they are viewing. It has equally been argued that the analysis of serially-sectioned material provides only a static view of events. This is, of course, in a narrow sense correct, but anyone attempting to understand the dynamic and sometimes rapid changes that are occurring during, for example, embryogenesis or organogenesis, is most unlikely to be satisfied with looking at serial sections through a single embryo, or even through serial sections of a small number of embryos. Where possible, other supplementary means of analysing a particular event should be studied using, for example, cell lineage markers if these are available, or by the production and analysis of interspecies chimeras. Even these seemingly more dynamic approaches can have their limitations, and it would still be essential to analyse the latter material, in particular, histologically to check that the end result of the exercise bears some resemblance to the normal architecture of the tissue or organ system under investigation. Undoubtedly, it would be invaluable if a wide variety of cell-specific lineage markers could be found, as this would provide an analytical approach which would be of immense value in facilitating our understanding of the complex interactions that take place during embryonic development. Certain cell-specific markers are already available, for example, for the demonstration of neural crest-derived tissue, and it is likely that other similar cell-specific markers will soon appear.

Over the last few years, an explosion of information has emerged as a result of the use of molecular biological probes to investigate the genetic control of (particularly) pattern formation during early mouse embryonic development. Even though this field is still in its infancy, it seems certain that it will provide new insights into the morphogenetic interactions that occur during the pre- and early postimplantation period when the overall pattern for all subsequent development is laid down. A descriptive account of the gross changes that take place during the early postimplantation period, during embryogenesis and organogenesis, has largely evolved from the analysis of serially-sectioned material. The various, and almost entirely descriptive, accounts of the early development of the mouse, including that presented in this Atlas, now form one type of data base. This baseline information has been supplemented over the years by the findings from appropriately designed experimental studies, and from the analysis of spontaneously or induced genetic mutations, and from teratological studies. It now seems likely, however, that in order to gain a more detailed understanding of the fundamental mechanisms that control how cells and tissues interact during early embryonic development, we will additionally require the results obtained from an altogether different source of information, namely from the use of genetic probes. Even the initial information that has emerged as a result of the use of this relatively new experimental approach suggests that the interpretation of the findings is likely to be extremely complex, as even single genes or gene complexes not infrequently appear to act in more than one location in the developing organism at any one time, and over variable periods of time. Despite the fact that the interpretation of these findings is likely to be several orders of magnitude more complex than was initially envisaged, this is the direction that must be explored, and will almost certainly produce findings of the most fundamental importance in relation to increasing our understanding of the morphogenetic factors that influence the development of both mice and men.

Even with the emergence of these new technologies, it will nevertheless inevitably still be necessary to establish at the gross histological level the final common pathways by which individual cells or tissues are controlled by individual genes or groups of related, or even unrelated, genes. This type of genetically derived fate map and the morphogenetic maps that will emerge in the not too distant future, are likely to require very sophisticated computer technology to make the maximum sense of the four-dimensional informational input. While the latter may provide a novel perspective from which to view early mammalian development, any attempt at interpretation of this material will inevitably require a very thorough understanding of the events displayed in this Atlas. Old-fashioned and descriptive it might be, but the amount of information that can be obtained from the analysis of serially-sectioned paraffin-embedded material is directly related to the degree of perseverance, perspicacity and intelligence of the observer, who must be just as aware of the limitations of this material as the observer is of its exceptional potential.

To gain the maximum possible information, paraffin histology should, where possible, be complemented by a range of other analytical approaches that are appropriate to the questions that are being addressed. While scanning electron microscopy, in isolation, may be of limited value, it can occasionally provide invaluable information which can greatly facilitate the interpretation of "difficult" histological sections. Similarly, though beyond the scope of this Atlas, transmission electron microscopy can often provide invaluable information in relation to inter- and intracellular structure, and thereby shed important light on the complexities of cellular and tissue function which can only be hinted at from the analysis of paraffin sections. With the recent availability of gene probes and other sophisticated molecular biological methodology, the use of these techniques can often give the researcher the impression that all other approaches are old-fashioned, and consequently incapable of providing new insights into complex problems. This is clearly wishful thinking on their part, and is unfortunately not infrequently propagated by those grant-giving bodies that give the impression that if a technique is novel or trendy, it must be worth supporting, even though at least a proportion of those that use the new methodologies are clearly incapable of fully interpreting their findings.

It is probably worth briefly indicating here why this monograph has had such a long gestation period. The original plan was to produce an Atlas covering only the period from implantation up to about day 12 of gestation. This exercise was almost completed, when the author decided that a less detailed Atlas covering the period from implantation to day 17.5 of gestation would be of greater value to a wider audience. The duration of the project has additionally been increased by a period of at least a year or two, and possibly even more, because all of the work involved in the preparation and labelling of the plates has been undertaken by the author. While the latter was an exceedingly onerous component of the effort involved in the preparation of the Atlas, it was undertaken by the author in order to try to minimize labelling errors which could easily arise had the work been carried out by individuals unfamiliar with the material being labelled.

While every attempt has been made by the author to ensure the accurate identification of structures, only the plates and associated figure legends of the cephalic region of the more advanced embryos have been subjected to detailed scrutiny by others. Dr K.S. O'Shea and Dr Sam Hicks (Department of Anatomy and Cell Biology, University of Michigan Medical School, Ann Arbor), because of their familiarity with the features of the embryonic mouse brain, kindly volunteered to look through this material. The latter exercise resulted in an amended, improved and slightly more detailed labelling of this material, though, of course, the author accepts full responsibility for all errors of labelling or identification that may appear both in these plates and elsewhere in the Atlas. For the sake of accuracy, the author would be grateful if all errors in labelling or points requiring clarification could be addressed to him, so that if the Atlas were to be revised at some future date, appropriate changes could be made at that time.

A careful selection has had to be made to keep the total number of plates down to a minimum, in order to ensure that publication costs could be kept as low as possible. This has inevitably meant that certain areas have only received a superficial coverage, while other equally important topics may have been omitted altogether. As with every monograph, the final product is heavily coloured by the interests and expertise of the author. Clearly, with the eventual appearance of this Atlas, other individuals with more specialist interests can now produce their own atlases to fill in these deficiencies.

I would like to take this opportunity to acknowledge those individuals initially in the Department of Anatomy, University of Cambridge, and more recently in the Department of Anatomy, University of Edinburgh, who provided me with technical assistance essential during the preparation of this Atlas. I am particularly indebted to Mr Jack Cable (Edinburgh) for the meticulous care he has taken in the printing of the vast majority of the micrographs that appear in the plates, Miss Susan Bates (Cambridge) who sectioned most of the pre- and early somite stage embryos and the early limb-bud stage material, and Miss Corinne Arnott (Edinburgh) who sectioned all of the more advanced embryos, as well as for providing practical advice regarding the preparation of embryos for histological examination. Thanks are also due to Mrs (now Dr) Sheila Webb for her invaluable support and assistance in maintaining the author's research activities in Edinburgh while he was not infrequently distracted preparing material for the Atlas. The dissections of the palate were the work of Mr Peter AhPin, and diagrammatic reconstructions illustrating the differentiation of the branchial arch arterial system are based on the findings from a research project undertaken by Mr Paul MacKenney. Mr (now Dr) Jeremy Skepper (Cambridge) and Mr Bob Shields (Edinburgh) assisted in the preparation of the scanning electron micrographs, and Mr Raith Overhill (Cambridge) and Mr Ian Lennox (Medical Illustrations, Edinburgh) assisted in the preparation of the diagrammatic illustrations that accompany the plates, though all of the reconstructions necessary to produce these illustrations and their labelling was undertaken by the author. The typing of the manuscript material was undertaken by Miss (now Dr) Helen Dingwall, Miss S. Macdonald and Mrs June Gillam. I also wish to express my indebtedness to Academic Press, but particularly to Dr Susan King, for their patience and encouragement over the years, and to IRL Press, Oxford, for allowing me to reproduce a limited number of tables, diagrams and micrographs that were used to illustrate my contribution to *Postimplantation Mammalian Embryos* (Copp and Cockcroft, 1990). I also wish to acknowledge the patience of my scientific colleagues (but particularly Dr Anne McLaren) for their unfailing optimism over the years that this Atlas would eventually see the light of day, and Action Research for the Crippled Child who provided microscope equipment to support this project. Finally, I wish to record my indebtedness to my wife and sons for their unfaltering loyalty, patience and consideration during this seemingly endless exercise.

Matthew H. Kaufman

1

Introduction

1 INTRODUCTORY REMARKS

In order to understand the events that occur during the early postimplantation period in any mammalian species, it is essential that intact embryos at different developmental stages, carefully removed from the uterus and dissected from within their extra-embryonic membranes, are studied in isolation, as this provides the simplest means of analysing the gross morphological and conformational changes that take place during the period under investigation. This material may be readily isolated and examined in the living state under a dissecting microscope in either phosphate-buffered saline or tissue culture medium, both of which should preferably be supplemented with protein (e.g. 4 mg/ml bovine serum albumin). In recently isolated embryos with more than about 6 pairs of somites present, the heart should show regular contractions, but the rate of the latter may be considerably slower than occurs *in vivo* if the temperature of the solution is substantially below 37°C.

If the embryonic material is to be examined in more detail, it must be "fixed" in one of several ways, depending on the analytical technique to be employed. If the material is to be examined by standard light microscopy, a wide range of fixatives may be employed, though for most purposes either Bouin's solution or 10% neutral buffered formalin is usually adequate. If, however, more sophisticated histological techniques are to be employed, for example if the sectioned material is to be examined to investigate intracellular enzyme activity, then different methods of fixation and/or sectioning (e.g. the use of a cryostat) will almost certainly be required.

In order to gain a detailed understanding of the three-dimensional relationships of the various components of the embryo, it is essential that serially-sectioned material is analysed. For most purposes, the material is embedded in paraffin wax, and sections cut at a nominal thickness of either 5, 7, 8 or 10 μm, whichever is the most convenient. Normally the sections are stained with haematoxylin and eosin, though a wide variety of other stains may be used, for example, if it is necessary to investigate the disposition of connective tissue, or to analyse in detail the various components of the nervous system, when one of the silver staining techniques may be employed.

To complement the analysis of the living embryos and serially-sectioned material, embryos at similar developmental stages may also be analysed by scanning electron microscopy. This latter technique has the principal advantage that it provides an excellent means of viewing the surface contours of an embryo in great detail. However, this approach is only of relatively limited value in itself in the analysis of embryonic material, as under normal circumstances a detailed knowledge of the subsurface structures is also required. However, as amply demonstrated in several of the plates, dissected material may also be scanned, and the resultant products can be extremely valuable for facilitating the interpretation of serially-sectioned embryos. Should the need arise, it is also feasible for material that has previously been examined by scanning electron microscopy to be embedded in epoxy resin (e.g. Araldite, (Ciba–Geigy (UK) Ltd) and subsequently serially-sectioned. While there are certain technical difficulties involved in this manoeuvre, this dual approach has much to commend it. It is of particular value, for example, in the analysis of embryos with gross morphological abnormalities such as of the cephalic region which are difficult to describe adequately or illustrate by any means other than by scanning electron microscopy because of the small size of the specimens.

Similar fixatives to those used for scanning electron microscopy may also be used if the material is to be examined by transmission electron microscopy. While the latter technique may provide unique information on the ultrastructure of early embryonic tissues, particularly with respect to their subcellular components, any consideration of this technique is unfortunately beyond the scope of this Atlas. However, it is appropriate to mention at this stage that material previously embedded in epoxy resin for ultrathin sectioning for transmission electron microscopy, may also be used to provide individual sections or even serial sections cut at a nominal thickness of from about 0.5 to 1.0 μm, or even up to 2–3 μm in thickness. This approach can provide information with respect to cellular detail, which is not readily seen when similar material is examined by conventional paraffin histology.

A selection of scanning electron micrographs is provided in the Atlas in order to illustrate the gross conformational changes that take place in the external appearance of the embryo, and of certain of its internal organs. This material is supplemented with intermittent serial sections from a representative selection of embryos which have been chosen because they most clearly illustrate the principal features to be seen at each stage of development. In most instances, the embryos have been sectioned in the transverse plane, because these sections are relatively easy to interpret. Furthermore, it will also be rapidly apparent that this plane of sectioning provides maximum information of the type most often required in developmental studies. For certain special requirements, however, material may be optimally sectioned in another plane. A number of representative examples are therefore illustrated where embryos have been sectioned sagittally, principally, though not exclusively, in order to investigate developmental events that occur in the median plane. In nearly all of the material illustrated here that has been sectioned in the sagittal plane, selected intermittent sections are provided which pass parasagittally from one side to the other side of the embryo, through the median plane. These sections particularly emphasize the progressive degree of asymmetry that develops very soon after the early somite "headfold stage" is achieved. In the examination of the cephalic region of mid- and late-gestational stages of embryonic development, for example, it may be of value to section material in the coronal plane, as this provides a particularly useful means of analysing the development of the orofacial structures such as the tongue or palate. This also tends to be the plane of section favoured by the neuroanatomists/neurophysiologists for their analyses of the early post-natal and adult rodent brain.

For the benefit of the reader, and for obvious aesthetic reasons, every effort has been made to illustrate, in relation to the intermittent serially-sectioned material, bilaterally symmetrical histological sections, as these substantially facilitate all forms of analysis. While this can usually be achieved in "unturned" embryos, in partially "turned" and fully "turned" embryos this is not technically feasible, since the lower half of the trunk and the proximal part of the tail region are invariably deflected away from the median plane. The situation is further complicated by the fact that the internal organs and the vascular system rapidly lose their bilateral symmetry. Nevertheless, in all of the embryos illustrated, the sections that pass through the cephalic region retain their bilateral symmetry, and this symmetrical arrangement is still clearly apparent as far as the external features of these embryos are concerned. The material illustrated therefore demonstrates what can be achieved if adequate care is taken in orientating the material before any sectioning is undertaken.

2. METHODOLOGY

Stages in the preparation of early embryos for histological examination

In order to obtain histological sections of the highest possible quality, extreme care must be taken in the handling of the material. The rules that apply for the handling of adult tissues have invariably to be modified, since early embryonic material is extremely delicate and intolerant of extreme changes in its external environment. Consequently, the tissues will not remain intact if they are not handled with great care.

There are clearly many means that have been developed over the years for preparing serial sections of early mammalian embryonic material. The instructions that follows are not absolute, and should be modified according to local conditions. Such variables as the air temperature and humidity, as well as the purity of the chemicals used, can have a dramatic effect during the critical processing stages. In the author's laboratory, the timings of the various stages have been modified over the years by trial and error, so that it is now possible for serial sections of the highest quality to be reliably achieved.

Table 1 Preparation of paraffin-embedded embryos

Process	Solution	Crown–rump length of embryo (mm) <2	3–5	>6
Fixation	Bouin's fluid (saturated picric acid 75 ml glacial acetic acid 5 ml 40% formaldehyde 25 ml)	2–24 h	2–24 h	2–24 h
Storage after fixation	70% Ethanol	Variable period	Variable period	Variable period
Dehydration and clearing	80% Ethanol	1 min	3–5 min	30 min
	90% Ethanol	1 min	3–5 min	30 min
	96% Ethanol I	1 min	3–5 min	60 min
	96% Ethanol II	–	–	30 min
	100% Ethanol I	1 min	3–5 min	60 min
	100% Ethanol II	–	–	30 min
	(1:1) 100% Ethanol:benzene[a]	1 min	3–5 min	15 min
	Benzene[a]	Until cleared	Until cleared	—
	Xylene	–	–	Until cleared
Embedding	Paraplast I	1 min	3–5 min	Vacuum embed
	Paraplast II	1 min	3–5 min	Vacuum embed
	Paraplast III	1 min	3–5 min	Vacuum embed

[a] With embryos of up to about 5 mm crown–rump length, benzene is preferable to alternatives such as xylene, toluene, inhibisol, because it facilitates visualizing when the end point of "clearing" has occurred. The other agents appear to cause increased hardening of the tissues, and possible problems with processing because the material is more brittle and consequently difficult to handle. With larger embryos these other agents should be employed instead of benzene for the "clearing" stage. Since benzene is a particularly dangerous chemical, appropriate safety measures must be taken when this agent is being used.

From Kaufman, 1990, with permission.

Fixation

For most purposes, exposure to Bouin's fixative (saturated aqueous picric acid, 75 ml; 40% formaldehyde, 25 ml; glacial acetic acid, 5 ml) for 2–24 hours, depending on the size of the embryo, and long-term storage in 70% alcohol at room temperature until needed, has proved to be entirely satisfactory. This particular fixative produces acceptable levels of artefactual damage. If, however, the degree of damage to a particular tissue is unacceptable, then a wide variety of other simple fixatives (such as 10% neutral buffered formalin) are available, and their effects on the tissue being analysed should be investigated. In general, the time that the material spends in 70% alcohol is relatively unimportant, though the volume of this solution should be large compared to the size of the embryo. The 70% alcohol will inevitably remove some of the characteristic yellow picric acid staining of the material, and it is advisable that this solution should be changed on at least several occasions. The yellow coloration produced can be useful when the embryo is being orientated in the paraffin wax. The addition of a few drops of eosin to the 90% alcohol stage during the dehydration sequence stains the embryo pink, and also facilitates its orientation in the paraffin wax.

As a general guide, the smaller the embryo, the shorter the duration in Bouin's fixative. Thus embryos with a crown–rump (C–R) length of about 2 mm or less need only about 1 hour in this fixative; 3–5 mm C–R length embryos require only about 2 hours, while 6–10 mm C–R length embryos require about 6 hours, and 11–15 mm C–R length embryos would require about 10 hours up to a maximum of about 24 hours – but certainly no longer in Bouin's solution, as excessive hardening of the tissue occurs which inevitably leads to problems during the sectioning stage. A general guide to the duration of the various stages involved in the processing of embryos of different gestational ages is presented in Table 1.

If special histochemical studies are to be undertaken, then the appropriate reference books should be consulted to establish in the first instance the optimal fixative which should be employed. To demonstrate the presence of *alkaline phosphatase* activity, for example, we have found that early embryonic material is best fixed in 80% alcohol at 4°C for a maximum period of about 2 hours. The duration of exposure to the fixative should of course be modified according to the size of the material to be analysed.

Dehydration, clearing and embedding of embryos in paraffin wax

The timing of the individual steps during the dehydration sequence has been found to be of critical importance, since overexposure during processing can result in excessive hardening of the tissue and disintegration of the material during sectioning. Extreme care must also be taken during the clearing and mounting of the sections onto slides, otherwise damage will inevitably occur. To facilitate the processing of this material, a fine wire mesh cassette or carrier may be employed (e.g. wire mesh thin bar grids No. G206, Agar Aids, Stansted, Essex, UK). The different alcohols may be retained in glass Petri dishes, rather than small bottles, as this allows more room for manoeuvre. As the number of changes of solution is considerable, the use of a wire mesh carrier reduces the necessary handling to a minimum, and consequently decreases the risk of damaging the material.

During the dehydration sequence the embryo has to progress from 70% alcohol, through successively 80%, 90%, 96% and subsequently absolute alcohol, then into a 50:50 mixture of absolute alcohol:benzene. For embryos of about 2 mm C–R length or less, it will be apparent from reference to Table 1 that the duration of exposure to each stage should be for 1 minute, and certainly for no longer than 1.5 minutes. For embryos of 3–5 mm C–R length, the duration of each step should be between 3 and 5 minutes. For embryos of 6–10 mm C–R length, the duration of exposure of each step should be as follows: 80% and 90% alcohol, 30 minutes, followed by two changes, each of 30 minutes duration, and finally exposure for 15 minutes in the 50:50 mixture of absolute alcohol:benzene.

Embryos of less than 5 mm C–R length should be retained in benzene until they clear, and then removed immediately into the first wax stage. Since benzene is a particularly dangerous chemical, appropriate safety measures must be taken when this agent is being used. For the larger embryos, xylene appears to be a more satisfactory clearing agent, and material should be retained in this agent until it clears, then immediately transferred into the first wax stage. Paraplast (Monoject Scientific Inc., Athy, Co. Kildare, Ireland) has been found to be a suitable embedding agent.

For embryos of less than 5 mm C–R length it has been observed that adequate penetration with paraffin wax occurs if, during the embedding procedure, the material is left in either a hot oven (at about 60°C) or preferably retained under a hot lamp. Three changes of wax are required. For embryos of 2 mm C–R length or less, they only require to be retained for 1 minute in each wax step. For 3–5 mm C–R length embryos, the duration of each wax step should be 3–5 minutes. The larger embryos require to be processed using a vacuum embedding oven, otherwise incomplete penetration of the wax into the embryo inevitably occurs.

As indicated earlier, the addition of a few drops of eosin into the 90% alcohol stage stains the embryo pink, and greatly facilitates the next step which involves the accurate orientation of the embryo in the paraffin wax.

Orientation of embryos in paraffin wax prior to sectioning

In the embedding of particularly small embryos, it is advantageous to use a dissecting microscope as this greatly helps in the visualizing of the material. In addition, the availability of a pair of *heated* forceps (e.g. heated forceps with control unit, from Histolab-Cytolab, PO Box 101, Hemel Hempstead, Herts, UK) will also prove to be invaluable if minor adjustments are to be made to the position of an embryo within wax. It has been found that much less damage may be done to the embryo if its orientation is carried out under a heated lamp whose temperature is sufficient to allow the wax to be retained in a melted state throughout these critical stages of the embedding procedure. If the wax cools down and solidifies, then has to be melted down once more (and this may have to be repeated on several further occasions) until the embryo is satisfactorily orientated within the wax block, then damage inevitably results. The rule, then, is that the embryo should be manoeuvred as little as possible both prior to and once it is in the wax.

Rather conveniently, on most occasions, early "turned" embryos that have been isolated from within their extra-embryonic membranes automatically fall onto their side in the wax. It is then only necessary to establish in which direction the embryo is facing, and the location of its head and tail regions. Pins with different coloured heads may be used for delineating these three critical landmarks. When smaller embryos are being sectioned, similar principles apply, so that specific landmarks should be located and delineated as described above.

The cutting of serial sections

Having carefully orientated the embryo in the paraffin wax, it is next necessary to mount the wax block onto an appropriate "chuck". It is firstly necessary to ensure that the long axis of the embryo is exactly horizontal when the block is mounted in the chuck, so that the sections through the cephalic and trunk regions will be symmetrical. When transverse sections are required, the embryo should be located in the block so that the cephalic region is sectioned first, with subsequent sections being directed towards the thoracic and finally the lower trunk region. When analysing individual sections, but more particularly when photomicrographs are published, it is the convention that on transverse sections the neural axis and dorsal part of the embryo always appears towards the top of the section. In order to achieve this end, it is convenient if the embryo is sectioned on its side in the "block" with its neural axis facing towards the operator, and ventral surface facing away from the operator. The right side of the embryo therefore is directed towards the upper surface of the block. The block should be trimmed to reduce the width of the wax ribbon and reduce the distance between consecutive sections, but taking care to ensure that an adequate gap still remains both between the sections on the ribbon, and between the rows of sections on the slide.

For most purposes, sections are cut at a nominal thickness of 7 or 8 μm. The ribbons of sections are mounted by floating them onto glass slides previously coated with glycerine/albumin placed on a hot plate. The latter may be prepared by saturating 10 ml of

Table 2 Staining paraffin sections of embryos

Dewax	Xylene	5 min
Rehydration	100% Ethanol	5 min
	96% Ethanol	5 min
	90% Ethanol	5 min
	70% Ethanol	5 min
	Running tap water	Wash
Staining	Haematoxylin (Delafield's or Ehrlich's)	10 min
Differentiate	Acid–alcohol (1% HCl in 70% ethanol)	15–30 s
	Running tap water	Wash
Counterstain	Eosin (aqueous)	5 min
Dehydration	Ascending alcohols to 100% ethanol	Rapid changes
Clearing	Xylene	2 changes of 5 min
Mounting	Dammar xylene mounting medium	–

From Kaufman, 1990, with permission.

distilled water with flaked egg albumin. When this has been dissolved, add an equal volume of glycerine, and 0.1–0.2 g of sodium salicylate. A small drop of the albumin/glycerine is put on the slide with an equal volume of distilled water, and these are then mixed well together. This is smeared over the surface of the slide to produce a thin coating, and the excess removed. It is usual to place the first sections to come off the microtome at the top left-hand corner of the slide, with the last sections of the final ribbon towards the bottom right. This convention facilitates the subsequent examination of the tissues. To reduce the total number of slides required, particularly for the larger embryos, precleaned 76 × 39 × 1.2 mm glass slides are usually employed. While sections for histochemistry are usually dried in a 37°C incubator, all other slides are dried in an incubator at 60°C.

The staining of paraffin sections with haematoxylin and eosin

Prior to staining, the sections are dewaxed in xylene for 5 minutes, then rehydrated through absolute alcohol, 96%, 90% and finally 70% alcohol (5 minutes each stage), then washed in running tap water. The sections should then be stained for 10 minutes in either Delafield's or Ehrlich's haematoxylin, washed in running tap water, and differentiated in acid–alcohol for 15–30 seconds. The sections are then washed once more in running tap water, then counterstained with eosin for 5 minutes. Following a final wash in running tap water, the sections are then dehydrated through ascending alcohols to absolute alcohol, care being taken to make sure that the eosin is not lost into the alcohol solutions. The sections are then cleared in xylene and finally cover slipped using Dammar xylene mounting medium (Raymond A Lamb, UK). The steps involved in staining paraffin sections with haematoxylin and eosin are summarized in Table 2. Although haematoxylin and eosin are probably the most popular staining combination for the analysis of serially-sectioned mammalian embryonic material, a wide range of alternative staining methods is available, particularly for the analysis of specific tissues, such as connective tissue or nervous tissue. For further details, the reader should consult one of the standard manuals of histological techniques (see, for example, Bancroft and Stevens (1982), Culling *et al.* (1985)). For histochemical stains, water-based mountants generally need to be employed.

3 GENERAL OBSERVATIONS ON THE STAGING OF POSTIMPLANTATION MOUSE EMBRYOS

Terminology

The overall changes that take place during the early postimplantation period of all mammalian species are fairly similar, and a knowledge of the embryology of one should be an invaluable guide to the events that occur in other mammalian species. In the mouse, and in other rodents, but not in other mammals, inversion of the germ layers occurs, and attention will be drawn in Chapter 5 to the configurational changes that take place during the process of "turning" in these species. Similarly, substantial differences occur in the fine structure of the placenta between different mammalian species. This also applies to the relationship which exists between the extra-embryonic tissues and the embryo, and their individual functional roles. These aspects of comparative placentation, however, are beyond the scope of this Atlas, and the interested reader should refer in the first instance to one of the specialist texts which deals specifically with this topic (see, for example, Amoroso, 1952; Hamilton and Mossman, 1972; Steven, 1975).

The few reviews, books, and chapters in others, that deal specifically with the embryology of the mouse, namely Otis and Brent (1954), Snell and Stevens (1966), Rugh (1968, reprinted 1990) and Theiler (1972, reprinted with minor changes, 1989) and the rat (Keibel, 1937; Witschi, 1962; Hebel and Stromberg, 1986), are all useful to some degree, but are of only limited value when it is necessary to recognize specific structures, though they are more helpful when it is of value to assess the exact stage of development of an embryo. For this reason, it is hoped that sufficient information will be available in this Atlas for those who need to have a reasonably detailed knowledge of the development and histological features of the mouse, and with certain reservations, those of other related species. In Table 3, a comparison is made between the timing of equivalent stages of *early* embryogenesis in the mouse, rat and human embryos (see also Butler and Juurlink, 1987). A similar staging system also exists for the chick embryo (Hamburger and Hamilton, 1951).

If the staging system proposed by Theiler (1972, reprinted with minor changes, 1989) for the mouse is adopted, which is largely based on the classification of Streeter (1942) in his Developmental Horizons of Human Embryos, a classification which itself has recently been updated by O'Rahilly and Müller (1987), it is then possible to assign all pre- and early postimplantation mouse embryos to specific stages with relative ease. For mouse embryos, beyond about 14–15 days of gestation, and approximately equivalent to human fetal material of 8 or more weeks of gestation, the gestational age alone has been used as the simplest means of staging this material, until a more comprehensive classification system is developed for the more advanced stages of this species.

As will be clearly apparent from the brief synopsis of Stages 6–19 of Theiler's staging system provided in Table 3, all of the critical events described that occur up to about 11.5 days of gestation in the mouse may be readily recognized both in intact embryos and in sectioned material. No account, however, is taken of the changes that occur during, for example, cardiogenesis, nor of the differentiation of other organ systems such as the kidney or the lung, presumably because these are less readily apparent from a superficial examination of the embryo. In practice, it is probably more important to establish whether an embryo is morphologically normal or otherwise, and whether its degree of development is consistent with its gestational age, than assigning it to one specific stage or another. In order to make such an assessment, it is essential to have a thorough knowledge (i) of the normal morphological appearance (in gross terms, and at the histological level) of embryos at the various stages studied, and (ii) to be aware when specific developmental features are first likely to be encountered. The detailed 17-point scoring system that is used in teratological studies for analysing the morphological differentiation of early rat embryos devised by Brown and Fabro (1981; see also Brown, 1990) provides an objective means of evaluating this material, and could easily be adapted for the analysis of similar somite-stage mouse embryos. Similarly, it is also helpful to acquire an appreciation of the normal range of size both of intact embryos and of their component parts at different stages of embryogenesis. While the *crown–rump length* would appear to be a rather crude measurement, it does in fact provide a particularly useful means of evaluating whether a "turned" embryo, which may be morphologically normal or abnormal, displays evidence of moderate or severe degrees of growth retardation. Early "turned" diploid parthenogenetic and diandric triploid mouse embryos, for example, appear on gross inspection to be morphologically normal, but are invariably substantially smaller than normal (diploid) fertilized embryos at similar stages of development (Kaufman, 1983a; Kaufman *et al.*, 1989).

As in all biological systems, a limited degree of variability exists even in embryos encountered within a single litter (possibly by the equivalent of up to 6–12 h of growth between the least and the most advanced embryo). Similarly, embryos from different females that have been isolated at an identical gestational time may vary quite considerably in the stage of development achieved. An appreciation of the normal range is therefore of particular importance, and this can only be achieved from the analysis of large numbers of embryos.

In order to achieve a reasonable degree of uniformity, all of the serially-sectioned embryos analysed in the present study have been isolated from spontaneously cycling (C57BL × CBA)F1 hybrid females that had previously been mated to genetically similar F1 hybrid males. While autopsies were carried out at a variety of different times on days 4.5–12 of gestation, in order to obtain a representative selection of embryos which adequately demonstrate the major morphological features likely to be encountered during embryogenesis and the early stages of organogenesis, the isolation of developmentally more advanced embryos was usually carried out during the morning on each successive day of gestation. Similarly, while the embryos isolated from randomly bred mice or from different inbred or F1 hybrid strains of mice may vary slightly in the degree of development achieved at a specific time postconception, for most purposes such differences are of little consequence during the second half of gestation. As indicated above, this degree of interstrain variation is usually no greater than the difference observed between the least and the most advanced embryos present within a single litter, or between litters isolated from different females of the same strain isolated at a similar time postconception. Furthermore, since the morphological differences observed between individual strains are generally so small, no difficulties should be encountered in the analysis of embryonic material isolated from genetically dissimilar strains of mice to those analysed in the present study.

Whereas in the analysis of human and other mammalian embryos, the *crown–rump* length is a commonly employed parameter, it has recently been

Table 3 Timing of appearance of principal features in mouse, rat, and human embryos during the early postimplantation period

Principal features in rodents	Mouse[a]				Rat[b]			Human[c]			
	Theiler's stage	Age (days)	Size (mm) (unfixed)	Pairs of somites	Age (days)	Size (mm)	Pairs of somites	Carnegie stage	Age (days)	Size (mm) (fixed)	Pairs of somites
Attaching blastocyst	6	4.5			5–6			4	5–6	0.1–0.2	
Early egg cylinder. Ectoplacental cone appears. Embryo implanted, although previllous (human)	7	5						5	7–12	0.1–0.2	
Differentiation of egg cylinder. Proamniotic cavity appears	8	6			7–8						
Advanced endometrial reaction. Formation of mesoderm. Primitive villi (human, stage 6), branching villi (human, stage 7)	9	6.5			8–9			6, 7	13–16	0.2–0.4	
Amnion formation, primitive groove, allantois	10	7						8	18	1.0–1.5	
Neural plate, presomite stage	11	7.5									
First somites. Neural folds begin to close in occipital/cervical region	12	8–8.5		1–7	9–10		1–3	9	20	1.5–2.5	1–3
"Turning". Two pharyngeal bars, optic sulcus	13	8.5–9		8–14	10–10.5	2.0	4–13	10	22	2–3.5	4–12
Elevation of cephalic neural folds, formation of rostral neuropore. Optic vesicle formation	14	9–9.5		13–20	10.5	2.2–4.0	13–28	11	24	2.5–4.5	13–20
Formation of caudal neuropore. Three pharyngeal bars. Optic pit. Forelimb buds appear	15	9.5–10.25	1.8–3.3	21–29	11–11.5			12	26	3–5	21–29
Closure of caudal neuropore. Four branchial bars. Hindlimb buds appear. Lens placode, otic vesicle olfactory placode/pit	16	10.25–10.75	3.1–3.9	30–34	11.5–12	4–6	28–40	13	27–29	4–6	30+
Deep lens indentation, optic cup	17		3.5–4.9	35–39	12–12.5			14	32	5–7	
Formation of lens vesicle, deep nasal pit. Cervical somites becoming indistinct. Hand plate (human)	18	11	5–6	40–44	12.5–13	6–8.5	40–48	14,15	32–33	7–9	
Lens vesicle completely separates from surface. Cervical somites indistinct. Distal part of forelimb bud paddle-shaped. Foot plate (human)	19	11.5	6–7	>45				16	37	8–11	

[a] Data from Theiler (1989). [b] Data from Witschi (1962). [c] Data from O'Rahilly and Müller (1987). From Kaufman, 1990, with permission.

proposed (see O'Rahilly and Müller, 1987) that this measurement should be replaced by the *greatest* length (GL), being measured without any attempt to straighten the curvature of the specimen. For most purposes, particularly in any analysis of early mouse embryos, the differences in length obtained by these two means are likely to be insignificant. While it is recognized that it is generally only practicable to measure early human embryos following their fixation, this is not necessarily the case in experimental animal studies, where recently isolated and consequently unfixed material is readily available. The degree of shrinkage obtained following fixation may be considerable, and values of from 5–10% to as much as 30% are commonly quoted, though the higher figure generally only relates to the smallest specimens. The degree of shrinkage of the larger specimens is generally at the lower end of the range.

Most embryologists who work with human material consider that the embryonic period proper is completed by the end of the 7th week of gestation, and that this is followed by the fetal period. While an enormous volume of literature has accumulated with regard to the first of these periods, which has enabled a standardized staging system to be devised for human embryos, no satisfactory staging system is yet available for the fetal period. In the human, it was arbitrarily decided that the time of onset of bone marrow formation in the humerus should be adopted as the conclusion of the embryonic and beginning of the fetal period of prenatal life. This event occurs in fixed human specimens of about 30 mm crown–rump length. If this Carnegie staging system is transferred directly to the mouse (see Theiler, 1989), then the dividing line between the embryonic and fetal period occurs in this species somewhere between the 14th and 15th days of gestation. Whereas in the human it may occasionally be of critical importance to distinguish between the individual stages, in the mouse, with its much shorter gestation period, these finer points of classification seem less important. Indeed, in the mouse there seems little justification for separating the postimplantation period into these two arbitrary stages, namely, the *embryonic* and the *fetal* stages, for the simple reason that it serves no particularly useful purpose. For this reason, throughout this Atlas, the term "embryo" has been used for all stages of development between fertilization and birth, though it has to be admitted that a case could be made for retaining the equivalence with the human (principally Carnegie) terminology, so that the term *embryo* is used for conceptuses up to and including Theiler Stage 23, while developmentally more advanced conceptuses are designated as *fetuses*. However, what is of critical importance is that in any account the *developmental age postconception* is clearly and unambiguously stated.

Standardized timing schemes

A variety of different conventions are used in the literature for assessing gestational age in the mouse, and this has undoubtedly led to a considerable degree of confusion, unless the timing scheme or system is clearly and unambiguously explained. While the morning on which a vaginal plug is found is often referred to as the morning of the *first day of pregnancy*, this is now more commonly referred to as *0.5 days of gestation* (or 0.5 days p.c., i.e. postconception). This latter convention assumes that during the normal oestrus cycle in mice housed under standardized laboratory conditions, both ovulation and conception normally occur at about the mid-point of the dark phase. In most animal houses, the artificial light regime is controlled, for example, to provide a 12 h light:12 h dark cycle. Since timing of ovulation and mating may vary considerably, and ovulation even in individual females probably takes place over a period of one or more hours, these factors may partly explain the considerable degree of variability seen in embryonic development both within and between litters, as indicated above. Even taking these factors into account, this convention still provides the most accurate statement of the gestational age of the conceptus, and has therefore been adopted in this Atlas as well as in most recent reference works on this species (see, for example, Hogan *et al.*, 1986; Copp and Cockroft, 1990). Moderate variations may also be observed in the degree of development achieved when embryos are isolated from particular strains or strain combinations at specific times after conception. For this reason it is difficult to provide here more than a general guide to the developmental stages that are likely to be seen at particular times after conception. Similarly, some of the events observed in these strains may therefore be slightly prolonged or foreshortened compared to those described here.

Since most staging systems rely on the recognition of a limited number of easily observed developmental parameters, they tend to suggest that early embryos evolve in a rather intermittent or erratic way, rather than smoothly progressing towards the fully differentiated state. For this reason, while the earlier embryos selected for illustration in this Atlas are based largely on Theiler's useful staging system, additional material has been inserted where the transition from one stage to the next is too great to allow a full understanding of the intermediate events. At later stages of gestation, the Theiler staging system is fairly rigidly adhered to, though the timing of the first appearance of particular features is occasionally at variance with those described in his monograph.

4 GENERAL OBSERVATIONS ON THE PLATES

A fairly detailed series of histological sections is presented in this Atlas covering the important developmental events that may be observed between the early headfold stage, observed at 7.5–8 days p.c. to the early forelimb bud stage, observed at 9.5–10 days p.c. Because of the rapid changes that are occurring during early organogenesis, associated with the increasing complexity of these embryos, larger numbers of representative histological sections have had to be presented (which individually also necessarily occupy an increased area), in order to allow the detailed morphology of these embryos to be accurately represented. As a result, the total number of plates required to adequately illustrate representative sections from embryos that are more advanced than 13.5 days of gestation increases dramatically, and is far more than that required to adequately illustrate embryos at earlier developmental stages. Whereas it is possible in the less complex embryos to provide a virtually complete histological account of their component parts in only one or two plates, as many as 12 plates have had to be devoted to each of the developmentally more advanced embryos, though the total number of representative sections required to provide a reasonably comprehensive series may not be significantly greater. Even when about 50 representative sections are used to illustrate the morphology of the more advanced embryos, while all of the larger structures are well represented, some of the smaller, but nevertheless equally critical structures may occasionally not be illustrated in any of the sections. This, of course, particularly applies in relation to the developmentally most advanced embryos studied, where the representative sections can only hope to give a reasonable overview, rather than a complete account, of their morphological features.

While a reasonable number of representative examples of embryos isolated at approximately half-daily intervals between about 7.5 days and 9.5 days of gestation are presented here, since each occupies from only one to a maximum of three plates, because of obvious limitations of space, during the succeeding days of gestation, it has generally only been possible to provide single representative transverse and sagittally sectioned embryos that are typical of those isolated at daily intervals between 10 and 17.5 days of gestation. Two representative series of coronally sectioned heads have also been included in order to illustrate the usefulness of this plane of section when studying this particular region of the body.

The approximate location of the individual sections illustrated, as determined from an analysis of the location of these sections on the original slides, with regard to each of the embryos in each intermittent serially-sectioned series, is presented in diagrammatic form on the page facing the relevant plate(s). It should also be noted that in each of the series of intermittent serially-sectioned embryos, all of the sections illustrated are shown at the same magnification. While this has presented certain technical problems with regard to the preparation of the plates, it has the considerable advantage that it provides the reader with an accurate representation of the size and disposition of individual component parts in each of the embryos studied. Moreover, once the overall size and magnification of an individual embryo is taken into account, it allows the proportionate size of its component parts and their degree of differentiation to be compared with similar structures or tissues in embryos at different stages of development.

In order to provide an accurate indication of the size of the embryos studied and that of their component parts, a uniform system has been employed throughout the Atlas. In the case of the complete "staged" embryos where representative intermittent serial sections have been provided, a scale bar is provided in association with the diagram of the intact (fixed) embryo. In the other plates in which isolated sections are presented to illustrate the features of a few of the earliest embryos studied, for example, to show their relationship to their extraembryonic tissues, but more specifically to illustrate the differentiation of a variety of organ systems, the magnification of the individual micrographs is incorporated into the figure legend. In the case of a few of the scanning electron micrographs, either a scale bar or details of the magnification used are provided, though in all instances the overall size of these embryos, or of their component parts, can be accurately assessed from a knowledge of the developmental age of each specimen analysed. The approximate developmental stage, as indicated in Theiler's classification of mouse development, or gestational age in the case of the more advanced embryos, is also provided for each of the representative "staged" embryos studied. This information is complemented by the nearest equivalent stage of rat development, based on Witschi's staging system (Witschi, 1962), and the nearest equivalent stage of human embryonic development (in the case of mouse embryos up to 13.5–14 days p.c.), based on the Carnegie staging system, as recently updated by O'Rahilly and Müller (1987). In the case of the plates of scanning electron micrographs, and the supplementary plates of histological sections which illustrate the sequential changes that may be observed in the development and differentiation of a selected number of organs and organ systems, the appropriate developmental stage of the material is also provided.

Care has been taken throughout the text to maintain a degree of uniformity in the terminology employed both in relation to the naming of individual morphological entities, as well as in the abbreviated accounts of the underlying developmental processes taking place. While every effort has been made to use terms in accordance with the English language version of current anatomical/embryological nomenclature (Hamilton and Mossman, 1972; Warwick and Williams, 1973; Gasser, 1975; O'Rahilly and Müller, 1987), alternative terms are used (or occasionally provided in parentheses) if these are the commonest terms used in the current embryological literature, or would assist in the interpretation of some of the earlier publications in this and closely related fields.

In most instances, identical terms are used in the anatomical/embryological literature to describe similar entities, whether they occur in the human or in the rodent embryo. This has the advantage of providing a uniformity of nomenclature, whichever the mammalian species studied. However, since fundamental anatomical differences may exist between even the closest of species, this has inevitably influenced the terminology used in relation to the detailed description of certain organ systems in the mouse which are in fact quite dissimilar to their human counterparts. The lungs may be cited as an interesting example which stresses the difficulties which can be encountered if the terminology used to describe the human embryo is directly applied to the mouse. In the human, the left lung is divided into two lobes: an upper (superior) and a lower (inferior) lobe; while the right lung is divided into three lobes: an upper (superior), a middle and a lower (inferior) lobe. In the mouse (as in the rat) a different arrangement exists: the left lung is smaller than the right, and is not separable into lobes, while the right lung is clearly divided into four lobes: a cranial, middle, caudal and an accessory lobe (which may also be subdivided into a larger and a smaller component). Similarly, the subdivisions of the liver are quite different in the rodent from those present in man (see Hebel and Stromberg, 1986), and the terminology is accordingly dissimilar.

Occasionally, a single pair of organs may be present in rodents, as occurs in the case of the parathyroid glands, whereas in man two pairs of parathyroid glands are present, one pair derived from the dorsal part of the third, and the other pair derived from the dorsal part of the fourth pharyngeal pouches. In rodents, the parathyroids are believed to be derivatives of the dorsal part of the third pharyngeal pouch, though accessory glands are commonly found either within the thymus or dorsolateral to the oesophagus at the level of the larynx (Hebel and Stromberg, 1986). In one of the embryos illustrated in the Atlas, the right parathyroid gland is located within the carotid sheath, close to the bifurcation of the common carotid artery to form the internal and external carotid arteries (see *Plate 40d*, d). This is, however, explicable on the grounds that both the parathyroid gland and this region of the carotid artery are of third branchial (pharyngeal) arch origin.

In some instances, each of the various reference books consulted has used a different terminology, and in these cases one system has been selected, and the alternative term(s) given in parentheses when the situation is first encountered, but not generally thereafter.

Every attempt has been made to see that each plate, whether from one of the "staged" series of embryos, or from those used to illustrate the differentiation of a particular organ system, can be understood in isolation, and without the need for consulting previous or succeeding plates. In the case of both the histological sections as well as the scanning electron micrographs all of the important structures present in an individual plate have been labelled to facilitate their immediate identification. While this has enormously increased the labelling exercise, this feature of the Atlas should greatly increase its usefulness in providing an immediate and relatively simple means of identifying specific tissues and structures.

There are certain structures that have quite specifically not been labelled, principally because the author did not feel confident that they could be accurately identified. This applies, for example, to the rhombomeres, which are in any case not as clearly defined in the mouse as they are in other species, such as in the chick. These structures appear in an approximately rostrocaudal sequence, and it is not possible from an analysis of conventionally stained material to accurately determine which particular rhombomere appears on any of the sections illustrated. Recent studies, however, indicate that certain genes have roles in the formation and differentiation of the rhombomeres during the early development of the hindbrain, and that their expression is often segment restricted. Moreover, these studies have indicated that the expression of these genes may be established before the morphological appearance of these segments. This is clearly an area where new techniques may assist in the accurate labelling of structures that may otherwise be extremely difficult to identify, when conventional labelling techniques are employed.

Recent studies indicate that various cranial nerves/cranial ganglia are associated with specific rhombomeres. Thus the trigeminal (V) nerve is associated with rhombomere 2. The facio-acoustic (VII/VIII) ganglion complex is associated with rhombomere 4, though rhombomere 5 (which is at about the level of the optic pit/otocyst) may contribute to the acoustic (VIII) nerve. The glossopharyngeal (IX) nerve is associated with rhombomere 6, while the vagus (X) nerve is associated with rhombomere 7. Rhombomere 7 also contributes to the *cranial* part of the accessory (XI) nerve, while rhombomere 8 gives rise to the *spinal* part of the accessory nerve. Recent information from the analysis of chick embryos suggests that rhombomeres 3 and 5 are not involved in neural crest cell formation, while findings in the mouse suggest that neural crest cells are probably produced by all of the rhombomeres, but that the crest cells derived from rhombomeres 3 and 5 may have a very limited migration capability. The derivatives of rhombomeres 2, 4 and 6, also migrate into the first, second and third branchial arches, respectively (see Hunt *et al.*, 1991; Nieto *et al.*, 1992).

Since any individual who is likely to gain maximal use from this Atlas is likely to be reasonably familiar with the general features of their own material, it was decided that little value would be served by the inclusion of a conventional index, as most of the important structures are present in each of the embryos studied, though their degree of differentiation might vary widely at succeeding stages of development. In order to emphasise the chronology of development, the index is given in terms of the timing of first appearance of specific tissues, organs, etc., as indicated by Theiler Stage. A guide to facilitate usage of the index is also provided.

In order to facilitate the recognition of a particular structure, or to fully understand the relationship between different structures, the reader is advised to initially consult the abbreviated list of plates to establish which is likely to illustrate material most similar to that under analysis. If the item(s) under investigation are not immediately recognized and a positive identification made, then the corresponding region in either a more or a less developmentally advanced embryo should be consulted. In most instances, this approach should provide the desired information, though, exceptionally, relatively small structures in the more advanced embryos may not be seen for the reasons indicated previously.

Similarly, no attempt has been made to provide an exhaustive literature review. As the total number of publications relating to all aspects of mouse development must now run into the tens of thousands, and hundreds if not thousands of additional items are appearing each year, it was decided that, in general terms, only those key items that were used to assist in the identification exercise would be cited. These days, with the increasing use of computer-aided literature searches, it was considered that such information could be legitimately omitted from the Atlas, which in any case is likely to be most effectively used in association with the literature relevant to the problem under investigation, with the one complementing the other.

5 EMBRYOGENESIS

Before drawing attention to the principal features that may be recognized in serial sections of embryos at various stages between implantation and full term, it is of value to briefly consider the characteristic conformational changes that take place in rodent embryos as they progress from the egg cylinder stage to the primitive streak stage. It is only with this background information to hand that the subsequent events, particularly with regard to the changing relationship which develops between the embryo and its extra-embryonic membranes, can readily be understood. This necessarily involves an analysis of the process of "turning", whereby the primitive streak/early somite stage mouse embryo "reverses" the initial location of its germ layers. During this process, the embryonic endoderm, which is initially on the outer aspect of the embryo, comes to line the gut tube, while the embryonic ectoderm, which initially appears to be located on the inner aspect of the embryo, eventually gives rise to its skin and neural tube. Possibly even more complex are the changes that take place in relation to the extra-embryonic membranes which eventually completely surround the embryo. To facilitate an understanding of these events, the sequential changes that take place over this critical period of embryogenesis are illustrated in simplified form diagrammatically and the latter are complemented with appropriate histological sections.

The critical configurational changes that take place in the early postimplantation period, particularly in relation to the process of "turning" and associated "inversion" of the germ layers

During the early postimplantation period, the cells of the inner cell mass region of the mouse embryo proliferate rapidly, and the embryonic mass thus formed expands into the blastocoelic (subsequently to become the yolk sac) cavity to form a structure known as the egg cylinder. The latter is initially a double-layered structure enclosing a narrow lumen (termed the proamniotic cavity or canal), consisting of an inner layer of ectoderm cells (equivalent to the epiblast of the chick embryo) and an outer layer of endoderm cells (equivalent to the chick hypoblast).

A similar relationship is also found in the rat, rabbit, guinea pig and their close relatives, and is the reverse of that found in other chordates. The type of configuration observed in the mouse is therefore referred to as *inversion of the germ layers*. Snell and Stevens (1966) have suggested that one useful consequence of the inversion of the germ layers is that the embryo is very compact, and much of the space normally occupied by the yolk sac cavity in other mammals is eliminated. The gross morphological changes involved in the conversion of the 5.5 day mouse egg cylinder into the 7.5 day primitive streak stage embryo are illustrated in Fig. 1. The rapid expansion of the posterior amniotic fold and its apposition and eventual fusion with the considerably smaller anterior amniotic fold results in the formation of the chorion and amnion, which divide the proamniotic cavity into the ectoplacental, exocoelomic and amniotic cavities, respectively.

Up to the stage when the embryo possesses about 6–8 pairs of somites, the relationship between the embryonic ectoderm and endoderm is similar to that observed at the egg cylinder stage. However, from the primitive streak stage to the stage when the embryo has about 6–8 pairs of somites present, the embryo, on sagittal section, has the form of a "U". The cephalic region is located at the periphery of one of the arms of the "U", while the caudal region (primitive streak, tail and allantois) is located at the periphery of the other arm. The lower trunk and proximal part of the tail

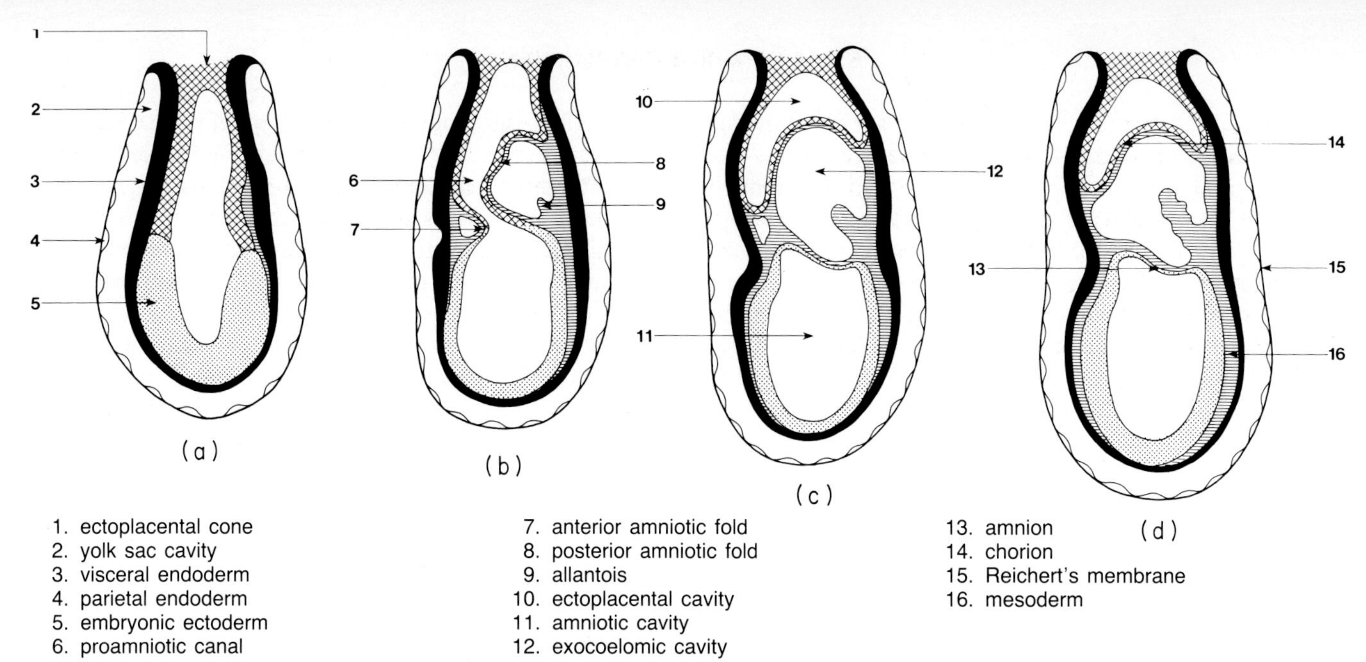

Figure 1 Stages in the conversion of the egg cylinder into the primitive streak stage embryo. (a) Egg cylinder stage; (b) proamniotic canal present; (c) during closure of proamniotic canal; (d) after closure of proamniotic canal. (From Kaufman, 1990, with permission.)

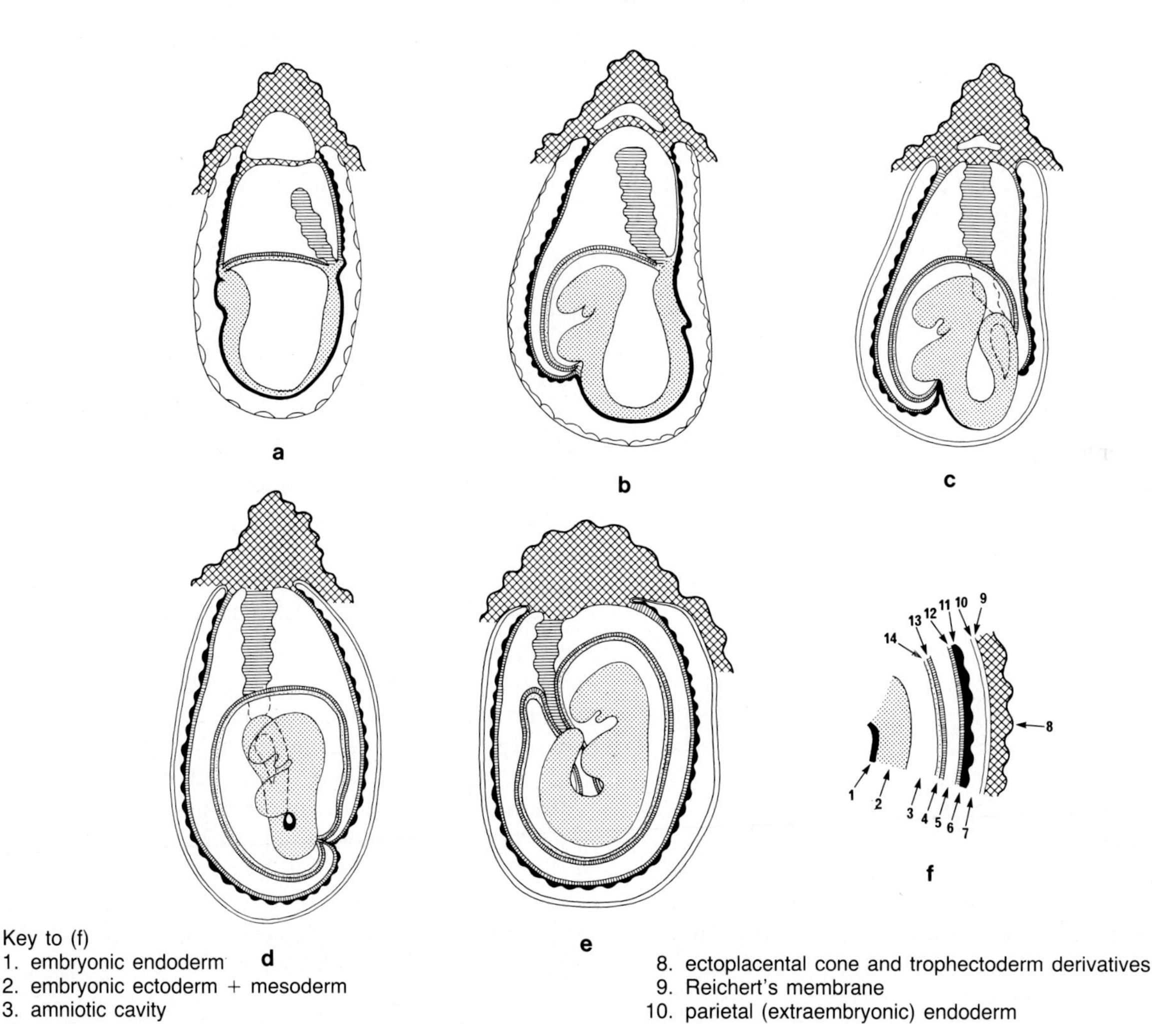

Figure 2 Simplified diagrammatic sequence to illustrate changes in the conformation of the embryo and the way in which the extraembryonic membranes surround it as it undergoes the process of "turning". The following stages are illustrated: (a) presomite headfold stage, 7.5–8 days; (b) embryo with 8–10 pairs of somites, about 8.5 days; (c) embryo with 10–12 pairs of somites, about 8.75 days; (d) embryo with 12–14 pairs of somites, about 9 days; (e) embryo with 15–20 pairs of somites, about 9.5 days; (f) embryonic layers, extraembryonic tissues, and cavities encountered in the embryo illustrated in (e). (From Kaufman, 1990, with permission.)

region are located in the middle section of the "U". Because of the inversion of the germ layers, the surface and neural ectoderm are consequently initially located within the concavity of the "U", while the exposed part of the gut (in the prospective midgut region) is located on the outer aspect of the convexity of the "U".

In order to achieve the characteristic "fetal" position, and thus conform with the configuration adopted by all other chordates, the curvature of the whole trunk region needs to be reversed. This manoeuvre is achieved in the mouse and in the other related species by the process of *"turning"* (or *axial rotation*), and is usually initiated when the embryo possesses about 6–8 pairs of somites, and is normally completed by the time that the embryo possesses about 14–16 pairs of somites. Once the embryo has completed the "turning" sequence, the dorsal surface of the embryo and its subjacent neural ectoderm are principally located on the outer aspect of the convexity of the "U", while the midgut region becomes located within the concavity of the "U". At the same time that the embryo is "turning" and inverting its germ layers, it invaginates into and consequently becomes surrounded by its amnion and yolk sac. This is a consequence of the fact that the extra-embryonic membranes are attached along the borders of the umbilical ring (the boundary zone between the body wall and the future site of attachment of the umbilical cord to the embryo), and during the course of rotation the embryo simply rolls into its membranes. These events and their approximate timing are illustrated in Fig. 2, which also identifies the various embryonic and extra-embryonic layers present after "turning".

The diagrammatic and associated histological sequences presented here, enable the entire process of "turning" to be seen as a smooth transition from one stable morphological state – as observed in the advanced "unturned" embryo, to another equally stable configuration – as observed after the completion of the "turning" sequence.

If the locations of several landmarks on the surface of the embryo are followed at regular intervals during the "turning" sequence, it will be seen that the solution adopted by the mouse to achieve the "fetal" position is virtually the only way that this particular problem could be solved. The method employed is seen to be both extremely simple and elegant, involving as it does the "rolling" of the central region of the embryo through 180° in an anticlockwise direction when the embryo is viewed towards its caudal pole. Apart from rotating clockwise (which apparently does very occasionally occur, and may be associated with malrotation of the heart and/or viscera) instead of anticlockwise, no other simple mechanism enables the embryo to adopt the "fetal" position while at the same time allowing it to become surrounded by its extra-embryonic membranes.

The corresponding five stages in the "turning" sequence represented diagrammatically in Fig. 2, are also shown in a much simplified form in Fig. 3, where the "embryo" has been isolated from its extra-embryonic membranes. The first stage shown represents the situation before the "turning" sequence has been initiated, and the fifth stage shown represents the situation observed after the completion of the "turning" sequence. Three intermediate stages are also illustrated. Between each stage, the proximal part of the tail region rotates through an arc of 45° with respect to the lower (distal) trunk region, while the most distal region of the tail with its wedge-shaped open neural segment is seen to rotate through 90°.

In order to simplify the overall sequence of events still further, a corresponding simplified diagrammatic series is presented in Fig. 4, in which the embryo is viewed from above. Such a view has the advantage that it enables the relationship which exists between the cephalic and caudal extremities of the embryo to be clearly established. In this diagrammatic sequence, the caudal region is shown as though it migrates anticlockwise around half the circumference of a circle, at the

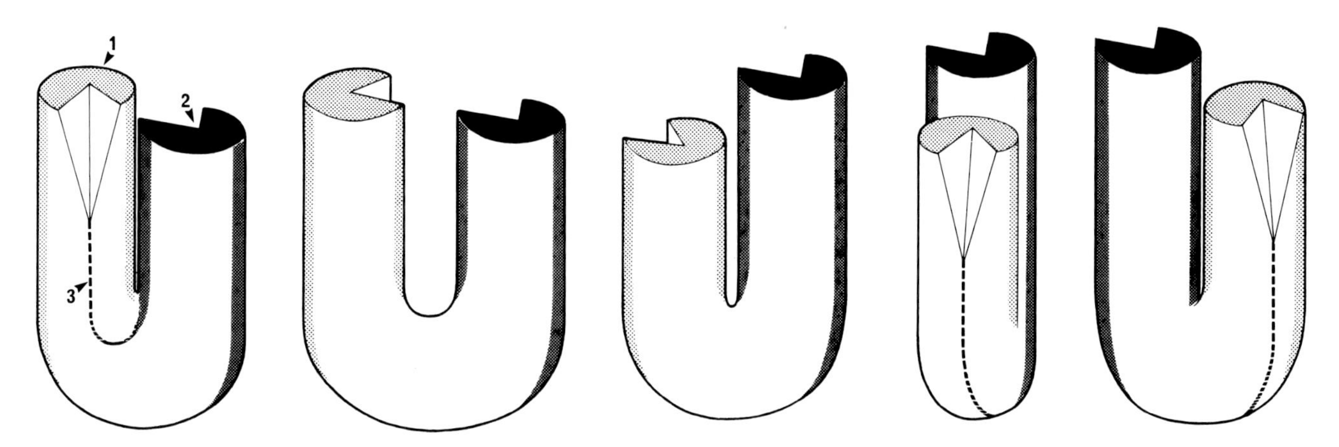

1. caudal extremity of embryo
2. cephalic extremity of embryo
3. neural exis.

Figure 3 Simplified sequence, divided into five stages, illustrating the overall configuration of the embryo as it proceeds through the 'turning' sequence. (From Kaufman, 1990, with permission.)

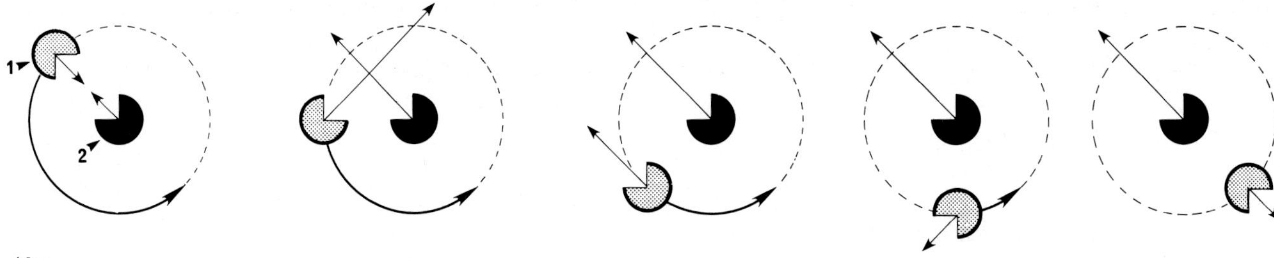

Key
1. transverse section through caudal part of embryo
2. transverse section through cephalic part of the embryo

Figure 4 Corresponding diagrammatic sequence to that presented in Fig. 3, in which the embryo, which is proceeding through the "turning" sequence, is now viewed from above.

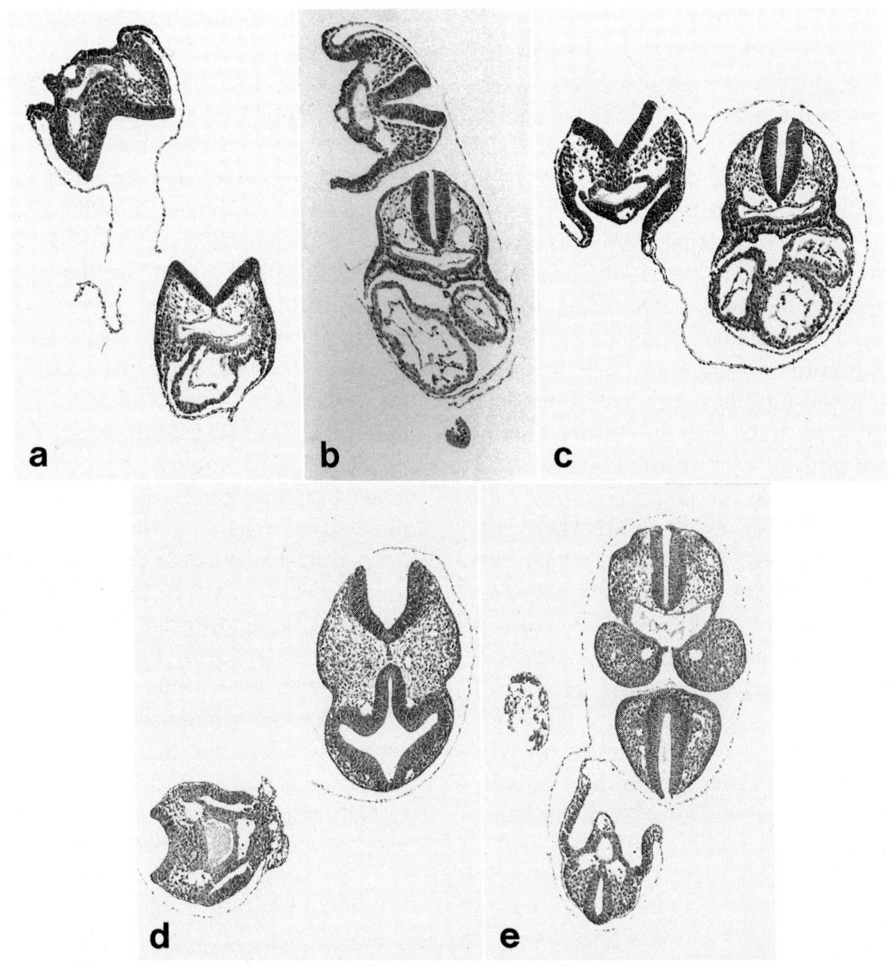

Figure 5 A series of representative transverse histological sections through embryos of increasing gestational ages which are proceeding through the "turning" sequence. These sections correspond almost exactly with the five stages of the diagrammatic sequence illustrated in Fig. 4 and have been labelled accordingly. (From Kaufman, 1990, with permission.)

centre of which is the stationary cephalic region. During the actual "turning" sequence as it occurs *in vivo*, the cephalic and caudal extremities would of course both move in relation to each other to achieve the same end result. While the proximal part of the tail region rotates through 180°, its most caudal segment containing the caudal neuropore in fact rotates through 360°.

A series of representative transverse histological sections through embryos of increasing gestational age which are proceeding through the "turning" sequence is illustrated in Fig. 5. Each section corresponds almost exactly with the five stages of the diagrammatic sequence illustrated in Figs 2, 3 and 4.

If only histological sections are available for analysis, I believe that it would be extremely difficult to determine the exact sequence of events that occurs during "turning". However, reference to the diagrammatic sequence illustrated in Figs 3 and 4 quite clearly reveals that the overall configurations observed in the sections shown in Fig. 5 are in their correct order.

The changes observed in the relationship between the embryo and its extra-embryonic membranes when sequential vertical sections through the mid-trunk region are viewed during the five stages of the "turning" sequence indicated above are also illustrated diagrammatically in Fig. 6. All of the sections are viewed as though looking towards the caudal region of

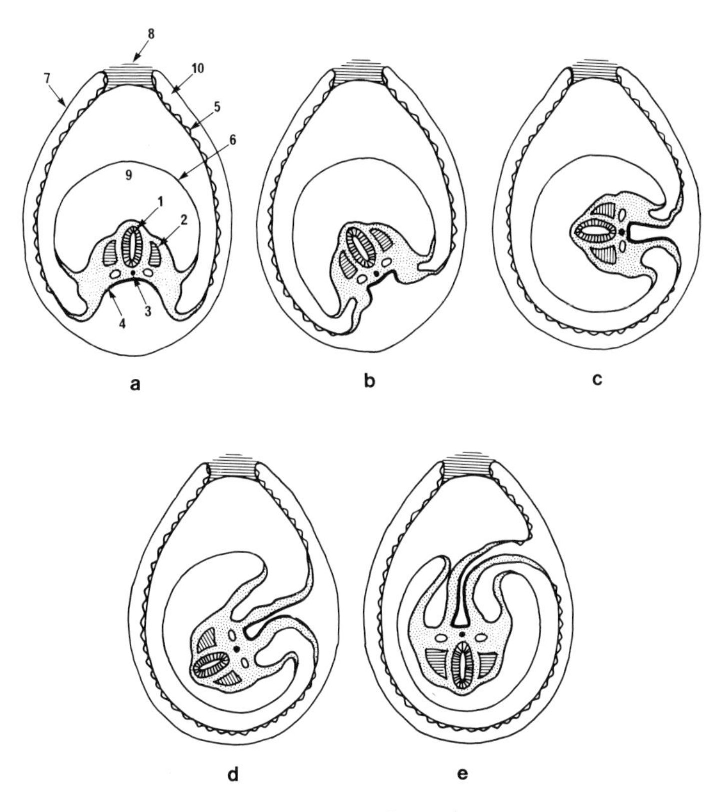

1. neural tube
2. somite
3. notochord
4. prospective midgut endoderm
5. yolk sac
6. amnion
7. parietal endoderm
8. ectoplacental cone
9. amniotic cavity
10. yolk sac cavity

Figure 6 Simplified diagrammatic series illustrating the changes in the relationship between the embryo and its extra-embryonic membranes when sequential vertical sections through the mid-trunk region are viewed during the five stages of the "turning" sequence illustrated in Fig. 3. All of the sections are viewed as though looking towards the caudal region of the embryo.

the embryo. Whereas in the first section in this series (section a) the left side of the embryo is shown on the right side of the diagram, because the sections are all viewed as though looking towards the caudal region of the embryo, in the final diagram (section e) the left side of the embryo now appears on the left side of the diagram. By following this diagrammatic sequence, the rotation that occurs in relation to the prospective mid-gut endoderm, from its initial location on the outer aspect of the mid-"U" section of the embryo, to its final destination within the concavity of the embryo is clearly seen. The lateral borders of the prospective midgut region coincide approximately with the sites of attachment of the amnion (dorsally) and the yolk sac (ventrally). During the "turning" sequence, these points also migrate anticlockwise through 180° around the circumference of a circle which represents the surface of the embryo in the mid-trunk region.

Relationship between Theiler Stage and Developmental age (days p.c.)

Theiler Stage	Days p.c.
1	0–1
2	1
3	2
4	3
5	4
6	4.5
7	5
8	6
9	6.5
10	7
11	7.5
12	8–8.5
13	8.5–9
14	9–9.5
15	9.5–10.25
16, 17	10.25–10.75
18	11
19	11.5
20	12
21	13
22	14
23	15
24	16
25	17
26	18

2

Assessment of development stage of pre- and postimplantation mouse embryos based on the staging system of Theiler (1989)

A. PREIMPLANTATION PERIOD

(Plates 1a, b)

Stage 1 Developmental age, 0–1 days p.c.

Fertilized one-cell stage egg (embryo). The embryo is located in the ampullary region of the oviduct.

Stage 2 Developmental age, 1 day p.c.

Two-cell stage embryo. The embryo is seen to be passing down the oviduct, beyond its ampullary region.

Stage 3 Developmental age, 2 days p.c.

Four- to 16-cell stage embryos, which may range from early to fully compacted morulae. Embryos are seen to be passing down the oviduct towards the utero-tubal junction.

Stage 4 Developmental age, 3 days p.c.

During this stage, embryos progress from the morula to the blastocyst stage (zona-intact), and possess a distinct inner cell mass (ICM) and an outer layer of trophectoderm cells, as a consequence of the process of polarization. Such embryos are usually located in the uterine lumen.

Stage 5 Developmental age, 4 days p.c.

Embryos at this stage are invariably zona-free blastocysts and are located within the uterine lumen.

B. POSTIMPLANTATION PERIOD

Stage 6 Developmental age, 4.5 days p.c. Plate 1c

During this stage the blastocyst implants. The first evidence of embryonic (proximal or visceral) endoderm cells is seen at this stage, as a distinct layer covering the blastocoelic surface of the inner cell mass. At a slightly later stage, the endoderm cells migrate to eventually cover (during Stage 7) the entire blastocoelic surface of the mural trophectoderm. This second group of cells constitutes the extra-embryonic distal (or parietal) layer of endoderm.

Stage 7 Developmental age, 5 days p.c. Plate 2, advanced Stage 7/early Stage 8 embryos

A substantial increase occurs in the number of inner cell mass cells present, and this results in their

(*continued on page 38*)

PLATE 1a

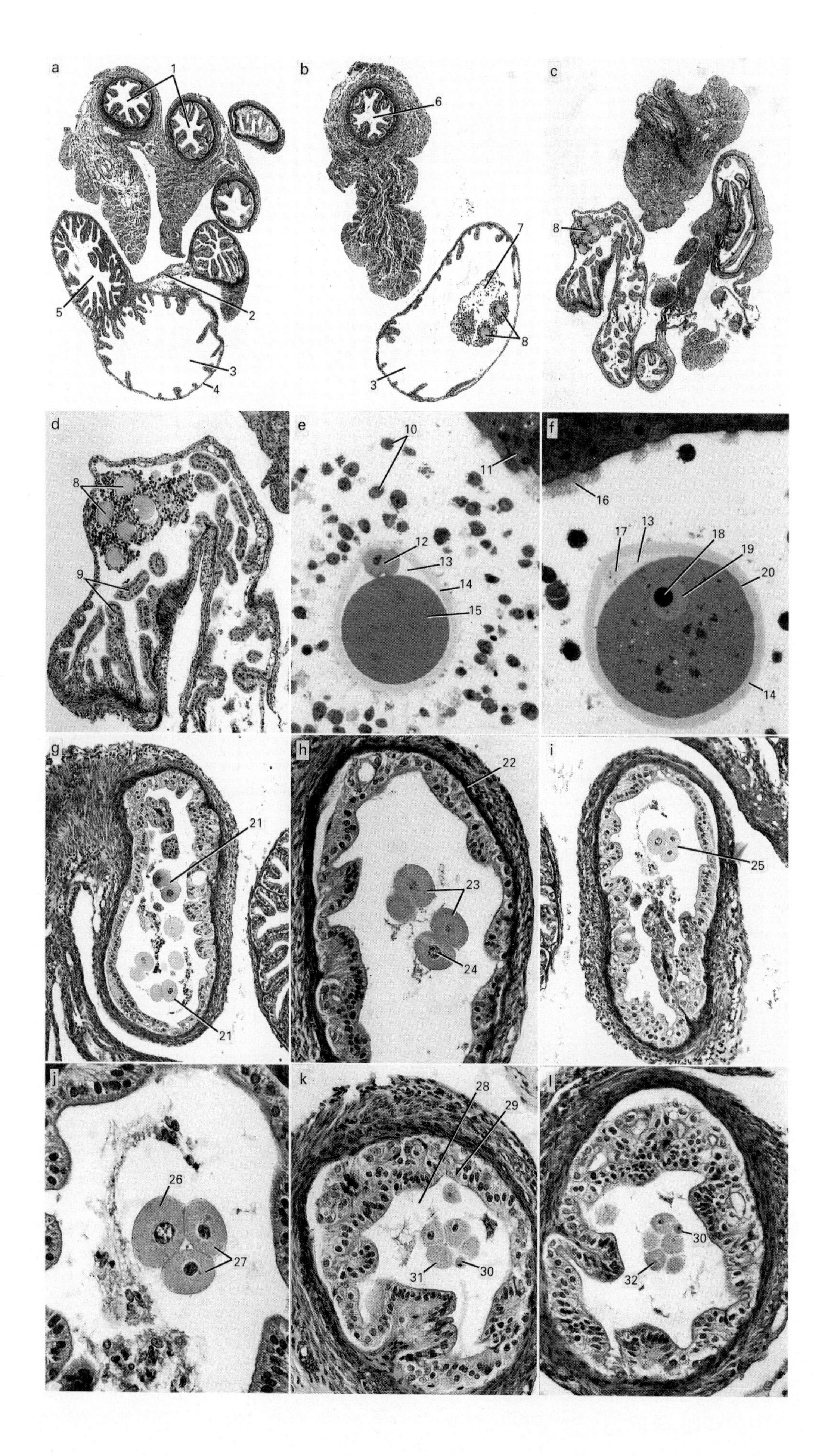
a
b
c
d
e
f
g
h
i
j
k
l
1
2
3
4
5
6
7
8
9
10
11
12
13
14
15
16
17
18
19
20
21
22
23
24
25
26
27
28
29
30
31
32

Plate 1a (Preimplantation period, I)

Embryonic development during the preimplantation period, I

a–f unfertilized eggs and recently fertilized 1-cell stage embryos located within the ampullary region of the oviduct.
a. ×40
b. ×40
c. ×40
d. ×100
e. ×310
f. ×400
(e, f) plastic-embedded material.

g, h 2-cell stage embryos located within the lumen of the oviduct just distal to the ampullary region (g. ×100; h. ×250).

i, j 3-cell stage embryo located within the lumen of the oviduct just distal to the ampullary region (i. ×100; j. ×400).

k 4- to 8-cell stage precompacted embryo located in the distal part of the oviduct (×250).

l 8-cell stage precompacted embryo located in the distal part of the oviduct (×250).

1. distal part of oviduct (uterine tube, Fallopian tube)
2. lateral part of broad ligament
3. lumen of ampullary region of oviduct
4. wall of ampullary region of oviduct
5. lumen of oviduct just distal to ampullary region
6. lumen in region of utero-tubal junction
7. peripheral part of cumulus mass
8. unfertilized eggs each surrounded by corona radiata
9. mucosal folds which extend into lumen of oviduct
10. follicle cells (cumulus cells) at an early stage of dispersal of the cumulus mass
11. mucous membrane in ampullary region of oviduct
12. first polar body
13. perivitelline space
14. zona pellucida
15. cytoplasm of unfertilized egg
16. ciliary border of epithelium lining the oviduct
17. degenerating first polar body
18. nucleolus
19. pronucleus (haploid, probably female)
20. plasma membrane of fertilized egg
21. 2-cell stage embryo
22. muscular layer of wall of oviduct consisting of external longitudinal layer and internal circular layer of non-striated muscle (in some parts of the oviduct there are in addition internal longitudinal groups of fibres)
23. blastomere of 2-cell stage embryo
24. nucleus (diploid) of 2-cell stage embryo
25. 3-cell stage embryo
26. blastomere of 3-cell stage embryo (this blastomere is one of the products of the first cleavage division, and has yet to divide to allow the embryo to achieve the 4-cell stage)
27. the two cells that result from the division of the first of the 2-cell stage blastomeres to divide (they are therefore the first two products of the second cleavage division to form)
28. lumen of distal part of oviduct
29. mucosal lining of distal part of oviduct
30. second polar body
31. blastomere of 4- to 8-cell stage precompacted embryo
32. blastomere of 8-cell stage precompacted embryo

PLATE 1b

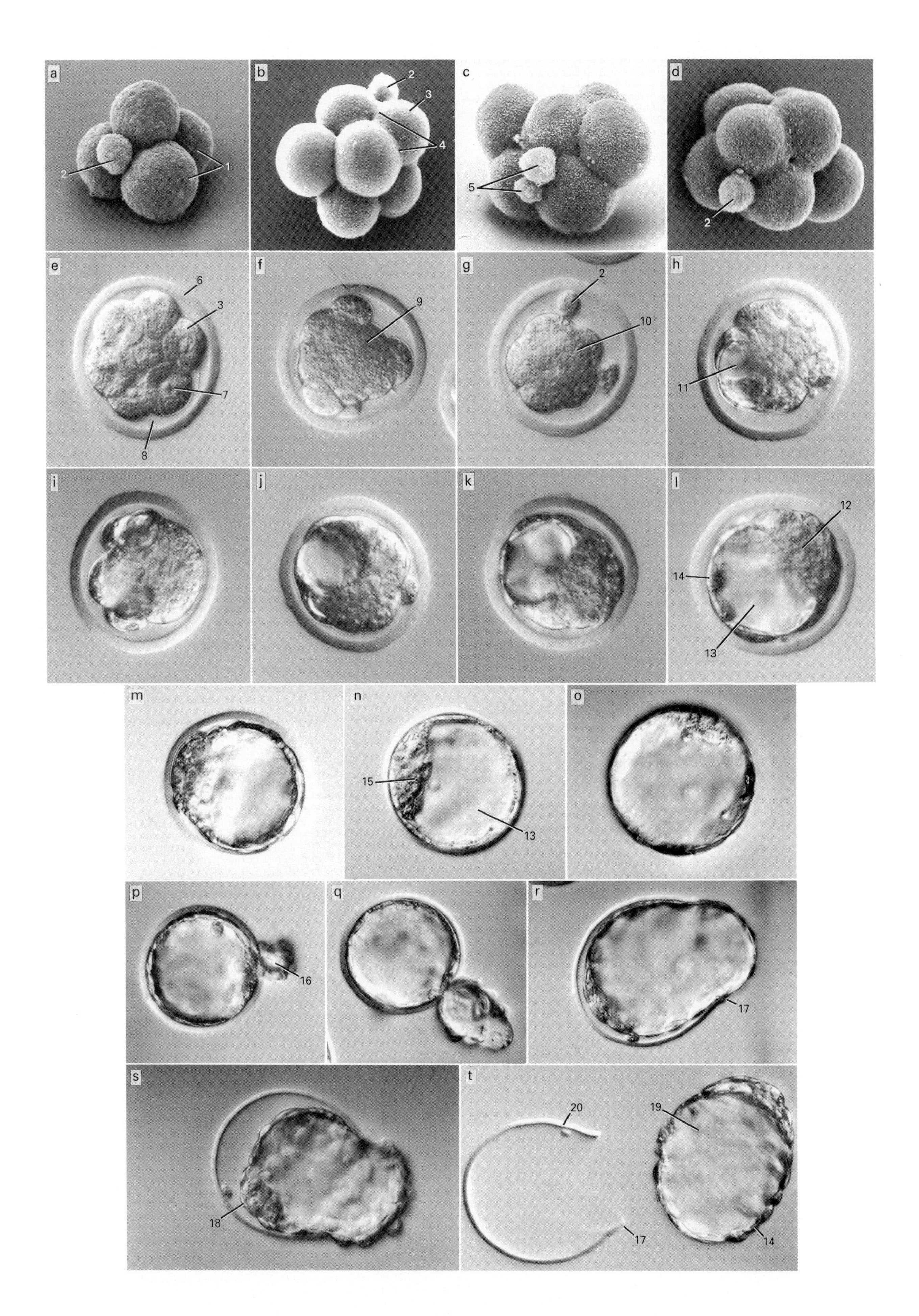
a
b
c
d
e
f
g
h
i
j
k
l
m
n
o
p
q
r
s
t
1
2
3
4
5
6
7
8
9
10
11
12
13
14
15
16
17
18
19
20

Plate 1b (Preimplantation period, II)

Embryonic development during the preimplantation period, II

a–d scanning electron micrographs of 4- to 8-cell stage embryos showing pre- and early compaction stages. (a) 4-cell stage embryo, (b) 8-cell stage embryo showing very early evidence of compaction, (c, d) 8-cell stage embryos at early stages of compaction in which boundaries of individual blastomeres are no longer distinct.

e–l embryos viewed by interference contrast optics showing normal sequence of events observed when embryos progress from the 8-cell to the blastocyst stage.

m–t embryos viewed by interference contrast optics illustrating "hatching" of blastocysts *in vitro*, though on time-lapse cinephotography the embryo is seen to rhythmically expand and contract over this period of time (Cole, 1967). As the blastocyst gradually increases in volume, this results in the progressive thinning of the zona pellucida (m–o). An area of weakness eventually leads to a breach in the integrity of the zona pellucida (usually at the site of sperm penetration) through which the trophectoderm escapes (p–s). The first region of trophectoderm to escape is usually at the abembryonic pole of the blastocyst (r, s). Shortly after the blastocyst has escaped from the zona, the embryo re-expands and reverts to its characteristic spherical/ovoid form (t).

1. blastomeres of 4-cell stage embryo
2. second polar body
3. blastomere of 8-cell stage embryo
4. early evidence of compaction in which the boundaries between the individual blastomeres are no longer distinct
5. the two division products when the second polar body divides (usually shortly after its extrusion) at the completion of the second meiotic division
6. zona pellucida
7. nucleus of 8-cell stage embryo
8. perivitelline space
9. early morula stage embryo where most (but not yet all) of the boundaries between individual blastomeres are indistinct
10. fully compacted morula stage embryo
11. early evidence of cavity formation (with accumulation of blastocoelic fluid)
12. location of inner cell mass
13. blastocoelic (blastocystic) cavity
14. mural trophectoderm cell
15. inner cell mass
16. trophectoderm escaping through breach in zona pellucida
17. boundary of breach in zona pellucida
18. polar trophectoderm cell
19. fully expanded zona-free blastocyst
20. empty zona pellucida

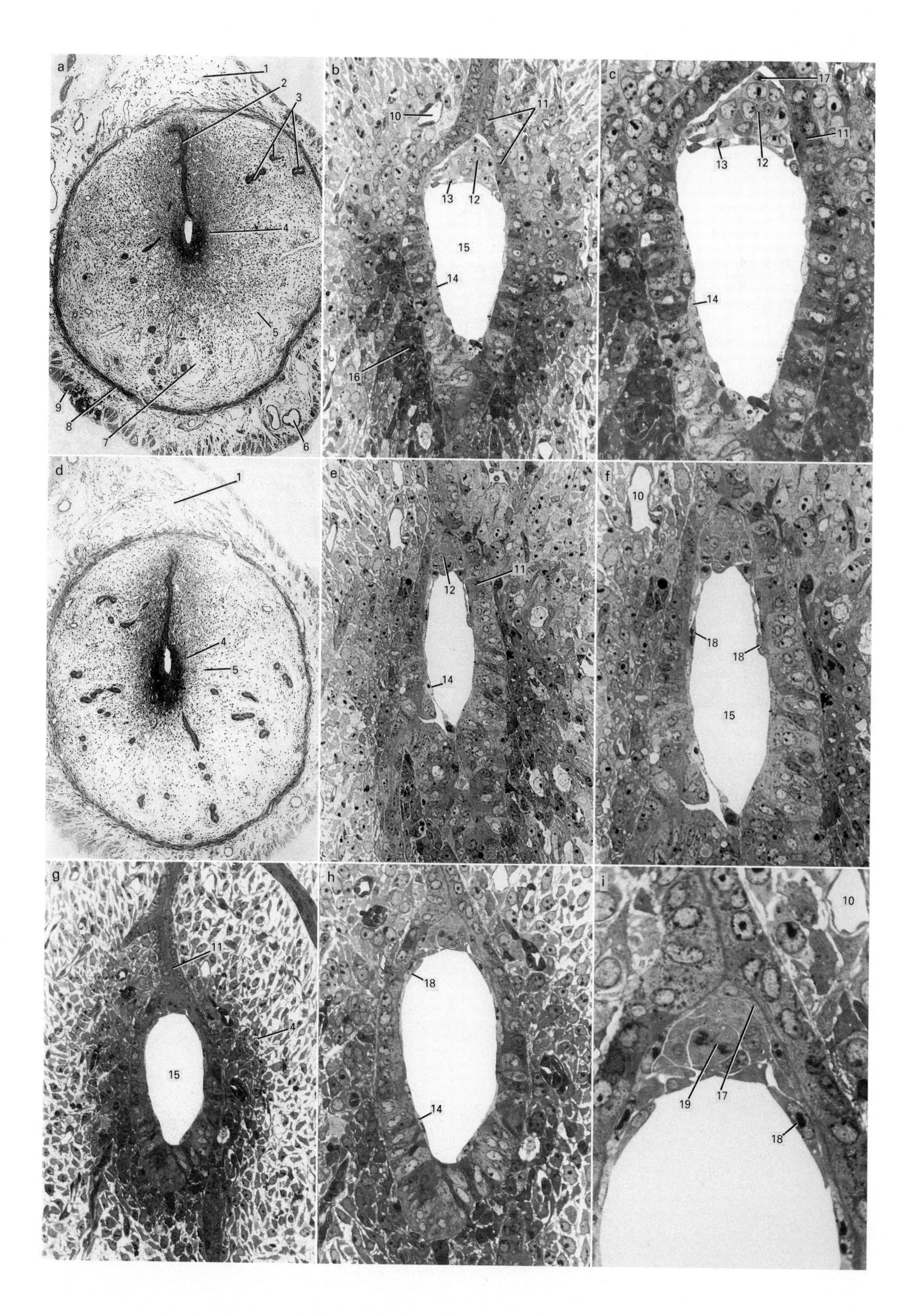
a
1
2
3
4
5
6
7
8
9
b
10
11
12
13
14
15
16
c
17
11
12
13
14
d
1
4
5
e
12
11
14
f
10
18
18
15
g
11
4
15
h
18
14
i
10
19
17
18

Plate 1c (4.5 days p.c.)

4.5 days p.c. Three implanting blastocysts. Sagittal sections. (Theiler Stage 6; rat, Witschi Stage 8; human, Carnegie Stage 4.) Semi-thin sections (1 μm) stained with toluidine blue.

a ×50
b ×312
c ×500
d ×50
e ×312
f ×500
g ×250
h ×400
i ×1000

1. mesometrium
2. remnant of uterine lumen
3. uterine glands
4. boundary of inner zone of decidual reaction
5. boundary of outer zone of decidual reaction
6. dilated blood vessels (branches of uterine artery) within uterine wall
7. area of stromal oedema
8. inner circular layer of myometrial smooth muscle
9. outer longitudinal layer of myometrial smooth muscle
10. dilated blood vessel in close proximity to endometrium
11. endometrial layer
12. inner cell mass (ICM)
13. visceral (proximal) layer of primary embryonic endoderm cells
14. layer of mural trophectoderm (trophoblast) cells
15. blastocoelic cavity (blastocoele)
16. zone of glycogen-containing decidual cells
17. polar trophectoderm cell
18. parietal (distal) layer of primary extra-embryonic endoderm cells
19. inner cell mass cell in mitosis

PLATE 2

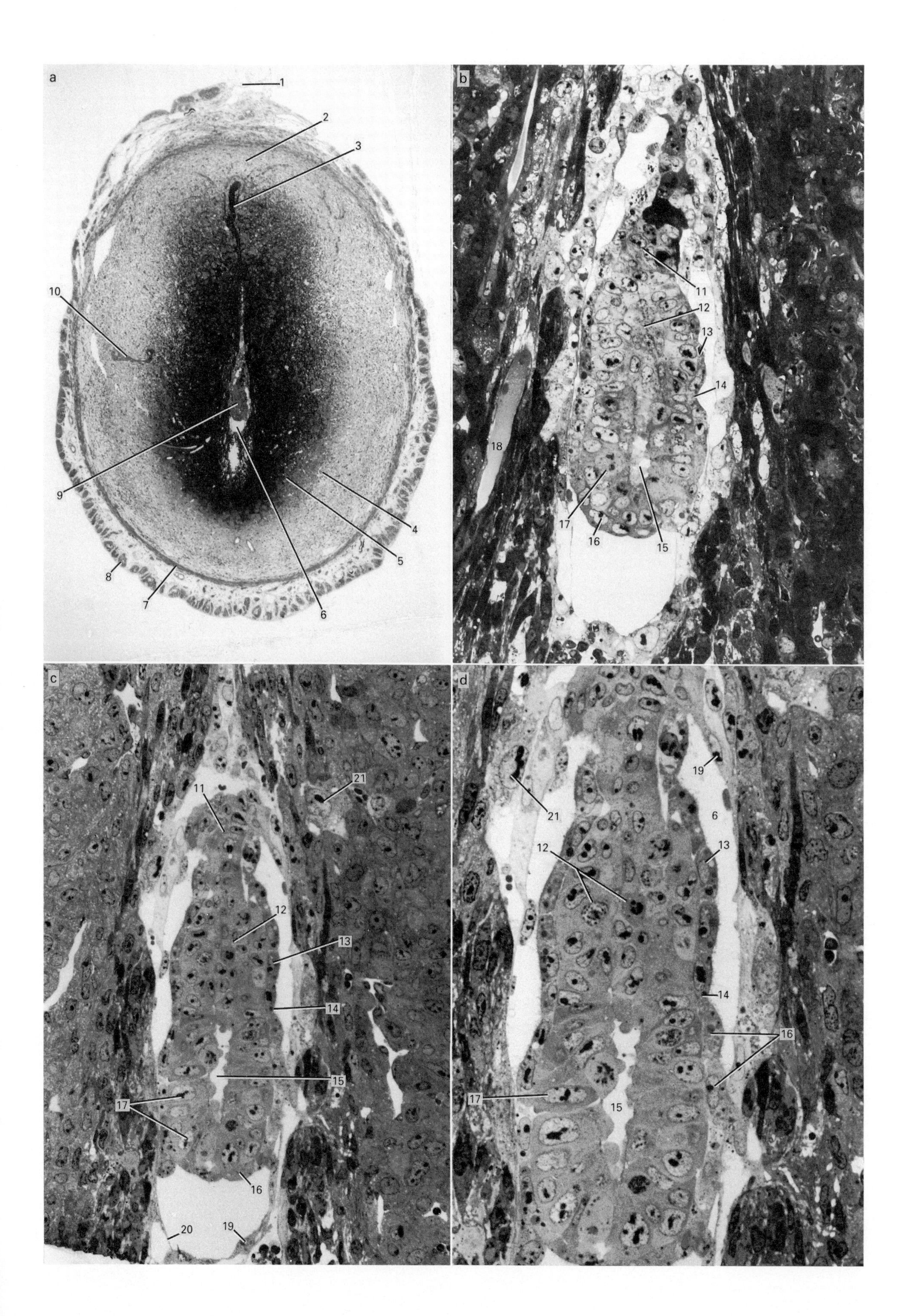

a
1
2
3
4
5
6
7
8
9
10
b
11
12
13
14
15
16
17
18
c
11
12
13
14
15
16
17
19
20
21
d
12
13
14
15
16
17
19
21
6

Plate 2 (5.5 days p.c.)

5.5 days p.c. Early stage in differentiation of egg cylinder. Sagittal sections. (Theiler Stages 7–8; rat, Witschi Stage 10; human, Carnegie Stage 5.) Semi-thin sections (1 μm) stained with toluidine blue.
a ×31
b ×250
c ×250
d ×400

1. mesometrium
2. area of stromal oedema subjacent to inner circular layer of myometrial smooth muscle
3. remnant of uterine lumen
4. boundary of outer zone of decidual reaction
5. boundary of inner zone of decidual reaction
6. yolk sac cavity
7. inner circular layer of myometrial smooth muscle
8. outer longitudinal layer of myometrial smooth muscle
9. egg cylinder stage embryo at centre of decidual reaction
10. uterine glands
11. ectoplacental cone
12. extra-embryonic ectoderm
13. visceral layer of extra-embryonic endoderm
14. approximate line of demarcation between extra-embryonic and embryonic endoderm
15. proamniotic cavity
16. visceral (proximal) layer of primary embryonic endoderm cells
17. embryonic ectoderm cells
18. dilated blood vessel (branch of uterine vessel) within decidual reaction
19. parietal (distal) layer of primary extra-embryonic endoderm cells
20. Reichert's membrane
21. trophoblast giant cell

PLATE
3

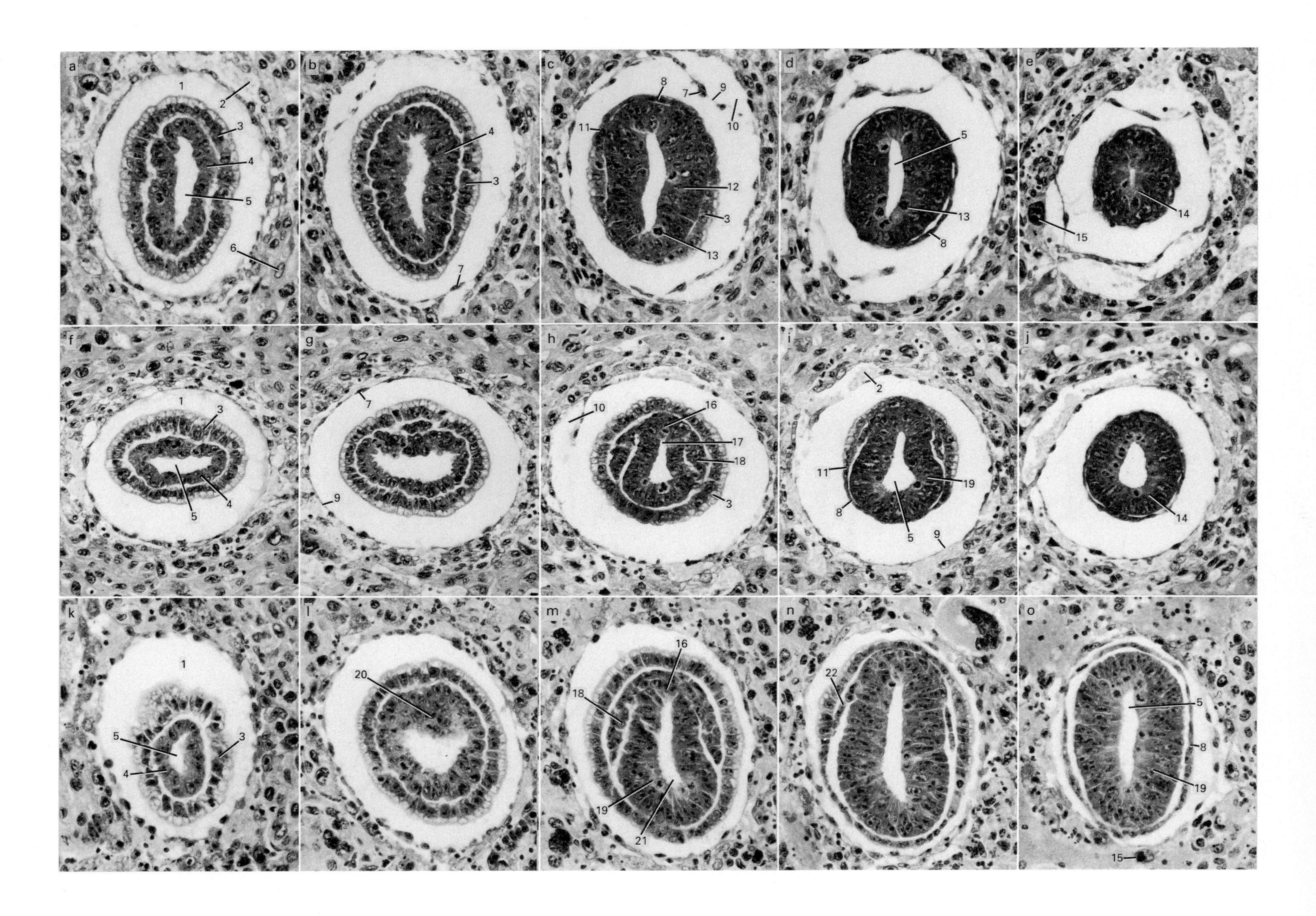

Plate 3 (6.5 days p.c.; 7 days p.c.)

a–e 6.5 days p.c. Advanced egg cylinder stage embryo (Theiler Stage 9; rat, Witschi Stage 11; human, Carnegie Stage 6).

f–j, k–o 7 days p.c. Two advanced egg cylinder/early primitive streak stage embryos (Theiler Stage 10; rat, Witschi Stage 12; human, Carnegie Stages 6–7). These embryos all sectioned transversely from the region of the ectoplacental cone towards the distal region of the embryonic pole.

1. yolk sac cavity
2. maternal red blood cells within maternal sinusoid
3. primitive extra-embryonic endoderm cells with characteristic columnar/cuboidal shape, apical vacuoles and microvillous "brush" border
4. extra-embryonic ectoderm cells
5. proamniotic canal
6. endometrial decidual cell
7. parietal endoderm cell attached to inner (i.e. embryonic) aspect of Reichert's membrane
8. embryonic endoderm-type cells with characteristic squamous morphology
9. Reichert's membrane
10. artefactual space due to shrinkage away of the embryonic and extra-embryonic tissues from the surrounding (maternal) endometrial tissue
11. cells with transitional morphology at boundary zone between embryonic and extra-embryonic endoderm
12. embryonic ectoderm
13. embryonic ectoderm cell in mitosis
14. embryonic pole of egg cylinder
15. primary giant cell of mural trophectodermal origin
16. primitive streak region
17. primitive groove
18. intra-embryonic mesoderm
19. neuroepithelium of future headfold region
20. mass of extra-embryonic ectoderm cells in region of future posterior amniotic fold
21. neural groove
22. polar extent of intra-embryonic mesoderm, which has yet to separate the ectoderm from the endoderm in the embryonic pole of the egg cylinder

PLATE 4

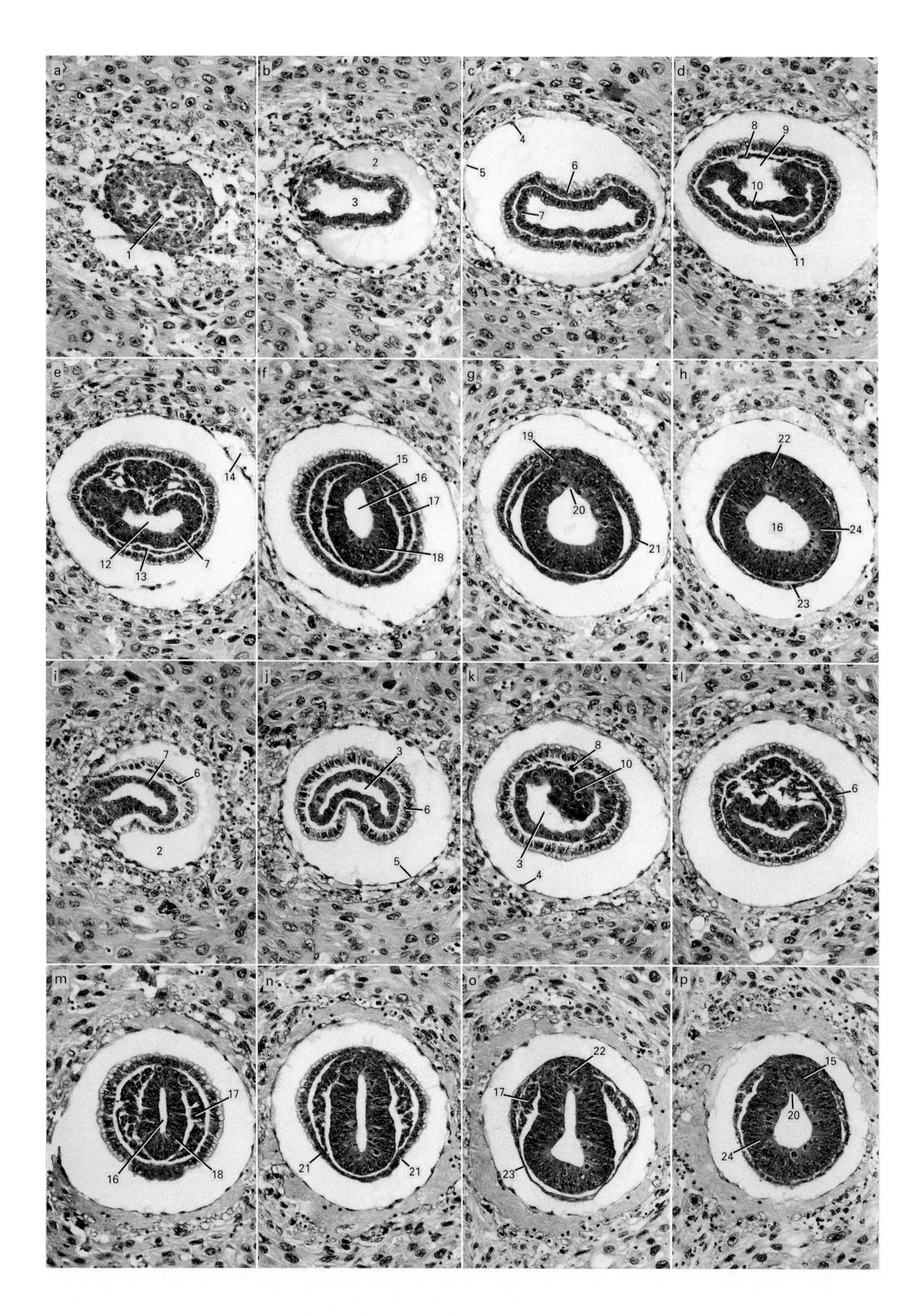
a
1
b
2
3
c
4
5
6
7
d
8
9
10
11
e
14
12
13
7
f
15
16
17
18
g
19
20
21
h
22
16
24
23
i
7
6
2
j
3
6
5
k
8
10
3
4
l
6
m
17
16
18
n
21
21
o
22
17
23
p
15
20
24

Plate 4 (7 days p.c.)

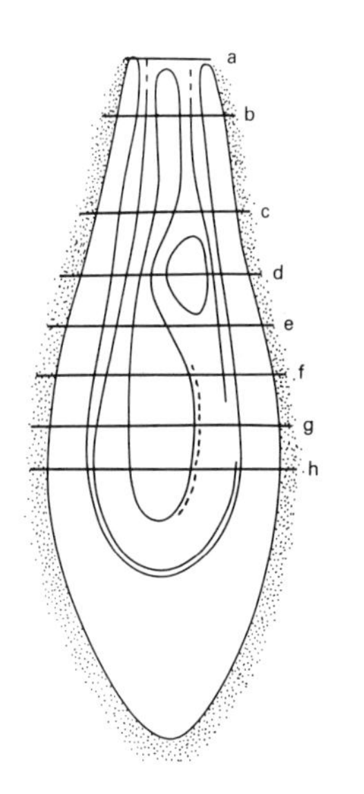

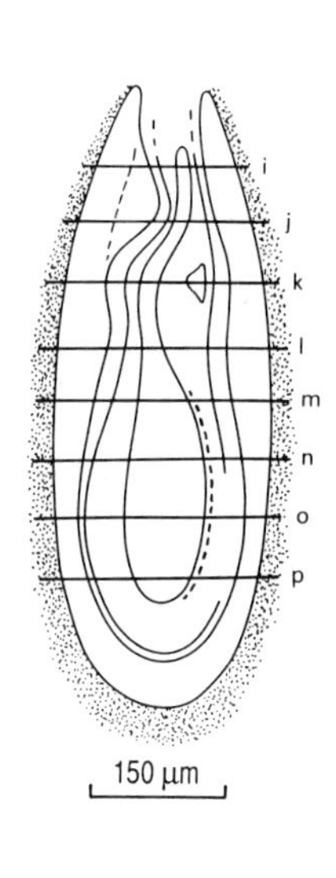

7 days p.c. Two early primitive streak stage embryos (a–h; i–p), sectioned transversely from the region of the ectoplacental cone towards the distal region of the embryonic pole (Theiler Stage 10; rat, Witschi Stage 12; human, Carnegie Stages 6–7).

1. ectoplacental cone
2. yolk sac cavity
3. abembryonic part of proamniotic canal
4. parietal endoderm cell attached to inner (i.e. embryonic) aspect of Reichert's membrane
5. Reichert's membrane
6. extra-embryonic endoderm
7. extra-embryonic ectoderm
8. mesothelial lining (extra-embryonic mesodermal cells) of posterior amniotic fold (future extra-embryonic coelomic (exocoelomic) cavity)
9. extra-embryonic coelomic (exocoelomic) cavity
10. posterior amniotic fold
11. proamniotic canal (ectoplacental duct)
12. site of communication between proamniotic canal and amniotic cavity
13. extra-embryonic mesoderm
14. artefactual space due to shrinkage away of the embryonic and extra-embryonic tissues from the surrounding (maternal) endometrial tissue
15. primitive streak region
16. amniotic cavity
17. intra-embryonic mesoderm
18. neural ectoderm (neuroepithelium) in future headfold region
19. region where intra-embryonic mesoderm merges into caudal region of notochordal plate subjacent to neuroepithelium of primitive streak
20. primitive groove
21. transitional zone between extra-embryonic endoderm cells (with characteristic columnar/cuboidal morphology), and embryonic endoderm cells (with characteristic squamous morphology)
22. condensation of cells (poorly defined) subjacent to primitive streak (caudal region of notochordal plate)
23. embryonic endoderm
24. lower pole of embryo, where intra-embryonic mesoderm has yet to colonize and consequently separate its ectodermal from its endodermal components

PLATE 5

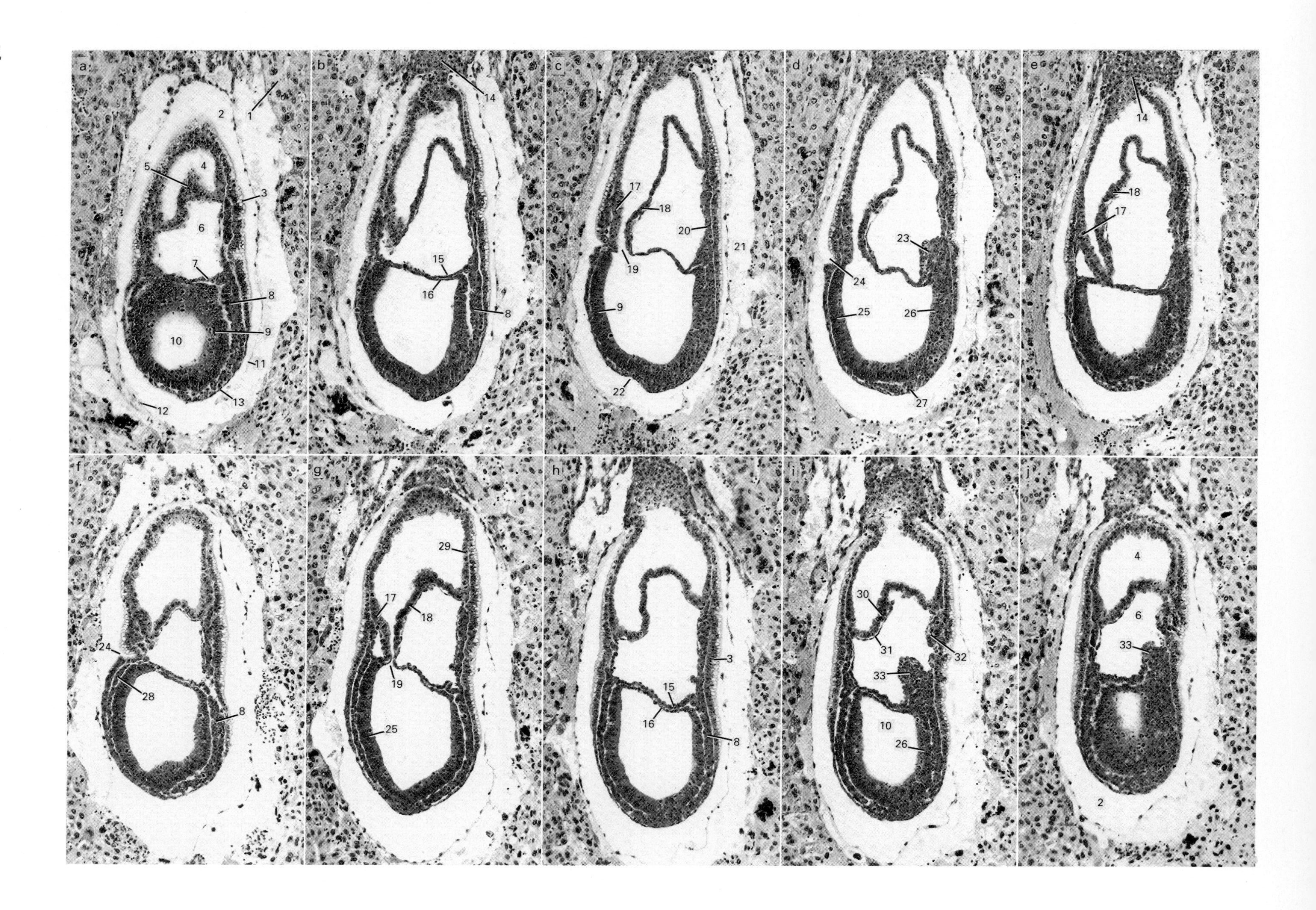
a
b
c
d
e
f
g
h
i
j
1
2
3
4
5
6
7
8
9
10
11
12
13
14
15
16
17
18
19
20
21
22
23
24
25
26
27
28
29
30
31
32
33

Plate 5 (7 days p.c.)

7 days p.c. Two primitive streak stage embryos (a–e; f–j), sectioned sagittally (Theiler Stages advanced 10–early 11; rat, Witschi Stage 12; human, Carnegie Stages 6–7).

1. spongy layer of endometrial tissue in decidua basalis in abembryonic region of implantation site
2. yolk sac cavity
3. extra-embryonic endodermal component of visceral yolk sac – the cells are cuboidal or columnar with large numbers of apical vacuoles and a pronounced microvillous border
4. ectoplacental cavity
5. chorion
6. extra-embryonic coelomic (exocoelomic) cavity
7. amnion
8. intra-embryonic mesoderm in primitive streak region
9. embryonic ectoderm
10. amniotic cavity
11. Reichert's membrane
12. parietal endoderm cell attached to inner (i.e. embryonic) aspect of Reichert's membrane
13. embryonic endoderm consisting principally of cells with a squamous morphology
14. ectoplacental cone
15. mesodermal component of amnion
16. ectodermal component of amnion
17. anterior amniotic fold
18. posterior amniotic fold
19. proamniotic canal (also termed ectoplacental duct) allowing communication between future ectoplacental and amniotic cavities
20. extra-embryonic mesodermal component of visceral yolk sac
21. artefactual space due to shrinkage away of the embryonic and extra-embryonic tissues from the surrounding (maternal) endometrial tissue
22. indentation indicating site of future foregut diverticulum
23. allantoic rudiment
24. endodermal furrow representing a transverse fold often seen in this location
25. neural ectoderm (neuroepithelium) in future headfold region
26. neural ectoderm (neuroepithelium) in primitive streak region
27. notochordal plate
28. intra-embryonic mesoderm subjacent to future headfold region of embryo
29. extra-embryonic ectoderm
30. extra-embryonic ectodermal component of chorion
31. extra-embryonic mesodermal component of chorion
32. early stage in differentiation of primitive "blood island" within extra-embryonic mesodermal component of visceral yolk sac
33. allantois

PLATE 6

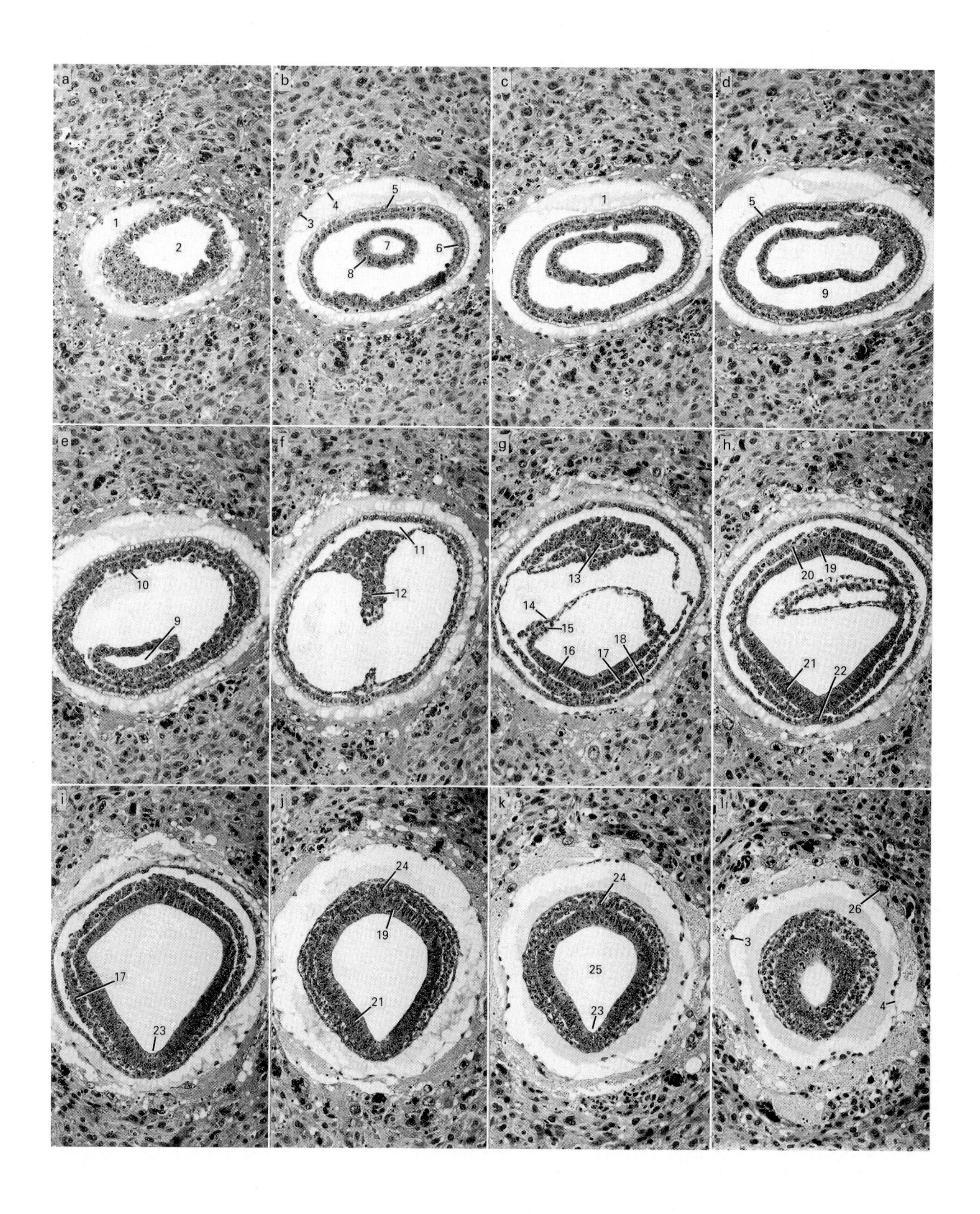
a
1
2
b
3
4
5
6
7
8
c
1
d
5
9
e
10
9
f
11
12
g
13
14
15
16
17
18
h
20
19
21
22
i
17
23
j
24
19
21
k
24
25
23
l
26
3
4

Plate 6 (7 days p.c.)

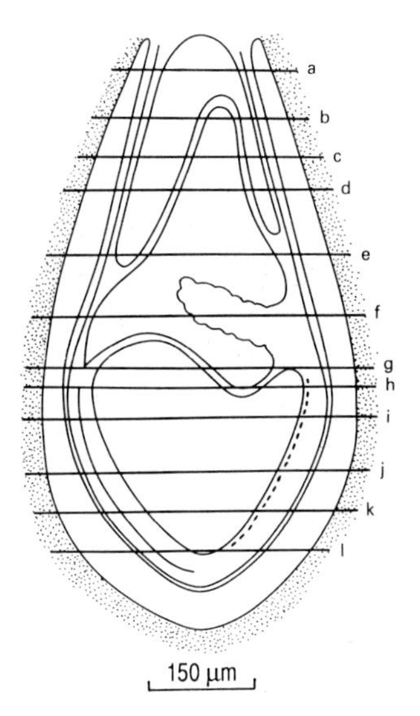

7 days p.c. Primitive streak stage embryo, sectioned transversely from the region of the ectoplacental cone towards the distal region of the embryonic pole (Theiler Stage 11; rat, Witschi Stage 12; human, Carnegie Stages 6–7).

1. yolk sac cavity
2. ectoplacental cavity
3. parietal endoderm cell attached to inner (i.e. embryonic) aspect of Reichert's membrane
4. Reichert's membrane
5. extra-embryonic endodermal component of visceral yolk sac
6. extra-embryonic mesodermal component of visceral yolk sac
7. extra-embryonic coelomic (exocoelomic) cavity within posterior amniotic fold
8. prospective chorion (abembryonic component of posterior amniotic fold)
9. ventral extension of ectoplacental cavity
10. mesothelial lining (extra-embryonic mesodermal cells) of extra-embryonic coelomic (exocoelomic) cavity
11. caudal diverticulum of amniotic cavity into base of allantois
12. allantois
13. base of allantois
14. mesodermal component of amnion
15. ectodermal component of amnion
16. neural ectoderm (neuroepithelium)
17. intra-embryonic mesoderm subjacent to primitive headfold region
18. embryonic endoderm
19. neural ectoderm (neuroepithelium) in primitive streak region
20. intra-embryonic mesoderm subjacent to neural ectoderm in primitive streak region
21. primitive headfold
22. notochordal plate
23. neural groove
24. condensation of tissue (poorly defined) in caudal region of notochordal plate subjacent to primitive streak
25. amniotic cavity
26. trophoblast giant cell

PLATE 7

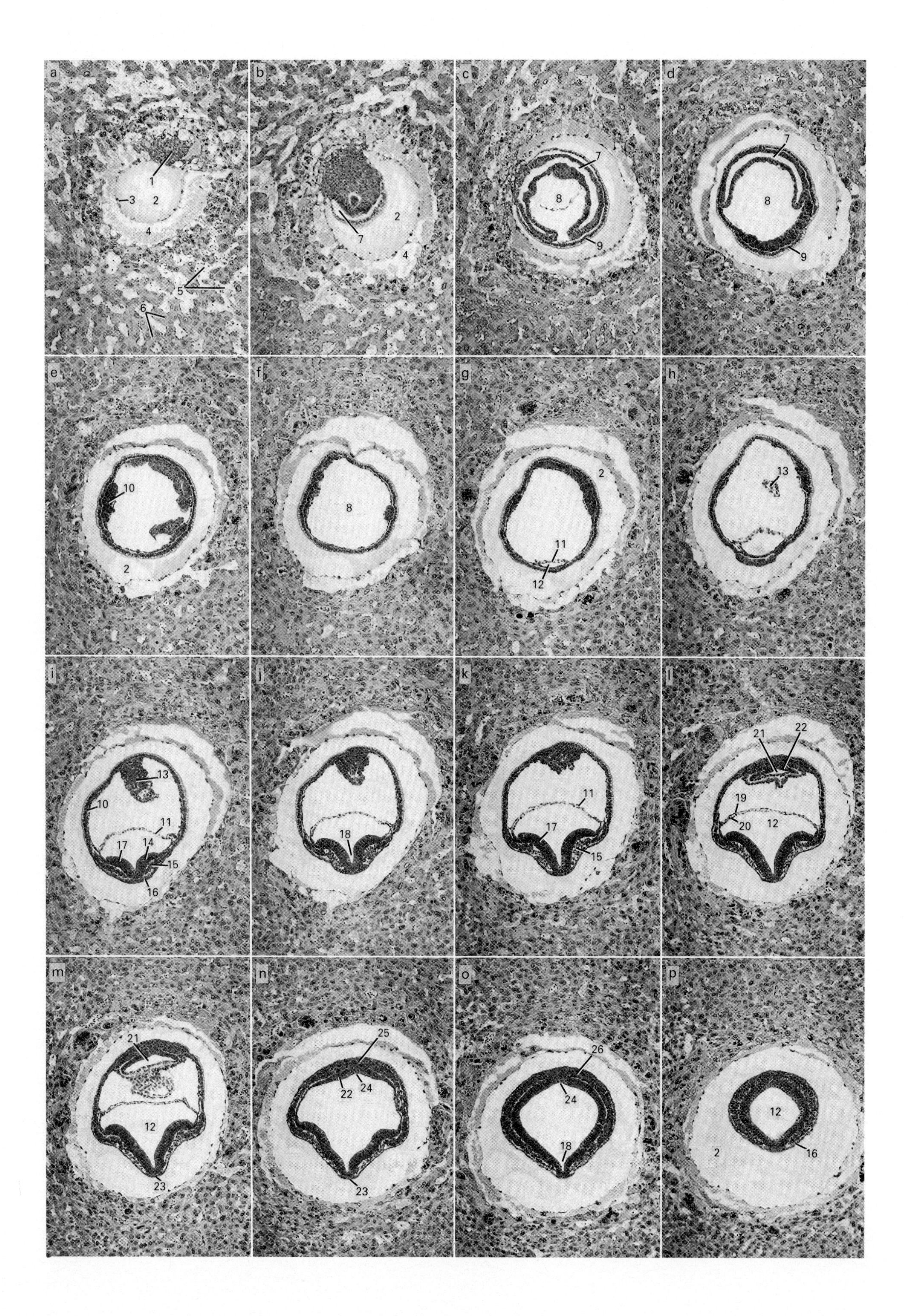
a
1
3
2
4
5
6
b
7
2
4
c
7
8
9
d
7
8
9
e
10
2
f
8
g
2
11
12
h
13
i
13
10
11
17
14
15
16
j
18
k
11
17
15
l
21
22
19
20
12
m
21
12
23
n
25
22
24
23
o
26
24
18
p
12
2
16

Plate 7 (7.5 days p.c.)

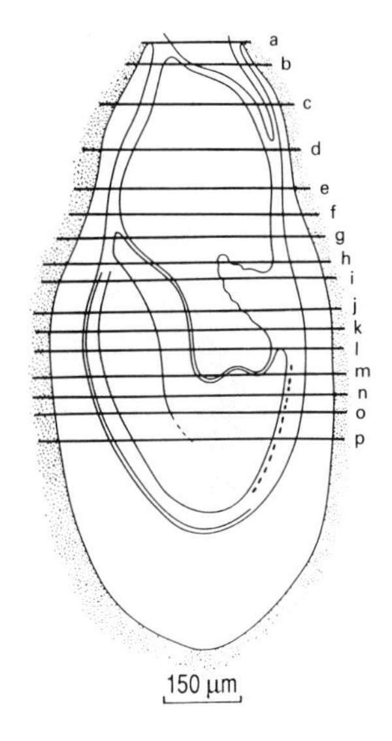

7.5 days p.c. Advanced primitive streak stage embryo, sectioned transversely from the region of the ectoplacental cone towards the distal region of the embryonic pole (Theiler Stage 11; rat, Witschi Stages 12–13; human, Carnegie Stages 7–8).

1. ectoplacental cone
2. yolk sac cavity
3. parietal endoderm cells attached to inner (i.e. embryonic) aspect of Reichert's membrane
4. artefactual space due to shrinkage away of the embryonic and extra-embryonic tissues from the surrounding (maternal) endometrial tissue
5. sinusoid containing maternal blood
6. spongy layer of endometrial tissue in decidua basalis in abembryonic region of implantation site
7. ectoplacental cavity
8. extra-embryonic coelomic (exocoelomic) cavity
9. extra-embryonic endodermal component of visceral yolk sac
10. extra-embryonic mesodermal component of visceral yolk sac (aggregation of tissue which will subsequently form "blood island" – source of primitive erythroblast cells)
11. amnion
12. amniotic cavity
13. allantois
14. neural ectoderm (neuroepithelium)
15. intra-embryonic mesoderm subjacent to headfold
16. embryonic endoderm
17. neural fold (headfold)
18. neural groove
19. mesodermal component of amnion
20. ectodermal component of amnion
21. caudal diverticulum of amnion into base of allantois
22. neuroepithelium in primitive streak region
23. notochordal plate
24. primitive groove
25. condensation of tissue (poorly defined) in caudal region of notochord subjacent to primitive streak
26. intra-embryonic mesoderm subjacent to neuroepithelium of primitive streak region

PLATE 8

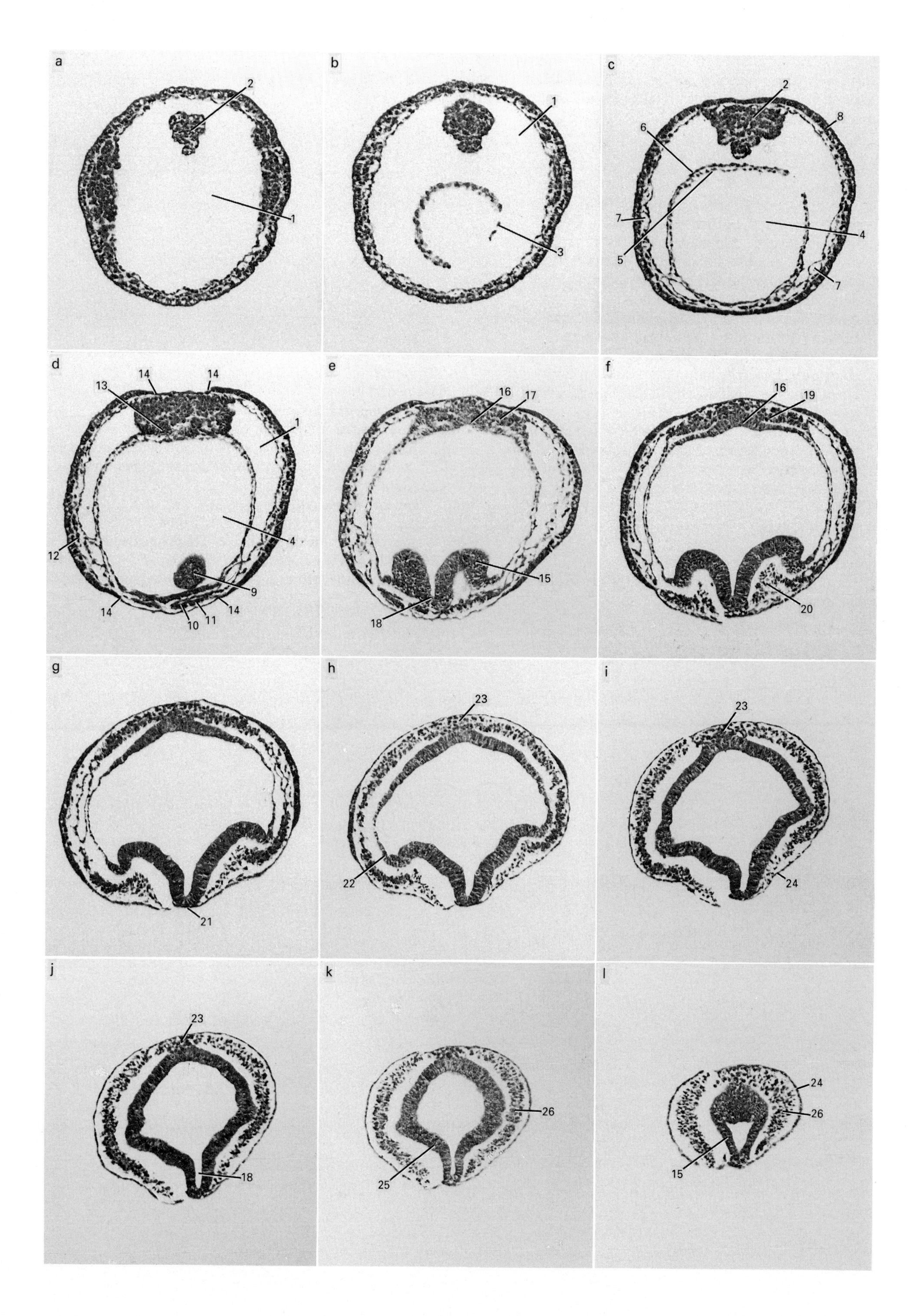

a
2
1
b
1
3
c
2
8
6
5
7
4
7
d
14
14
13
1
12
4
9
14
10
11
14
e
16
17
15
18
f
16
19
20
g
21
h
23
22
i
23
24
j
23
18
k
26
25
l
24
26
15

Plate 8 (7.5 days p.c.)

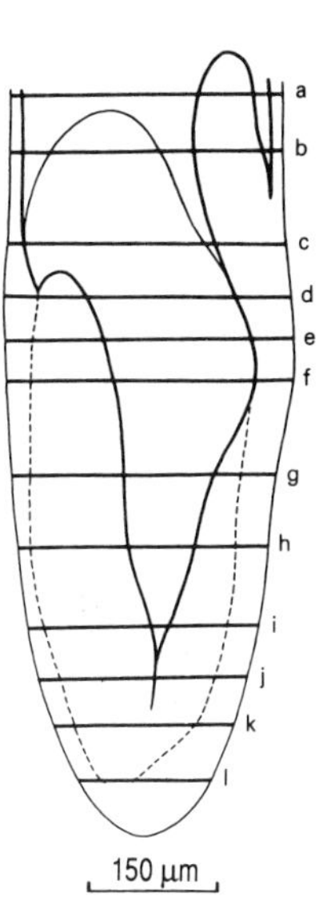

7.5 days p.c. Early headfold presomite stage embryo, sectioned transversely from the region of the ectoplacental cone towards the distal region of the embryonic pole (Theiler Stage 11; rat, Witschi Stage 13; human, Carnegie Stage 8).

1. extra-embryonic coelomic (exocoelomic) cavity
2. allantois
3. amnion
4. amniotic cavity
5. amnion – ectodermal component
6. amnion – mesodermal component
7. blood island within mesodermal component of yolk sac
8. extra-embryonic endodermal component of yolk sac
9. cephalic neural fold (headfold)
10. intra-embryonic coelomic cavity (region of future pericardial cavity)
11. thickened mesothelial cells – future cardiogenic (myocardial) plate
12. nucleated red blood cells within yolk sac blood island
13. base of allantois
14. junction between yolk sac endoderm (columnar) and embryonic endoderm (squamous)
15. neuroepithelium (neural ectoderm)
16. primitive streak region
17. region of continuity between tissue of allantois and mesoderm subjacent to primitive streak
18. neural groove
19. mesoderm subjacent to primitive streak
20. intra-embryonic mesoderm – future cephalic mesenchyme tissue
21. rostral extent of notochordal process/plate
22. junction between neural and surface ectoderm
23. condensation of tissue (poorly defined) in caudal region of notochord
24. embryonic endoderm
25. caudal extremity of cephalic neural fold
26. intra-embryonic mesoderm

subsequent growth towards the abembryonic pole of the implanting blastocyst, to form the egg cylinder. The proximal endoderm cells are cuboidal in appearance.

Stage 8 Developmental age, 6 days p.c.

The differentiation of the egg cylinder occurs during this stage. The proamniotic cavity forms within the mass of primitive ectoderm cells, initially in the future embryonic pole, and then progressively towards the abembryonic pole. The first evidence of Reichert's membrane (secreted by distal endoderm cells) is seen at this stage. Similarly, the formation of the ectoplacental cone occurs at this time, and this is associated with the invasion of maternal tissue by trophoblast (primary) giant cells. The latter originate from the polar region of the trophectoderm.

Stage 9 Developmental age, 6.5 days p.c.
Plate 3,embryo illustrated a–e (see pp. 26–27)

The embryo is at the advanced egg cylinder stage. The first evidence of an embryonic axis is seen at this stage. A clear morphological distinction is apparent between the embryonic and extraembryonic ectoderm, and between the embryonic endoderm cells (which have a largely squamous morphology) and extra-embryonic endoderm cells (which have a characteristic columnar/cuboidal shape, apical vacuoles and a microvillous "brush" border). The ectoplacental cone region becomes invaded by maternal blood.

Stage 10 Developmental age, 7 days p.c.
Plate 3, embryos illustrated f–j, k–o; Plate 4, embryos illustrated a–h, i–p; Plate 5, a–e (see pp. 26–31)

Early to mid-primitive streak stage. The first appearance of the intraembryonic mesoderm is observed at this stage (*Plate 3*, embryos illustrated *f–j*, *k–o*). These cells migrate forward from the region of the primitive streak to separate the neuroepithelium of the future headfolds from the subjacent extra-embryonic endoderm distally, and the embryonic endoderm proximally. Extra-embryonic ectoderm and subjacent mesoderm cells proliferate to form the posterior amniotic fold. The latter is first evident at the early primitive streak stage (*Plate 3,l*), and is more pronounced at the mid-primitive streak stage (*Plate 4,d,k*). The fully differentiated posterior and anterior amniotic folds are seen towards the end of this stage, and these are associated with the narrowing (*Plate 5,c,d,g*) and eventual obliteration of the proamniotic canal when they eventually fuse together at the beginning of the next stage. The first evidence of the allantoic rudiment is seen at the end of this stage (*Plate 5,d*)

Stage 11 Developmental age, 7.5 days p.c.
Plate 5, f–j; Plates 6, 7 and 8 (see pp. 30–37)

Late primitive streak stage. More extensive growth of the allantois occurs into the cavity of the posterior amniotic fold. Shortly after the onset of this stage, the anterior and posterior amniotic folds amalgamate, and this results in the obliteration of the proamniotic canal and the formation of the exocoelomic cavity. As a consequence of these events, the conceptus becomes divided into three separate compartments which consist of the ectoplacental cavity, the exocoelomic cavity and the amniotic cavity, being separated by the chorion and amnion, respectively (*Plate 5,h,i*). Discrete blood islands make their first appearance in the (extra-embryonic) mesodermal component of the visceral yolk sac (*Plate 5,i*; *Plate 7,e*).

The neural plate is clearly defined anteriorly and laterally, but merges into the primitive streak region posteriorly. In the midline subjacent to the neural groove the notochordal plate is clearly evident, though it is only poorly defined in the region subjacent to the primitive streak (*Plate 6,j*; *Plate 7,n*). Towards the anterior end of the primitive streak, an indentation may be observed on its ventral aspect – this is sometimes referred to as the "archenteron", but for convenience should be referred to as the "node", as its function is believed to be equivalent to that of Hensen's node region of the chick embryo, and is the site where the basic organization of the primitive body plan is believed to be formulated (the latter site is more clearly defined in the chick than in the mouse embryo).

The primitive endoderm is gradually displaced by the definitive endoderm (and in the midline by the notochord), which are derived from the epiblast cells which migrate from the anterior part of the primitive streak. A smaller proportion of the definitive endoderm cells are believed to be recruited directly from the overlying embryonic ectoderm cells (as well as being recruited directly into the notochord) without their first having to migrate through the primitive streak (Tam and Beddington, 1992). The neural groove is well defined in the future headfold region, but becomes gradually less well defined posteriorly, in the region of the primitive streak. A slight indentation in the ventral midline in the region of the embryonic pole is sometimes seen at this stage, and represents an early stage in the differentiation of the foregut pocket. All of the definitive endoderm cells that line the region of the primitive gut, are believed to be derived from the anterior end of the primitive streak (Tam and Beddington, 1992).

In the more advanced embryos observed at this stage, the ectoplacental cavity diminishes in volume due to the expansion of the exocoelomic and amniotic cavities. Similarly at this time, the headfolds become much more marked, and the first evidence is seen of the intra-embryonic coelomic cavity which, in the ventral midline (region of future pericardial cavity), is lined by thickened mesothelial cells with a columnar rather than a squamous morphology, representing the future cardiogenic or myocardial plate (*Plate 8,d*).

For a comparison of various gastrulation and early neurulation staging systems, see Fujinaga *et al.*, (1992).

Stage 12 Developmental age, 8 days p.c. Plates 9,10 and 11

EXTERNAL FEATURES

"Unturned" headfold stage embryos with 1–7 pairs of somites. The allantois extends further into the exocoelom, and grows towards the chorion with which it eventually makes contact towards the end of this stage, or during the early part of the next stage of development. The maxillary components of the first branchial arches also become prominent at this time. In intact embryos, the somites are particularly prominent in the mid-lordotic region of the embryo.

THE HEART AND VASCULAR SYSTEM

Relatively few blood islands are present in the mesoderm component of the visceral yolk sac of embryos with only 1 or 2 pairs of somites, but by the end of this stage the blood islands have amalgamated together to form the yolk sac vasculature. This contains large numbers of primitive nucleated red blood cells, a few of which are seen to be in division. Most embryos with 3–5 pairs of somites show little development of the cardiogenic plate beyond that seen in presomite headfold stage embryos. Endocardial cells, however, proliferate subjacent to the cardiogenic plate across the ventral midline in the pericardial region of the intraembryonic coelom (*Plate 9,d–f*), and extend from here into the right and left horns of the intra-embryonic coelomic channels (the future pericardio-peritoneal channels) (*Plate 9,g–k*), where they will form the endothelial lining of the two horns of the sinus venosus (*Plate 10,n,o*) (Kaufman and Navaratnam, 1981). Elsewhere, angiogenetic (endothelial) elements amalgamate to form the embryonic vasculature, with the paired dorsal aortae and first branchial (pharyngeal) arch arteries being clearly seen in embryos towards the end of this stage of development.

The primitive blood vessels which extend from the cephalic region to the most caudal region of these embryos, contain no primitive nucleated red blood cells at this stage, as the embryonic and extra-embryonic circulations have yet to amalgamate (this occurs during Stage 13, in embryos with about 8–10 pairs of somites, e.g. *Plate 13a,e*; *Plate 14a,e,f*). Towards the end of this stage, the heart differentiates to form a substantial broad median tubular mass which is suspended from a wide dorsal mesentery (mesocardium) (*Plate 10,i–l*; *Plate 11,g,h*). The myocardial tissue which forms the outer wall of the primitive heart tube bulges prominently on either side of a deep ventral median sulcus. The outflow tract of the heart leads directly into the first branchial arch arteries. The first, initially irregular, but subsequently more regular contractions of the heart are observed towards the end of this stage of development.

THE PRIMITIVE GUT

In early somite stage embryos, the foregut pocket is clearly evident and extends rostrally from a wide entrance portal where it is in continuity with the future midgut region. More caudally, the endodermal lining of the future midgut is in continuity with the future hindgut region, though the entrance to the hindgut diverticulum is only clearly seen in embryos towards the end of this stage of development (*Plate 10,c*; *Plate 11,h,i*). The foregut diverticulum extends rostrally into the headfold region and its endodermal lining is only separated from the neuroepithelial cells lining the neural groove by the cells of the notochordal plate (*Plate 9,f–i*). In more advanced embryos, the rostral part of the foregut diverticulum widens from side to side to form the first branchial (pharyngeal) pouches (*Plate 10,g*; *Plate 11,e–g*), though the caudal extremities of these pouches are not well defined as the distal region of the foregut in these embryos is significantly wider from side to side than formerly. The foregut pocket characteristically contains acellular debris (*Plate 11,d,e*), and similar material is also observed in the hindgut pocket at later stages of development (e.g. *Plate 15b,a*).

THE NERVOUS SYSTEM

The headfolds become particularly prominent during this stage, and the boundary zone which separates the neural ectoderm (neuroepithelium) from the surface ectoderm becomes progressively clearer both in the cephalic region (cf. *Plate 9,f–h*; *Plate 11,e–h*) as well as elsewhere along the embryonic axis. Initially apposition, and later fusion of the neural folds occurs in the region of the 4th and 5th somites, and extends rostrally and caudally from this site. Optic placodes are first evident during the early part of this stage, and become indented in their central regions to form the optic pits (*Plate 10,d*). Towards the end of this stage, the increasing depth of the optic evaginations results in the formation of the optic eminences which are located on either side of the future forebrain region (*Plate 11,e*).

In conventionally sectioned material, it is difficult to establish the caudal extent of the future hindbrain region. In intact embryos, two rhombomeres (A and B) are first evident in early headfold stage embryos, and become increasingly more prominent throughout this stage of development, rhombomere B being at the level of the future otic placode which is first clearly seen during the early part of the next stage of development (*Plate 13a, h*). Originating from the junctional zone between the neural and surface ectoderm in the future mid-pontine region of the hindbrain, the trigeminal (V) neural crest cells emerge and migrate into the maxillary component of the first branchial arches (*Plate 11,d,e*; *Plate 12,c*), while the facial (VII) neural crest cells migrate from the future caudal region of the pons into the second branchial arches in slightly more advanced embryos (*Plate 12,e,f*).

(*continued on page 46*)

PLATE 9

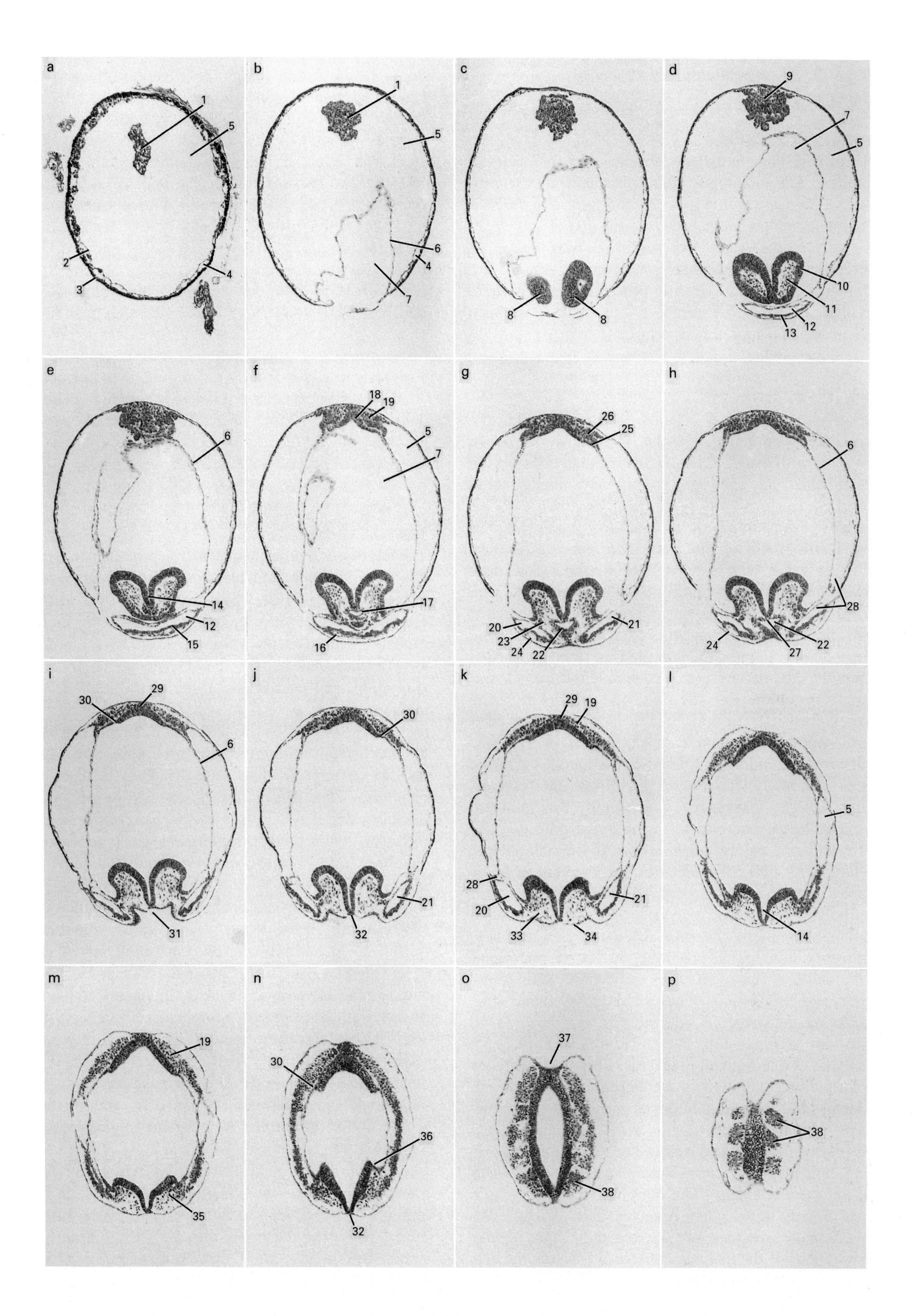
a
1
5
2
3
4
b
1
5
6
4
7
c
8
8
d
9
7
5
10
11
12
13
e
6
14
12
15
f
18
19
5
7
17
16
g
26
25
20
23
24
22
21
h
6
28
24
27
22
i
30
29
6
31
j
30
21
32
k
29
19
28
20
21
33
34
l
5
14
m
19
35
n
30
36
32
o
37
38
p
38

Plate 9 (7.5–8 days p.c.)

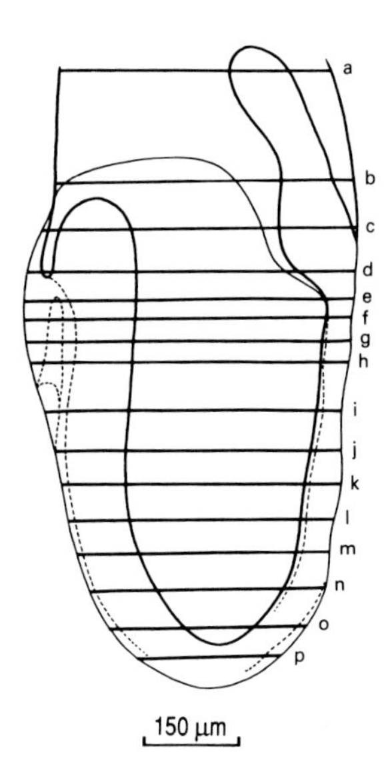

7.5–8 days p.c. Early headfold unturned embryo with 4–5 pairs of somites. Transverse sections. (Theiler Stage 12; rat, Witschi Stage 14; human, Carnegie Stage 9).

1. allantois
2. yolk sac – mesodermal component
3. yolk sac – endodermal component
4. blood island within mesodermal component of yolk sac
5. extra-embryonic coelomic (exocoelomic) cavity
6. amnion
7. amniotic cavity
8. cephalic neural fold (headfold)
9. base of allantois
10. neuroepithelium (neural ectoderm)
11. cephalic mesenchyme cells
12. intra-embryonic coelomic cavity (prospective pericardial cavity)
13. mesothelial cells lining intra-embryonic coelomic cavity
14. neural groove
15. cardiogenic plate
16. endocardial cells subjacent to cardiogenic plate
17. rostral extension of foregut diverticulum (pocket)
18. primitive streak region
19. mesoderm subjacent to primitive streak
20. right horn of intra-embryonic coelomic channel
21. left horn of intra-embryonic coelomic channel
22. foregut diverticulum (pocket)
23. rostral extension of dorsal aorta
24. endothelial lining of prospective right horn of sinus venosus
25. body wall – surface ectoderm and somatopleure (parietal mesodermal layer)
26. splitting of lateral plate mesoderm to form splanchnopleure and somatopleure
27. rostral extension of notochordal plate
28. continuity between intra- and extra-embryonic coelomic cavities
29. condensation of tissue (poorly defined) in caudal region of notochord
30. junction between surface and neural ectoderm at lateral boundary of primitive streak region
31. entrance to foregut diverticulum
32. notochord/notochordal plate
33. dorsal aorta
34. endoderm cells lining prospective midgut region
35. intra-embryonic mesoderm subjacent to caudal part of headfold
36. boundary zone between surface and neural ectoderm in caudal extremity of headfold – location of prospective neural crest tissue
37. shallow depression representing the future site of entrance into the hindgut diverticulum
38. somite derived from condensation of paraxial mesoderm

PLATE 10

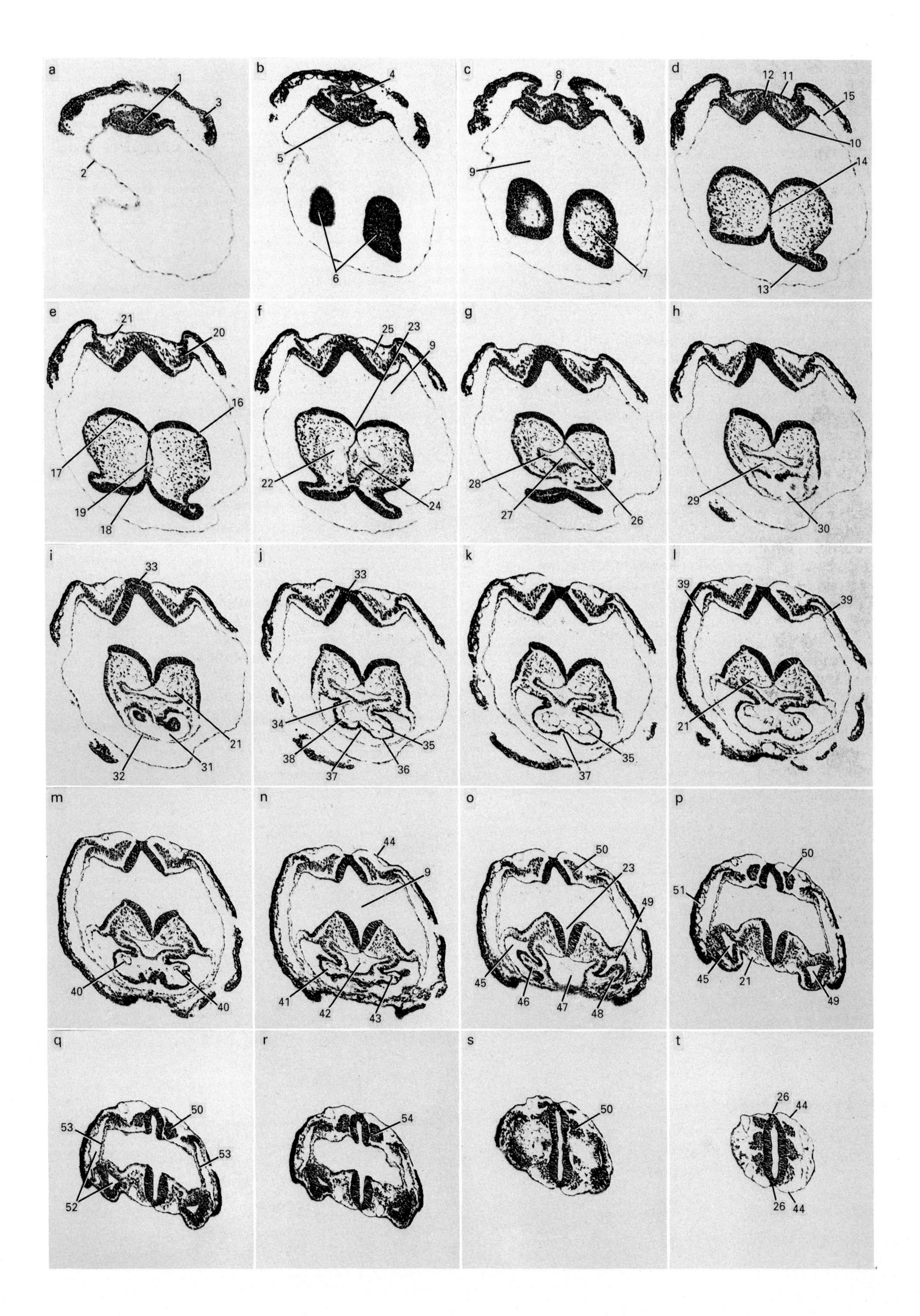

a
1
3
2
b
4
5
6
c
8
9
7
d
12 11
15
10
14
13
e
21
20
16
17
19
18
f
25
23
9
22
24
g
28
27
26
h
29
30
i
33
21
32
31
j
33
34
38
37
36
35
k
37
35
l
39
39
21
m
40
40
n
44
9
41
42
43
o
50
23
49
45
46
47
48
p
50
51
45
21
49
q
50
53
53
52
r
54
s
50
t
26
44
26
44

Plate 10 (8 days p.c.)

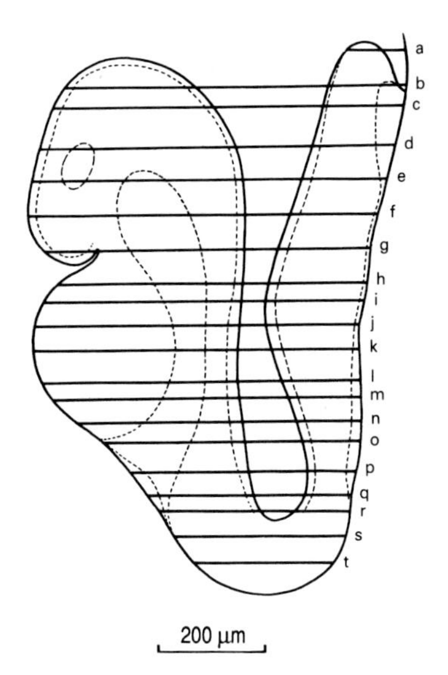

8 days p.c. Unturned embryo with 5–7 pairs of somites. Transverse sections (Theiler Stage 12; rat, Witschi Stage 15; human, Carnegie Stage 9).

1. base of allantois
2. amnion
3. visceral yolk sac
4. caudal extremity of hindgut diverticulum
5. primitive streak region (caudal extent of neural plate)
6. cephalic neural folds
7. cephalic mesenchyme tissue
8. entrance to hindgut diverticulum
9. amniotic cavity
10. junctional zone between surface and neural ectoderm
11. endodermal lining of prospective hindgut – midgut junction
12. condensation of tissue (poorly defined) in caudal region of notochord
13. central region of optic placode
14. rostral extremity of notochord
15. extra-embryonic coelomic (exocoelomic) cavity
16. junctional zone between surface and neural ectoderm in prospective hindbrain region
17. neuroepithelium of cephalic neural fold in prospective hindbrain region
18. prospective forebrain region
19. rostral extremity of foregut diverticulum
20. splitting of lateral plate mesoderm to form splanchnopleure and somatopleure
21. dorsal aorta
22. cephalic extension of dorsal aorta (prospective internal carotid artery)
23. neural groove
24. foregut diverticulum
25. paraxial mesoderm
26. notochord
27. pharyngeal region of foregut diverticulum
28. first branchial (pharyngeal) pouch
29. coalescence of endothelial elements to form the first branchial arch artery
30. rostral extremity of pericardial cavity
31. pericardial cavity
32. body wall in future thoracic region with outer layer of surface ectoderm and inner layer of intra-embryonic coelomic mesothelial cells
33. notochordal plate
34. broad dorsal mesentery of heart (dorsal mesocardium)
35. endocardial tissue lining the primitive heart tube
36. myocardial tissue forming outer wall of primitive heart tube
37. ventral median sulcus
38. gap between endocardial and myocardial tissue filled with cardiac jelly
39. lateral extent of body wall, in continuity with amnion
40. endothelial continuity between sinus venosus and common atrial chamber of primitive heart tube
41. right horn of sinus venosus
42. caudal region of foregut diverticulum
43. left horn of sinus venosus
44. endodermal lining of prospective midgut region
45. intra-embryonic coelomic cavity (future right pericardio-peritoneal canal)
46. right vitelline (omphalomesenteric) vein
47. entrance to foregut diverticulum
48. left vitelline (omphalomesenteric) vein
49. intra-embryonic coelomic cavity (future left pericardio-peritoneal canal)
50. somite
51. yolk sac blood island
52. communication between intra- and extra-embryonic coelomic cavities
53. body wall in mid-lordotic region of embryo
54. myocoele

PLATE 11

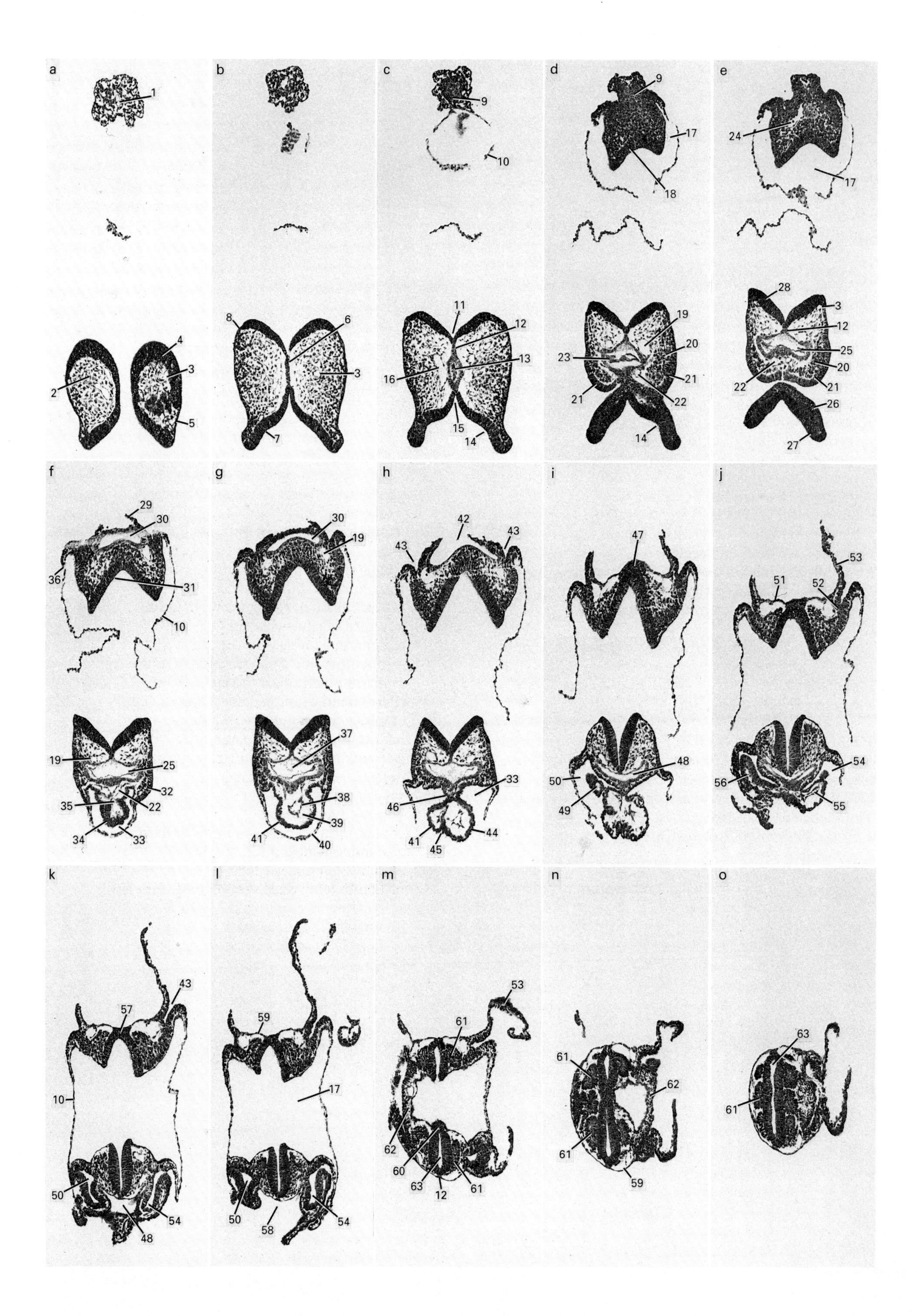

a
1
4
3
2
5
b
8
6
3
7
c
9
10
11
12
13
16
15
14
d
9
17
18
19
20
23
21
22
14
e
24
17
28
3
12
25
20
22
21
26
27
f
29
30
36
31
10
19
25
32
35
22
34
33
g
30
19
37
38
39
41
40
h
42
43
43
33
46
44
41
45
i
47
48
50
49
j
53
51
52
54
56
55
k
43
57
10
50
54
48
l
59
17
50
58
54
m
53
61
62
60
63
12
61
n
61
62
61
59
o
63
61

Plate 11 (8 days p.c.)

8 days p.c. Unturned embryo with 6–8 pairs of somites. Transverse sections (Theiler Stage 12; rat, Witschi Stage 15; human, Carnegie Stage 10).

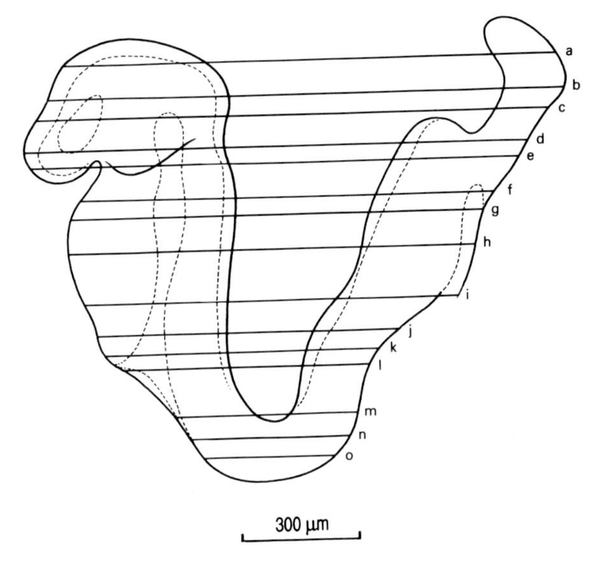

1. allantois
2. cephalic mesenchyme tissue in prospective midbrain region
3. branch of primary head vein
4. neural ectoderm (neuroepithelium)
5. surface ectoderm
6. rostral extremity of notochord
7. optic placode
8. junctional zone between surface (squamous) and neural (columnar) ectoderm
9. base of allantois
10. amnion
11. neural groove in region of prospective hindbrain (rhombencephalon)
12. notochord
13. rostral extremity of foregut diverticulum
14. optic pit (optic evagination)
15. neural groove in prospective diencephalic region of forebrain
16. communication between first branchial arch artery and dorsal aorta
17. amniotic cavity
18. neuroepithelium in caudal extremity of neural plate (caudal neuropore)
19. dorsal aorta
20. trigeminal (V) neural crest
21. first branchial arch (future maxillary component)
22. first branchial arch artery
23. acellular debris characteristically observed within cephalic extension of foregut diverticulum
24. nucleated primitive red blood cells within embryonic vasculature
25. first branchial (pharyngeal) pouch
26. optic eminence
27. cephalic neural fold in region of prospective forebrain
28. neuroepithelium in prospective hindbrain region
29. torn edge of vitelline (omphalomesenteric) artery
30. caudal extension of hindgut diverticulum
31. neural plate in caudal part of tail region
32. caudal part of first branchial arch (future maxillary component)
33. pericardial cavity
34. myocardial wall in bulbo-truncal junction region of primitive heart tube
35. aortic sac
36. lateral extent of body wall in continuity with amnion
37. pharyngeal region of foregut diverticulum
38. endocardial lining of primitive heart tube
39. gap between endocardial and myocardial tissue filled with cardiac jelly
40. body wall overlying pericardial cavity
41. bulbus cordis region of primitive heart tube
42. entrance to hindgut diverticulum
43. extra-embryonic coelomic cavity (exocoelom)
44. myocardium in ventricular region of primitive heart tube
45. bulbo-ventricular groove
46. mesentery of heart (dorsal mesocardium)
47. condensation of tissue (poorly defined) in caudal region of notochord
48. foregut diverticulum
49. right horn of sinus venosus
50. intra-embryonic coelomic cavity (future right pericardio-peritoneal canal)
51. endodermal lining of gut in region of prospective midgut–hindgut junction
52. splitting of lateral plate mesoderm to form splanchnopleure and somatopleure
53. visceral yolk sac
54. intra-embryonic coelomic cavity (future left pericardio-peritoneal canal)
55. left horn of sinus venosus
56. right vitelline (omphalomesenteric) vein
57. notochordal plate
58. entrance to foregut diverticulum
59. endodermal lining of prospective midgut region
60. rostral extremity of neural tube closure
61. somite
62. body wall in lordotic region of embryo
63. neural lumen

THE UROGENITAL SYSTEM
Histochemical staining of embryonic and extra-embryonic tissues for the presence of intracellular alkaline phosphatase enzyme activity, revealed that in embryos with 1–10 pairs of somites, 16% of primordial germ cells were located near the base of the allantois and in the yolk sac, while about 84% were located in association with the hindgut endoderm (see Table 4).

Table 4 Location of origin and timing of migration of primordial germ cells in the mouse

Developmental stage	Time (days p.c.)	Yolk sac mesoderm/ epiblast	Base of allantois/ caudal end of primitive streak	Hindgut endoderm	Hindgut mesentery/ coelomic wall	Gonadal ridge	Human (days p.c.)
Early primitive streak (amnion present)	7–7.5	0–50 alk. phos. +ve cells					18
Advanced primitive streak (early allantois	7.5–8	50–150 alk. phos. +ve cells					
1–10 somites	8–8.5		16%	84%			20
11–20 somites	8.5–9.5		1%	92%	7%		24
21–30 somites	9.5–10		1%	67%	30%	2%	26
31–36 somites	10–10.5			6%	86%	8%	27–29

Mouse data: Tam and Snow (1981); Snow and Monk (1983); Ginsburg *et al.* (1990).
Alk. phos.: alkaline phosphatase.

Stage 13 Developmental age, 8.5 days p.c. Plates 12, 13a, 13b; 14a, 14b; Plates 15a, 15b and 16a, 16b illustrate advanced Stage 13/early Stage 14 embryos (see pp. 60–67)

EXTERNAL FEATURES

Embryos with 8–12 pairs of somites. The process of "turning" is initiated in embryos with about 6–8 pairs of somites, and the process is normally completed by the time the embryo possesses about 14–16 pairs of somites (a detailed description of the process of "turning" and associated inversion of the germ layers is presented in pages 11–15 of the Introduction). A second branchial arch is now evident, and separated from the maxillary component of the first branchial arch by the first branchial groove (cleft) (*Plate 13a,g*; *Plate 14a,g,h*). The mandibular component of the first branchial arch is first evident in embryos with about 8–10 pairs of somites, and appears initially to be continuous with the maxillary component of the first branchial arch (*Plate 12,d*; *Plate 13a,g*; *Plate 14a,f–h*). The otic placodes are first evident as circumscribed regions of columnar ectoderm located on either side of the developing hindbrain region, at the level of rhombomere B, in embryos with about 9–11 pairs of somites (*Plate 13a,h*). In more advanced embryos, the central region of the placode indents to form the otic pit (*Plate 14a,j,k*; *Plate 15a,j,k*). In advanced Stage 13 embryos, a ridge runs along the lateral part of the body wall, and is subsequently associated with the development of the limb buds (extremitätenleiste (*Plate 15b,g*)).

THE HEART AND VASCULAR SYSTEM

The heart is now an asymmetrical globular structure in embryos with 8–10 pairs of somites, and subsequently becomes more S-shaped with clear evidence of regionalization along its length. It is divided into a common atrial chamber (*Plate 13b,c*; *Plate 14b,c*) which receives blood from the right and left horns of the sinus venosus (*Plate 13b,d*; *Plate 14b,d*), a common ventricular chamber which is in direct continuity with the bulbus cordis region of the primitive heart (*Plate 14b,a*), and this in turn is in direct continuity with the outflow tract of the heart. The latter consists of the truncus arteriosus and aortic sac, though it is not possible to distinguish on histological analysis between these regions of the outflow tract. The right and left components of the primitive common atrial chamber are initially bilaterally symmetrical, but this region of the primitive heart becomes progressively directed towards the left side (*Plate 14b,a–c*; *Plate 15a,m,n*). The primitive ventricle (the future left ventricle) also becomes directed towards the left side of the pericardial cavity, while the bulbus cordis region (the future right ventricle) becomes located on the right side of the pericardial cavity. Since the proximal part of the outflow tract (the bulbo-truncal junction) is to the right of the midline, and the aortic sac in the midline, the outflow tract is therefore directed rostrally and medially. A small segment of the dorsal mesocardium breaks down between the inflow and outflow regions of the primitive heart tube, thus creating the transverse pericardial sinus (*Plate 14b,a,b*; *Plate 15a,l–n*). The heart is the first organ system to differentiate and to function, and at this stage of development is the most prominent of all of the organ systems present in the embryo. Since the embryonic and extra-embryonic circulations amalgamate in embryos with about 8–10 pairs of somites, the embryonic vasculature is now seen to contain primitive nucleated red blood cells (*Plate 13a,e*; *Plate 14a,e,f*). It is also possible to recognize both the umbilical and vitelline vessels in the more advanced embryos at this stage of development (*Plate 15b,c*; *Plate 16a,e*).

THE PRIMITIVE GUT

The hindgut diverticulum is well defined in embryos with 8–10 pairs of somites, and extends almost to the caudal extremity of the tail (*Plate 12,f*; *Plate 13a,c*). In the primitive foregut region, the buccopharyngeal membrane is first evident in embryos with about 8–10 pairs of somites (*Plate 12,d*). The first branchial pouches become enormously dilated and their lateral extremities become separated from the first branchial clefts by the differentiation of the first branchial membranes (*Plate 13a,g*; *Plate 14a,h*). The first indication of the thyroid rudiment is seen at this stage of development, and is in the form of a thickened region of cells on the ventral aspect of the floor of the future pharynx (*Plate 13a,i–k*). The first evidence of the septum transversum is also seen in embryos with 8–10 pairs of somites (*Plate 12,n*), though it is more clearly evident at slightly later stages of development (*Plate 13b,f*; *Plate 14b,e*) in close proximity to the entrance to the foregut diverticulum. The hepatic diverticulum, which originates at the foregut–midgut junction, grows into the septum transversum (*Plate 14b,e*; *Plate 15a,p*; *Plate 15b,a*), and is destined to give rise to the intra- and extra-hepatic components of the biliary system.

THE NERVOUS SYSTEM

Two principal changes occur during this relatively short period, namely in relation to the postcephalic neural axis, and in relation to the rapid changes that occur in the cephalic region. An increase in the extent of neural tube closure along the neural axis is evident, extending caudally from a level opposite the outflow tract of the heart (*Plate 13a,k*; *Plate 14a,k*) to the proximal part of the tail (*Plate 13b,c,d*; *Plate 14b,f*). The profile of the closed segment of the neural tube is conspicuously flattened from side to side, and the lumen consequently has along the majority of its length the appearance of a narrow slit (eg *Plate 13b,a–c*). The notochord extends caudally from the mid-cephalic region (*Plate 13a,d,f*). Along the majority of its length it appears to have a relatively small profile, but it suddenly becomes quite diffuse in the

(*continued on page 58*)

PLATE 12

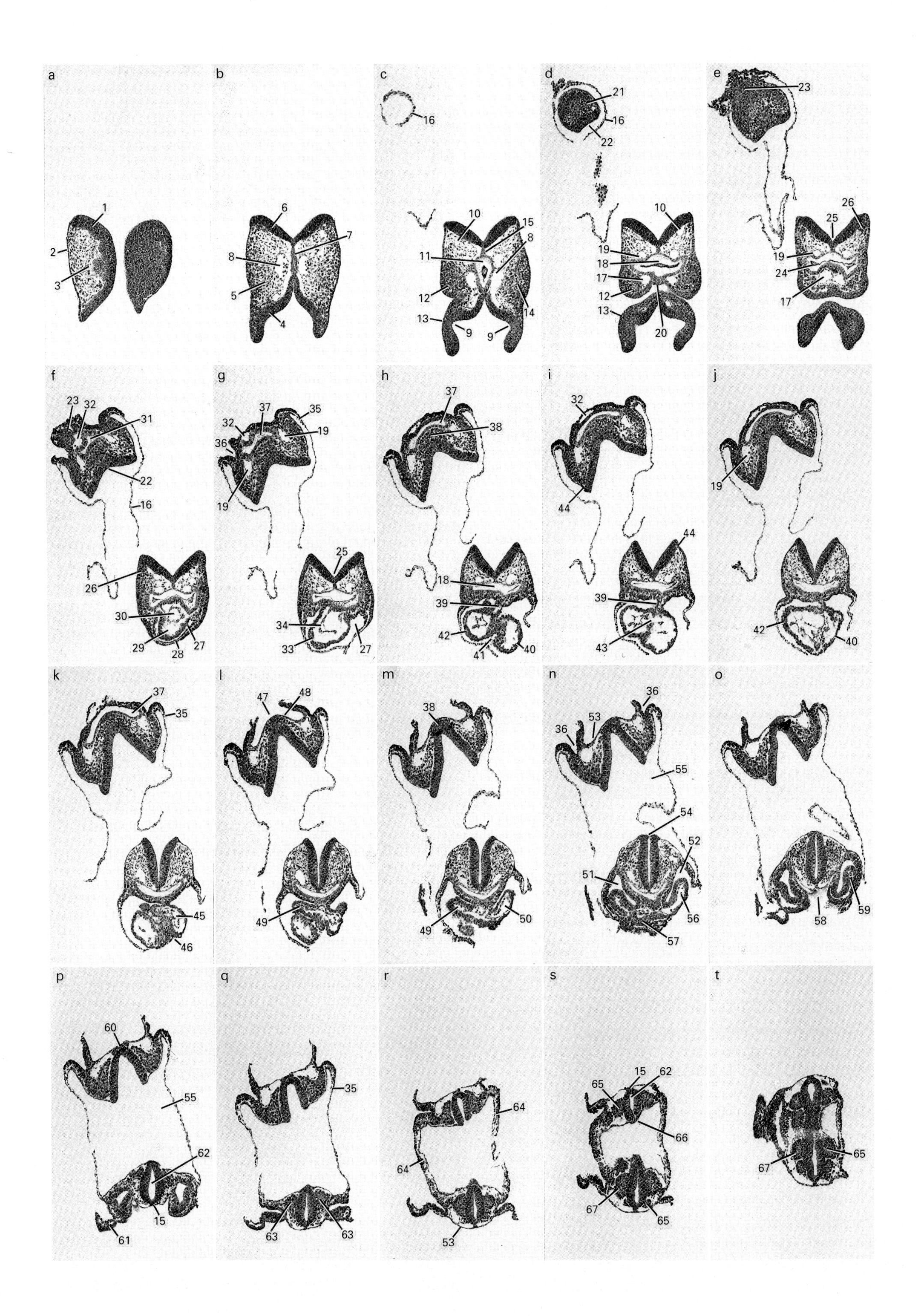
a
1
2
3
b
6
7
8
5
4
c
16
10
15
11
8
12
14
13
9
9
d
21
16
22
10
19
18
17
12
13
20
e
23
26
25
19
24
17
f
23
32
31
22
16
26
30
29
28
27
g
37
35
32
19
36
19
25
34
33
27
h
37
38
18
39
42
41
40
i
32
44
44
39
43
j
19
42
40
k
37
35
45
46
l
47
48
49
m
38
49
50
n
36
36
53
55
54
52
51
56
57
o
58
59
p
60
55
62
15
61
q
35
63
63
r
64
64
53
s
15
62
65
66
67
65
t
65
67

Plate 12 (8–8.5 days p.c.)

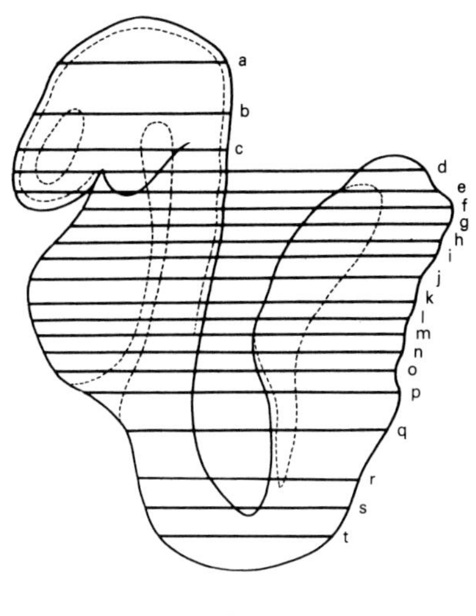

8–8.5 days p.c. Embryo with 8–10 pairs of somites, which has just initiated the turning sequence. Transverse sections (Theiler Stage 13; rat, Witschi Stage 15; human, Carnegie Stage 10).

1. neuroepithelium
2. surface ectoderm
3. cephalic mesenchyme tissue
4. prospective forebrain region (prosencephalon)
5. cephalic mesenchyme tissue in prospective midbrain region (mesencephalon)
6. neuroepithelium in prospective hindbrain region (rhombencephalon)
7. rostral extremity of notochord
8. communication between first branchial arch artery and dorsal aorta
9. optic pit (optic evagination)
10. primary head vein
11. rostral extremity of foregut diverticulum
12. first branchial arch (future maxillary component which is continuous anteroventrally with the mandibular component)
13. optic eminence
14. trigeminal (V) neural crest
15. notochord
16. amnion
17. first branchial (pharyngeal) arch artery
18. pharyngeal region of foregut diverticulum
19. dorsal aorta
20. buccopharyngeal membrane (intact)
21. caudal extremity of embryo (future tail region)
22. caudal extremity of neural plate (caudal neuropore)
23. base of allantois
24. first branchial (pharyngeal) pouch
25. neural groove
26. facial (VII) neural crest
27. pericardial cavity
28. body wall overlying pericardial cavity
29. truncus arteriosus region of outflow tract of primitive heart
30. aortic sac
31. caudal extremity of hindgut diverticulum
32. vitelline (omphalomesenteric) artery
33. myocardium in junctional zone between bulbus cordis and truncus arteriosus
34. endocardial tissue
35. lateral extent of body wall, in continuity with amnion
36. extra-embryonic coelom in region of caudal part of future peritoneal cavity
37. hindgut diverticulum
38. condensation of tissue (poorly defined) in caudal region of notochord
39. mesentery of heart (dorsal mesocardium)
40. primitive ventricle (future left ventricle)
41. bulbo-ventricular groove
42. bulbus cordis (future right ventricle)
43. bulbo-ventricular canal
44. boundary zone between surface and neural ectoderm-location of pre-neural crest tissue
45. common atrial chamber of primitive heart
46. atrio-ventricular groove
47. entrance to hindgut diverticulum
48. endodermal lining of gut in region of prospective hindgut–midgut junction
49. right horn of sinus venosus
50. left horn of sinus venosus
51. intra-embryonic coelomic cavity (future right pericardio-peritoneal canal)
52. intra-embryonic coelomic cavity (future left pericardio-peritoneal canal)
53. endodermal lining of prospective midgut region
54. rostral extremity of neural tube closure
55. amniotic cavity
56. nucleated primitive red blood cells within embryonic vasculature
57. septum transversum
58. entrance to foregut diverticulum
59. right vitelline (omphalomesenteric) vein
60. notochordal plate
61. visceral yolk sac
62. neural lumen
63. most rostral clearly defined somite (first somite)
64. body wall in lordotic region of embryo
65. somite
66. caudal extremity of neural tube closure
67. myocoele

PLATE 13a

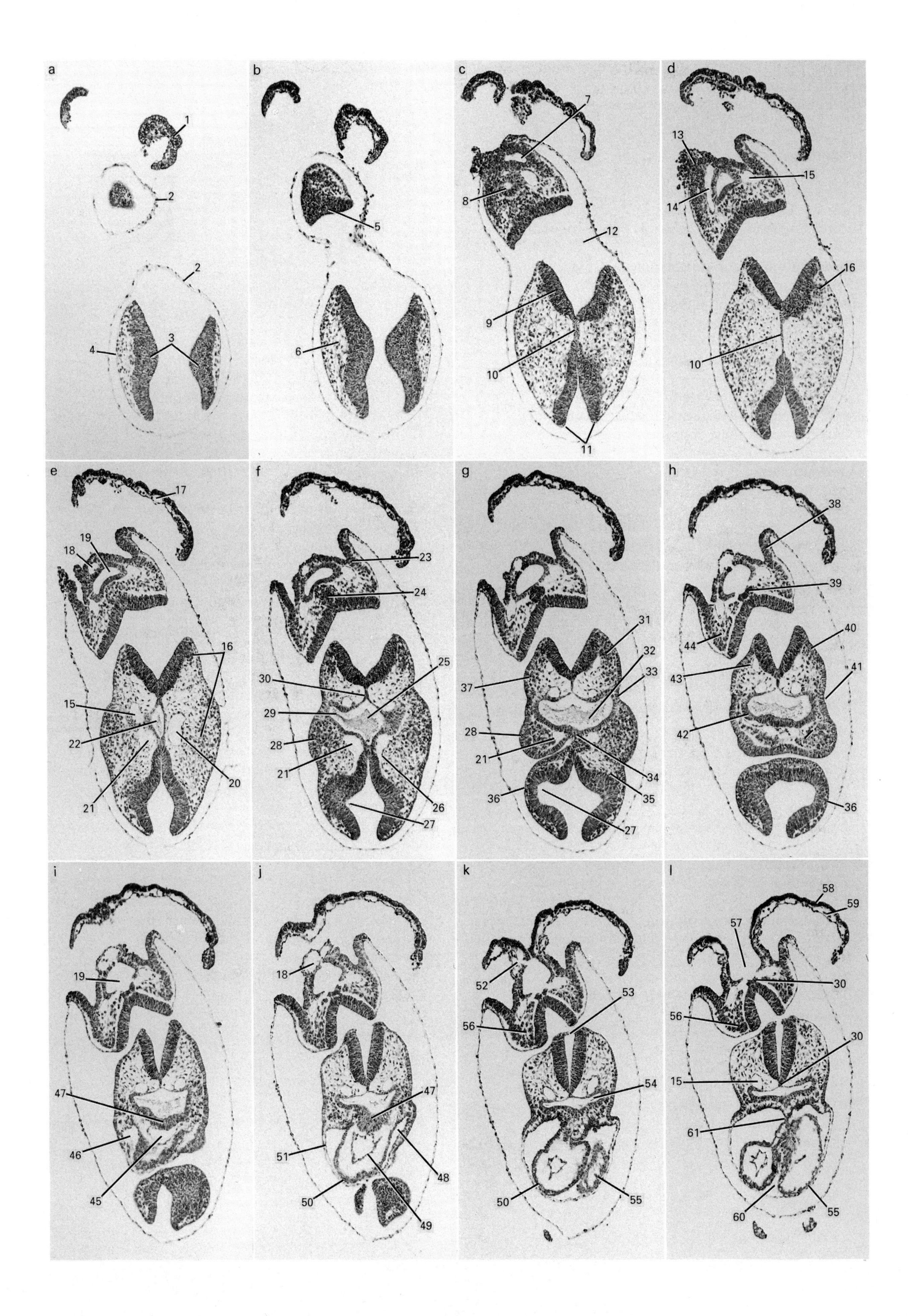

a
1
2
2
3
4
b
5
6
c
7
8
12
9
10
11
d
13
15
14
16
10
e
17
19
18
16
15
22
21
20
f
23
24
25
30
29
28
21
26
27
g
31
32
33
37
28
21
34
35
36
27
h
38
39
40
44
43
41
42
36
i
19
47
46
45
j
18
47
51
48
50
49
k
52
53
56
54
50
55
l
58
59
57
30
56
30
15
61
60
55

Plate 13a (8–8.5 days p.c.)

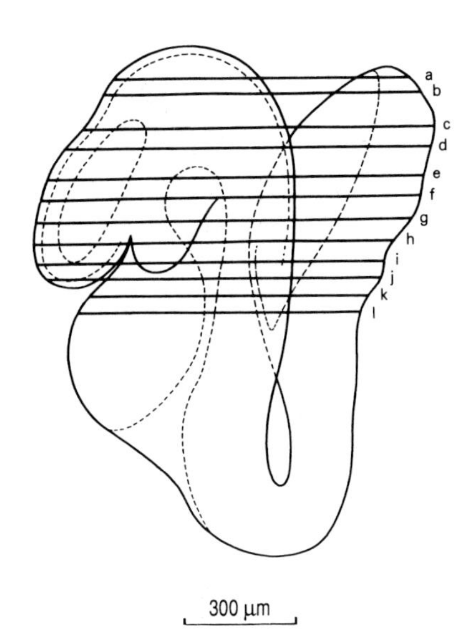

8–8.5 days p.c. Embryo with 9–11 pairs of somites, which has just initiated the turning sequence. Transverse sections (Theiler Stage 13; rat, Witschi Stage 15; human, Carnegie Stage 10).

1. visceral yolk sac
2. amnion
3. neuroepithelium in prospective midbrain region
4. surface ectoderm
5. neuroepithelium (neural plate) in primitive streak region
6. cephalic mesenchyme tissue
7. caudal extension of intra-embryonic coelomic cavity
8. caudal extension of hindgut diverticulum
9. neuroepithelium in prospective hindbrain region
10. rostral extemity of notochord
11. forebrain neural folds
12. amniotic cavity
13. base of allantois
14. communication between dorsal aorta and vitelline (omphalomesenteric) artery
15. dorsal aorta
16. trigeminal (V) neural crest
17. vascular plexus (blood island) within mesodermal component of visceral yolk sac
18. vitelline (omphalomesenteric) artery
19. hindgut diverticulum
20. communication between dorsal aorta and first branchial arch artery
21. first branchial arch artery
22. rostral extremity of foregut diverticulum
23. splitting of lateral plate mesoderm to form splanchnopleure and somatopleure
24. condensation of tissue (poorly defined) in caudal region of notochord
25. foregut diverticulum
26. branch from first branchial arch artery to perioptic vascular plexus
27. optic pit (optic evagination)
28. first branchial arch (future maxillary component which is continuous anteroventrally with the mandibular component)
29. first branchial pouch
30. notochord
31. facio-acoustic (VII–VIII) neural crest
32. eosin-positive non-cellular material characteristically found within the lumen of the foregut diverticulum
33. first branchial membrane (future tympanic membrane)
34. intact buccopharyngeal membrane
35. perioptic neural crest tissue
36. optic eminence
37. second branchial arch
38. body wall in tail region of embryo
39. notochordal plate
40. otic placode
41. first branchial cleft
42. layer of cuboidal endodermal cells lining the first branchial pouch
43. primary head vein (rostral extension of anterior cardinal vein)
44. paraxial (presomite) mesoderm
45. aortic sac giving rise to first branchial arch arteries
46. rostral extremity of pericardial cavity
47. endodermal thickening, location of thyroid primordium
48. pericardial cavity
49. endocardial lining of truncal region of primitive heart
50. wall of bulbus cordis region of primitive heart
51. thoracic wall overlying pericardial cavity
52. communication between vitelline artery and yolk sac circulation
53. rostral extremity of neural tube closure
54. second branchial pouch
55. wall of common ventricular chamber of primitive heart
56. condensation of paraxial mesoderm to form somite
57. entrance to hindgut diverticulum
58. extra-embryonic endodermal component of visceral yolk sac
59. extra-embryonic mesodermal component of visceral yolk sac with associated yolk sac vascular plexus and blood islands
60. bulbo-ventricular sulcus
61. mesentery of heart (dorsal mesocardium)

PLATE 13b

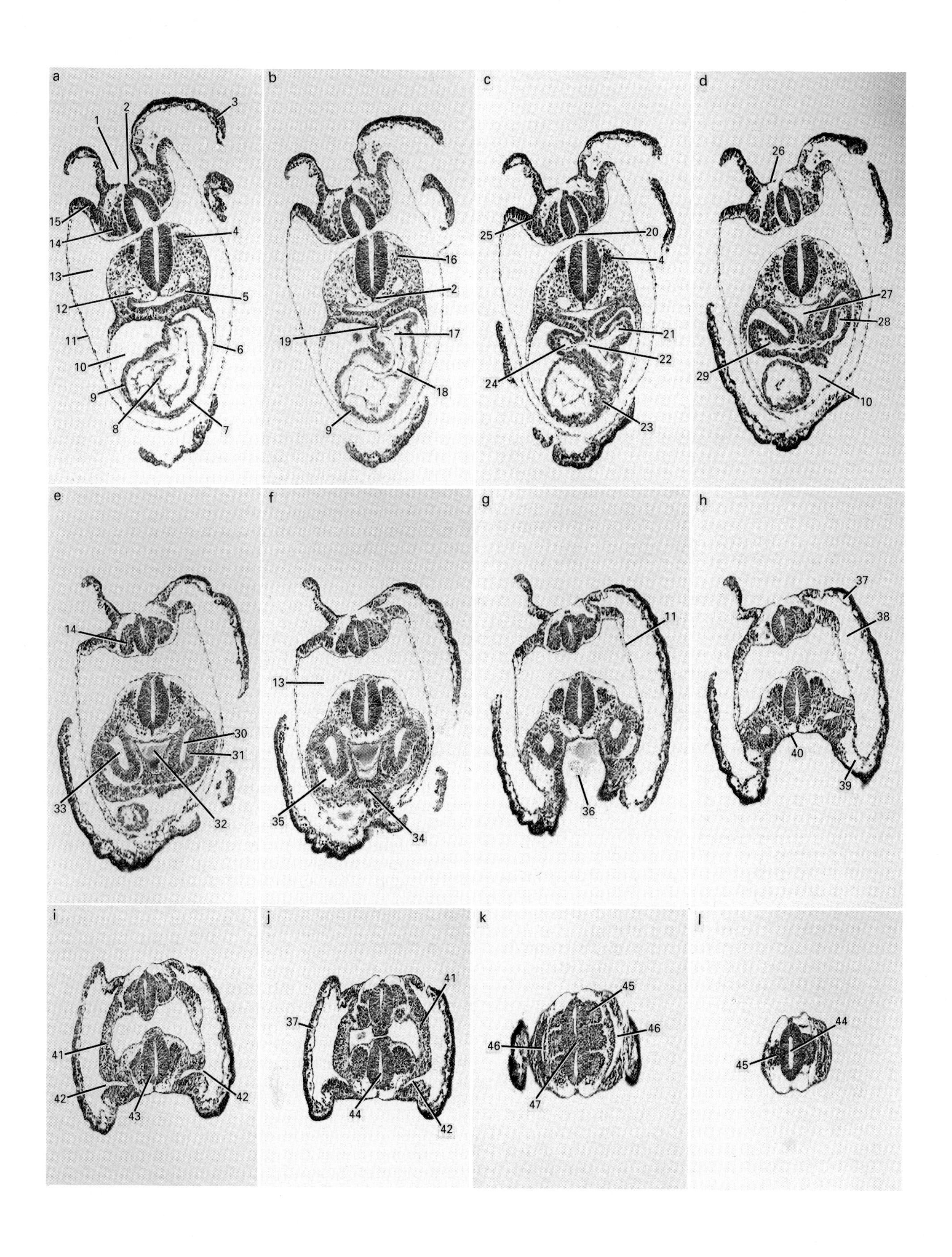

a
b
c
d
e
f
g
h
i
j
k
l
1
2
3
4
5
6
7
8
9
10
11
12
13
14
15
16
17
18
19
20
21
22
23
24
25
26
27
28
29
30
31
32
33
34
35
36
37
38
39
40
41
42
43
44
45
46
47

Plate 13b (8–8.5 days p.c)

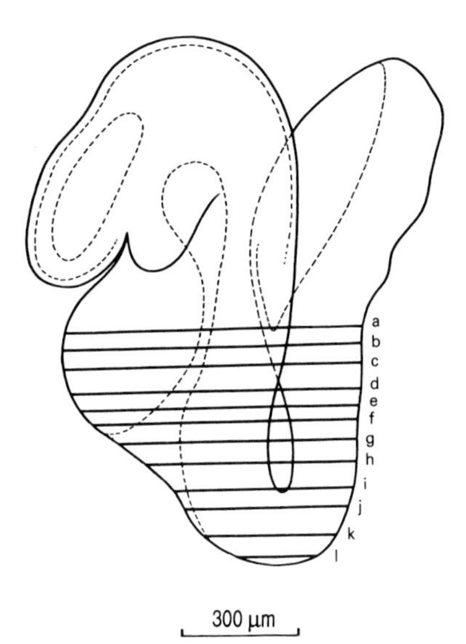

8–8.5 days p.c. Embryo with 9–11 pairs of somites, which has just initiated the turning sequence. Transverse sections (Theiler Stage 13; rat, Witschi Stage 15; human, Carnegie Stage 10).

1. entrance to hindgut diverticulum
2. notochord
3. visceral yolk sac
4. presomite aggregation of paraxial mesoderm
5. second branchial pouch
6. body wall overlying pericardial cavity
7. myocardium in region of common ventricular chamber of primitive heart
8. endocardium in region of bulbo-ventricular canal
9. myocardium in bulbus cordis region of primitive heart
10. pericardial cavity
11. amnion
12. dorsal aorta
13. amniotic cavity
14. condensation of paraxial mesoderm to form caudally located somite
15. body wall in mid-tail region of embryo
16. paraxial mesoderm in interzone region between somite condensations
17. left side of common atrial chamber of heart
18. atrio-ventricular canal
19. mesentery of heart (dorsal mesocardium)
20. caudal extremity of neural tube closure
21. opening of left horn of sinus venosus into common atrial chamber of heart
22. common atrial chamber of heart
23. trabeculation in myocardial wall of common ventricular chamber of primitive heart
24. opening of right horn of sinus venosus into common atrial chamber of heart
25. splitting of lateral plate mesoderm to form splanchnopleure and somatopleure
26. endodermal lining of gut at midgut–hindgut junction
27. caudal region of foregut diverticulum
28. left horn of sinus venosus
29. right horn of sinus venosus
30. pericardio-peritoneal canal
31. thickened (columnar) layer of coelomic mesothelial cells lining left pericardio-peritoneal canal region of intra-embryonic coelom
32. eosin-positive non-cellular material characteristically found within the foregut diverticulum
33. right pericardio-peritoneal canal
34. septum transversum
35. right umbilical vein
36. entrance to foregut diverticulum
37. vascular plexus (blood island) within mesodermal component of visceral yolk sac
38. extra-embryonic coelomic cavity
39. communication between vitelline vein and yolk sac circulation
40. endodermal lining of future midgut region
41. body wall in mid-lordotic region of embryo
42. site of communication between intra- and extra-embryonic coelomic cavities
43. neural tube in mid-trunk region of embryo
44. neural lumen
45. somite
46. intra-embryonic coelomic cavity (region of future peritoneal cavity)
47. mid-lordotic region of neural tube

PLATE 14a

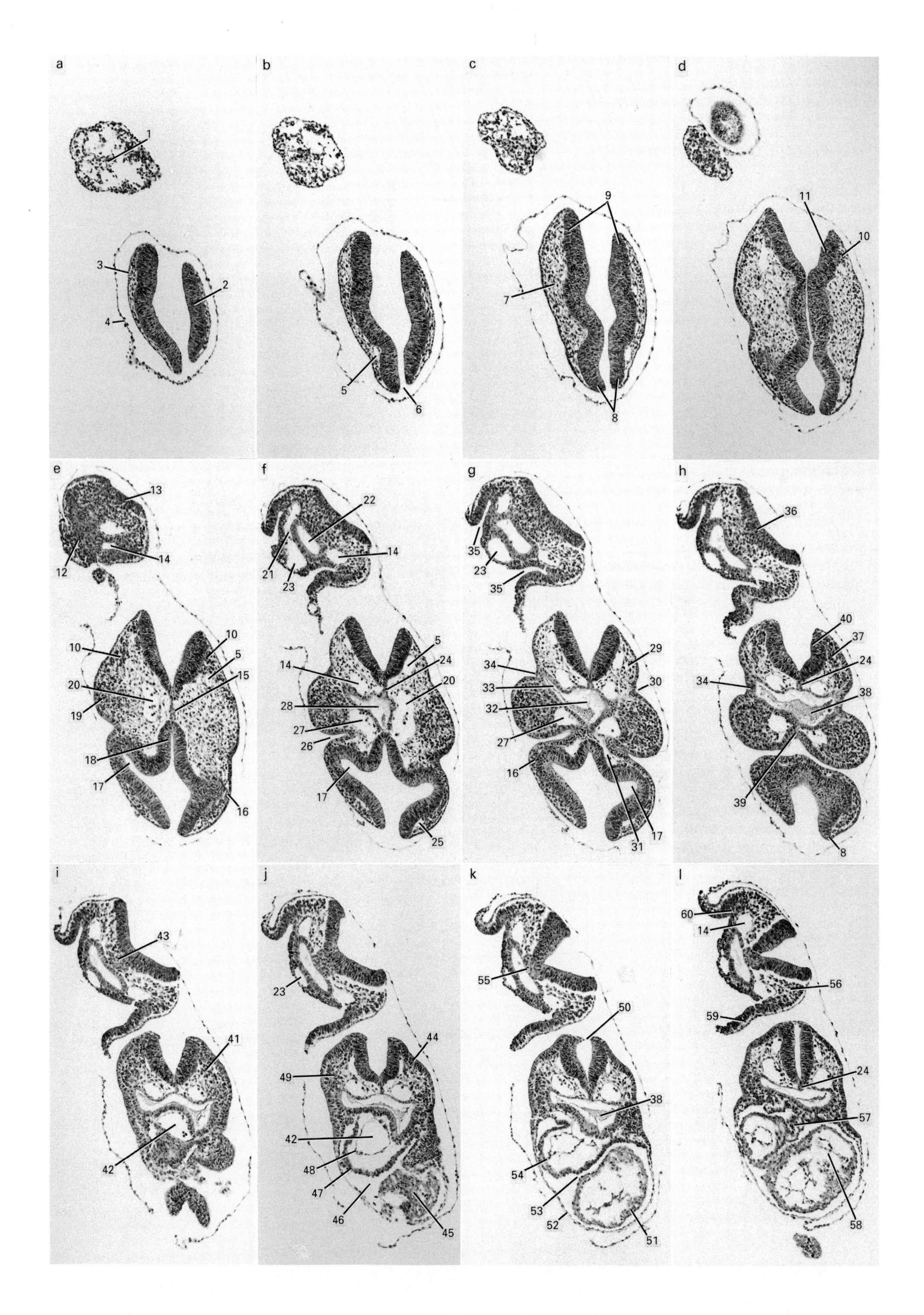

a
1
3
2
4
b
5
6
c
9
7
8
d
11
10
e
13
12
14
10
10
5
20
15
19
18
17
16
f
22
21
14
23
14
5
24
20
28
27
26
17
25
g
35
23
35
29
34
33
30
32
27
16
31
17
h
36
40
37
34
24
38
39
8
i
43
41
42
j
23
44
49
42
48
47
46
45
k
55
50
38
54
53
52
51
l
60
14
56
59
24
57
58

Plate 14a (8–8.5 days p.c.)

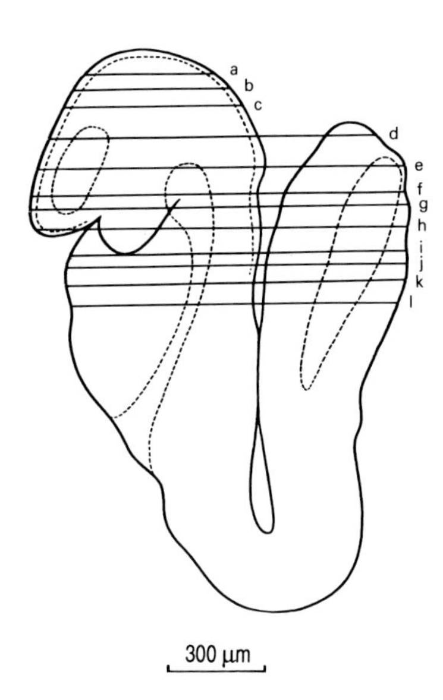

8–8.5 days p.c. Embryo with 10–12 pairs of somites, which is in the middle of the turning sequence. Transverse sections (Theiler Stage 13; rat, Witschi Stage 15; human, Carnegie Stage 10).

1. allantois
2. neuroepithelium in prospective midbrain region
3. surface ectoderm
4. amnion
5. primary head vein (one branch of anterior plexus of veins)
6. junctional zone between prospective forebrain and midbrain regions
7. cephalic mesenchyme tissue
8. forebrain neural fold
9. hindbrain neural fold
10. trigeminal (V) neural crest cells
11. neuroepithelial prominence overlying trigeminal (V) neural crest tissue
12. base of allantois
13. neuroepithelial tissue (neural plate) at caudal extremity of primitive streak region
14. dorsal aorta
15. rostral extremity of notochord
16. optic eminence
17. optic pit (optic evagination)
18. prospective diencephalon
19. first branchial arch (future maxillary component which is continuous anteroventrally with the mandibular component)
20. communication between dorsal aorta and first branchial arch artery
21. communication between dorsal aorta and vitelline (omphalomesenteric) artery
22. caudal extremity of hindgut diverticulum
23. vitelline (omphalomesenteric) artery
24. notochord
25. branch of perioptic vascular plexus
26. groove between first branchial arch and optic eminence which extends rostrally from oropharyngeal region
27. first branchial arch artery
28. rostral extremity of foregut diverticulum
29. second branchial arch
30. first branchial groove (cleft)
31. branch from first branchial arch artery to perioptic vascular plexus
32. foregut diverticulum
33. first branchial pouch
34. first branchial membrane (future tympanic membrane)
35. communication between intra- and extra-embryonic coelomic cavities
36. neuroepithelium (neural plate) in primitive streak region
37. facio-acoustic (VII–VIII) neural crest tissue
38. pharyngeal region of foregut diverticulum containing characteristic eosin-staining non-cellular material
39. intact buccopharyngeal membrane
40. neuroepithelial prominence overlying facio-acoustic (VII–VIII) neural crest tissue
41. mesenchyme tissue within second branchial arch
42. aortic sac
43. condensation of tissue (poorly defined) in caudal region of notochord
44. otic placode
45. trabeculation in wall of common ventricular chamber of primitive heart
46. pericardial cavity
47. bulbo-truncal junction region of outflow tract of primitive heart
48. endocardial tissue
49. facial (VII) neural crest tissue
50. rostral extremity of neural tube closure
51. wall of common ventricular chamber of primitive heart
52. body wall overlying pericardial cavity
53. bulbo-ventricular groove
54. bulbus cordis region of primitive heart
55. notochordal plate
56. condensation of paraxial mesoderm to form somite
57. mesentery of heart (dorsal mesocardium)
58. atrio-ventricular canal
59. body wall in middle part of tail region
60. splitting of lateral plate mesoderm to form splanchnopleure and somatopleure

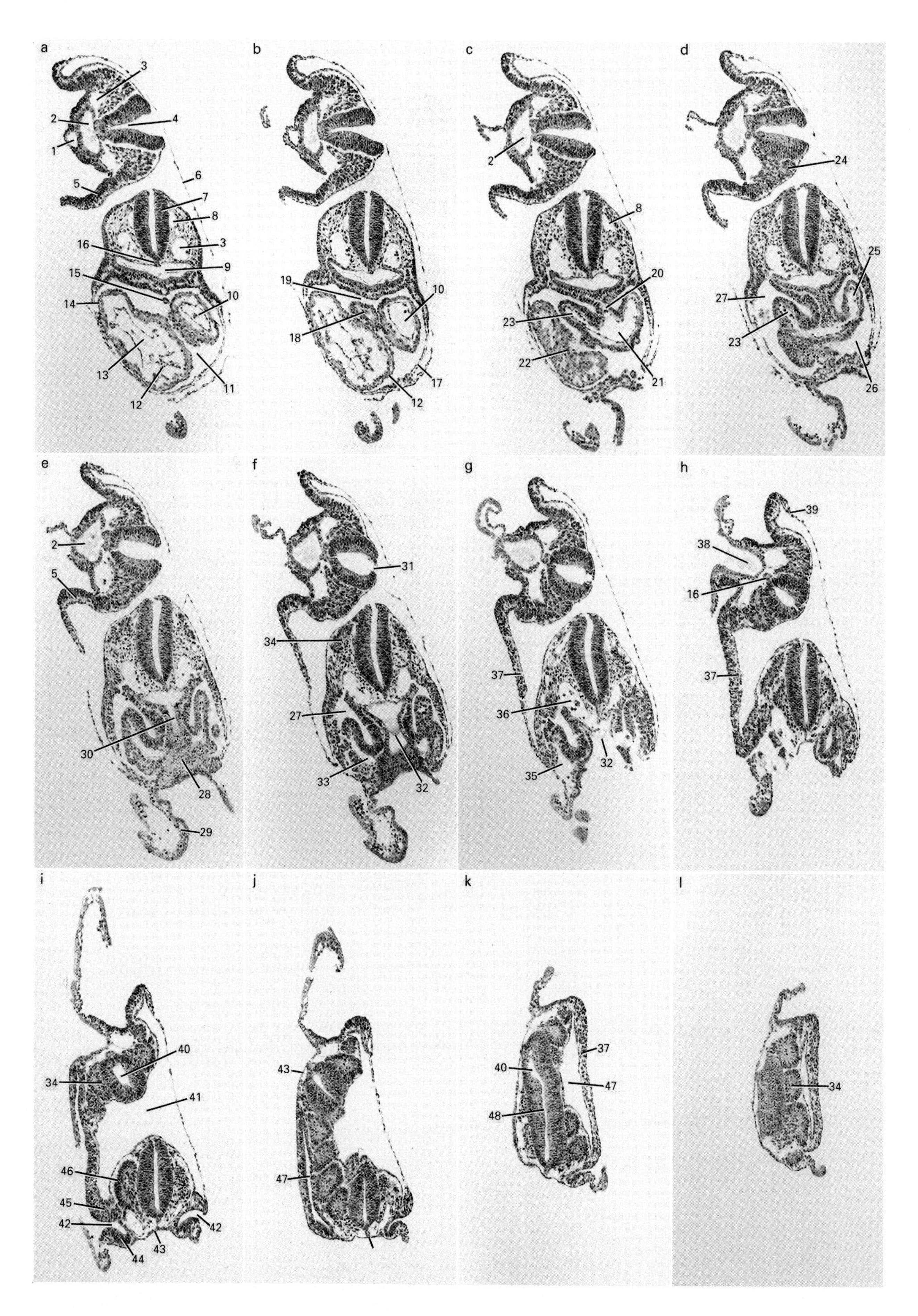
a
b
c
d
e
f
g
h
i
j
k
l
1
2
3
4
5
6
7
8
9
10
11
12
13
14
15
16
17
18
19
20
21
22
23
24
25
26
27
28
29
30
31
32
33
34
35
36
37
38
39
40
41
42
43
44
45
46
47
48

Plate 14b (8–8.5 days p.c)

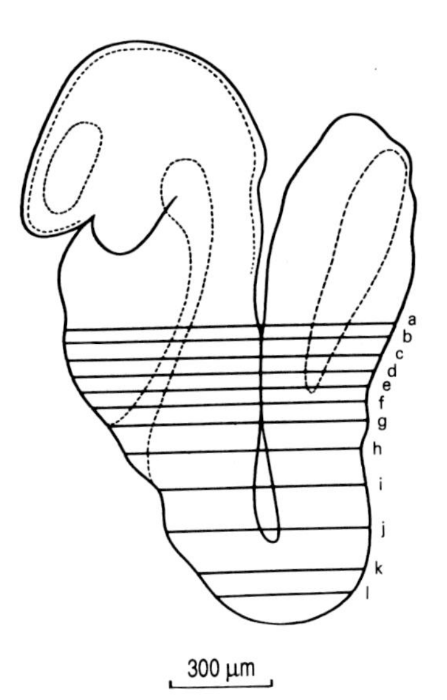

8–8.5 days p.c. Embryo with 10–12 pairs of somites, which is in the middle of the turning sequence. Transverse sections (Theiler Stage 13; rat, Witschi Stage 15; human, Carnegie Stage 10).

1. vitelline (omphalomesenteric) artery
2. hindgut diverticulum
3. dorsal aorta
4. neural groove in region of caudal neuropore
5. body wall in middle part of tail region
6. amnion
7. neural tube
8. anterior cardinal vein
9. pharyngeal region of foregut diverticulum
10. left side of common atrial chamber of primitive heart
11. pericardial cavity
12. common ventricular chamber of primitive heart
13. bulbo-ventricular canal
14. bulbus cordis region of primitive heart
15. small component of dorsal mesocardium
16. notochord
17. body wall overlying pericardial cavity
18. right side of common atrial chamber of primitive heart
19. transverse pericardial sinus (discontinuity in dorsal mesocardium)
20. ventral (intact) component of dorsal mesocardium
21. common atrial chamber of primitive heart
22. trabeculated myocardial wall in region of bulbo-ventricular canal
23. right horn of sinus venosus
24. condensation of paraxial mesoderm
25. left horn of sinus venosus
26. left pericardio-peritoneal canal
27. right pericardio-peritoneal canal
28. septum transversum just ventral to entrance to foregut diverticulum
29. visceral yolk sac
30. hepatic diverticulum (from region of foregut–midgut junction)
31. caudal extremity of neural tube closure
32. entrance to foregut diverticulum
33. right vitelline (omphalomesenteric) vein
34. somite
35. communication between right vitelline and right umbilical veins
36. nucleated primitive red blood cells within embryonic vasculature
37. body wall in mid-lordotic region of embryo
38. entrance to hindgut diverticulum
39. lateral extent of body wall, in continuity with amnion
40. neural lumen
41. amniotic cavity
42. site of communication between intra- and extra-embryonic coelomic cavities
43. endodermal lining of prospective midgut region
44. lateral plate mesoderm – splanchnic component (splanchnopleure)
45. lateral plate mesoderm – somatic component (somatopleure)
46. outer (dermatomyotome) component of somite
47. intra-embryonic coelomic cavity (future peritoneal cavity)
48. mid-lordotic region of neural tube

caudal region of the tail (*Plate 13a,f–h*). Similarly, along its entire length, it appears to be attached dorsally to the basement membrane of the neural plate/neural tube, and ventrally to the basement membrane of the dorsal aspect of the endodermal wall of the primitive gut tube (e.g. *Plate 13a,f–l*).

A conspicuous increase occurs in the overall volume of the cephalic region during this stage of development, particularly in relation to the maxillary component of the first branchial arch, and as a consequence of the first appearance of the second branchial arch. Substantial changes also occur in the shape of the forebrain neural folds, concomitant with an increase in the depth and volume of the optic pits (*Plate 13a,g,h*; *Plate 14a,e–g*). Towards the end of this stage of development, the cephalic neural folds approach the ventral midline at a site approximately overlying the junctional zone between the future forebrain and midbrain (*Plate 14a,d,e*; *Plate 15a,a–e*). The neural folds overlying the majority of the future forebrain region subsequently become apposed and fuse across the ventral midline towards the end of this stage, or during the early part of the next stage of development, to form the forebrain vesicle with its two principal evaginations, namely the (primary) optic vesicles (*Plate 16a,d,e*). The neural folds overlying the future midbrain region become apposed and subsequently fuse in embryos with about 14–15 pairs of somites, and the neural folds in the region overlying the future hindbrain become apposed and fuse shortly thereafter. The neural folds overlying the most inferior part of the ventral aspect of the future forebrain close in embryos with about 15–18 pairs of somites, and with their closure the last of the three primary brain vesicles, namely the forebrain vesicle, the midbrain (or mesencephalic) vesicle and the hindbrain vesicle, are now formed, and their walls form the prosencephalon, mesencephalon and rhombencephalon, respectively. As the overall pattern of closure of the cephalic neural folds in the mouse is so different from that observed in the human embryo, there is no exact equivalent in the mouse of the rostral neuropore described in the human embryo, and it is probably inappropriate to use this term in the mouse.

Stage 14 Developmental age, 9 days p.c. Plates 15a, 15b and 16a, 16b illustrate advanced Stage 13/early Stage 14 embryos; Plate 17, Plates 18a, 18b)

EXTERNAL FEATURES

Embryos with 13–20 pairs of somites. The principal difference between embryos assigned to Stage 14 compared with those in Stage 13 relates to the overall configuration of the body, so that during the early part of Stage 14 embryos complete the process of "turning". Equally important is the fact that the rostral extremity of the neural tube closes in embryos with about 15–18 pairs of somites (for further discussion on the pattern of closure of the cephalic neural folds, see Stage 13, *the nervous system*). However, the caudal extremity of the neural tube (i.e. the caudal neuropore), in the tail region, is still in the process of closing (*Plate 18a,a–e*), and only closes at about the time when the embryo possesses about 30–35 pairs of somites (*Plate 23c,a–h*).

During Stage 14, the otic pit becomes progressively more deeply indented (*Plate 17,g*; *Plate 18a,k–m*), but only becomes completely closed to form the otocyst (otic vesicle) and separated from the overlying surface ectoderm in embryos with about 25–30 pairs of somites (see *Plate 19a,l*, for illustration of almost completely formed otocyst, and *Plate 20b,d–f*, for fully formed otocyst). The first indication of a third branchial arch is evident towards the end of this stage (*Plate 18a,o*), and both a first and second branchial membrane are now clearly evident (see *Plate 18a,h–j*, and *Plate 18a,n,o*, for sections illustrating first and second branchial membranes, respectively). The mandibular component of the first branchial arch is particularly prominent during this stage of development (*Plate 18a,h–j*). The first evidence of a forelimb bud is seen in embryos with about 15–18 pairs of somites (*Plate 17,a–d*; *Plate 18b,n*), as an increasingly prominent ridge on the lateral part of the body wall in the mid-trunk region at about the level of somites 8–12.

THE HEART AND VASCULAR SYSTEM

In embryos with about 8–10 pairs of somites, the heart is seen to contract irregularly, but at this stage of development the heart beats regularly and powerfully. The functional activity of the heart is directly related to the degree of myofibrillogenesis seen at the ultrastructural level (Kaufman, 1981; Navaratnam *et al.*, 1986). With the amalgamation of the embryonic and yolk sac vascular systems (e.g. *Plate 16b,g,h*), and a significant increase in the number of primitive nucleated red blood cells, it is now possible to see red blood circulating throughout the embryonic and yolk sac vasculature. While it is possible to recognize small branches of the primary head veins in embryos with about 6–8 pairs of somites (e.g. *Plate 11,a–e*), towards the end of this stage of development these are now seen to drain into the quite substantial anterior cardinal veins (*Plate 17,h*; *Plate 18b, a*). The latter then drain into the common cardinal veins (ducts of Cuvier) (*Plate 18b,g–i*) which then flow into the left and right horns of the sinus venosus (*Plate 18b,f–j*). The umbilical veins also drain into the common cardinal veins (*Plate 18b,h,i*).

The most obvious change that occurs in the arterial system at this stage of development is the localized fusion of the right and left dorsal aortae in the caudal part of the trunk region of the embryo (the intervening endothelial walls of the two vessels break down) to form a single midline dorsal aorta in this location (*Plate 18b,l*). In more advanced embryos, the fusion process extends both rostrally and caudally from this initial site, and in early Stage 15 embryos, the midline dorsal aorta is seen to extend from the mid-trunk region (*Plate 19c,g*) caudally into the proximal part of the tail region (*Plate 19b,c*). At this stage, the heart appears to be the only region within the embryo where the endothelial elements of the circulation are surrounded by a vessel wall. While the thickness of the myocardial wall is fairly uniform in the previous stage, throughout this stage of development the walls of the common ventricular chamber in particular, and those of the bulbus cordis and common atrial chamber to a lesser extent, display an increasing degree of trabeculation. The lumen of the endothelial (i.e. endocardial) tube in the region of the bulbus cordis and outflow tract is particularly narrow (*Plate 18a, l-n*), and both here and elsewhere in the heart, the substantial is between the endothelial and myocardial elements are filled with very loose mesenchyme or reticulum, the so-called cardiac jelly.

THE PRIMITIVE GUT

Significant changes occur in the configuration of the primitive gut tube during this stage of development, principally as a consequence of the completion of "turning", so that the peripheral margin of the wide communication between the midgut and the yolk sac (in the region of the umbilical ring), seen in "unturned" embryos towards the end of the previous stage of development (*Plate 14b,h–j*), narrows dramatically, with the formation of the vitelline duct. In the rostral extremity of the foregut, the buccopharyngeal membrane breaks down to allow continuity to be established between the pharynx and the amniotic cavity (*Plate 17,a–c*; *Plate 18a,h–j*).

In the superior aspect of the oral cavity, just in front of the remnants of the buccopharyngeal membrane, a thickened and indented region of the ectoderm represents the first evidence of Rathke's pouch (*Plate 17,a*; *Plate 18a,g*). Towards the end of this stage, but particularly during the early part of the next stage of development, the midgut region has a uniform lumen throughout its length (*Plate 18b,l*; *Plate 19c,d,e*).

Just caudal to the bulbo-ventricular region of the primitive heart, the tissue of the septum transversum appears to be proliferating (*Plate 18b,f–i*), as does the wall of the hepatic diverticulum (*Plate 18b,g–i*). The caudal extremity of the hindgut diverticulum becomes increasingly dilated throughout this stage (*Plate 16a,c–f*;

(*continued on page 74*)

PLATE 15a

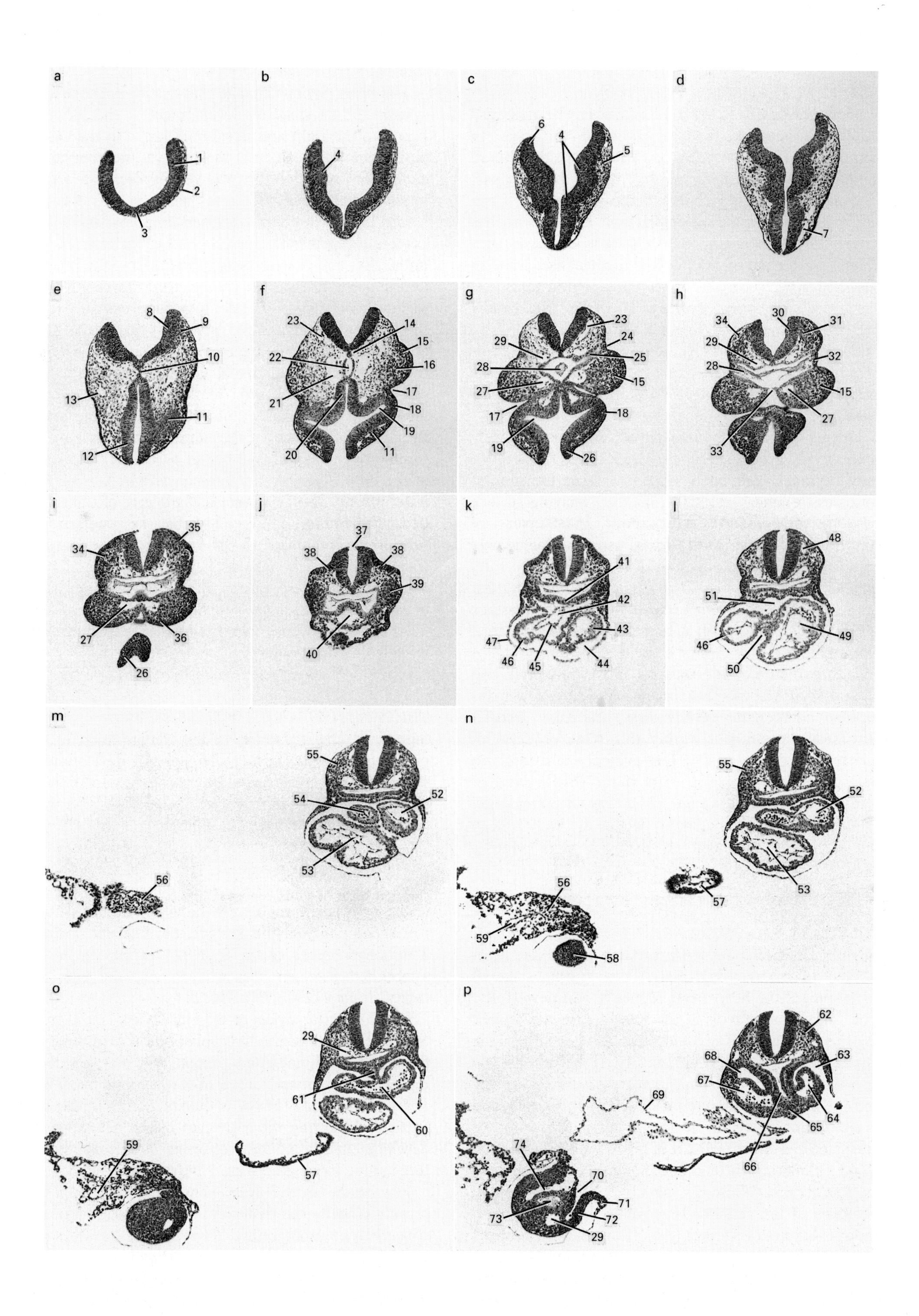

Plate 15a (8.5 days p.c.)

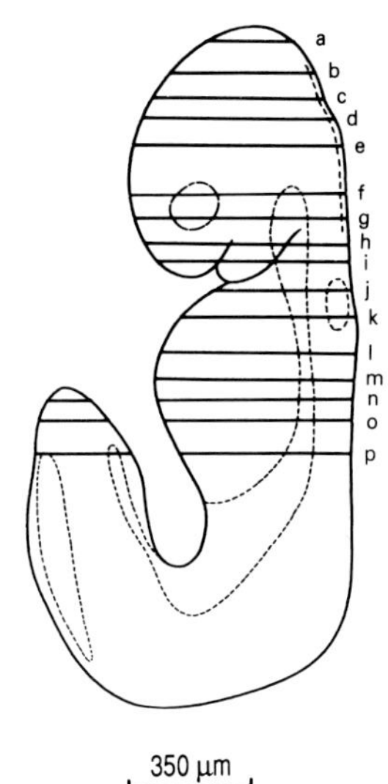

8.5 days p.c. Embryo with 11–13 pairs of somites, which has almost completed the process of turning. Transverse sections (Theiler Stages 13–14; rat, Witschi Stage 16; human, Carnegie Stages 10–11).

1. neuroepithelium in prospective midbrain region
2. surface ectoderm
3. site of fusion of neural folds at forebrain–midbrain junction
4. rhombomere
5. cephalic mesenchyme tissue
6. hindbrain neural fold
7. anterior branch of primary head vein (will give rise to anterior plexus of veins)
8. prominence overlying origin of trigeminal (V) neural crest
9. trigeminal (V) neural crest tissue
10. rostral extremity of notochord
11. perioptic neural crest tissue of trigeminal (V) neural crest origin
12. forebrain
13. medial branch of primary head vein (will give rise to medial plexus of veins)
14. notochord
15. first branchial arch (future maxillary component)
16. trigeminal (V) neural crest tissue within first branchial arch
17. groove between first branchial arch and optic eminence
18. optic eminence
19. optic pit (optic evagination)
20. diencephalon
21. communication between dorsal aorta and first branchial arch artery
22. rostral extension of foregut diverticulum
23. posterior branch of primary head vein (will give rise to posterior plexus of veins)
24. first branchial cleft (groove)
25. first branchial pouch
26. forebrain neural fold
27. first branchial arch artery
28. eosin-positive non-cellular material characteristically found within the lumen of the foregut diverticulum
29. dorsal aorta
30. prominence overlying origin of facio-acoustic (VII–VIII) neural crest
31. facio-acoustic (VII–VIII) neural crest tissue
32. first branchial membrane (future tympanic membrane)
33. buccopharyngeal membrane
34. second branchial arch
35. otic placode tissue at peripheral margin of otic pit
36. first branchial arch (future mandibular component)
37. rostral extremity of neural tube closure
38. otic pit
39. facial (VII) neural crest tissue within second branchial arch
40. aortic sac giving origin to first branchial arch arteries
41. location of thyroid primordium
42. truncus arteriosus/aortic sac region of outflow tract of primitive heart
43. trabeculation in myocardial wall of common ventricular chamber of primitive heart
44. pericardial cavity
45. endocardial tissue lining primitive heart
46. bulbus cordis region of primitive heart
47. body wall overlying pericardial cavity
48. anterior cardinal vein
49. atrio-ventricular canal
50. bulbo-ventricular groove
51. transverse pericardial sinus
52. left side of common atrial chamber of primitive heart
53. bulbo-ventricular canal
54. right side of common atrial chamber of primitive heart
55. origin of third branchial arch
56. allantois
57. visceral yolk sac
58. caudal extremity of primitive streak region
59. part of allantoic vascular plexus
60. common atrial chamber of primitive heart
61. mesentery of heart (dorsal mesocardium) caudal to transverse pericardial sinus
62. presomite condensation of paraxial mesoderm
63. left pericardio-peritoneal canal
64. left horn of sinus venosus
65. septum transversum
66. hepatic diverticulum (from region of foregut–midgut junction)
67. right horn of sinus venosus
68. right pericardio-peritoneal canal
69. amnion
70. communication between extra- and intra-embryonic coelomic cavities
71. left umbilical vein
72. body wall in distal part of tail region
73. caudal extension of hindgut diverticulum
74. communication between dorsal aorta and vitelline (omphalomesenteric) artery

PLATE 15b

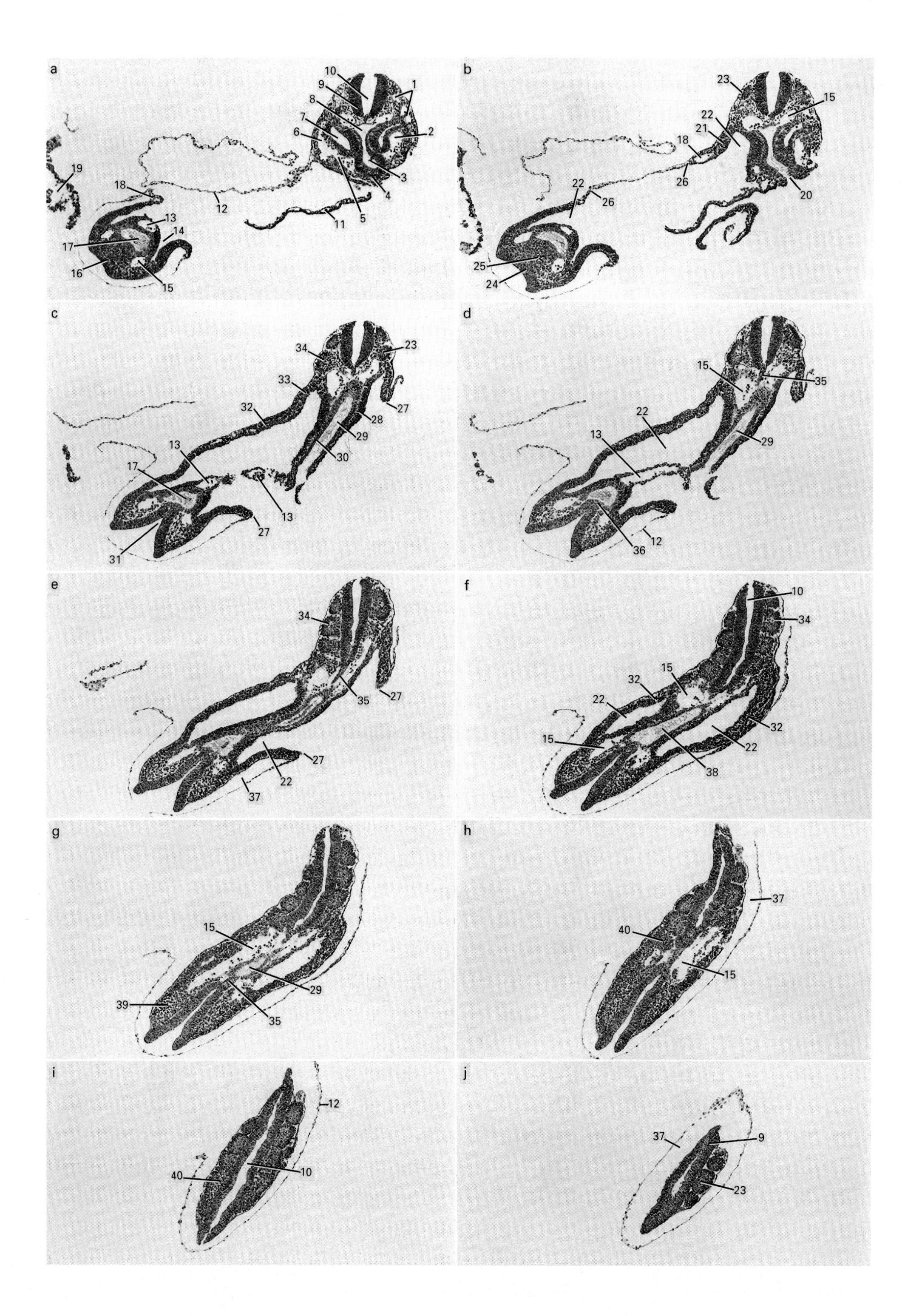
a
10
9
1
8
7
2
6
3
4
5
19
18
12
11
13
14
17
16
15
b
23
15
22
21
18
26
20
22
26
25
24
c
34
23
33
27
32
28
29
30
13
17
13
27
31
d
15
35
22
29
13
36
12
e
34
35
27
27
22
37
f
10
34
15
32
22
32
15
22
38
g
15
29
39
35
h
37
40
15
i
12
40
10
j
37
9
23

Plate 15b (8.5 days p.c.)

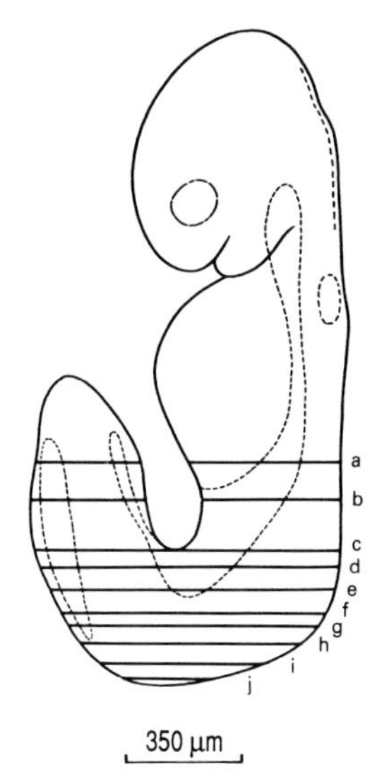

8.5 days p.c. Embryo with 11–13 pairs of somites, which has almost completed the process of turning. Transverse sections (Theiler Stages 13–14; rat, Witschi Stage 16; human, Carnegie Stages 10–11).

1. left anterior cardinal vein
2. left horn of sinus venosus
3. hepatic diverticulum from region of foregut–midgut junction
4. septum transversum
5. site of entry of common cardinal vein (duct of Cuvier) into right horn of sinus venosus
6. right anterior cardinal vein at site of communication with right common cardinal vein
7. right pericardio-peritoneal canal
8. foregut diverticulum
9. neural tube
10. neural lumen
11. visceral yolk sac
12. amnion
13. vitelline (omphalomesenteric) artery
14. communication between intra- and extra-embryonic coelomic cavities in primitive streak/tail region
15. dorsal aorta
16. neuroepithelium (neural plate) in primitive streak region
17. caudal extension of hindgut diverticulum
18. right umbilical vein
19. allantois
20. entrance to foregut diverticulum
21. body wall in mid-trunk region
22. intra-embryonic coelomic cavity (peritoneal cavity)
23. somite
24. neural plate (caudal part of caudal neuropore)
25. condensation of tissue (poorly defined) in caudal region of notochord
26. junction between body wall and amnion in middle region of embryo
27. boundary of umbilical ring
28. splanchnic component of lateral plate mesoderm (i.e. splanchnopleure) forming wall of midgut
29. lumen of the midgut
30. endodermal lining of midgut
31. neural groove in mid-tail region
32. body wall in mid-trunk region of the embryo
33. somatic component of lateral plate mesoderm (i.e. somatopleure) forming part of body wall musculature
34. myocoele
35. notochord
36. notochordal plate
37. amniotic cavity
38. midgut–hindgut junction
39. prominent lateral ridge – to be associated with development of the limb buds
40. presomite paraxial mesoderm

PLATE 16a

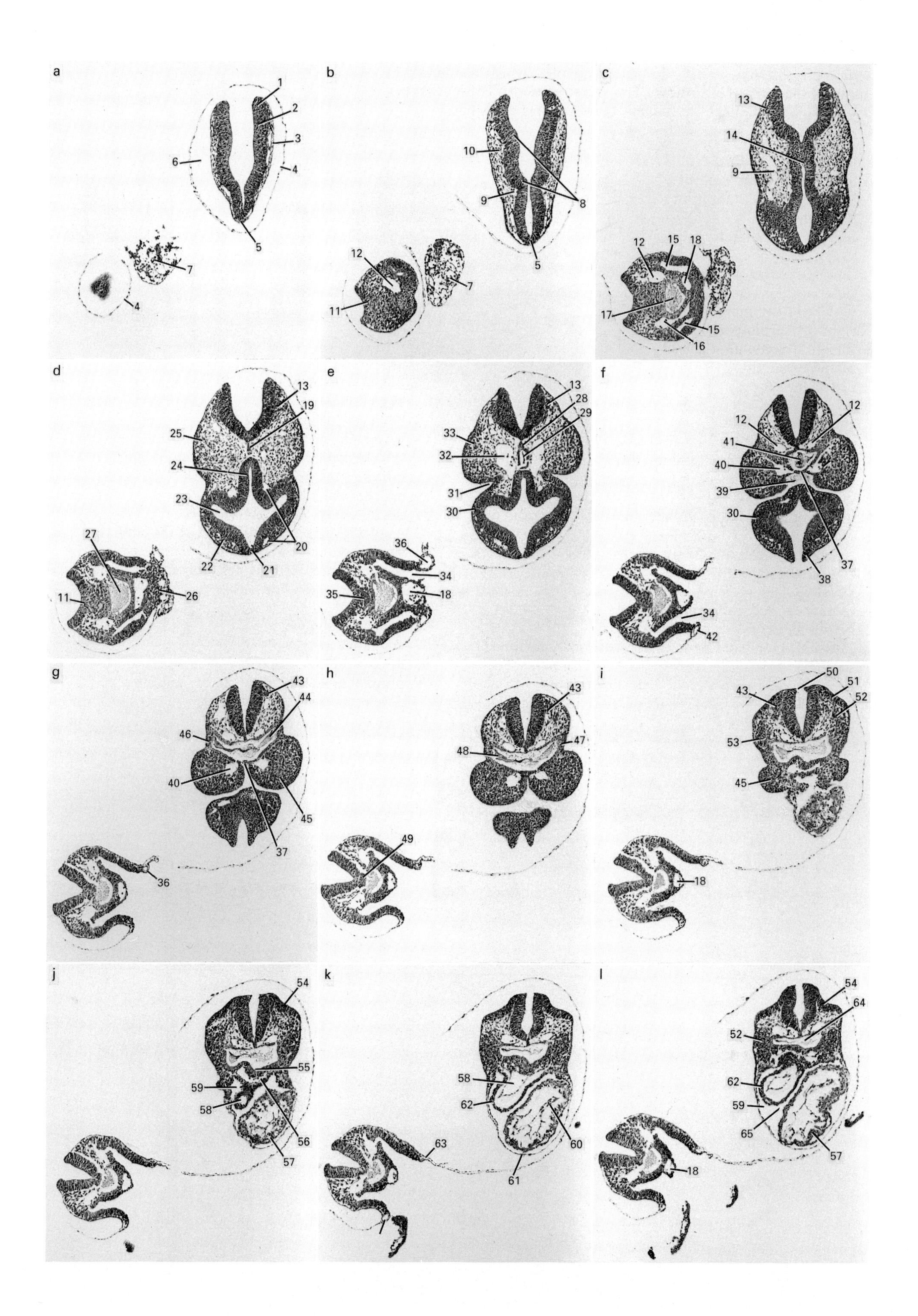
a
1
2
3
4
5
6
7
4
b
8
9
10
5
12
7
11
c
13
14
9
12
15
18
17
15
16
d
13
19
25
24
23
20
22
21
27
11
26
e
13
28
29
33
32
31
30
36
34
35
18
f
12
12
41
40
39
30
37
38
34
42
g
43
44
46
40
45
37
36
h
43
47
48
49
i
50
51
43
52
53
45
18
j
54
55
59
58
56
57
k
58
62
60
61
63
l
54
64
52
62
59
65
57
18

Plate 16a (8.5–9 days p.c.)

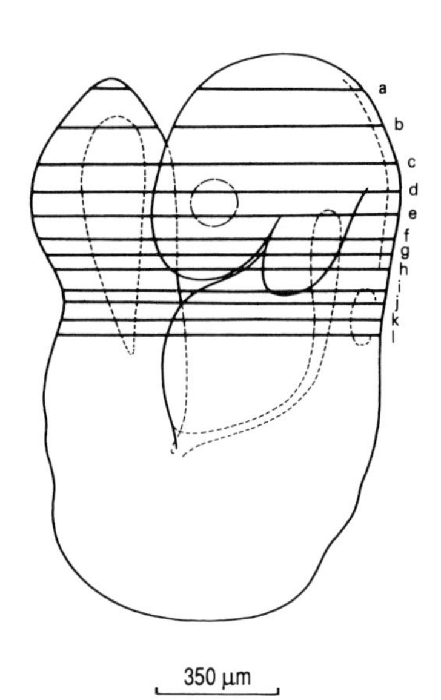

8.5–9 days p.c. Embryo with 12–14 pairs of somites, which has almost completed the process of turning. Transverse sections (Theiler Stages 13–14; rat, Witschi Stage 16; human, Carnegie Stages 10–11).

1. neural fold in prospective hindbrain region
2. neuroepithelium (neural ectoderm)
3. surface ectoderm
4. amnion
5. site of fusion of cephalic neural folds at junction between prospective forebrain and midbrain regions
6. amniotic cavity
7. allantois
8. rhombomere
9. part of venous plexus, branch of primary head vein
10. cephalic mesenchyme tissue
11. neuroepithelium (neural plate) in primitive streak region
12. dorsal aorta
13. trigeminal (V) neural crest cells
14. diencephalon (future thalamic region)
15. caudal diverticulum of intra-embryonic coelomic cavity
16. communication between dorsal aorta and vitelline artery
17. caudal extremity of hindgut diverticulum
18. vitelline (omphalomesenteric) artery
19. rostral extremity of notochord
20. perioptic vascular plexus
21. site of fusion of cephalic neural folds overlying the forebrain region
22. perioptic neural crest tissue
23. optic vesicle
24. diencephalon
25. origin of first branchial arch
26. base of allantois
27. hindgut diverticulum
28. notochord
29. rostral extremity of foregut diverticulum
30. optic eminence
31. groove between first branchial arch and optic eminence
32. communication between dorsal aorta and first branchial arch artery
33. first branchial arch (future maxillary component)
34. communication between intra- and extra-embryonic coelomic cavities
35. condensation of tissue (poorly defined) in caudal region of notochord
36. right umbilical vein
37. buccopharyngeal membrane
38. forebrain neural fold
39. oropharynx
40. first branchial arch artery
41. foregut diverticulum
42. left umbilical vein
43. facio-acoustic (VII–VIII) neural crest
44. first branchial pouch
45. first branchial arch (future mandibular component)
46. first branchial membrane (future tympanic membrane)
47. first branchial cleft
48. eosin-staining non-cellular material characteristically found in foregut diverticulum
49. notochordal plate
50. rostral extremity of neural tube closure
51. periphery of otic placode
52. facial (VII) neural crest tissue migrating into second branchial arch
53. origin of second branchial arch
54. otic pit
55. location of thyroid primordium
56. origin of first branchial arch artery from aortic sac
57. trabeculated myocardial wall of common ventricular chamber of heart
58. aortic sac
59. pericardial cavity
60. endocardial lining of atrio-ventricular canal
61. body wall overlying pericardial cavity
62. bulbo-truncal junction region of outflow tract of primitive heart
63. junctional zone between body wall and amnion
64. pharyngeal region of foregut diverticulum
65. rostral extremity of bulbo-ventricular groove

PLATE 16b

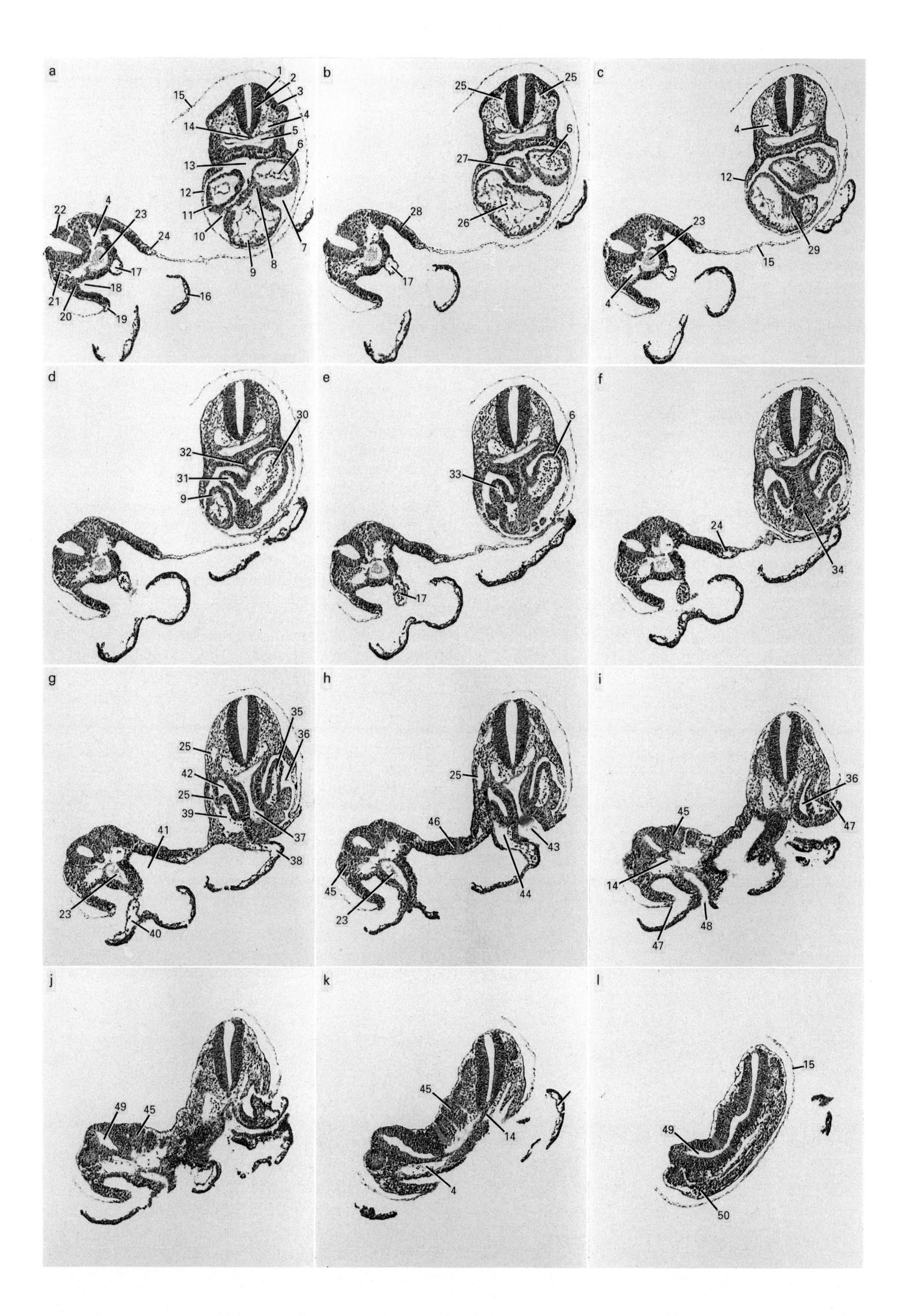

a
1
2
3
4
5
6
7
8
9
10
11
12
13
14
15
16
17
18
19
20
21
22
23
24
b
25
25
6
27
26
28
17
c
4
12
23
15
29
4
d
30
32
31
9
e
6
33
17
f
24
34
g
35
36
25
42
25
39
37
38
41
23
40
h
25
46
43
44
45
23
i
36
45
47
14
48
47
j
49
45
k
45
14
4
l
15
49
50

Plate 16b (8.5–9 days p.c.)

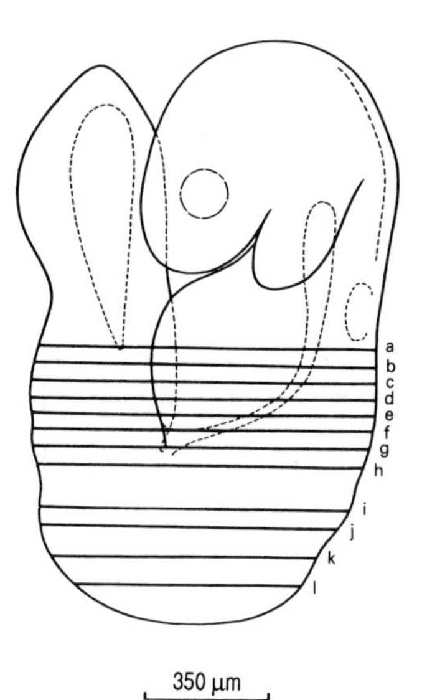

8.5–9 days p.c. Embryo with 12–14 pairs of somites, which has almost completed the process of turning. Transverse sections (Theiler Stages 13–14; rat, Witschi Stage 16; human, Carnegie Stages 10–11).

1. periphery of otic placode
2. neural tube in upper trunk region
3. otic pit
4. dorsal aorta
5. pharyngeal region of foregut diverticulum
6. endocardial lining of left side of common atrial chamber of heart
7. pericardial cavity
8. atrio-ventricular canal
9. myocardial wall of common ventricular chamber of heart
10. bulbo-ventricular groove
11. bulbus cordis region of heart
12. body wall overlying pericardial cavity
13. transverse pericardial sinus
14. notochord
15. amnion
16. visceral yolk sac
17. vitelline (omphalomesenteric) artery
18. entrance to intra-embryonic coelomic cavity
19. left umbilical vein
20. splitting of lateral plate mesoderm to form splanchnopleure and somatopleure
21. paraxial mesoderm (presomite)
22. neural tube in tail region
23. hindgut diverticulum
24. right umbilical vein located at junctional zone between body wall and amnion
25. anterior cardinal vein
26. bulbo-ventricular canal
27. right side of common atrial chamber of heart
28. body wall in mid-tail region of embryo
29. trabeculation of myocardial wall
30. primitive nucleated red blood cells within lumen of common atrial chamber of heart
31. site of entrance of right horn of sinus venosus into common atrial chamber of heart
32. mesentery of heart (dorsal mesocardium) caudal to transverse pericardial sinus
33. right horn of sinus venosus
34. septum transversum
35. left horn of sinus venosus
36. left pericardio-peritoneal canal
37. hepatic diverticulum from region of foregut–midgut junction
38. vascular plexus (blood island) within visceral yolk sac
39. common cardinal vein (duct of Cuvier)
40. communication between vitelline artery and yolk sac circulation
41. intra-embryonic coelomic cavity
42. right pericardio-peritoneal canal
43. entrance to foregut diverticulum
44. communication between right umbilical vein and yolk sac circulation
45. somite
46. body wall in mid-trunk region of embryo
47. boundary of umbilical ring
48. entrance to hindgut diverticulum
49. neural lumen
50. prominent lateral ridge – to be associated with development of the forelimb bud

PLATE 17

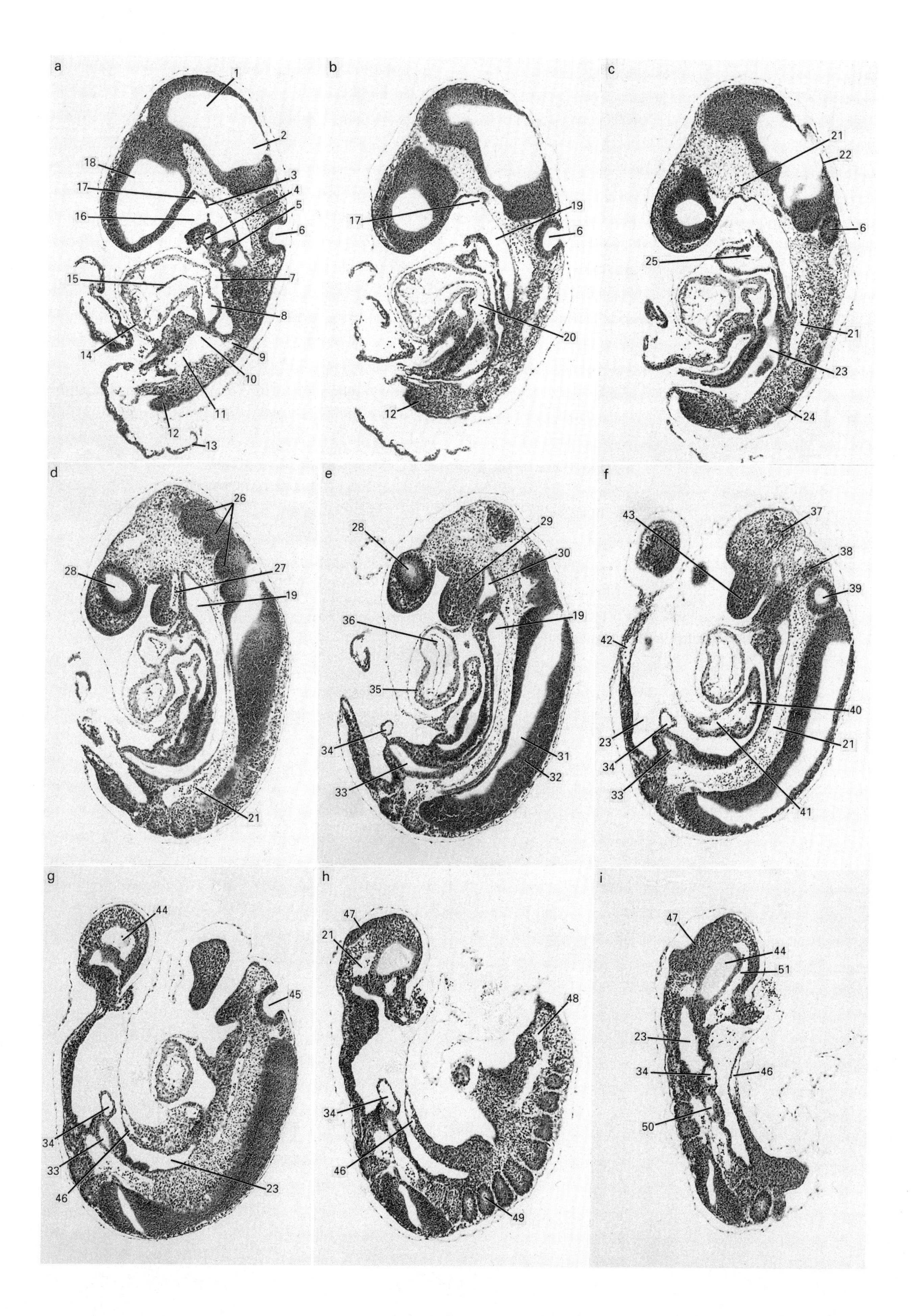
a
b
c
d
e
f
g
h
i
1
2
3
4
5
6
7
8
9
10
11
12
13
14
15
16
17
18
19
20
21
22
23
24
25
26
27
28
29
30
31
32
33
34
35
36
37
38
39
40
41
42
43
44
45
46
47
48
49
50
51

Plate 17 (9 days p.c.)

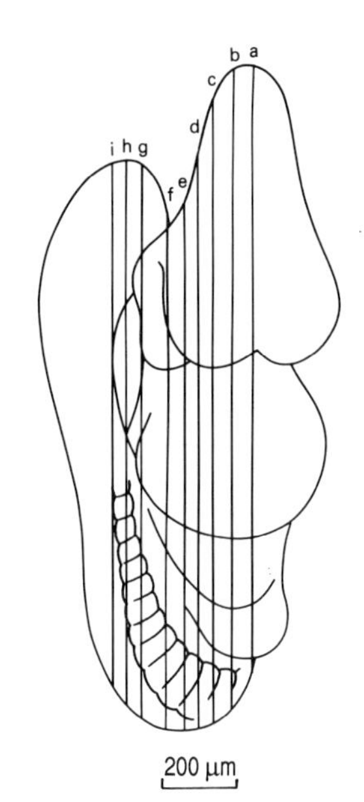

9 days p.c. Embryo with 15–18 pairs of somites, which has almost completed the process of turning. Crown–rump length 1.7 mm (fixed). Sagittal sections (Theiler Stage 14; rat, Witschi Stage 16; human, Carnegie Stage 11).

1. midbrain
2. hindbrain
3. remains of bucco- (oro-) pharyngeal membrane
4. first branchial arch artery
5. second branchial arch artery
6. left otic pit
7. pericardio-peritoneal canal
8. left part of common atrial chamber of heart
9. left anterior cardinal vein
10. left common cardinal vein
11. left posterior cardinal vein
12. left forelimb bud
13. visceral yolk sac
14. trabeculated myocardial wall of common ventricular chamber of heart
15. atrio-ventricular canal
16. stomatodaeum (oropharyngeal region)
17. entrance to Rathke's pouch (ectodermally lined)
18. forebrain
19. pharyngeal region of foregut diverticulum
20. left horn of sinus venosus
21. dorsal aorta
22. roof of hindbrain
23. intra-embryonic coelomic cavity (peritoneal cavity)
24. somite
25. aortic sac giving origin to first and second branchial arch arteries
26. rhombomere
27. communication between dorsal aorta and first branchial arch artery
28. right optic vesicle
29. first branchial arch mesoderm and trigeminal (V) neural crest tissue
30. first branchial pouch
31. lumen of neural tube
32. neuroepithelium of neural tube
33. midgut–hindgut junction
34. vitelline (omphalomesenteric) artery
35. bulbus cordis
36. truncus arteriosus/aortic sac
37. trigeminal (V) neural crest tissue
38. facial (VII) neural crest tissue
39. medial part of right otic pit
40. right part of common atrial chamber of heart
41. right horn of sinus venosus
42. left umbilical vein
43. first branchial arch (mandibular component)
44. caudal extremity of hindgut diverticulum
45. right otic pit
46. right umbilical vein
47. neuroepithelium (neural plate) in region of caudal neuropore
48. right anterior cardinal vein
49. myocoele
50. hindgut diverticulum
51. communication between dorsal aorta and vitelline artery

PLATE 18a

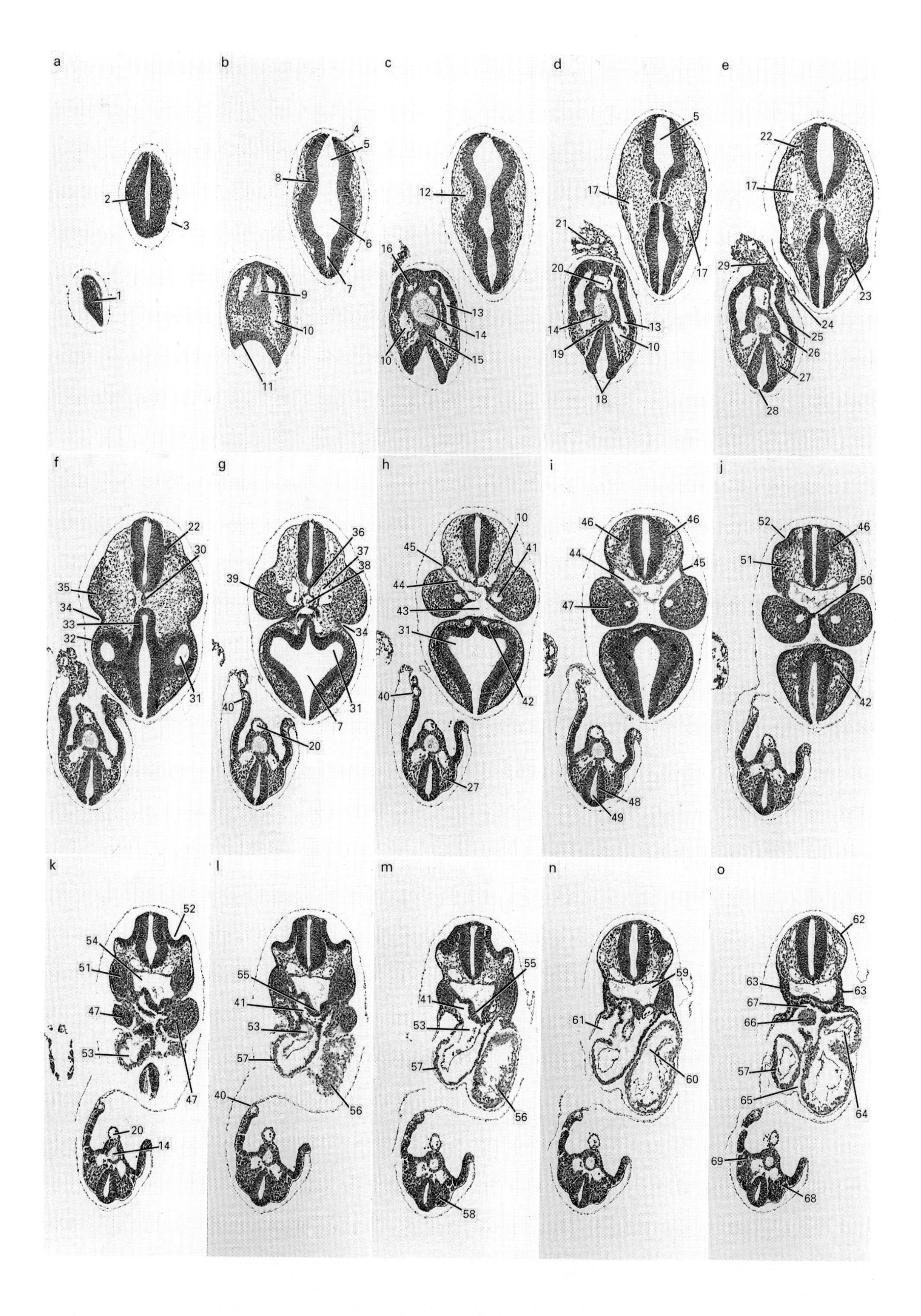

Plate 18a (9 days p.c.)

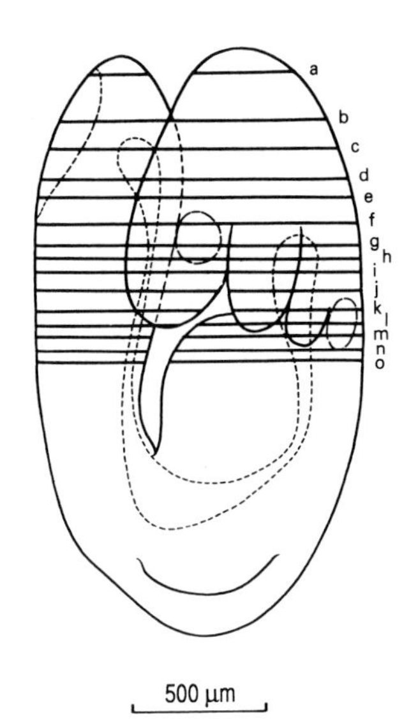

9 days p.c. Embryo with 15–20 pairs of somites, which has just completed the process of turning. Crown–rump length 2.1 mm (fixed). Transverse sections (Theiler Stage 14; rat, Witschi Stage 16; human, Carnegie Stage 11).

1. caudal region of embryo
2. cephalic region of embryo
3. amnion
4. surface ectoderm
5. lumen of prospective hindbrain
6. lumen of prospective midbrain
7. lumen of prospective forebrain
8. neuroepithelium (neural ectoderm)
9. caudal extremity of hindgut diverticulum
10. dorsal aorta
11. neural plate of caudal neuropore
12. cephalic mesenchyme tissue
13. caudal extremity of intraembryonic coelomic cavity
14. hindgut diverticulum
15. condensation of tissue (poorly defined) in caudal region of notochord
16. distal region of right dorsal aorta just proximal to site of fusion with left dorsal aorta to form the (midline) vitelline artery
17. primary head vein (part of anterior plexus of veins)
18. neural folds in region of caudal neuropore
19. caudal extremity of notochord
20. vitelline (omphalomesenteric) artery
21. component of allantoic vasculature
22. trigeminal (V) neural crest cells
23. perioptic neural crest tissue of trigeminal (V) crest origin
24. left umbilical vein
25. somatic component of lateral plate mesoderm (i.e. somatopleure) forming part of the body wall musculature
26. splanchnic component of lateral plate mesoderm (i.e. splanchnopleure) forming wall of the hindgut
27. paraxial mesoderm (presomite)
28. caudal extent of neural tube closure
29. base of allantois
30. rostral extremity of notochord
31. optic vesicle
32. optic eminence
33. diencephalon
34. groove between first branchial arch and optic eminence
35. first branchial arch
36. notochord
37. communication between dorsal aorta and first branchial arch artery
38. Rathke's pouch
39. trigeminal (V) neural crest tissue within maxillary component of first branchial arch
40. right umbilical vein
41. first branchial arch artery
42. perioptic vascular plexus
43. oropharynx
44. first branchial pouch
45. first branchial membrane (future tympanic membrane)
46. facio-acoustic (VII–VIII) neural crest tissue
47. first branchial arch (mandibular component)
48. neural tube in tail region
49. neural (spinal) lumen
50. tag of ruptured buccopharyngeal membrane
51. facial (VII) neural crest tissue within second branchial arch
52. otic pit
53. aortic sac
54. pharyngeal region of foregut diverticulum
55. location of thyroid primordium
56. trabeculation in myocardial wall of common ventricular chamber of heart
57. bulbus cordis region of heart
58. somite
59. second branchial pouch
60. atrio-ventricular canal
61. pericardial cavity
62. origin of third branchial arch
63. second branchial membrane
64. left component of common atrial chamber of heart
65. bulbo-ventricular groove
66. wall of right side of common atrial chamber of heart
67. transverse pericardial sinus
68. pronephric duct (yet to canalize)
69. intermediate plate mesoderm (nephrogenic cord)

PLATE 18b

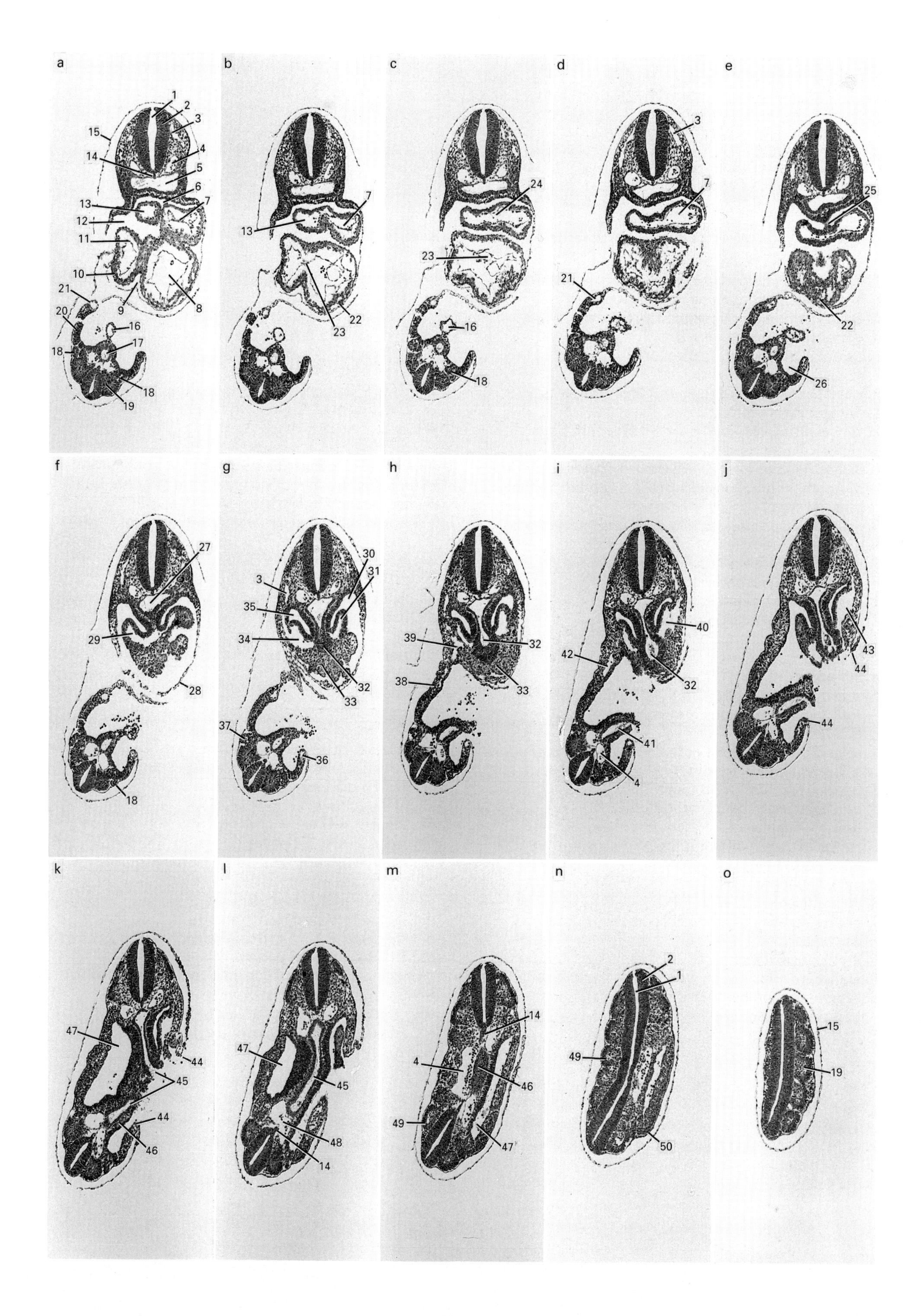

a
b
c
d
e
f
g
h
i
j
k
l
m
n
o
1
2
3
4
5
6
7
8
9
10
11
12
13
14
15
16
17
18
19
20
21
22
23
24
25
26
27
28
29
30
31
32
33
34
35
36
37
38
39
40
41
42
43
44
45
46
47
48
49
50

Plate 18b (9 days p.c.)

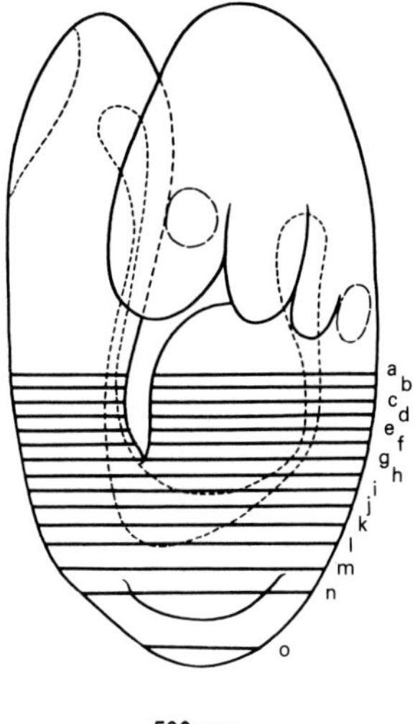

9 days p.c. Embryo with 15–20 pairs of somites, which has just completed the process of turning. Crown–rump length 2.1 mm (fixed). Transverse sections (Theiler Stage 14; rat, Witschi Stage 16; human, Carnegie Stage 11).

1. neural lumen
2. neuroepithelium of neural tube
3. anterior cardinal vein
4. dorsal aorta
5. pharyngeal region of foregut
6. transverse pericardial sinus
7. left component of common atrial chamber of heart
8. common ventricular chamber of heart
9. bulbo-ventricular groove
10. bulbus cordis region of heart
11. bulbo-truncal region
12. pericardial cavity
13. right component of common atrial chamber of heart
14. notochord
15. amnion
16. vitelline (omphalomesenteric) artery
17. hindgut diverticulum
18. rudiments of nephric duct and blastema
19. somite
20. body wall in tail region
21. right umbilical vein
22. trabeculation in myocardial wall of common ventricular chamber of heart
23. bulbo-ventricular canal
24. common atrial chamber of heart
25. caudal (intact) component of mesentery of heart (dorsal mesocardium)
26. intra-embryonic coelomic cavity (caudal part of peritoneal cavity)
27. foregut
28. body wall overlying pericardial cavity
29. right horn of sinus venosus
30. left pericardio-peritoneal canal
31. left horn of sinus venosus
32. hepatic diverticulum from region of foregut–midgut junction
33. septum transversum
34. communication between right common cardinal vein and right horn of sinus venosus
35. right pericardio-peritoneal canal
36. left umbilical vein
37. intermediate plate mesoderm (nephrogenic cord)
38. body wall in mid-trunk region of embryo
39. communication between right umbilical and right common cardinal veins
40. communication between left anterior and left common cardinal veins
41. midgut–hindgut junction
42. right umbilical vein where it passes rostrally and ventrally just before it joins the right common cardinal vein
43. left common cardinal vein
44. boundary of umbilical ring
45. midgut
46. wall of midgut of lateral plate mesodermal (splanchnopleuric) origin
47. intra-embryonic coelomic cavity (peritoneal cavity)
48. site of union of right and left dorsal aortae
49. myocoele
50. prominent lateral ridge to be associated with the development of the forelimb bud

Plate 18a,c–e), and has a significantly greater diameter than the rest of the hindgut and midgut regions (cf. hindgut diameter in *Plate 18a,k–m* and midgut diameter in *Plate 18b,i–l*).

THE NERVOUS SYSTEM

As in the previous stage, the most obvious changes that occur in relation to the developing nervous system are observed in the cephalic region, principally as a consequence of the closure of the rostral extremity of the neural tube to form the three primitive brain vesicles. In the forebrain region, the (primary) optic vesicles become particularly prominent (*Plate 18a,f–h*), and a deep groove separates the overlying optic eminence thus created from the maxillary and mandibular components of the first branchial arch (*Plate 18a,f–h*). The optic stalks, which connect the optic vesicles to the lumen of the primitive forebrain (the prosencephalic vesicle), have a wide diameter at this stage (*Plate 18a,g*), but this diminishes rapidly during the early part of the next stage of development. This coincides with the dramatic increase in the volume of the forebrain vesicle that occurs at that time (*Plate 19a,k,l*), with early evidence of its subdivision into a third ventricle with two associated telencephalic vesicles.

Even in the most advanced embryos at this stage of development, the neuroepithelium of the optic vesicles is still separated from the overlying surface ectoderm by cephalic mesenchyme (*Plate 18a,g*). Direct contact between the two is first evident in embryos with about 20–25 pairs of somites (*Plate 19a,h–k*) and is a necessary prerequisite for the induction of the lens placode by the subjacent neuroepithelium of the optic vesicle.

In the dorsal part of the floor of the forebrain vesicle a small evagination appears, and represents the first evidence of the infundibular recess (*Plate 18a,g*), though this is more clearly seen in slightly more advanced embryos (*Plate 19a,k,l*), where the infundibular recess is seen to make direct contact with the posterosuperior wall of Rathke's pouch (*Plate 19a,k,l*).

On histological sections, no clear demarcation is seen between the rostral part of the future spinal cord and the caudal part of the hindbrain. The profile of the neural lumen is, as in the previous stage, still conspicuously flattened from side to side throughout the majority of its length, except in its most caudal region where it has yet to close (the caudal neuropore) (*Plate 18a,b–e*).

Cells of the trigeminal (V) neural crest are clearly evident as they migrate into the first branchial arch (*Plates 15a,e*; *16a,d,e*; *17,f*; *18a,d–f*), and the facial (VII) neural crest is seen to migrate into the second branchial arch (*Plates 15a,h–j*; *16a,g–i*; *17,f*; *18a,i–l*). The facio-acoustic (VII–VIII) neural crest complex is first evident at this stage, though only the cells of its facial (VII) component migrate into the second branchial arch (see above), the facio-acoustic (VII–VIII) (or acoustico-facial) primordium is developed in part from neural crest and in part (probably) from cells contributed from the wall of the otocyst, and from a small additional ectodermal placode related to the second arch (Hamilton and Mossman, 1972). While it is difficult to appreciate them on transverse sections, between four and six rhombomeres may be observed in sagittal sections though the hindbrain region of the most advanced embryos isolated at this stage of development (*Plate 17,d*), though a total of eight rhombomeres are eventually present in this species.

THE UROGENITAL SYSTEM

In the most advanced embryos at this stage of development, the pronephric duct (which has yet to canalize) and nephric vesicle (or primordium), which are of intermediate plate mesodermal origin, are first clearly evident (*Plate 18a,m–o*; *Plate 18b,a–c*) in the caudal part of the trunk/proximal part of the tail region. In sections stained histochemically to demonstrate the presence of intracellular alkaline phosphatase enzyme activity, over 90% of the primordial germ cells are seen to be located in association with the wall of the hindgut, a few are still located at the base of the allantois, while about 7% have already migrated into the hindgut mesentery (see Table 4).

Stage 15 Developmental age, 9.5 days p.c. Plates 19a–c, 20a–c, 21a, 21b

EXTERNAL FEATURES

Embryos with 21–29 pairs of somites. The major difference between the external appearance of embryos at this stage compared with that observed at the previous stage of development, relates to the gradual but substantial changes that occur in the cephalic region, principally as a consequence of the subdivision of the primitive forebrain vesicle to form the third ventricle and the two telencephalic vesicles. The volume of these derivatives of the primitive forebrain, as well as the mesencephalic vesicle and, but more particularly, the hindbrain vesicle increases markedly throughout this stage (*Plate 21a,c,d*), and consequently plays an important part in increasing the overall volume of the cephalic region. A significant proliferation is also evident in the amount of trigeminal (V) neural crest tissue seen to be migrating from its source of origin towards the maxillary component of the first branchial arch (*Plate 19a,g–l*; *Plates 20a,i–l, 20b,a–c*), sufficient to create a diffuse prominence overlying this region which is distinct from the maxillary component of the first branchial arch. The migrating trigeminal (V) neural crest tissue is characteristically located posteriorly and lateral to the large primary head vein (*Plate 19a,h*; *Plate 20a,k,l*).

Since the volume of the optic vesicles does not increase to the same extent as that of the telencephalic vesicles, the optic eminences become less prominent than previously and displaced more dorsally, so that they appear to become wedged between the telencephalic vesicles (which are anterior to them) and the maxillary components of the first branchial arches (which are located inferiorly and slightly dorsal to them). A distinct groove is seen to separate the two prominences produced by the telencephalic vesicles and the maxillary components of the first branchial arches, and represents the first evidence of a nasolacrimal groove (which will eventually form the nasolacrimal duct).

The further degree of differentiation of the otic pits also influences the external appearance of the dorsal part of the cephalic region, since these structures are located on either side of the caudal part of the hindbrain vesicle (or fourth ventricle). In embryos with about 20–25 pairs of somites, the otic pit is seen to be deeply indented (*Plate 19a,l*), and this structure only becomes completely separated from the overlying surface ectoderm (to form the otocyst) in embryos with about 25–30 pairs of somites (*Plate 20b,d–f*; *Plates 21a,b, 21b,c*). Over this relatively short period, the volume of the otic vesicles also increases substantially.

While the tail does not increase in length to any great extent, it becomes displaced laterally, more frequently towards the right side of the cephalic region than towards the left side, and in more advanced embryos at this stage of development, the distal region of the tail may become recurved on itself. In embryos with about 25–30 pairs of somites, the caudal neuropore is still seen to be in the process of closing (*Plate 20b,d–f*), and closure is only finally achieved in embryos which possess about 30–35 pairs of somites. In embryos with about 25–30 pairs of somites, the forelimb buds are becoming increasingly prominent and are now clearly seen to be located at about the level of somites 8–12 (*Plate 20c,g–j*; *Plate 21b,a–d*). Two rather indistinct condensations, being the first evidence of the hindlimb buds, appear towards the end of this stage, and extend caudally from the middle part of the tail region (*Plates 20a,j–l, 20b,a–h*). The first evidence of a third branchial arch is also seen towards the end of this stage (*Plate 20b,i*).

THE HEART AND VASCULAR SYSTEM

No fundamental changes occur in the overall configuration of the heart during this stage of development, though its component parts become more clearly delineated. The latter is associated with a substantial increase in the overall volume of the common atrial chamber, and a concomitant reduction in the space occupied by the cardiac jelly in this region (*Plate 20c,a–c*). Prominent ridges also develop during this stage at the sites of entry of the two horns of the sinus venosus into the common atrial chamber of the heart (*Plate 21b,a*). Despite the substantial increase that occurs in the volume of the primitive ventricle (*Plate 19b,g–j*; *Plate 20b,g–l*), the atrio-ventricular channel appears to have a relatively narrow luminal diameter (*Plate 19b,g–i*; *Plate 20b,k*), and this is principally due to an increase in the volume of the bulbar cushion tissue in this location at this time (*Plate 19b,h,i*; *Plate 20b,k*). The bulbo-ventricular groove tends to deepen, and this helps to define the boundary between the primitive ventricle and the bulbus cordis region of the heart (*Plate 19b,j*; *Plate 20b,i,j*). The degree of trabeculation in the primitive ventricle is now quite marked, and serves also to delineate this region of the primitive heart.

The luminal diameter of the outflow tract of the primitive heart is also relatively narrow at this stage (*Plate 19b,g,h*; *Plate 20b,h,i*). This is where the aortico-pulmonary spiral septum forms, the first definitive evidence of which is seen in this location in embryos with about 30–35 pairs of somites (*Plate 22b,a*; *Plate 23c,a,b*) (Fananapazir and Kaufman, 1988). The distal part of the outflow tract of the heart opens into the region of the aortic sac (*Plate 20b,i,j*). The blood from the aortic sac then flows into the paired dorsal aortae via the aortic (or pharyngeal or branchial) arch arteries. Of those present at this stage, the first has a relatively narrow luminal diameter, while that of the second and third arch arteries is substantially greater (*Plate 20b,f–j*; *Plate 21a,a,b*).

A more extensive degree of fusion of the paired dorsal aortae is seen at this stage of development than previously, and towards the end of Stage 15, the single midline dorsal aorta is seen to extend from the mid-trunk region (*Plate 20c,i*) caudally into the mid-tail region (*Plate 20b,d*) of the embryo. The posterior cardinal veins are clearly seen in embryos with 25–30 pairs of somites, and drain into the common cardinal

(*continued on page 92*)

PLATE 19a

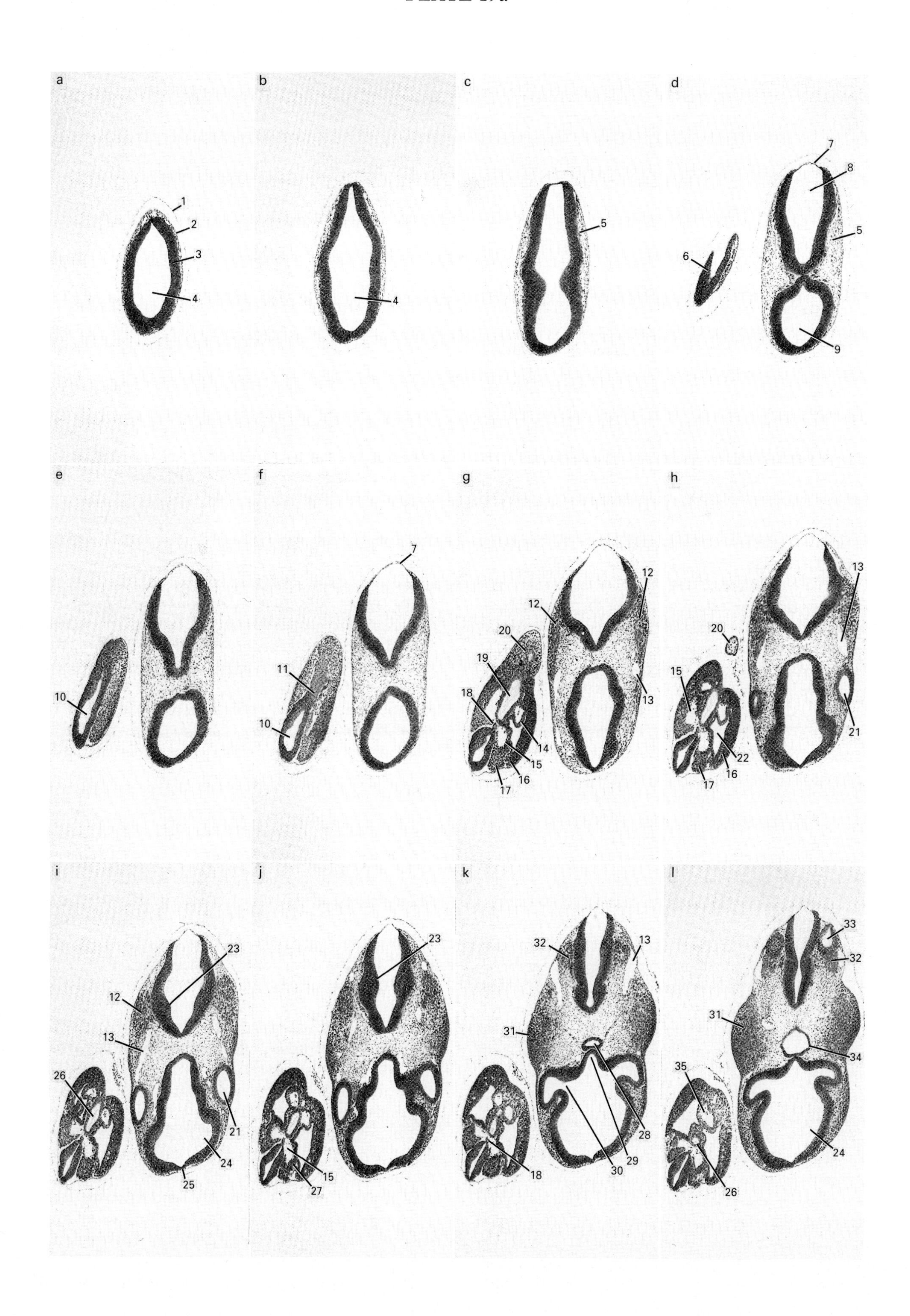
a
b
c
d
e
f
g
h
i
j
k
l
1
2
3
4
5
6
7
8
9
10
11
12
13
14
15
16
17
18
19
20
21
22
23
24
25
26
27
28
29
30
31
32
33
34
35

Plate 19a (9.5 days p.c)

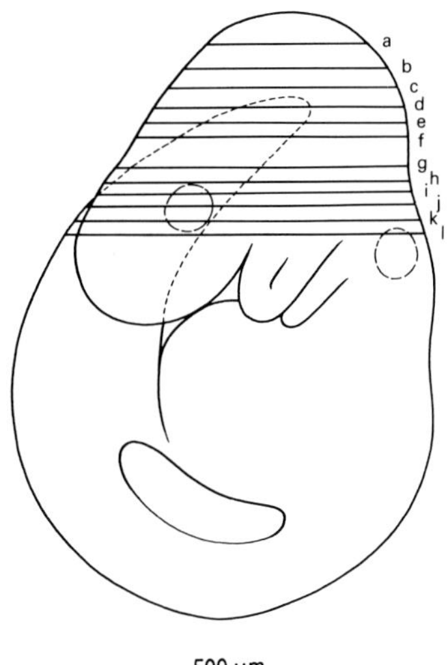

9.5 days p.c. Embryo with 20–25 pairs of somites, which has completed the process of turning. Crown–rump length 2.5 mm (fixed). Transverse sections (Theiler Stage 15; rat, Witschi Stage 17; human, Carnegie Stage 12).

1. amnion
2. surface ectoderm
3. neuroepithelium of midbrain region
4. mesencephalic vesicle
5. cephalic mesenchyme tissue
6. neuroepithelium of caudal neuropore
7. roof of hindbrain
8. fourth ventricle
9. forebrain vesicle
10. dilated region of neural tube in caudal region of the tail
11. tangential section through wall of neural tube in region overlying the caudal extremity of the hindgut diverticulum
12. trigeminal (V) neural crest tissue
13. branch of primary head vein
14. distal recurved portion of dorsal aorta
15. proximal part of dorsal aorta in tail region
16. early stage in differentiation of hindlimb bud
17. somite
18. notochord
19. caudal dilated extremity of hindgut diverticulum
20. recurved most distal portion of tail region
21. optic vesicle
22. caudal extremity of intra-embryonic coelomic cavity (peritoneal cavity)
23. rhombomere
24. telencephalic vesicle
25. lamina terminalis
26. hindgut diverticulum
27. pronephric (nephrogenic cord) tissue
28. Rathke's pouch
29. infundibular recess of diencephalon
30. lumen of optic stalk
31. origin of maxillary component of first branchial arch containing trigeminal (V) neural crest tissue
32. facio-acoustic (VII–VIII) neural crest tissue
33. otocyst (otic vesicle) (the otic pit has almost completely separated from the overlying surface ectoderm)
34. peripheral boundary of entrance to Rathke's pouch
35. site of amalgamation of distal trunks of dorsal aortae to form the vitelline (omphalomesenteric) artery

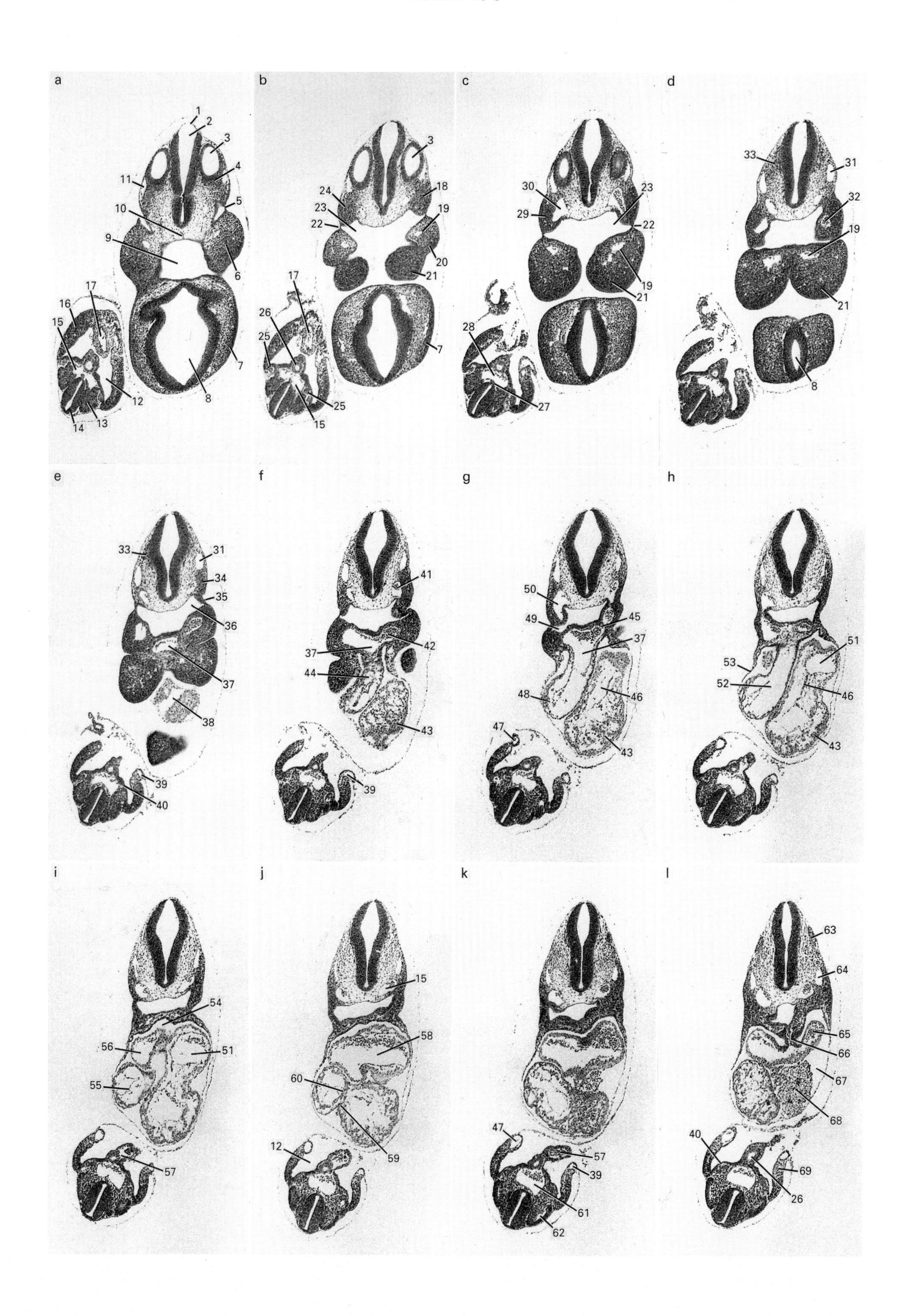
a
b
c
d
e
f
g
h
i
j
k
l

Plate 19b (9.5 days p.c)

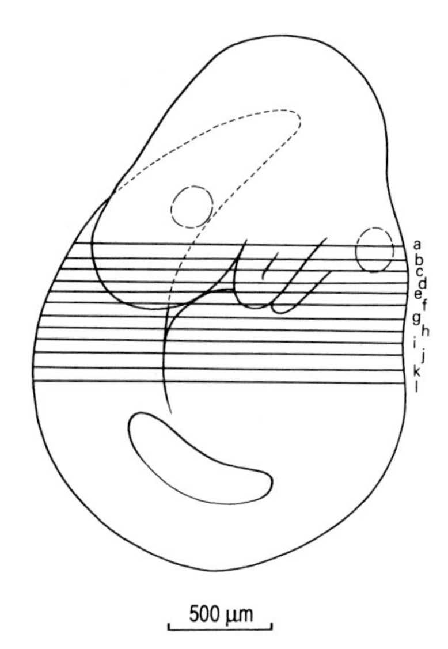

9.5 days p.c. Embryo with 20–25 pairs of somites, which has completed the process of turning. Crown–rump length 2.5 mm (fixed). Transverse sections (Theiler Stage 15; rat, Witschi Stage 17; human, Carnegie Stage 12).

1. roof of hindbrain
2. fourth ventricle
3. otocyst (otic vesicle)
4. facio-acoustic (VII–VIII) neural crest complex
5. first branchial cleft
6. maxillary component of first branchial arch containing trigeminal (V) neural crest
7. olfactory placode
8. forebrain vesicle
9. oropharynx
10. rostral extremity of notochord
11. branch of primary head vein
12. intra-embryonic coelomic cavity (caudal part of peritoneal cavity)
13. somite
14. neural tube in proximal part of tail region
15. dorsal aorta
16. right hindlimb bud
17. vitelline (omphalomesenteric) artery
18. second branchial arch containing facial (VII) neural crest
19. first branchial arch artery
20. maxillary component of first branchial arch
21. mandibular component of first branchial arch
22. first branchial membrane (future tympanic membrane)
23. first branchial pouch
24. origin of second branchial arch
25. nephrogenic cord tissue including nephric duct and vesicle
26. hindgut diverticulum
27. notochord
28. caudal extremity of midline dorsal aorta – more distally, paired dorsal aortae
29. second branchial arch
30. communication between second branchial arch artery and dorsal aorta
31. primary head vein/cephalic extension of anterior cardinal vein
32. second branchial arch artery
33. glossopharyngeal–vagal (IX–X) neural crest complex
34. origin of third branchial arch
35. second branchial cleft and membrane
36. second branchial pouch (dorsal component)
37. aortic sac
38. cephalic extremity of pericardial cavity
39. left umbilical vein
40. urogenital ridge
41. third branchial arch artery
42. origin of second branchial arch artery from aortic sac
43. trabeculation in myocardial wall of common ventricular chamber of heart
44. truncus arteriosus region of outflow tract of heart
45. origin of third branchial arch artery from aortic sac
46. atrio-ventricular canal
47. right umbilical vein
48. bulbus cordis region of heart
49. second branchial pouch (ventral component)
50. communication between third branchial arch artery and dorsal aorta
51. left component of common atrial chamber of heart
52. bulbo-truncal junction
53. body wall overlying pericardial cavity
54. transverse pericardial sinus
55. endocardium (endothelial lining of heart)
56. right component of common atrial chamber of heart
57. vitelline vein
58. common atrial chamber of heart
59. bulbo-ventricular groove
60. bulbo-ventricular canal
61. midline dorsal aorta
62. myocoele
63. condensation of paraxial mesoderm (cervical myotome)
64. anterior cardinal vein
65. left horn of sinus venosus
66. caudal component of dorsal mesocardium
67. pericardial cavity
68. septum transversum containing elements of hepatic primordia
69. body wall in caudal region of trunk

PLATE 19c

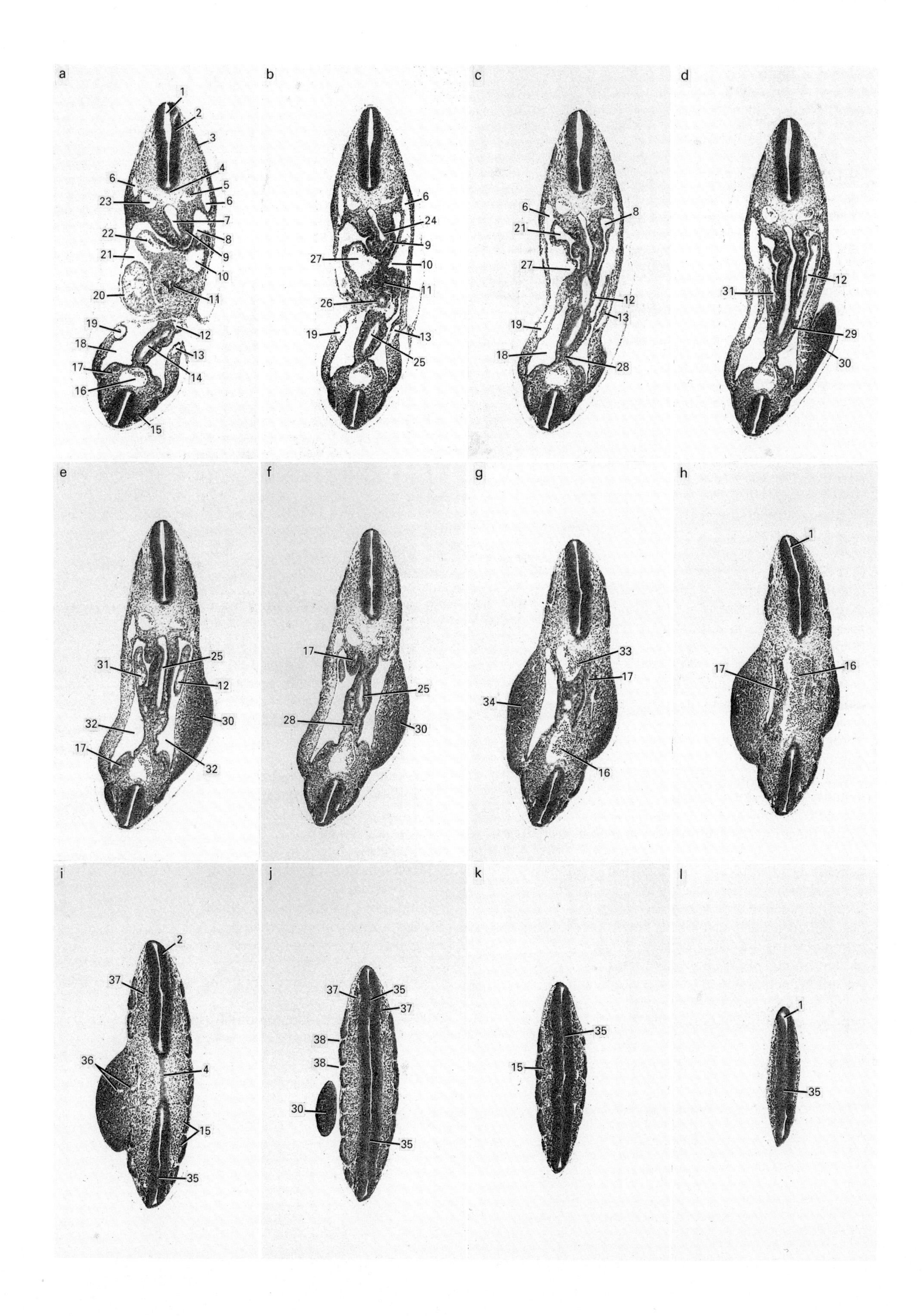
a
1
2
3
4
5
6
6
23
7
22
8
9
21
10
20
11
19
12
18
13
17
14
16
15
b
6
24
9
27
10
11
26
19
13
25
c
6
8
21
27
12
13
19
18
28
d
12
31
29
30
e
25
31
12
30
32
17
32
f
17
25
28
30
g
33
17
34
16
h
1
17
16
i
2
37
36
4
15
35
j
37
35
37
38
38
30
35
k
35
15
l
1
35

Plate 19c (9.5 days p.c.)

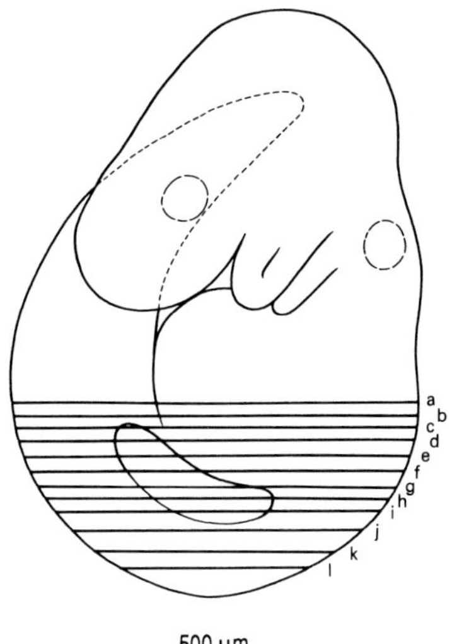

9.5 days p.c. Embryo with 20–25 pairs of somites, which has completed the process of turning. Crown–rump length 2.5 mm (fixed). Transverse sections (Theiler Stage 15; rat, Witschi Stage 17; human, Carnegie Stage 12).

1. neural lumen
2. neuroepithelium of neural tube
3. condensation of paraxial mesoderm (cervical myotome)
4. notochord
5. left dorsal aorta
6. anterior cardinal vein
7. ventral bifurcation of foregut to produce tracheal diverticulum (laryngo-tracheal groove)
8. left pericardio-peritoneal canal
9. early stage in differentiation of lung bud
10. left common cardinal vein
11. septum transversum containing elements of hepatic primordia
12. left vitelline vein
13. left umbilical vein
14. hindgut–midgut junction
15. somite
16. midline dorsal aorta
17. nephrogenic cord tissue (nephric duct and vesicle) subjacent to urogenital ridge
18. intra-embryonic coelomic cavity (caudal part of peritoneal cavity)
19. right umbilical vein
20. trabeculation in myocardial wall of common ventricular chamber of heart
21. right pericardio-peritoneal canal
22. right horn of sinus venosus
23. right dorsal aorta
24. tracheal diverticulum
25. midgut
26. cystic (gall bladder) primordium
27. right common cardinal vein
28. dorsal mesentery of gut
29. wall of gut tube containing vitelline venous plexus
30. left forelimb bud
31. right vitelline vein
32. peritoneal cavity
33. rostral site of fusion of right and left dorsal aortae
34. right forelimb bud
35. localized site of neural luminal occlusion
36. axial vessels to and from right forelimb bud
37. neural crest derived spinal primordia (will give rise to sensory cells of dorsal (posterior) root ganglia)
38. intersegmental zone of decreased mesodermal cell packing density

PLATE 20a

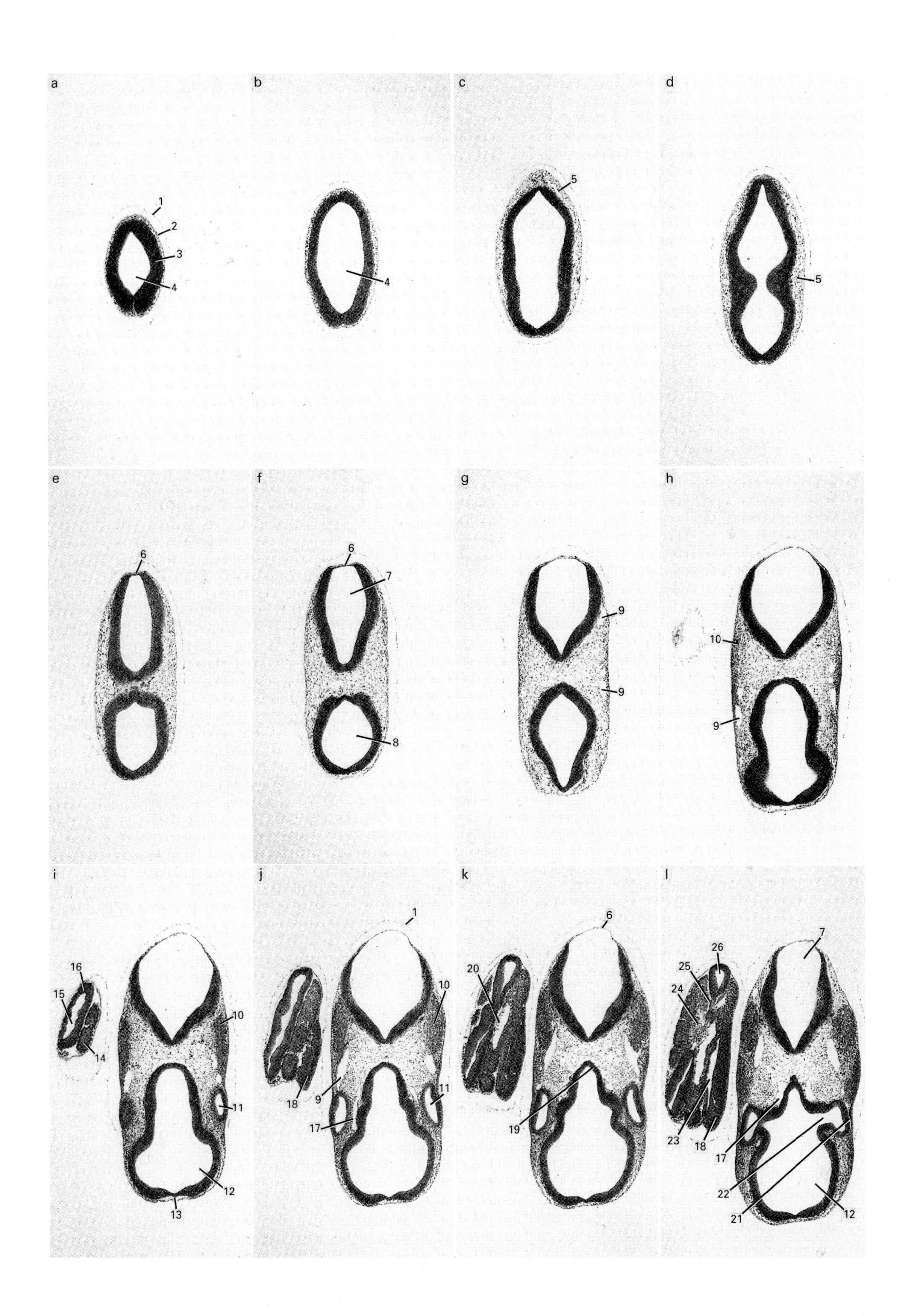

a
b
c
d
e
f
g
h
i
j
k
l
1
2
3
4
5
6
7
8
9
10
11
12
13
14
15
16
17
18
19
20
21
22
23
24
25
26

Plate 20a (10–10.25 days p.c.)

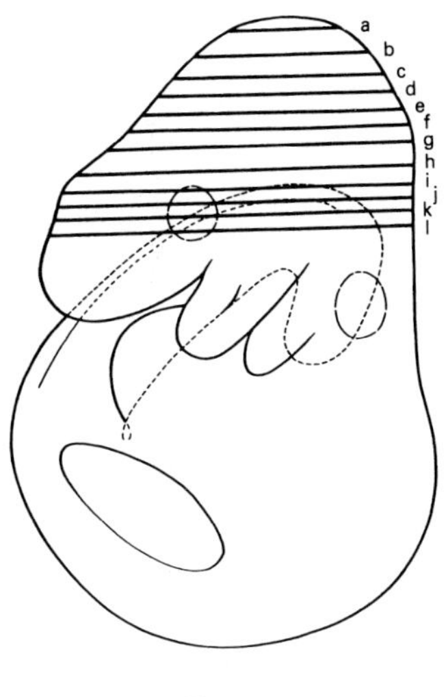

10–10.25 days p.c. Embryo with 25–30 pairs of somites. Crown–rump length 2.6 mm (fixed). Transverse sections (Theiler Stage 15; rat, Witschi Stages 18–19; human, Carnegie Stage 12).

1. amnion
2. surface ectoderm
3. neuroepithelium of midbrain region
4. mesencephalic vesicle
5. cephalic mesenchyme tissue
6. roof of hindbrain
7. fourth ventricle
8. forebrain vesicle
9. branch of primary head vein
10. trigeminal (V) neural crest tissue
11. optic vesicle/early stage in development of optic cup
12. telencephalic vesicle
13. lamina terminalis
14. somite
15. neural lumen in tail region
16. neuroepithelium of neural tube in tail region
17. perioptic vascular plexus
18. hindlimb bud
19. diencephalon
20. left dorsal aorta in caudal part of tail region
21. lens placode
22. optic stalk
23. intra-embryonic coelomic cavity (caudal extension of peritoneal cavity)
24. paired dorsal aortae, only separated by their endothelial walls
25. notochord
26. dilated region of neural tube in caudal region of the tail just proximal to the caudal neuropore

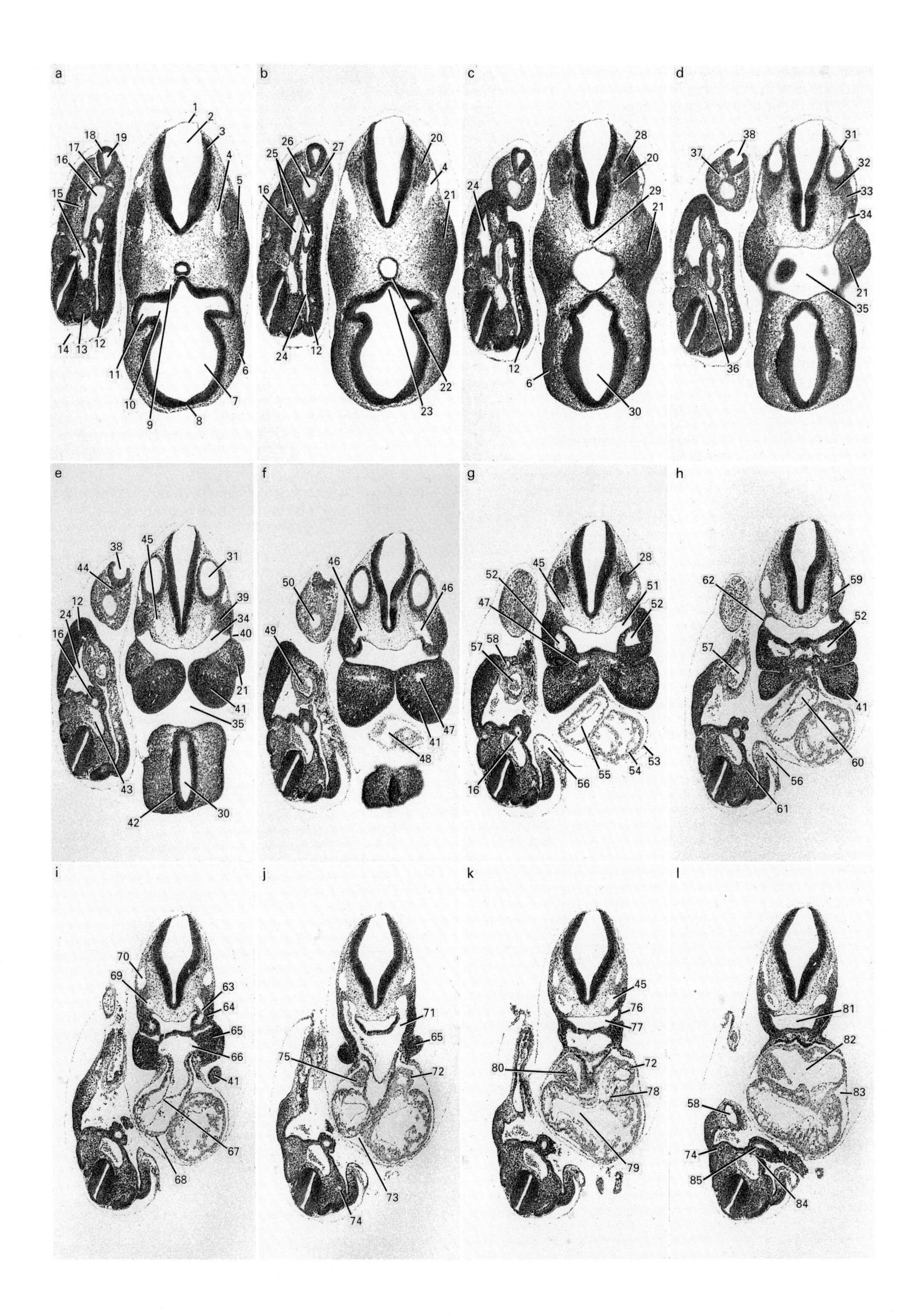
a
1 2 3 18 19 17 16 4 5 15 14 13 12 11 10 9 8 7 6
b
26 27 25 20 4 16 21 24 12 22 23
c
28 20 24 29 21 12 6 30
d
38 31 37 32 33 34 21 35 36
e
38 45 31 44 39 12 34 24 40 16 21 41 35 43 42 30
f
46 50 46 49 47 41 48
g
45 28 52 51 47 52 57 58 55 54 53 16 56
h
62 59 52 57 41 60 56 61
i
70 69 63 64 65 66 41 67 68
j
71 65 75 72 73 74
k
45 76 77 80 72 78 79
l
81 82 83 58 74 85 84

Plate 20b (10–10.25 days p.c.)

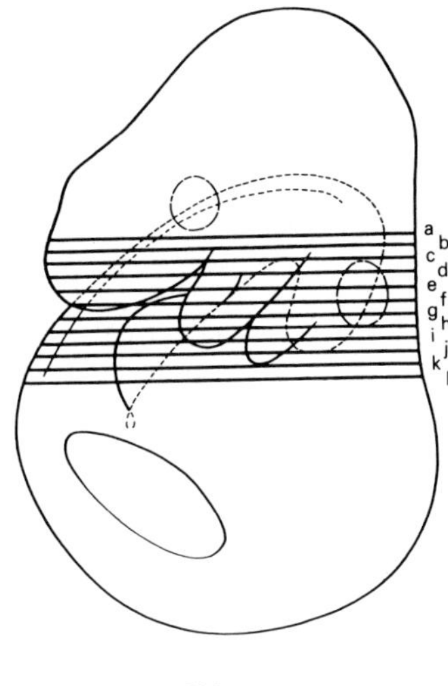

10–10.25 days p.c. Embryo with 25–30 pairs of somites. Crown–rump length 2.6 mm (fixed). Transverse sections (Theiler Stage 15; rat, Witschi Stages 18–19; human, Carnegie Stage 12).

1. roof of hindbrain
2. fourth ventricle
3. neuroepithelium of hindbrain
4. primary head vein
5. trigeminal (V) neural crest cells within origin of maxillary component of first branchial arch
6. olfactory placode
7. telencephalic vesicle
8. lamina terminalis
9. perioptic vascular plexus
10. optic stalk
11. optic vesicle/early stage in development of optic cup
12. hindlimb bud
13. somite
14. amnion
15. paired dorsal aortae, only separated by their endothelial walls
16. hindgut diverticulum
17. notochord
18. dilated region of neural tube in caudal region of the tail just proximal to the caudal neuropore
19. neural lumen
20. facio-acoustic (VII–VIII) neural crest complex
21. maxillary component of first branchial arch
22. peripheral boundary of entrance to Rathke's pouch
23. caudal margin of infundibular recess of diencephalon
24. intra-embryonic coelomic cavity (caudal extension of peritoneal cavity)
25. distal recurved parts of left and right dorsal aortae
26. distal dilated part of hindgut diverticulum
27. dorsal aorta in distal region of the tail
28. wall of otocyst (otic vesicle)
29. rostral extremity of notochord
30. forebrain vesicle
31. otocyst (otic vesicle)
32. acoustic (auditory, VIII) component of facio-acoustic (VII–VIII) neural crest complex
33. facial (VII) component of facio-acoustic (VII–VIII) neural crest complex
34. first branchial pouch
35. oropharynx
36. caudal extremity of midline dorsal aorta – more distally, paired dorsal aortae
37. caudal extremity of dorsal aorta
38. caudal neuropore
39. facial (VII) neural crest cells within second branchial arch
40. first branchial membrane (future tympanic membrane)
41. mandibular component of first branchial arch
42. neuroepithelium of forebrain
43. midline dorsal aorta
44. condensation of tissue (poorly defined) in caudal region of notochord subjacent to caudal neuropore
45. dorsal aorta
46. communication between second branchial arch artery and dorsal aorta
47. first branchial arch artery
48. rostral extremity of pericardial cavity
49. site of communication between paired dorsal aortae and umbilical artery
50. caudal extremity of hindgut diverticulum
51. second branchial pouch (dorsal component)
52. second branchial arch artery
53. body wall (thoracic) overlying pericardial cavity
54. trabeculation in myocardial wall of common ventricular chamber of heart
55. truncus arteriosus region of outflow tract of heart
56. left umbilical vein
57. umbilical artery
58. right umbilical vein
59. glossopharyngeal (IX) neural crest tissue in third branchial arch
60. aortic sac
61. nephrogenic cord tissue subjacent to urogenital ridge – caudal extremity of nephric duct (not yet canalized in this location)
62. second branchial membrane
63. communication between third branchial arch artery and dorsal aorta
64. third branchial arch
65. second branchial arch
66. origin of first branchial arch artery from aortic sac
67. endothelial lining of bulbo-truncal canal
68. bulbus cordis region of heart
69. third branchial arch artery
70. right anterior cardinal vein
71. communication between aortic sac and third branchial arch artery
72. wall of left component of common atrial chamber of heart
73. bulbo-ventricular groove
74. nephric duct
75. wall of right component of common atrial chamber of heart
76. third branchial membrane
77. third branchial pouch
78. atrio-ventricular canal
79. bulbo-ventricular canal
80. right component of common atrial chamber of heart
81. pharyngeal region of foregut
82. common atrial chamber of heart
83. pericardial cavity
84. communication between vitelline (omphalomesenteric) artery and midline dorsal aorta
85. hindgut

PLATE 20c

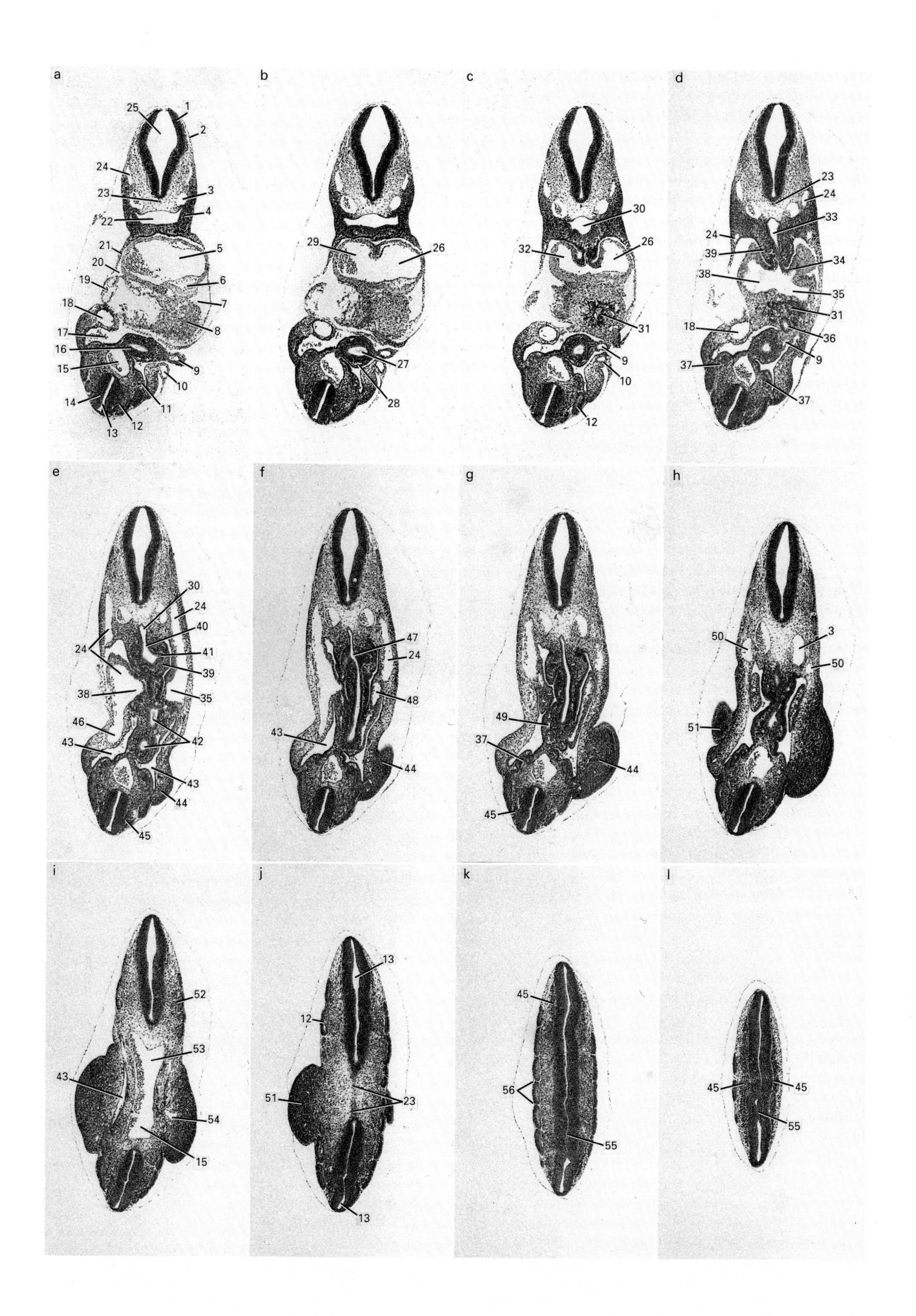

a
25
1
2
24
23
3
22
4
21
5
20
6
19
7
18
8
17
16
9
15
10
11
14
12
13
b
29
26
27
28
c
30
32
26
31
9
10
12
d
23
24
33
24
39
34
38
35
31
18
36
9
37
37
e
30
24
40
24
41
39
38
35
46
42
43
43
44
45
f
47
24
48
43
44
g
49
37
44
45
h
50
3
50
51
i
52
53
43
54
15
j
13
12
51
23
13
k
45
56
55
l
45
45
55

Plate 20c (10–10.25 days p.c.)

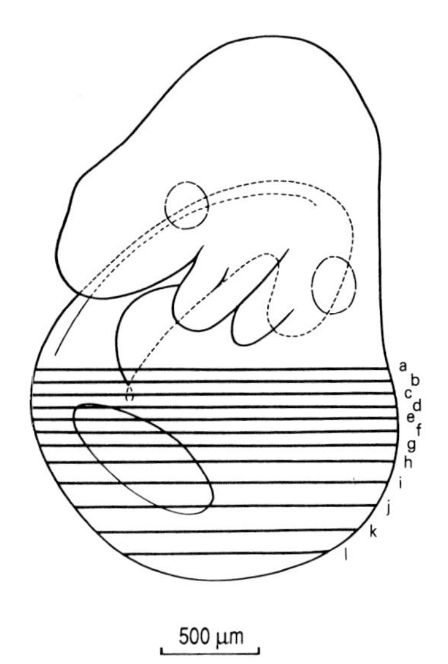

10–10.25 days p.c. Embryo with 25–30 pairs of somites. Crown–rump length 2.6 mm (fixed). Transverse sections (Theiler Stage 15; rat, Witschi Stages 18–19; human, Carnegie Stage 12).

1. neuroepithelium of caudal part of hindbrain
2. surface ectoderm
3. dorsal aorta
4. glossopharyngeal (IX) neural crest tissue within third branchial arch
5. common atrial chamber of heart
6. endocardial cushion tissue associated with wall of atrio-ventricular canal
7. left pericardio-peritoneal canal
8. septum transversum
9. vitelline artery
10. left umbilical vein
11. nephric duct (canalized at this level)
12. somite
13. neural lumen
14. neural tube
15. midline dorsal aorta
16. hindgut
17. intra-embryonic coelomic cavity (caudal extension of peritoneal cavity)
18. right umbilical vein
19. wall of bulbus cordis region of heart
20. right pericardio-peritoneal canal
21. body wall (thoracic) overlying pericardial cavity
22. pharyngeal region of foregut
23. notochord
24. anterior cardinal vein
25. caudal region of fourth ventricle
26. left component of common atrial chamber of heart
27. hindgut–midgut junction
28. vitelline vessels within wall of gut tube
29. right component of common atrial chamber of heart
30. oesophageal region of foregut
31. elements of hepatic/biliary primordia within septum transversum
32. right horn of sinus venosus
33. ventral bifurcation of foregut to produce tracheal diverticulum (laryngo-tracheal groove)
34. left horn of sinus venosus
35. left common cardinal vein
36. cystic (gall bladder) primordium
37. nephrogenic cord tissue (nephric duct and vesicle) subjacent to urogenital ridge
38. right common cardinal vein
39. lung bud
40. trachea
41. left main bronchus
42. midgut
43. intra-embryonic coelomic (peritoneal) cavity
44. left forelimb bud
45. neural crest derived spinal primordia (will give rise to sensory cells of dorsal (posterior) root ganglia)
46. communication between right umbilical vein and right posterior cardinal vein
47. foregut–midgut junction
48. communication between vitelline vein and left common cardinal vein
49. vitelline venous plexus within wall of midgut
50. posterior cardinal vein
51. right forelimb bud
52. condensation of paraxial (presomite) mesoderm
53. rostral site of fusion of right and left dorsal aortae
54. principal (axial) artery to left forelimb bud
55. localized region of neural luminal occlusion
56. intersegmental zone of decreased mesodermal cell packing density

PLATE 21a

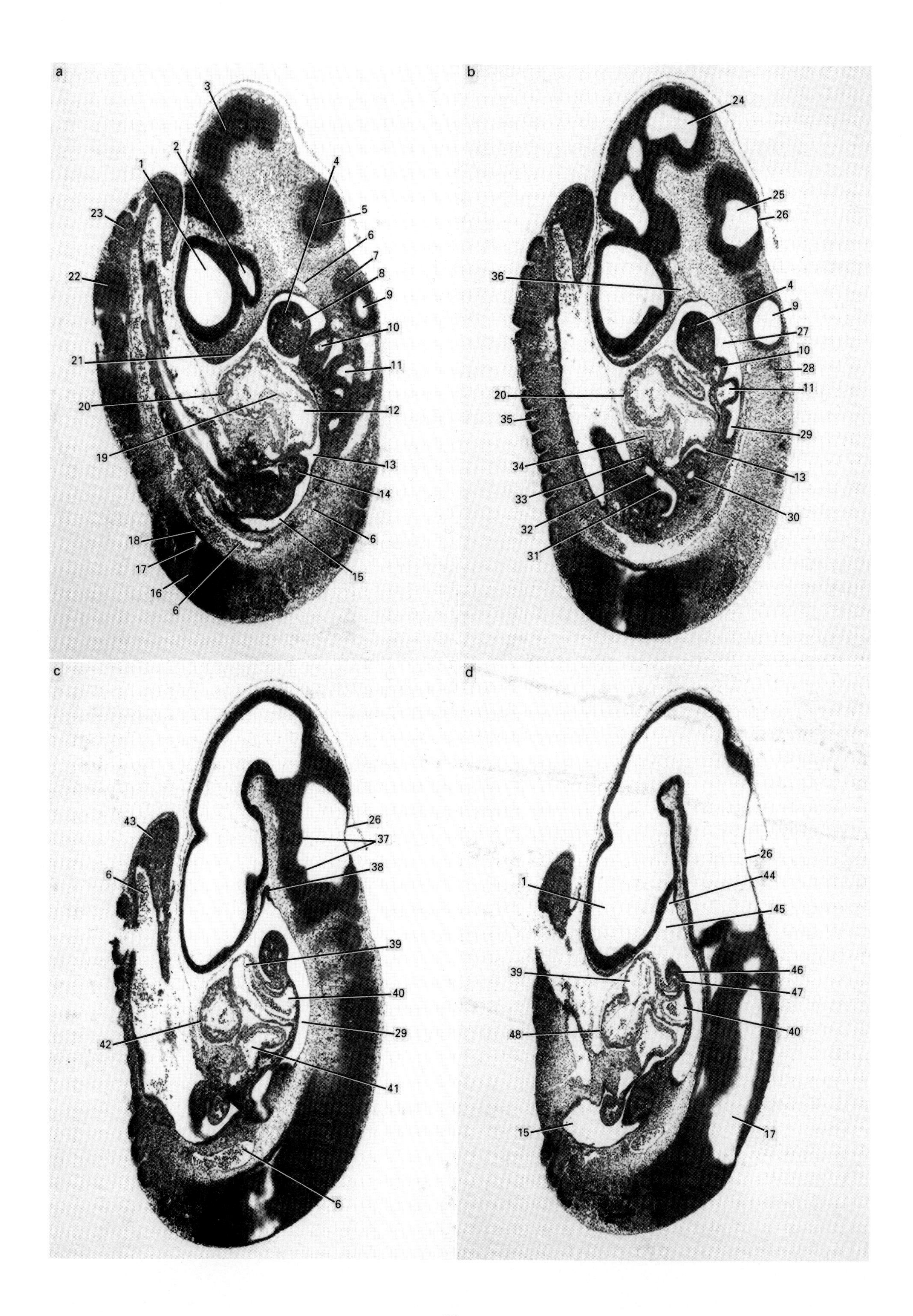
a
1
2
3
4
5
6
7
8
9
10
11
12
13
14
15
16
17
18
19
20
21
22
23
b
24
25
26
4
9
27
10
28
11
29
13
30
31
32
33
34
35
20
36
c
26
37
38
39
40
29
41
42
43
6
d
26
44
45
46
47
40
48
39
1
15
17

Plate 21a (10–10.25 days p.c.)

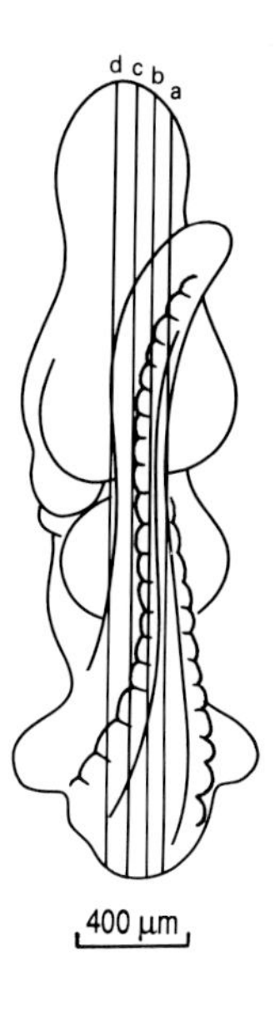

10–10.25 days p.c. Embryo with 25–30 pairs of somites. Crown–rump length 2.8 mm (fixed). Sagittal sections (Theiler Stage 15; rat, Witschi Stages 18–19; human, Carnegie Stage 12).

1. forebrain (telencephalic vesicle)
2. optic stalk/optic vesicle
3. neuroepithelial lining of midbrain
4. mandibular component of left first branchial arch
5. neuroepithelial lining of hindbrain
6. dorsal aorta
7. facio-acoustic (VII–VIII) neural crest complex
8. first branchial arch artery
9. left otic vesicle (otocyst)
10. second branchial arch artery
11. third branchial arch artery
12. left component of common atrial chamber
13. pericardio-peritoneal canal
14. lung bud
15. intra-embryonic coelomic cavity (peritoneal cavity)
16. neural tube (cut obliquely)
17. neural lumen
18. notochord
19. atrio-ventricular canal
20. trabeculation in myocardial wall of common ventricular chamber of heart
21. olfactory placode
22. neural tube in tail region
23. somite
24. lumen of midbrain (mesencephalic vesicle)
25. lumen of hindbrain (fourth ventricle)
26. roof of hindbrain
27. pharyngeal region of foregut
28. glossopharyngeal (IX) neural crest
29. oesophageal region of foregut
30. left main bronchus
31. midgut
32. hepatic diverticulum from foregut–midgut junction
33. hepatic primordia within septum transversum
34. tissue of septum transversum
35. myocoele
36. perioptic branches from primary head vein
37. rhombomere
38. Rathke's pouch
39. truncus arteriosus
40. aortic sac giving origin to branchial arch arteries
41. right horn of sinus venosus
42. bulbo-ventricular junction
43. neuroepithelium in caudal region of tail
44. entrance to Rathke's pouch
45. oropharynx
46. site of fusion in the midline of the two mandibular components of the first branchial arch
47. thyroid primordium
48. bulbus cordis

PLATE 21b

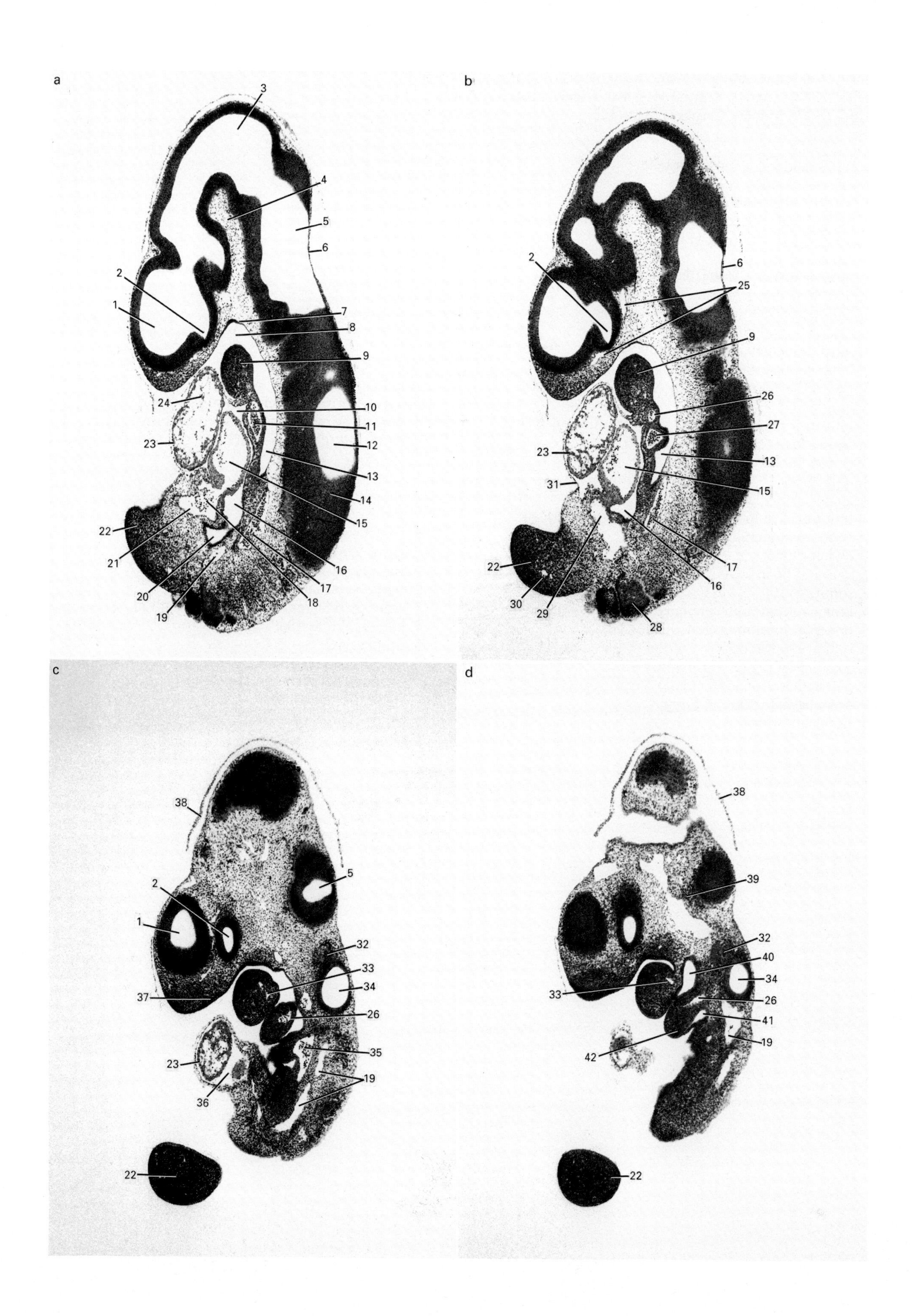
a
b
c
d
1
2
3
4
5
6
7
8
9
10
11
12
13
14
15
16
17
18
19
20
21
22
23
24
25
26
27
28
29
30
31
32
33
34
35
36
37
38
39
40
41
42

Plate 21b (10–10.25 days p.c.)

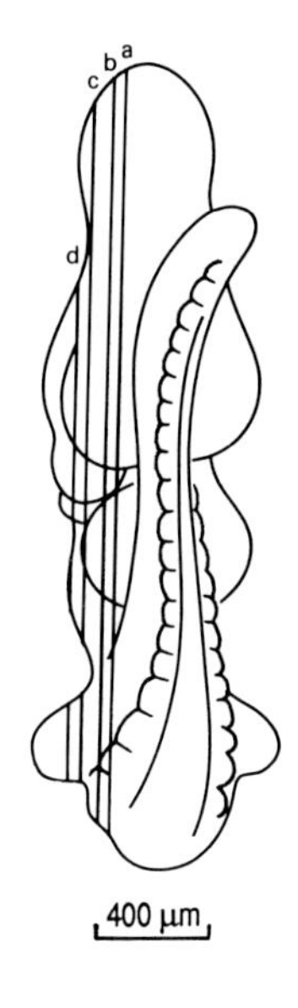

10–10.25 days p.c. Embryo with 25–30 pairs of somites. Crown–rump length 2.8 mm (fixed). Sagittal sections (Theiler Stage 15; rat, Witschi Stages 18–19; human, Carnegie Stage 12).

1. forebrain (telencephalic vesicle)
2. optic stalk/optic vesicle
3. lumen of midbrain (mesencephalic vesicle)
4. cephalic mesenchyme tissue
5. fourth ventricle
6. roof of hindbrain
7. peripheral margin of Rathke's pouch
8. oropharynx
9. mandibular component of first branchial arch
10. transverse pericardial sinus
11. origin of second and third branchial arch arteries from the aortic sac
12. neural lumen
13. oesophagus
14. neuroepithelium of neural tube
15. right component of common atrial chamber
16. pericardio-peritoneal canal
17. dorsal aorta
18. right horn of sinus venosus
19. anterior cardinal vein
20. peritoneal cavity
21. umbilical vein
22. right forelimb bud
23. bulbus cordis
24. truncus arteriosus
25. perioptic branches from primary head vein
26. second branchial arch artery
27. third branchial arch artery
28. somite
29. common cardinal vein
30. axial limb vessels
31. body wall overlying pericardial cavity
32. facio-acoustic (VII–VIII) neural crest complex
33. first branchial arch artery
34. otic vesicle (otocyst)
35. communication between third branchial arch artery and dorsal aorta
36. pericardial cavity
37. olfactory placode
38. amnion
39. trigeminal (V) neural crest
40. first branchial pouch
41. second branchial pouch
42. second branchial arch

veins, while the umbilical veins, which may be observed within the lateral body wall in the mid-trunk region, drain either into the posterior cardinal veins (*Plate 20c,e,f*), or directly into the common cardinal veins. When embryos at this stage of development are maintained *in vitro* under optimal culture conditions, the heart rate is about 100–110 beats/minute.

THE PRIMITIVE GUT

As in the previous stage, the pharynx remains the most impressive region of the primitive gut. In the roof of the oropharynx, the entrance to Rathke's pouch becomes wider throughout this stage (*Plate 19a,k,l*; *Plate 20b,a,b*), while its lining cells become progressively more prominent. The basal layers of the dorsal wall of Rathke's pouch and that of the ventral wall of the infundibular recess of the third ventricle make direct contact (*Plate 19a,k*; *Plate 20b,a*; *Plate 21a,d*), with no intervening cephalic mesenchyme.

The gradual increase in size of the third branchial arch, allows the second branchial pouch to differentiate, so that it develops both a dorsal (*Plate 19b,e*; *Plate 20b,g*) and a ventral (*Plate 19b,g*) component. In embryos with 25–30 pairs of somites, a third branchial pouch is also evident (*Plate 20b,k*). The first indication of a fourth pouch may also be observed at this time (*Plate 20c,b*). Slightly caudal to this region, the luminal diameter of the primitive pharynx narrows significantly, and this region represents the upper part of the future oesophagus (*Plate 20c,c*). On the ventral aspect of this region, a deep groove is first apparent (*Plate 19c,a*), termed the laryngo-tracheal groove, which indents further to form the lumen of the tracheal diverticulum of the foregut (*Plate 20c,d*). Its distal part is seen to bifurcate to form the left and right main (or principal) bronchi (*Plate 20c,e*), and it is around the latter that the two lung buds differentiate (*Plate 20c,d–f*).

The septum transversum is now seen to contain an increasing amount of hepatic/biliary primordia (*Plate 19b,k,l*; *19c,a,b*; *Plate 20c,a–d*), and a gall bladder (or cystic) primordium is now also clearly seen (*Plate 19c,b*; *Plate 20c,d*). The biliary system originates from an outgrowth (the biliary bud) at the foregut–midgut junction (*Plate 20c,f*), and this is also the level from which the pancreatic primordia will shortly emerge. The vitelline duct closes in embryos with about 25 pairs of somites. A marked increase also occurs in the vascularity of the wall of the midgut and hindgut during this stage of development, and the vitelline venous plexus drains to the common cardinal veins via the right and left vitelline veins (*Plate 19c,c–e*; *Plate 20c,e–g*).

THE NERVOUS SYSTEM

Apart from the changes occurring in the prosencephalon to form the third ventricle and the two telencephalic vesicles (described above), the most impressive features observed during this stage also relate to this region. A deepening of the infundibular recess of the third ventricle occurs, and its basal lamina makes direct contact with the basal lamina of the dorsal wall of Rathke's pouch (see above). In the ventral midline, the wall of the prosencephalon between the two telencephalic vesicles thins to form the lamina terminalis (*Plate 20a,i–l*; *Plate 21a,c*).

The primitive optic apparatus also undergoes marked changes during this stage of development, and this is particularly evident in relation to their gradual change in location, which is principally due to the increase in the volume of the telencephalic vesicles. The optic stalks are now seen to have a considerably smaller luminal diameter than that of the optic vesicles (*Plate 20a,l*). The outer surface of the optic vesicles shows the first evidence of collapsing inwards to form the (secondary) optic cups (*Plate 20a,j–l*), while the surface ectoderm in the localized regions overlying the central part of the optic cups is induced to form the lens placodes (*Plate 20a,l*).

A change is also evident in the surface ectoderm overlying the ventrolateral aspect of the telencephalic vesicles, with the differentiation of the olfactory placodes. However, unlike the situation described above in relation to the lens placode, a considerable volume of cephalic mesenchyme intervenes between the surface and neural ectoderm in this region (*Plate 20b,a–d*), and the peripheral boundaries of the olfactory placodes are only poorly defined at this stage. In the hindbrain region, the volume of the fourth ventricle increases markedly throughout this stage, and its roof plate becomes stretched and thinned (which is a characteristic feature of this region) (*Plate 19a,d–l*; *19b,a*; *Plate 20a,e–l*, *20b,a–l*; *Plate 21a,d*, *21b,a,b*).

In embryos with 20–25 pairs of somites, the otocysts have completely separated from the overlying surface ectoderm (*Plate 20b,d–f*; *Plate 21a,b*). The trigeminal (V) neural crest tissue that does not migrate into the first branchial arch has progressively condensed to form a major (the sensory) component of the trigeminal (V) ganglion (*Plate 20a,i–l*; *20b,a*). The facio-acoustic (VII–VIII) neural crest complex also becomes progressively more prominent, and is located in close association with the superior and anterior surfaces of the otocyst (*Plate 19a,k,l*, *19b,a*; *Plate 20b,b–d*; *Plate 21a,b*, *21b,c,d*).

In the mid-trunk region of the embryo, it is possible to recognize condensations of neural crest-derived cells located on either side of the neural tube (*Plate 19c,j–l*; *Plate 20c,k,l*), and these will give rise to the sensory cells of the dorsal (posterior) root ganglia. As indicated above, the caudal neuropore is seen to be in the process of closing throughout this stage (*Plate 20b,d–f*), and closure is only finally achieved in embryos with about 30–35 pairs of somites.

In paraffin-embedded material, only a moderate degree of residual neural luminal occlusion is seen (e.g. *Plate 19c,i–l*; *Plate 20c,k,l*), particularly in the lower trunk region of the embryo. However, when comparable material is sectioned, having previously been embedded in plastic, this phenomenon is first seen in embryos with about 10–12 pairs of somites. In embryos with about 20 pairs of somites, almost complete neural luminal occlusion may extend caudally (*Plate 20b*) from the region of the otic pits to the proximal part of the tail region, and this process is believed to play a critical role in facilitating the dilatation of the primary brain vesicles (Kaufman, 1983b, 1986).

THE UROGENITAL SYSTEM

While nephrogenic cord tissue and the nephric ducts are first seen during the latter part of the previous stage (*Plate 18a,o*), during this stage of development, the urogenital ridges enlarge on either side of the dorsal mesentery of the hindgut, being most pronounced in the mid-trunk region (*Plate 19c,a–f*; *Plate 20c,b–h*), where they protrude into the dorsal part of the peritoneal cavity. The urogenital ridges extend caudally from this region to the mid-tail region (*Plate 19a,g–l*; *Plate 20b,a–g*). It is also just possible to discern the presence of nephric vesicles, and note that the nephric ducts are now canalized along much of their length (*Plate 20c,a–g*), and are directed caudally towards the region of the cloaca.

Stage 16 Developmental age, 10 days p.c. Plates 22a–c; Plates 23a–d illustrate an advanced Stage 16/early Stage 17 embryo

EXTERNAL FEATURES

Embryos with 30–34 pairs of somites. As in previous stages of development, the cephalic region is in developmental terms significantly more advanced than the rest of the embryo, and appears to be disproportionately enlarged compared to the situation observed at later stages of development. The most obvious new external features that differentiate in the cephalic region occur in relation to the optic and olfactory systems. While the overall size of the optic cup is similar to that seen in the previous stage, its outer surface collapses further, and this allows the lens pit to deepen (*Plate 23b,b,c*).

During the early part of this stage, the central region of the olfactory placodes becomes indented, with the consequent formation of the medial and lateral nasal processes (*Plate 22a,b–d*; *Plate 23b,h–l*). On each side, the lateral parts of the medial and lateral nasal processes are in contact with the anteroventral part of the maxillary component of the first arch, and are only separated from it by a shallow groove (the naso-lacrimal groove). The olfactory pit is now seen to be lined by cuboidal/columnar cells, and its peripheral boundaries are more clearly evident than previously.

A shallow and fairly wide groove develops between the two medial nasal processes, and this extends upwards between the two telencephalic vesicles; the groove, however, only extends as far dorsally as the region overlying the ventral part of the midbrain, and coincides with the location of the (subjacent) lamina terminalis (*Plate 23b,a–j*).

In some embryos at this stage, the roof of the oropharyngeal region may be exposed to view, and this allows the entrance to Rathke's pouch to be seen. While the third branchial arch is first evident in embryos with about 20 pairs of somites (*Plate 18a,o*), its overall volume as well as that of the fourth arch, which is just seen in embryos with about 35 pairs of somites, even when fully differentiated, is substantially less than that of the second arch (*Plate 47,e,g*). The third and fourth arches are soon overgrown by the second arches to enclose an ectodermal (retro-hyoid) depression known as the cervical sinus (*Plate 23c,a*; *Plate 47,g*).

Towards the end of this stage, or during the early part of the following stage, the caudal neuropore finally closes (*Plate 23c,f–h*). Just proximal to its caudal extremity, the tail is seen to be flattened and laterally expanded, possibly due to a localised dilatation of the hindgut diverticulum in this region (*Plate 23c,a*). The hindlimb buds, though significantly less well differentiated than the forelimb buds, are quite substantial structures, and now have reasonably well-defined proximal and distal margins (*Plate 22a,a–d, 22b,a,b*; *Plate 23c,a–l*), and are located opposite somites 23–28.

THE HEART AND VASCULAR SYSTEM

The most important vascular event that takes place in embryos with about 30–35 pairs of somites relates to the differentiation of the outflow tract of the heart, with the first definitive evidence of aortico-pulmonary spiral septum formation (*Plate 22b,a*; *Plate 23c,a,b*). Initially, randomly orientated mesenchyme cells with large intercellular spaces appear between the outer myoepicardial layer and the inner layer of endocardial cells. Slightly later, in embryos with about 35–40 pairs of somites, an increase in mesenchyme is seen in this location, and two principal ridges become apparent that are clearly aligned in an oblique direction in relation to the main direction of the outflow tract (*Plate 24c,a–d*) (Fananapazir and Kaufman, 1988). At the same time that the outflow tract is showing early evidence of differentiation to form the spiral septum, an increase is also apparent in the volume of endocardial cushion tissue present, particularly in the region of the atrio-ventricular canal (*Plate 23c,d–g*), so much so, in fact, that it almost obliterates its lumen.

The proximal axial arterial branches from the dorsal aorta to the forelimb buds are now clearly seen (*Plate 23d,d*), as are the principal axial veins of the forelimb bud that drain, at this stage of development, into the posterior cardinal veins (*Plate 23d,b,c*). At the peripheral margins of the forelimb buds, the marginal veins are readily recognized (*Plate 23d,b–e*). In addition to the posterior cardinal veins, it is also possible to recognize the subcardinal veins (*Plate 23c,a,b*), as well as the quite substantial intersubcardinal venous anastomosis ventral to the aorta (*Plate 23c,d*). This system of veins principally drains the mesonephroi and the germinal epithelium of the gonadal ridges that extend along the medial aspects of each mesonephros, and they terminate cranially and caudally in the corresponding posterior cardinal veins, to which they are also joined by various transverse anastomoses.

In the cephalic region, the internal carotid arteries are now recognized (*Plate 23b,c,d*), and represent the rostral extensions of the dorsal aortae. The umbilical vessels within the proximal part of the umbilical cord are particularly well seen, as is their relationship to the intra-embryonic vasculature (*Plates 23c,k,l, 23d,a*).

THE PRIMITIVE GUT

In the floor of the pharyngeal region of the foregut, the thyroid primordium is now clearly seen as a diverticulum of pharyngeal endoderm whose outer surface appears to be in direct contact with the wall of the aortic sac (*Plate 23b,k,l*). In the oesophageal region of the foregut, the laryngo-tracheal groove gives origin to the trachea (*Plate 23c,d–g*), and this in turn gives rise to the left and right main bronchi (*Plate 23c,h*). The two lung buds are somewhat larger than previously, and extend laterally into the pericardio-peritoneal canals (*Plate 23c,i,j*). The lung buds are now located on the ventral aspect of the gastric dilatation of the foregut (future stomach region) (*Plate 23c,k,l*), which is also first evident at this stage of development.

(*continued on page 110*)

PLATE 22a

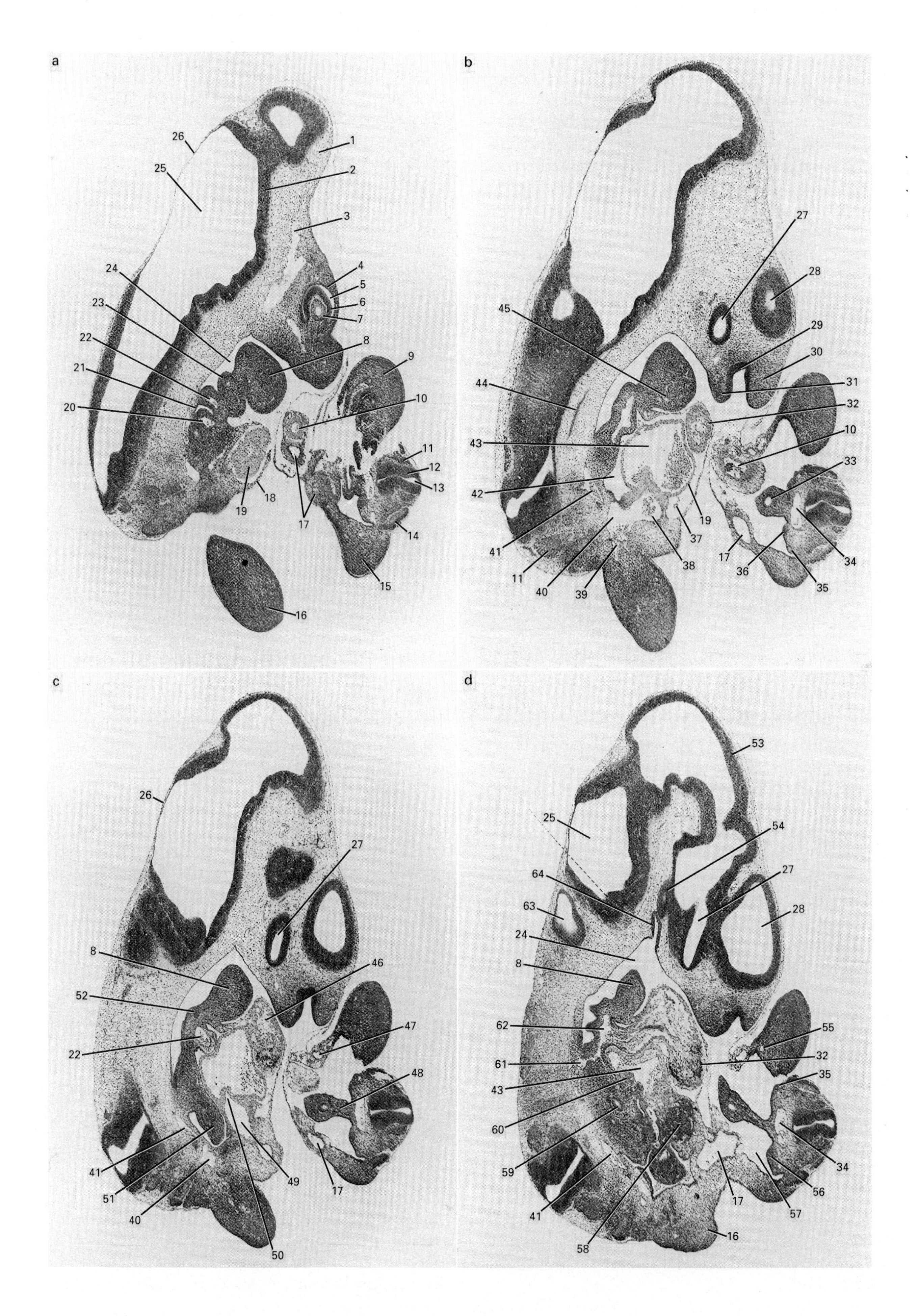
a
1
2
3
4
5
6
7
8
9
10
11
12
13
14
15
16
17
18
19
20
21
22
23
24
25
26
b
27
28
29
30
31
32
10
33
34
35
36
17
37
38
19
39
40
11
41
42
43
44
45
c
26
27
8
46
52
47
22
48
41
51
40
50
49
17
d
53
25
54
64
27
63
28
24
8
62
55
61
32
43
35
60
59
34
41
58
17
56
57
16

Plate 22a (10.25–10.5 days p.c.)

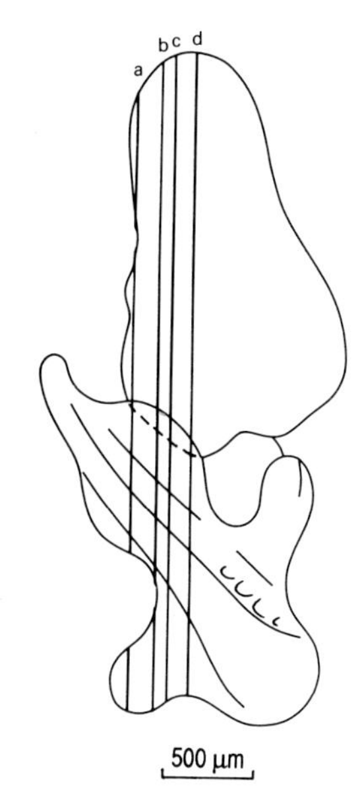

10.25–10.5 days p.c. Embryo with 30–35 pairs of somites. Crown–rump length 3.6 mm (fixed). Sagittal sections (Theiler Stage 16; rat, Witschi Stages 20–21; human, Carnegie Stage 13).

1. cephalic mesenchyme tissue
2. neuroepithelium of hindbrain
3. primary head vein
4. outer layer of optic cup (future pigment layer of retina)
5. intra-retinal space, between inner and outer layers of optic cup
6. inner (neural) layer of optic cup (future nervous layer of retina)
7. lens vesicle
8. mandibular component of first branchial arch
9. left hindlimb bud
10. umbilical artery
11. dorsal (posterior) root ganglion
12. neural tube in tail region
13. neural lumen
14. myotome component of somite
15. right hindlimb bud
16. right forelimb bud
17. right umbilical vein
18. body wall (thoracic) overlying pericardial cavity
19. wall of right component of common atrial chamber of heart
20. fourth branchial arch artery
21. third branchial pouch
22. third branchial arch artery
23. second branchial arch
24. pharyngeal region of foregut
25. fourth ventricle
26. roof of hindbrain
27. optic stalk
28. telencephalic vesicle
29. olfactory epithelium lining olfactory pit
30. lateral nasal process
31. medial nasal process
32. wall of bulbus cordis
33. hindgut
34. midline dorsal aorta
35. mesonephric duct
36. mesonephric tubule
37. pericardial cavity
38. common cardinal vein
39. principal (axial) vein from forelimb bud draining into anterior cardinal vein
40. anterior cardinal vein
41. right dorsal aorta
42. transverse pericardial sinus
43. common atrial chamber of heart
44. notochord
45. first branchial arch artery
46. bulbo-truncal junction
47. left umbilical vein
48. dorsal mesentery of hindgut
49. right horn of sinus venosus
50. venous valves where sinus venosus enters common atrial chamber of heart
51. lung bud
52. thyroid primordium
53. neuroepithelial wall of midbrain
54. neural component of pituitary (infundibulum) arising from the floor of the diencephalon – this gives origin to the pituitary stalk and pars nervosa of the hypophysis
55. principal (axial) artery to the left hindlimb bud
56. urogenital ridge
57. intra-embryonic coelomic cavity (peritoneal cavity)
58. elements of hepatic/biliary primordia within septum transversum
59. main bronchus within lung bud
60. mesentery of heart (dorsal mesocardium)
61. communication between right fourth branchial arch artery and right dorsal aorta
62. origin of the second, third and fourth branchial arch arteries from the aortic sac
63. otocyst (otic vesicle)
64. Rathke's pouch, extending rostrally from the roof of the primitive buccal cavity, which subsequently principally develops into the anterior lobe of the hypophysis

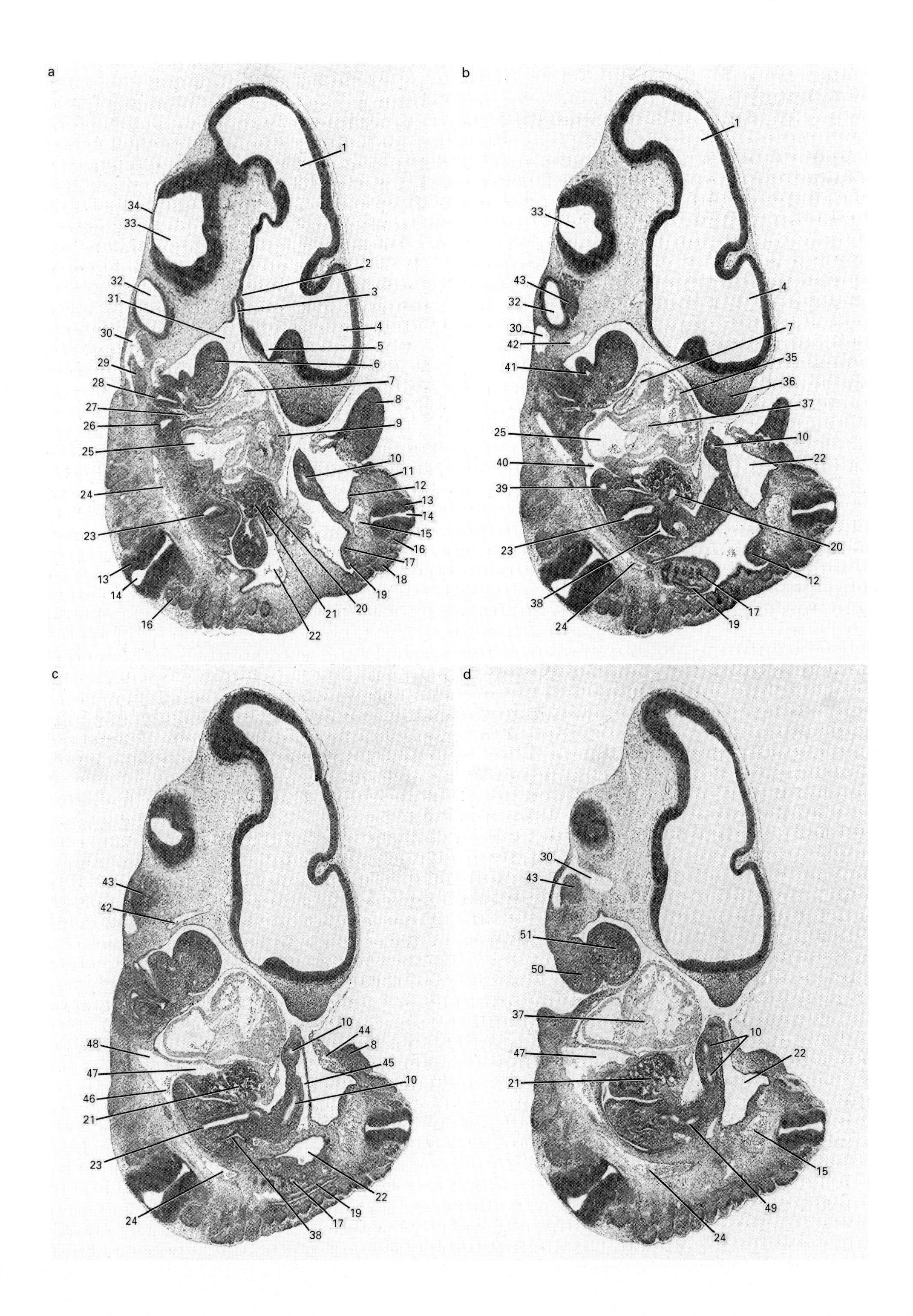
a
1
34
33
2
32
3
31
4
30
5
6
29
7
28
8
27
26
9
25
10
11
24
12
13
14
23
15
16
17
13
18
14
19
16
20
21
22
b
1
33
43
4
32
30
42
7
35
41
36
37
25
10
22
40
39
23
20
12
38
17
24
19
c
43
42
10
44
48
8
45
47
10
46
21
23
22
24
19
17
38
d
30
43
51
50
37
10
47
22
21
15
49
24

Plate 22b (10.25–10.5 days p.c.)

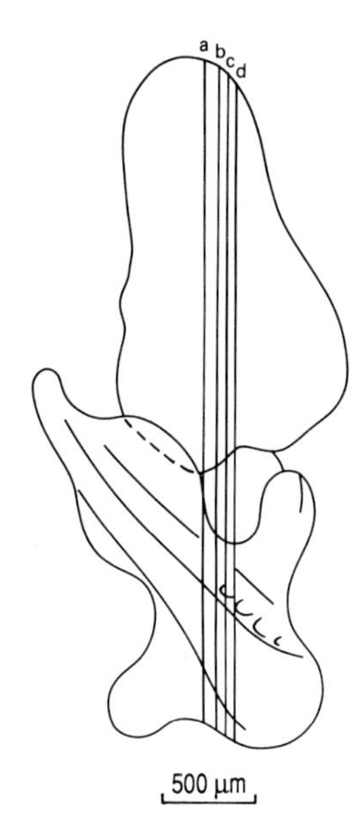

10.25–10.5 days p.c. Embryo with 30–35 pairs of somites. Crown–rump length 3.6 mm (fixed). Sagittal sections (Theiler Stage 16; rat, Witschi Stages 20–21; human, Carnegie Stage 13).

1. mesencephalic vesicle
2. infundibular recess
3. Rathke's pouch
4. telencephalic vesicle
5. entrance to optic stalk (optic recess)
6. mandibular component of first branchial arch
7. truncus arteriosus region with early evidence of aortico-pulmonary spiral septum formation
8. left hindlimb bud
9. bulbus cordis
10. midgut with its associated dorsal mesentery
11. myotome component of somite
12. urogenital ridge
13. neuroepithelium of neural tube
14. neural lumen
15. midline dorsal aorta
16. dorsal (posterior) root ganglion
17. mesonephric tubule
18. sclerotome component of somite
19. mesonephric duct
20. cystic primordium
21. elements of hepatic/biliary primordia within septum transversum
22. intra-embryonic coelomic cavity (peritoneal cavity)
23. lumen of stomach
24. dorsal aorta
25. common atrial chamber of heart
26. fourth branchial pouch
27. communication between aortic sac and third branchial arch artery
28. third branchial pouch
29. glossopharyngeal (IX) neural crest
30. primary head vein
31. oropharyngeal region
32. otocyst (otic vesicle)
33. fourth ventricle
34. roof of hindbrain
35. trabeculated wall of common ventricular chamber of heart
36. medial nasal process
37. endocardial cushion tissue associated with wall of atrio-ventricular canal
38. omental bursa
39. main bronchus within lung bud
40. transverse pericardial sinus
41. second branchial arch artery
42. internal carotid artery (rostral extension of dorsal aorta)
43. facio-acoustic (VII–VIII) preganglion complex
44. left umbilical vein
45. vitelline vein
46. posterior cardinal vein
47. common cardinal vein
48. anterior cardinal vein
49. duodenum
50. left second branchial arch
51. mandibular component of left first branchial arch

PLATE 22c

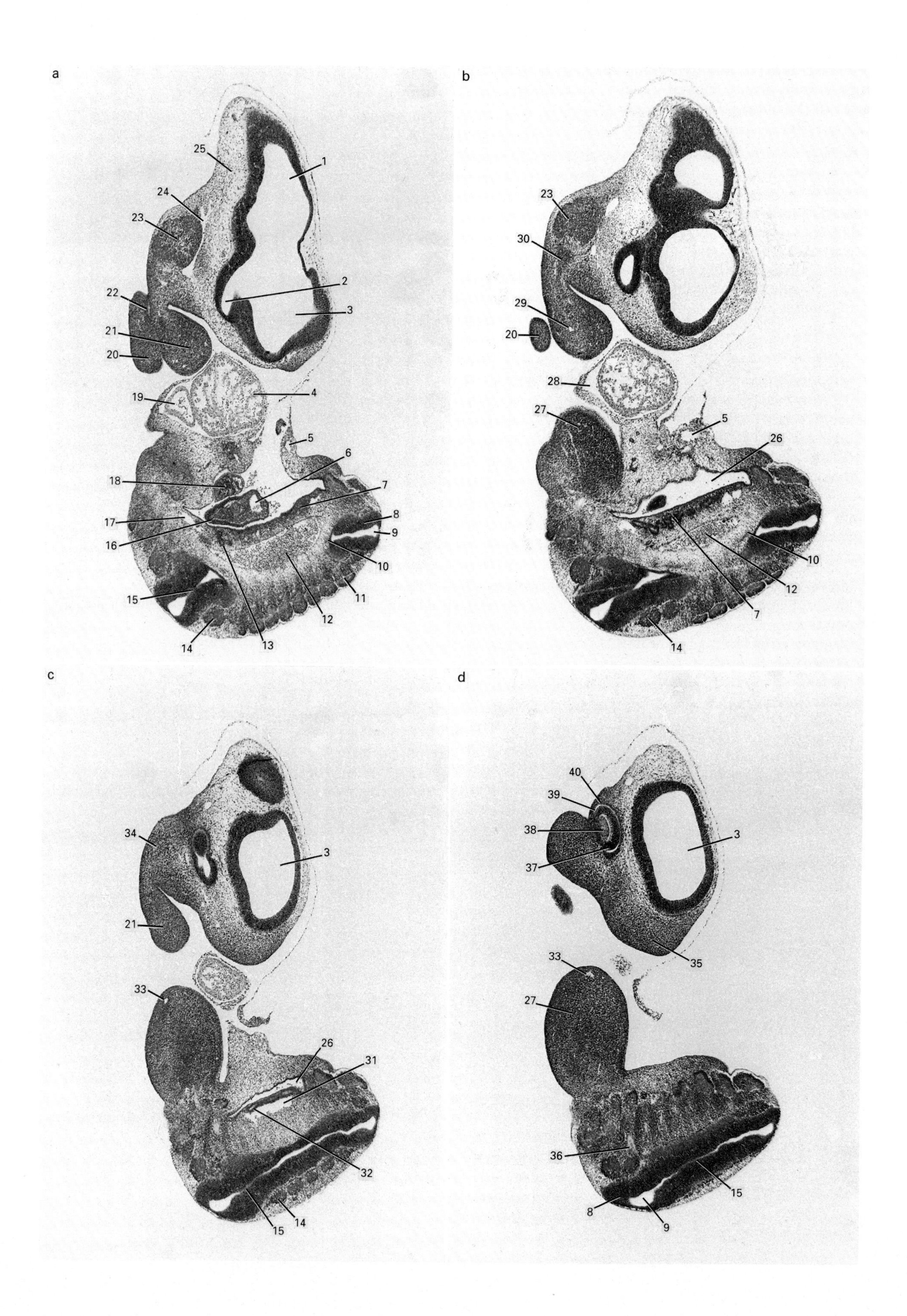
a
b
c
d
1
2
3
4
5
6
7
8
9
10
11
12
13
14
15
16
17
18
19
20
21
22
23
24
25
26
27
28
29
30
31
32
33
34
35
36
37
38
39
40

Plate 22c (10.25–10.5 days p.c.)

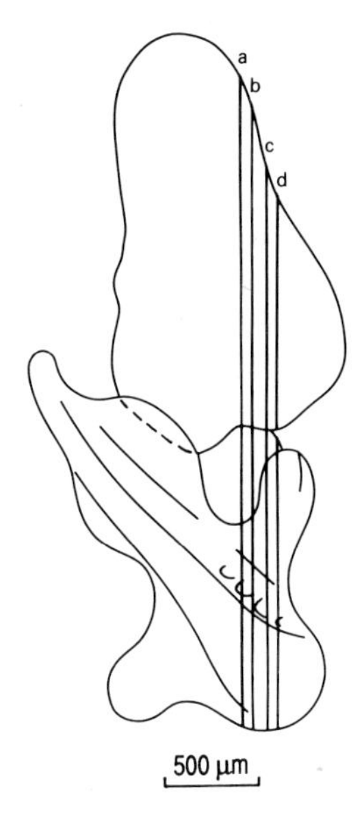

10.25–10.5 days p.c. Embryo with 30–35 pairs of somites. Crown–rump length 3.6 mm (fixed). Sagittal sections (Theiler Stage 16; rat, Witschi Stages 20–21; human, Carnegie Stage 13).

1. mesencephalic vesicle
2. optic stalk
3. telencephalic vesicle
4. trabeculated wall of common ventricular chamber of heart
5. left umbilical vein
6. vitelline vein
7. urogenital ridge showing early stage of differentiation of the mesonephros
8. neuroepithelium of neural tube
9. neural lumen
10. notochord
11. somite
12. midline dorsal aorta
13. mesonephric tubule (vesicle)
14. dorsal (posterior) root ganglion
15. region of neural luminal occlusion
16. mesentery of midgut
17. intersubcardinal venous anastomosis
18. elements of hepatic/biliary primordia within septum transversum
19. left component of common atrial chamber of heart
20. second branchial arch
21. mandibular component of first branchial arch
22. first branchial cleft (groove)
23. trigeminal (V) ganglion
24. primary head vein
25. cephalic mesenchyme tissue
26. intra-embryonic coelomic cavity (peritoneal cavity)
27. left forelimb bud
28. pericardial cavity
29. first branchial arch artery
30. trigeminal (V) neural crest tissue migrating into first branchial arch
31. subcardinal vein
32. mesonephric duct
33. marginal vein
34. maxillary component of first branchial arch
35. lateral nasal process
36. cervical nerve trunk directed towards forelimb bud, and destined to form part of the brachial plexus
37. inner (neural) layer of optic cup (future nervous layer of retina)
38. lens vesicle
39. intra-retinal space
40. outer layer of optic cup (future pigment layer of retina)

PLATE 23a

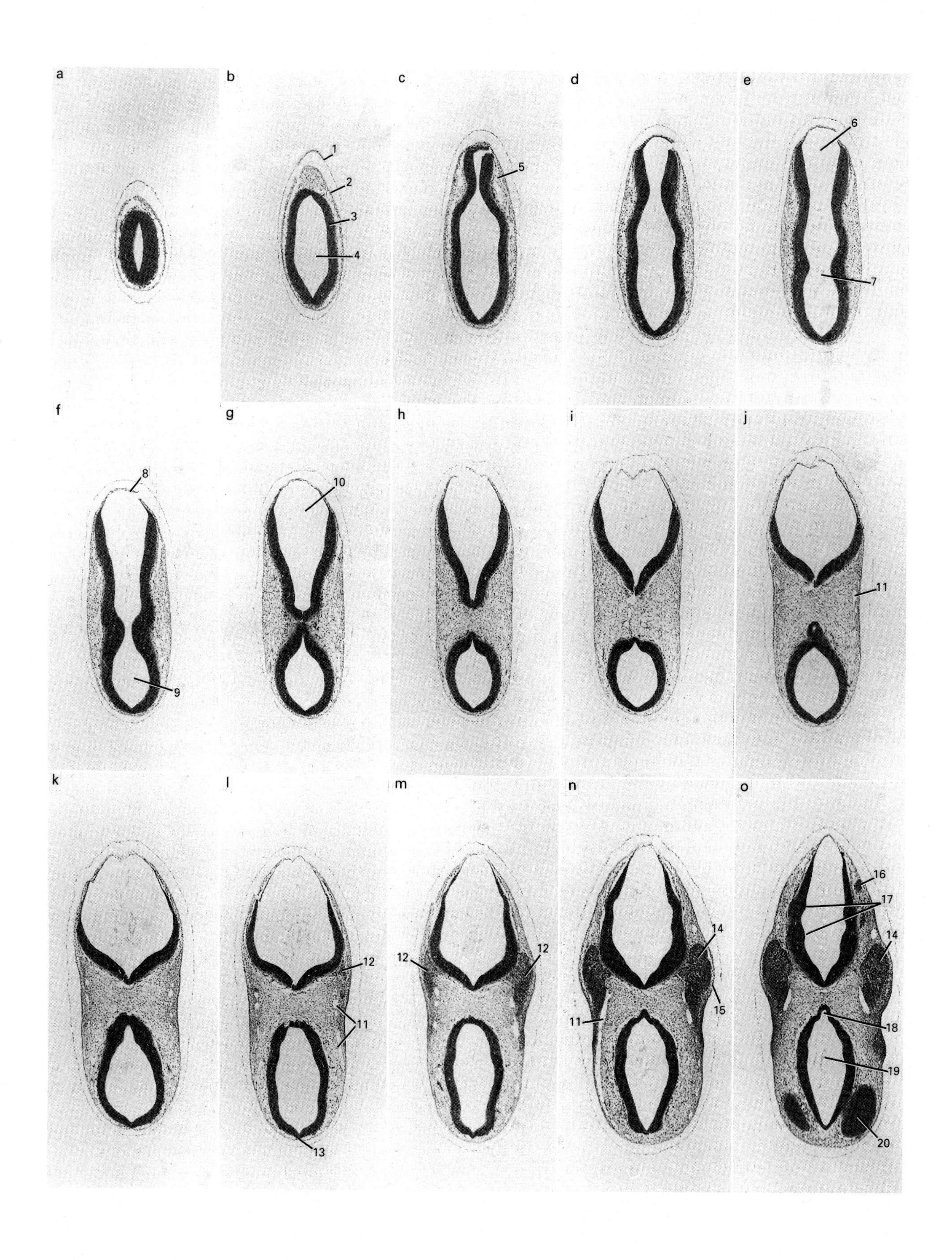

a
b
c
d
e
1
2
3
4
5
6
7
f
g
h
i
j
8
9
10
11
k
l
m
n
o
12
11
13
12
12
14
15
11
16
17
14
18
19
20

Plate 23a (10.25–10.5 days p.c.)

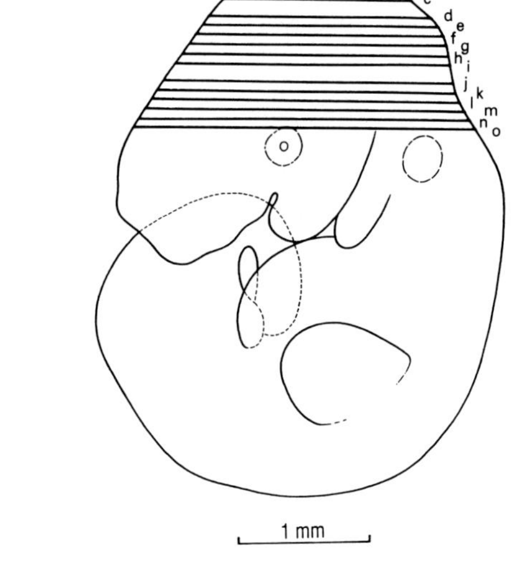

10.25–10.5 days p.c. Embryo with 30–35 pairs of somites. Crown–rump length 3.9 mm (fixed). Transverse sections (Theiler Stages 16–17; rat, Witschi Stages 20–22; human, Carnegie Stage 13).

1. amnion
2. surface ectoderm
3. neuroepithelium
4. mesencephalic vesicle
5. cephalic mesenchyme tissue
6. hindbrain
7. forebrain–midbrain junction
8. roof of hindbrain
9. forebrain
10. fourth ventricle
11. branch of primary head vein
12. trigeminal (V) neural crest tissue
13. lamina terminalis
14. trigeminal (V) ganglion
15. prominence overlying trigeminal (V) ganglion
16. wall of otocyst (otic vesicle)
17. rhombomere
18. diencephalon
19. third ventricle
20. wall of left telencephalic vesicle

PLATE 23b

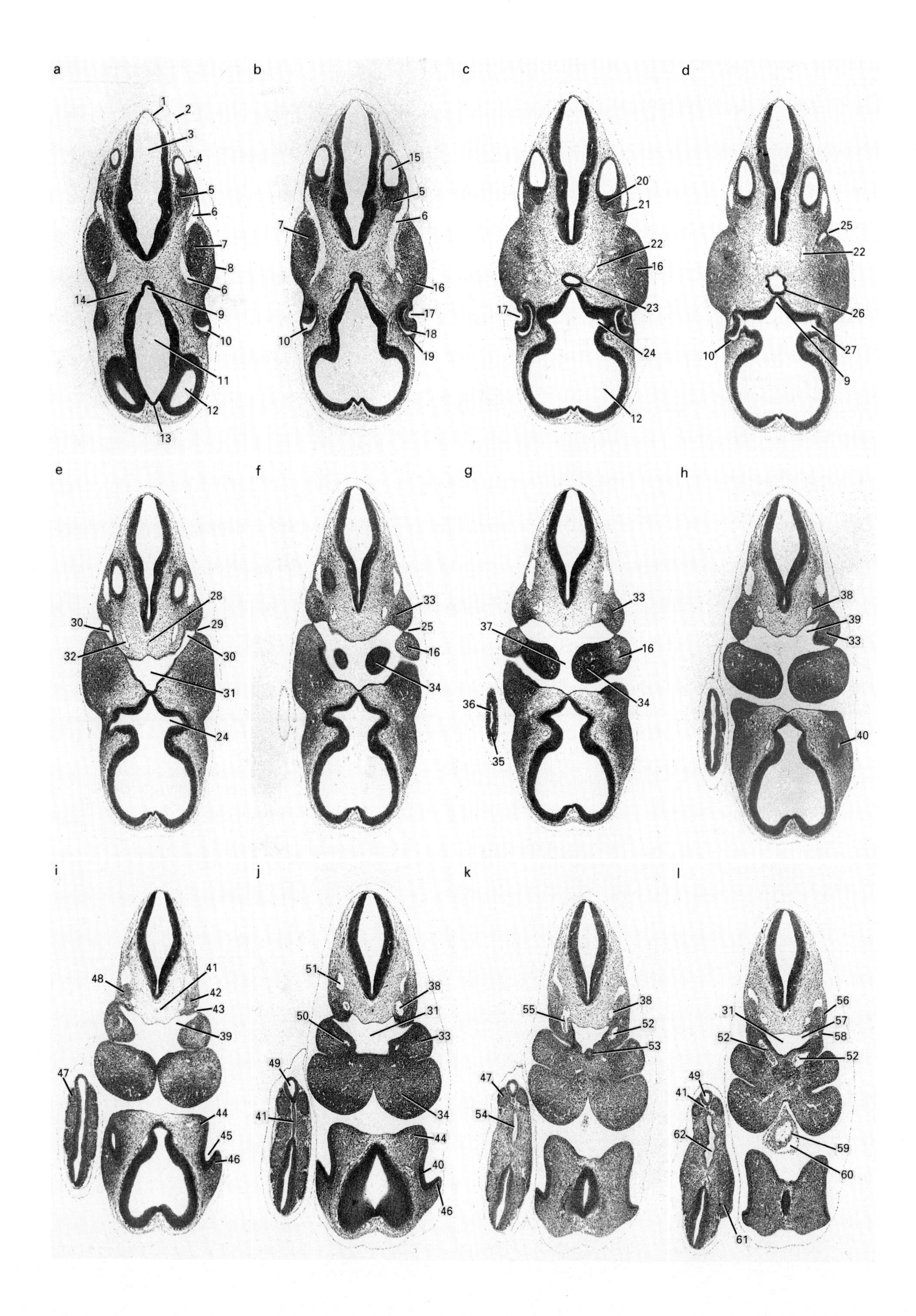
a
b
c
d
e
f
g
h
i
j
k
l
1
2
3
4
5
6
7
8
9
10
11
12
13
14
15
16
17
18
19
20
21
22
23
24
25
26
27
28
29
30
31
32
33
34
35
36
37
38
39
40
41
42
43
44
45
46
47
48
49
50
51
52
53
54
55
56
57
58
59
60
61
62

Plate 23b (10.25–10.5 days p.c.)

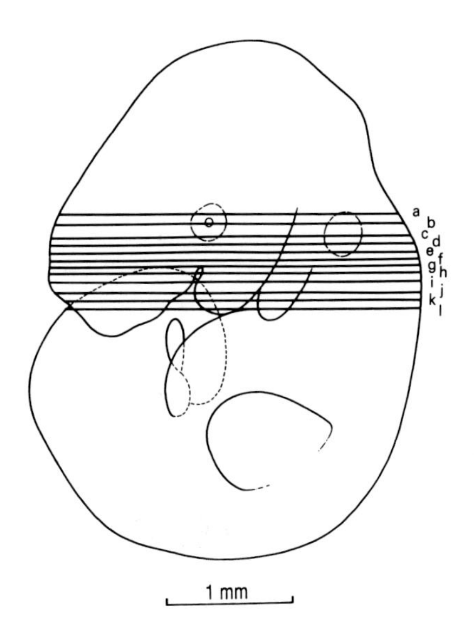

10.25–10.5 days p.c. Embryo with 30–35 pairs of somites. Crown–rump length 3.9 mm (fixed). Transverse sections (Theiler Stages 16–17; rat, Witschi Stages 20–22; human, Carnegie Stage 13).

1. roof of hindbrain
2. amnion
3. fourth ventricle
4. wall of otocyst (otic vesicle)
5. facio-acoustic (VII–VIII) neural crest complex (facio-acoustic (VII–VIII) preganglion)
6. primary head vein
7. trigeminal (V) ganglion
8. prominence overlying trigeminal (V) ganglion
9. infundibular recess of diencephalon
10. intra-retinal space, located between inner and outer layers of optic cup
11. third ventricle
12. telencephalic vesicle
13. lamina terminalis
14. cephalic mesenchyme tissue
15. otocyst (otic vesicle)
16. maxillary component of first branchial arch
17. lens pit
18. inner (neural) layer of optic cup (future nervous layer of retina)
19. outer layer of optic cup (future pigment layer of retina)
20. acoustic (auditory, VIII) component of facio-acoustic (VII–VIII) neural crest complex
21. facial (VII) component of facio-acoustic (VII–VIII) neural crest complex
22. left internal carotid artery (rostral extension of left dorsal aorta)
23. Rathke's pouch
24. optic stalk
25. first branchial cleft (groove)
26. peripheral boundary of entrance to Rathke's pouch
27. communication between lumen of optic stalk and intra-retinal space
28. rostral extremity of notochord
29. first branchial membrane (future tympanic membrane)
30. first branchial pouch
31. pharyngeal region of foregut
32. right internal carotid artery (rostral extension of right dorsal aorta)
33. second branchial arch containing facial (VII) neural crest tissue
34. mandibular component of first branchial arch
35. neural tube in caudal region of the tail
36. neural lumen
37. stomatodaeum
38. dorsal aorta
39. second branchial pouch
40. epithelial lining of olfactory (nasal) pit
41. notochord
42. third branchial arch containing glossopharyngeal (IX) neural crest tissue
43. second branchial membrane
44. medial nasal process
45. olfactory (nasal) pit
46. lateral nasal process
47. somite
48. glossopharyngeal (IX) preganglion
49. dilated region of neural tube in caudal region of the tail
50. second branchial arch artery
51. anterior cardinal vein
52. third branchial arch artery
53. thyroid primordium (thyroid diverticulum)
54. caudal extension of hindgut diverticulum into caudal region of tail (postanal component)
55. communication between right third branchial arch artery and right dorsal aorta
56. origin of fourth branchial arch
57. third branchial pouch
58. third branchial membrane
59. truncus arteriosus/aortic sac region of outflow tract of heart
60. rostral extremity of pericardial cavity
61. cloaca

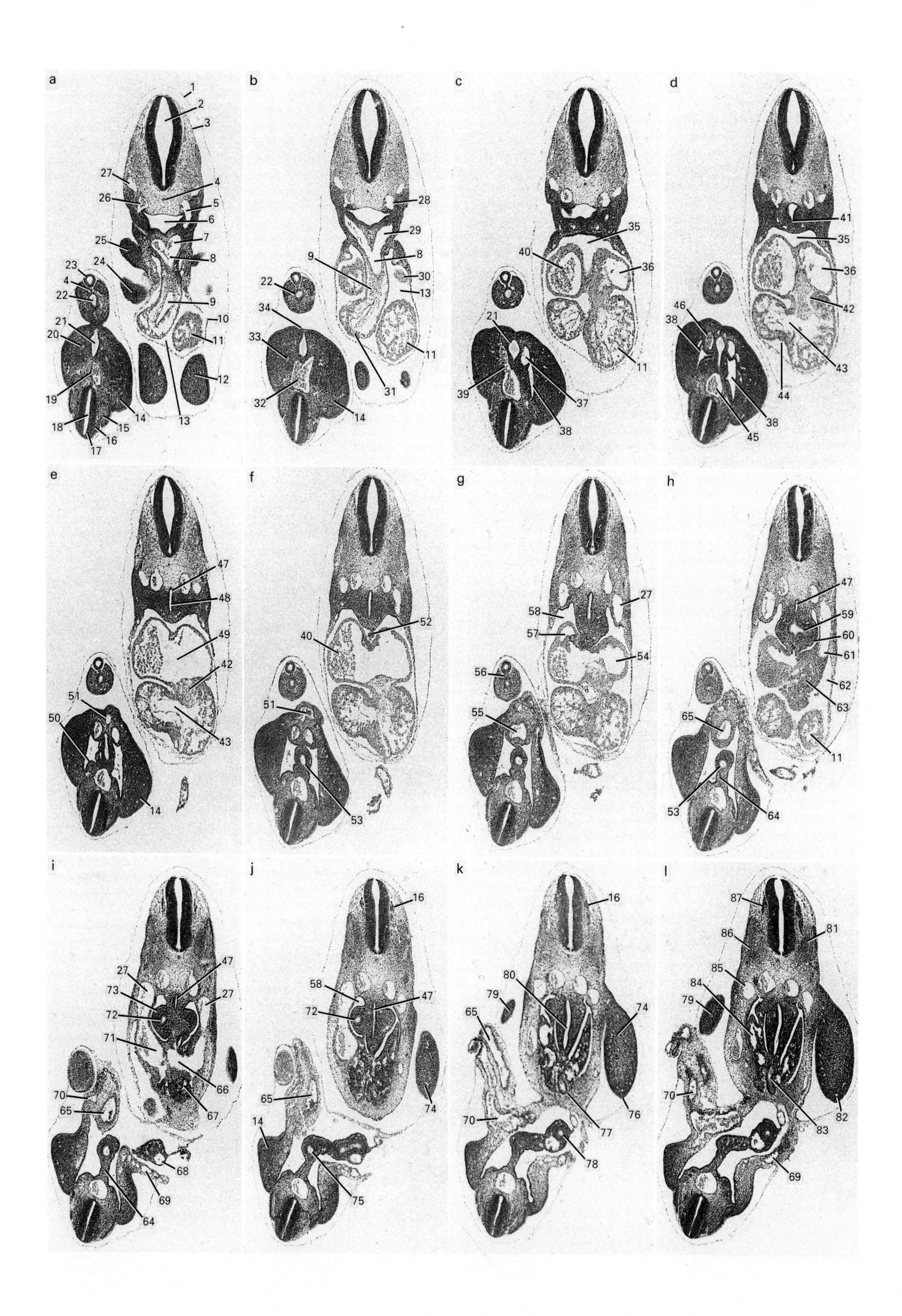
a
1 2 3 27 4 26 5 6 25 7 8 23 24 4 22 9 10 21 20 11 12 19 14 13 18 15 16 17
b
28 29 9 8 22 30 13 34 33 11 32 31 14
c
35 40 36 21 11 39 37 38
d
41 35 36 46 42 38 43 44 45 38
e
47 48 49 42 51 50 43 14
f
52 40 51 53
g
27 58 57 54 56 55
h
47 59 60 61 62 63 65 11 53 64
i
27 47 73 27 72 71 66 70 67 65 68 69 64
j
16 58 47 72 74 65 14 75
k
16 80 79 74 65 70 77 76 78
l
87 81 86 85 84 79 70 82 83 69

Plate 23c (10.25–10.5 days p.c.)

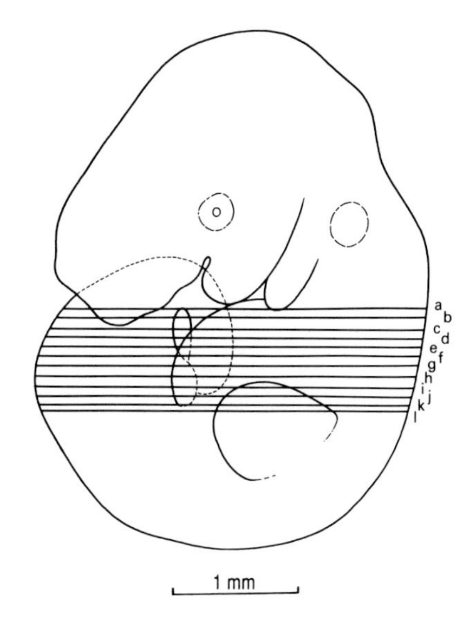

10.25–10.5 days p.c. Embryo with 30–35 pairs of somites. Crown–rump length 3.9 mm (fixed). Transverse sections (Theiler Stages 16–17; rat, Witschi Stages 20–22; human, Carnegie Stage 13).

1. amnion
2. caudal region of fourth ventricle
3. surface ectoderm
4. notochord
5. communication between left third branchial arch artery and left dorsal aorta
6. pharyngeal region of foregut
7. origin of third branchial arch artery from aortic sac
8. aortic sac
9. truncus arteriosus region of outflow tract of heart with early evidence of aortico-pulmonary spiral septum formation
10. body wall (thoracic) overlying pericardial cavity
11. trabeculation in myocardial wall of common ventricular chamber of heart
12. medial nasal process
13. pericardial cavity
14. hindlimb bud
15. somite
16. neural crest derived spinal primordia (will give rise to sensory cells of posterior (dorsal) root ganglia)
17. neural tube
18. neural lumen
19. paired dorsal aortae separated by their endothelial walls
20. common excretory duct (mesonephric duct) just proximal to site of entry into cloaca
21. cloaca
22. caudal extension of hindgut diverticulum into caudal region of tail (postanal component)
23. neural tube in caudal region of tail
24. mandibular component of first branchial arch
25. second branchial arch
26. communication between right third branchial arch artery and right dorsal aorta
27. anterior cardinal vein
28. dorsal aorta
29. third branchial arch artery
30. wall of left component of common atrial chamber of heart
31. bulbus cordis region of heart
32. caudal extremity of midline dorsal aorta
33. common excretory duct (mesonephric duct)
34. cloacal membrane
35. transverse pericardial sinus
36. left component of common atrial chamber of heart
37. recurved distal part of left dorsal aorta (left umbilical artery)
38. intra-embryonic coelomic cavity (caudal extension of peritoneal cavity into tail region)
39. communication between midline dorsal aorta and recurved distal part of right dorsal aorta (right umbilical artery)
40. right component of common atrial chamber of heart
41. ventral bifurcation of foregut to produce tracheal diverticulum (laryngo-tracheal groove)
42. endocardial cushion tissue associated with wall of atrio-ventricular canal
43. bulbo-ventricular canal
44. bulbo-ventricular groove
45. midline dorsal aorta
46. recurved distal part of right dorsal aorta (right umbilical artery)
47. oesophageal region of foregut
48. trachea
49. common atrial chamber of heart
50. urogenital ridge
51. midline umbilical vein
52. mesentery of heart (dorsal mesocardium)
53. hindgut
54. left horn of sinus venosus
55. site of communication between paired dorsal aortae and umbilical artery
56. caudal extremity of notochord
57. right horn of sinus venosus
58. right pericardio-peritoneal canal
59. left main bronchus
60. left lung bud
61. communication between left anterior and left common cardinal veins
62. left pericardio-peritoneal canal
63. septum transversum
64. dorsal mesentery of hindgut
65. umbilical artery
66. left common cardinal vein
67. elements of hepatic/biliary primordia within septum transversum
68. vitelline vein
69. left umbilical vein
70. right umbilical vein
71. right common cardinal vein
72. right main bronchus
73. right lung bud
74. left forelimb bud
75. hindgut–midgut junction
76. apical ectodermal ridge
77. cystic (gall bladder) primordium
78. midgut
79. right forelimb bud
80. gastric dilatation of foregut (future stomach region)
81. condensation of paraxial mesoderm (presomite)
82. marginal vein
83. cystic duct
84. right subcardinal vein
85. posterior cardinal vein
86. cervical nerve trunk, destined to form component of brachial plexus
87. most rostral differentiated dorsal (posterior) root ganglion

PLATE 23d

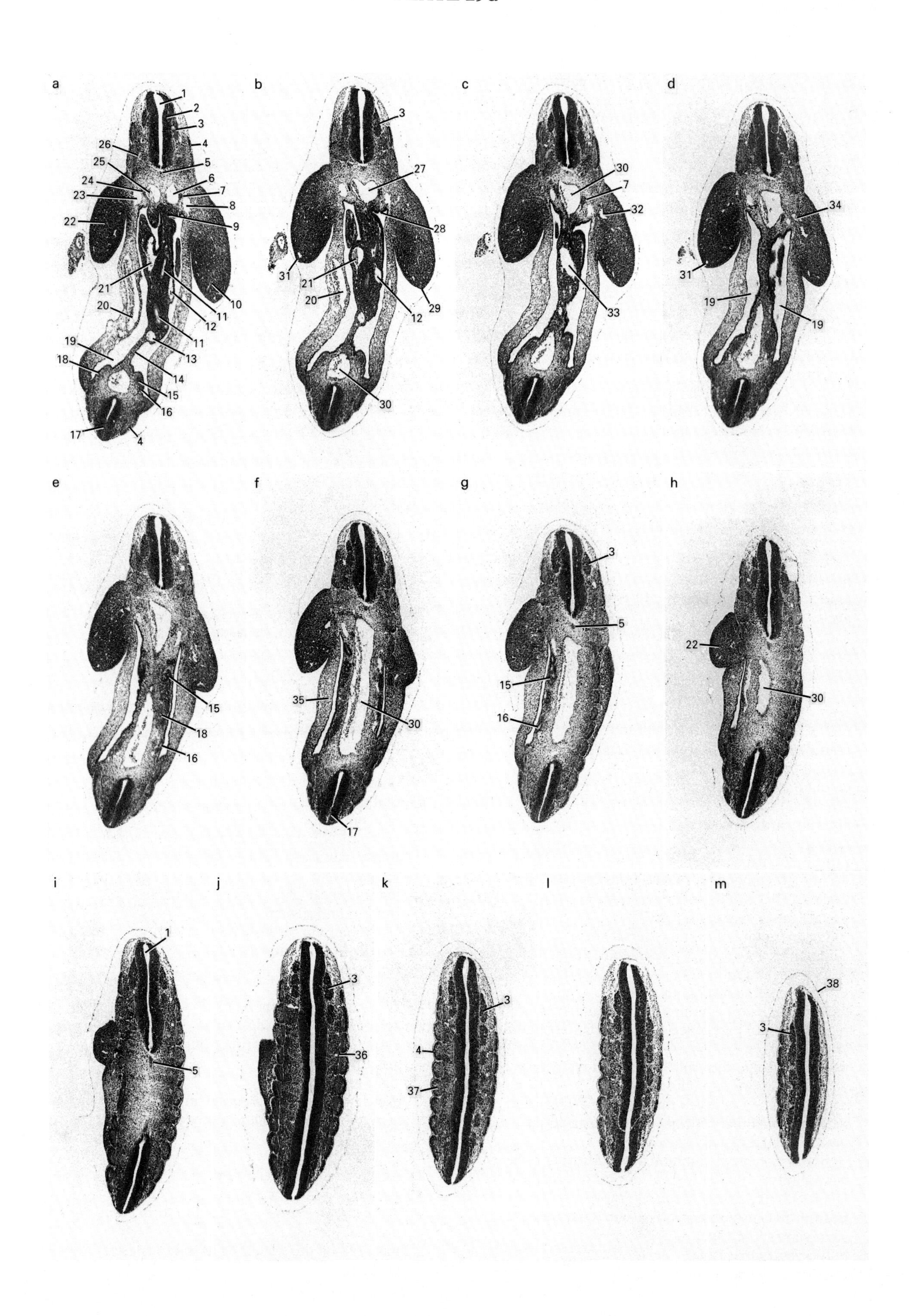
a
b
c
d
e
f
g
h
i
j
k
l
m
1
2
3
4
5
6
7
8
9
10
11
12
13
14
15
16
17
18
19
20
21
22
23
24
25
26
27
28
29
30
31
32
33
34
35
36
37
38

Plate 23d (10.25–10.5 days p.c.)

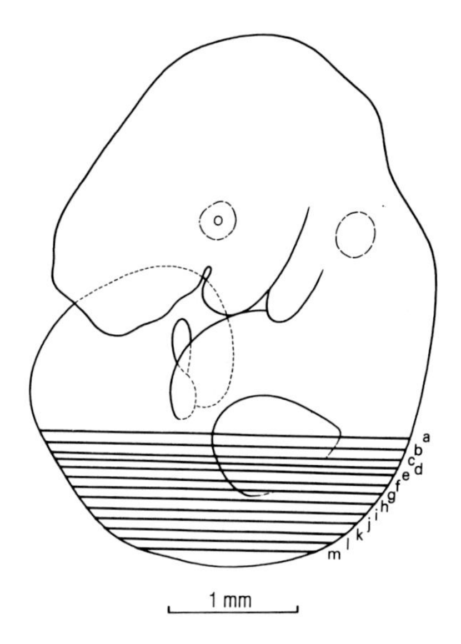

10.25–10.5 days p.c. Embryo with 30–35 pairs of somites. Crown–rump length 3.9 mm (fixed). Transverse sections (Theiler Stages 16–17; rat, Witschi Stages 20–22; human, Carnegie Stage 13).

1. neural lumen
2. cervical region of neural tube
3. dorsal (posterior) root ganglion
4. dermomyotome component of somite
5. notochord
6. left dorsal aorta
7. left sympathetic trunk
8. left anterior cardinal vein
9. dorsal mesentery of gut in foregut–midgut junction
10. left forelimb bud
11. lumen of midgut
12. left subcardinal vein
13. vitelline vein
14. dorsal mesentery of hindgut
15. mesonephric tubule (vesicle)
16. mesonephric duct
17. localized segment of neural luminal occlusion
18. urogenital ridge
19. intra-embryonic coelomic cavity (peritoneal cavity)
20. right umbilical vein
21. right subcardinal vein
22. right forelimb bud
23. right posterior cardinal vein
24. right sympathetic trunk
25. right dorsal aorta
26. cervical nerve trunk destined to form component of brachial plexus
27. rostral site of fusion of right and left dorsal aortae
28. rostral extremity of left urogenital ridge
29. apical ectodermal ridge
30. midline dorsal aorta
31. marginal vein
32. principal (axial) vein from forelimb bud
33. intersubcardinal venous anastomosis
34. principal (axial) arterial branch to forelimb bud from dorsal aorta
35. body wall in mid-trunk region
36. intersegmental artery
37. sclerotome component of somite
38. amnion

An increasing degree of differentiation of the septum transversum is also apparent, and the parenchymal tissue is seen to be invaded by large numbers of venous channels, principally of vitelline venous origin (*Plate 23c,j–l*), to form the hepatic sinusoids. Both the cystic (gall bladder) primordium and its (cystic) duct are clearly seen (*Plate 23c,k,l*). In the most caudal region of the hindgut, the cloaca is now seen as an elongated dilated region (*Plate 23b,l; 23c,a–d*), the endodermal lining of which makes contact with surface ectoderm at the cloacal membrane (*Plate 23c,b*).

THE NERVOUS SYSTEM

In the cephalic region, the two telencephalic vesicles show an increased degree of differentiation than observed previously, and this is particularly well seen on either side of the superior part of the third ventricle (*Plate 23b,a,b*). The trigeminal (V) ganglia are readily recognized because of their enormous size and location (being a reflection of their future functional importance) (*Plates 23a,m–o, 23b,a,b*), and it is now possible in developmentally more advanced specimens at this stage to clearly distinguish between the facial (VII) and acoustic (VIII) components of the facio-acoustic (VII–VIII) ganglion complex (*Plate 23b,c*). It is also possible to recognize the glossopharyngeal (IX) neural crest which is seen to be migrating into the third branchial arch (*Plate 22b,a*), and the glossopharyngeal (IX) preganglion (*Plate 23b,i*). The increased cellularity observed in the fourth branchial arch (*Plate 23b,l*) probably represents the first indication of the vagal (X) neural crest in this location. The neural crest-derived spinal primordia are also clearly seen, and the most rostral of the differentiated dorsal (posterior) root ganglia are now located at a level opposite the proximal part of the forelimb buds (*Plate 23c,l*). The infundibular recess of the third ventricle continues to enlarge, and now makes contact with the rostral extremity as well as part of the posterior surface of Rathke's pouch (*Plates 22a,d, 22b,a*).

With the increasing degree of invagination of the optic cup, it is now possible to distinguish between the histological features of its inner layer of cells which are destined to form the neural part of the retina, and the outer layer of cells which will form the pigment layer of the retina (*Plates 22a,a, 22c,d; Plates 23b,c*), as well as the intra-retinal space located between them. Continuity between the latter space and the third ventricle is via the lumen of the optic stalk.

Localized segments of neural luminal occlusion are evident in the caudal part of the trunk region (*Plates 22b,d, 22c,a–d*), and may extend into the proximal region of the tail (*Plate 23d,a–f*). In paraffin-embedded embryos, it only appears to involve either the middle segment or the dorsal half or one-third of the lumen, whereas in resin-embedded material, the degree and extent of neural luminal occlusion is seen to be considerably more marked (Kaufman, 1983b, 1986).

THE UROGENITAL SYSTEM

A substantial degree of differentiation occurs particularly in the lateral parts of the urogenital ridges, with the development of large numbers of mesonephric vesicles, with several forming at each segmental level (*Plates 22b,b,c, 22c,a,b; Plate 23d,e–g*). The mesonephric ducts are now patent along their entire length (*Plate 22b,b,c; Plate 23d,e,g*), but do not yet drain into the cloaca (*Plate 23c,a*). Analysis of sections stained histochemically to demonstrate the presence of intracellular alkaline phosphatase enzyme activity reveals that while a few primordial germ cells are still located in the dorsal mesentery of the hindgut (*Plate 67,n*), the majority are to be found in the gonadal component of the urogenital ridges, on the medial aspects of the mesonephroi (*Plate 67,m*) (see Table 4).

Stage 17 Developmental age, 10.5 days p.c. Plates 23a–d illustrate an advanced Stage 16/early Stage 17 embryo (see pp. 102–109). Plates 24a–d illustrate an advanced Stage 17/early Stage 18 embryo, and the descriptive account relates principally to the features of this second embryo

EXTERNAL FEATURES

Embryos with 35–39 pairs of somites. The two most obvious distinguishing features of embryos at this stage of development relate to the degree of differentiation of the lens vesicle, and the fact that there is the first evidence of the physiological umbilical hernia containing, initially, only a small segment of the primitive midgut (see below). Concerning the differentiation of the lens, it is possible to see on histological sections that the lens pit has deepened (*Plate 24b,b,c*; *Plate 51,m,n*) compared with the situation observed in previous stages (e.g. *Plate 51,k,l*), and that the diameter of the outer pore-like opening is substantially reduced. The lens pore will eventually close completely, and the outer surface of the lens vesicle thus formed will then lose its contact with the overlying surface ectoderm. This occurs in embryos with about 40–44 pairs of somites (*Plate 25c,d*; *Plate 51,o,p*). Scanning electron micrographs of the cephalic region are particularly instructive in revealing the appearance of the lens pore in embryos at this stage of development (*Plate 47,g*). Another feature of these embryos relates to the deepening of the olfactory pits (*Plate 24b,e–h*), and the marked degree of eversion of the lateral nasal processes that may be observed at this time (*Plate 47,g,h*).

As in the previous stage, it is often possible to view the roof of the oropharyngeal region and the entrance to Rathke's pouch (*Plate 47,g,h*), and in doing so observe its gradual constriction, and consequent reduction in diameter, in intact embryos between about 30–35 and 35–39 pairs of somites (*Plate 25b,b*; *Plate 26b,c*). Its final closure and separation from the roof of the oropharynx may be observed in embryos with about 45–50 pairs of somites.

The other change that may be observed in the external appearance of the cephalic region relates to the different roles that will shortly be subsumed by the maxillary and mandibular components of the first branchial arch. Towards the end of this stage, and during the next stage, a deep groove develops on the anterolateral surface in this region, so that the maxillary component of the first branchial arch gradually becomes incorporated into the developing face (it is one of the upper facial processes), while the mandibular component becomes involved in the differentiation of the lower jaw (*Plate 47,i,j*).

During this stage, the limb buds become increasingly prominent, and begin their rostral "ascent". The sharp apical ectodermal ridge, which runs along the leading edge (i.e. the outer border) of the limb buds, is more clearly seen in the forelimb buds than in the hindlimb buds at this stage. In more advanced embryos, the apical ectodermal ridge will be confined to the region of the handplate and footplate (*Plate 43,i*), whereas in developmentally less advanced embryos, the rostral and caudal extents of these ridges are much less well defined (*Plate 43,f,g*).

Throughout this stage and during the succeeding stage, the tail elongates and its distal one-third becomes markedly thinner than previously. Its subterminal region continues to be somewhat flattened in appearance, and its distal extremity is often slightly upturned (*Plate 43,f–h*).

THE HEART AND VASCULAR SYSTEM

The most obvious events that occur at this stage of development relate to the changes that are taking place in the heart and in the vessels that emerge from it, and drain into it. With regard to the outflow tract, further differentiation of the aortico-pulmonary spiral septum takes place during this stage, as a prelude to its division into two distinct outflow vessels, namely the aortic and pulmonary trunks. This is the earliest stage that the sixth branchial (aortic) arch arteries (the future pulmonary arteries) may be unequivocally recognized (*Plate 24c,j–n*), though the first indication of a pair of vessels (with a relatively small diameter) which, in all probability, are the sixth arch arteries, are seen in embryos with about 30–35 pairs of somites (*Plate 23c,c,d*). Unlike the first four branchial arch arteries that tend to be directed posterolaterally around the foregut, the sixth arch arteries are aimed almost directly posteriorly from the dorsal region of the aortic sac.

Of the vessels flowing into the heart, it is now possible to observe that the primitive symmetrical arrangement of the right and left horns of the sinus venosus (which formerly drained into the right and left halves of the common atrial chamber of the heart (e.g. *Plate 18b,f,g*) is no longer seen, so that while the right horn of the sinus venosus drains directly into the right side of the common atrial chamber (its entrance into that chamber being guarded by a venous valve (*Plate 24c,l*), the left horn of the sinus venosus now passes across the midline to enter the right side of the common atrial chamber at approximately the same site (*Plate 24c,k,l*).

Within the heart, important changes are also taking place, being the earliest events associated with the process of septation. In embryos at slightly earlier stages of development, a relatively broad thickened region may be observed in the midline in the dorsal part of the wall of the common atrial chamber, directly subjacent to the caudal part of the transverse pericardial sinus (*Plate 19b,k*; *Plate 20c,b*; *Plate 23c,e,f*). However, it is unclear whether this actually represents the primordium of the septum primum. The first clear indication of the septum primum is seen at this stage of development, and constitutes the initial thin crescent-shaped dorsal/postero-inferior component of the septum primum. This appears to descend close to the median plane from the caudal part of the broad elevation indicated above.

The degree of trabeculation observed in the bulbus cordis region is now quite marked, and clearly

(*continued on page 120*)

PLATE 24a

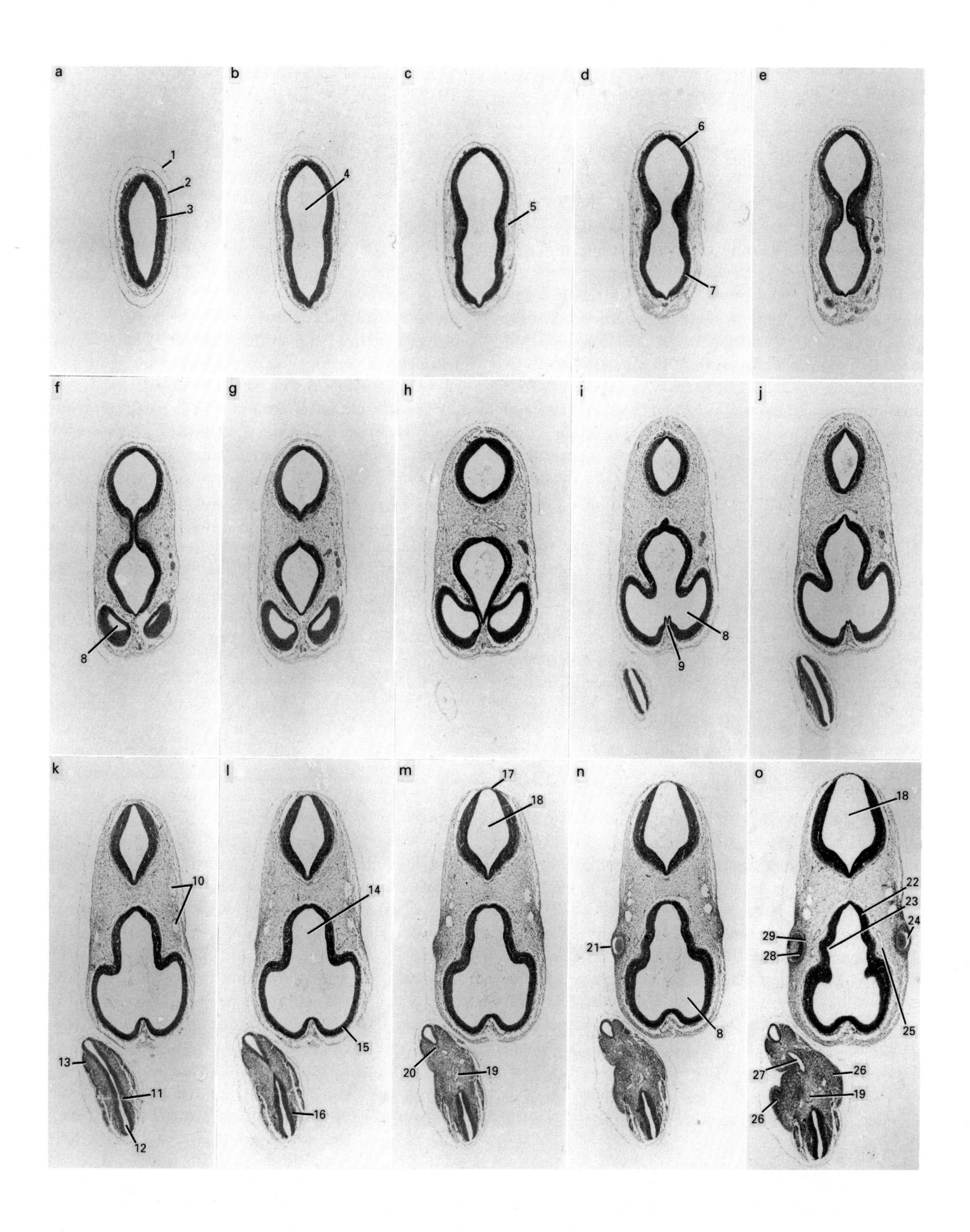
a
1
2
3
b
4
c
5
d
6
7
e
f
8
g
h
i
8
9
j
k
10
13
11
12
l
14
15
16
m
17
18
20
19
n
21
8
o
18
22
23
24
29
28
25
27
26
19
26

Plate 24a (10.5–11 days p.c.)

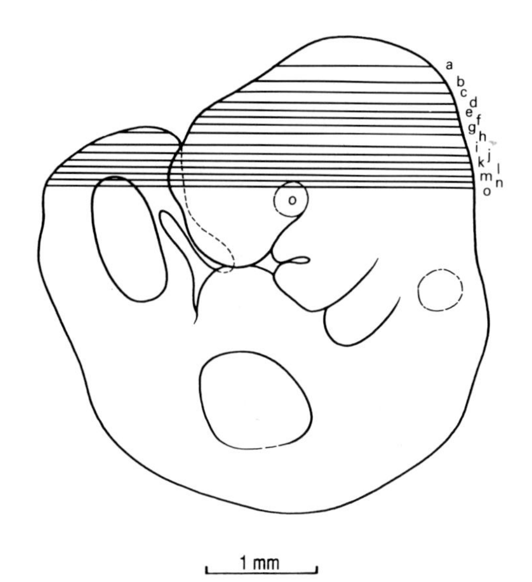

10.5–11 days p.c. Embryo with 35–40 pairs of somites. Crown–rump length 4.1 mm (fixed). Transverse sections (Theiler Stages 17–18; rat, Witschi Stages 24–25; human, Carnegie Stages 14–15).

1. amnion
2. surface ectoderm
3. neuroepithelium
4. mesencephalic vesicle
5. cephalic mesenchyme tissue
6. wall of hindbrain
7. forebrain–midbrain junction
8. telencephalic vesicle
9. lamina terminalis
10. primary head vein
11. neuroepithelium of neural tube in mid-tail region
12. neural lumen
13. somite
14. third ventricle
15. wall of telencephalic vesicle
16. dorsal (posterior) root ganglion
17. roof of hindbrain
18. fourth ventricle
19. midline dorsal aorta
20. notochord
21. optic eminence
22. diencephalon
23. entrance to optic stalk
24. intra-retinal space
25. perioptic vascular anastomosis
26. hindlimb bud
27. hindgut diverticulum (postanal component)
28. inner (neural) layer of optic cup (future nervous layer of retina)
29. outer layer of optic cup (future pigment layer of retina)

PLATE 24b

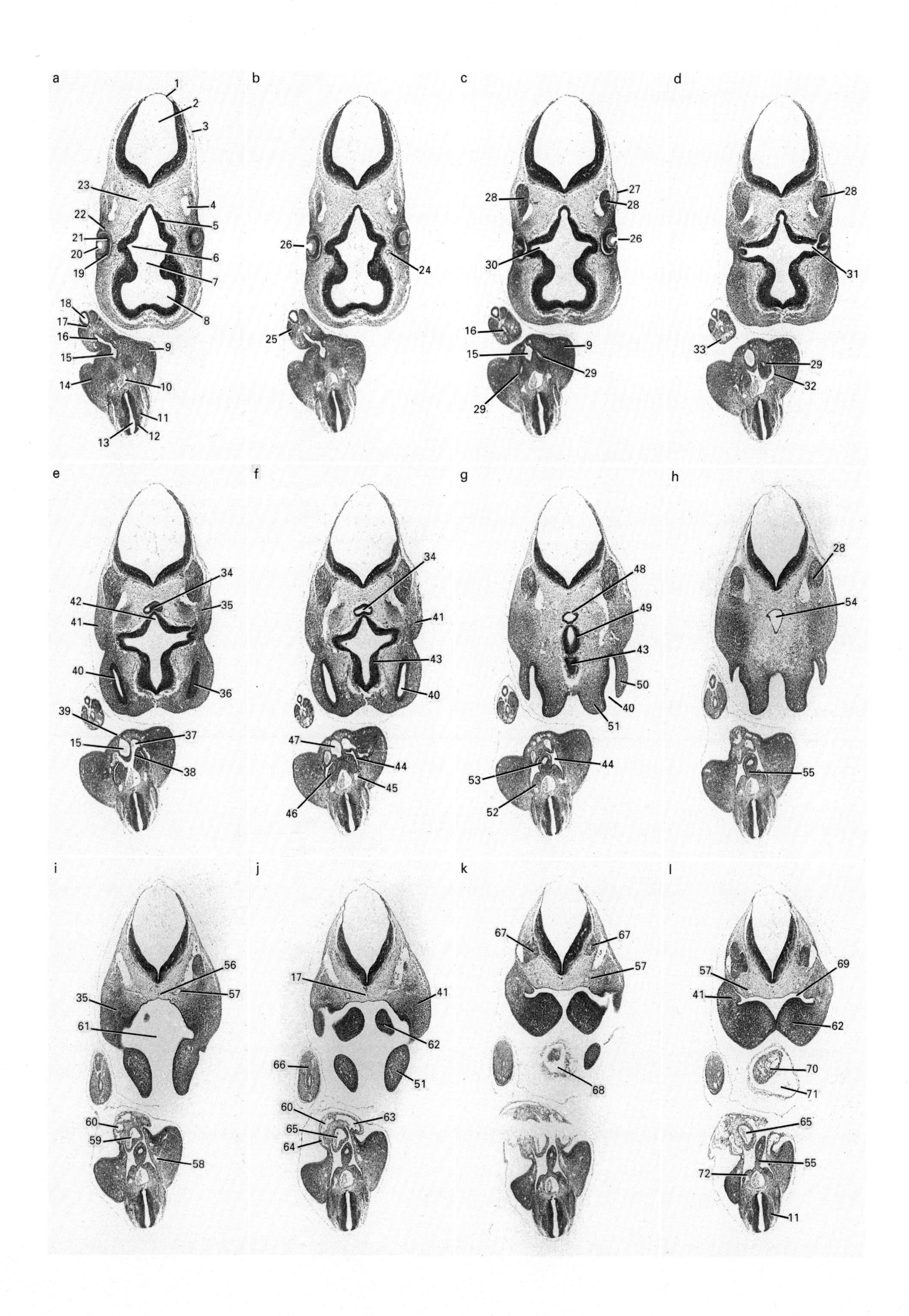
a
1
2
3
4
5
6
7
8
9
10
11
12
13
14
15
16
17
18
19
20
21
22
23
b
24
25
26
c
26
27
28
28
29
29
30
16
15
9
d
28
29
31
32
33
e
34
35
36
37
38
39
40
41
42
15
f
34
41
43
40
44
45
46
47
g
48
49
43
50
40
51
44
52
53
h
28
54
55
i
56
57
35
61
60
59
58
j
17
41
62
66
51
60
63
65
64
k
67
67
57
68
l
57
69
41
62
70
71
65
55
72
11

Plate 24b (10.5–11 days p.c.)

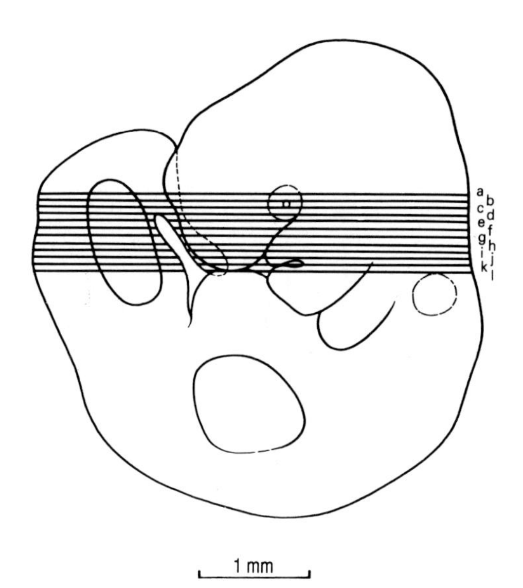

10.5–11 days p.c. Embryo with 35–40 pairs of somites. Crown–rump length 4.1 mm (fixed). Transverse sections (Theiler Stages 17–18; rat, Witschi Stages 24–25; human, Carnegie Stages 14–15).

1. roof of hindbrain
2. fourth ventricle
3. surface ectoderm
4. primary head vein
5. wall of diencephalon
6. entrance to optic stalk (optic recess)
7. third ventricle
8. telencephalic vesicle
9. left hindlimb bud
10. midline dorsal aorta
11. dorsal (posterior) root ganglion
12. neuroepithelium of neural tube in mid-tail region
13. neural lumen
14. right hindlimb bud
15. cloaca
16. hindgut diverticulum (postanal component)
17. notochord
18. neural tube in caudal region of tail
19. intra-retinal space
20. wall of lens vesicle
21. inner (neural) layer of optic cup (future nervous layer of retina)
22. outer layer of optic cup (future pigment layer of retina)
23. cephalic mesenchyme tissue
24. perioptic vascular anastomosis
25. somite
26. lens pit
27. prominence overlying trigeminal (V) ganglion
28. trigeminal (V) ganglion, characteristically located lateral to primary head vein
29. mesonephric duct just proximal to its entry into the posterolateral surface of cloaca
30. optic stalk
31. communication between optic stalk and intra-retinal space
32. principal (axial) arterial branch to forelimb bud from dorsal aorta
33. right dorsal aorta in caudal region of tail
34. Rathke's pouch
35. trigeminal (V) neural crest cells migrating into maxillary component of first branchial arch
36. olfactory epithelium
37. left dorsal aorta
38. communication between midline dorsal aorta and recurved part of left dorsal aorta (left umbilical artery)
39. cloacal membrane
40. olfactory pit
41. maxillary component of first branchial arch
42. infundibular recess of diencephalon
43. floor of forebrain
44. caudal extremity of intra-embryonic coelomic cavity (peritoneal cavity)
45. posterior cardinal vein
46. caudal extremity of urogenital ridge
47. right recurved part of dorsal aorta (right umbilical artery)
48. peripheral boundary of entrance to Rathke's pouch
49. floor of diencephalon
50. lateral nasal process
51. medial nasal process
52. mesonephric duct
53. hindgut
54. communication between rostral part of buccal cavity and Rathke's pouch
55. dorsal mesentery of hindgut
56. rostral extremity of notochord
57. internal carotid artery (rostral extension of dorsal aorta)
58. principal (axial) vein draining the left hindlimb bud
59. site of communication between paired recurved parts of dorsal aortae (right and left umbilical arteries) and midline umbilical artery
60. umbilical venous anastomosis
61. oropharyngeal region
62. mandibular component of first branchial arch
63. left umbilical vein
64. right umbilical vein
65. midline umbilical artery
66. caudal extremity of notochord
67. facio-acoustic (VII–VIII) ganglion complex characteristically located medial to the primary head vein
68. rostral extremity of pericardial cavity
69. first branchial pouch
70. wall of truncus arteriosus region of outflow tract of heart
71. pericardial cavity
72. urogenital ridge

PLATE 24c

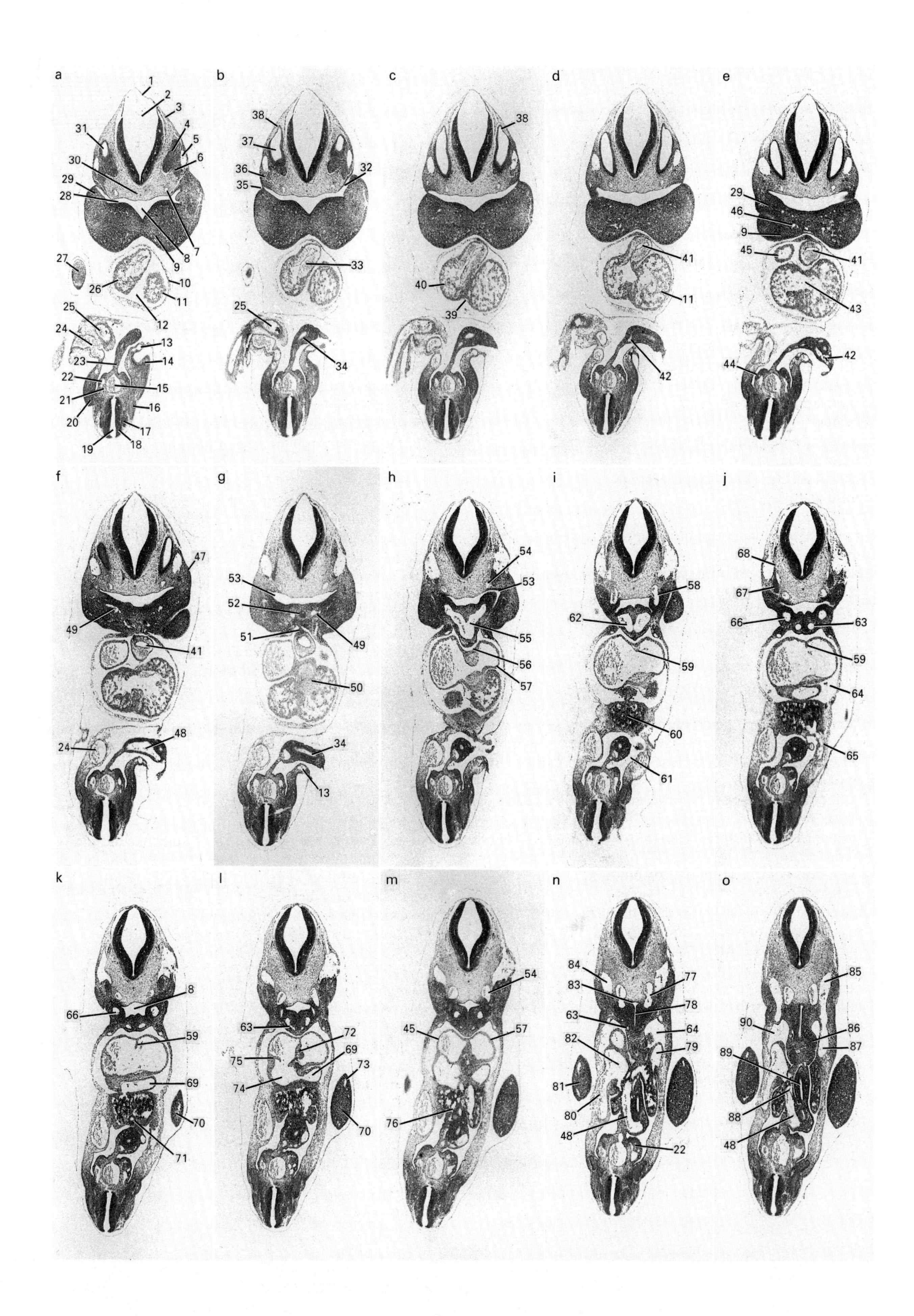

a
1 2 3 4 5 6 7 8 9 10 11 12 13 14 15 16 17 18 19 20 21 22 23 24 25 26 27 28 29 30 31
b
38 37 36 35 32 33 25 34
c
38 40 39
d
41 11 42
e
29 46 9 45 41 43 44 42
f
47 49 41 24 48
g
53 52 51 49 50 34 13
h
54 53 55 56 57
i
58 62 59 60 61
j
68 67 66 63 59 64 65
k
8 66 59 69 70 71
l
63 72 75 69 73 74 70
m
54 45 57 76
n
84 77 83 78 63 64 82 79 81 80 48 22
o
85 90 86 87 89 88 48

Plate 24c (10.5–11 days p.c.)

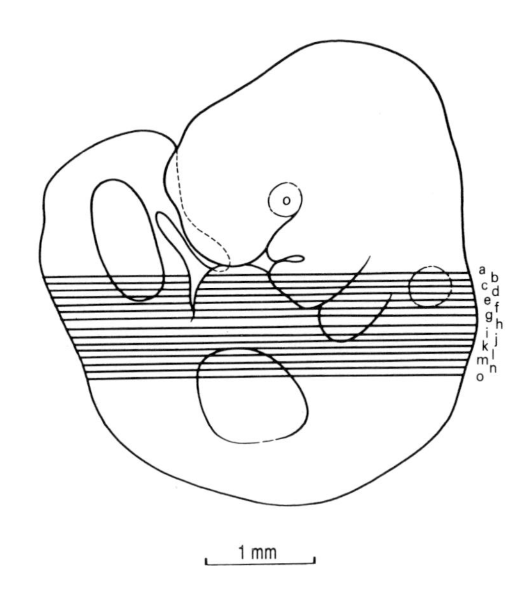

10.5–11 days p.c. Embryo with 35–40 pairs of somites. Crown–rump length 4.1 mm (fixed). Transverse sections (Theiler Stages 17–18; rat, Witschi Stages 24–25; human, Carnegie Stages 14–15).

1. roof of hindbrain
2. fourth ventricle
3. surface ectoderm
4. acoustic (VIII) component of facio-acoustic (VII–VIII) ganglion complex
5. primary head vein
6. facial (VII) component of facio-acoustic (VII–VIII) ganglion complex
7. internal carotid artery (rostral extension of dorsal aorta)
8. pharyngeal region of foregut
9. mandibular component of first branchial arch
10. body wall (thoracic) overlying pericardial cavity
11. trabeculated wall of common ventricular chamber of heart
12. pericardial cavity
13. left umbilical vein
14. left hindlimb bud
15. midline dorsal aorta
16. somite
17. dorsal (posterior) root ganglion
18. neuroepithelium of neural tube
19. neural lumen
20. right hindlimb bud
21. posterior cardinal vein
22. urogenital ridge
23. dorsal mesentery of gut
24. right umbilical vein
25. midline umbilical artery
26. wall of truncus arteriosus region of outflow tract of heart
27. caudal extremity of tail
28. first branchial pouch
29. maxillary component of first branchial arch
30. notochord
31. otocyst (otic vesicle)
32. dorsal portion of first branchial pouch (future tubo-tympanic recess)
33. endocardial cushion tissue (future spiral septum) within bulbo–truncal region of outflow tract of heart
34. midgut
35. first branchial membrane (future tympanic membrane) and cleft (groove)
36. facial (VII) neural crest tissue migrating into second branchial arch
37. main (vestibular) portion of otic vesicle
38. endolymphatic appendage of otic vesicle
39. bulbo-ventricular groove
40. wall of bulbus cordis region of heart
41. aortic sac
42. vitelline venous plexus within wall and dorsal mesentery of midgut
43. bulbo-ventricular canal
44. mesonephric duct
45. wall of right atrial chamber of heart
46. first branchial arch artery
47. second branchial arch
48. vitelline vein
49. second branchial arch artery
50. endocardial cushion tissue in atrio-ventricular canal
51. rostral component of heart mesentery (dorsal mesocardium)
52. thyroid primordium
53. second branchial pouch
54. left dorsal aorta
55. communication between aortic sac and third branchial arch artery
56. transverse pericardial sinus
57. wall of left atrial chamber of heart
58. communication between third branchial arch artery and dorsal aorta
59. dorsal component of septum primum
60. elements of hepatic/biliary primordia within septum transversum
61. intra-embryonic coelomic cavity (future peritoneal cavity)
62. origin of fourth branchial arch artery from aortic sac
63. sixth branchial arch artery (future pulmonary artery)
64. left pericardio-peritoneal canal
65. communication between left umbilical vein and intra-hepatic venous sinusoids
66. fourth branchial arch artery
67. glossopharyngeal (IX) ganglion
68. glossopharyngeal (IX) nerve
69. left horn of sinus venosus
70. left forelimb bud
71. cystic primordium
72. caudal component of septum primum
73. marginal vein
74. right horn of sinus venosus
75. leaf of venous valve where right horn of sinus venosus enters right atrial chamber of heart
76. vitelline venous plexus entering dorsal aspect of hepatic primordium
77. communication between sixth branchial arch artery and dorsal aorta
78. ventral bifurcation of foregut to produce tracheal diverticulum (laryngo-tracheal groove)
79. left common cardinal vein
80. communication between right umbilical vein and common cardinal vein
81. right forelimb bud
82. right common cardinal vein
83. oesophagus
84. right anterior cardinal vein
85. left anterior cardinal vein
86. left lung bud
87. communication between left anterior and left common cardinal veins
88. communication between vitelline vein and intra-hepatic venous sinusoids
89. lumen of stomach
90. right pericardio-peritoneal canal

PLATE 24d

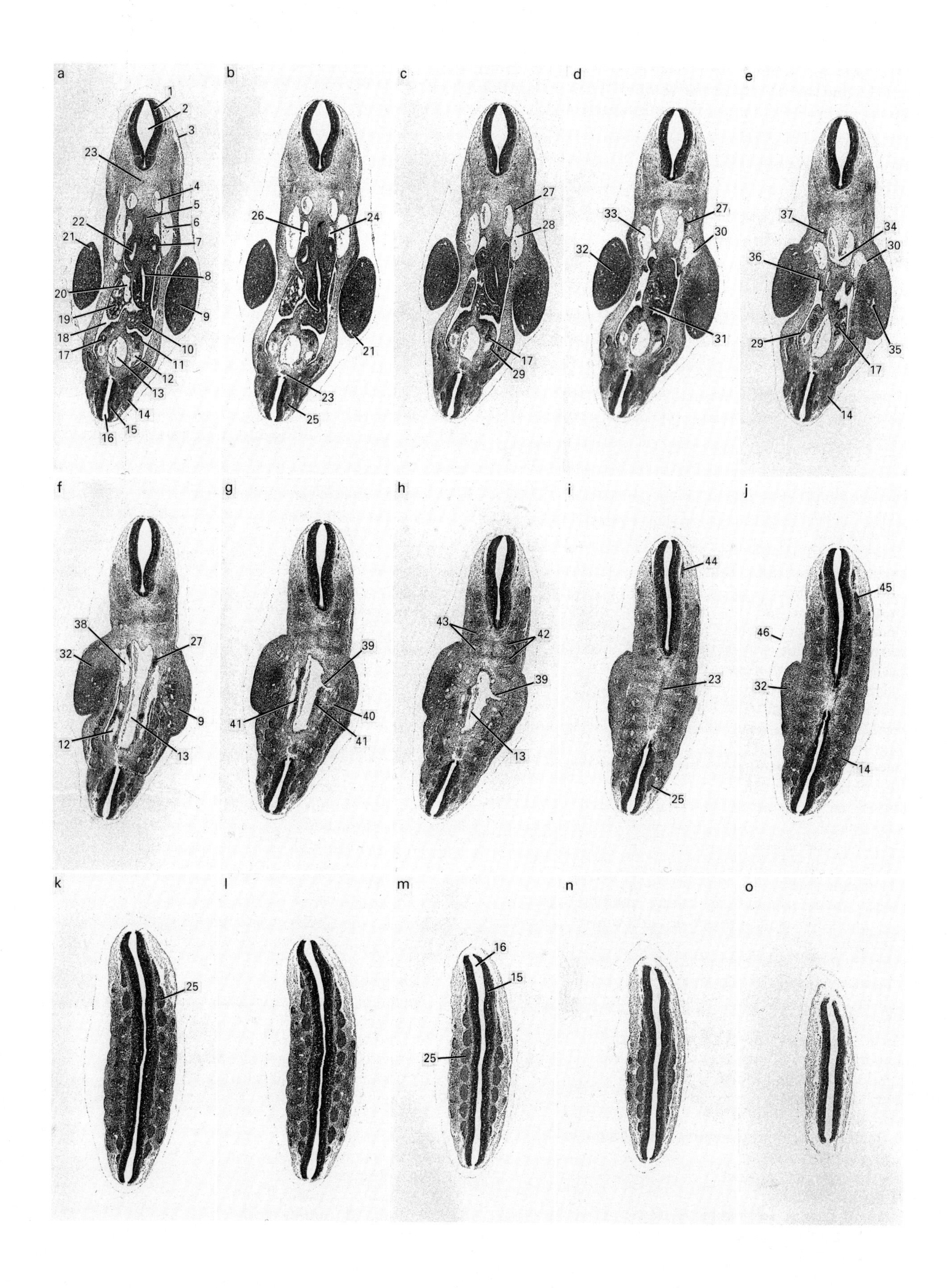

a
1
2
3
23
4
5
6
22
7
21
8
20
9
19
18
10
17
11
12
13
14
15
16
b
26
24
21
23
25
c
27
28
17
29
d
33
27
30
32
31
e
37
34
30
36
29
35
17
14
f
38
27
32
9
12
13
g
39
41
40
41
h
43
42
39
13
i
44
23
25
j
45
46
32
14
k
25
l
m
16
15
25
n
o

Plate 24d (10.5–11 days p.c.)

10.5–11 days p.c. Embryo with 35–40 pairs of somites. Crown–rump length 4.1 mm (fixed). Transverse sections (Theiler Stages 17–18; rat, Witschi Stages 24–25; human, Carnegie Stages 14–15).

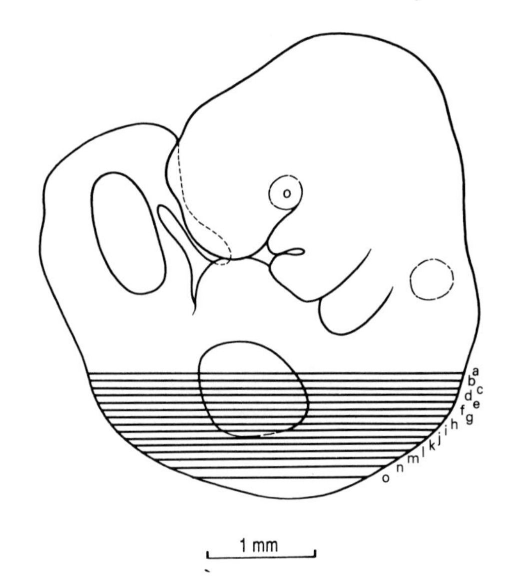

1. neuroepithelium of caudal part of hindbrain
2. caudal extremity of fourth ventricle
3. surface ectoderm
4. left dorsal aorta
5. oesophagus
6. anterior cardinal vein
7. left lung bud
8. lumen of stomach
9. left forelimb bud
10. mesentery of gut containing vitelline venous plexus
11. urogenital ridge
12. subcardinal vein
13. midline dorsal aorta
14. somite
15. neuroepithelium of neural tube
16. neural lumen
17. mesonephric tubule (vesicle)
18. intra-embryonic coelomic cavity (peritoneal cavity)
19. hepatic primordium containing biliary primordia and venous sinusoids
20. vitelline vein within hepatic primordium
21. marginal vein
22. right main bronchus
23. notochord
24. left pericardio-peritoneal canal
25. dorsal (posterior) root ganglion
26. right pericardio-peritoneal canal
27. left sympathetic trunk
28. left posterior cardinal vein
29. mesonephric duct
30. communication between principal (axial) vein from left forelimb bud and left posterior cardinal vein
31. dorsal mesentery of midgut
32. right forelimb bud
33. right posterior cardinal vein
34. rostral site of fusion of right and left dorsal aortae
35. principal (axial) artery within forelimb bud
36. rostral extremity of urogenital ridge
37. right sympathetic trunk
38. communication between right subcardinal and right posterior cardinal veins
39. principal (axial) artery from dorsal aorta to left forelimb bud
40. cervical nerve trunk destined to form part of brachial plexus
41. sympathetic trunk with associated segmental enlargements (ganglia)
42. segmental zone of increased cell packing density
43. segmental interzone of decreased cell packing density
44. segmental nerves with closely associated intersegmental arteries
45. cervical (posterior) root ganglion
46. amnion

distinguishes this region from the proximal part of the outflow tract, where the space between the myoepicardial wall and the endocardial cells is filled with differentiating mesenchyme which, in this location, will form the right and left bulbar ridges.

It is also of interest to note that the diameters of the right and left umbilical veins, which were formerly of fairly similar dimensions (see *Plate 18b,e–g*), are now quite dissimilar, so that the diameter of the right vessel is now significantly greater than that of the equivalent vessel on the left side (*Plate 24c,f–m*). This change from a symmetrical to an asymmetrical arrangement is clearly apparent from an analysis of embryos at intermediate stages of development (cf. *Plate 20b,f–h*; *Plate 23c,i,l*).

THE PRIMITIVE GUT

In the cephalic region, the relationship between the infundibular recess of the third ventricle and Rathke's pouch is gradually changing. Whereas at earlier stages of development, the ventral wall of the infundibular recess was in close proximity principally to the superior and dorsal walls of Rathke's pouch (*Plate 21a,d*; *Plate 22a,d*; *Plate 23b,a–d*), the infundibular recess is now almost exclusively a superior relation of Rathke's pouch (*Plate 24b,a–d*). Furthermore, the inferior border of the infundibular recess is seen to indent the rostral part of Rathke's pouch, so that this region eventually bifurcates to form two wing-like extensions (*Plate 56,d,h,m,n*). In the pharyngeal region, the thyroid primordium is relatively easily seen (*Plate 24c,g*), and shows the first evidence of canalization to form the proximal part of the thyroglossal duct (cf. *Plate 22a,c*; *Plate 25b,a,b*).

The location of the lung buds at this stage of development is of interest, in that they are quite clearly inferior to the common atrial chamber of the heart, and still retain their very close relationship to the oesophageal region of the foregut (*Plate 24c,o*; *Plate 24d,a–c*).

A considerable degree of differentiation is also seen in relation to the primitive liver, with in addition a substantial degree of enlargement apparent in its overall volume compared to the situation observed in previous stages (*Plate 24c,i–m*). The latter is principally due to the invasion of the mesenchyme of the septum transversum by the hepatic (biliary) cords, and the initiation of haematopoietic activity in this organ. The right vitelline vein (later to form the portal vein) appears to have a surprisingly large diameter considering its limited function at this stage of development, and is seen to drain into the deep surface of the liver (*Plate 24c,n,o*). The left umbilical vein meanwhile drains into the left side of the upper part of this organ (*Plate 24c,j*), and will in due course form the ductus venosus. The latter vessel allows the oxygenated blood from the placenta to largely bypass the liver as it is directed towards the right atrium via the posthepatic part of the inferior vena cava.

A characteristic feature of this stage of development, is the first indication of the physiological umbilical hernia which initially contains only a relatively small region of the midgut loop (*Plate 24c,b–h*). The obliterated vitelline duct (formerly the neck of the middle region of the midgut) may also be seen (*Plate 24c,d–f*).

THE NERVOUS SYSTEM

As indicated above, the most characteristic feature of this stage of development relates to the appearance of the lens, which has almost completed its initial differentiation to form the lens vesicle (*Plate 24b,b,c*; *Plate 51,m,n*). A substantial difference is now apparent between the thickness of the inner and outer layers of the optic cup, with the inner layer being three to four times thicker than the outer layer. On a more detailed inspection of the eye, it is also possible to recognize the first evidence of the primitive hyaloid vessels which are located within the hyaloid cavity (*Plate 51,n*).

The fourth ventricle appears to be more dilated than previously, and its roof is characteristically extremely thin. As in earlier stages of development, it is not possible to distinguish between the caudal extremity of the primitive hindbrain and the rostral extremity of the primitive spinal cord. The otocysts are now more ovoid than spherical, and characteristically have a small posteromedial endolymphatic appendage (*Plate 24c,b–d*). The glossopharyngeal (IX) ganglion (*Plate 24c,j*) is also more easily recognized than previously (cf. *Plate 23b,i*).

The lower cervical dorsal (posterior) root ganglia are also becoming increasingly well defined, and this is associated with the earliest stage in the differentiation of the central components of the brachial plexus (*Plate 23c,i–l*). From the mid-trunk region caudally to the mid-tail region, the dorsal (posterior) root ganglia are also becoming particularly prominent structures (*Plate 24d,k–n*). Another feature of this stage of development is the first appearance of the sympathetic trunks, and these are located initially between the dorsal aortae and the anterior cardinal veins at the level of the rostral border of the forelimb buds (*Plate 23c,k,l*; *Plate 24d,c–e*). Only in the developmentally more advanced embryos at this stage is it also possible to recognize the sympathetic ganglia (*Plate 24d,f–h*).

THE UROGENITAL SYSTEM

The urogenital ridges are now more prominent than previously, and numerous mesonephric vesicles are present (*Plate 24d,a–e*) which drain via tubules into the mesonephric ducts. The mesonephric ducts pass caudally and make contact with the wall of the cloaca in the region of the urogenital sinus (*Plate 23c,a*; *Plate 24b,c,d*), but do not drain into it until about 14 days p.c. For this reason, and because no well-differentiated mesonephric glomeruli are seen in mice, it appears unlikely that the mesonephros ever functions as an effective excretory organ in this species.

Stage 18 Developmental age, 11 days p.c.
Plates 24a–d illustrate an advanced Stage 17/early Stage 18 embryo (see pp. 112–119). Plates 25a–c, and the descriptive account relates principally to the features of this second embryo

EXTERNAL FEATURES

Embryos with 40–44 pairs of somites. Apart from the progressive degree of closure of the lens vesicle which occurs during this stage of development, no other stage-specific externally visible morphological features characterize this stage of development. It is also possible to observe, however, that the margins around the olfactory pits are beginning to fuse (*Plate 47,j*). A gradual increase in the degree of differentiation of the limb buds is also seen, as well as a slight increase in the volume of the physiological umbilical hernia compared to the situation observed previously. The majority of the features which characterize this stage therefore are most clearly evident from an analysis of histological sections of this material.

With the increased degree of flexion of these embryos, it is no longer possible to view the roof of the oral cavity in intact specimens. However, analysis of this region in decapitated specimens reveals that there is now only a narrowed entrance to Rathke's pouch visible at this stage (*Plate 25b,a–c*; *Plate 54,i–l*). The latter will eventually close completely by about 12–12.5 days p.c. (*Plate 56,g–l*), when the stalk loses its contact with the roof of the oral cavity, so that usually no evidence of the subjacent structures is seen (see *Plate 59,b*). Occasionally, however, analysis of sagittal sections through this region reveals the presence of remnants of the neck and/or connecting stalk of Rathke's pouch that have failed to separate from the roof of the oral cavity with the ascent of this structure (see *Plate 55,g–i*). If a gross palatal defect is present, for example, this may be associated with the presence of a "pharyngeal" pituitary (of Rathke's pouch origin) that develops independently, and is quite distinct from the pars nervosa tissue which is located in the sella turcica (hypophyseal fossa). On other occasions, remnants of Rathke's pouch origin may be located in association with a craniopharyngeal canal.

The second branchial arch also increases in size during this stage, and progressively comes to overly the third and fourth arches (*Plate 47,i,j*). The cervical somites are no longer clearly visible, though from the region of the forelimb caudally, but particularly in the tail region, the somites are generally very easily discerned. During this stage, the tail gradually narrows and elongates, and its caudal part is particularly slender. The latter tends to pass on one side of the neck region, and it is often in close proximity to the margins of the nasal pit (similar to the situation illustrated in *Plate 43,h*).

THE HEART AND VASCULAR SYSTEM

There are no specific features that characterize this stage of development, though an increased degree of differentiation is evident in relation to the aortico-pulmonary spiral septum (*Plate 25a,d*; *Plate 25b,a*), the caudal part of which is seen to extend well into the bulbus cordis region (*Plate 25b,a*).

THE PRIMITIVE GUT

In the cephalic region, the relationship between the roof of the oral cavity and Rathke's pouch has been detailed above, though the relationship between the inferior part of the wall of the infundibulum and the superior part of Rathke's pouch is similar to that seen during the previous stage of development. In appropriate sagittal sections, it is possible to see that the thyroid primordium is still connected to the floor of the pharynx in the midline by the thyroglossal duct, whereas in developmentally slightly more advanced embryos, the thyroid primordium may have lost its connection with the floor of the pharynx (this is at the site of the future foramen caecum).

While the oesophagus (*Plate 25a,d*) and duodenum (*Plate 25b,a–c*) are seen to have a relatively narrow lumen, that of the stomach is seen to be considerably dilated. Furthermore, the stomach is now seen to be principally located on the left side of the midline, and its longitudinal axis is now almost horizontal (*Plate 25a,a–c*). This change in the orientation of the stomach compared to the situation observed in previous stages of development, where its longitudinal axis is seen to be almost vertical and close to the median plane (*Plate 24c,o*; *Plate 24d,a–c*), is largely brought about by the rotation of the distal part of the primitive foregut and its derivatives. The latter also facilitates the formation of the omental bursa (*Plate 25a,c,d*), which is first clearly seen in sagittal sections of embryos with about 30–35 pairs of somites (*Plate 22b,b,c*).

The close relationship between the dorsal mesentery of the stomach and the primitive lung buds is also clearly seen at this stage of development (*Plate 25a,b,c*). The lung buds, though relatively small in volume at this stage, bulge into the pericardio-peritoneal canals (*Plate 24c,n,o*; *Plate 24d,a,b*; *Plate 25a,b,c*; *Plate 25b,a*), and are considerably more caudally situated than in subsequent stages of development. Similarly, the left lobes of the liver are both anterior and superior relations of the stomach (*Plate 25a,a–c*), while the right lobes of the liver occupy a substantial volume of the upper part of the peritoneal cavity (*Plate 25b,a–d*). In the median plane, the cystic primordium (gall bladder) is now reasonably well differentiated (*Plate 25b,a,b*).

The volume of the physiological umbilical hernia is enlarged (*Plate 25a,a–d*; *Plate 25b,a,b*) compared with the situation observed previously (*Plate 24c,c–g*), and the hernial sac is directed principally towards the left of the midline, whereas the umbilical cord and its associated vessels are principally located towards the right of the midline (*Plate 24c,a–e*; *Plate 25c,a–d*). It

(*continued on page 128*)

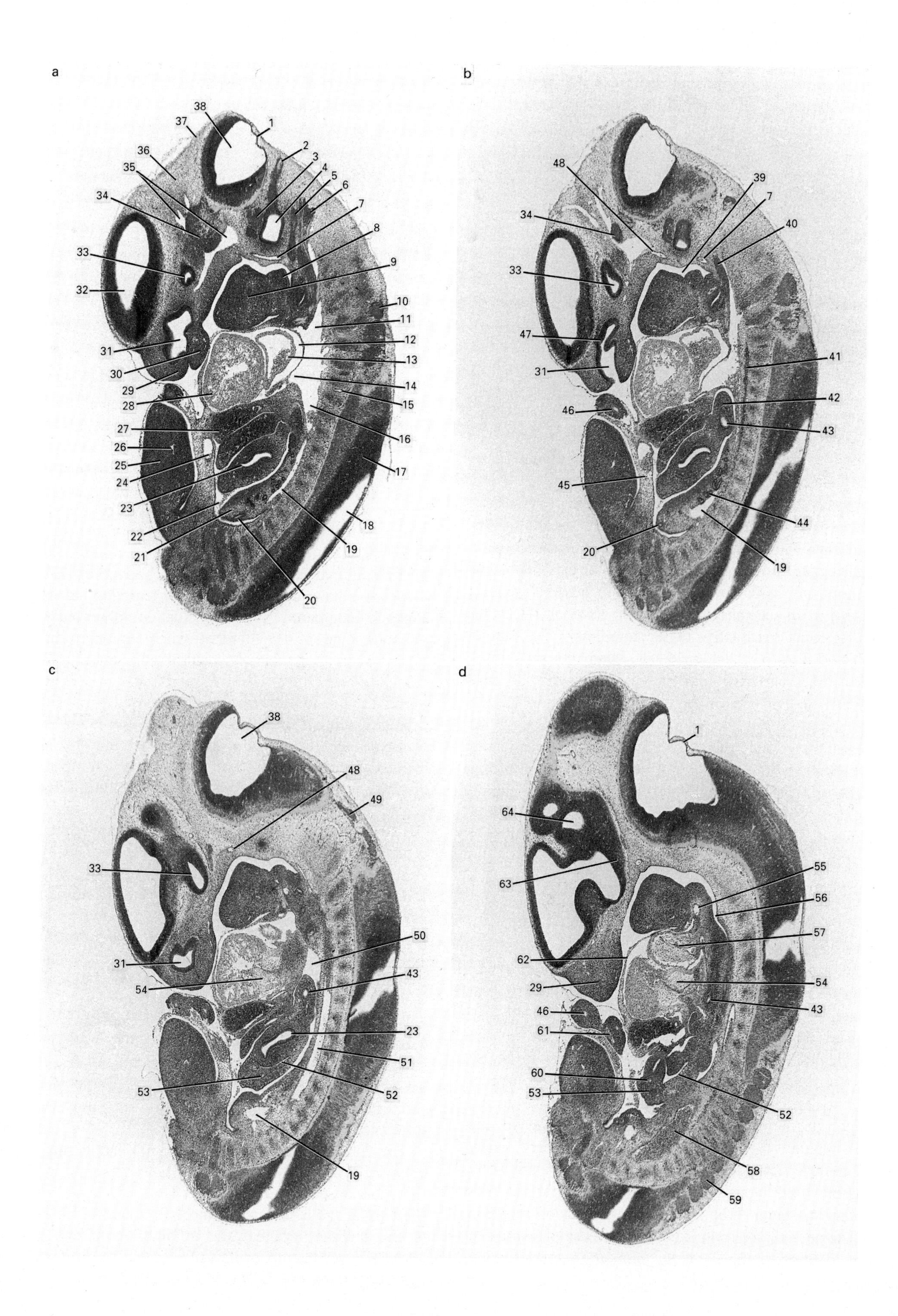
a
b
c
d
1
2
3
4
5
6
7
8
9
10
11
12
13
14
15
16
17
18
19
20
21
22
23
24
25
26
27
28
29
30
31
32
33
34
35
36
37
38
39
40
41
42
43
44
45
46
47
48
49
50
51
52
53
54
55
56
57
58
59
60
61
62
63
64

Plate 25a (11–11.5 days p.c.)

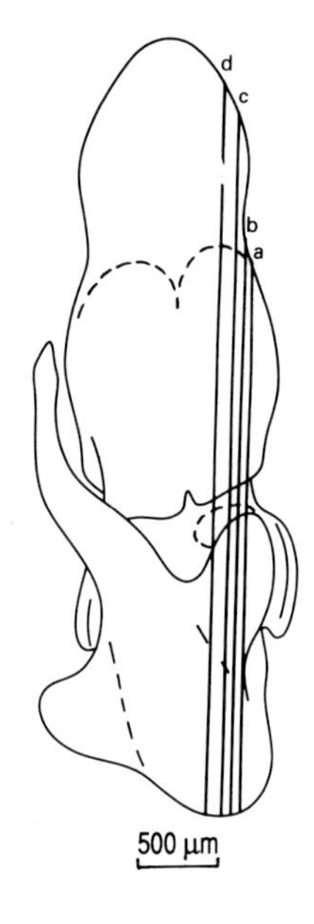

11–11.5 days p.c. Embryo with 40–45 pairs of somites. Crown–rump length 4.6 mm (fixed). Sagittal sections (Theiler Stage 18; rat, Witschi Stages 25–26; human, Carnegie Stage 15).

1. roof of hindbrain
2. wall of endolymphatic diverticular appendage of otic vesicle
3. facio-acoustic (VII–VIII) ganglion complex
4. otocyst (otic vesicle)
5. superior and inferior glossopharyngeal (IX) ganglia
6. superior ganglion of vagus (X) nerve
7. rostral extremity of dorsal aorta
8. second branchial arch
9. mandibular component of first branchial arch
10. first cervical dorsal (posterior) root ganglion
11. anterior cardinal vein
12. pericardial cavity
13. wall of left atrial chamber of heart
14. common cardinal vein
15. intersegmental artery from dorsal aorta
16. posterior cardinal vein
17. neuroepithelium of neural tube
18. neural lumen
19. subcardinal vein
20. mesonephric duct
21. urogenital ridge
22. peritoneal cavity
23. lumen of stomach
24. umbilical vein
25. left hindlimb bud
26. principal (axial) artery of hindlimb bud
27. hepatic primordium (liver)
28. trabeculated wall of common ventricular chamber of heart (future left ventricle)
29. medial nasal process
30. lateral nasal process
31. olfactory (nasal) pit
32. left telencephalic vesicle
33. lumen of optic stalk with its characteristic ventral choroidal (fetal) fissure
34. trigeminal (V) ganglion
35. primary head vein
36. cephalic mesenchyme
37. surface ectoderm
38. fourth ventricle
39. buccal cavity (oropharynx)
40. inferior ganglion of vagus (X) nerve
41. sympathetic trunk
42. lung bud
43. main bronchus
44. mesonephric tubule (vesicle)
45. body wall in abdominal region
46. midgut loop within physiological umbilical hernia
47. olfactory epithelium in region of vomeronasal organ (Jacobson's organ)
48. internal carotid artery (rostral extension of dorsal aorta)
49. cranial/spinal accessory (XI) nerve
50. pericardio-peritoneal canal
51. dorsal aorta
52. lesser sac (omental bursa)
53. pancreatic primordium
54. endocardial cushion tissue lining the atrio-ventricular canal
55. second branchial arch artery
56. oesophagus
57. endocardial cushion tissue in outflow tract of heart (future aortico-pulmonary spiral septum)
58. midline dorsal aorta
59. dorsal (posterior) root ganglion
60. lumen of duodenum
61. vitelline vein within wall of midgut loop
62. body wall (thoracic) overlying pericardial cavity
63. diencephalon
64. mesencephalic vesicle

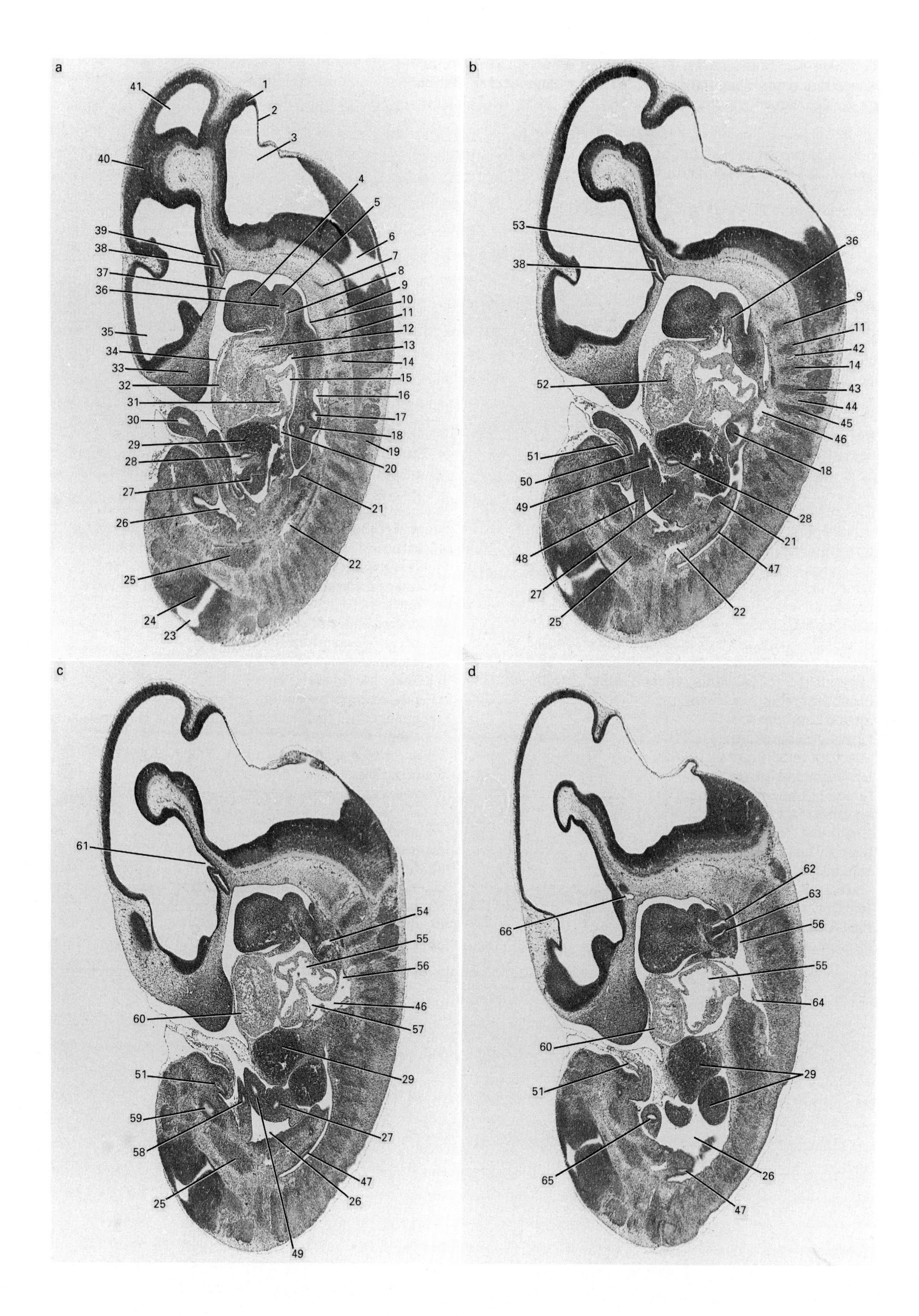
a
1
2
3
4
5
6
7
8
9
10
11
12
13
14
15
16
17
18
19
20
21
22
23
24
25
26
27
28
29
30
31
32
33
34
35
36
37
38
39
40
41
b
53
38
36
9
11
42
14
52
43
44
45
46
18
51
50
49
48
27
25
28
21
47
22
c
61
54
55
56
46
57
60
29
51
59
58
27
47
25
26
49
d
62
63
56
66
55
64
60
29
51
65
26
47

Plate 25b (11–11.5 days p.c.)

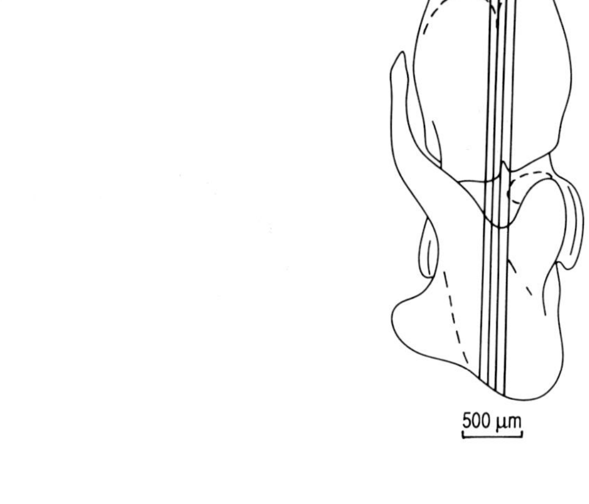

11–11.5 days p.c. Embryo with 40–45 pairs of somites. Crown–rump length 4.6 mm (fixed). Sagittal sections (Theiler Stage 18; rat, Witschi Stages 25–26; human, Carnegie Stage 15).

1. cerebellar primordium (cerebellar rudiment derived from dorsal part of the alar lamina of the metencephalon)
2. roof of hindbrain
3. fourth ventricle
4. mandibular component of first branchial arch
5. second branchial arch
6. caudal extremity of fourth ventricle
7. notochord
8. second branchial arch artery
9. condensation of sclerotomic material forming centrum of atlas (C1)
10. caudal region of pharynx
11. centrum of axis (C2)
12. endocardial cushion tissue in outflow tract of heart (future aortico-pulmonary spiral septum)
13. transverse pericardial sinus
14. centrum of C3
15. left atrial chamber of heart
16. pericardio-peritoneal canal
17. main bronchus
18. lung bud
19. dorsal (posterior) root ganglion
20. site of entrance of common cardinal vein into sinus venosus
21. urogenital ridge
22. subcardinal vein
23. neural lumen
24. neuroepithelium of neural tube
25. midline dorsal aorta
26. intra-embryonic coelomic cavity (peritoneal cavity)
27. lumen of duodenum
28. cystic primordium (gall bladder)
29. hepatic primordium (liver)
30. lumen of midgut
31. venous valves at site of entrance of sinus venosus into atrial chamber of heart
32. wall of heart at bulbo-ventricular junction
33. medial nasal process
34. body wall (thoracic) overlying pericardial cavity
35. third ventricle
36. thyroid primordium (thyroglossal duct)
37. buccal cavity (oropharynx)
38. Rathke's pouch
39. diencephalon in region of infundibular recess
40. wall of midbrain
41. mesencephalic vesicle
42. intersegmental artery
43. centrum of C4
44. cervical nerve (C5)
45. centrum of C5
46. common cardinal vein
47. mesonephric duct
48. dorsal mesentery of distal loop of midgut
49. proximal loop of midgut
50. distal loop of midgut
51. umbilical vein
52. bulbo-ventricular canal
53. infundibulum (pars nervosa)
54. fourth arch artery
55. right atrial chamber of heart
56. anterior cardinal vein
57. right horn of sinus venosus
58. midgut–hindgut junction
59. urogenital sinus
60. wall of bulbus cordis
61. infundibular recess
62. third branchial arch artery
63. entrance to third branchial pouch
64. posterior cardinal vein
65. lumen of hindgut
66. arterial supply to pituitary from internal carotid artery

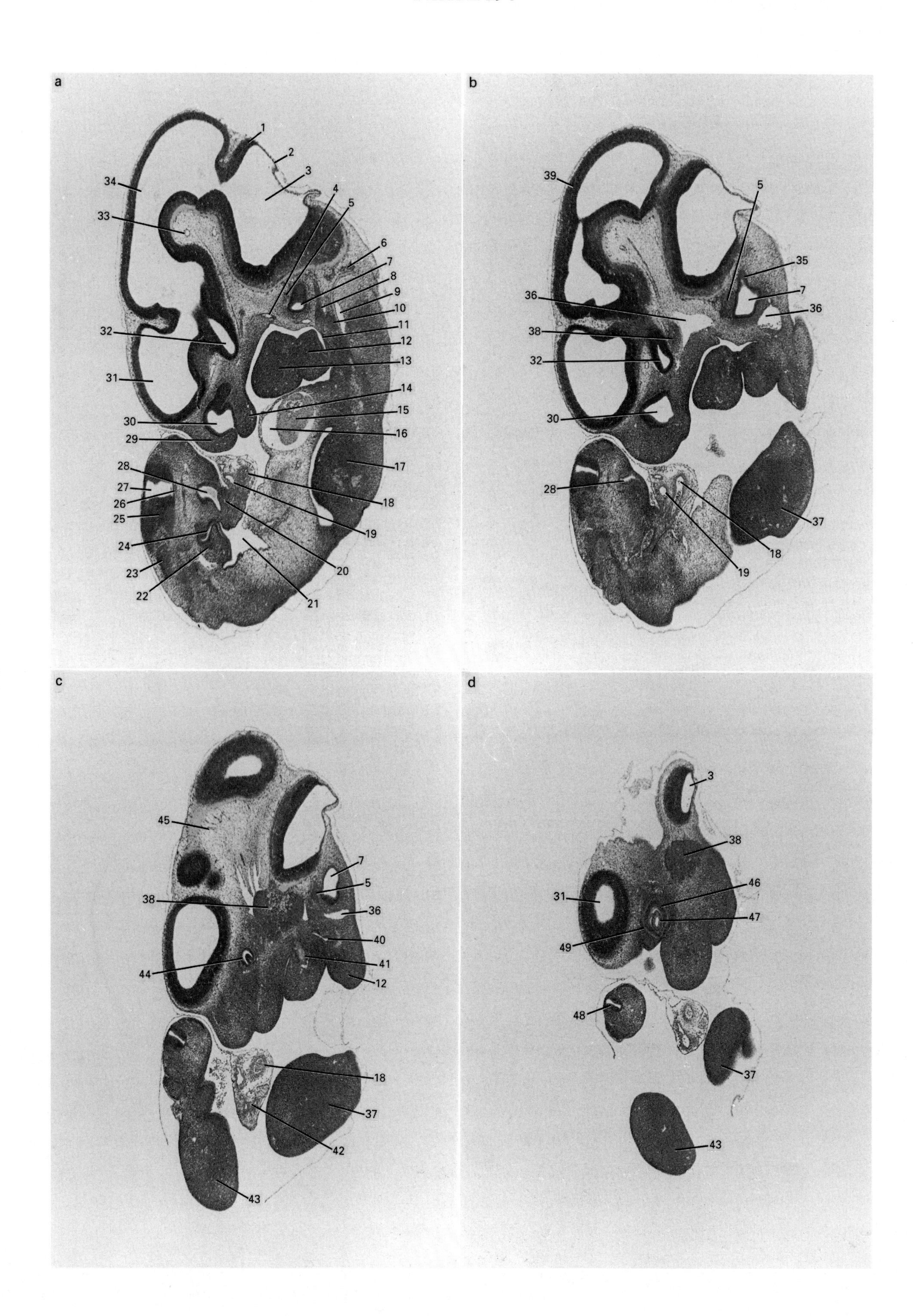
a
1
2
3
4
5
6
7
8
9
10
11
12
13
14
15
16
17
18
19
20
21
22
23
24
25
26
27
28
29
30
31
32
33
34
b
39
5
35
7
36
36
38
32
30
28
37
18
19
c
45
7
5
38
36
40
44
41
12
18
37
42
43
d
3
38
31
46
47
49
48
37
43

Plate 25c (11–11.5 days p.c.)

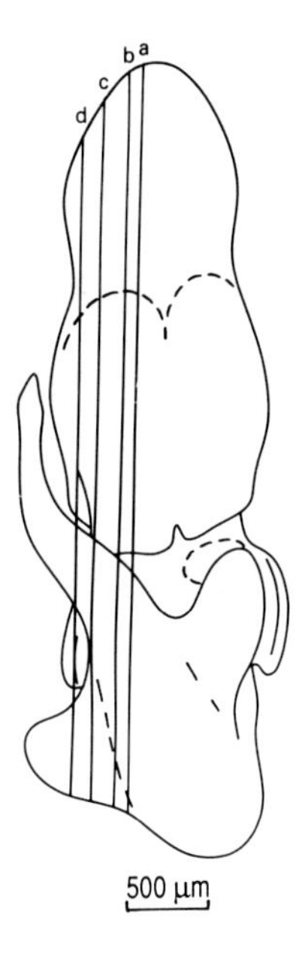

11–11.5 days p.c. Embryo with 40–45 pairs of somites. Crown–rump length 4.6 mm (fixed). Sagittal sections (Theiler Stage 18; rat, Witschi Stages 25–26; human, Carnegie Stage 15).

1. cerebellar primordium
2. roof of hindbrain
3. fourth ventricle
4. internal carotid artery (rostral extension of dorsal aorta)
5. facio-acoustic (VII–VIII) ganglion complex
6. cranial/spinal accessory (XI) nerve
7. otocyst (otic vesicle)
8. glossopharyngeal (IX) nerve and associated ganglion
9. anterior cardinal vein
10. inferior ganglion of vagus (X) nerve
11. oropharynx
12. second branchial arch
13. mandibular component of first branchial arch
14. lateral nasal process
15. wall of right atrium
16. pericardial cavity
17. right forelimb bud
18. umbilical artery
19. umbilical vein
20. mesonephric duct (common excretory duct) just proximal to its site of entry into the posterolateral surface of the cloaca
21. intra-embryonic coelomic cavity (peritoneal cavity)
22. metanephric blastema
23. dorsal (posterior) root ganglion
24. ureteric bud (metanephric duct)
25. neuroepithelium of neural tube
26. notochord
27. neural lumen
28. lumen of cloaca
29. medial nasal process
30. olfactory (nasal) pit
31. right telencephalic vesicle
32. optic stalk
33. branch of primary head vein
34. mesencephalic vesicle
35. proximal part of endolymphatic diverticular appendage of otic vesicle
36. primary head vein
37. right forelimb bud
38. right trigeminal (V) ganglion
39. roof of midbrain
40. second branchial pouch
41. maxillary division of trigeminal (V) nerve passing into maxillary component of first branchial arch
42. umbilical cord
43. left hindlimb bud
44. lumen of optic stalk with its characteristic ventral choroidal (fetal) fissure
45. cephalic mesenchyme
46. inner (neural) layer of optic cup
47. lens vesicle
48. neural tube in tail region
49. outer layer of optic cup (future pigment layer of retina)

is believed that the principal reason why the physiological umbilical hernia develops (it initially appears during the previous stage of development) is because there is insufficient space within the peritoneal/coelomic cavity to accommodate the rapidly elongating midgut loop. The space is limited due to the presence of the liver, which occupies most of the upper part of the cavity, and because of the substantial volume occupied by the mesonephroi which, though retroperitoneal structures, impinge to a considerable degree on the dorsal region of the peritoneal/coelomic cavity.

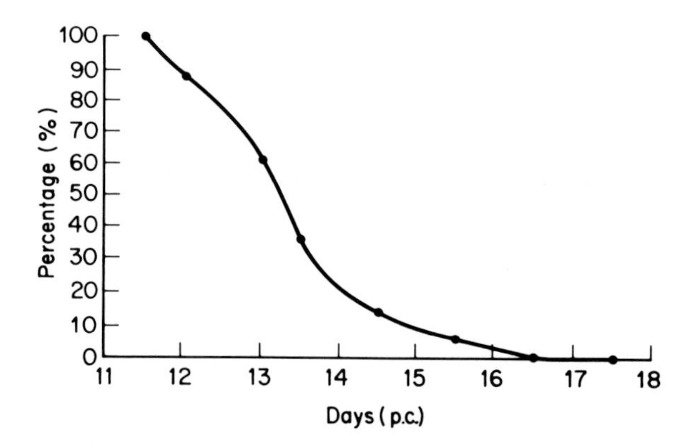

Figure 7 Incidence of primitive nucleated red blood cells in the embryonic circulation between 11.5 and 17.5 days p.c.

The liver rapidly enlarges at this time because it takes over from the yolk sac as the principal source of (definitive) haematopoietic activity, though yolk sac-derived nucleated (primitive) red blood cells persist for several more days before they eventually disappear from the circulation (see Fig. 7). It should also be noted that both the cellular and nuclear volumes of the primitive red blood cells closely correlate with the developmental age of normal diploid and tetraploid mouse embryos during the duration of their presence in both the extra-embryonic and embryonic circulations (see Figs 8 and 9). Indeed, the morphometric analysis of their respective nuclear and cellular

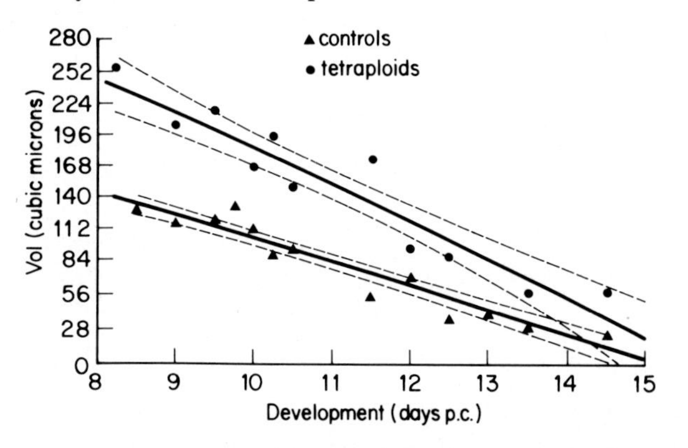

Figure 8 Graph in which the primitive red blood cell *nuclear* volume is plotted against developmental age of carefully staged embryos. Each point represents the mean value obtained from the analysis of at least 100 cells (Henery and Kaufman, 1991). 95% confidence limits for each curve are also illustrated (dotted lines).

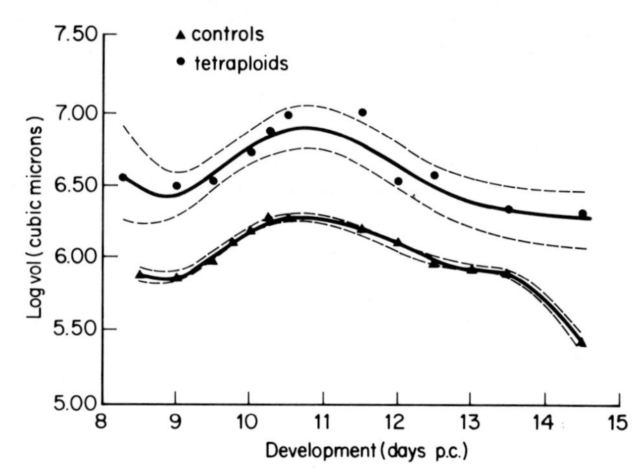

Figure 9 Graph in which the log of primitive red blood *cellular* volume is plotted against developmental age of carefully staged embryos. Each point represents the mean value obtained from the analysis of at least 100 cells (Henery and Kaufman, 1991). 95% confidence limits for each curve are also illustrated (dotted lines).

volumes may be used as one of a number of accurate means for assessing developmental age of embryos, if this information is not available from other sources.

The other intraperitoneal organs that are enlarging at this time are the stomach (*Plate 25a,a–c*), and the pancreatic primordia. The latter develop in close association with the foregut/midgut junction, in the region of the second part of the duodenum (*Plate 25a,a–d*), and are, at this early stage in their differentiation, more easily recognized in sagittal than in transverse sections.

THE NERVOUS SYSTEM

A characteristic feature of this stage of development is the progressive closure of the lens pit to form the lens vesicle. With the formation of the latter the entire vesicle eventually loses contact with the overlying surface ectoderm (*Plate 51,o,p*). An increase in the thickness of the inner (neural) layer of the optic cup also occurs during this stage (*Plate 51,o,p*). With the formation of the optic cup from the invagination of the distal (outer) part of the optic vesicle, continuity between the third ventricle and the intra-retinal space is maintained through the lumen of the optic stalk. On the ventral (inferior) aspect of the distal part of the latter, an indentation termed the choroidal (or "fetal") fissure (*Plate 25a,a*) forms. This is continuous with an indentation on the inferior part of the optic cup, and accommodates the hyaloid artery and vein (*Plate 52,a,c*). The artery breaks up into a plexus of small vessels which supply the inner layer of the optic cup, the lens vesicle and the intervening mesenchyme. Subsequently, the choroidal fissure becomes narrowed by the growth and later fusion of its inferior margins. The latter process extends distally, so that the fetal fissure eventually becomes completely obliterated. If fusion fails to occur in this location, or is incomplete, the condition known as coloboma results, with a deficiency of the inferomedial part of the iris and (often) the choroid.

The otocyst gradually elongates (*Plate 24c,b–f*), and a diverticulum grows out from its superomedial border (*Plate 24c,b–d*), to form the endolymphatic duct (*Plate 25a,a*; *Plate 25c,b*). A number of cranial nerve ganglia are also particularly prominent at this stage of development, attention being especially drawn to the relatively enormous trigeminal (V) ganglion (*Plate 25a,a*; *Plate 25c,c,d*), the facio-acoustic (VII–VIII) ganglion complex (*Plate 25a,a*; *Plate 25c,b,c*), the superior and inferior glossopharyngeal (IX) ganglia (*Plate 25a,a*; *Plate 25c,a*), and the superior and inferior vagal (X) ganglia (*Plate 25a,a,b*; *Plate 25c,a*). As a more general feature, the marginal layer is more easily seen in the region of the developing brain and spinal cord (*Plate 25a,a–d*; *Plate 25b,a–d*; *Plate 25c,a,b*) than in previous stages of development, and this is particularly obvious in sagittally sectioned material.

THE UROGENITAL SYSTEM

Despite a slight increase in the volume of the urogenital ridges compared to that observed in the previous stage, little difference is seen in the histological appearance of either the mesonephric ducts (*Plate 25a,a,b*; *Plate 25b,b–d*) or the mesonephric tubules/vesicles (*Plate 25a,a,b*; *Plate 25b,b,c*). The ureteric bud is first seen at this stage, being a diverticulum from the mesonephric duct just proximal to where it makes contact with the urogenital sinus (*Plate 25b,c*). It grows towards, and its tip soon becomes surrounded by, metanephric blastemal tissue (*Plate 25c,a*).

Stage 19 Developmental age, 11.5 days p.c.
Plates 26a–e

EXTERNAL FEATURES

Embryos with about 43–48 pairs of somites. The most obvious changes that occur at this stage of development relate to the external morphological features of the cephalic region, particularly in the area around the developing eyes. The lens vesicles become completely separated from the surface ectoderm, and the latter differentiates to form the corneal epithelium. The peripheral margins of the eye become increasingly well defined (*Plate 49,a*; *Plate 52,a,c*), so that over the next few days these will differentiate to form the upper and lower eyelids (*Plate 49,b,c*; *Plate 52,e,i*). The surface ectoderm between the peripheral margin of the cornea and the under surface of the eyelids will also subsequently differentiate to form the conjunctival epithelium (*Plate 52,e,i*).

The naso-lacrimal grooves deepen (*Plate 47,k*; *Plate 49,a*), while the maxillary component of the first branchial arches becomes more prominent and develops more sharply defined borders (*Plate 47,k*) than previously (*Plate 47,g,i*). At about this time, the medial and lateral margins of the olfactory pits tend to migrate towards each other, and consequently reduce their entrance to a relatively narrow slit (*Plate 47,l*; *Plate 59,b*). The medial and lateral nasal processes also become more prominent (*Plate 47,l*) than previously (*Plate 47,j*). The maxillary and mandibular components of the first branchial arch are more readily distinguished, particularly because of the change in the shape and greater degree of prominence of the maxillary component (see above).

On either side of the first branchial groove (cleft), a series of about six small tubercles (the auditory hillocks) appear (*Plate 47,k*), which will eventually enlarge and coalesce to give rise to the various components, but principally the pinna, of the external ear (*Plate 49,c*), while the first branchial groove gives rise to the external auditory (acoustic) meatus (*Plate 49,c*).

With the rapid increase in the volume of the telencephalic vesicles, the prominences overlying them tend to occupy a greater proportion of the cephalic region than previously, though the transverse groove between them and the rostral part of the midbrain, which is clearly seen in the previous stage of development (*Plate 47,i,j*), becomes much less well defined (*Plate 47,k,l*).

Elsewhere in the embryo, the limb buds show early evidence of differentiation. In developmentally less well advanced embryos at this stage, only the forelimbs are seen to be divided into two distinct regions, namely into a proximal part consisting of the future limb-girdle region and the arm region, and the more peripherally located almost circular or paddle-shaped hand plate (*Plate 43,h–j*). In these embryos, the hindlimb buds have yet to differentiate into these various regions. In developmentally more advanced embryos, however, the hindlimb buds are clearly distinguishable into a proximally located part consisting of the future limb-girdle and leg regions, and the more peripherally located paddle-shaped foot plate (*Plate 43,h–j*). At the peripheral margins of both the fore- and hindlimb buds, the well-defined apical ectodermal ridges are also clearly seen, in both regions running craniocaudally (*Plate 43,h–j*). On sections through this region, it is possible to see that the marginal veins are located just subjacent to the apical ectodermal ridges (*Plate 26d,d–g*), and that in appropriately stained material, the cells within the ridges have high levels of intracellular alkaline phosphatase enzyme activity (*Plate 67,i–k*).

Between the caudal part of the hindlimb buds, a prominent transversely running genital ridge (or tubercle) is also clearly seen (*Plate 43,j*; *Plate 70,a*). The latter overlies the cloacal region (and urogenital sinus) of the hindgut diverticulum (*Plate 26d,d–f*).

THE HEART AND VASCULAR SYSTEM

This is a critical stage in the differentiation of the vascular system, in that it is the first stage when evidence of asymmetry is seen in relation to the branchial (pharyngeal or aortic) arch arterial system. Whereas during the 9th and 10th days p.c., successively the first, second, third, fourth and sixth arch arteries differentiate in relation to their respective branchial arches, as the caudal vessels become apparent, the most rostral vessels largely disappear (the sequential appearance and evolution of these vessels in the mouse is illustrated in Fig. 10). The first arch artery largely disappears between 10.5 and 11 days p.c., and its only named derivative is the maxillary artery. The second arch artery also largely disappears at about this time, and its only named derivatives are the stapedial and the hyoid arteries. While the former vessel is not easily located at this stage of embryonic development, it is, however, recognizable after about 13.5 days p.c., because of its close relationship to the stapes (see, for example, *Plate 32c,b*; *Plate 39c,c*). The third arch arteries are first evident in embryos at about 9.5 days p.c. (*Plate 19b,g*), and these make contact with the dorsal aorta by about 10–10.25 days p.c. (*Plate 20b,j*). The third branchial arch arteries give rise to the common carotid arteries in the region of the carotid body and carotid sinus which receive their nerve supply from the glossopharyngeal (IX) nerve, which is the nerve supply to the derivatives of the third branchial arch. The majority of the internal carotid artery is derived from the rostral extremity of the dorsal aorta. The fourth branchial arch arteries are also first evident in embryos at about 10.25–10.5 days p.c. (*Plate 22a,a*), but are more easily recognized in developmentally slightly more advanced embryos, at about 10.5–11 days p.c. (*Plate 24c,j,k*). At this stage, it is also possible to recognize the sixth arch arteries (*Plate 24c,j–n*). Similarly, the pulmonary arteries are also easily recognized at this stage as a pair of narrow diameter vessels which pass directly posteriorly from the middle region of the sixth arch arteries, and are directed caudally towards the two primitive lung buds (*Plate 24c,l–o*).

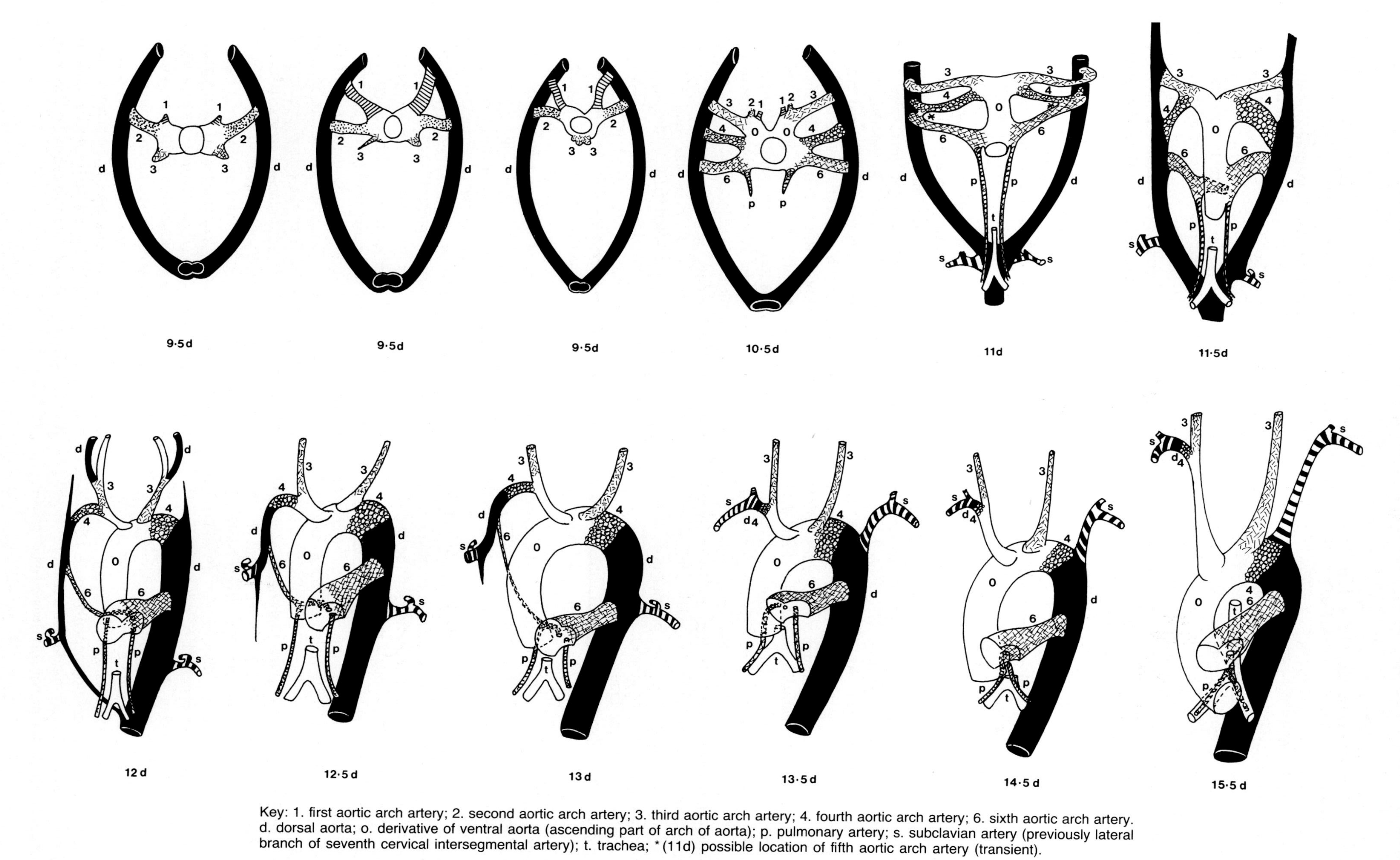

Key: 1. first aortic arch artery; 2. second aortic arch artery; 3. third aortic arch artery; 4. fourth aortic arch artery; 6. sixth aortic arch artery. d. dorsal aorta; o. derivative of ventral aorta (ascending part of arch of aorta); p. pulmonary artery; s. subclavian artery (previously lateral branch of seventh cervical intersegmental artery); t. trachea; * (11d) possible location of fifth aortic arch artery (transient).

Figure 10 Sequential changes that occur during the differentiation of the aortic (pharyngeal) arch arteries.

By about 11.5 days p.c., the right fourth and sixth arch arteries are seen to have a slightly narrower diameter than the corresponding vessels of the left side. By 12–12.5 days p.c., the difference is particularly marked, especially in relation to the right sixth arch artery in its course between the origin of the right pulmonary artery and the dorsal aorta, which is now seen to have only a very narrow lumen (*Plate 27b,h*). The right sixth arch artery in this location is no longer seen after about 13–13.5 days p.c., and it can only be assumed that its cellular components are removed by the process of phagocytosis.

The most important change that occurs in the venous system at this time is a direct consequence of the events that are occurring in relation to the liver. Even in the relatively short period between 10.5–11 days and 11.5–12 days p.c., the overall size and complexity of the liver has increased dramatically. At 10.5–11 days p.c., the liver is a relatively small organ, which still shows clear evidence of its origin within the septum transversum (*Plate 24c,i–o*), whereas by 11.5–12 days p.c., it is not only a substantially larger organ than previously, but its peripheral boundaries are much more clearly defined (*Plate 26d,a–h*). While previously its principal vascular supply was from the left umbilical vein (*Plate 24c,j*) and from the vitelline veins (*Plate 24c,l–n*), both of which contributed to the hepatic sinusoidal system, by 11.5–12 days p.c., the left umbilical vein appears principally to drain into the ductus venosus (*Plate 26d,b–e*). The latter forms a large diameter channel that effectively bypasses most of the liver substance, while the hepatic sinusoids are now principally filled with blood which drains into them from the portal vein (*Plate 26d,f*). This vessel is the principal derivative of the right vitelline vein. The disparity in size between the left and right umbilical veins previously noted (*Plate 24c,f–m*), is now less marked with the differentiation of the ductus venosus, and may even be reversed, so that the diameter of the left umbilical vein may even be greater than that of the right umbilical vein (*Plate 26d,f–i*). Within the liver, foci of haematopoiesis may be observed.

The ductus venosus drains into the right hepatocardiac vein (the primitive inferior vena cava) (*Plate 26d,a,b*), and this vessel largely drains into the right horn of the sinus venosus (*Plate 26c,i*), and the largely oxygenated blood within this channel then drains into the right atrium (*Plate 26c,h,i*). At the site of entrance of the right horn of the sinus venosus into the right atrium, a venous valve with several leaflets may be clearly discerned (*Plate 26c,f,g*).

Within the common atrial chamber of the heart, increasing evidence of septation may be seen, with the growth of the septum primum towards the atrioventricular bulbar cushion tissue (*Plate 26c,d–g*). At this stage, the peripheral margin of the septum primum appears to be quite bulbous (*Plate 26c,e–g*). During the next stage of development, by the time that the septum primum makes contact with the atrioventricular cushion tissue, the ostium secundum is already present and acts as the channel between the right and left atria (*Plate 27c,b*). The two bulbar ridges within the outflow tract of the heart, which form the aortico-pulmonary spiral septum, are also clearly seen (*Plate 26b,i*; *Plate 26c,a,b*), and are directed caudally, so that they are soon to form the membranous component of the interventricular septum, while the muscular component is largely formed from the ventricular wall subjacent to the bulboventricular groove.

THE PRIMITIVE GUT AND ITS DERIVATIVES

Scanning electron microscopy of the oropharyngeal region of embryos at this stage of development reveals the first indication of the lateral lingual swellings. These develop on the dorsal (oral) surface of the first branchial (pharyngeal) arch (*Plate 58,a,b*), the two swellings being separated at this stage by a deep groove (median sulcus), located at the site of fusion in the midline of the left and right mandibular processes. A smaller slightly more posteriorly located median swelling probably represents the tuberculum impar (*Plate 58,b*), which is also of first arch origin. These structures are more difficult to recognize on transverse histological sections (*Plate 26b,d*), but the lingual vessels are relatively easily located (*Plate 26b,e*).

The thyroid gland differentiates from a midline thickening of the endoderm in the floor of the pharynx just caudal to the tuberculum impar. The thickening progresses to form a diverticulum (*Plate 23b,k*; *Plate 24c,g*), which is attached to the floor of the pharynx by the thyroglossal duct (*Plate 25b,a*), though at this stage the thyroid primordium may appear to have lost contact with the floor of the pharynx (*Plate 26b,f*).

In the roof of the pharynx, scanning electron micrographs of this region reveal the first evidence of the palatal shelves of the maxillae and the entrances to the primitive posterior naris. Only at about 12.5–13 days p.c. is it first possible to recognize the ventral surface of the primitive nasal septum. The site of attachment of Rathke's pouch is now quite constricted (*Plate 26b,c*; *Plate 56,f*), but complete separation from the roof of the oropharynx is not usually seen before 12–12.5 days p.c.

In the upper pharyngeal region, the laryngo-tracheal groove delineates the rostral extremity of the trachea (*Plate 26b,i*). The latter courses through the thoracic region (*Plate 26c,b–f*), until it bifurcates to form the left and right main bronchi. By this stage of development, both primary and secondary (lobar) bronchi are present (*Plate 26c,h,i*; *Plate 26d,a–c*; *Plate 66,b*). In the mid-abdominal region, the two pancreatic primordia are now seen to be in close proximity to the wall of the duodenum (*Plate 26d,g,h*). More caudally, the urorectal septum is forming, and will separate the urogenital sinus anteriorly from the hindgut posteriorly (*Plate 26d,f–h*). The caudal (post-anal) extension of the hindgut diverticulum terminates near the tip of the tail (*Plate 26c,i*; *Plate 26d,a*).

THE NERVOUS SYSTEM

The most obvious features at this stage of development are again associated with the cephalic region, so that, for example, the lens vesicle is now seen to be

(*continued on page 144*)

PLATE 26a

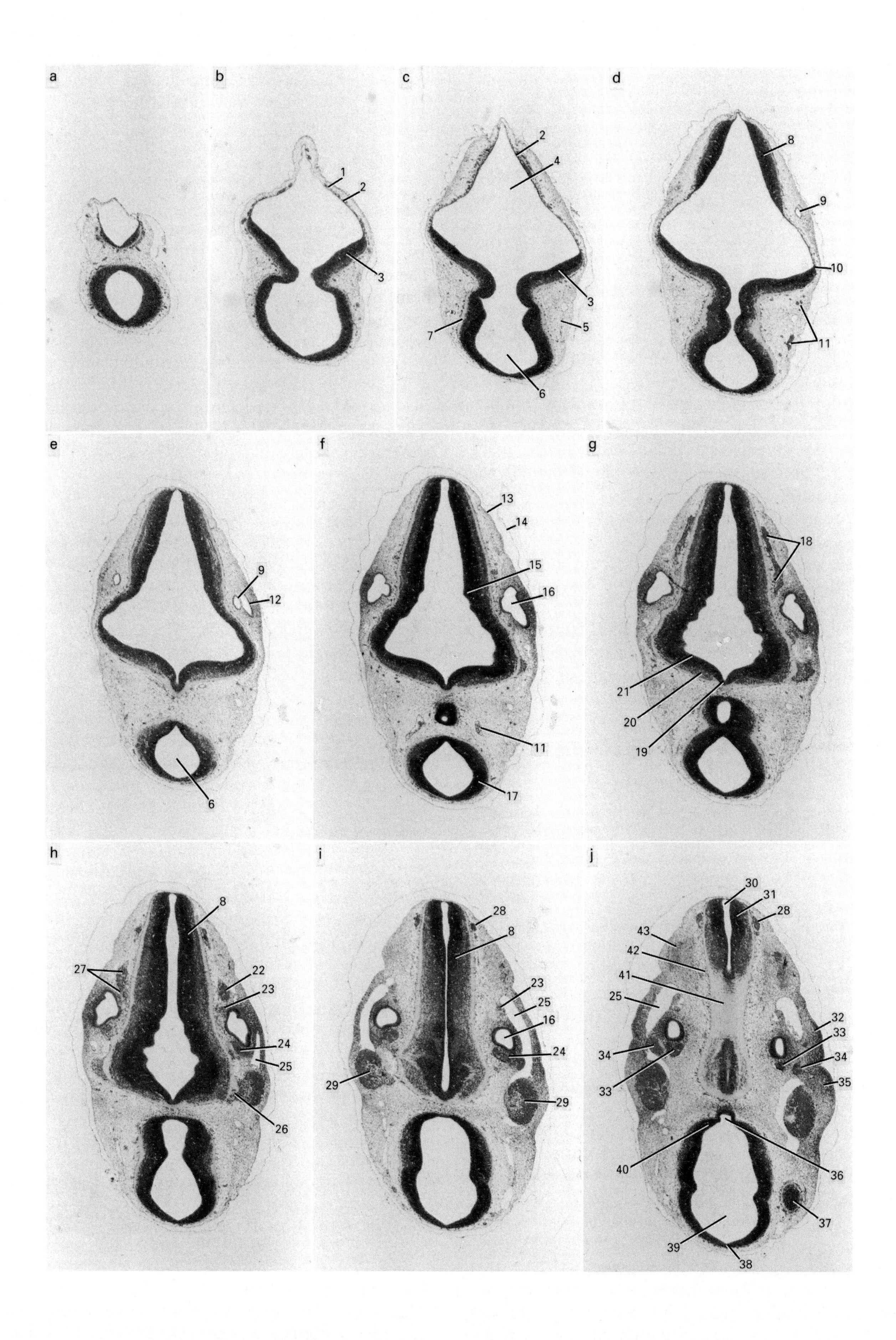

a
b
c
d
e
f
g
h
i
j
1
2
3
4
5
6
7
8
9
10
11
12
13
14
15
16
17
18
19
20
21
22
23
24
25
26
27
28
29
30
31
32
33
34
35
36
37
38
39
40
41
42
43

Plate 26a (11.5–12 days p.c.)

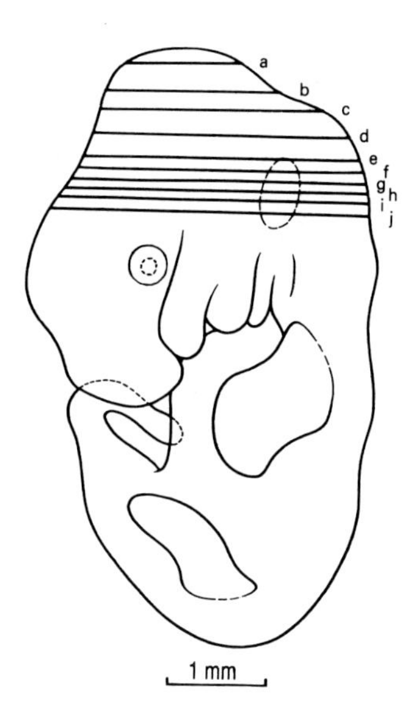

11.5–12 days p.c. Embryo with 43–48 pairs of somites. Crown–rump length 5.8 mm (fixed). Transverse sections (Theiler Stage 19; rat, Witschi Stages 26–27; human, Carnegie Stage 16).

1. posterior cerebral plexus of veins
2. roof of fourth ventricle
3. metencephalic part of rhombencephalon
4. fourth ventricle
5. cephalic mesenchyme
6. mesencephalic vesicle
7. marginal layer
8. myelencephalic part of rhombencephalon (future medulla oblongata)
9. endolymphatic diverticular appendage of otic vesicle (otocyst)
10. rhombic lip (dorsal part of alar lamina of metencephalon – the cerebellar primordium)
11. small arterial branches from internal carotid artery
12. main (vestibular) portion of otic vesicle
13. surface ectoderm
14. amnion
15. neuromere
16. otic vesicle (otocyst)
17. wall of mesencephalon (midbrain)
18. cranial part of accessory (XI) nerve
19. median sulcus
20. mantle layer
21. ependymal layer
22. superior ganglion of vagus (X) nerve
23. superior ganglion of glossopharyngeal (IX) nerve
24. facio-acoustic (VII–VIII) ganglion complex
25. primary head vein
26. rootlets of trigeminal (V) ganglion
27. glossopharyngeal–vagal (IX–X) ganglion complex
28. spinal part of accessory (XI) nerve
29. trigeminal (V) ganglion
30. central canal
31. neural tube at level of caudal hindbrain/upper cervical region
32. second branchial arch
33. acoustic (VIII) component of facio-acoustic (VII–VIII) ganglion complex
34. facial (VII) component of facio-acoustic (VII–VIII) ganglion complex sending branches into second branchial arch
35. maxillary component of first branchial arch
36. infundibular recess (future pars nervosa)
37. wall of telencephalic vesicle
38. lamina terminalis
39. third ventricle
40. wall of diencephalon
41. floor of myelencephalon
42. rootlets of hypoglossal (XII) nerve
43. occipital myotome

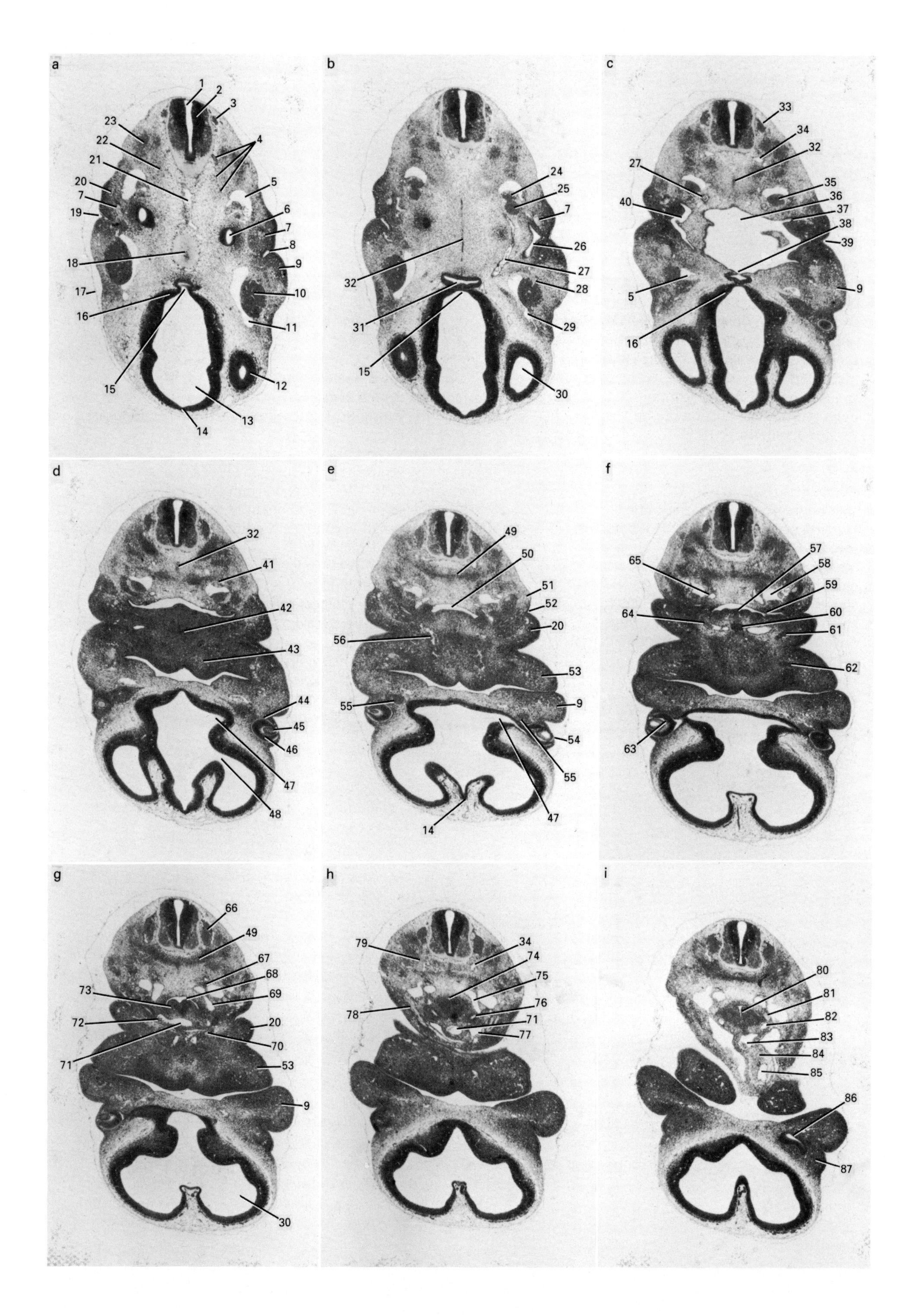
a
1
2
3
23
22
4
21
5
20
7
6
19
7
8
18
9
17
10
16
11
15
12
13
14
b
24
25
7
32
26
27
28
31
29
15
30
c
33
34
32
27
35
40
36
37
38
39
5
9
16
d
32
41
42
43
44
45
46
47
48
e
49
50
51
52
20
56
53
55
9
54
55
47
14
f
57
65
58
59
64
60
61
62
63
g
66
49
67
68
73
69
72
20
70
71
53
9
30
h
79
34
74
75
78
76
71
77
i
80
81
82
83
84
85
86
87

Plate 26b (11.5–12 days p.c.)

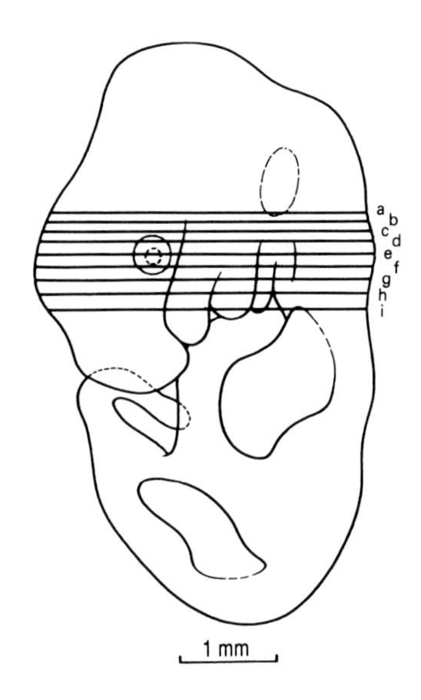

11.5–12 days p.c. Embryo with 43–48 pairs of somites. Crown–rump length 5.8 mm (fixed). Transverse sections (Theiler Stage 19; rat, Witschi Stages 26–27; human, Carnegie Stage 16).

1. central canal
2. neural tube at level of caudal hindbrain/upper cervical region
3. spinal part of accessory (XI) nerve
4. rootlets of hypoglossal (XII) nerve
5. primary head vein
6. otic vesicle (otocyst)
7. branches of facial (VII) nerve passing into second branchial arch
8. first branchial pouch
9. maxillary component of first branchial arch
10. trigeminal (V) ganglion
11. site of communication between primary head vein and anterior cerebral plexus of veins
12. wall of telencephalic vesicle (future lateral ventricle)
13. third ventricle
14. lamina terminalis
15. infundibular recess (neurohypophyseal bud, future pars nervosa)
16. wall of diencephalon
17. amnion
18. ventral region of pons
19. first branchial membrane
20. second branchial arch
21. basilar artery
22. occipital sclerotome
23. occipital myotome premuscle mass
24. united ganglion of spinal accessory (XI) and vagus (X) nerves
25. inferior ganglion of glossopharyngeal (IX) nerve
26. tubo-tympanic recess
27. internal carotid artery
28. branches of trigeminal (V) nerve passing into maxillary component of first branchial arch
29. anterior cerebral plexus of veins
30. telencephalic vesicle
31. Rathke's pouch
32. notochord
33. dorsal (posterior) root ganglion of C1
34. vertebral artery
35. inferior ganglion of vagus (X) nerve
36. oropharyngeal region (buccal cavity)
37. entrance to second branchial pouch
38. constriction at site of attachment of Rathke's pouch to roof of pharynx
39. first branchial cleft (groove)
40. second branchial pouch
41. hypoglossal (XII) nerve
42. tuberculum impar
43. premuscle mass of lateral lingual swelling (mandibular swelling)
44. outer layer of optic cup (future pigment layer of retina)
45. lens vesicle
46. inner (neural) layer of optic cup (future nervous layer of retina)
47. optic recess (entrance to optic stalk)
48. interventricular foramen
49. cervical sclerotome
50. pharyngeal region of foregut
51. third branchial arch
52. second branchial cleft (groove)
53. mandibular component of first branchial arch
54. corneal ectoderm
55. extra-ocular premuscle mass
56. lingual vessels
57. epiglottic swelling
58. anterior cardinal vein
59. entrance into second branchial pouch
60. thyroid primordium (thyroglossal duct)
61. precursor of second branchial arch cartilage (hyoid cartilage)
62. precursor of first branchial arch cartilage (Meckel's cartilage)
63. intra-retinal space
64. second branchial arch artery
65. dorsal aorta
66. dorsal (posterior) root ganglion of C3
67. sympathetic chain
68. laryngeal orifice
69. communication between dorsal aorta and second branchial arch artery
70. superior laryngeal branch of vagus (X) nerve
71. aortic sac
72. origin of third branchial arch artery
73. arytenoid swelling
74. oesophagus
75. communication between dorsal aorta and fourth branchial arch artery
76. fourth branchial pouch
77. rostral extremity of pericardial cavity
78. phrenic nerve
79. anterior primary division of C4
80. laryngo-tracheal groove
81. vagus (X) nerve
82. sixth branchial arch artery
83. communication between aortic sac and sixth branchial arch artery
84. cranial ridge of mesenchymatous tissue of aortico-pulmonary spiral septum
85. pulmonary component of outflow tract of heart
86. olfactory epithelial lining of nasal pit
87. rostral part of lateral nasal process

PLATE 26c

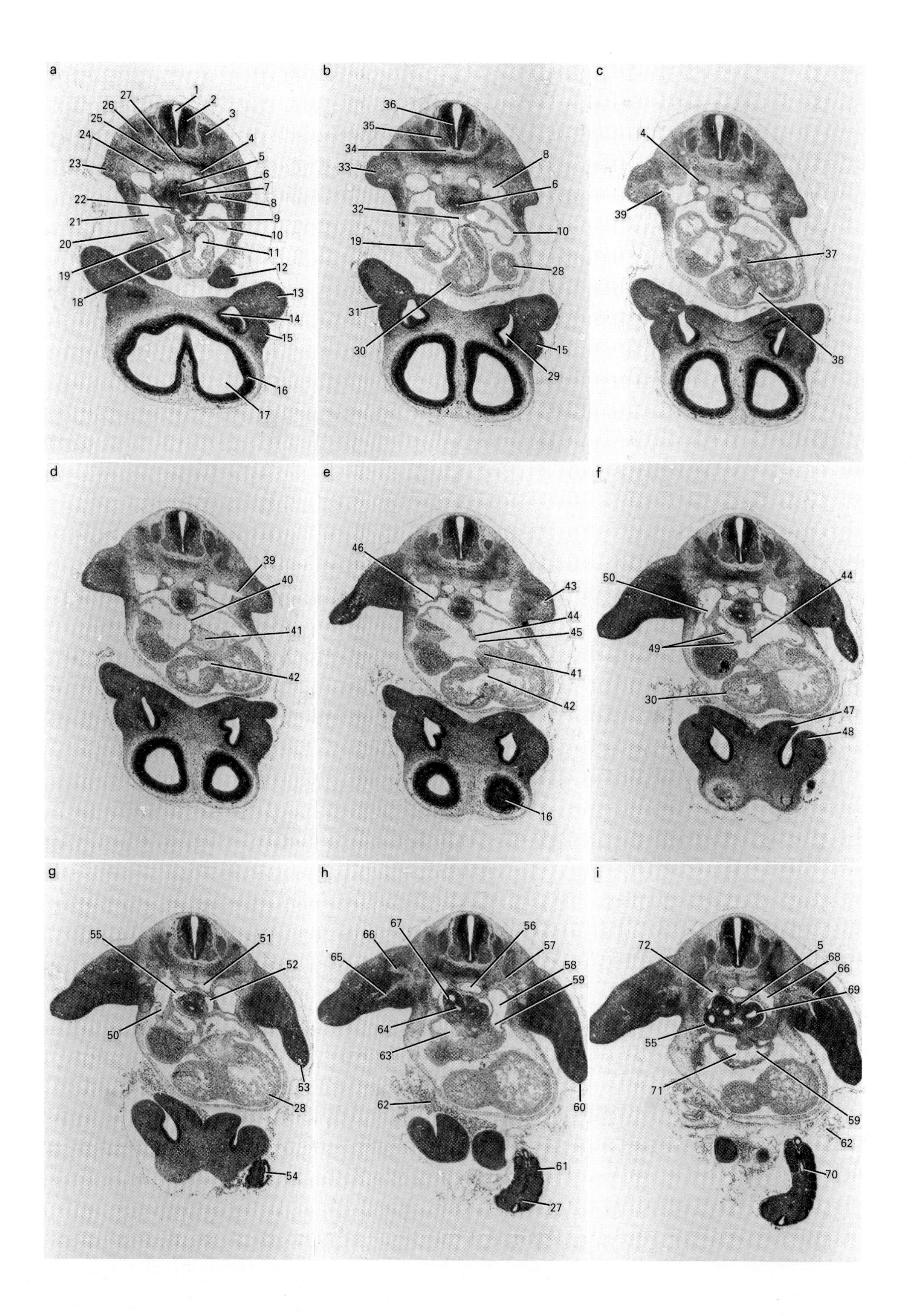

a
1
2
3
4
5
6
7
8
9
10
11
12
13
14
15
16
17
18
19
20
21
22
23
24
25
26
27
b
36
35
34
33
8
6
32
10
19
28
31
30
15
29
c
4
39
37
38
d
39
40
41
42
e
46
43
44
45
41
42
16
f
50
44
49
30
47
48
g
55
51
52
50
53
28
54
h
67
66
56
57
65
58
59
64
63
62
60
61
27
i
72
5
68
66
69
55
71
59
62
70

Plate 26c (11.5–12 days p.c.)

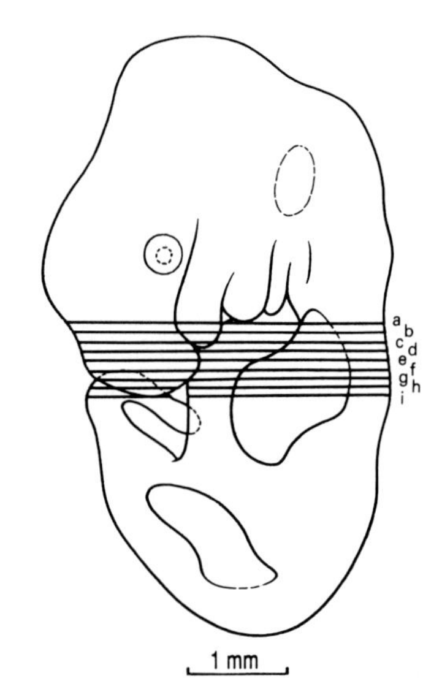

11.5–12 days p.c. Embryo with 43–48 pairs of somites. Crown–rump length 5.8 mm (fixed). Transverse sections (Theiler Stage 19; rat, Witschi Stages 26–27; human, Carnegie Stage 16).

1. central canal
2. neural tube in cervical region
3. dorsal (posterior) root ganglion
4. sympathetic chain
5. oesophagus
6. trachea
7. vagus (X) nerve
8. anterior cardinal vein
9. aortic sac
10. wall of left atrium
11. pulmonary trunk
12. mandibular component of first branchial arch
13. maxillary component of first branchial arch
14. olfactory epithelium in region of vomeronasal organ (Jacobson's organ)
15. rostral part of lateral nasal process
16. wall of telencephalic vesicle (future lateral ventricle)
17. telencephalic vesicle
18. cranial ridge of mesenchymatous tissue of aortico-pulmonary spiral septum
19. wall of right atrium
20. thoracic wall
21. pericardial cavity
22. sixth branchial arch artery
23. rostral extremity of forelimb bud
24. dorsal aorta
25. cervical sclerotome (dense caudal component)
26. cervical myotome premuscle mass
27. notochord
28. trabeculated wall of common ventricular chamber (future left ventricle) of heart
29. nasal (olfactory) pit
30. wall of bulbus cordis (future right ventricle) region of heart
31. maxillo-nasal groove
32. transverse pericardial sinus
33. forelimb bud
34. marginal layer
35. mantle layer
36. ependymal layer
37. caudal component of bulbar ridge
38. bulbo-ventricular groove
39. principal (axial) vein from forelimb bud
40. rostral component of septum primum
41. endocardial cushion tissue in atrio-ventricular canal
42. bulbo-ventricular canal
43. principal (axial) artery of forelimb bud
44. lower bulbous edge of septum primum
45. foramen primum
46. phrenic nerve
47. medial nasal process
48. lateral nasal process
49. leaflets of venous valve at entrance of sinus venosus into right atrium
50. right common cardinal vein
51. rostral site of amalgamation of right and left dorsal aortae
52. left pericardio-peritoneal canal
53. marginal vein
54. neural tube in tail region
55. right pericardio-peritoneal canal
56. midline dorsal aorta
57. cervical nerve trunk (part of brachial plexus) passing into forelimb bud
58. left common cardinal vein
59. left horn of sinus venosus
60. apical ectodermal ridge
61. somite
62. extravasated blood
63. right horn of sinus venosus
64. right main bronchus
65. median nerve
66. radial nerve
67. right lung bud
68. left posterior cardinal vein
69. left main bronchus
70. caudal extension of hindgut diverticulum (postanal component)
71. site of amalgamation of right and left horns of sinus venosus
72. right posterior cardinal vein

PLATE 26d

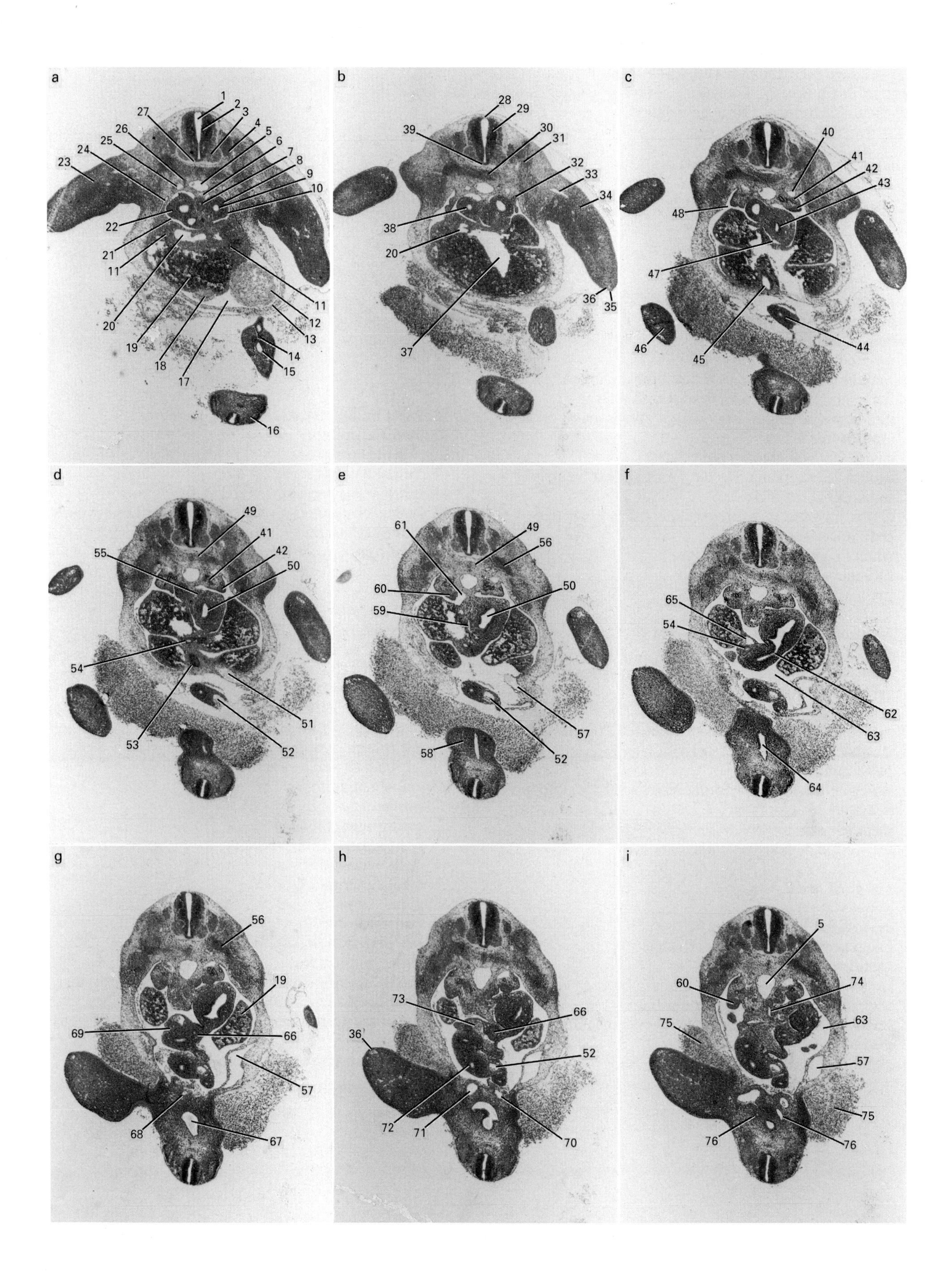
a
b
c
d
e
f
g
h
i
1
2
3
4
5
6
7
8
9
10
11
12
13
14
15
16
17
18
19
20
21
22
23
24
25
26
27
28
29
30
31
32
33
34
35
36
37
38
39
40
41
42
43
44
45
46
47
48
49
50
51
52
53
54
55
56
57
58
59
60
61
62
63
64
65
66
67
68
69
70
71
72
73
74
75
76

Plate 26d (11.5–12 days p.c.)

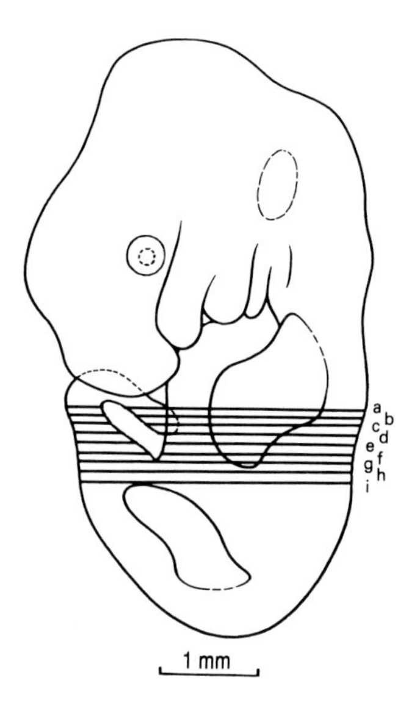

11.5–12 days p.c. Embryo with 43–48 pairs of somites. Crown–rump length 5.8 mm (fixed). Transverse sections (Theiler Stage 19; rat, Witschi Stages 26–27; human, Carnegie Stage 16).

1. central canal
2. ependymal layer
3. mantle layer (basal plate region)
4. dorsal (posterior) root ganglion
5. midline dorsal aorta
6. dorsal meso-oesophagus
7. oesophagus
8. left lung bud
9. left main bronchus
10. left pericardio-peritoneal canal
11. septum transversum
12. wall of common ventricular chamber of heart (future left ventricle)
13. thoracic wall
14. notochord
15. dilated caudal extension of hindgut diverticulum (postanal component)
16. somite
17. pericardial cavity
18. wall of bulbus cordis
19. hepatic sinusoids within liver
20. right hepato-cardiac vein
21. right pericardio-peritoneal canal
22. right lung bud
23. right forelimb bud
24. rostral extension of peritoneal cavity
25. right posterior cardinal vein
26. sympathetic chain
27. marginal layer
28. roof plate
29. neural tube in mid-cervical region
30. sclerotome (caudal condensed portion)
31. cervical myotome premuscle mass
32. mesonephric duct
33. principal (axial) artery of forelimb bud
34. left forelimb bud
35. apical ectodermal ridge
36. marginal vein
37. ductus venosus
38. right lobar bronchus
39. floor plate
40. left posterior cardinal vein
41. mesonephric tubules
42. rostral extremity of urogenital ridge
43. gastro-oesophageal junction
44. loop of midgut within physiological umbilical hernia
45. gall bladder
46. right hindlimb bud
47. pneumo-enteric recess
48. site of communication between pericardio-peritoneal canal and peritoneal cavity
49. sclerotome (rostral less condensed portion)
50. lumen of stomach
51. site of entrance of left umbilical vein into liver prior to joining ductus venosus
52. vitelline vein
53. wall of gall bladder
54. duodenum
55. dorsal mesogastrium
56. rib primordium
57. left umbilical vein
58. genital tubercle
59. lesser sac of omental bursa
60. gonadal ridge
61. medial coelomic bay
62. gastroduodenal junction
63. peritoneal cavity
64. hindgut diverticulum (cloacal region)
65. portal vein
66. primordium of left (dorsal) lobe of the pancreas
67. urogenital sinus
68. midline umbilical artery
69. primordium of right (ventral) lobe and body of the pancreas
70. left umbilical artery
71. right umbilical artery
72. distal part of midgut
73. superior mesenteric vein
74. superior mesenteric artery
75. extravasated blood
76. metanephric duct (ureteric bud)

PLATE 26e

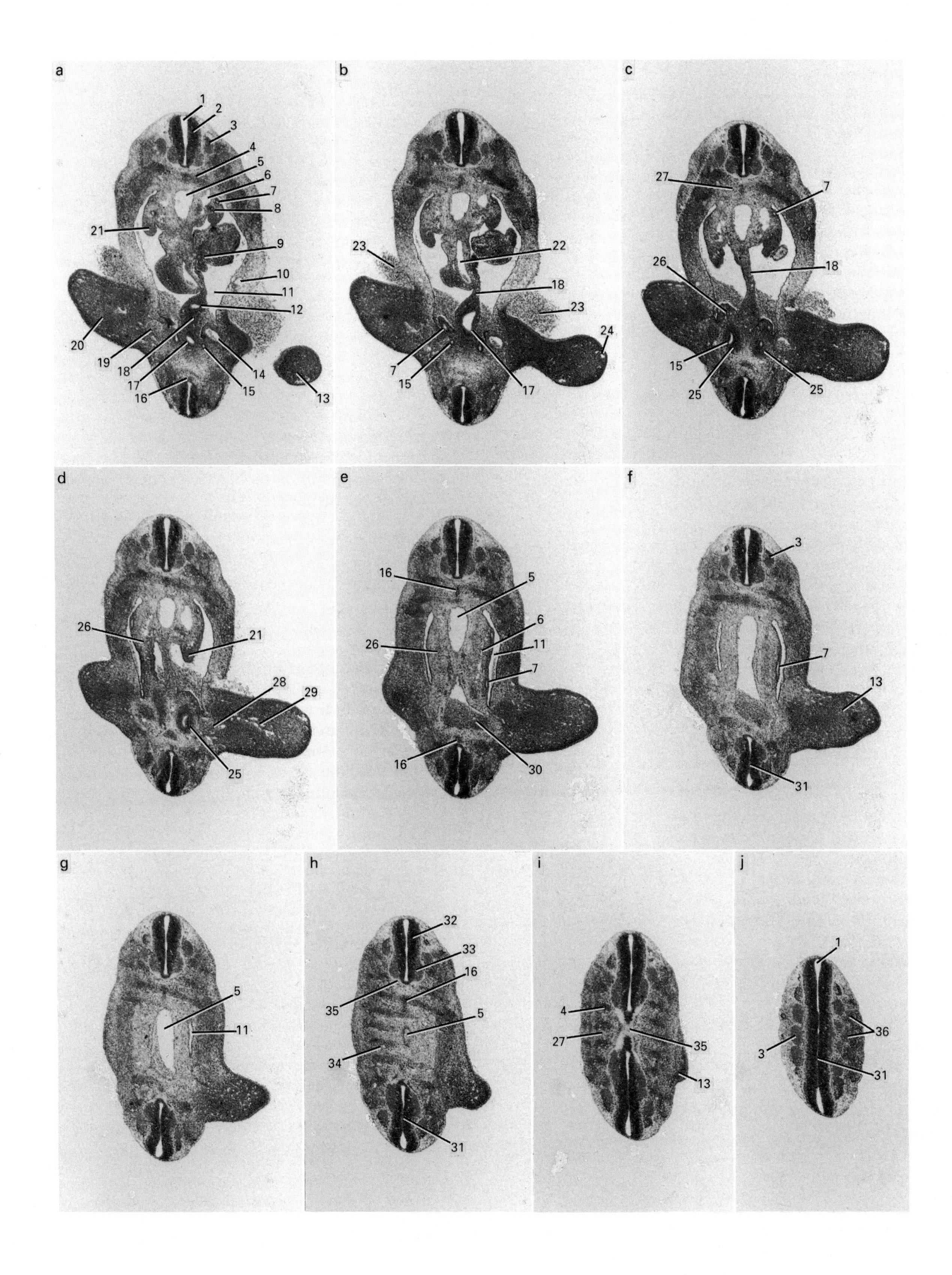

a
1
2
3
4
5
6
7
8
9
10
11
12
13
14
15
16
17
18
19
20
21
b
22
23
18
23
24
7
15
17
c
27
7
18
26
15
25
25
d
26
21
28
29
25
e
16
5
6
26
11
7
16
30
f
3
7
13
31
g
5
11
h
32
33
16
35
5
34
31
i
4
27
35
13
j
1
36
3
31

Plate 26e (11.5–12 days p.c.)

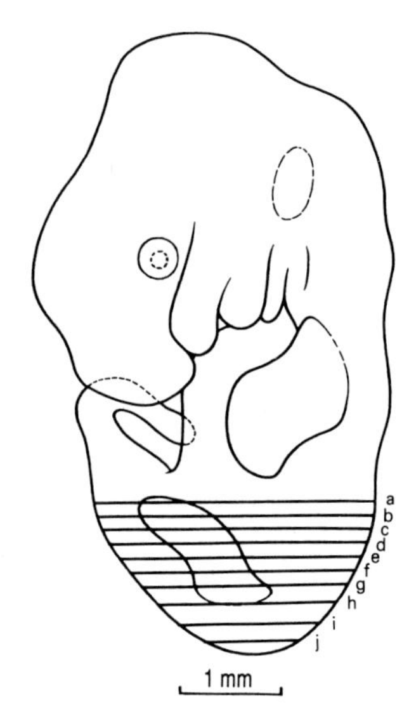

11.5–12 days p.c. Embryo with 43–48 pairs of somites. Crown–rump length 5.8 mm (fixed). Transverse sections (Theiler Stage 19; rat, Witschi Stages 26–27; human, Carnegie Stage 16).

1. central canal
2. spinal cord
3. dorsal (posterior) root ganglion
4. sclerotome (caudal condensed portion)
5. midline dorsal aorta
6. subcardinal vein
7. mesonephric duct
8. mesonephric tubule
9. primordium of left lobe of the pancreas
10. left umbilical vein
11. peritoneal cavity
12. hindgut
13. left hindlimb bud
14. left umbilical artery
15. ureteric bud (metanephric duct)
16. notochord
17. urogenital sinus
18. dorsal mesentery of hindgut
19. right umbilical artery
20. right hindlimb bud
21. gonadal ridge
22. inferior mesenteric artery
23. extravasated blood
24. marginal vein
25. metanephric blastema
26. urogenital ridge
27. sclerotome (rostral less condensed portion)
28. principal (axial) artery to hindlimb bud (branch of common iliac artery)
29. principal (axial) vein draining the hindlimb bud
30. common iliac artery
31. localized site of neural luminal occlusion
32. mantle layer (alar plate)
33. mantle layer (basal plate)
34. segmental artery
35. marginal layer
36. presomite condensation of paraxial mesoderm

completely detached from the overlying surface ectoderm, as indicated earlier. On histological examination of the eye of more advanced embryos at this stage of development, it is possible to observe that the cells of the posterior wall of the lens vesicle are elongated, and that their nuclei are mostly located towards the periphery of the lens vesicle (*Plate 52,a,b*). The hyaloid vessels are also more prolific than previously. In appropriate strains of mice, the first evidence of pigment granules may be seen in the cells of the outer (pigment) layer of the retina.

The otocysts have a more irregular shape than previously, and the endolymphatic ducts are slightly more elongated (*Plate 26a,d–j*). The tubo-tympanic recesses are also first evident at this stage (*Plate 26b,b,c*), and are believed to be formed from the dorsal portion of the first pharyngeal pouch, the adjacent pharyngeal wall and possibly from the dorsal part of the second pouch (Hamilton and Mossman, 1972). The precartilage precursors of the first and second arch cartilages (the Meckel's and the hyoid (Reichert's) cartilages, respectively) are also first seen at this stage (*Plate 26b,f*).

As noted previously, many of the cranial ganglia are well differentiated by 11–11.5 days p.c., with the elements of the spinal and cranial parts of the accessory (XI) nerves being particularly well seen at this stage of development (*Plate 26a,g–j*; *Plate 26b,a*). In the neck region, the rostral parts of the sympathetic chains are now present (*Plate 26b,g*; *Plate 26c,a–e*), as are the vagal (X) ganglia, and the vagal (X) trunks and their branches (*Plate 26b,b–d,i*; *Plate 26c,a*). Along the spinal axis, the dorsal (posterior) root ganglia are particularly prominent. Not infrequently, it is possible to see localized regions of neural luminal occlusion (Kaufman, 1983b, 1986). The latter are particularly clearly seen in the mid-trunk region of the embryo (*Plate 26e,e–j*), but may also be seen as far rostrally as the lower cervical/upper thoracic region (*Plate 26c, b–d,g–i*), and usually involve the ventral half of the developing spinal cord.

The spinal cord shows increasing evidence of differentiation, with the presence of a broad ependymal layer as well as a broad mantle layer, especially in the region of the basal plate (future ventral horn). The marginal layer (future white matter) is also clearly seen, particularly in the ventral and lateral parts of the cord. In the mid-thoracic region, the shape of the cord at this stage is characteristically still narrower from side to side than dorsiventrally, though this situation gradually changes over the next 1–2 days (see, for example, *Plate 63,c*), so that the spinal cord in this region eventually becomes wider from side to side than dorsiventrally.

THE UROGENITAL SYSTEM

The urogenital ridges extend caudally from the upper abdominal region, from about the level of the caudal part of the forelimbs (*Plate 26d,c–i*; *Plate 26e,a–e*). The mesonephros is seen to be situated on the lateral aspect of the urogenital ridge, while the gonadal primordium is seen to be located on its medial aspect (*Plate 68,c*). The urogenital mesentery at this stage is quite broad. Within the mesonephroi, the tubules are particularly prominent (*Plate 26e,a–e*; *Plate 68,c*), and more laterally, the mesonephric ducts are clearly seen (*Plate 26e,a–f*). Despite its size, the gonad is still at the so-called "indifferent" stage, and it is not possible to determine the sex of the embryo on purely morphological or histological analysis at this stage (*Plate 68,a*). The gonads contain substantial numbers of primordial germ cells. The distal part of the ureteric buds is dilated, and this region is surrounded by the metanephric blastema (*Plate 26e,c–e*). The genital ridge is also particularly prominent at this stage (*Plate 70*,a).

Stage 20 Developmental age, 12 days p.c.
Plates 27a–d

EXTERNAL FEATURES

Embryos with about 48–52 pairs of somites. An analysis of the state of differentiation of the limbs in embryos at this stage of development is particularly instructive. By 11.5–12 days p.c., the handplate is no longer paddle-shaped but has angular contours which correspond to the location of the future digits (*Plate 44,a*), and these are separated by four indented "rays" which correspond to the digital interzones. All of the latter features are accentuated by 12–12.5 days p.c. (*Plate 44,b,c*). By 11.5–12 days p.c., the footplate is still paddle-shaped, and quite clearly demarcated from the more proximal part of the hindlimb. By 12–12.5 days p.c., however, the footplate is now also seen to be polygonal in shape, and only minimally less differentiated than the handplate (cf. *Plate 44,b,c*). In both the handplate and the footplate, the remodelling process has been initiated, so that eventually the sites of programmed (physiological) cell death will be observed within the mesenchyme of the digital interzones as the process of digit formation proceeds. However, at this stage of development, no evidence of programmed cell death is seen in the digital interzones.

In the cephalic region, since the precursors of the eyelids are only minimally in evidence at this stage, it is possible to see the pigmentation in the exposed peripheral part of the outer (pigmented) layer of the retina through the almost transparent cornea, even in Bouin's fixed material (*Plate 46,a–d*; *Plate 52,a–d*). With regard to the primitive pituitary gland, it is possible to see that Rathke's pouch has already lost its original site of contact with the roof of the oropharynx (*Plate 55,a*; *Plate 56,1*), and scanning electron micrographs of this region in decapitated specimens clearly reveal that no evidence of the entrance site remains (*Plate 59,b*).

A substantial degree of growth occurs in the tongue rudiment during this stage (*Plate 58,c–e,f*). In addition, the tuberculum impar appears to have completely merged with the posteromedial parts of the two lateral lingual swellings, all three being derivatives of the first branchial arch (*Plate 58,c–e,f*). These derivatives of the first arch subsequently fuse with the derivatives of the second, third and fourth arches, but the latter structures are not readily recognized even when this region is studied by scanning electron microscopy, as they do not appear to be in the form of discrete entities as might have been expected from the standard description of comparable events observed during the development of the human tongue (see, for example, Hamilton and Mossman, 1972).

With regard to the palatal shelves, these are seen to be slightly less widely separated than the situation observed in embryos at 11.5–12 days p.c. (cf. *Plate 59,b*; *Plate 59,c,d*). The ventral (lower) surface of the primitive nasal septum is also reasonably well defined (*Plate 59,c*), and in the more advanced embryos at this stage of development, the first evidence of rugae formation is seen on the palatal shelves (*Plate 59,f*). It is also possible to see that the width of the tongue is now almost exactly the same as the distance between the two palatal shelves (*Plate 59,e,f,*). The latter feature is particularly well seen in transverse or coronal sections of this region (*Plate 61,a–c*), where it is also apparent that the medial borders of the two palatal shelves are, in fact, directed vertically.

The overall appearance of the distal part of the tail at this stage of development is similar to that seen in embryos at about 11.5 days p.c., where its caudal half is seen to gradually narrow towards the blunt-ended tip (cf. *Plate 46,a,b*). By 13.5 days p.c., the overall shape of the tail has changed quite dramatically, so that its distal part is seen to narrow to a relatively sharp point (*Plate 46,e,f*). Similarly, the segmental arrangement of the somites in the tail and lower trunk regions, which is so clearly seen at 12.5 days p.c. (*Plate 46,a–d*), is no longer seen by 13.5 days p.c. (*Plate 46,e,f*).

THE HEART AND VASCULAR SYSTEM

Critical events are occurring at this stage of development in the outflow tract of the primitive heart, as well as in the relationship between this region and the branchial (pharyngeal) arch arteries. The first and second arch arteries are not seen in embryos after about 10.5–11 days p.c. The symmetrical arrangement of the arch arteries seen up to about 11–11.5 days p.c. has gradually evolved and changed, so that by 12 days p.c. the arrangement is clearly asymmetrical, particularly as far as the fourth and sixth arch arteries and the dorsal aortae are concerned. Because of the relatively low magnification and limited number of sections in this series, it is extremely difficult (from an analysis of these sections in isolation) to fully understand the detailed relationship that exists between the arch arteries and the outflow tract of the primitive heart, and for this reason it is essential that reference be made to Fig. 10, so that the changes that are taking place in this location may be more clearly followed. Despite these reservations, the third, fourth and sixth arch arteries on the left side are readily recognized in these sections (*Plates 27b,f,g,h*, respectively).

The principal features of note are as follows: in the region of the outflow tract of the primitive heart, it is possible to observe that aortico-pulmonary spiral septum formation is now progressing apace, so that in the more distal part of this region two distinct channels are now seen (*Plate 27b,g*), which will form the proximal part of the ascending (thoracic) aorta, and the origin of the pulmonary trunk. The former channel will form the outlet of the primitive ventricle (the future left ventricle), while the latter channel will form the outlet of the bulbus cordis region of the primitive heart (the future right ventricle).

In the majority of the outflow tract, the lumen is much less than the overall diameter of the outflow tract. The latter feature is a direct consequence of the increased cellular activity in the mesenchyme tissue constituting the bulbar ridges in this location. In developmentally more advanced embryos, the lumen of the outflow is now seen to be divided into two almost completely separate channels. The bulbar ridges are aligned in an oblique direction in relation to the main direction of the outflow tract, so that the cranial ridge runs obliquely backwards and dorsally from right to left, while the more caudally located ridge, which also runs obliquely, runs ventrally from left to right. The direction of the two bulbar (spiral) ridges therefore tend to cross at right angles to each other.

While the cranial part of the two dorsal aortae beyond the site of fusion with the third arch arteries are retained and form the internal carotid arteries, the segments of the dorsal aortae between the points at which they are joined by the third and fourth arch arteries (termed the *ductus caroticus*) gradually disappear. This event considerably alters the original arch arterial pattern, so that subsequent changes are often referred to as postbranchial (see Hamilton and Mossman 1972). Similarly, and at about the same time, the right dorsal aorta between its connection with the right fourth arch artery and its connection with the left dorsal aorta, also gradually disappears (see Fig. 10). The distal part of the right sixth aortic arch artery (just beyond the site of origin of the right pulmonary artery) also disappears a day or so later, though evidence of this vessel may still be seen in embryos of about 13–13.5 days p.c.

As a consequence of the gradual rostral movement of the forelimbs, the lateral branch of the seventh cervical intersegmental arteries gradually enlarges to form the axial artery of the forelimb, while the branches of the fifth lumbar intersegmental arteries enlarge to form the axial arteries of the hindlimb. At this stage of development, the upper limbs are in an intermediate position, having yet to achieve their definitive vertebral level, and the two axial vessels to the forelimbs still arise from the dorsal aortae (*Plate 27c,c*). It is also possible to see that the diameter of the left dorsal aorta, at the level of origin of the left seventh intersegmental (future subclavian) artery, is substantially greater than the diameter of the right dorsal aorta at a comparable level (*Plate 27c,c*).

Within the common ventricular chamber of the primitive heart, it is possible to see that atrial septation is progressing apace, so that by this stage of development the lower border of the septum primum has now contacted and fused with the atrio-ventricular cushion tissue. The only channel that now allows communication between the right and left atria (as they will now be designated) is via the foramen secundum (*Plate 27c,b*).

The components of the postcardiac (i.e. caudal) parts of the venous system are also seen to be evolving, and, as in the previous stage, this is largely a consequence of the changes that occur in the relationship between these vessels and the liver. The ductus venosus, for example, which in the previous stage of development was in an early stage of its differentiation (*Plate 26d,b,c*), is now seen to be a somewhat smaller but more distinct vessel than previously (*Plate 27c,e*). This vessel is almost exclusively formed from the left umbilical vein, there being little evidence at this stage of development of the right umbilical vein which, though a very significant vessel up to about 11 days p.c. (*Plate 24c,f–n*), has now almost completely regressed. Similarly, with the differentiation of the midgut and hindgut, their original venous drainage via the right and left vitelline veins has become considerably modified, and is now via the portal vein, which is principally derived from the right vitelline vein (*Plate 27c,f*; *Plate 27d,a*).

Less significant changes are observed in the rostral part of the cardinal venous system, as these vessels (in the mouse, but not in man) largely retain their original symmetrical arrangement. The most rostral part of each anterior cardinal vein forms the internal jugular veins (*Plate 27b,a*), while the more caudal part of each of these vessels differentiates to form the right and left superior venae cavae (*Plate 27b,h*).

THE PRIMITIVE GUT AND ITS DERIVATIVES

The most obvious structure in the oropharyngeal region at this stage of development is the tongue (*Plate 27a,i*; *Plate 27b,a,b*). This organ continues to grow forwards, but the anterior part of its lower surface becomes separated from most of the floor of the mandibular component of the first branchial arch by a deep transversely running groove (*Plate 58,f*). The posterior half of the tongue is still very closely related to the vertically directed palatal shelves (*Plate 61,a–c*), while the anterior half of the tongue is more rounded and only poorly applied to the medial borders of the palatal shelves (*Plate 61,d–f*). Only in the latter region is the dorsal surface of the tongue related to the ventral surface of the primitive nasal septum. An oblique groove is characteristically present (located about one-third of the way along the palatal shelves) behind which the medial borders of the palatal shelves are directed vertically (*Plate 61,a–c*), whereas in front of this oblique groove, the palatal shelves are more bulbous and their medial borders are less well defined (*Plate 59,d*; *Plate 61,d–f*).

Within the lateral walls of the lower anterior part of the developing nasal septum, the two vomeronasal organs (Jacobson's organs) are first clearly seen at this stage of development, as invaginations of the olfactory epithelium (*Plate 27b,b–d*). The vomeronasal organs are subsequently supplied by the vomeronasal branch of the olfactory (I) nerve (this organ develops maximally in man by about the fifth month of gestation, then generally completely regresses, though occasionally remnants may be found in adults).

The thyroid is at a similar stage of development to that described in the previous stage, though its connection with the floor of the pharynx via the thyroglossal duct is often severed. By this stage of development, Rathke's pouch has now also lost its tenuous connection with the roof of the oropharynx (see above). Within the anterior wall of Rathke's pouch, active cellular proliferation is seen (*Plate 55,a*), and this is destined to form the pars anterior of the pituitary gland. As a result of further cellular activity in this region, the original lumen of Rathke's pouch eventually becomes reduced to a narrow cleft (*Plate 55,g–i*). A lumen is still present at this stage in the region of the infundibular recess of the third ventricle (*Plate 55,a*), though this is fairly rapidly obliterated with the proliferative activity that occurs in its wall (*Plate 55,d–f*), with the formation of the pars nervosa and stalk regions of the neurohypophysis.

In the pharyngeal region of the foregut, while the first branchial pouch is clearly seen (*Plate 27b,a–d*), the second, third and fourth pouches, though previously easily recognized (e.g. *Plate 21b,d*; *Plate 22a,a*; *Plate 22b,a,b*), gradually decrease in size and differentiate to form their various derivatives. At this stage of development these derivatives are not easily recognized because of their small size, though the thymus (of third pouch origin) is readily recognized at 12.5–13 days p.c. The thyroid lobes are in close association with the laryngeal cartilage by about 13.5 days p.c. (*Plate 30e,a*), and the parathyroid glands (of third pouch origin) are found as posterior relations of the latter by 14–14.5 days p.c. (*Plate 32e,a*). The fourth pouches give rise to the ultimobranchial bodies.

The lumen of the pharyngeal region of the foregut narrows at the entrance to the oesophagus, and at its anterior bifurcation where it forms the laryngotracheal groove (*Plate 27b,f*). Just proximal to this site, the arytenoid swellings guard the entrance to the primitive larynx (*Plate 27b,e*). More distally, the oesophagus and trachea are surrounded by dense splanchnic mesenchyme tissue which will subsequently contribute to the formation of the connective tissue and smooth muscle of the larynx and the laryngeal cartilages (*Plate 27b,g,h*; *Plate 27c,a–e*). The two vagal (X) trunks, initially located on the lateral side of the fourth and sixth arch arteries (*Plate 27b,g,h*), gradually migrate medially, so that slightly more caudally they come to lie (on either side) between the oesophagus and the trachea (*Plate 27c,a*). The trachea bifurcates (*Plate 27c,c*) to form the two main bronchi (*Plate 27c,d*), and these then successively branch to form the secondary and tertiary bronchi within the lung buds (*Plate 27d,a–c*; *Plate 66,c,d*). The latter continue to expand into the pericardio-peritoneal canals.

The liver is seen to be a well differentiated organ at this stage of development (*Plate 27c,c–f*; *Plate 27d,a–c*), and it is clear that less of its structure is occupied by venous sinusoids and more by functional hepatic parenchymal tissue than previously (cf. *Plate 26d,a–h*), and as a direct consequence of this, the liver appears to have a denser architecture than previously. In addition, the liver now occupies a substantially greater proportion of the abdominal cavity than previously. The right half of the liver has a slightly greater volume than that of the left side, and is divided by a fissure into an upper and a lower lobe (this feature is more clearly seen during the next stage of development (see *Plate 28d,f–i*), and the lower lobe is further subdivided by a vertical fissure (*Plate 28e,a,b*).

The volume of the stomach is now considerable (*Plate 27d,b–d*), and this organ also occupies a substantial part of the left side of the abdominal cavity. The first evidence of the "compound" stomach that occurs in this species is evident at this stage. This organ is divided roughly equally into two parts (proventricular and glandular). On the left side, the wall is thin and almost transparent, while on the right side, and leading to the pyloric region, the wall is considerably thicker due to the presence of the fundic glandular mucous membrane. The line of transition from "cutaneous" (or proventricular) to glandular mucous membrane is more clearly seen during the next stage of development (*Plate 27d,b–d*; cf. *Plate 28e,a,b*). The two pancreatic primordia (of the future right and left lobes) are in close proximity to the wall of the duodenum, but rostrally they are seen to be separated by the portal vein (*Plate 27d,a*).

THE NERVOUS SYSTEM

The various regions of the brain are now more clearly defined than previously, and the two telencephalic vesicles have expanded upwards and are now both dorsolateral as well as inferolateral relations of the third ventricle (*Plate 27a,b–i*), whereas previously they were principally inferolateral relations of the third ventricle (*Plate 26b,a–i*; *Plate 26c,a–e*). In addition to the infundibular recess (*Plate 27a,f*), the third ventricle also has an epithalamic recess (*Plate 27a,c*), though the first evidence of the pineal primordium is not seen in this location until about 13.5–14.5 days p.c. (*Plate 32a,a,b*). The hypothalamic and thalamic regions are also differentiating and enlarging, and the volume of the third ventricle becomes correspondingly diminished. The corpus striatum in particular undergoes a rapid degree of expansion at this time (*Plate 27a,f–h*).

Possibly the most impressive degree of differentiation occurs in the otocyst at this time. In the previous stage, while the endolymphatic duct was readily recognized (*Plate 26a,d–f*), the rest of the otocyst was otherwise fairly featureless. By this stage, however, the vestibular component is seen to contain the first indication of the semicircular canals, which initially tend to be flattened rather than tubular in profile (*Plate 27a,f–h*). The latter develop as diverticula from the saccule (*Plate 27a,h*), and this is seen to be in continuity caudally with the primitive cochlea (*Plate 27a,i,j*). All of the latter develop in close proximity to the vestibulocochlear (VIII) ganglion complex (*Plate 27a,f–i*), which in developmentally more advanced embryos at this stage is seen to contain two discrete components, namely the vestibular and cochlear ganglia. It is also possible to distinguish the mesenchyme pre-cartilage condensation of the otic capsule, particularly around the saccule (*Plate 27a,j*), which in due course forms the petrous part of the temporal bone.

The posterior cells of the lens vesicle are progressively increasing in length to form the so-called lens "fibres", though at this relatively early stage in their differentiation, their nuclei are principally located towards the periphery of these cells (*Plate 52,c,d*). By

(*continued on page 156*)

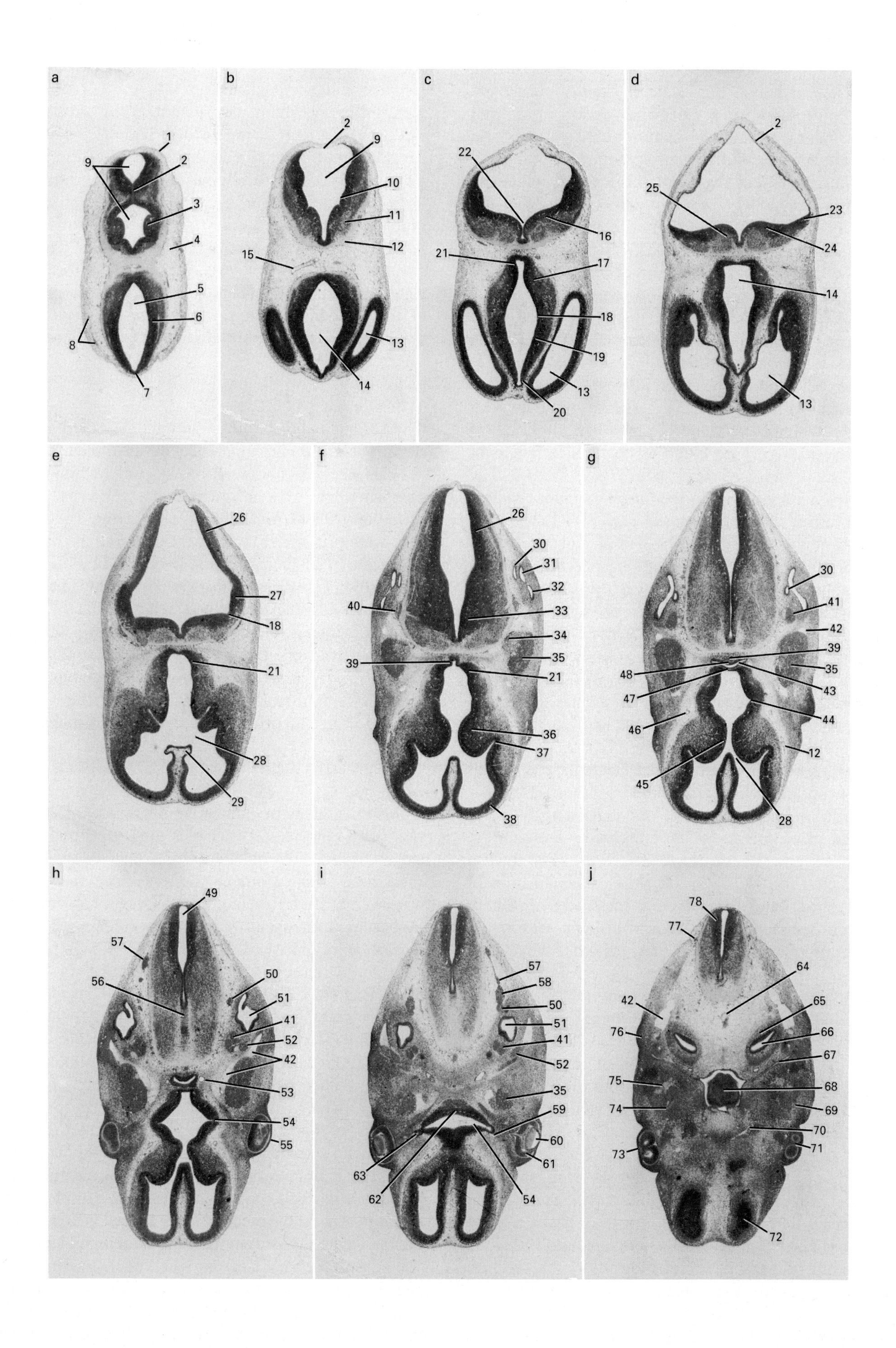
a
b
c
d
e
f
g
h
i
j
1 2 3 4 5 6 7 8 9 10 11 12 13 14 15 16 17 18 19 20 21 22 23 24 25 26 27 28 29 30 31 32 33 34 35 36 37 38 39 40 41 42 43 44 45 46 47 48 49 50 51 52 53 54 55 56 57 58 59 60 61 62 63 64 65 66 67 68 69 70 71 72 73 74 75 76 77 78

Plate 27a (12–12.5 days p.c.)

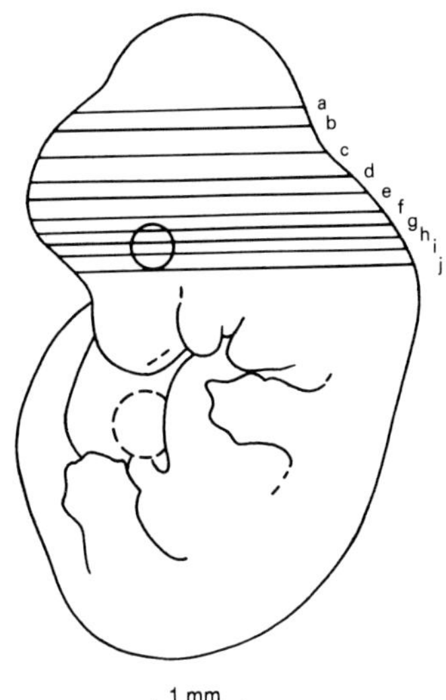

12–12.5 days p.c. Embryo with 48–52 pairs of somites. Crown–rump length 7.0 mm (fixed). Transverse sections (Theiler Stage 20; rat, Witschi Stage 28; human, Carnegie Stage 17).

1. surface ectoderm
2. roof of hindbrain
3. neuromere
4. cephalic mesenchyme
5. mesencephalic vesicle
6. wall of midbrain
7. lamina terminalis
8. peripheral branches of primary head vein
9. fourth ventricle
10. ependymal layer
11. mantle layer
12. marginal layer
13. telencephalic vesicle (lateral ventricle)
14. third ventricle
15. posterior cerebral artery
16. metencephalon
17. diencephalon (ventral thalamus)
18. sulcus limitans
19. diencephalon (dorsal thalamus)
20. epithalamus
21. diencephalon (hypothalamus)
22. median sulcus
23. rhombic lip (dorsal part of alar lamina of metencephalon – the cerebellar primordium)
24. tegmentum of pons
25. basal plate
26. alar plate of myelencephalon (medulla oblongata)
27. alar plate of metencephalon (future pons/cerebellar junction)
28. interventricular foramen
29. choroid invagination
30. endolymphatic sac
31. posterior semicircular canal
32. lateral semicircular canal
33. basal plate of pons
34. rootlets of trigeminal (V) nerve
35. trigeminal (V) ganglion
36. thalamus
37. epithalamic sulcus
38. neopallial cortex
39. infundibulum (future pars nervosa)
40. rootlets of vestibulocochlear (VIII) nerve
41. vestibulocochlear (VIII) ganglion complex
42. primary head vein
43. lumen of anterior lobe of pituitary (previously Rathke's pouch)
44. hypothalamic sulcus
45. corpus striatum
46. middle cerebral artery
47. pars anterior of pituitary
48. pars intermedia of pituitary
49. caudal part of fourth ventricle
50. superior ganglion of glossopharyngeal (IX) nerve
51. saccule
52. facial (VII) ganglion
53. internal carotid artery
54. optic recess
55. corneal ectoderm
56. medullary raphe
57. cranial accessory (XI) nerve
58. superior ganglion of vagus (X)–cranial accessory (XI) ganglion complex
59. ophthalmic artery passing distally towards fetal fissure of optic stalk
60. lens vesicle
61. inner (neural) layer of retina
62. region of optic chiasma
63. optic stalk (optic (II) nerve)
64. basilar artery
65. mesenchymal precartilage condensation of otic capsule (future petrous part of temporal bone)
66. cochlea
67. dorsal component of first branchial pouch
68. dorsal surface of tongue mass
69. maxillary component of first branchial arch
70. anterior cerebral vein
71. fetal fissure
72. wall of telencephalic vesicle
73. optic eminence
74. oculomotor (III) nerve
75. maxillary division of trigeminal (V) nerve
76. primordium of external ear
77. spinal root of accessory (XI) nerve
78. caudal part of medulla oblongata

PLATE 27b

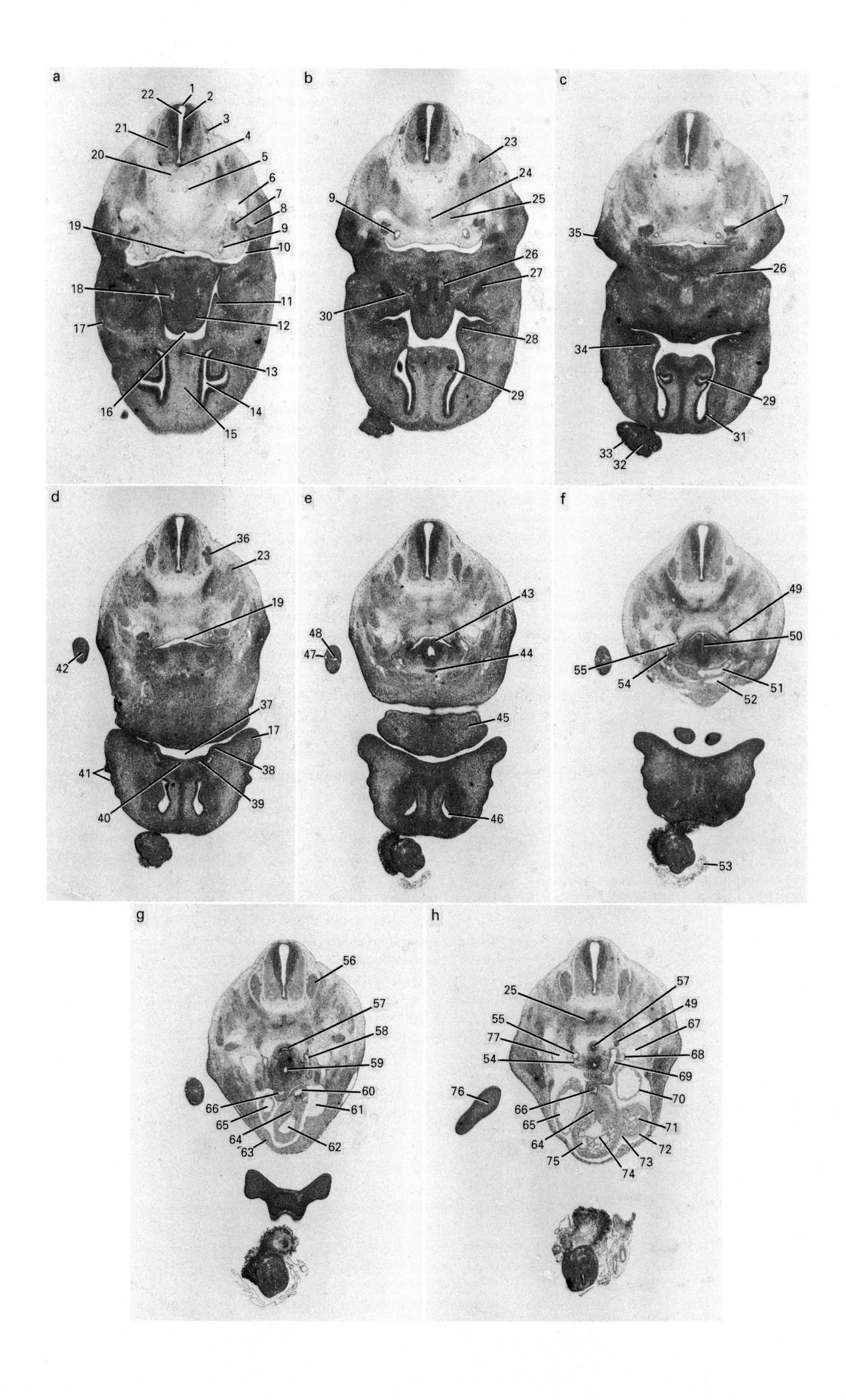

a
b
c
d
e
f
g
h
1
2
3
4
5
6
7
8
9
10
11
12
13
14
15
16
17
18
19
20
21
22
23
24
25
26
27
28
29
30
31
32
33
34
35
36
37
38
39
40
41
42
43
44
45
46
47
48
49
50
51
52
53
54
55
56
57
58
59
60
61
62
63
64
65
66
67
68
69
70
71
72
73
74
75
76
77

Plate 27b (12–12.5 days p.c.)

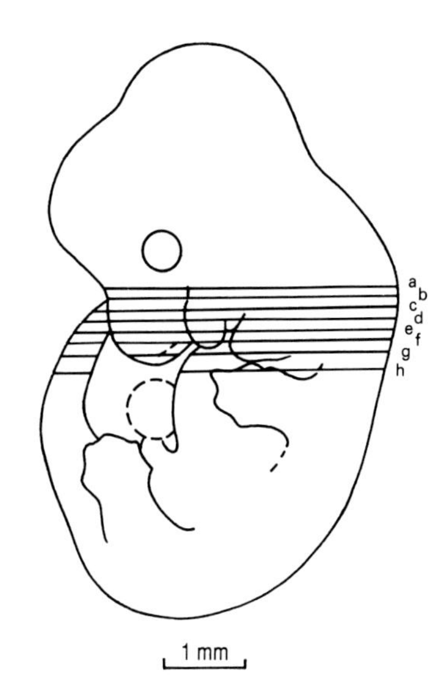

12–12.5 days p.c. Embryo with 48–52 pairs of somites. Crown–rump length 7.0 mm (fixed). Transverse sections (Theiler Stage 20; rat, Witschi Stage 28; human, Carnegie Stage 17).

1. roof plate
2. ependymal layer
3. spinal accessory (XI) nerve
4. floor plate
5. vertebral artery
6. anterior cardinal vein (internal jugular vein)
7. inferior ganglion of vagus (X) nerve
8. facial (VII) nerve
9. internal carotid artery
10. first branchial pouch
11. vertically directed palatal shelf of maxilla
12. tongue
13. precartilage mass of nasal septum
14. primitive nasal cavity
15. nasal septum
16. median sulcus
17. maxillary process
18. lingual vessels
19. pharyngeal region of foregut
20. marginal layer
21. mantle layer
22. central canal
23. premuscle mesodermal condensation
24. notochord
25. sclerotomal vertebral condensation
26. hypoglossal (XII) nerve
27. precartilage condensation of first branchial arch (Meckel's precartilage mass)
28. lateral palatine process
29. vomeronasal organ (Jacobson's organ)
30. lingual nerve
31. olfactory epithelium
32. proximal region of tail
33. somite
34. greater palatine artery
35. primordium of pinna of ear
36. dorsal (posterior) root ganglion of C1
37. stomodaeum (stomatodaeum)
38. lateral nasal process
39. medial nasal process
40. primary palate
41. elevations associated with the formation of follicles of vibrissae
42. distal extremity of tail
43. arytenoid swelling
44. thyroid primordium
45. mandibular process
46. entrance to primitive nasal cavity (anterior naris)
47. caudal extremity of neural tube
48. caudal extremity of hindgut diverticulum (postanal component)
49. sympathetic chain
50. laryngo-tracheal groove
51. left third branchial arch artery
52. rostral extremity of pericardial cavity
53. amnion
54. right vagus (X) nerve
55. right dorsal aorta
56. dorsal (posterior) root ganglion
57. oesophagus
58. left fourth branchial arch artery
59. trachea
60. pulmonary trunk
61. pericardial cavity
62. outflow tract of heart (region of truncus arteriosus, just distal to the bulbo-truncal junction)
63. thoracic wall overlying pericardial cavity
64. endocardial cushion tissue (bulbar ridges) associated with formation of the aortico-pulmonary spiral septum
65. right atrium
66. origin of thoracic aorta (aortic sac)
67. anterior cardinal vein (region of future left superior vena cava)
68. left vagus (X) nerve
69. left sixth branchial arch artery
70. wall of left atrium
71. trabeculated wall of left ventricle
72. mesothelial covering of heart (epicardium/visceral pericardium)
73. interventricular groove
74. endothelial lining of heart (endocardium)
75. bulbus cordis (future right ventricle) region of heart
76. right forelimb bud
77. anterior cardinal vein (region of future right superior (cranial) vena cava)

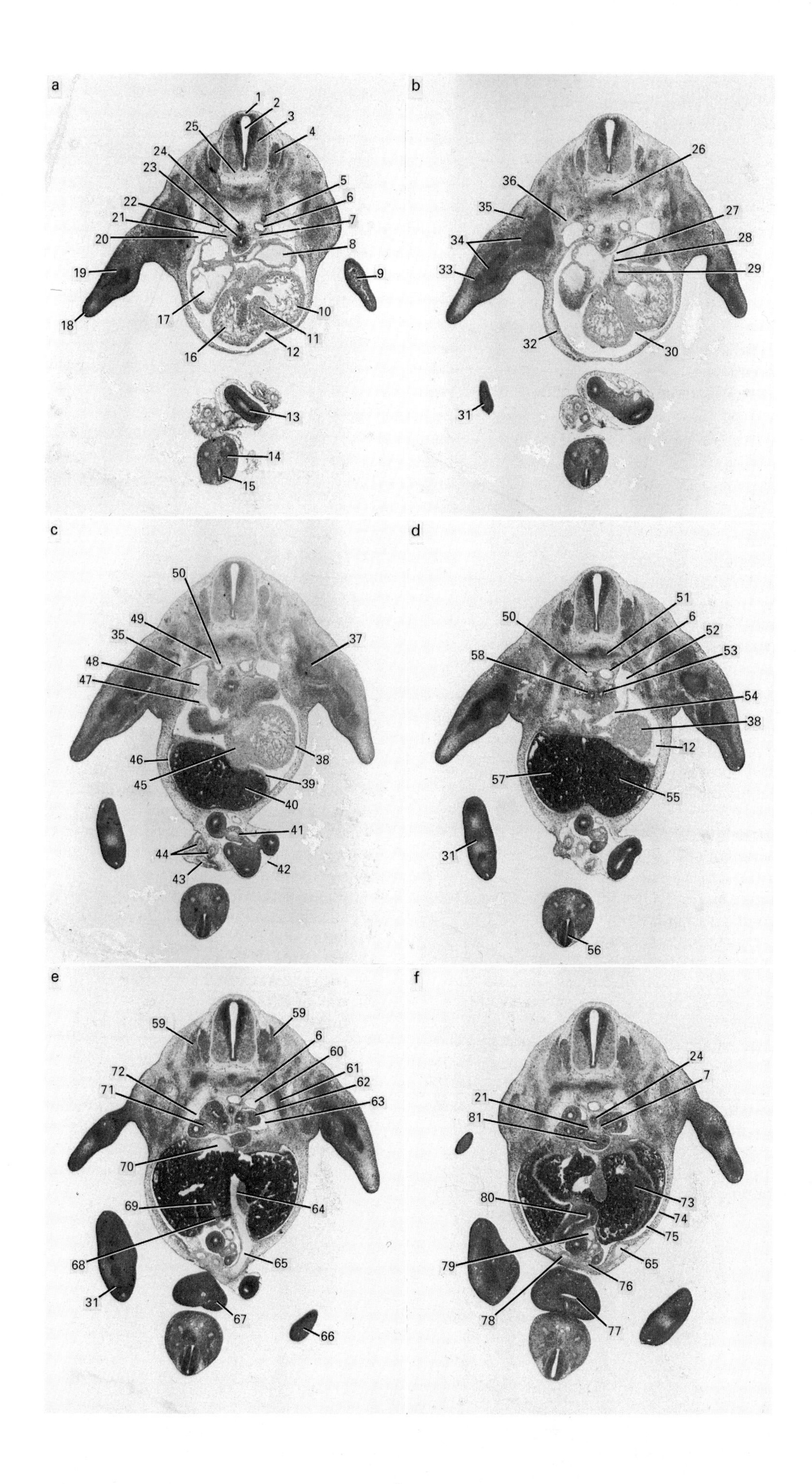
a
1
2
3
4
5
6
7
8
9
10
11
12
13
14
15
16
17
18
19
20
21
22
23
24
25
b
26
27
28
29
30
31
32
33
34
35
36
c
35
37
38
39
40
41
42
43
44
45
46
47
48
49
50
d
50
51
6
52
53
54
38
12
55
56
57
58
31
e
59
59
6
60
61
62
63
64
65
66
67
68
69
70
71
72
31
f
24
7
21
81
73
74
75
65
76
77
78
79
80

Plate 27c (12–12.5 days p.c.)

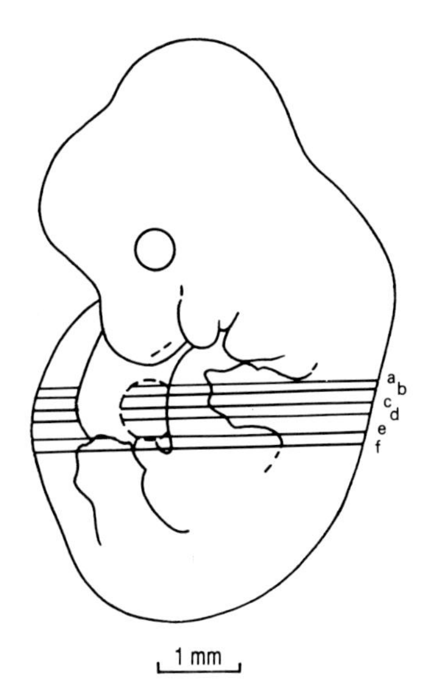

12–12.5 days p.c. Embryo with 48–52 pairs of somites. Crown–rump length 7.0 mm (fixed). Transverse sections (Theiler Stage 20; rat, Witschi Stage 28; human, Carnegie Stage 17).

1. roof plate of neural tube
2. central canal
3. neural tube in mid-thoracic region
4. dorsal (posterior) root ganglion
5. anlage of cervical sympathetic ganglion
6. descending (thoracic) aorta
7. left vagal (X) trunk
8. left atrium
9. left forelimb bud
10. wall of left ventricle
11. muscular part of interventricular septum
12. pericardial cavity
13. loop of midgut within physiological umbilical hernia
14. proximal part of tail region
15. neural tube in tail region
16. right ventricle (bulbus cordis)
17. right atrium
18. marginal vein
19. right forelimb bud
20. right anterior cardinal vein (future right superior vena cava)
21. right vagal (X) trunk
22. right common carotid artery
23. trachea
24. oesophagus
25. marginal layer
26. notochord surrounded by sclerotomal condensation
27. foramen secundum
28. septum primum
29. atrio-ventricular endocardial cushion tissue
30. interventricular groove
31. right hindlimb bud
32. thoracic wall overlying pericardial cavity
33. proximal radio-ulnar mesenchymal condensation
34. right humeral mesenchymal condensation
35. dorsal premuscle mesenchymal condensation
36. cervical nerve trunk (part of brachial plexus)
37. precartilage primordium of left humerus
38. wall of left ventricle (trabeculated myocardium with outer endocardial layer first clearly evident at this stage)
39. mesenchymal condensation of septum transversum forming major component of diaphragm
40. liver
41. mesentery of midgut loop containing vitelline (superior mesenteric) artery
42. wall of proximal part of umbilical cord covering physiological umbilical hernia
43. umbilical vein
44. paired umbilical arteries
45. mesenchymal condensation of septum transversum (future fibrous pericardium)
46. peritoneal cavity (subdiaphragmatic recess)
47. venous valves at entrance of sinus venosus/common cardinal vein into right atrium
48. common cardinal vein
49. right subclavian artery
50. right brachiocephalic (innominate) artery
51. sclerotomal condensation for vertebral body
52. site of communication between left anterior cardinal vein (future left superior vena cava) and left common cardinal vein
53. left main bronchus
54. left horn of sinus venosus
55. left lobe of liver
56. localized region of neural luminal occlusion
57. right lobe of liver
58. right main bronchus
59. myotome-derived premuscle mass
60. left posterior cardinal vein
61. mesenchymal condensation for rib primordium
62. left lung bud
63. left pericardio-peritoneal canal
64. ductus venosus
65. left umbilical vein
66. left hindlimb bud
67. genital tubercle
68. wall of duodenum
69. common bile duct
70. origin of post hepatic inferior vena cava
71. right lung bud
72. right pericardio-peritoneal canal
73. hepatic sinusoids
74. abdominal wall
75. peritoneal cavity
76. left umbilical artery
77. urogenital sinus
78. right umbilical artery
79. portal vein/superior mesenteric vein
80. lumen of duodenum
81. right lung bud (accessory lobe)

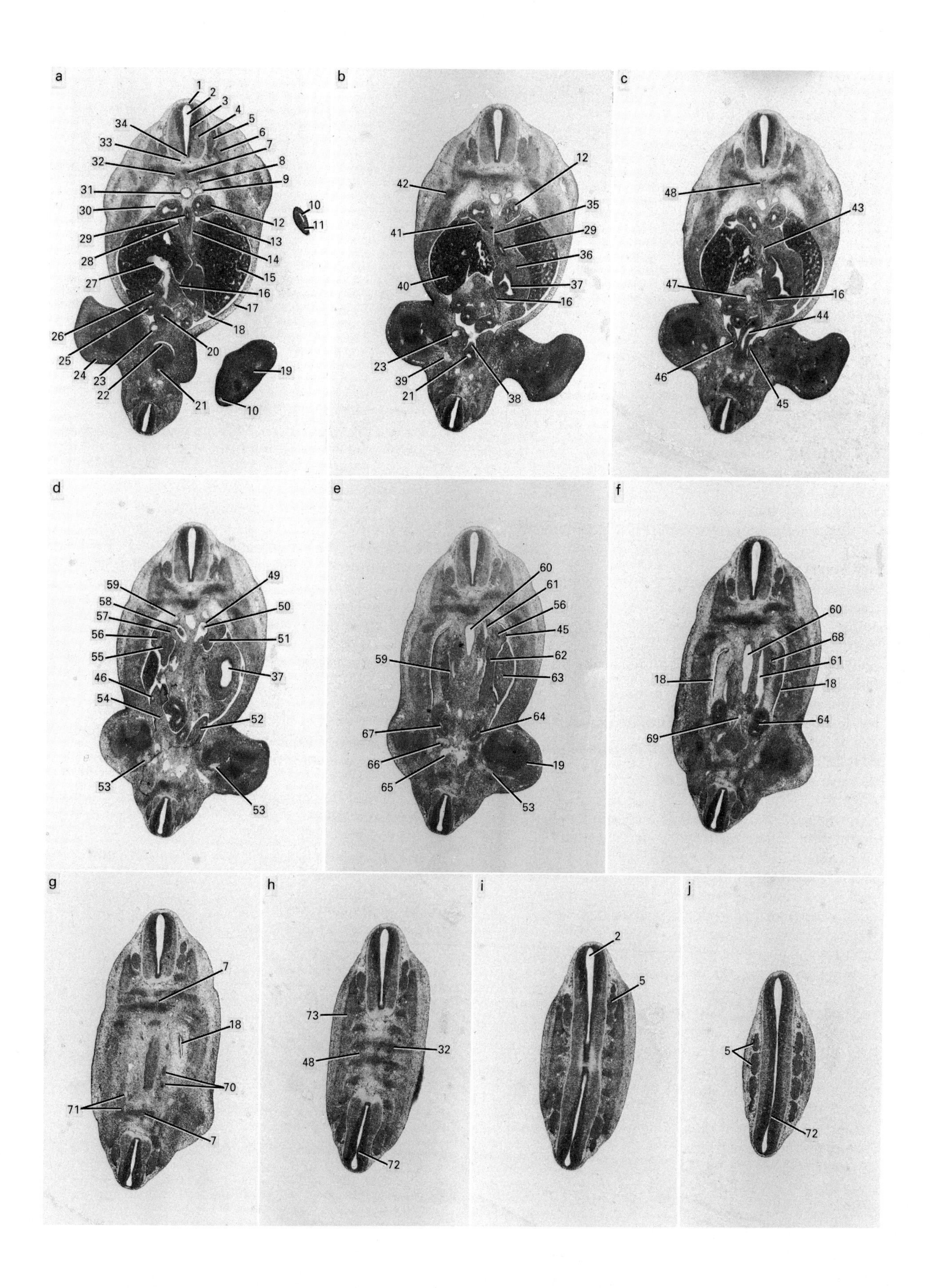
a
1
2
3
4
5
6
7
8
9
10
11
12
13
14
15
16
17
18
19
20
21
22
23
24
25
26
27
28
29
30
31
32
33
34
b
35
36
37
38
39
40
41
42
c
43
44
45
46
47
48
d
49
50
51
52
53
54
55
56
57
58
59
e
60
61
62
63
64
65
66
67
f
68
69
g
70
71
h
72
73
i
j

Plate 27d (12–12.5 days p.c.)

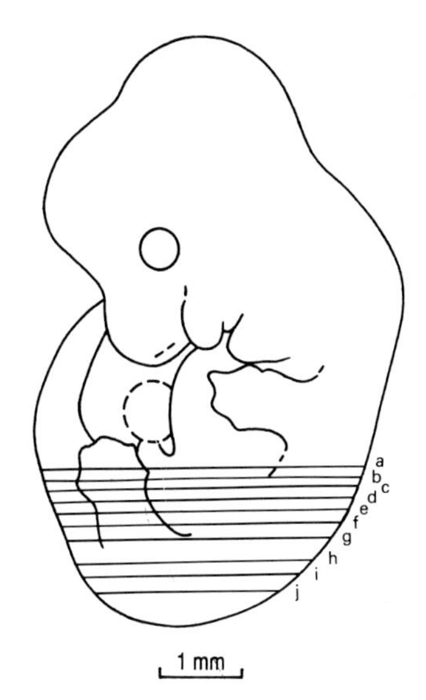

12–12.5 days p.c. Embryo with 48–52 pairs of somites. Crown–rump length 7.0 mm (fixed). Transverse sections (Theiler Stage 20; rat, Witschi Stage 28; human, Carnegie Stage 17).

1. roof plate
2. central canal
3. ependymal layer
4. mantle layer
5. dorsal (posterior) root ganglion
6. premuscle mesodermal condensation
7. notochord
8. sympathetic chain
9. left posterior cardinal vein
10. marginal vein
11. left forelimb bud
12. left lung bud
13. left pericardio-peritoneal canal
14. left vagal (X) trunk
15. sinusoids within left lobe of liver
16. primordium of left lobe of the pancreas
17. abdominal wall
18. peritoneal cavity
19. left hindlimb bud
20. midgut
21. hindgut
22. caudal extension of peritoneal cavity
23. right umbilical artery
24. right hindlimb bud
25. duodenum
26. primordium of right lobe and body of the pancreas
27. right vitelline vein/portal vein
28. right vagal (X) trunk
29. oesophagus
30. right pericardio-peritoneal canal
31. descending aorta (caudal part of thoracic aorta)
32. sclerotomal condensation for vertebral body (caudal condensed portion)
33. marginal layer
34. floor plate
35. right vagal (X) trunk becoming posterior vagal (X) trunk
36. wall of stomach
37. lumen of stomach
38. mesentery of hindgut
39. external iliac artery
40. right lobe of liver
41. mesodermal condensation of septum transversum (future central tendon of diaphragm)
42. rib primordium
43. posterior vagal (X) trunk
44. lumen of hindgut
45. left mesonephric duct
46. right mesonephric duct within caudal extremity of mesonephric ridge
47. superior mesenteric artery
48. sclerotomal condensation for vertebral body (rostral less condensed portion)
49. medial coelomic bay
50. caudal extremity (lower pole) of left lung bud
51. rostral extremity of left gonadal ridge (“indifferent” gonad stage)
52. caudal extremity of left mesonephric ridge
53. nerve trunk from lumbo-sacral plexus entering hindlimb bud
54. caudal extremity of right mesonephric ridge
55. rostral extremity of urogenital ridge
56. mesonephric tubules
57. urogenital mesentery
58. caudal extremity (lower pole) of right lung bud
59. right subcardinal vein
60. abdominal aorta
61. left subcardinal vein
62. left gonadal ridge
63. posterior (dorsal) wall of stomach
64. metanephric blastema
65. common iliac vein
66. common iliac artery
67. ureteric bud
68. left mesonephric ridge
69. bifurcation of the aorta to give rise to the common iliac arteries
70. segmental branches of left subcardinal vein draining the body wall
71. segmental nerves
72. localized region of neural luminal occlusion
73. mesodermal premuscle condensation (dermatomyotome)

12.5–13 days p.c., the lumen of the lens vesicle is completely obliterated due to the elongation of the lens fibres. The nuclei of the lens fibres characteristically become located towards the middle region of these cells (*Plate 52,e,f*). The hyaloid vessels are much in evidence (*Plate 52,d*), and in appropriate sections, the ventrally located "fetal" fissure is seen (*Plate 27a,j*). The intra-retinal space is also seen to be diminished in well-fixed specimens (*Plate 52,c*).

In the spinal cord, the ependymal layer (*Plate 63,b*) is less pronounced than previously (*Plate 63,a*), and by 12.5–13 days p.c. is even further reduced in volume (*Plate 63,c*). Minimal regional differences are observed in the morphology of the spinal cord along the neural axis at this time.

THE UROGENITAL SYSTEM

The mesonephros contains tubular tissue, but no obvious glomeruli at this stage, and appears to be regressing (*Plate 68,d,e*), and the urogenital mesentery is slightly narrower than previously (cf. *Plate 68,b,c*). The gonad is still at the "indifferent" stage, though it has been suggested that a prominent blood vessel is commonly observed that runs along the length of the peripheral part of the genital ridge of the presumptive testis, which is not seen in association with the presumptive ovary (Lovell-Badge, 1992). Indeed, it has recently been demonstrated that increased gonadal vascularity is the first sign of testicular differentiation, and is seen before the first appearance of the testicular cords (Mackay *et al.*, 1993).

In typical specimens at this stage, the metanephros is located within the caudal part of the urogenital ridge (*Plate 27d,e,f*). The ureteric bud is seen to have expanded within the tissue of the metanephric blastema to form the primitive renal pelvis, with one or several buds which represent the primordia of the major calyces (*Plate 27d,e,f*). By 12.5–13 days p.c., and in the most advanced embryos at this stage, the metanephros is seen to be divided into an outer cortical zone in which numerous primitive vesicles may be seen, and an inner medullary region in which tubules at various stages of differentiation may be found (*Plate 72,a,b*). The cloacal region is in the process of becoming divided by the downward growth (towards the cloacal membrane) of the urorectal septum, to form the urogenital sinus anteriorly (*Plate 27c,e,f*), and the hindgut posteriorly (*Plate 27d,a,b*).

Stage 21 Developmental age, 13 days p.c. Plates 28 a–e; Plates 29a, b

EXTERNAL FEATURES

Embryos with about 52–55 pairs of somites. The overall shape of the embryo is beginning to change, so that the pronounced "hump" characteristically present in the caudal hindbrain/upper cervical region of embryos at about 11.5–12.5 days p.c. (*Plate 46,b,d*) has almost completely disappeared, while the face is seen to take on more "adult" features. No clear line of demarcation is seen between the caudal part of the hindbrain and the spinal cord in the cervical and upper thoracic region. Since the latter is subcutaneous at this stage, it produces a prominent dorsal "bulge" in this location (*Plate 46,f*). More caudally, because of the greater width of the lower thoracic and lumbar regions of the spinal cord, the prominence overlying the dorsal part of the spinal cord is substantially less marked. Whereas previously the external ear was only poorly differentiated, being formed by the amalgamation of the auditory hillocks located around the first branchial groove (cleft) (*Plate 46,c,d*), by 13–13.5 days p.c., a well-formed anteriorly directed pinna has formed, and the external acoustic meatus is also seen to be directed anteriorly (*Plate 49,i*). Similarly, possibly as a result of the increasing length of the cervical region, the external ear appears to be located more rostrally than previously (cf. *Plate 46,d*; *Plate 46,f*).

In the front (maxillary part) of the facial region, and associated with the upper lips, the precursors of the vibrissae are now clearly seen to be arranged in five almost parallel rows (*Plate 49,i*; *Plate 28b,b–i*). The unerupted follicles are first seen at about 15.5 days p.c. (*Plate 49,g*), and the latter rupture through the skin (leaving a characteristic mound of cellular debris (see *Plate 49,o*) at about 16–16.5 days p.c. On the upper lips, the vibrissae do not extend forwards beyond the lateral margins of the nostrils (superiorly), so that the region on either side of the philtrum is devoid of vibrissae (*Plate 49,i*). In addition to the vibrissae, early evidence of the single tactile or sinus hair follicles characteristically located just caudal to the angle of the mouth (*Plate 49,i*), and in an infraorbital position, are first seen at this stage, as well as the primordia of the pair of tactile hair follicles characteristically located in the supraorbital region (*Plate 49,n*; *Plate 28a,d–f*).

The distal borders of the handplates and footplates of the limbs are now indented, and the definitive location and width of the digits are clearly seen. The webbing seen in the region of the digital interzones is a characteristic feature of this stage of development (*Plate 44,d–f*). The degree of indentation is slightly more pronounced in the forelimbs than in the hindlimbs (*Plate 44,d*), and more in evidence on the dorsal than on the plantar surface of the footplates (*Plates 44,d,f*). At this stage, the digits of both the handplate and the footplate are splayed out, and almost symmetrical, with the first and fifth digits being similar in size. On the upper limb, the location of the elbow and wrist can just be recognized (*Plate 44,d*), though in the lower limb, only the ankle region is distinguishable at this stage of development (*Plate 44,d*). While the upper limbs tend to be directed laterally on either side of the trunk region, the palmar surfaces of the handplates are directed inferomedially. The hindlimbs as well as the plantar surfaces of the footplates are now seen to be directed medially, and are often in close contact with the tail (*Plate 46,e*).

The shape of the tail gradually changes between the previous stage and the end of this stage of development, so that instead of being elongated and gradually tapering towards its tip (*Plate 46,c,d*), the tail becomes proportionately shorter and blunter, and ends in a relatively sharp tip (*Plate 46,e,f*). Whereas at slightly earlier stages of development the most caudal of the somites are clearly visible in the tail region, this is no longer the case (cf. *Plate 46,c,d*; *Plate 46,e,f*). The full length of the tail at this stage tends to be located in the median plane (*Plate 46,e,f*), and is no longer directed to one or other side of the neck region as observed in previous stages (*Plate 46,a–d*).

THE HEART AND VASCULAR SYSTEM

Fundamental changes have occurred between the previous stage and this stage of development, with regard to both the outflow tract of the heart and the configuration of the branchial arch arteries. Similarly, important changes are apparent within the heart with regard to both inter-atrial and inter-ventricular septation, and the first evidence of the cardiac valves is also seen at this stage. In addition, important changes are observed in the more peripheral parts of the vascular system.

The most important event that occurs in relation to the outflow tract is the complete separation of this region into two distinct channels by the aortico-pulmonary spiral septum to form the ascending (thoracic) aorta (*Plate 28c,c*), which becomes the outlet of the left ventricle (*Plate 28c,e*), and the pulmonary trunk (*Plate 28c,e*), which is the outlet of the right ventricle (*Plate 28c,c*). The arch of the aorta (which develops from the left fourth arch artery) (*Plate 28c,c*) is seen to be a superior relation of the left sixth aortic arch artery (*Plate 28c,e*). The principal derivative of the latter vessel is the ductus arteriosus which has a similar luminal diameter to that of the arch of the aorta, and allows the largely deoxygenated (venous) blood from the head and neck region, and (at this stage) also from the upper limbs (see Fig. 10), to largely bypass the lungs and flow into the postductal part of the descending aorta (*Plate 28c,e–g*). The origins of the two pulmonary arteries are located in the distal part of the pulmonary trunk (*Plate 28c, e*), and are vessels with a relatively narrow luminal diameter. The latter vessels descend anteriorly and on either side of the trachea (*Plate 28c,f,g*), and enter the lung buds at the level of the bifurcation of the trachea (*Plate 28d,a*).

All components of the aortic arch vasculature are significantly more asymmetrical than previously, so that while the left sixth arch artery is a very substantial vessel, that on the right has progressively regressed, and will have completely disappeared by about 13.5 days p.c. The subclavian arteries are also interesting, in that the origin of the left subclavian artery is now at approximately the same level as the site of entrance of the ductus arteriosus into the aorta (*Plate 28c,f*), while the right subclavian artery is slightly more rostrally located (*Plate 28c,d*), and is seen to be a branch of the innominate artery (see Fig. 10). The relationship between the right and left vagal (X) trunks and their recurrent laryngeal branches is now approaching the adult arrangement, this being already seen on the left side (*Plate 28c,e*), where the vagal (X) trunk is seen to be closely related to the lateral wall of the ductus arteriosus.

All of the cardiac valves are recognizable at this stage of development, so that the pulmonary valve is located at the junction between the right ventricle and origin of the pulmonary trunk (*Plate 28c,c*), the proximal part of the latter vessel being directed towards the left side of the thorax. The aortic valve is located at the junction between the outlet of the left ventricle and the proximal part of the ascending aorta (*Plate 28c,e*), and the proximal part of this vessel is directed towards the right. The mitral (bicuspid) valve is located in the left atrio-ventricular region (*Plate 28c,g*), while the tricuspid valve is located in a comparable position on the right side (*Plate 28c,g*). The leaflets of the venous valve at the site of entrance of the posthepatic part of the inferior vena cava into the floor of the right atrium are also readily recognized (*Plate 28d,a*), and this represents the earlier site of entrance of the right horn of the sinus venosus into this chamber of the heart (*Plate 26c,e–g*). Valve leaflets are also seen to be associated with the site of entrance of the right superior (cranial) vena cava (previously the right common cardinal vein in this location) into the posterior part of the right atrium (*Plate 28d,a*), while the left superior (cranial) vena cava (previously the left anterior cardinal vein) amalgamates with the left posterior cardinal vein to form the left common cardinal vein (*Plate 28d,b*). The latter vessel crosses the midline posterior to the atrial region of the heart before it also enters the right atrium.

Both inter-atrial and inter-ventricular septation are progressing apace. Within the atrial region of the heart, the septum primum obstructs blood flow through the caudal half of the inter-atrial channel, though the ostium secundum is now a particularly wide channel (*Plate 28c,g*). No evidence of the septum secundum is yet evident. In the ventricular region of the heart, the walls of both the right and left ventricles are of similar thickness, and are largely composed of trabeculae carneae (*Plate 28c,c–g*; *Plate 28d,a*). The muscular part of the interventricular septum is composed largely of non-trabeculated muscle (*Plate 28c,e*), and extends towards the atrio-ventricular cushion tissue from the interventricular (previously bulbo-ventricular) groove. A small channel is still present between the rostral part of the muscular component of the interventricular septum and the atrio-ventricular cushion tissue (*Plate 28c,f*), and this communication between the two ventricles will soon be closed by the downgrowth of the aortico-pulmonary spiral ridges. The latter form the membranous part of the interventricular septum.

Rostrally, the two vertebral arteries pass cranially through the foramina transversaria of the cervical vertebrae (mostly incomplete at this stage) (*Plate 28b,f*), and then pass medially to enter the subarachnoid space, there to amalgamate with the corresponding vessel from the other side to form the basilar artery (*Plate 28b,c*). This vessel gives origin to the medullary arteries and, in due course, the anterior and posterior cerebellar arteries. At the same time, the internal carotid arteries each give rise to an anterior cerebral artery, a choroidal artery and the posterior cerebral artery. The anterior cerebral artery then gives origin to the middle cerebral artery and the anterior communicating artery, while the posterior cerebral artery gives origin to the posterior communicating artery. Thus the basilar artery and the two internal carotid arteries give origin to all of the components of the arterial circle of Willis.

In the trunk region of the embryo, the two umbilical arteries, which take origin from the common iliac arteries (*Plate 28e,f*), pass forwards to reach the anterior abdominal wall, then pass rostrally on either side of the median plane towards the physiological umbilical hernia (*Plate 28d,f–i*), where they enter the umbilical cord. These two vessels may be of similar luminal diameter, as seen in the representative embryo of 13.5 days p.c. (*Plate 30k,a–c*), the right vessel may have a significantly larger diameter than the left, as seen in the representative embryos of 12–12.5, 12.5–13, 14.5 and 15.5 days p.c. (*Plate 27c,e,f*; *Plate 28d,f*; *Plate 32j,a*; *Plate 35k,a*), or the left vessel may have a significantly greater diameter than the right, as seen in the representative embryos of 16.5 and 17.5 days p.c. (*Plate 37j,b*; *Plate 40j,d*). The significance of the latter is unclear, as the left umbilical vein is usually the dominant vessel (*Plate 27c,e,f*).

THE PRIMITIVE GUT AND ITS DERIVATIVES

Scanning electron micrographs of decapitated specimens clearly reveal that, as observed in the previous stage, the palatal shelves are widely separated due to the intervention of the tongue (*Plate 59,e*). The posterior two-thirds of their medial borders are directed vertically, while their anterior one-third tends to be more rounded, and not in such close contact with the dorsal and lateral surfaces of the tongue (*Plate 59,c–g*; *Plate 61,d–f*). At this stage, the width of the nasal septum is fairly similar to that of the anterior half of the tongue, the lumen of the primitive nasal (olfactory) cavity is small, and the olfactory epithelium relatively undifferentiated. The primitive oropharynx, which has yet to be subdivided into nasal and oral components by the intervention of the palate, is in direct continuity anteriorly with the external naris via the primary choana (primitive posterior naris). These features are also clearly seen in sagittally sectioned material, particularly the continuity between the oropharynx and the external naris (*Plate 29a,a*; *Plate 29b,b*). Similarly, median sections through the tongue demonstrate the substantial increase that occurs in its volume during the 24 h period between 11.5–12 and 12.5–13 days p.c. (*Plate 25b,a*; *Plate 29a,a*), as well as confirming that the components of the various branchial arches that are involved in its formation, are no longer individually recognizable.

A diffuse swelling represents the primordium of the epiglottis at this stage (*Plate 29a,a*), but this is more clearly seen at about 13–13.5 days p.c. (*Plate 31a,a*), where it is seen to guard the entrance to the laryngeal aditus. This is also the first stage that Meckel's cartilage is recognizable, and this is seen to extend forwards from its dorsal end, which is located close to the tubo-tympanic recess (*Plate 28b,c,d*), towards the tip of the mandible (*Plate 28b,g*).

This is also the first stage that the dental laminae (tooth primordia) are seen, and are formed by an infolding and ingrowth of the surface stratified epithelium (*Plate 28b,c–h*). The dental laminae initially appear as discrete fairly homogeneous condensations of tissue, but by 13.5 days p.c. it is possible to see that the tooth primordia consist of an inner (less dense) core of the dental lamina, surrounded by an outer epithelial layer which represents the enamel organ (*Plate 30e,a,b*) (the dental lamina is the anlage of the whole of the ectodermal part of the dentition, whereas the dental papilla, which is mesenchymal in origin, and neural crest-derived, differentiates to form the inner component of the tooth germ (or bud)) (*Plate 40c,a*). The dorsal component of the first branchial pouch (the pharyngo-tympanic or Eustachian tube), possibly in association with the dorsal part of the second pouch, forms the tubo-tympanic recess (*Plate 28b,d*), which will later form the middle ear cavity.

The thymic primordia have lost their connection with the pharynx, and have descended to the upper mediastinum (*Plate 28c,a,b*), but the other derivatives of the third pouch (from its dorsal part), namely the parathyroid glands, are not readily distinguishable before about 14.5–15.5 days p.c., when they are usually, but not invariably, seen to be embedded in the middle of the posterolateral part of the lobes of the thyroid gland (*Plate 35e,a*). While the entrance into the oesophagus from the pharynx is clearly seen (*Plate 28b,h,i*), throughout its entire intrathoracic course, the lumen of the oesophagus is only minimally patent (*Plate 28c,a–g*). The first indication of a lumen is only clearly seen at this stage at about the level of the bifurcation of the trachea (*Plate 28d,b–d*). By contrast, the trachea and main bronchi are seen to have a substantially wider luminal diameter than that of the oesophagus, and are patent throughout their length (*Plate 28c,b–g*; *Plate 28d,a–c*).

The lungs are now subdivided into definitive lobes, each with their own primary and segmental bronchi (*Plate 28d,b–i*; *Plate 66,e,f*). The left lung is considerably smaller than the right, and is not divided into lobes (*Plate 28d,c–i*), whereas the right lung is divided into four lobes, a cranial, middle and caudal lobe that are located on the right side of the thorax (*Plate 28d,b–i*), and an accessory lobe that takes origin on the right side, but is predominately located to the left of the midline (*Plate 28d,d,e*). The first evidence of differentiation of the epithelial lining of both the oesophagus and the lower respiratory tree is seen at this stage, as well as the condensation of splanchnic mesenchyme tissue that will subsequently form the connective tissue and smooth muscle in both systems, as well as the C-shaped cartilaginous "rings" characteristically found in the wall of the trachea and main bronchi.

The pericardial cavity is now a discrete entity with the closure of the pleuro-pericardial canals by the growth and apposition of the pleuro-pericardial folds (shelves) to the dorsal mesentery of the oesophagus medially, and to the septum transversum anteriorly (*Plate 28d,a*). The majority of the components of the diaphragm are also present at this stage, the largest of which is the central tendon which overlies the "bare" area of the liver (*Plate 28d,b*). With the eventual closure of the two pleuro-peritoneal canals (*Plate 28e,a,b*) by the pleuro-peritoneal folds at about 13.5–14 days p.c., this completes the formation of the (thoracic) diaphragm. The latter is then seen to be derived from (i) the dorsal mesentery of the oesophagus, (ii) the part of the septum transversum that forms the central tendon of the diaphragm, (iii) the two pleuro-peritoneal folds (these become invaded by myoblasts which develop from C3–C5 cervical somites, and are supplied by the phrenic (motor) nerve), and (iv) a peripheral component from the body wall which is innervated by intercostal nerves (sensory only).

Where the oesophagus pierces the diaphragm, the left vagal (X) trunk is now seen to be anterior to the oesophagus, while the right vagal (X) trunk is now seen to be a posterior relation of the oesophagus (*Plate 28d,g–i*). As discussed in the appropriate section of the previous stage, the stomach is seen to be divided into two roughly equal parts (the proventricular and glandular), its wall being considerably thicker in the pyloric region than in the left half of the stomach due to the presence of the fundic glandular mucous membrane. The difference in the thickness of the wall between the region of the pyloric antrum and "body" of the stomach is particularly clearly seen in the sagittal sections illustrated (*Plate 29b,a,b*).

The relationship between the pancreatic primordia and the first and third parts of the duodenum is also clearly seen in sagittal sections (*Plate 29a,a*), as is the relatively small volume of the peritoneal cavity at this stage of development (*Plate 29a,b*; *Plate 29b,a*). The liver is seen to occupy the majority of the upper part, the stomach occupies most of the left side (*Plate 29b,b*), while the liver and pancreas occupy much of the right side (*Plate 29a,a*). Despite the fact that the mesonephroi have virtually completely regressed and are replaced by the somewhat smaller metanephros (*Plate 29b,a*), little space is available at this time for the midgut to elongate and differentiate. The majority of this region of the gut is consequently located within the physiological umbilical hernia (*Plate 28d,a–g*), and attached to the posterior abdominal wall by its elongated dorsal mesentery. The two pancreatic primordia have substantially increased in size over the previous 24 h, and show early evidence of differentiation. The primordium of the body and right lobe of the pancreas is embedded in the dorsal mesoduodenum and the beginning of the mesojejunum (*Plate 28d,h,i*; *Plate 28e,a,b*), whereas the left lobe is initially a medial relation, but more distally is a dorsal relation of the stomach (*Plate 28e,a–e*). The primordium of the right lobe of the pancreas is slightly more rostrally located than that of the left lobe. Sagittal sections also reveal that the two components of the cloaca are now separated by the downgrowth of the urorectal septum, to form the distal part of the rectum posteriorly and the urogenital sinus anteriorly (*Plate 29a,b*).

THE NERVOUS SYSTEM

The first evidence of choroid plexus is present at this stage of development. The latter is seen in two locations, namely a relatively small amount that arises from the medial wall of the lateral ventricles (*Plate 28a,b,c*; *Plate 29a,a*; *Plate 29b,a,b*), and a rather more voluminous amount that arises from the roof and extends into the central part of the cavity of the fourth ventricle (*Plate 28a,d,e*; *Plate 29a,a*). All of the major components of the forebrain are now expanding and differentiating, and this is particularly evident in relation to the hypothalamus and thalamus (*Plate 28a,b–e*). The lateral ventricles continue to expand, and are now principally dorsolateral relations of the third ventricle (*Plate 28a,g,h*). The neopallial cortex expands forwards to overly the olfactory epithelium in the roof of the primitive nasal cavity, and this region represents the future olfactory lobe of the brain (*Plate 28a,g,h*). Numbers of olfactory (I) nerves pierce the nasal capsule and the mesenchymal pre-cartilage precursor of the ethmoid bone to pass upwards towards the olfactory cortex (*Plate 28a,h*).

The pituitary gland also continues to differentiate, so that there is a progressive narrowing of the neck of the infundibular recess (region of future stalk of the pituitary) compared to the situation seen previously (*Plate 55,a,b*). The remnants of the original stalk connecting the roof of the oropharynx and Rathke's pouch have now disappeared in most specimens, though occasionally remnants of tissue may be encountered along this pathway (e.g. *Plate 55,h,i*), but more particularly there is increasing evidence of active cellular proliferation in the region of the anterior wall of Rathke's pouch (region of future pars anterior), as well as an increasingly rich vascular network principally derived from the plexus on the floor of the diencephalon.

The eye also shows increasing evidence of differentiation with regard to the changes that occur in the lens, as well as those seen in relation to the neural part of the retina. The lumen of the lens vesicle is no longer present, due to the elongation of the lens fibres (*Plate 52,e,f*), and this is a characteristic feature of this stage of development. The lens fibre nuclei have migrated away from the basal region, so that they are now located towards the middle of these cells, and it is now possible to see that new lens fibres are added from the cuboidal cells at the equatorial region of the lens (*Plate 52,f*). There is a slight suggestion that the nuclei of the lens fibres located towards the centre of the lens may be less densely staining than the more peripheral nuclei, and this tendency increases, so that by 18.5 days p.c. the central "nucleus" of the lens is almost completely translucent (*Plate 53,i*).

The most significant change observed in relation to the neural layer of the retina at this stage is the first evidence of differentiation of its various layers, though these changes are not readily recognized in conventionally (ie. haematoxylin and eosin) stained sections. The silver stained sections illustrated (*Plate 52,e,f*), clearly demonstrate the presence of nerve fibres (that constitute the innermost layer of the neural retina) that pass from the ganglion cells towards the region of the future optic disc, and thence centrally along the optic stalk. However, the various nuclear layers recognizable at later stages of development are not yet distinguishable at this stage. Despite the fact that the lumen of the lens vesicle has disappeared, the overall volume of the lens remains fairly constant, as does the thickness of the neural retina. The overall volume of the globe of the eye increases, however, and the enlargement observed is principally due to an increase in the volume of the vitreous (hyaloid body). An increase is also observed at this time in the hyaloid vascular plexus, branches of which contact the posterior surface of the lens capsule to form the tunica vasculosa lentis.

While the first evidence of the sympathetic trunks was seen in the trunk region in embryos at about 10.25–10.5 days p.c. (*Plate 23d,a–c*), and suggestions of sympathetic ganglia in the same region at about 10.5–11 days p.c. (*Plate 24d,f–h*), the sympathetic trunks are first clearly seen in the cervical region (where they are attached to the dorsal surface of the common carotid arteries) by about 11.5–12 days p.c. (*Plate 26b,g–i*; *Plate 26c,a–f*). The anlage of the cervical sympathetic ganglia are not observed, however, before about 12–12.5 days p.c. (*Plate 27c,a*). By this stage, the superior, middle and cervico-thoracic (stellate) sympathetic ganglia are recognizable (*Plate 28b,e,i*; *Plate 28c,e,g*). In the mid- to lower thoracic region, while the sympathetic trunks are readily located, the ganglia are difficult to recognize before about 13.5 days p.c. (*Plate 30h,a*).

The typical shape of a cross-section through the mid-thoracic region of the spinal cord changes between 12 and 13 days p.c., principally because formerly the cord was elongated anteroposteriorly, whereas the cord is now seen to be wider from side to side (cf. *Plate 63,b,c*). In addition, the width of the ependymal layer has substantially diminished, while that of the mantle and marginal layers have proportionately increased. On cross-section, the central canal is now shaped like an elongated rhomboid, and between 13.5 and 15.5 days p.c. will progressively diminish in size (*Plate 63,d–f*), so that by about 16.5 days p.c. the central canal has been all but obliterated (*Plate 63,h*). It is also just possible to discern that the alar and basal plates of grey matter correspond to the dorsal and ventral grey horns of the spinal cord, respectively.

THE UROGENITAL SYSTEM

This stage is of critical importance, in that it is the earliest stage that the sex of an embryo can be determined from the histological analysis of its gonads, and (but with less certainty) from the analysis of its internal genital duct system. The mesonephros has all but regressed by this stage in the female, leaving a few degenerating tubules (*Plate 68,d–f*). The relationship between the mesonephric ducts and the more laterally located paramesonephric ducts is also clearly seen (*Plate 68,f*). In females, the ovary, while initially similar in shape to the testis (both are elongated and sausage-shaped at this stage), has a quite dissimilar histological appearance, the former being relatively homogeneous but having a "spotty" appearance (*Plate 28e,c,d*; *Plate 68,f*), whereas the testis is seen to contain wide regularly arranged bands of tissue (the testicular cords, which are destined to form the seminferous tubules), and consequently has a characteristic "striped" appearance (*Plate 68,g,h*).

If gonads are isolated from embryos at this stage of development and examined by transmitted light under a dissecting microscope, the morphological features indicated above (namely that the ovary appears to be homogeneous in consistency, while the testis appears to be "striped") provide a simple and almost foolproof means of distinguishing between the gonads of the two sexes. Before this stage, since the gonads as well as the internal genital duct system appear in all respects to be identical, the (genetic) sex of an embryo can only be determined cytogenetically, for example, from an analysis of its extra-embryonic membranes, if the embryo is to be retained intact for histological analysis. Detailed analysis of both ovaries and testes reveals the presence of considerable numbers of primordial germ cells, some of which are seen to be in division.

The gonads at this stage are suspended by a wide urogenital mesentery, but this gets progressively narrower over the next 24–36 h, the ovary being suspended by the mesovarium and the testis by the mesorchium. In the female, the paramesonephric ducts are destined to give rise to the uterine horns and the ovarian capsule (*Plate 71,f–h*). Sections through the latter reveal the presence within it of the distal part of the paramesonephric duct, which is destined to form the oviduct, though it is uncoiled, and in a relatively undifferentiated state even at 17.5 days p.c. (*Plate 69,h*). More caudally, the paramesonephric ducts make contact with the wall of the urogenital sinus (*Plate 29a,a*; *Plate 29b,a*). In the male, the paramesonephric duct is believed to give rise to the appendix testis and part of the utriculus prostaticus. In both sexes, the genital tubercle is quite large, but in neither the male nor the female have they any distinguishing features at this stage.

Now that the mesonephros has virtually completely regressed, its only functional remnant in the two sexes, namely the ureteric bud (which is intitially a diverticulum of the mesonephric duct), has been taken over by its successor, the metanephros. In the male, however, the mesonephric duct becomes the "drainage system" for the contents of the seminiferous tubules, and forms (i) the rete testis (possibly), (ii) the efferent ducts, (iii) the epididymis, (iv) the ductus deferens, (v) the ejaculatory ducts, and (vi) the seminal vesicles. The blind cranial end of the mesonephric duct persists as the appendix epididymis. In the female, the mesonephric duct is believed to give rise to a few vestigial structures such as the epoophoron, paroophoron etc.

The metanephros (the definitive kidney) continues to differentiate, and enlarge, gradually "ascending" from its original pelvic site of origin. It is unclear whether the kidney actually ascends, or only appears to do so, due to the differential growth of the caudal half of the embryo. At this stage, the kidney is at a very early stage in its "ascent", its upper pole being slightly below the level of the 13th rib (*Plate 29b,a,b*). The metanephric cap tissue is becoming broken up into relatively small fragments, each of which becomes localized around diverticular derivatives of the ureteric bud. The latter are induced to form ampullae, and these subsequently divide to produce the primitive intrarenal collecting system. At this stage, some of the metanephric blastema cells differentiate to form nephrogenic vesicles, which are an early stage in the histogenesis of glomeruli, and these are first seen at about 14.5 days p.c. (*Plate 72,e,f*). The ureters are clearly seen though (*Plate 28e,f*), and they make contact with, but do not yet open into, the urogenital sinus.

SKELETAL SYSTEM

In the cranial vault, condensations of mesenchyme tissue are seen which subsequently ossify to form the frontal, parietal and parts of the temporal and occipital bones, as well as the majority of the facial bones. In the region of the base of the skull, precartilaginous condensations initially appear which differentiate to form the various components of the chondrocranium. The latter subsequently ossify to form the bony elements of the cranial base. Along the postcranial axis, the somite-derived segmentally arranged sclerotomes appear, initially as precartilaginous condensations. The latter migrate medially to surround the notochord. Each sclerotome becomes subdivided into a caudal condensed portion, and a rostral (cranial) less-condensed portion. The condensed portion then tends to move caudally, and fuses with the rostral part of its immediately succeeding neighbour to form the precartilaginous vertebral body, while the lower part of the less-condensed portion/upper part of the condensed portion differentiates to form the intervertebral disc (for fuller discussion, see Hamilton and Mossman, 1972).

In the skull at this stage, parts of the chondrocranium are present in the form of cartilaginous elements, though other parts are still present in the form of precartilaginous condensations. The largest of these

(*continued on page 176*)

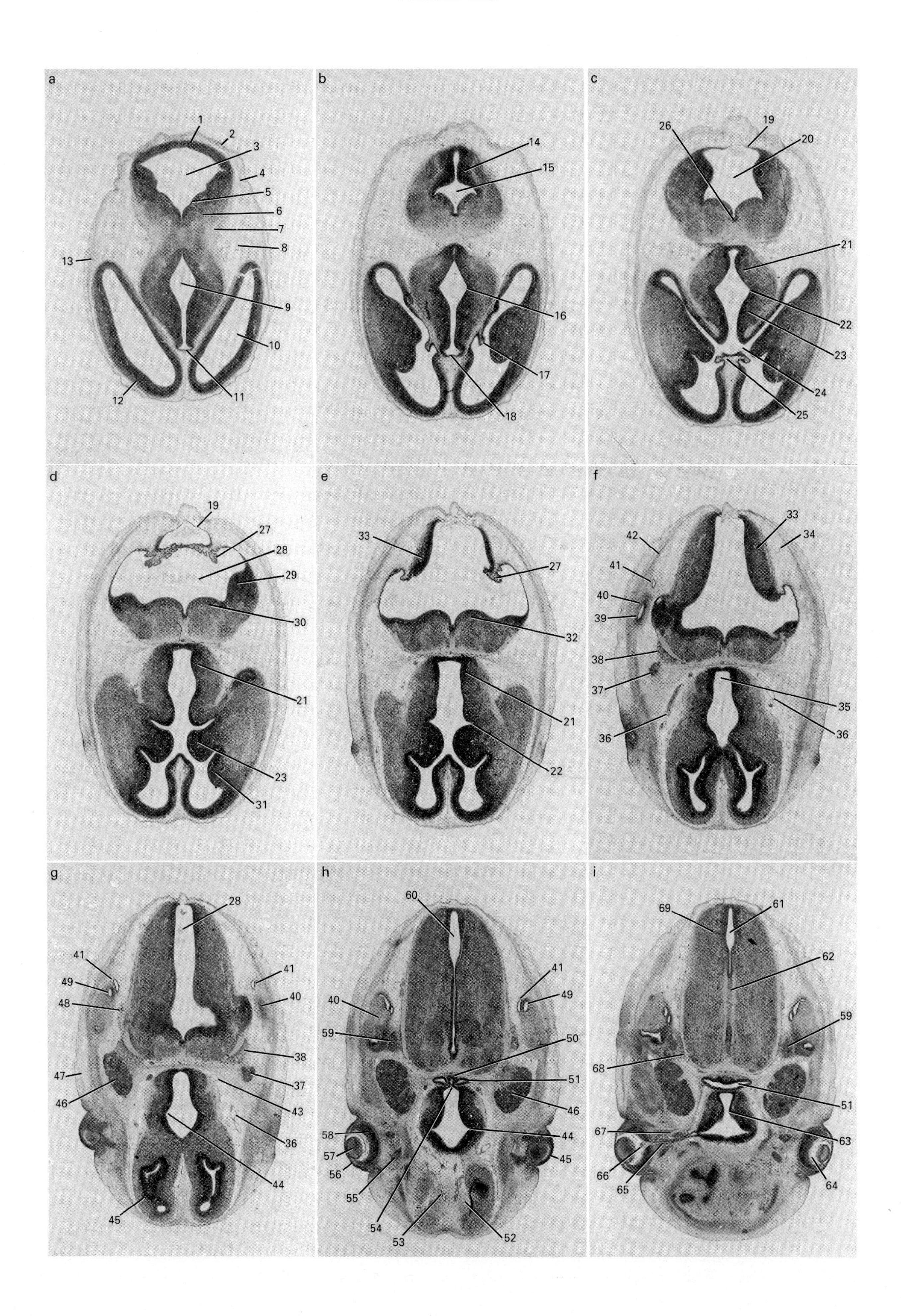
a
b
c
d
e
f
g
h
i

Plate 28a (12.5–13 days p.c.)

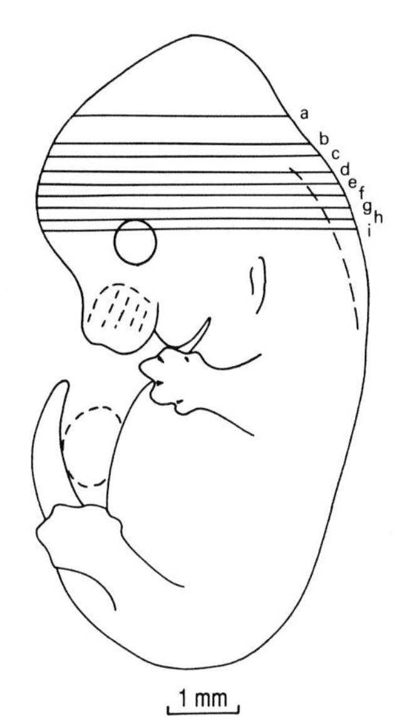

12.5–13 days p.c. Female embryo with 52–55 pairs of somites. Crown–rump length 8.8 mm (fixed). Transverse sections (Theiler Stage 21; rat, Witschi Stages 29–30; human, Carnegie Stage 18).

1. roof of dorsal part of midbrain
2. surface ectoderm
3. mesencephalic vesicle
4. part of mesencephalic venous plexus
5. ependymal layer
6. mantle layer
7. marginal layer
8. loosely packed cephalic mesenchyme
9. dorsal part of third ventricle
10. telencephalic vesicle (future lateral ventricle)
11. lamina terminalis
12. roof of neopallial cortex (future cerebral cortex)
13. condensation of mesenchyme tissue forming the outer table of the primitive ectomeninx
14. junction between caudal midbrain and rostral hindbrain
15. cerebral aqueduct
16. sulcus limitans
17. first evidence of choroid plexus in telencephalic vesicle
18. roof of third ventricle
19. roof of fourth ventricle
20. rostral part of fourth ventricle
21. diencephalon (hypothalamus)
22. hypothalamic sulcus
23. diencephalon (thalamus)
24. interventricular foramen
25. choroidal fissure
26. median sulcus
27. first evidence of choroid plexus in roof of fourth ventricle
28. fourth ventricle
29. alar plate (region of interventricular portion of future cerebellum)
30. basal plate of metencephalon
31. epithalamus
32. tegmentum of pons
33. alar plate of myelencephalon (medulla oblongata)
34. branches of posterior cerebral venous plexus
35. ventral part of third ventricle
36. middle cerebral artery
37. dorsal part of trigeminal (V) ganglion
38. rootlets of trigeminal (V) nerve from basal region of pons
39. superior semicircular canal
40. precartilage primordium of petrous part of temporal bone
41. rostral extremity of endolymphatic sac
42. primitive ectomeninx here seen to be divided into an outer thinner portion and an inner thicker and more condensed portion, separated by a narrow relatively acellular zone
43. internal carotid artery
44. optic recess
45. neopallial cortex in region of future olfactory lobe
46. trigeminal (V) ganglion
47. maxillary process
48. rootlets of vestibulocochlear (VIII) nerve
49. posterior semicircular canal
50. infundibulum (future pars nervosa)
51. residual lumen of anterior lobe of pituitary (previously Rathke's pouch)
52. olfactory (I) nerves
53. branches of anterior cerebral artery
54. infundibulum
55. extrinsic ocular muscle
56. corneal ectoderm (epithelium)
57. lens
58. inner (neural) layer of retina
59. vestibulocochlear (VIII) ganglion
60. caudal extremity of fourth ventricle
61. junction between fourth ventricle and central canal of spinal cord
62. medullary raphe
63. region of optic chiasma
64. hyaloid cavity
65. optic stalk (optic (II) nerve)
66. hyaloid artery
67. branch of oculomotor (III) nerve to extrinsic ocular muscle
68. rootlets of facial (VII) nerve
69. caudal part of medulla oblongata

PLATE 28b

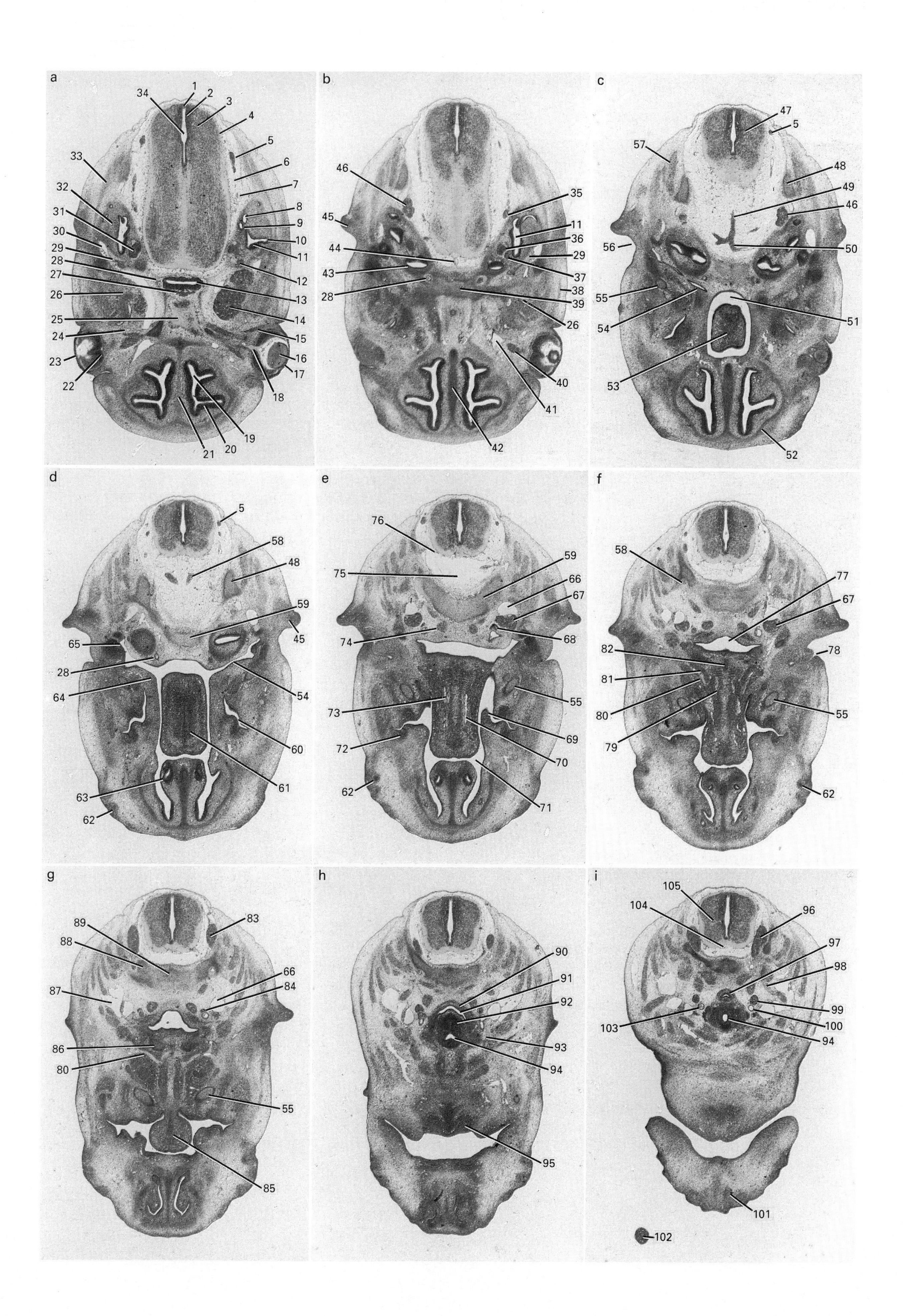

Plate 28b (12.5–13 days p.c.)

12.5–13 days p.c. Female embryo with 52–55 pairs of somites. Crown–rump length 8.8 mm (fixed). Transverse sections (Theiler Stage 21; rat, Witschi Stages 29–30; human, Carnegie Stage 18).

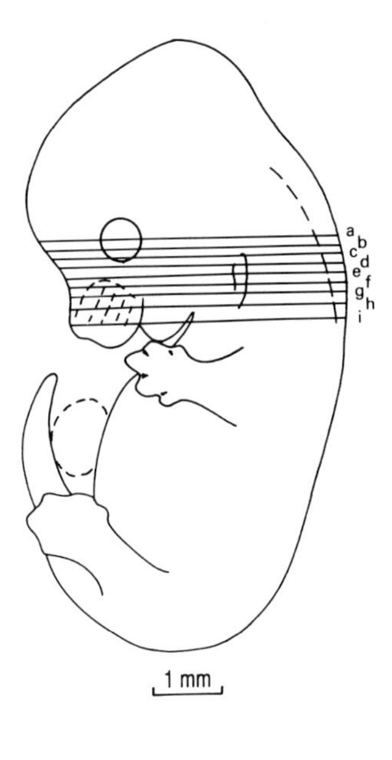

1. roof plate
2. ependymal layer
3. mantle layer
4. marginal layer
5. spinal accessory (XI) nerve
6. loose mesenchyme
7. cranial rootlets of accessory (XI) nerve
8. posterior semicircular canal
9. caudal extremity of endolymphatic sac
10. lateral semicircular canal
11. saccule
12. rootlets of facial (VII) nerve
13. residual lumen of anterior lobe of pituitary (previously Rathke's pouch)
14. trigeminal (V) ganglion
15. extrinsic ocular muscle (lateral rectus)
16. lens
17. corneal ectoderm
18. optic (II) nerve
19. lumen of primitive nasal cavity
20. olfactory epithelium
21. nasal septum
22. peripheral extremity of fetal fissure
23. optic eminence
24. ophthalmic division of trigeminal (V) nerve
25. floor of third ventricle
26. maxillary division of trigeminal (V) nerve
27. pars nervosa of pituitary
28. internal carotid artery
29. facial (VII) nerve
30. temporal branch of cerebral venous plexus
31. vestibulocochlear (VIII) nucleus
32. precartilage primordium of petrous part of temporal bone
33. condensation of mesenchyme tissue forming the outer table of the primitive ectomeninx
34. caudal lumen of medulla oblongata (caudal extremity of fourth ventricle)
35. cranial accessory (XI) ganglion
36. vestibular (VIII) ganglion
37. cochlear (VIII) ganglion
38. maxillary process
39. precartilage primordium of sphenoid bone
40. extrinsic ocular muscle (inferior rectus)
41. ophthalmic vein
42. precartilage primordium of nasal septum
43. lumen of cochlea
44. basilar artery
45. primordium of pinna
46. superior ganglion of vagus (X)–cranial accessory (XI) ganglion complex
47. rostral extremity of spinal cord
48. precartilage primordium of exoccipital bone
49. union of the two vertebral arteries to form the basilar artery
50. posterior cerebral artery
51. oropharyngeal region of foregut
52. precartilage primordium of nasal capsule
53. dorsum of tongue
54. dorsal component of first branchial pouch (pharyngo-tympanic tube)
55. precartilage primordium of Meckel's cartilage
56. external acoustic meatus
57. myotome-derived premuscle mass
58. vertebral artery
59. precartilage primordium of basioccipital bone
60. ventral component of first branchial pouch
61. premuscle precursors of intrinsic muscles of tongue
62. elevations associated with formation of follicles of vibrissae
63. vomeronasal organ (Jacobson's organ)
64. dorsal extension of palatal shelf of maxilla
65. dorsal component of first branchial pouch (location of tubo-tympanic recess)
66. internal jugular vein
67. inferior ganglion of glossopharyngeal (IX) nerve
68. superior cervical sympathetic ganglion
69. vertically directed palatal shelf of maxilla
70. lingual vessels
71. primitive nasopharynx
72. dental lamina (tooth primordium) of first upper molar tooth
73. lingual nerve
74. prevertebral premuscle mass
75. loosely packed mesenchyme within future subarachnoid space
76. vascular elements of pia mater
77. pharyngeal region of foregut
78. entrance to first branchial cleft
79. glossopharyngeal (IX) nerve
80. hypoglossal (XII) nerve
81. premuscle precursors of extrinsic muscle of tongue
82. thyroid primordium
83. dorsal (posterior) root ganglion of C1
84. sympathetic trunk
85. tip of tongue
86. precartilage primordium of laryngeal cartilage
87. external jugular vein
88. precartilage primordium of atlas vertebra (C1)
89. notochord
90. premuscle condensation of pharyngeal constrictor muscle
91. laryngeal region of pharynx
92. arytenoid swelling
93. superior laryngeal branch of vagus (X) nerve
94. trachea
95. dental lamina (tooth primordium) of lower incisor tooth
96. dorsal (posterior) root ganglion of C2
97. oesophagus
98. C2 spinal nerve
99. middle cervical sympathetic ganglion
100. left common carotid artery
101. ventral extremity of upper jaw
102. caudal extremity of tail
103. right common carotid artery
104. anterior spinal artery
105. upper cervical region of spinal cord

PLATE 28c

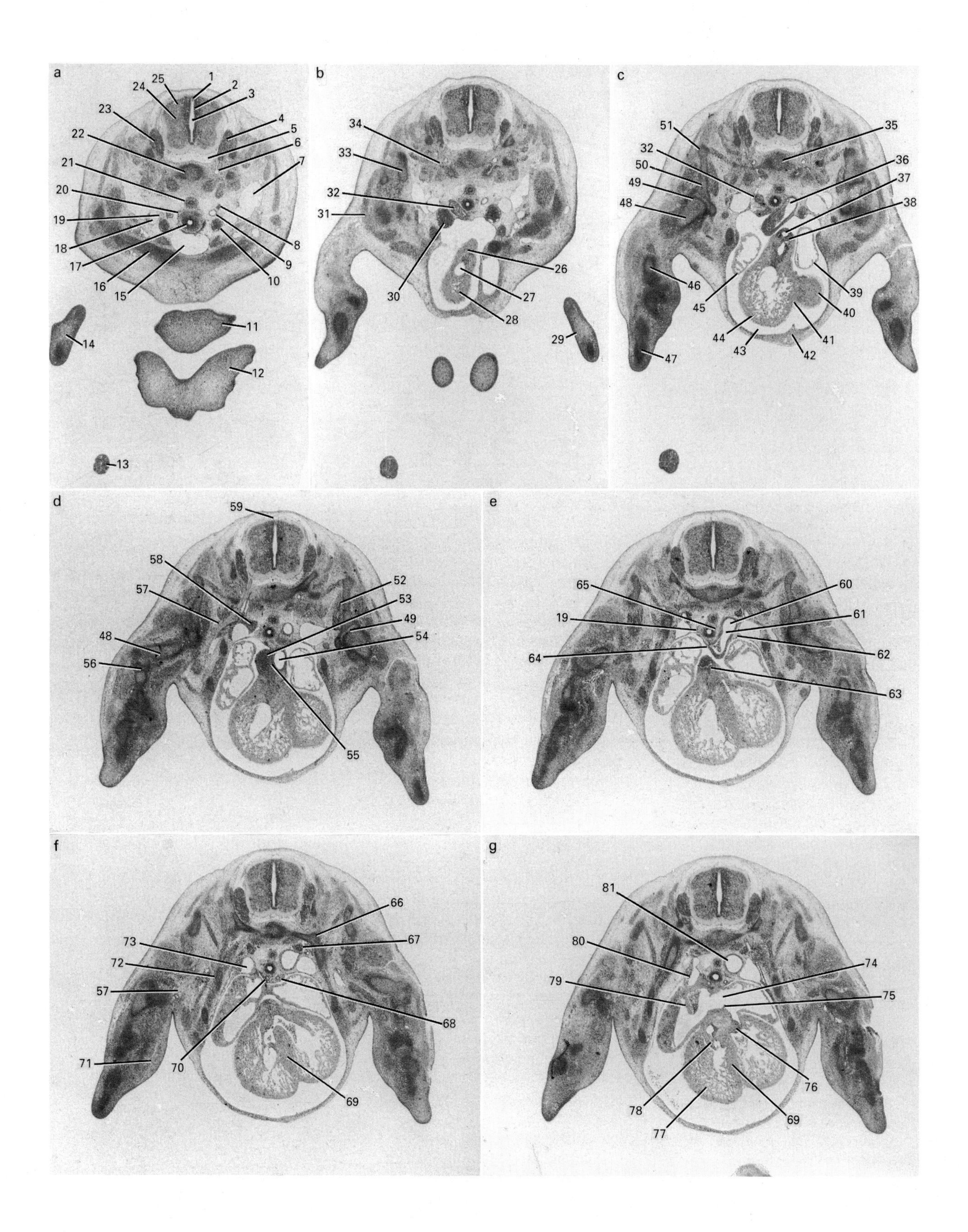

a
b
c
d
e
f
g
1
2
3
4
5
6
7
8
9
10
11
12
13
14
15
16
17
18
19
20
21
22
23
24
25
26
27
28
29
30
31
32
33
34
35
36
37
38
39
40
41
42
43
44
45
46
47
48
49
50
51
52
53
54
55
56
57
58
59
60
61
62
63
64
65
66
67
68
69
70
71
72
73
74
75
76
77
78
79
80
81

Plate 28c (12.5–13 days p.c.)

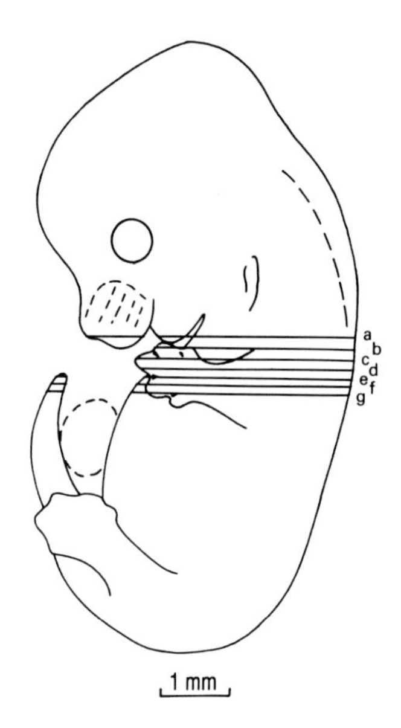

12.5–13 days p.c. Female embryo with 52–55 pairs of somites. Crown–rump length 8.8 mm (fixed). Transverse sections (Theiler Stage 21; rat, Witschi Stages 29–30; human, Carnegie Stage 18).

1. roof plate
2. ependymal layer
3. central canal
4. marginal layer
5. loosely packed mesenchyme within future subarachnoid space
6. vertebral artery
7. external jugular vein
8. sympathetic chain
9. common carotid artery
10. left thymic primordium
11. ventral tip of mandibular process (lower jaw)
12. tip of maxillary process (upper jaw)
13. caudal extremity of tail
14. "hand plate" of right forelimb bud
15. rostral extremity of pericardial cavity
16. mesenchymal primordium of clavicle
17. lumen of trachea
18. internal jugular vein
19. right vagal (X) trunk
20. recurrent laryngeal branch of right vagus (X) nerve
21. oesophagus
22. precartilage primordium of C3 vertebral body (centrum)
23. dorsal (posterior) root ganglion of C3
24. mantle layer
25. upper cervical region of spinal cord
26. wall of truncus arteriosus
27. lumen of outflow tract of right ventricle
28. trabeculated wall of outflow tract of right ventricle
29. "hand plate" of left forelimb bud
30. right thymic primordium
31. premuscle mass of deltoid
32. right fourth aortic arch artery (right brachiocephalic (innominate) artery)
33. precartilage primordium of rostal part of scapula
34. cervical spinal nerve (mixed, motor and sensory)
35. notochord
36. left fourth aortic arch artery (arch of the aorta, preductal component of thoracic aorta)
37. origin of pulmonary trunk
38. leaflets of pulmonary valve
39. wall of left atrium
40. wall of left ventricle
41. interventricular (previously bulbo-ventricular) groove
42. thoracic body wall
43. pericardial cavity
44. wall of right ventricle
45. wall of right atrium
46. precartilage primordium of ulna/radius
47. precartilage primordium of digital bones
48. precartilage primordium of humerus
49. primordium of shoulder (gleno-humeral) joint
50. precartilage primordium of blade of scapula
51. medial border of scapula
52. premuscle mass of subscapularis
53. wall of proximal (ascending) part of aorta
54. lumen of proximal part of pulmonary trunk
55. aortico-pulmonary spiral septum
56. primordium of elbow joint
57. component of brachial plexus passing into the right forelimb
58. origin of right subclavian artery
59. posterior spinal vein
60. arch of the aorta at site of entrance of the ductus arteriosus
61. ductus arteriosus (left sixth aortic arch artery)
62. left vagal (X) trunk
63. leaflets of aortic valve
64. origin of right pulmonary artery
65. left recurrent laryngeal branch of vagus (X) nerve
66. precartilage primordium of tubercle and neck region of rib
67. left cervico-thoracic (stellate) sympathetic ganglion
68. left pulmonary artery
69. muscular part of interventricular septum
70. right pulmonary artery
71. ulnar nerve
72. precartilage primordium of shaft of rib
73. right anterior cardinal vein (future right superior (cranial) vena cava)
74. inter-atrial communication via ostium secundum
75. septum primum
76. leaflets of left atrio-ventricular (mitral) valve
77. trabeculae carneae of right ventricular wall
78. leaflets of right atrio-ventricular (tricuspid) valve
79. leaflet of venous valve at site of entrance of inferior vena cava into right atrium
80. common cardinal vein
81. postductal part of thoracic (descending) aorta

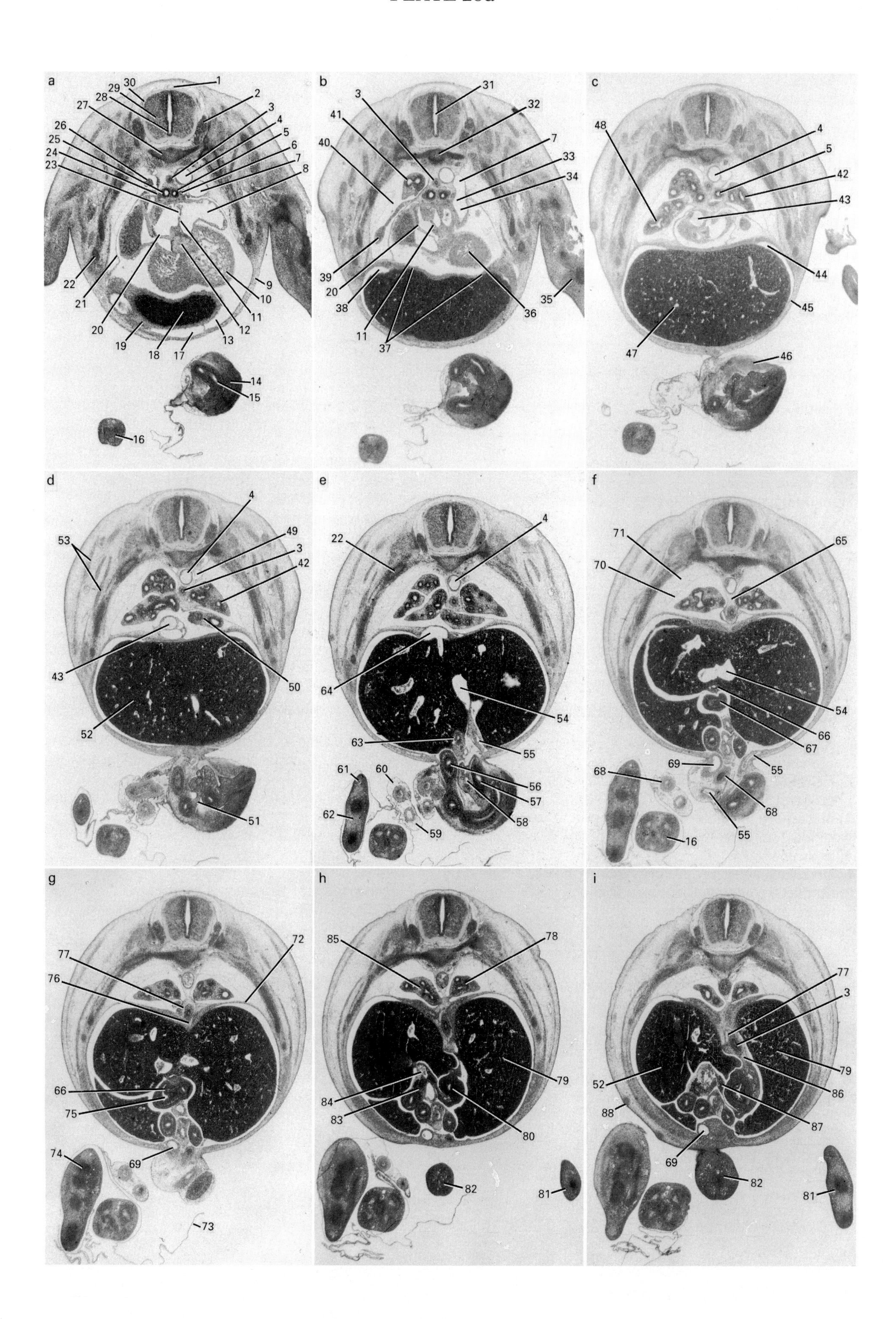
a
1 2 3 4 5 6 7 8 9 10 11 12 13 14 15 16 17 18 19 20 21 22 23 24 25 26 27 28 29 30
b
3 31 32 41 40 7 33 34 39 20 38 11 37 36 35
c
48 4 5 42 43 44 45 47 46
d
4 49 3 42 53 43 50 52 51
e
22 4 64 54 63 55 61 60 56 57 62 59 58
f
71 65 70 54 66 67 69 55 68 68 55 16
g
72 77 76 66 75 74 69 73
h
85 78 79 84 83 80 82 81
i
77 3 79 52 86 88 87 69 82 81

Plate 28d (12.5–13 days p.c.)

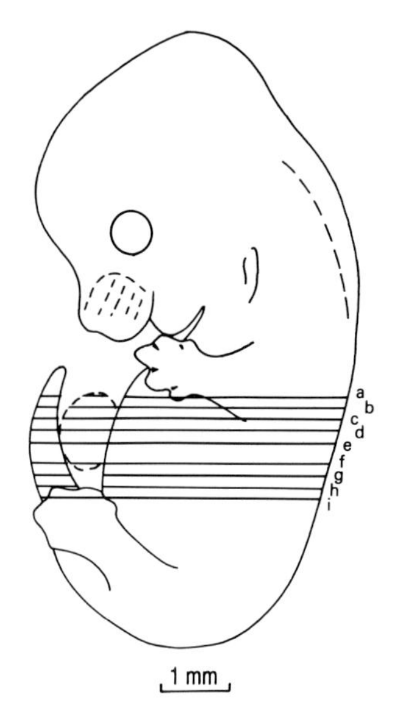

12.5–13 days p.c. Female embryo with 52–55 pairs of somites. Crown–rump length 8.8 mm (fixed). Transverse sections (Theiler Stage 21; rat, Witschi Stages 29–30; human, Carnegie Stage 18).

1. posterior spinal vein
2. dorsal (posterior) root ganglion
3. oesophagus
4. thoracic (descending) aorta
5. left main bronchus
6. left pulmonary artery
7. left anterior cardinal vein
8. lumen of left atrium
9. thoracic wall
10. trabeculated wall of left ventricle
11. septum primum
12. atrio-ventricular endocardial cushion tissue
13. pericardial cavity
14. extravasated blood within tissues of physiological umbilical hernia
15. midgut loop of bowel within physiological umbilical hernia
16. caudal extremity of tail
17. rostral extremity of peritoneal cavity
18. rostral extremity of liver
19. condensation of mesenchyme of septum transversum associated with differentiation of central tendon of the diaphragm
20. leaflet of venous valve at site of entrance of inferior vena cava into right atrium
21. pleuro-pericardial membrane
22. precartilage primordium of shaft of rib
23. ostium secundum
24. right pulmonary artery
25. right main bronchus
26. apex of upper lobe of right lung
27. precartilage primordium of vertebral body (centrum)
28. ependymal layer
29. mantle layer
30. marginal layer
31. central canal
32. notochord
33. communication between left anterior and left common cardinal veins
34. left common cardinal vein
35. left forelimb
36. caudal extremity of wall of left ventricle
37. central tendon of the diaphragm with subjacent "bare" area of the liver
38. right subphrenic region of peritoneal cavity
39. rostral extremity of middle lobe of right lung
40. pericardio-peritoneal canal (future pleural cavity)
41. segmental bronchus (right lung – cranial lobe)
42. segmental bronchus (left lung)
43. posthepatic component of inferior vena cava
44. left subphrenic region of peritoneal cavity
45. abdominal wall
46. wall of physiological umbilical hernia
47. hepatic sinusoid
48. segmental bronchus (right lung – middle lobe)
49. left subcardinal vein
50. right lung – accessory lobe
51. dorsal mesentery of midgut loop
52. right lobe of liver
53. premuscle masses of body wall musculature
54. ductus venosus
55. left umbilical artery
56. proximal part of midgut loop
57. distal part of midgut loop
58. vitelline vessels (superior mesenteric arteries and veins) within dorsal mesentery of midgut
59. Wharton's jelly (connective tissue in the umbilical cord)
60. umbilical cord containing a single umbilical artery and umbilical vein
61. marginal vein
62. right hindlimb
63. wall of gall bladder
64. origin of posthepatic part of inferior vena cava
65. dorsal meso-oesophagus
66. common bile duct
67. wall of duodenum
68. umbilical vein
69. right umbilical artery
70. mesothelial lining of pericardio-peritoneal canal (future parietal pleura)
71. loose connective tissue forming wall of pericardio-peritoneal canal
72. diaphragm
73. amnion
74. precartilage condensation within hindlimb
75. lumen of duodenum at site of entrance of common bile duct
76. left (becoming anterior) vagal (X) trunk
77. right (becoming posterior) vagal (X) trunk
78. caudal region of left lung
79. left lobe of liver
80. pyloric region of stomach
81. left hindlimb
82. genital tubercle
83. primordium of right lobe and body of the pancreas
84. portal vein/superior mesenteric vein
85. right lung – caudal lobe
86. dorsal mesogastrium
87. primordium of proximal part of left lobe of the pancreas
88. mammary gland primordium

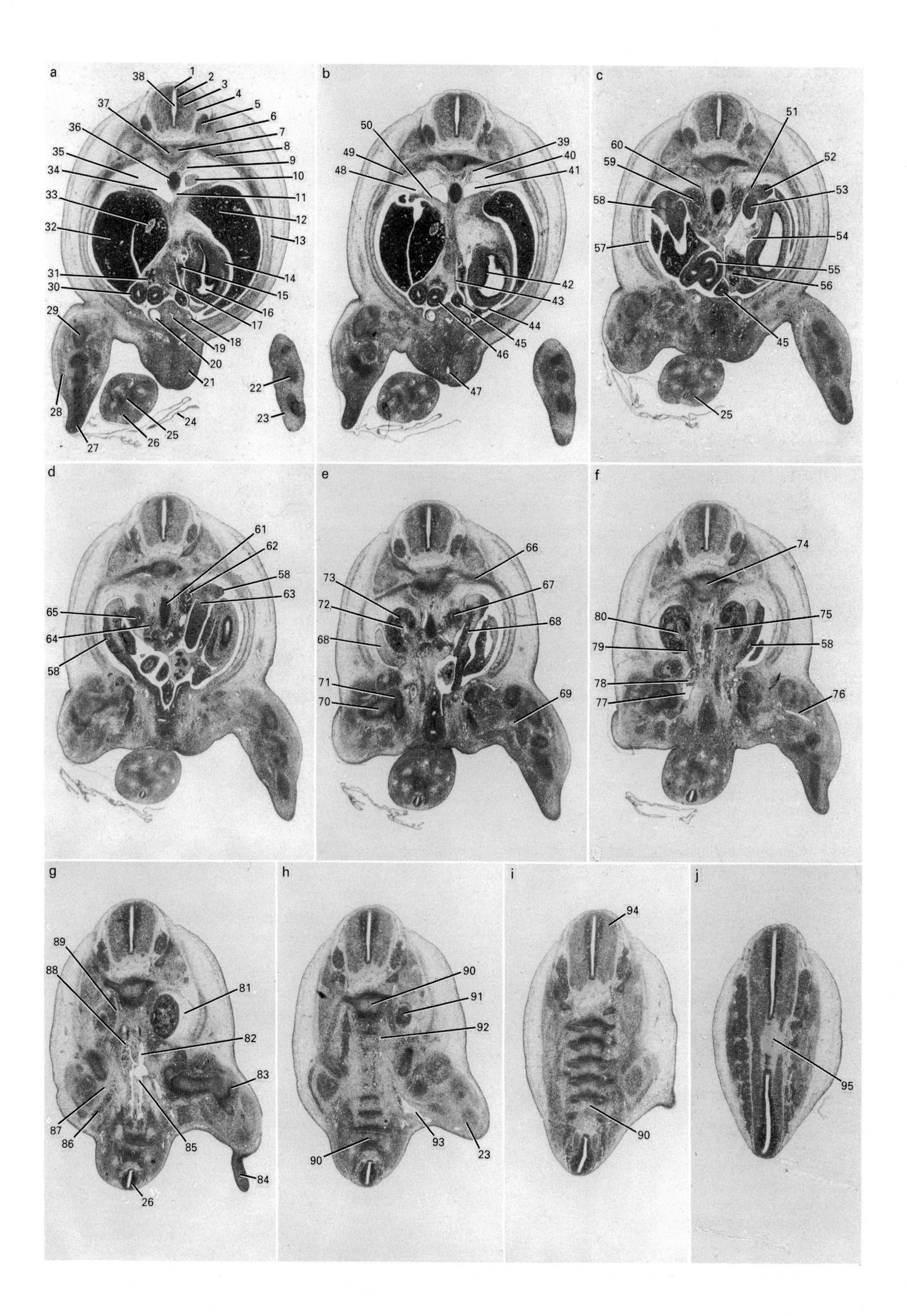
a
1
2
3
4
5
6
7
8
9
10
11
12
13
14
15
16
17
18
19
20
21
22
23
24
25
26
27
28
29
30
31
32
33
34
35
36
37
38
b
39
40
41
42
43
44
45
46
47
48
49
50
c
51
52
53
54
55
56
45
25
57
58
59
60
d
61
62
58
63
65
64
58
e
66
67
68
73
72
68
71
70
69
f
74
75
58
80
79
78
77
76
g
89
88
81
82
83
87
86
85
84
26
h
90
91
92
93
23
90
i
94
90
j
95

Plate 28e (12.5–13 days p.c.)

12.5–13 days p.c. Female embryo with 52–55 pairs of somites. Crown–rump length 8.8 mm (fixed). Transverse sections (Theiler Stage 21; rat, Witschi Stages 29–30; human, Carnegie Stage 18).

1. roof plate
2. ependymal layer
3. mantle layer
4. marginal layer
5. dorsal (posterior) root ganglion
6. myotome-derived epaxial (paravertebral) muscle mass
7. precartilage primordium of vertebral body (centrum)
8. precartilage primordium of shaft of rib
9. subcardinal vein
10. caudal extremity of left lung
11. dorsal mesogastrium
12. left lobe of liver
13. premuscle mass of outer layer of thoracic/abdominal musculature
14. ventral (anterior) wall of stomach
15. primordium of proximal part of left lobe of the pancreas
16. lumen of stomach
17. portal vein/superior mesenteric vein
18. left umbilical artery
19. urachus
20. right umbilical artery
21. genital tubercle
22. precartilage condensation within hindlimb
23. left hindlimb
24. amnion
25. proximal part of tail
26. neural tube within tail
27. right hindlimb
28. premuscle mass of proximal extensor group of muscles
29. precartilage primordium of distal region of femur
30. duodenum
31. primordium of right lobe and body of the pancreas
32. right lobe of liver
33. hepatic vein
34. right pleuro-peritoneal canal
35. mesothelial lining of pleuro-peritoneal canal (future parietal pleura)
36. distal part of thoracic (descending) aorta
37. notochord
38. central canal
39. left sympathetic trunk
40. left pleuro-peritoneal fold
41. left pleuro-peritoneal canal
42. gastric mucosa lining the stomach
43. dorsal mesentery of hindgut
44. lesser omentum (ventral mesentery of stomach)
45. hindgut
46. midgut
47. urethral lumen
48. right pleuro-peritoneal fold
49. segmental (intercostal) nerve
50. caudal extremity of diaphragmatic primordium
51. urogenital mesentery
52. mesonephros (regressing)
53. gonad (very early stage in differentiation of ovary)
54. omental bursa (lesser sac of peritoneal cavity)
55. lumen of midgut
56. primordium of body and proximal part of tail of left lobe of the pancreas
57. peritoneal cavity
58. paramesonephric duct
59. right adrenal primordium (initially principally cortical tissue)
60. branch from right sympathetic trunk to adrenal medullary primordium
61. descending (abdominal) aorta
62. left adrenal primordium
63. rete ovarii
64. prehepatic component of inferior vena cava (derived from right subcardinal vein)
65. rostral extremity (upper pole) of right metanephros (definitive renal (kidney) primordium)
66. precartilage primordium of caudal rib
67. upper pole of left metanephros (kidney)
68. urogenital ridge
69. component of lumbo-sacral plexus passing into the left hindlimb
70. precartilage primordium of proximal part of femoral shaft (diaphysis)
71. primordium of hip joint
72. upper branch of right metanephric (renal) artery
73. right metanephros (kidney)
74. precartilage primordium of lumbar vertebral body
75. branch of sympathetic chain to pre-aortic plexus
76. left axial vein (femoral vein)
77. right common iliac vein
78. origin of right umbilical artery from right common iliac artery
79. right ureter
80. pelvis of right metanephros (kidney)
81. loose mesenchyme tissue in prerenal region
82. left posterior cardinal vein (regressing) at site of amalgamation with left subcardinal vein
83. primordium of left knee joint
84. lateral border of left "foot plate"
85. sacral vein draining into common iliac vein
86. right axial artery (femoral artery)
87. right sciatic nerve
88. caudal extremity of right posterior cardinal vein (site of union of the two common iliac veins)
89. lumbar nerve descending to form part of the sciatic nerve
90. segmental interzone associated with the formation of the intervertebral disc with notochordal precursor of nucleus pulposus
91. lower pole of left metanephros (kidney)
92. segmental (lumbar) artery
93. left common iliac vein
94. neural tube (spinal cord) in lower lumbar region
95. marginal layer (cut obliquely)

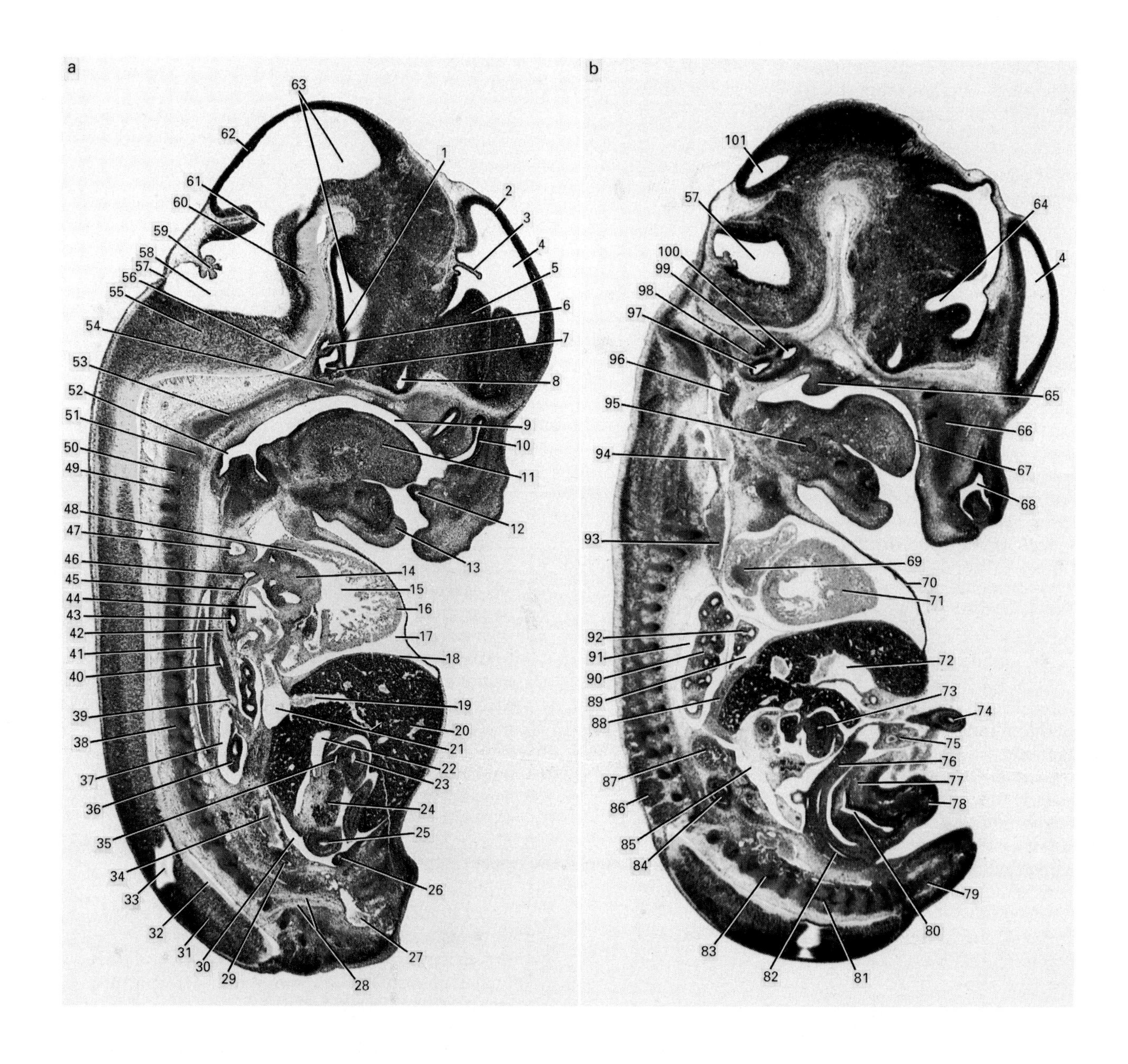
a
1
2
3
4
5
6
7
8
9
10
11
12
13
14
15
16
17
18
19
20
21
22
23
24
25
26
27
28
29
30
31
32
33
34
35
36
37
38
39
40
41
42
43
44
45
46
47
48
49
50
51
52
53
54
55
56
57
58
59
60
61
62
63
b
101
57
64
4
100
99
98
97
96
65
95
66
94
67
68
93
69
70
71
92
91
90
72
89
73
74
88
75
76
87
77
86
78
85
84
79
80
83
82
81

Plate 29a (12.5–13 days p.c.)

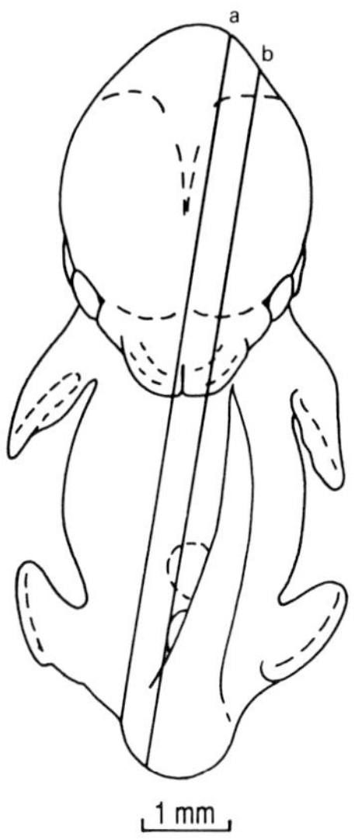

12.5–13 days p.c. Female embryo. Crown–rump length 8.2 mm (fixed). Sagittal sections (Theiler Stage 21; rat, Witschi Stages 29–30; human, Carnegie Stage 18).

1. wall of diencephalon (hypothalamus)
2. roof of neopallial cortex (future cerebral cortex)
3. choroid plexus extending into lateral ventricle
4. lateral ventricle
5. corpus striatum mediale (medial aspect of ganglionic eminence)
6. infundibulum of pituitary (future pars nervosa)
7. early evidence of vascular differentiation which develops in close association with anterior wall of Rathke's pouch (future anterior pituitary)
8. optic recess of diencephalon
9. oropharynx
10. olfactory epithelium
11. intrinsic muscles of tongue
12. primordium of primary palate
13. ventral extremity of lower jaw
14. right atrio-ventricular bulbar cushion tissue
15. lumen of right ventricle of heart
16. wall of right ventricle of heart
17. pericardial cavity
18. central tendon of diaphragm
19. right hepatic vein
20. right lobe of liver
21. origin of posthepatic part of inferior vena cava
22. portal vein
23. first part (i.e. transverse part) of duodenum
24. pancreatic primordium
25. third part of duodenum
26. caudal part of right paramesonephric duct where it approaches the urogenital sinus
27. right common iliac vein
28. components of lumbo-sacral plexus passing into the right hindlimb
29. right paramesonephric duct within caudal part of urogenital mesentery
30. caudal (pelvic) part of peritoneal cavity
31. mantle layer of spinal cord in lower lumbar region
32. marginal layer of spinal cord
33. central canal of spinal cord
34. prehepatic part of inferior vena cava
35. common bile duct
36. caudal lobe of right lung
37. right pleural cavity
38. cartilage primordium of lower thoracic vertebral body
39. posterior vagal (X) trunk
40. oesophagus
41. descending (thoracic) aorta
42. anterior vagal (X) trunk
43. right main bronchus
44. lumen of right atrium
45. right pulmonary artery
46. pulmonary trunk
47. arch of aorta
48. outflow tract of right ventricle (origin of pulmonary trunk)
49. cartilage primordium of body of C3 vertebra
50. cartilage primordium of body of C2 vertebra (axis)
51. cartilage primordium of anterior arch of C1 vertebra (atlas)
52. entrance to oesophagus
53. cartilage primordium of basioccipital bone (clivus)
54. cartilage primordium of basisphenoid bone
55. medulla oblongata
56. basilar artery
57. roof of fourth ventricle
58. fourth ventricle
59. choroid plexus differentiating from roof of fourth ventricle
60. region of pons–midbrain junction
61. cerebral aqueduct
62. roof of midbrain
63. third ventricle
64. interventricular foramen
65. vertically directed palatal shelf/process of maxilla
66. cartilage primordium of nasal septum
67. dorsum of tongue
68. entrance into nasopharynx from external naris
69. lumen of left atrium of heart
70. thoracic body wall
71. muscular part of interventricular septum
72. left umbilical vein passing dorsally to form the ductus venosus within the liver
73. pyloric sphincter
74. proximal part of midgut loop passing into the physiological umbilical hernia
75. mesenteric vessels within the dorsal mesentery of the midgut
76. distal (intra-abdominal) part of midgut loop
77. wall of urogenital sinus
78. tip of genital tubercle
79. middle region of tail
80. lumen of vesical part of urogenital sinus
81. notochord
82. lumen of hindgut (rectum)
83. cartilage primordium of lumbar vertebral body
84. medial border of left metanephros
85. peritoneal cavity
86. dorsal (posterior) root ganglion
87. left adrenal primordium
88. left crus of the diaphragm
89. right pleural cavity
90. left lung
91. left pleural cavity
92. accessory lobe of right lung
93. left common cardinal vein (distal part of internal jugular vein)
94. left anterior cardinal vein (proximal part of internal jugular vein)
95. Meckel's cartilage
96. inferior ganglion of vagus (X) nerve
97. cochlea
98. cartilage primordium of petrous part of temporal bone
99. vestibulocochlear (VIII) ganglion
100. saccule
101. caudal part of mesencephalic vesicle

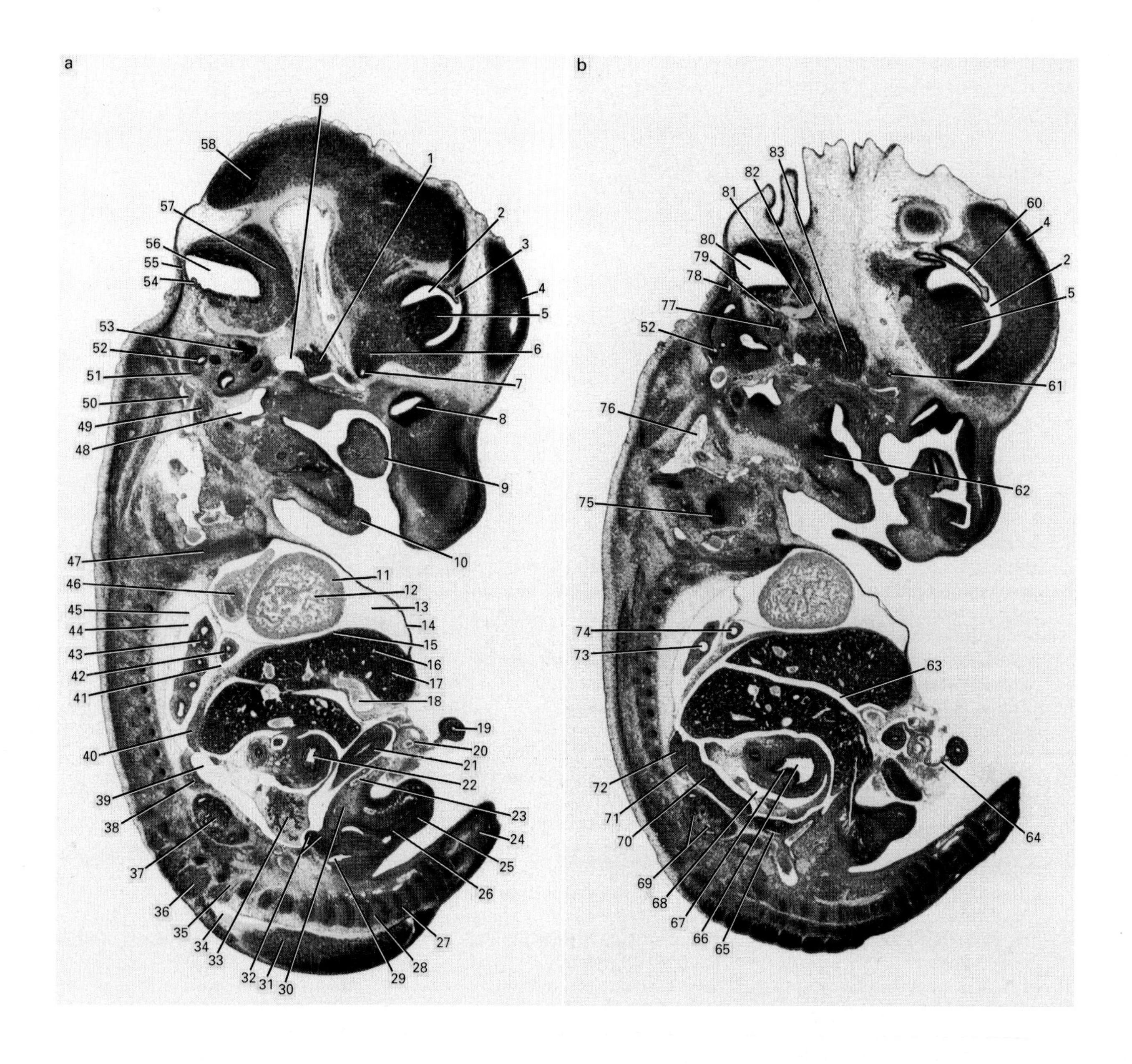
a
b
1
2
3
4
5
6
7
8
9
10
11
12
13
14
15
16
17
18
19
20
21
22
23
24
25
26
27
28
29
30
31
32
33
34
35
36
37
38
39
40
41
42
43
44
45
46
47
48
49
50
51
52
53
54
55
56
57
58
59
60
61
62
63
64
65
66
67
68
69
70
71
72
73
74
75
76
77
78
79
80
81
82
83

Plate 29b (12.5–13 days p.c.)

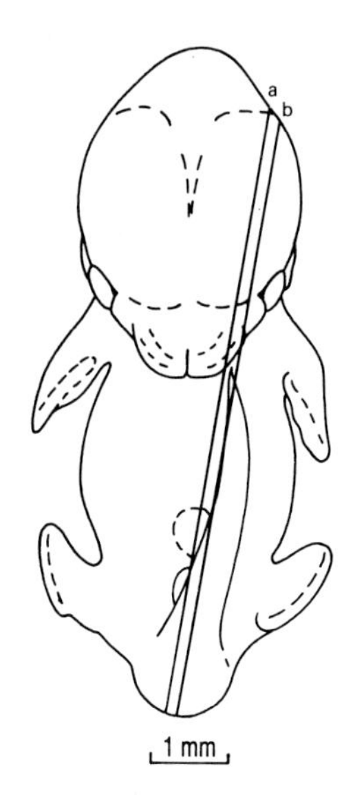

12.5–13 days p.c. Female embryo. Crown–rump length 8.2 mm (fixed). Sagittal sections (Theiler Stage 21; rat, Witschi Stages 29–30; human, Carnegie Stage 18).

1. medial margin of left trigeminal (V) ganglion
2. lateral ventricle
3. origin of choroid plexus, arising from medial wall (choroidal fissure) of lateral ventricle
4. roof of neopallial cortex (future cerebral cortex)
5. striatum
6. diencephalon (region of optic stalk)
7. optic stalk (optic (II) nerve)
8. vomeronasal organ (Jacobson's organ)
9. lateral part of tip of tongue
10. ventral extremity of lower jaw
11. wall of left ventricle of heart
12. trabeculated part of wall of left ventricle of heart
13. pericardial cavity
14. thoracic body wall
15. central tendon of diaphragm
16. hepatic sinusoids
17. left lobe of liver
18. left umbilical vein passing dorsally to form the ductus venosus within the liver
19. midgut loop within physiological umbilical hernia
20. mesenteric vessels within the dorsal mesentery of the midgut
21. lumen of distal part of midgut loop
22. lumen of pyloric antrum region of stomach
23. left umbilical artery
24. middle region of tail
25. tip of genital tubercle
26. phallic part of urethra
27. notochord
28. cartilage primordium of vertebral body in lumbo-sacral region
29. lumen of distal part of hindgut (future rectum)
30. wall of urogenital sinus (future bladder)
31. mantle region of spinal cord in lumbo-sacral region
32. caudal part of paramesonephric duct where it approaches the urogenital sinus
33. pancreatic primordium
34. marginal region of spinal cord in lumbar region
35. segmental lumbar nerve, being a component of the lumbo-sacral plexus passing into the left hindlimb
36. dorsal (posterior) root ganglion
37. left metanephros (definitive kidney)
38. left adrenal primordium
39. peritoneal cavity
40. left crus of the diaphragm
41. right pleural cavity surrounding accessory lobe of right lung
42. accessory lobe of right lung
43. left lung
44. left pleural cavity
45. mesothelial lining (parietal pleura) of left pleural cavity
46. lumen of left atrium of heart
47. primordium of left clavicle
48. first pharyngeal pouch
49. inferior ganglion of vagus (X) nerve
50. internal jugular vein
51. cartilage primordium of petrous part of temporal bone
52. posterior semicircular canal
53. vestibulocochlear (VIII) ganglion
54. origin of choroid plexus differentiating in roof of fourth ventricle
55. roof of fourth ventricle
56. fourth ventricle
57. region of pons/midbrain junction
58. wall of midbrain
59. branch of primary head vein – drains into internal jugular vein
60. choroid plexus extending into lateral ventricle
61. lumen of optic stalk
62. Meckel's cartilage
63. interlobar space
64. dorsal mesentery of midgut
65. paramesonephric duct within caudal part of urogenital mesentery
66. lumen of body of stomach
67. mucous membranous lining of the stomach
68. lesser sac (omental bursa) of peritoneal cavity
69. tubules within left metanephros
70. left ovary
71. urogenital mesentery (region of rete ovarii)
72. mesonephric tubules (mesonephros almost completely regressed at this stage)
73. segmental bronchus within left lung
74. segmental bronchus within accessory lobe of right lung
75. cartilage primordium of left humerus
76. external jugular vein
77. facial (VII) ganglion
78. rostral extremity of endolymphatic duct
79. rootlets of origin of facial (VII) ganglion
80. lateral part of fourth ventricle
81. basal region of pons
82. rootlets of trigeminal (V) ganglion
83. left trigeminal (V) ganglion

elements are destined to form the petrous part of the temporal bones (*Plate 28a,g–i*; *Plate 28b,a–d*), the exoccipital bones (*Plate 28b,b–d*), and the basioccipital bone (*Plate 28b,d,e*). In the cephalic region, the presence of Meckel's cartilage has already been noted (*Plate 28b,e–h*). The segmental precartilage/cartilaginous sclerotome-derived condensations along the vertebral axis are most readily seen in sagittal sections (*Plate 29a,a*). In developmentally more advanced embryos, the long bones may be chondrified, as may many of the vertebral bodies. The condensations for the ribs as well as the bones of the pectoral and pelvic girdles are also recognizable at this stage, as are the primordia of the shoulder and hip joints (*Plate 28c,c,d*; *Plate 28e,e,g*). At 12.5–13 days p.c., no evidence of ossification is seen.

Stage 22 Developmental age, 14 days p.c.
Plates 30 a–l; Plates 31a, b

EXTERNAL FEATURES

Embryos with about 60 pairs of somites. Between 13.5–14.5 days p.c., a marked change is observed in the morphology of the hand plate and foot plate. In the former region, the digital interzones, with their associated webbing, have retreated proximally, so that the digits are now present as completely separated units. This is clearly seen in the scanning electron micrographs of both the dorsal and palmar surfaces of the hand plate at this time (*Plate 44,g–i*). A dramatic change is also observed with regard to the appearance of the foot plate, though there is a substantially less marked degree of differentiation of the hindlimb digits (i.e. the toes), than that seen in the forelimb digits (*Plate 45,a,b*). In addition to the residual degree of webbing still seen in the hindlimbs, the foot plate shows the first evidence of asymmetry, which is a feature of this region at later stages of development. The digits of both the forelimb and hindlimb are seen to be fairly splayed out (*Plate 44,h,i*; *Plate 45,a,b*), though none of the digits show any distinguishing features at this stage. The differentiation of the nail beds is not evident until about 15–15.5 days p.c., though a slight flattening on the dorsum of the digits overlying the terminal phalanges of the forelimb is just discernible (*Plate 44,h*). The orientation of the hand plate and foot plate is similar to that seen at the previous stage. Within the limbs, the cartilaginous primordia of all of the long bones are present, but the more distal elements are still present in the form of condensed mesenchyme precartilage precursor elements (*Plate 30f,b–d*; *Plate 30g,a–d*).

Relatively few changes are observed in the external appearance of the cephalic region between 13.5 and 14.5 days p.c., so that the more adult features observed to be present in the previous stage are becoming more consolidated. The latter particularly applies to the morphological appearance of the ears, in which the pinnae are seen to have expanded and are now clearly seen to be turned forwards, and cover about half of the external acoustic meatus (*Plate 46,f,h*). The neck is also seen to have elongated slightly and narrowed. As indicated in the appropriate section of the previous stage, all of the primordia of the vibrissae, as well as those of the prominent tactile or sinus hair follicles that are characteristically present in specific locations in the cephalic region, are readily seen (*Plate 49,i*). At about 13.5–14 days p.c., the first evidence of hair follicle primordia is seen, and these spread caudally from the pectoral region. In the limbs, the primitive hair follicles do not extend distally beyond the skin overlying either the wrist or ankle joints (*Plate 44,g*).

The volume of the physiological umbilical hernia is proportionately somewhat larger at this stage than previously (*Plate 30i,a–d*; *Plate 30j,a,b*; cf. *Plate 28d,a–g*), and remains at about this size until about 14.5 days p.c. (*Plate 32h,a–d*). By about 15.5 days p.c. the midgut loops of bowel are beginning to return to the peritoneal cavity (*Plate 35h,d*; *Plate 35i,a–c*). By 16.5 days p.c., the physiological umbilical hernia is no longer evident, so that the only structure to be found in the region of the umbilicus is the umbilical cord and its contents, consisting of the umbilical arteries, the umbilical vein and the urachus.

THE HEART AND VASCULAR SYSTEM

In the region of the outflow tract, the ascending aorta and pulmonary trunk are now seen to be two quite distinct vessels, each with a similar luminal diameter (*Plate 30f,d*; *Plate 31a,a*). The ascending aorta flows directly into the arch of the aorta, and gives off a series of major branches, the first of which is the innominate artery (*Plate 30f,b*; *Plate 31a,a*). The principal branches of the latter are the right subclavian and right common carotid arteries (*Plate 30f,a,b*). The next major branches to emerge from the arch of the aorta are the left common carotid artery (*Plate 30f,a*), and the left subclavian artery. The latter vessel has almost reached its definitive position, and between 13 days p.c. and 14 days p.c. is seen to have "ascended" from a position directly opposite the site of entrance of the ductus arteriosus into the concavity of the arch of the aorta (ie. where the left sixth arch artery joins the left dorsal aorta) (see Fig. 10). The "ascent" of the subclavian vessels, of course, simply reflects the rostral ascent of the forelimbs that occurs during this period. More significant, however, from an embryological point of view, is the fact that between 13 days p.c. and 14 days p.c. the already tenuous connection which exists between the right sixth aortic arch artery and the right dorsal aorta finally disappears. As a result of the latter event, the right subclavian artery is now seen to take origin from the innominate artery (*Plate 30f,b*), while the right sixth arch artery loses its connection with the right dorsal aorta (see Fig. 10).

Apart from a limited degree of elongation of the proximal parts of the innominate artery and the left common carotid and left subclavian arteries, the vascular arrangements seen at this stage persist until birth. At the time of birth, however, major rearrangements within the vascular system are brought about as a direct consequence of altered demands on the circulation, due to the severing of the umbilical cord and the necessity of the fetus to lead an independent existence. The other change that occurs, which is also a direct consequence of the disappearance of the distal part of the right sixth arch artery, is the asymmetrical relationship that now develops between the recurrent laryngeal branches of the right and left vagus (X) nerves, and the modified branchial arch arterial system. The arrangement on the left side, where the recurrent laryngeal branch of the vagus (X) nerve hooks around the left sixth arch artery (the ductus arteriosus of the embryo, that is destined to become the ligamentum arteriosum of the adult circulation), retains the original "primitive" arrangement, whereas on the right side, as a consequence of the disappearance of the right sixth arch artery, the recurrent laryngeal branch of the right vagus (X) nerve hooks around the right subclavian artery. While the two vagal (X) trunks are easily seen (*Plate 30g,a*), the relatively low magnification of the sections illustrated here precludes the visualization of their recurrent laryngeal branches.

With the differentiation of the membranous part of the interventricular septum from the caudal part of the bulbar ridges (they descend towards the atrioventricular cushion tissue, and subsequently fuse with the muscular part of the interventricular septum) (*Plate 30g,c*), the communication between the cavities of the right and left ventricles is now closed. As a consequence of the latter, the heart has now achieved the definitive arrangement that persists until the birth of the individual. As far as inter-atrial septation is concerned, the septum primum has fully formed, the foramen primum has been obliterated, and the foramen secundum remains as a channel through the rostral part of the septum primum (*Plate 30g,c,d*). The septum secundum is now formed, and is located to the right of the septum primum. The latter forms a crescent-shaped and fairly inflexible membrane that is suspended from the posterosuperior part of the wall of the inter-atrial region of the heart (*Plate 30g,c,d*).

The presence of these two septa (namely the septum primum and septum secundum) and their relationship to each other allows the largely oxygenated blood from the placenta (that flows into the right atrium from the inferior vena cava, having previously reached this vessel via the left umbilical vein and the ductus venosus), to be directed from this chamber of the heart into the left atrium through the foramen ovale. The critical component in this exercise is the lower free edge of the septum secundum (also termed the crista dividens), which overrides the orifice of the inferior vena cava, and consequently directs a considerable proportion of the inferior vena caval blood through the foramen ovale into the left atrium, while only allowing a relatively small proportion of this blood stream to pass into the right atrium proper.

The pericardial cavity now extends rostrally to just above the level of the neck of the first rib (*Plate 30f,a,b*; *Plate 31a,a*), and caudally to the level of the central tendon of the diaphragm (*Plate 30h,c*; *Plate 31a,a,b*). The pericardial cavity is now separated from the two pleural cavities by the thin pleuro-pericardial membranes (*Plate 30g,a–d*; *Plate 30h,a,b*). The two pleuro-peritoneal channels have almost closed at this stage, by the fusion of the majority of the pleuro-peritoneal folds to the diaphragmatic component of the septum transversum (i.e. the central tendon of the diaphragm). A tag of tissue is clearly seen to arise from the inner aspect of the anterior wall of the pericardial cavity in the midline, and this may represent a vestigial remnant of a primitive ventral mesentery of the heart (*Plate 28c,c,d*; *Plate 30f,c,d*; *Plate 30g,a–d*; *Plate 30h,a,b*). However, a similar structure is not apparent in developmentally less advanced embryos (cf. *Plate 26c,b–i*; *Plate 27c,a,b*), nor is it seen in more advanced embryos (cf. *Plate 32f,b–d*; *Plate 32g,a–d*), so its aetiology must remain uncertain until more information is forthcoming.

The left and right superior (cranial) venae cavae form principally from their corresponding internal jugular veins (*Plate 30f,a,b*), and are quite symmetrical as they pass caudally into the upper part of the thoracic region. Here they are joined by the subclavian veins, formed by the confluence of the axillary and external jugular veins. The subclavian veins pass medially beneath the clavicle and over the first rib, and are then joined by the corresponding internal jugular veins (*Plate 30f,a*). The two superior (cranial) venae cavae pass further caudally until just below the level of the bifurcation of the trachea, where the left superior (cranial) vena cava receives the accessory (superior) hemiazygos vein (previously the rostral part of the left posterior cardinal vein) which, with the left superior intercostal vein, drains all of the upper left intercostal spaces (*Plate 30g,c*). The azygos vein, which drains the right intercostal spaces, has a smaller luminal diameter than that of the corresponding accessory hemiazygos and hemiazygos veins on the left side. The left superior (cranial) vena cava passes across the midline, to drain into the caudal part of the right atrium close to the entrance of the inferior vena cava, and these entrances are guarded by venous valves (*Plate 30h,a–c*). The right superior (cranial) vena cava enters the rostral part of the right atrium just above the entrance of the inferior vena cava, and this entrance is also guarded by a venous valve (*Plate 30g,d*).

The posthepatic inferior vena cava is a vessel of substantial luminal diameter (*Plate 30h,c,d*; *Plate 30i,a–c*). It is formed principally from the ductus venosus (*Plate 30i,b,c*), though it also receives contributions from the portal system (*Plate 30j,a*), via the hepatic venous system (*Plate 30i,a*), and from the prehepatic inferior vena cava (which provides the venous drainage of the lower limbs, kidneys and other paired abdominal organs) (*Plate 30k,c,d*; *Plate 30l,a,b*).

THE PRIMITIVE GUT AND ITS DERIVATIVES

Between 12.5–13 and 13.5–14 days p.c., the tongue elongates and increases in volume as well as developing a prominent intermolar eminence. The latter is believed to be derived from the tuberculum impar (*Plate 58,h–k*; cf. *Plate 58,f,g*). The anlage of the epiglottis also shows a substantial increase in size over this period (*Plate 58,j*; *Plate 30d,b*; *Plate 31a,a*). During the early part of this stage, the palatal shelves are clearly directed vertically (*Plate 59,h,i*), whereas in developmentally more advanced embryos, the medial margins of the palatal shelves are now seen to be horizontally directed (*Plate 59,j,k*), the change in direction of the medial borders of the shelves from vertical to horizontal represents a critical stage in palatal shelf formation, and is followed by a relatively rapid period of shelf growth, in which the distance between their medial margins steadily decreases (*Plate 59,l*; *Plate 60,a*). Others, however, have suggested that the palatal shelves achieve their full length prior to their elevation, as a result of mesenchymal proliferation (Ferguson, 1987). It is unclear at the present time whether the palatal shelf mesenchyme is of neural crest origin in mammalian embryos, though neural crest contributes significantly to the maxillary mesenchyme in avian embryos (Johnston, 1966).

During the early part of this stage, the first indication of reflex opening and closing movements of the mouth occurs, and this enables the (lower jaw and) tongue to descend below the lower margins of the vertically directed palatal shelves for the first time. Similarly, as a result of the so-called "internal shelf forces", in which the vascularity of the shelves increases, as well as its tissue turgidity (due to the hydration of extracellular matrix material), and other related factors, elevation of the shelves is facilitated (Ferguson, 1987, 1988). It has also been suggested that the underlying mechanisms may be different for the anterior and posterior parts of the palatal shelves, so that anteriorly an intrinsic system of contractile proteins may cause a rapid reorientation, whereas posteriorly a more fluid reorientation may be achieved due to internal "forces" created within mesenchymal gel matrices, possibly by hydration of glycosaminoglycans such as hyaluronic acid (Ferguson, 1987, 1988). This latter phase is of relatively brief duration, and shelf apposition is first evident by about 14–14.5 days p.c. (*Plate 62,c–f*). The first evidence of fusion occurs shortly afterwards, initially involving the anterior half of the palatal shelves, and gradually more posteriorly until about 15.5 days p.c. (*Plate 60,h,i*), by which time palatal shelf formation is all but complete.

This is the first occasion that the nasopharynx is separated from the oropharynx. During the same period over which palatal shelf formation occurs, the nasal septum grows down towards the palatal shelves, but does not fuse with them until about 15–15.5 days p.c. (*Plate 62,h–l*). The latter event results in the separation of the nasopharynx into the left and right nasal cavities. Each of the latter has an inlet, the external naris (*Plate 60,l*), through which the air passes via the definitive choana (*Plate 59,i*) into the corresponding nasal cavity, and thence into the posterior pharynx (*Plate 60,j*).

The primordia of the dentition are more clearly seen at this stage than previously (cf. *Plate 28b,c–h*), particularly the primordia of both the upper and lower first molar teeth (*Plate 30d,d*), as well as those of the upper and lower incisor teeth (*Plate 30f,a,b*; *Plate 31a,a,b*; *Plate 31b,a,b*).

With regard to the respiratory tract, mesenchymal condensations of precartilage are now apparent both in the region of the larynx and intermittently along the length of the trachea and main bronchi, and these are particularly well seen in sagittal sections of this region in developmentally slightly more advanced embryos (*Plate 33a,a*). The left lung and the various lobes of the right lung are now clearly demarcated, and each contains numerous primary, secondary and tertiary bronchi (*Plate 30h,a–d*; *Plate 66,g,h*). By about 13.5–14 days p.c., the two pleuro-peritoneal canals are closed by the growth of the pleuro-peritoneal folds when they make contact and eventually fuse with the dorsal mesentery of the oesophagus, and the central tendon of the diaphragm (*Plate 30h,c,d*; *Plate 30i,a–d*; *Plate 30j,a,b*; *Plate 31b,a,b*).

The thyroglossal duct has previously descended from the region of the foramen caecum (*Plate 22a,c*; *Plate 24c,g*), and bifurcated to give rise to the two lobes of the thyroid gland. The latter are located as two distinct entities on either side of the larynx (*Plate 30e,a,b*; *Plate 31a,b*). At this stage, the two lobes of the thymus gland are quite substantial structures. These have differentiated from the ventral part of the third pharyngeal pouches, and their upper borders have descended to the level of the upper part of the trachea (*Plate 30e,c,d*), while their lower borders have reached the upper mediastinum. The two lobes of the thymus gland now encroach on the upper part of the pericardial cavity (*Plate 30f,a,b*), at about the level of the superior part of the arch of the aorta (*Plate 30f,b*; *Plate 31a,a*). The parathyroid glands, though present at this stage, are not shown in the sections illustrated. They are, however, quite difficult to find at this stage, and are more readily recognized in developmentally slightly more advanced embryos, for example, at about 14.5–15.5 days p.c. (*Plate 35e,a*).

The oesophagus is only barely canalized below the level of the bifurcation of the trachea (*Plate 30h,b–d*), and where it passes through the diaphragm, the two vagal (X) trunks are seen to be anterior and posterior relations of it (*Plate 30j,a*). The oesophagus enters the stomach at the gastro-oesophageal junction (*Plate 30j,b*). The lumen of the stomach is extremely large at this stage (*Plate 30j,b–d*; *Plate 30k,a,b*), and the stomach occupies a proportionately greater volume of the peritoneal cavity than previously (cf. *Plate 28e,a–d*). The lesser sac of the peritoneal cavity (the omental bursa) is also seen to be a potentially large space which is closely related to the dorsal wall of the stomach (*Plate 30j,b–d*; *Plate 30k,a*; *Plate 31b,b*). The pyloric region of the stomach and the pyloric sphincter at the gastro-duodenal junction are particularly clearly seen in sagittal sections (*Plate 31a,b*; *Plate 31b,a*). At this stage of development, the stomach and the liver combined occupy the majority of the peritoneal cavity (*Plate 31a,a,b*; *Plate 31b,a,b*), and consequently leave relatively little room for the majority of the midgut to elongate and differentiate, so that the midgut is still required to occupy the physiological umbilical hernia for this purpose.

The pancreas is a large and rather diffuse organ in the mouse (*Plate 30j,c,d*; *Plate 30k,a–d*), and is consequently less easily subdivided into regions (such as the head, uncinate process, body and tail) than is the human pancreas. The pancreas is seen to contain numerous secondary hollow branchings derived from the two original endodermal outgrowths at the foregut–midgut junction, and these extend into the mesenchyme of the organ. Numerous sprouts arise from the solid ends of the epithelial buds, and these canalize to form the drainage ducts. Acini arise from the tips of the latter. The islets of Langerhans are first recognized at about 16 days p.c., though the typical beta-cells do not appear until about 17.5–18.5 days p.c.

The hindgut is seen to be suspended by an

elongated dorsal mesentery (*Plate 30k,a–c*), except for its most caudal part where the mesentery gradually disappears (*Plate 30k,d*), so that within the pelvis it has no mesentery (*Plate 30l,a,b*). In sagittal sections, the anal canal is seen to be canalized, and the anal pit seen to be perforated (*Plate 31a,b*) due to the rupture of the anal membrane.

THE NERVOUS SYSTEM

The principal difference observed between neural differentiation in embryos at this stage and those at previous stages of development, relates to the increased degree of differentiation of the wall of the telencephalic vesicles. Whereas at about 12.5–13 days p.c., the roof of the telencephalic vesicles is fairly homogeneous, with little evidence of stratification (cf. *Plate 28a,a–e*), the most marked change seen at about 13.5–14 days p.c. relates to the first clear evidence of cellular migration from the mantle zone into the overlying marginal zone, to form the primary (neopallial) cortex. This now consists of a superficial, and at this stage a relatively narrow, cortical layer, which is separated from the mantle layer by a narrow anuclear intermediate zone (*Plate 30a,c,d*). The latter becomes progressively wider over the next few days (cf. *Plate 35a,a–d*), and beyond the darkly staining nuclear layer of the neopallial cortex, it will also be possible to recognize a further narrow relatively anuclear layer (the marginal zone) (*Plate 35a,e*). Subsequently, other neuroblasts and glioblasts migrate from the mantle layer into the intermediate zone, and this layer then also shows evidence of stratification (cf. *Plate 39b,a–d*) (for terminology of the cortical layers, see Sidman, 1970).

A marked change is also observed in the volume of choroid plexus present, both in the lateral ventricles and in association with the roof of the fourth ventricle. Choroid plexus is first evident in both locations at about 12.5–13 days p.c. (*Plate 28a,b–e*), and becomes increasingly evident in both regions over the next few days. It is particularly florid in the fourth ventricle (*Plate 30a,c*; *Plate 31a,a,b*; *Plate 31b,a*), where it is seen to extend progressively into its lateral recesses (*Plate 30a,d*; *Plate 32b,a,b*). A small volume of choroid plexus appears in the pineal recess of the third ventricle at about 14.5 days p.c. (*Plate 32a,c,d*), but only by about 16.5 days p.c. does the volume here increase to any great extent (*Plate 39c,a–c*). Similarly, the choroid plexus in the lateral ventricles only gradually increases until about 16.5 days p.c., particularly in relation to the superior part of the body of the lateral ventricles (*Plate 39b,d*; *Plate 39c,a–c*), though that associated with the fourth ventricle remains extremely florid (*Plate 39d,c*), and is probably of considerably greater functional importance.

The principal difference observed in the morphological appearance of the eye at this stage of development (*Plate 30c,d*; *Plate 30d,a*; *Plate 52,g,h*), and that seen at about 13 days p.c. (*Plate 52,e,f*), relates to the progressive decrease in the volume of the lens compared to that of the overall volume of the globe of the eye. In the haematoxylin and eosin stained section which provides a close-up view of the eye (*Plate 52,h*), it is just possible to observe differences in the staining of the nuclei of the lens fibres between the cells at the periphery and those at the centre of the lens, with the latter tending to be progressively less densely staining. Similarly, as the lens increases in volume and differentiates, the thickness of the anterior part (or capsule) of the lens decreases compared to the overall diameter of the lens (*Plate 52,h*; cf. *Plate 52,f*).

Regrettably, in these sections, the haematoxylin and eosin has only minimally stained the nerve fibres that pass from the ganglion cells of the neural retina towards the region of the future optic disc, and thence centrally along the optic nerve. The diameter of the optic nerve at this stage is quite considerable, and its original lumen (previously the optic stalk, *Plate 51,f,j*; *Plate 25a,a,b*; *Plate 27a,h*; *Plate 28a,i*), is obliterated by the nerve fibres that are passing centrally along it towards the optic chiasma (*Plate 30c,d*; *Plate 30d,a*).

An increase in the volume of the vitreous (hyaloid body) is seen at this stage (*Plate 52,h*), compared to that observed at the previous stage (*Plate 52,f*). It is also possible to observe that between 11.5 days p.c. (*Plate 46,b*) and 13.5–14.5 days p.c. (*Plate 46,f,h*), the location of the eyes in relation to the whole head is gradually changing. Whereas at the beginning of this period, the eye is located on the lateral side of the head, and only barely seen on a frontal view of this region, towards the end of this period, the eye is now seen to be in a more anterior position. Furthermore, the interocular distance progressively diminishes over this period.

The location of the future eyelids is clearly seen at this stage (*Plate 30c,c,d*; *Plate 30d,a,b*), but their appearance is not very different from that observed at about 14.5 days p.c. (*Plate 49,b*). Indeed, fundamental changes are only seen during the following day or so, when the first evidence of cellular activity is seen at the peripheral margins of the orbit (*Plate 49,d*), as a prelude to eyelid closure, which occurs at about 15.5–16 days p.c. (*Plate 49,h*). On the inferior (ventral) surface of the globe of the eye, the "fetal" fissure is seen (*Plate 30d,b*, unlabelled). While the extrinsic ocular muscles were first seen at about 12.5–13 days p.c. (*Plate 28a,i*; *Plate 28b,a*), it is now possible to recognize most of them individually (*Plate 30c,d*; *Plate 30d,a*), and (particularly in silver stained material) to discern their individual nerve supply.

The components of the inner ear apparatus are showing increased evidence of differentiation. Whereas previously it was just possible to recognize their various principal components, and note that they were embedded in the precartilage primordium of the petrous part of the temporal bone (*Plate 28a,g–i*; *Plate 28b,a–d*), it is now possible to discern that the cochlear duct has elongated and curved to form the first part of a spiral (*Plate 30c,c,d*). Towards its periphery, the precartilage mass is now more condensed than previously, and seen to be replaced by a cartilaginous capsule, whereas in its more central part, the precartilage tissue is said to dedifferentiate to form the much more loosely arranged periotic connective tissue. A localized region of the cuboidal basal epithelial lining along the length of the cochlear duct elongates, and these cells are destined to form the specialized sensory spiral organ of Corti (*Plate 40b,d*; *Plate 40c,a*, unlabelled).

Between 12.5–13 and 13.5–14 days p.c., a marked change is observed in the histological appearance of the pituitary gland, which is particularly well seen in sagittal sections. Formerly, the remnant of the lumen of Rathke's pouch was quite voluminous (*Plate 55,a*), and a remnant of its original connecting stalk which linked it to the roof of the oropharynx was still present. However, during the period between 13.5 and 14.5 days p.c., a progressive narrowing of the lumen of Rathke's pouch occurs (*Plate 55,b,c*), and its connecting stalk completely disappears. By 17.5 days p.c., all that remains of the original lumen of Rathke's pouch is a small cleft which separates the narrow pars intermedia from the much more cellular and vascular pars anterior (*Plate 55,i*). Similarly, over the same initial period of time, the diameter of the entrance to the infundibular recess of the third ventricle diminishes to a narrow slit (*Plate 55a*; c.f. *Plate 55,c,d*), and by about 15.5 days p.c., the lumen of the pars nervosa eventually becomes completely obliterated (*Plate 55,f*). On transverse sections through this region, it is possible to see that the rostrolateral extensions of Rathke's pouch, which were first seen to be extending on either side of the infundibulum (neurohypophysis) at about 11.5–12.5 days p.c. (*Plate 56,c,d*; *Plate 56,h*), have now almost surrounded this region (*Plate 56,m,n*), to form the pars tuberalis of the pituitary.

A typical section through the mid-thoracic region of the spinal cord reveals that at 13.5–14 days p.c., it still displays certain features which reflect its relatively early state of differentiation. This is particularly the case with regard to the size and shape of the central canal, and the overall shape of the dorsal and ventral grey horns (columns). At 13.5–14 days p.c., the central canal is still elongated, has a narrow diamond-shaped profile, and extends for most of the vertical diameter of the spinal cord (*Plate 63,d*). This appearance is clearly a transitional form, and an intermediate shape between that seen at 12–12.5 days p.c. (*Plate 63,b*), and that observed at 14.5 days p.c. (*Plate 63,e*). The latter shape is itself still evolving towards the definitive arrangement seen at 16.5–17.5 days p.c. (*Plate 63,h,k*), where the lumen of the central canal is seen to be almost obliterated. The dorsal and ventral grey horns (columns) are well demarcated at this stage, and are seen to be surrounded by a fairly uniform marginal layer of white matter (*Plate 63,d*). As development proceeds, the dorsal horns spread rather further towards the periphery of the cord than do the ventral horns. The proportionate volume of the white matter correspondingly increases at this level, particularly in the ventral half of the cord and on either side of the dorsal (posterior) median septum and the ventral median fissure (*Plate 63,h,k*).

The adrenal gland, though a close anatomical relation of the kidney, is best considered within this section because of its complex origin, the medulla of which is ectodermal in origin and neural crest derived, while the cortex is mesodermal in origin and derived from the proliferation of coelomic epithelium. The adrenal primordium has a more clearly defined border than previously (*Plate 31b,a*; cf. *Plate 28e,c*; *Plate 29a,b*). Initially, the adrenal primordium consists principally of cortical tissue which is derived from the cellular proliferation of coelomic epithelium. The latter forms cords of tissue which invade the subjacent mesenchyme to form two large condensations in the retroperitoneal region These are located in close proximity to the upper poles of the kidneys, one on either side of the abdominal aorta. The medullary tissue of the adrenal consists of sympathetic nerve cells of neural crest origin, and this is initially principally located on the medial aspect of the gland. The various branches that pass from the sympathetic trunks to the mediodorsal aspect of the primordium of the adrenal medulla (*Plate 30j,b,c*), are particularly clearly seen in silver-stained material.

THE UROGENITAL SYSTEM

The principal difference observed between the histological appearance of the kidney at 12.5–13 days and 13.5–14 days p.c., relates to the increasing degree of differentiation of the cortical region. At 12.5–13 days p.c., the metanephros appears to be a fairly homogeneous structure, with relatively few poorly differentiated metanephric vesicles and collecting tubules present (*Plate 72,a,b*). By 13.5–14 days p.c., the kidney has increased in size, and moderate numbers of vesicles and tubules are now seen to be present. The latter are principally located towards the periphery of the organ, with the future medullary tissue at this stage being fairly devoid of features. By 14–14.5 days p.c., primitive glomeruli are seen for the first time, and this is associated with an increase in the number of collecting tubules, most of which have a relatively narrow luminal diameter (*Plate 72,e,f*). In the hilum of the kidney, the collecting ducts amalgamate to form a primitive renal pelvis which then drains into the ureter (*Plate 30k,b*) and this passes caudally (*Plate 30k,c,d*; *Plate 31a,b*), towards the vesical part of the urogenital sinus (*Plate 31a,b*) into which it opens.

The region of the primitive urogenital sinus where contact was first made by the mesonephric ducts, serves to divide the urogenital sinus into an upper and lower portion. The former is termed the vesico-urethral canal, and subsequently differentiates into the vesical part of the urogenital sinus, and ultimately forms the bladder. The latter region is in continuity rostrally with the urachus (*Plate 29a,b*; *Plate 31a,b*). The lower part is termed the definitive urogenital sinus. The latter region is initially markedly flattened from side to side (*Plate 26d,f–h*; *Plate 27c,f*; *Plate 27d,a*), and is itself subdivided into an upper pelvic part, and a lower phallic part which is located at the base of the genital tubercle (*Plate 29a,b*, unlabelled;

(*continued on page 210*)

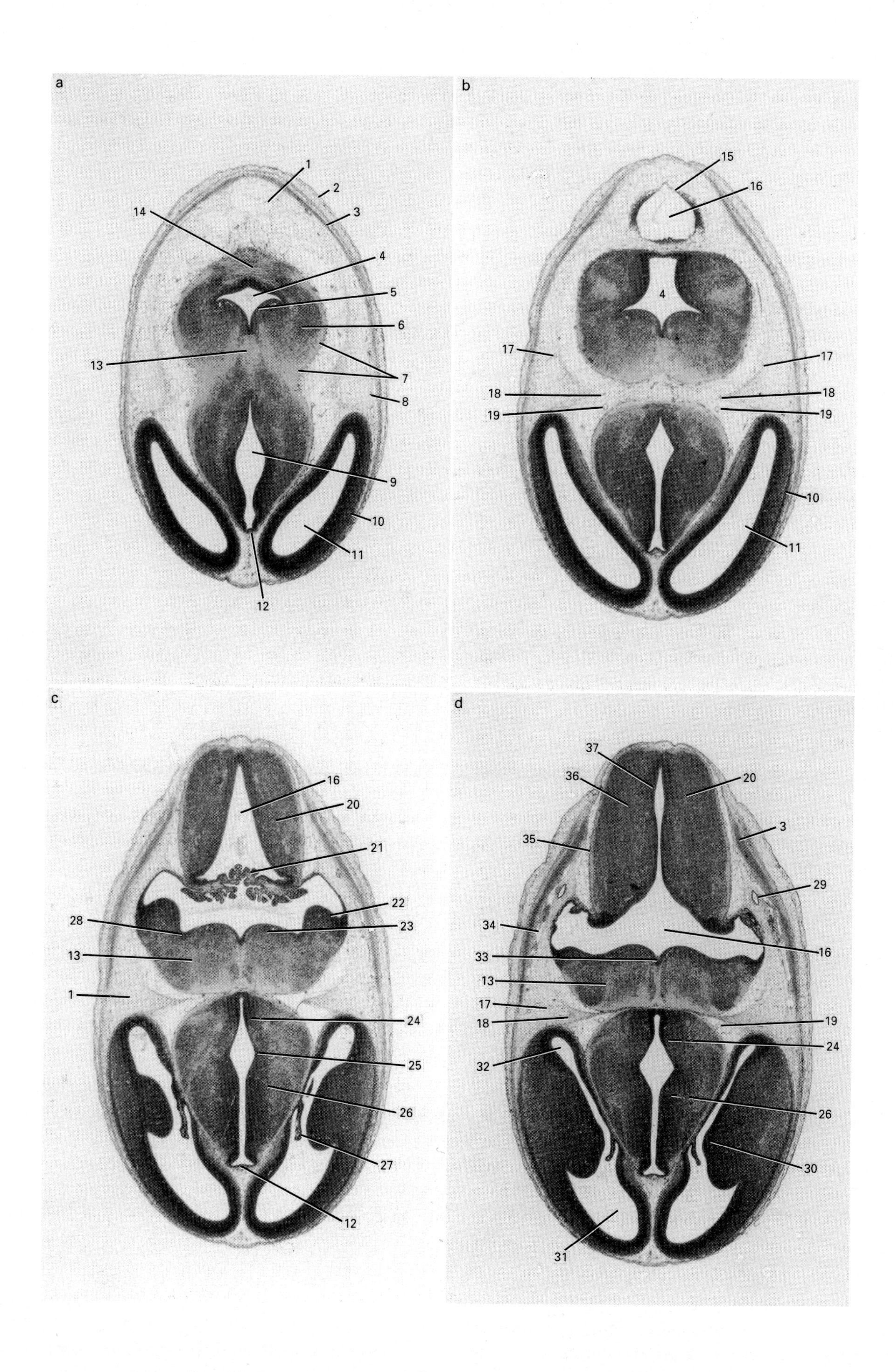
a
1
2
3
14
4
5
6
13
7
8
9
10
11
12
b
15
16
4
17
18
19
10
11
c
16
20
21
22
28
23
13
1
24
25
26
27
12
d
37
36
20
35
3
29
34
16
33
13
17
18
19
24
32
26
30
31

Plate 30a (13.5 days p.c.)

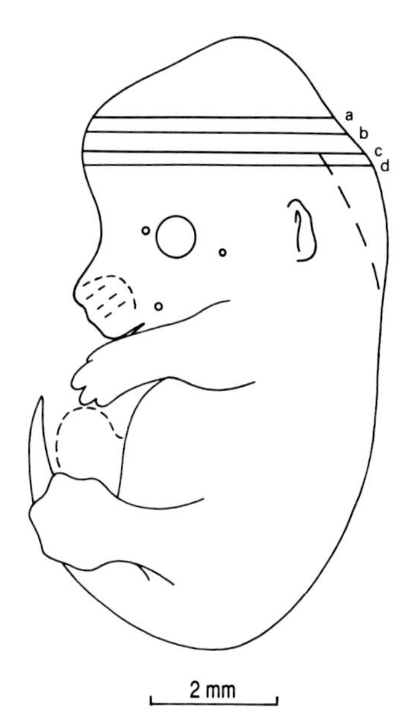

13.5 days p.c. Female embryo with about 60 pairs of somites. Crown–rump length about 9.1 mm (fixed). Transverse sections (Theiler Stage 22; rat, Witschi Stage 31; human, Carnegie Stages 20–23).

1. loosely packed cephalic mesenchyme tissue
2. skin
3. primitive ectomeninx here seen to be divided into an outer thinner portion and an inner thicker and more condensed portion separated by a narrow relatively acellular zone
4. cerebral aqueduct
5. ependymal layer of midbrain
6. mantle layer of midbrain
7. marginal layer of midbrain
8. primitive venous plexus, precursors of dural venous sinuses
9. dorsal part of third ventricle
10. roof of neopallial cortex (future cerebral cortex)
11. telencephalic vesicle (future lateral ventricle)
12. lamina terminalis
13. midbrain
14. junction between caudal midbrain and rostral hindbrain
15. roof of fourth ventricle
16. fourth ventricle
17. trochlear (IV) nerve
18. oculomotor (III) nerve
19. rostral extension of posterior communicating artery
20. alar plate of myelencephalon (medulla oblongata)
21. choroid plexus differentiating in roof of fourth ventricle
22. rhombic lip (dorsal part of alar lamina of metencephalon – location of the cerebellar primordium)
23. tegmentum of the pons (basal plate)
24. diencephalon (hypothalamus)
25. hypothalamic sulcus
26. diencephalon (thalamus)
27. first evidence of choroid plexus in medial wall of lateral ventricle
28. sulcus limitans
29. rostral extremity of endolymphatic sac
30. striatum
31. anterior portion of superior horn of lateral ventricle
32. posterior horn of lateral ventricle
33. median sulcus
34. primitive transverse dural venous sinus
35. marginal layer of medulla oblongata
36. mantle layer of medulla oblongata
37. ependymal layer of medulla oblongata

PLATE 30b

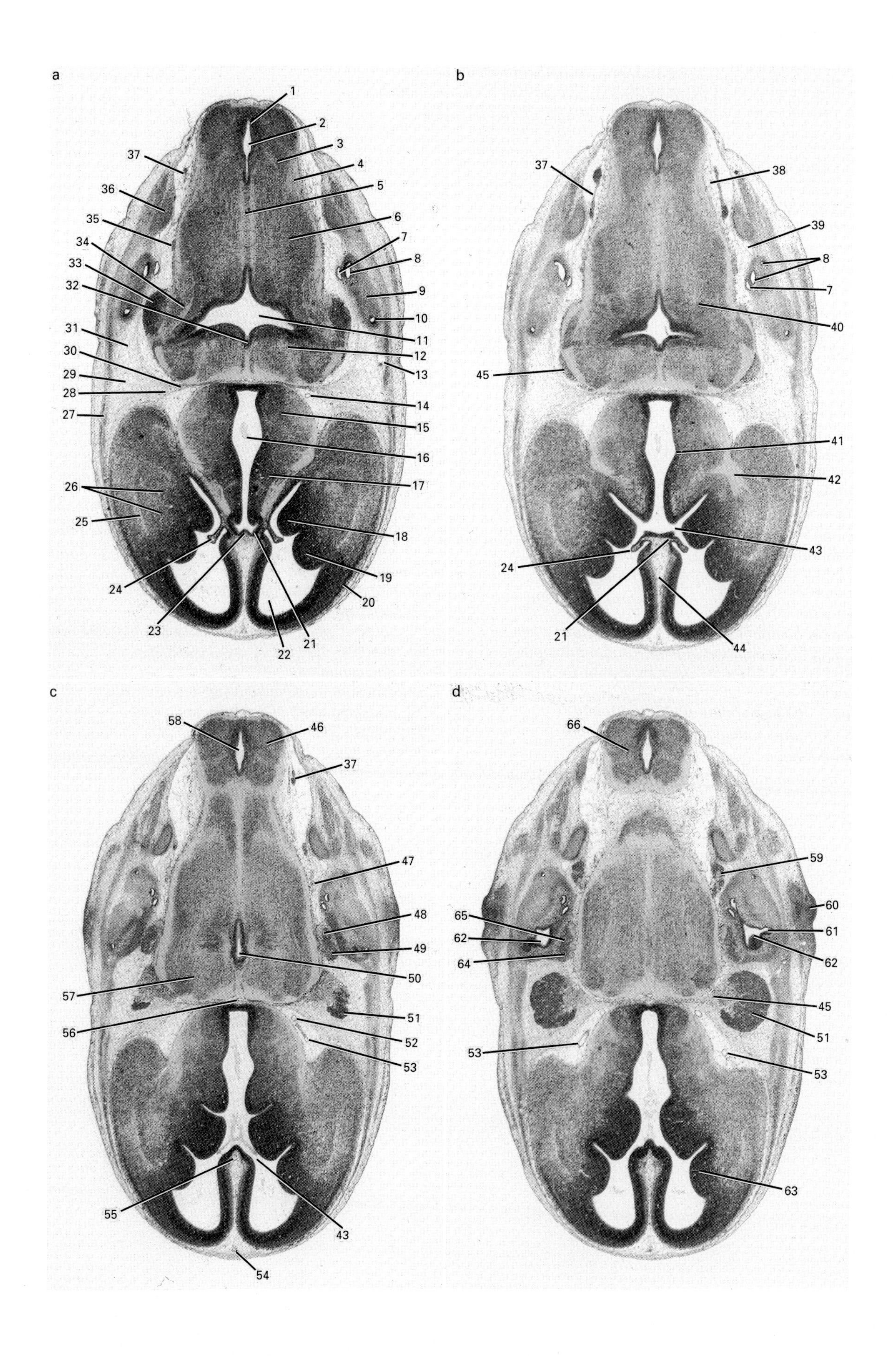
a
b
c
d
1
2
3
4
5
6
7
8
9
10
11
12
13
14
15
16
17
18
19
20
21
22
23
24
25
26
27
28
29
30
31
32
33
34
35
36
37
38
39
40
41
42
43
44
45
46
47
48
49
50
51
52
53
54
55
56
57
58
59
60
61
62
63
64
65
66

Plate 30b (13.5 days p.c.)

13.5 days p.c. Female embryo with about 60 pairs of somites. Crown–rump length about 9.1 mm (fixed). Transverse sections (Theiler Stage 22; rat, Witschi Stage 31; human, Carnegie Stages 20–23).

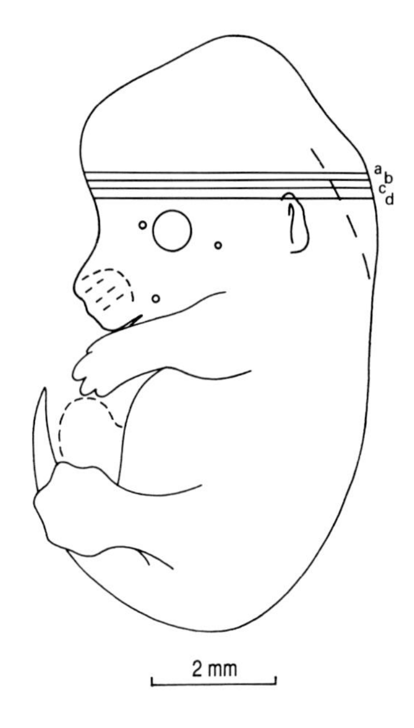

1. ependymal layer
2. caudal extremity of fourth ventricle
3. mantle layer
4. marginal layer
5. medullary raphe (location of decussation of the medial lemniscus)
6. medulla oblongata (myelencephalon)
7. endolymphatic sac
8. posterior semicircular canal
9. cartilage primordium of petrous part of the temporal bone
10. lateral semicircular canal
11. fourth ventricle
12. tegmentum of pons (basal plate)
13. primitive dural venous plexus
14. rostral extension of posterior communicating artery
15. diencephalon (hypothalamus)
16. third ventricle
17. diencephalon (thalamus)
18. corpus striatum mediale (medial aspect of ganglionic eminence)
19. corpus striatum laterale (lateral aspect of ganglionic eminence)
20. roof of neopallial cortex (future cerebral cortex)
21. choroidal fissure
22. anterior portion of superior horn of lateral ventricle
23. lamina terminalis (roof of third ventricle)
24. choroid plexus
25. internal capsule
26. caudate nucleus (caudate – putamen)
27. primitive ectomeninx
28. oculomotor (III) nerve
29. trochlear (IV) nerve
30. mesenchymal condensation forming tentorium cerebelli
31. cephalic mesenchyme in future subarachnoid space
32. median sulcus
33. region of inferior cerebellar peduncle
34. spinal tract of trigeminal (V) nerve
35. vascular elements of pia mater
36. cartilage primordium of exoccipital bone
37. spinal accessory (XI) nerve
38. rootlets of cranial accessory (XI) nerve
39. primitive petrosal dural venous sinus
40. region of rostral medulla oblongata
41. hypothalamic sulcus
42. fibres of internal capsule
43. interventricular foramen
44. mesenchymal condensation forming falx cerebri
45. rootlets of trigeminal (V) nerve
46. junctional zone between caudal part of medulla oblongata and rostral extremity of cervical region of spinal cord
47. rootlets of superior ganglion of vagus (X) nerve
48. rootlets of vestibulocochlear (VIII) nerve
49. vestibulocochlear (VIII) ganglion
50. floor of fourth ventricle
51. trigeminal (V) ganglion
52. internal carotid artery
53. middle cerebral artery
54. primitive superior sagittal dural venous sinus
55. primitive inferior sagittal dural venous sinus
56. basilar artery
57. basal region of the pons
58. rostral extremity of central canal
59. superior ganglion of vagus (X) nerve
60. rostral extremity of pinna of ear
61. origin of lateral semicircular canal from saccule
62. saccule
63. epithalamus/lateral ganglionic eminence
64. vestibular (VIII) ganglion
65. cochlear (VIII) ganglion
66. rostral extremity of cervical region of neural tube

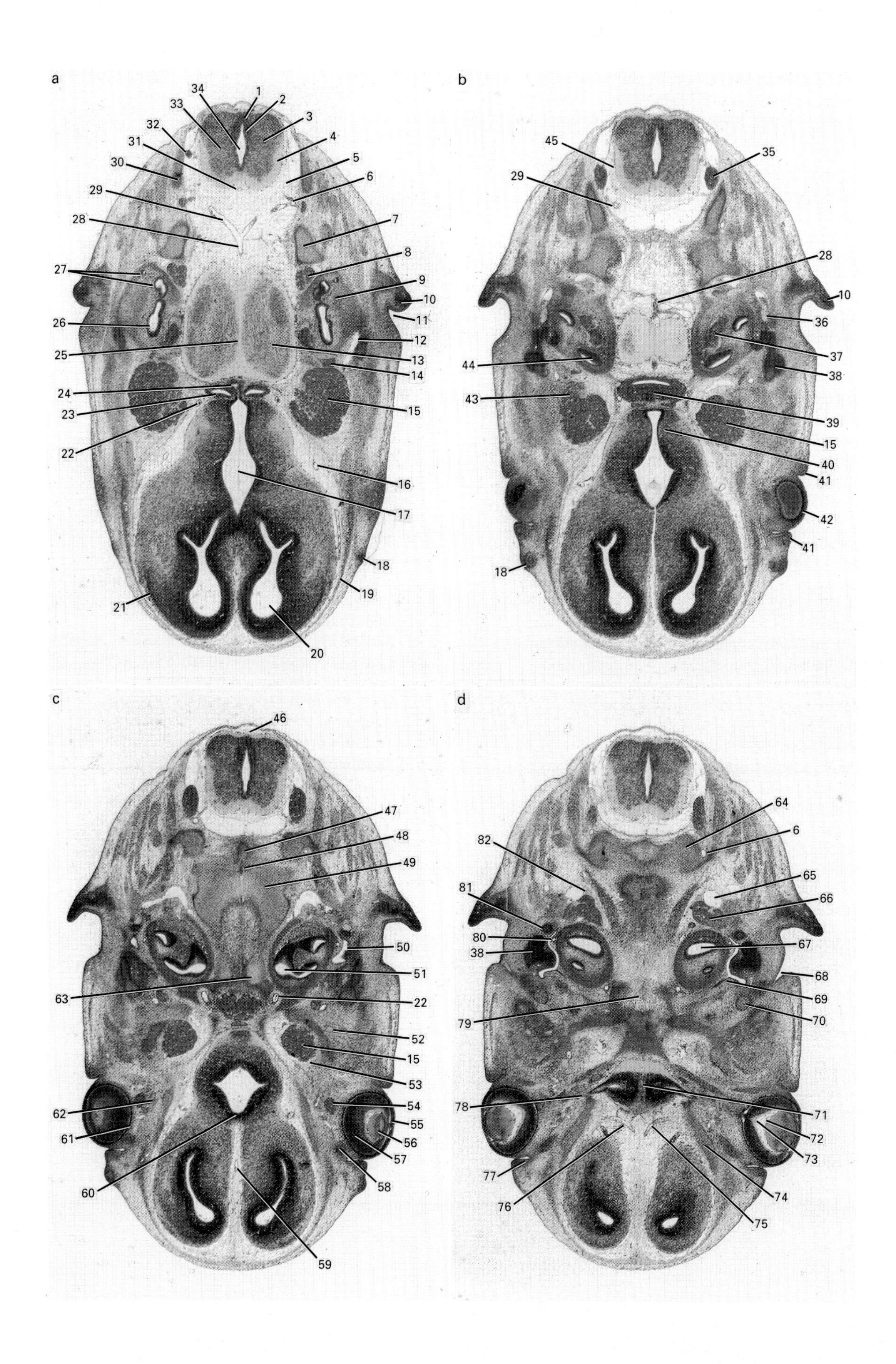
a
34
1
2
3
33
32
4
31
5
30
6
29
7
28
8
27
9
10
11
26
12
25
13
14
24
15
23
22
16
17
18
19
21
20
b
45
35
29
28
10
36
37
44
38
43
39
15
40
41
42
41
18
c
46
47
48
49
50
51
63
22
52
15
53
62
54
55
61
56
57
60
58
59
d
64
6
82
65
81
66
80
67
38
68
69
79
70
78
71
72
73
77
74
76
75

Plate 30c (13.5 days p.c.)

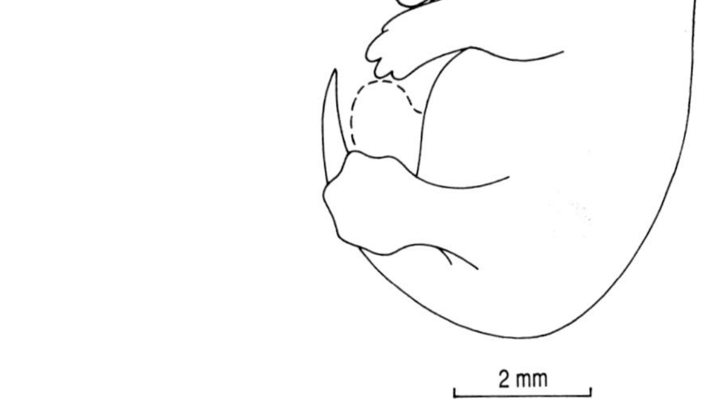

13.5 days p.c. Female embryo with about 60 pairs of somites. Crown–rump length about 9.1 mm (fixed). Transverse sections (Theiler Stage 22; rat, Witschi Stage 31; human, Carnegie Stages 20–23).

1. roof plate
2. ependymal layer
3. mantle layer
4. marginal layer
5. loosely packed mesenchyme tissue within future subarachnoid space
6. left vertebral artery
7. cartilage primordium of exoccipital bone
8. superior ganglion of glossopharyngeal (IX) nerve
9. cartilage primordium of petrous part of temporal bone
10. pinna of ear
11. external auditory meatus
12. branch of primary head vein draining to anterior cardinal vein (internal jugular vein)
13. basal portion of pons
14. facial (VII) ganglion
15. trigeminal (V) ganglion
16. middle cerebral artery
17. third ventricle
18. primordium of prominent tactile or sinus hair follicle characteristically found in this location
19. condensation of mesenchyme forming the outer table of the primitive ectomeninx
20. anterior portion of superior horn of lateral ventricle
21. roof of neopallial cortex (future cerebral cortex)
22. internal carotid artery
23. residual lumen of anterior lobe of pituitary (previously Rathke's pouch)
24. infundibulum (future pars nervosa) of pituitary
25. region of basilar sulcus
26. crus commune of the labyrinth
27. posterior semicircular canal
28. basilar artery
29. right vertebral artery
30. cartilage primordium of ventral part of neural arch of C1 vertebra (atlas)
31. anterior spinal artery
32. spinal accessory (XI) nerve
33. rostral extremity of cervical region of spinal cord
34. central canal
35. dorsal (posterior) root ganglion of C1 (unusually large – often considerably smaller)
36. chorda tympani branch of the facial (VII) nerve passing through the region of the future middle ear
37. vestibulocochlear (VIII) ganglion
38. primordia of two of the middle ear ossicles (malleus and incus) in close proximity to Meckel's cartilage
39. pars anterior of pituitary gland
40. hypothalamus
41. boundary of eyelid
42. eye
43. mandibular division of trigeminal (V) nerve
44. differentiating neuroepithelium in anterior part of wall of cochlear canal
45. vascular elements of pia mater
46. posterior spinal vein
47. cartilage primordium of dens
48. notochord
49. cartilage primordium of basioccipital bone
50. tubo-tympanic recess
51. inferomedial part of cochlear canal
52. maxillary division of trigeminal (V) nerve
53. ophthalmic division of trigeminal (V) nerve
54. extrinsic ocular muscle (superior/dorsal rectus)
55. corneal epithelium
56. lens
57. neural layer of retina
58. extrinsic ocular muscle (medial/nasal rectus)
59. median anterior cerebral artery (formed by amalgamation of right and left anterior cerebral arteries)
60. lamina terminalis
61. pigment layer of retina
62. oculomotor (III) nerve
63. cartilage primordium of basisphenoid bone
64. lateral mass/neural arch of C1 (atlas)
65. internal jugular vein
66. inferior ganglion of glossopharyngeal (IX) nerve
67. posterior part of cochlear canal
68. entrance to first branchial cleft
69. communication between pharyngo-tympanic tube (Eustachian tube) and tubo-tympanic recess
70. Meckel's cartilage
71. optic chiasma
72. hyaloid cavity
73. optic disc
74. nasal capsule
75. left anterior cerebral artery
76. right anterior cerebral artery
77. site of future conjunctival sac
78. optic (II) nerve passing through optic foramen
79. hypophyseal fossa (sella turcica)
80. stapedial artery
81. stapes
82. vagus (X) and spinal accessory (XI) nerves

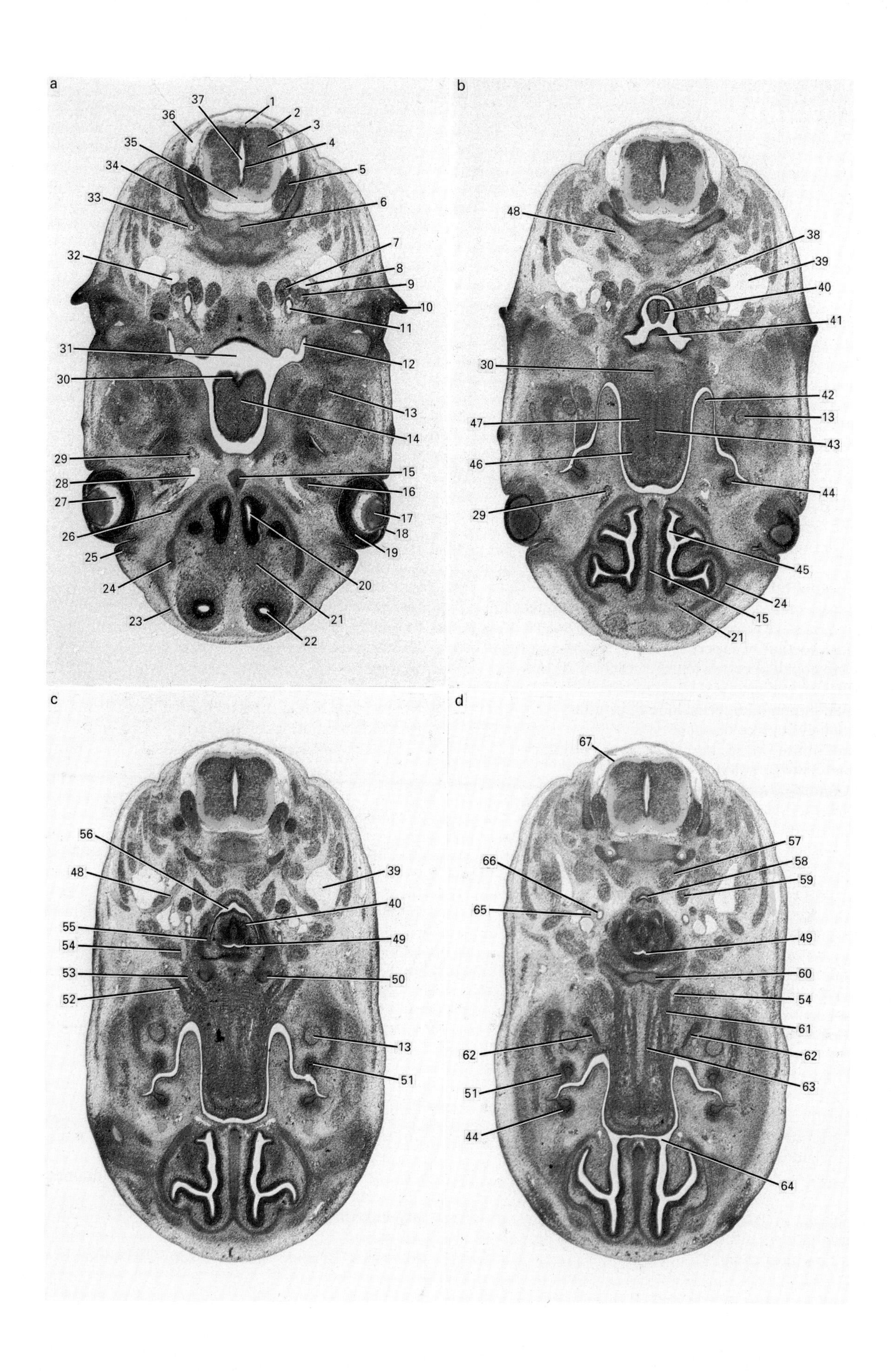
a
b
c
d
1
2
3
4
5
6
7
8
9
10
11
12
13
14
15
16
17
18
19
20
21
22
23
24
25
26
27
28
29
30
31
32
33
34
35
36
37
38
39
40
41
42
43
44
45
46
47
48
49
50
51
52
53
54
55
56
57
58
59
60
61
62
63
64
65
66
67

Plate 30d (13.5 days p.c.)

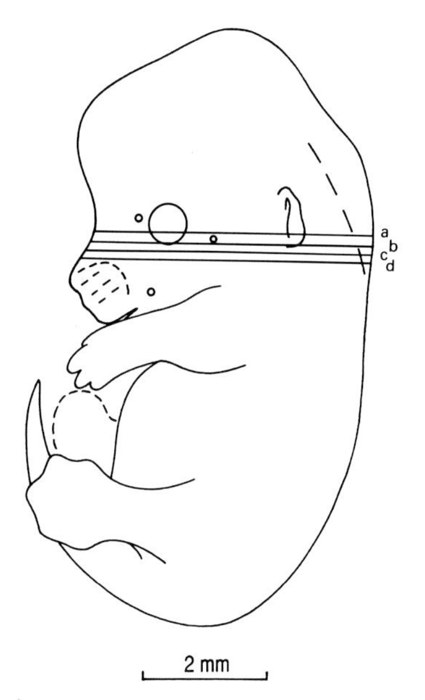

13.5 days p.c. Female embryo with about 60 pairs of somites. Crown–rump length about 9.1 mm (fixed). Transverse sections (Theiler Stage 22; rat, Witschi Stage 31; human, Carnegie Stages 20–23).

1. posterior spinal vein
2. marginal layer
3. mantle layer
4. ependymal layer
5. dorsal (posterior) root ganglion of C2
6. notochord
7. superior cervical sympathetic ganglion
8. left internal jugular vein
9. inferior ganglion of vagus (X) nerve
10. pinna of ear
11. carotid sinus
12. opening of pharyngo-tympanic (Eustachian) tube into pharynx
13. Meckel's cartilage
14. dorsum of tongue
15. cartilage primordium of nasal septum
16. optic (II) nerve (the lumen is obliterated by nerve fibres passing from the ganglion cells of the neural retina towards the optic chiasma)
17. lens
18. corneal epithelium
19. neural layer of retina
20. rostral extremity of primitive nasal cavity
21. olfactory (I) nerves
22. lateral ventricle (future olfactory lobe)
23. condensation of mesenchyme tissue forming the outer table of the primitive ectomeninx
24. nasal capsule
25. boundary of eyelid
26. extrinsic ocular muscle (medial/nasal rectus)
27. hyaloid cavity
28. ophthalmic vein
29. ciliary ganglion
30. circumvallate papilla
31. cavity of oropharynx
32. right internal jugular vein
33. right vertebral artery
34. cartilage primordium of neural arch of C2 vertebra (axis)
35. anterior spinal artery
36. subarachnoid space
37. central canal
38. pharyngeal constrictor muscle
39. jugular lymph sac
40. arytenoid swelling
41. epiglottis
42. vertically directed palatal shelf/process of maxilla
43. precursor of median fibrous septum of tongue
44. primordium of first upper molar tooth
45. primitive nasal cavity
46. intrinsic muscle of the tongue (longitudinal component)
47. intrinsic muscle of the tongue (transverse component)
48. ventral primary ramus of cervical nerve
49. laryngeal aditus
50. cartilage primordium of lesser horn of hyoid bone
51. primordium of lower molar tooth
52. extrinsic muscle of tongue (styloglossus)
53. extrinsic muscle of tongue (palatoglossus)
54. hypoglossal (XII) nerve
55. primordium of thyroid cartilage
56. entrance to the oesophagus
57. prevertebral muscles of the neck (longus capitis, longus cervicis)
58. oesophagus
59. cervical sympathetic ganglion
60. cartilage primordium of body of hyoid bone
61. extrinsic muscle of the tongue (hyoglossus)
62. submandibular duct
63. extrinsic muscle of the tongue (genioglossus)
64. primitive nasopharynx
65. vagal (X) nerve trunk
66. distal part of common carotid artery
67. vascular elements of pia mater

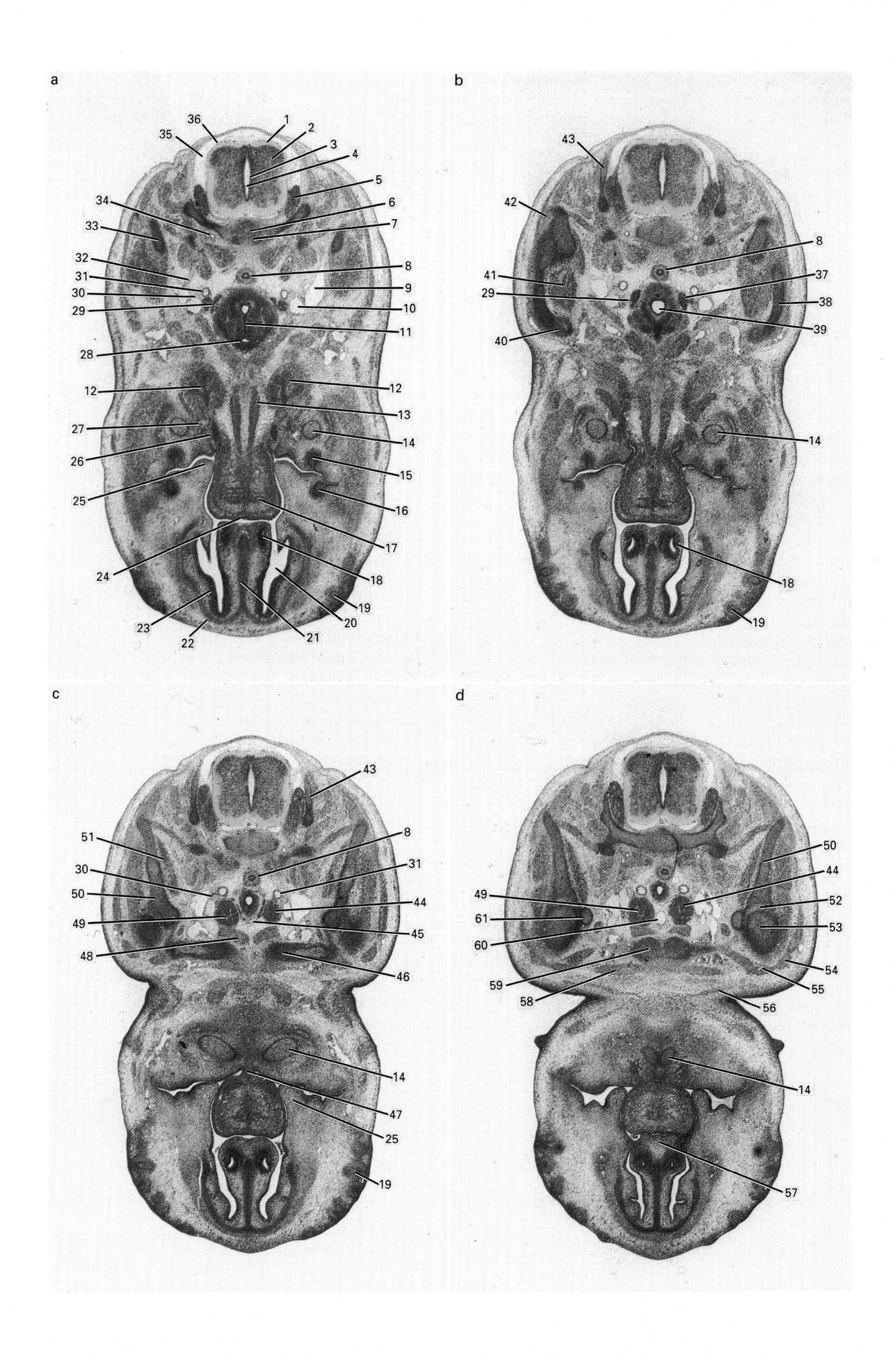
a
b
c
d
1
2
3
4
5
6
7
8
9
10
11
12
13
14
15
16
17
18
19
20
21
22
23
24
25
26
27
28
29
30
31
32
33
34
35
36
37
38
39
40
41
42
43
44
45
46
47
48
49
50
51
52
53
54
55
56
57
58
59
60
61

Plate 30e (13.5 days p.c.)

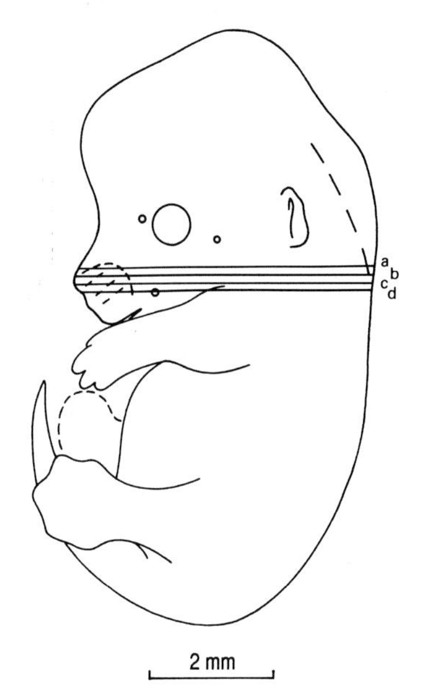

13.5 days p.c. Female embryo with about 60 pairs of somites. Crown–rump length about 9.1 mm (fixed). Transverse sections (Theiler Stage 22; rat, Witschi Stage 31; human, Carnegie Stages 20–23).

1. marginal layer
2. mantle layer
3. ependymal layer
4. central canal
5. dorsal (posterior) root ganglion
6. notochord
7. cartilage primordium of vertebral body (centrum)
8. oesophagus
9. jugular lymph sac
10. left internal jugular vein
11. arytenoid swelling
12. submandibular gland
13. extrinsic muscle of tongue (genioglossus)
14. Meckel's cartilage
15. primordium of lower molar tooth
16. primordium of first upper molar tooth
17. intrinsic muscle of tongue (transverse component)
18. vomeronasal organ (Jacobson's organ)
19. primordium of follicle of vibrissa
20. primitive nasopharynx
21. cartilage primordium of nasal septum
22. nasal capsule
23. epithelial lining of primitive nasal cavity
24. dorsum of tongue
25. vertically directed palatal shelf/process of maxilla
26. submandibular duct
27. hypoglossal (XII) nerve
28. primitive glottis
29. right thyroid lobe
30. right vagal (X) trunk
31. common carotid artery
32. ventral primary ramus of cervical nerve
33. cartilage primordium of upper angle of scapula
34. vertebral artery
35. subarachnoid space
36. vascular elements of pia mater
37. left thyroid lobe
38. spine of scapula
39. trachea
40. acromion of scapula
41. supraspinatus muscle
42. trapezius muscle
43. cartilage primordium of neural arch
44. left lobe of thymic rudiment
45. isthmus of thyroid gland
46. cartilage primordium of clavicle
47. frenulum of the tongue
48. sternomastoid muscle
49. right lobe of thymic rudiment
50. blade of scapula
51. subscapularis muscle
52. precursor of gleno-humeral joint
53. head of the humerus
54. deltoid muscle
55. axillary (circumflex) nerve supplying the deltoid muscle
56. platysma muscle
57. site of fusion between nasal septum and primary palate
58. clavicular head of pectoralis major muscle
59. medial end of clavicle
60. rostral extremity of pericardial cavity
61. cartilage primordium of coracoid process of scapula

PLATE 30f

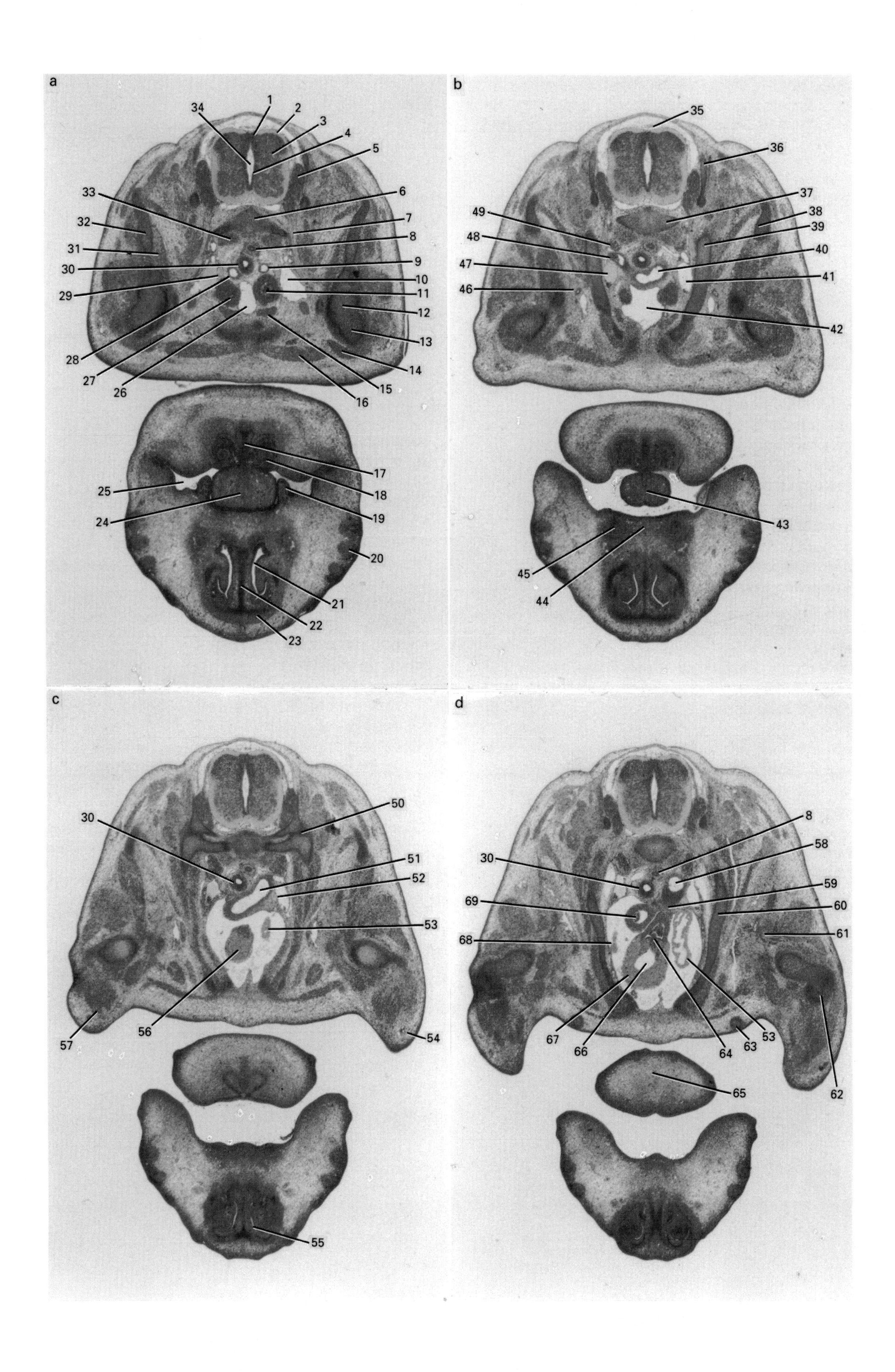

a
b
c
d
1
2
3
4
5
6
7
8
9
10
11
12
13
14
15
16
17
18
19
20
21
22
23
24
25
26
27
28
29
30
31
32
33
34
35
36
37
38
39
40
41
42
43
44
45
46
47
48
49
50
51
52
53
54
55
56
57
58
59
60
61
62
63
64
65
66
67
68
69

Plate 30f (13.5 days p.c.)

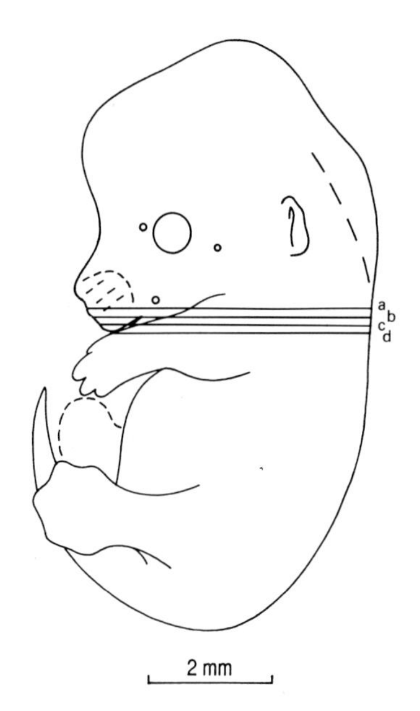

13.5 days p.c. Female embryo with about 60 pairs of somites. Crown–rump length about 9.1 mm (fixed). Transverse sections (Theiler Stage 22; rat, Witschi Stage 31; human, Carnegie Stages 20–23).

1. roof plate
2. marginal layer
3. mantle layer
4. ependymal layer
5. dorsal (posterior) root ganglion
6. notochord
7. ventral primary ramus of C8
8. oesophagus
9. proximal part of left common carotid artery
10. origin of left superior (cranial) vena cava from left internal jugular vein and left subclavian vein
11. left lobe of thymic rudiment
12. head of left humerus
13. upper region of shaft of humerus
14. deltoid muscle
15. sternomastoid muscle
16. clavicular head of pectoralis major muscle
17. site of apposition in the midline of Meckel's cartilages
18. primordium of lower incisor tooth
19. vertically directed palatal shelf/process of maxilla
20. primordium of follicle of vibrissa
21. primitive nasal cavity
22. cartilage primordium of nasal septum
23. nasal capsule
24. tongue
25. oral cavity
26. rostral extremity of pericardial cavity
27. right lobe of thymic rudiment
28. proximal part of right common carotid artery
29. right vagal (X) trunk
30. trachea
31. subscapularis muscle
32. blade of scapula
33. right inferior cervical sympathetic ganglion
34. central canal
35. posterior spinal vein
36. cartilage primordium of neural arch
37. cartilage primordium of T1 vertebral body (centrum)
38. medial (posterior) border of scapula
39. cartilage primordium of first rib
40. rostral extremity of arch of aorta and (in the midline) the origin of the innominate artery
41. left superior (cranial) vena cava
42. pericardial cavity
43. tip of tongue
44. primary palate
45. primordium of upper incisor tooth
46. components of brachial plexus passing into the forelimb
47. right superior (cranial) vena cava
48. right subclavian artery
49. right sympathetic trunk
50. dorsal articulation of first rib
51. arch of aorta
52. left vagal (X) trunk
53. wall of left atrium
54. left forelimb
55. entrance into primitive nasal cavity (external naris)
56. wall of distal part of pulmonary trunk
57. right forelimb
58. arch of aorta at site of entrance of ductus arteriosus
59. rostral part of wall of ductus arteriosus
60. cartilage primordium of second rib
61. radial nerve
62. primordium of elbow joint
63. mammary gland primordium
64. leaflet of pulmonary valve
65. tip of lower jaw
66. lumen of outlet of right ventricle
67. wall of right ventricle
68. wall of right atrium
69. proximal part of arch of aorta

PLATE 30g

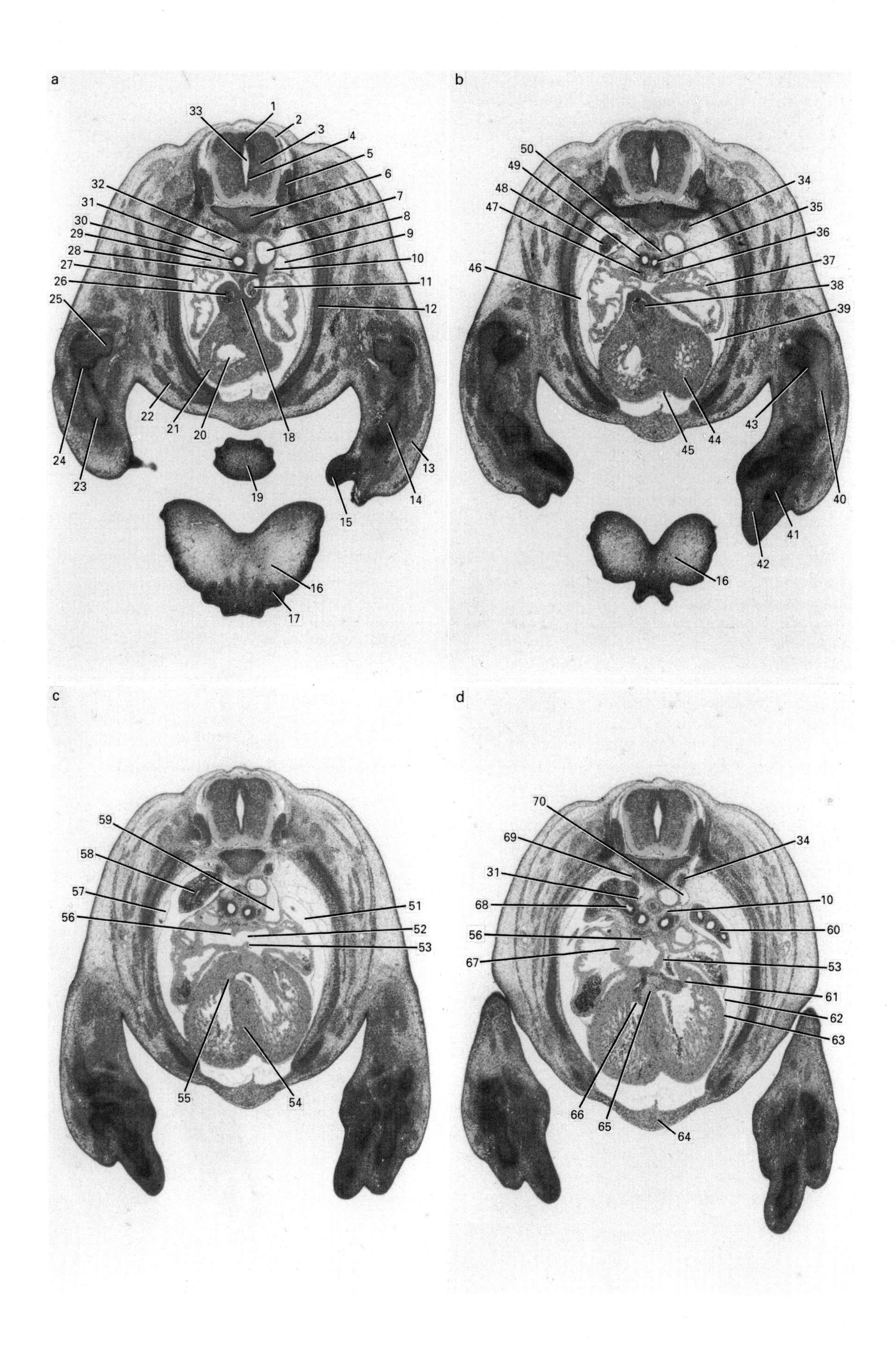

Plate 30g (13.5 days p.c.)

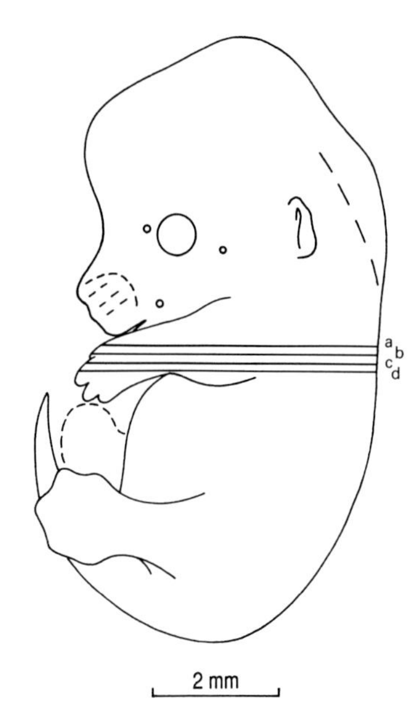

13.5 days p.c. Female embryo with about 60 pairs of somites. Crown–rump length about 9.1 mm (fixed). Transverse sections (Theiler Stage 22; rat, Witschi Stage 31; human, Carnegie Stages 20–23).

1. roof plate
2. marginal layer
3. mantle layer
4. ependymal layer
5. dorsal (posterior) root ganglion
6. notochord
7. thoracic sympathetic ganglion
8. arch of aorta just distal to site of entrance of ductus arteriosus
9. left superior (cranial) vena cava
10. left vagal (X) trunk
11. pulmonary trunk
12. cartilage primordium of third rib
13. left forelimb
14. cartilage primordium of left radius
15. first digit (lateral)
16. ventral extremity of upper jaw
17. primordium of follicle of vibrissa
18. aortico-pulmonary spiral septum
19. ventral extremity of lower jaw
20. cavity of right ventricle
21. wall of right ventricle
22. pectoralis major muscle (p. superficialis)
23. cartilage primordium of right radius
24. primordium of joint between head of the radius and capitulum of the humerus
25. cartilage primordium of distal part of shaft of humerus
26. origin of aortic trunk
27. wall of right atrium
28. caudal part of wall of ductus arteriosus (the recurrent laryngeal branch of the left vagus (X) nerve hooks around the ductus arteriosus just subjacent to its wall in this location)
29. right superior (cranial) vena cava
30. trachea
31. right vagal (X) trunk
32. oesophagus
33. central canal
34. left sympathetic trunk
35. left main bronchus
36. left pulmonary artery
37. wall of left atrium
38. leaflet of aortic valve
39. pericardial cavity
40. cartilage primordium of left ulna
41. precartilage primordium of metacarpal bone
42. "hand plate"
43. primordium of joint between proximal part of ulna and coronoid process of humerus
44. wall of left ventricle
45. interventricular groove
46. pleural component of right pericardio-peritoneal canal
47. right pulmonary artery
48. rostral extremity of apex of right lung bud
49. right main bronchus
50. left recurrent laryngeal nerve
51. pleural component of left pericardio-peritoneal canal
52. foramen secundum
53. septum primum
54. muscular component of interventricular septum
55. site of fusion of the bulbar ridges with the muscular part of the interventricular septum
56. septum secundum
57. mesothelial lining of pleuro-peritoneal canal (future parietal pleura)
58. tertiary bronchi within cranial lobe of right lung
59. site of communication between the accessory (superior) hemiazygos vein, the left superior intercostal vein and the left superior (cranial) vena cava
60. left lung
61. leaflets of left atrio-ventricular (mitral) valve
62. parietal pericardium (pleuro-pericardial membrane)
63. visceral pericardium
64. ventral part of thoracic body wall (no evidence of sternal precartilage condensation in this location)
65. atrio-ventricular bulbar cushion tissue
66. leaflets of right atrio-ventricular (tricuspid) valve
67. leaflet of venous valve at site of entrance of right superior (cranial) vena cava into right atrium
68. lobar bronchus to upper lobe of right lung
69. right sympathetic trunk
70. accessory (superior) hemiazygos vein (previously the left posterior cardinal vein)

PLATE 30h

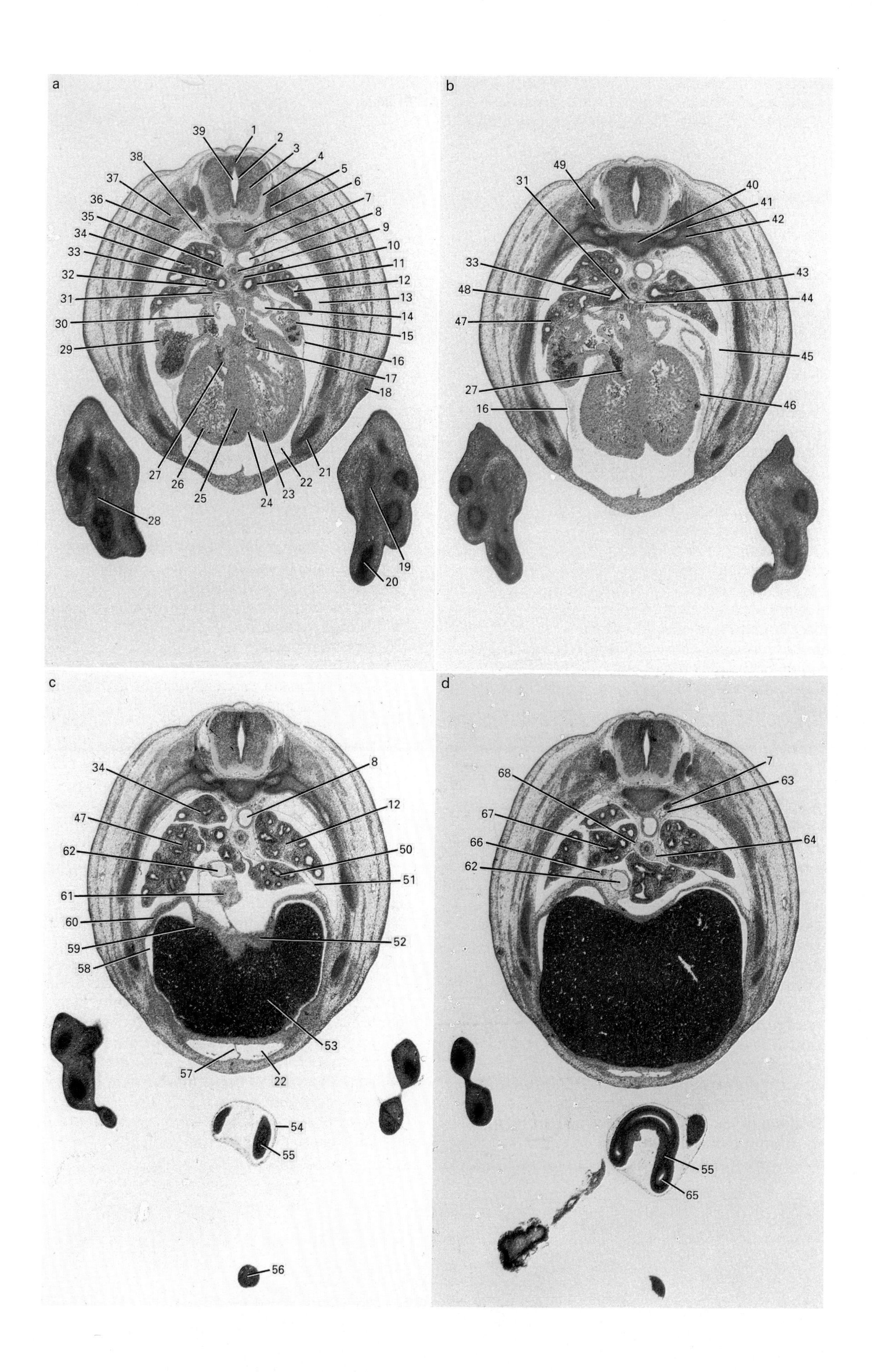

Plate 30h (13.5 days p.c.)

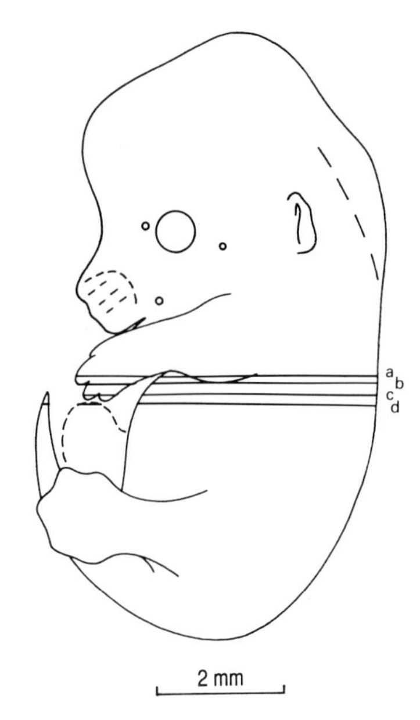

13.5 days p.c. Female embryo with about 60 pairs of somites. Crown–rump length about 9.1 mm (fixed). Transverse sections (Theiler Stage 22; rat, Witschi Stage 31; human, Carnegie Stages 20–23).

1. roof plate
2. ependymal layer
3. mantle layer
4. marginal layer
5. dorsal (posterior) root ganglion
6. thoracic vertebral body (centrum)
7. left sympathetic trunk
8. descending (thoracic) aorta
9. oesophagus
10. left vagal (X) trunk
11. left main bronchus
12. left lung
13. mesothelial lining of pleuro-peritoneal canal (future parietal pleura)
14. left superior (cranial) vena cava
15. wall of left atrium
16. parietal pericardium (pleuro-pericardial membrane)
17. leaflet of left atrio-ventricular (mitral) valve
18. mammary gland primordium
19. left "hand plate"
20. precartilage primordium of phalangeal bone
21. cartilage primordium of rib
22. pericardial cavity
23. wall of left ventricle
24. interventricular groove
25. muscular component of interventricular septum
26. wall of right ventricle
27. leaflet of right atrio-ventricular (tricuspid) valve
28. right "hand plate"
29. wall of right atrium
30. leaflet of venous valve at site of entrance of inferior vena cava into right atrium
31. right bronchial artery
32. right pulmonary artery
33. right main bronchus
34. cranial lobe of right lung
35. right vagal (X) trunk
36. anterior primary division of thoracic spinal nerve
37. posterior primary division of thoracic spinal nerve
38. white ramus communicans from spinal nerve to sympathetic ganglion
39. central canal
40. notochord
41. primordium of joint at location of rib articulation with its own vertebra
42. neck of rib
43. principal bronchus to upper part of left lung
44. left bronchial artery
45. pleural component of left pleuro-peritoneal canal
46. visceral pericardium
47. middle lobe of right lung
48. pleural component of right pleuro-peritoneal canal
49. cartilage primordium of neural arch
50. accessory lobe of right lung
51. fibrous intersection between left lung and accessory lobe of right lung
52. septum transversum
53. liver
54. wall of physiological umbilical hernia
55. wall of midgut loop within physiological umbilical hernia
56. distal extremity of tail
57. remnant of ventral mesentery
58. peritoneal cavity (right hepatic recess)
59. boundary of bare area of liver
60. pleuro-peritoneal membrane (right dome of the diaphragm)
61. caudal extremity of wall of atrium
62. posthepatic inferior vena cava
63. hemiazygos vein
64. left vagal (X) trunk becoming anterior vagal (X) trunk
65. lumen of midgut
66. right phrenic nerve
67. caudal lobe of right lung
68. right vagal (X) trunk becoming posterior vagal (X) trunk

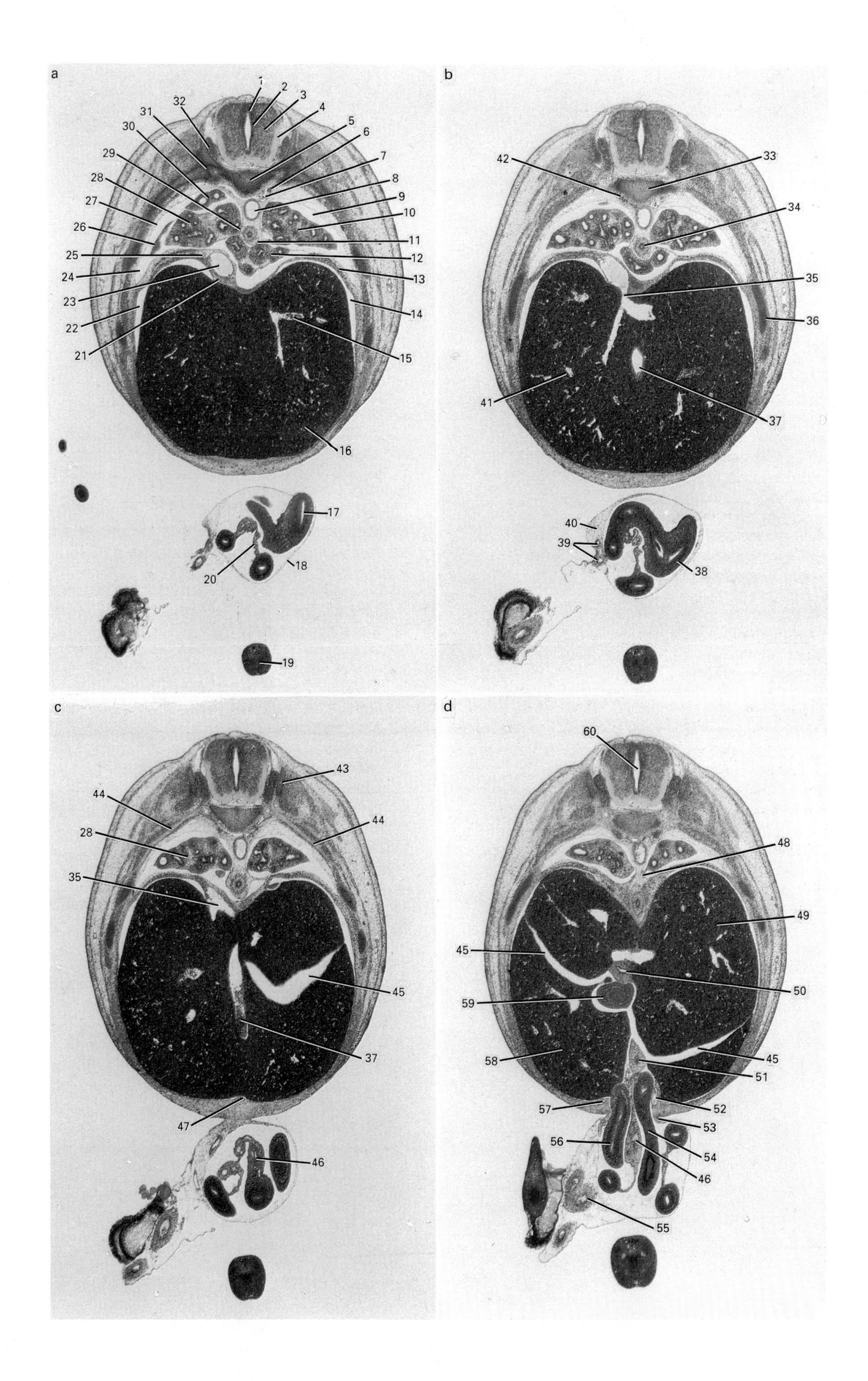
a
b
c
d
1
2
3
4
5
6
7
8
9
10
11
12
13
14
15
16
17
18
19
20
21
22
23
24
25
26
27
28
29
30
31
32
33
34
35
36
37
38
39
40
41
42
43
44
45
46
47
48
49
50
51
52
53
54
55
56
57
58
59
60

Plate 30i (13.5 days p.c.)

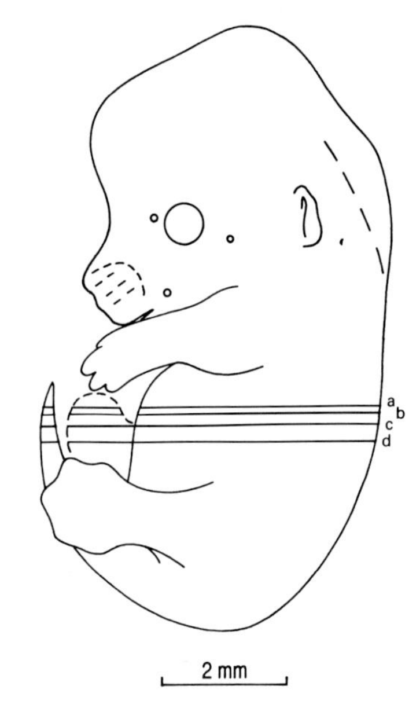

13.5 days p.c. Female embryo with about 60 pairs of somites. Crown–rump length about 9.1 mm (fixed). Transverse sections (Theiler Stage 22; rat, Witschi Stage 31; human, Carnegie Stages 20–23).

1. roof plate
2. ependymal layer
3. mantle layer
4. marginal layer
5. notochord
6. left sympathetic trunk
7. hemiazygos vein
8. descending (thoracic) aorta
9. pleural component of left pleuro-peritoneal canal
10. left lung
11. anterior (previously left) vagal (X) trunk
12. accessory lobe of right lung
13. left pleuro-peritoneal membrane (left dome of the diaphragm)
14. peritoneal cavity (left hepatic recess)
15. part of hepatic venous plexus
16. liver
17. lumen of midgut
18. wall of physiological umbilical hernia
19. distal extremity of tail
20. dorsal mesentery of midgut
21. boundary of bare area of liver
22. peritoneal cavity (right hepatic recess)
23. posthepatic inferior vena cava
24. pleural component of right pleuro-peritoneal canal
25. right phrenic nerve
26. caudal margin of middle lobe of right lung
27. mesothelial lining of right pleuro-peritoneal canal (future parietal pleura)
28. caudal lobe of right lung
29. posterior (previously right) vagal (X) trunk
30. cranial lobe of right lung
31. primordium of joint at location of rib articulation with its own vertebra
32. dorsal (posterior) root ganglion
33. cartilage primordium of vertebral body (centrum)
34. oesophagus
35. site of amalgamation of the ductus venosus and vitelline veins to form the posthepatic inferior vena cava
36. cartilage primordium of rib
37. ductus venosus
38. loop of midgut within physiological umbilical hernia
39. umbilical vessels
40. Wharton's jelly
41. hepatic sinusoid
42. right sympathetic trunk
43. cartilage primordium of neural arch
44. intercostal nerve
45. interlobar space
46. vitelline vessels within dorsal mesentery of midgut
47. ventral hepatic mesentery (falciform ligament)
48. dorsal meso-oesophagus
49. left lobe of liver
50. junction between right hepatic duct and common bile duct
51. mucous membrane lining the gall bladder
52. left umbilical artery
53. boundary of umbilical ring
54. distal part of midgut loop
55. umbilical artery embedded in Wharton's jelly within umbilical cord
56. proximal part of midgut loop
57. right umbilical artery
58. right lobe of liver
59. wall of duodenum
60. central canal

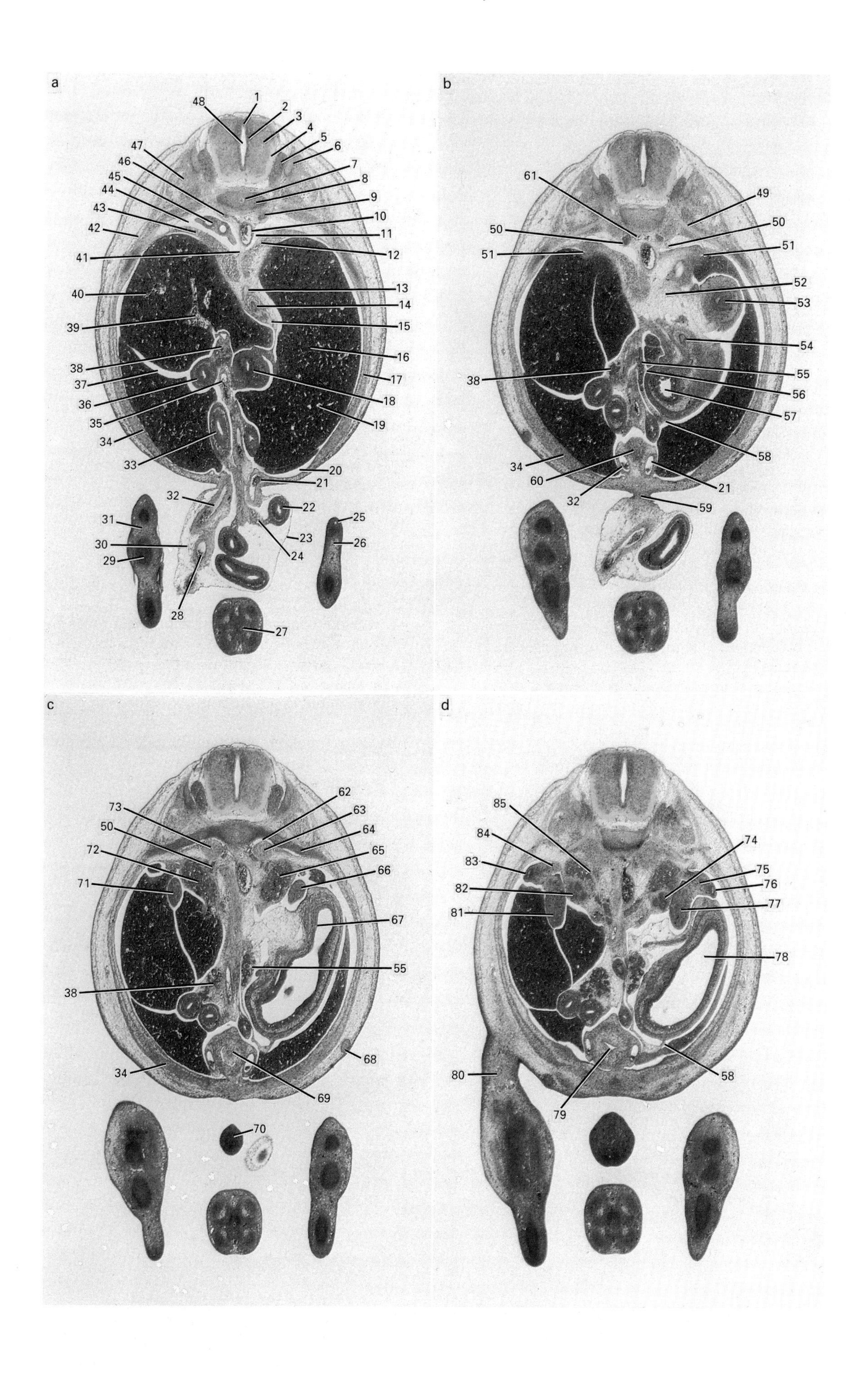
a
1
2
3
4
5
6
7
8
9
10
11
12
13
14
15
16
17
18
19
20
21
22
23
24
25
26
27
28
29
30
31
32
33
34
35
36
37
38
39
40
41
42
43
44
45
46
47
48
b
49
50
51
52
53
54
55
56
57
58
59
60
61
c
62
63
64
65
66
67
68
69
70
71
72
73
d
74
75
76
77
78
79
80
81
82
83
84
85

Plate 30j (13.5 days p.c.)

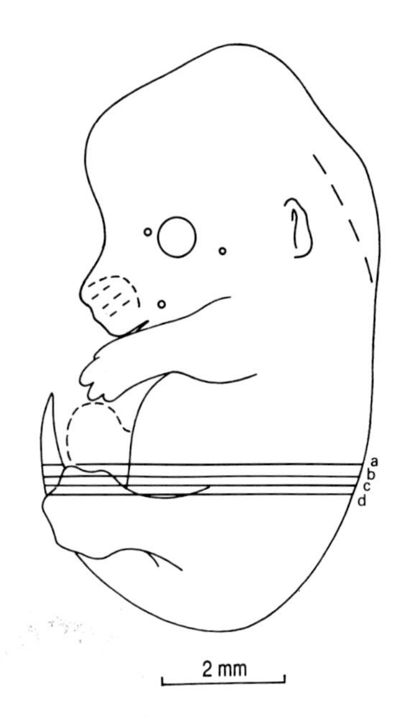

13.5 days p.c. Female embryo with about 60 pairs of somites. Crown–rump length about 9.1 mm (fixed). Transverse sections (Theiler Stage 22; rat, Witschi Stage 31; human, Carnegie Stages 20–23).

1. roof plate
2. ependymal layer
3. mantle layer
4. marginal layer
5. dorsal (posterior) root ganglion
6. cartilage primordium of neural arch
7. notochord
8. cartilage primordium of T11 vertebral body (centrum)
9. left sympathetic trunk
10. descending (thoracic) aorta
11. pleural component of left pleuro-peritoneal canal
12. pleuro-peritoneal membrane (future muscular component of diaphragm)
13. posterior vagal (X) trunk
14. oesophagus
15. anterior vagal (X) trunk
16. left lobe of liver
17. three layers of abdominal body wall musculature (external oblique, internal oblique and transversus abdominis muscles)
18. pyloric region of stomach
19. hepatic sinusoid
20. peritoneal cavity
21. left umbilical artery
22. midgut loop within physiological umbilical hernia
23. wall of physiological umbilical hernia
24. vitelline vessels within dorsal mesentery of midgut
25. marginal vein
26. "foot plate" region of left hindlimb
27. middle region of tail
28. umbilical artery in proximal part of umbilical cord
29. pre-cartilage primordium of phalangeal bone
30. Wharton's jelly
31. "foot plate" region of right hindlimb
32. right umbilical artery
33. proximal part of midgut (intra-abdominal component)
34. rectus abdominis muscle
35. portal vein (formed principally from right vitelline vein)
36. duodenum
37. interlobar space
38. primordium of right lobe and body of the pancreas
39. part of hepatic venous plexus
40. right lobe of liver
41. dorsal meso-oesophagus
42. right intercostal nerve
43. right pleuro-peritoneal membrane (boundary between caudal part of right dome of the diaphragm and right crus)
44. pleural component of right pleuro-peritoneal canal
45. caudal lobe of right lung
46. right sympathetic trunk
47. erector spinae muscle
48. central canal
49. anterior primary division of segmental nerve
50. branch from sympathetic trunk to adrenal medulla
51. rostral reflection of peritoneum (boundary of bare area of liver)
52. region of central tendon of the diaphragm
53. wall of fundus region of the stomach
54. gastro-oesophageal junction
55. primordium of left lobe of the pancreas
56. lesser sac (superior recess of omental bursa)
57. lumen of prepyloric region of the stomach
58. ventral mesentery of the stomach (lesser omentum)
59. caudal component of umbilical ring
60. urachus
61. azygos vein
62. lumbar (segmental) artery
63. left psoas major muscle
64. cartilage primordium T12 rib
65. left adrenal primordium
66. upper pole of left gonad (ovary)
67. mucosal lining of stomach
68. mammary gland primordium
69. wall of urogenital sinus
70. genital tubercle
71. upper pole of right gonad (ovary)
72. right adrenal primordium
73. right psoas major muscle
74. upper pole of left metanephros (definitive kidney)
75. mesonephric tubules (mesonephros almost completely regressed at this stage)
76. left paramesonephric duct (Müllerian duct)
77. left gonad (ovary)
78. lumen of stomach
79. urogenital sinus
80. right hindlimb
81. right gonad (ovary)
82. upper pole of right metanephros (definitive kidney)
83. right paramesonephric duct (Müllerian duct)
84. urogenital mesentery
85. hypogastric nerves

PLATE 30k

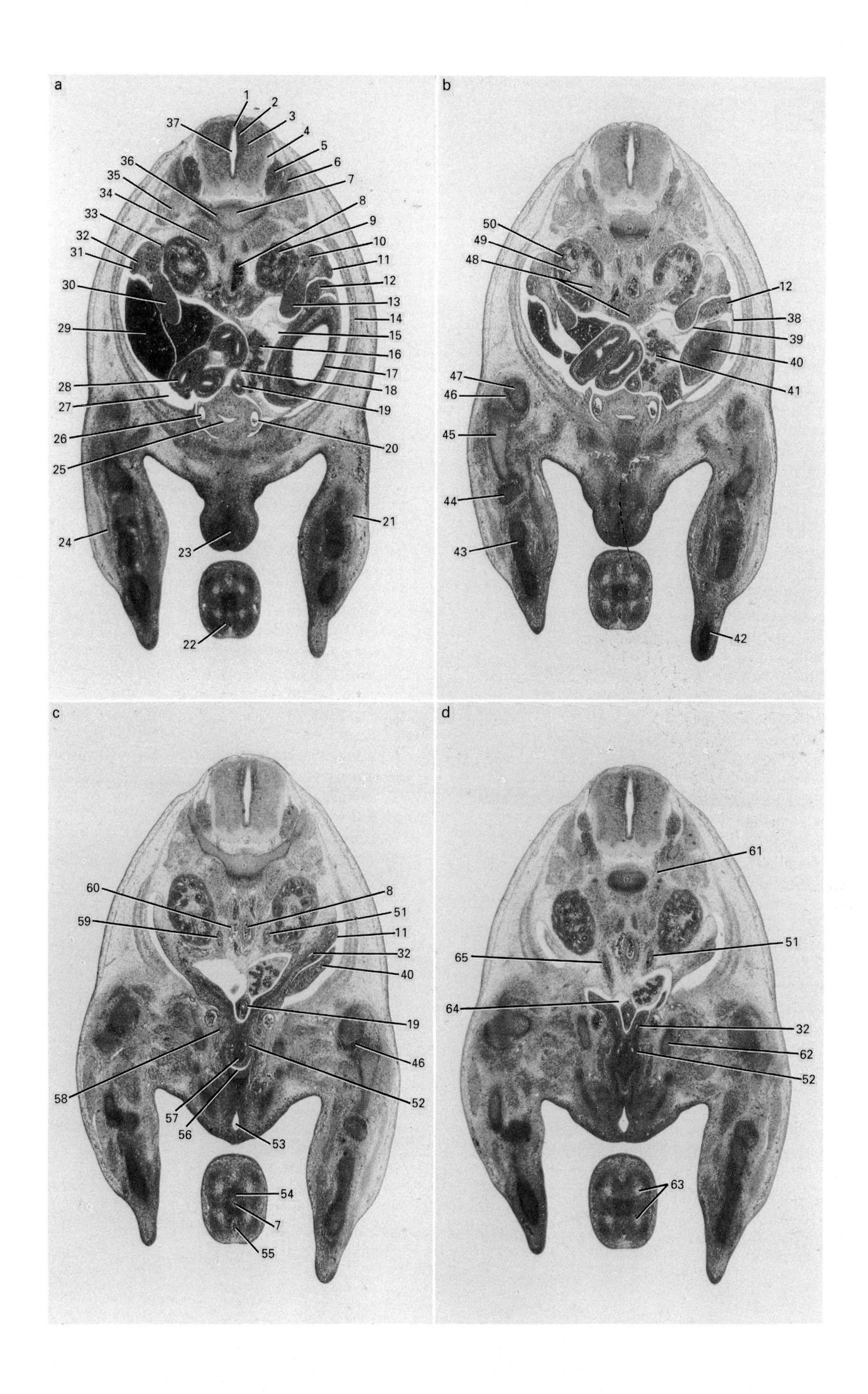

a
1
2
3
4
5
6
7
8
9
10
11
12
13
14
15
16
17
18
19
20
21
22
23
24
25
26
27
28
29
30
31
32
33
34
35
36
37
b
50
49
48
12
38
39
40
41
47
46
45
44
43
42
c
60
59
8
51
11
32
40
19
46
58
52
57
56
53
54
7
55
d
61
51
65
64
32
62
52
63

Plate 30k (13.5 days p.c.)

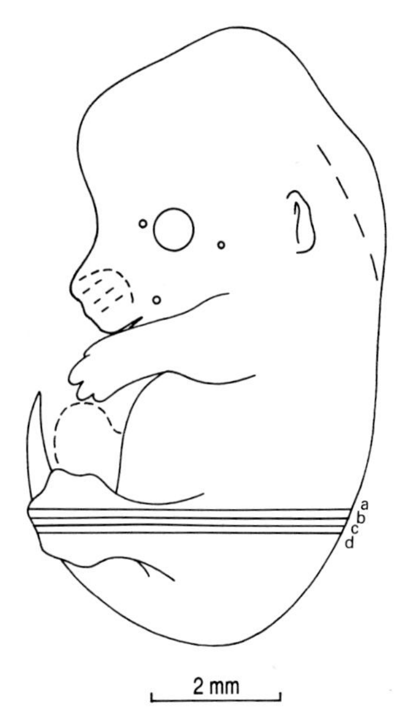

13.5 days p.c. Female embryo with about 60 pairs of somites. Crown–rump length about 9.1 mm (fixed). Transverse sections (Theiler Stage 22; rat, Witschi Stage 31; human, Carnegie Stages 20–23).

1. roof plate
2. ependymal layer
3. mantle layer
4. marginal layer
5. dorsal (posterior) root ganglion
6. cartilage primordium of neural arch
7. notochord
8. descending (abdominal) aorta
9. left metanephros
10. mesonephric tubules (mesonephros almost completely regressed at this stage)
11. left paramesonephric duct (Müllerian duct)
12. splenic primordium developing within dorsal mesogastrium
13. left gonad (ovary)
14. three layers of abdominal wall musculature (external oblique, internal oblique and transversus abdominis)
15. superior recess of omental bursa (lesser sac)
16. primordium of left lobe of the pancreas
17. mucosal lining of stomach
18. dorsal mesentery of hindgut
19. hindgut (rectum)
20. left umbilical artery
21. left hindlimb
22. middle region of tail
23. genital tubercle
24. right hindlimb
25. vesical region of urogenital sinus
26. right umbilical artery
27. peritoneal cavity
28. distal part of duodenum
29. right lobe of liver
30. right gonad (ovary)
31. right paramesonephric duct (Müllerian duct)
32. mesonephric duct (Wolffian duct)
33. urogenital mesentery
34. psoas major muscle
35. erector spinae muscle
36. cartilage primordium of lumbar vertebral body (centrum)
37. central canal
38. gastro-splenic ligament
39. lieno-renal ligament
40. wall of stomach
41. "body" of pancreatic primordium
42. left "foot plate"
43. cartilage primordium of metatarsal bone
44. cartilage primordia of talus and calcaneus
45. cartilage primordium of tibia
46. primordium of knee joint
47. cartilage primordium of distal part of femur
48. hilum of kidney with rostral part of right ureter and (medially) para-aortic bodies
49. metanephric tubule
50. metanephric glomerulus
51. proximal part of left ureter
52. distal part of left paramesonephric duct
53. definitive (phallic part) of urogenital sinus
54. cartilage primordium of tail vertebral body
55. filum terminale
56. pelvic part of urogenital sinus
57. Müllerian tubercle
58. distal part of right ureter where it approaches the vesical part of the urogenital sinus
59. right ureter
60. postrenal segment of inferior vena cava
61. lumbar plexus
62. cartilage primordium of iliac bone
63. segmental nerves
64. pelvic part of peritoneal cavity
65. wall of right ureter

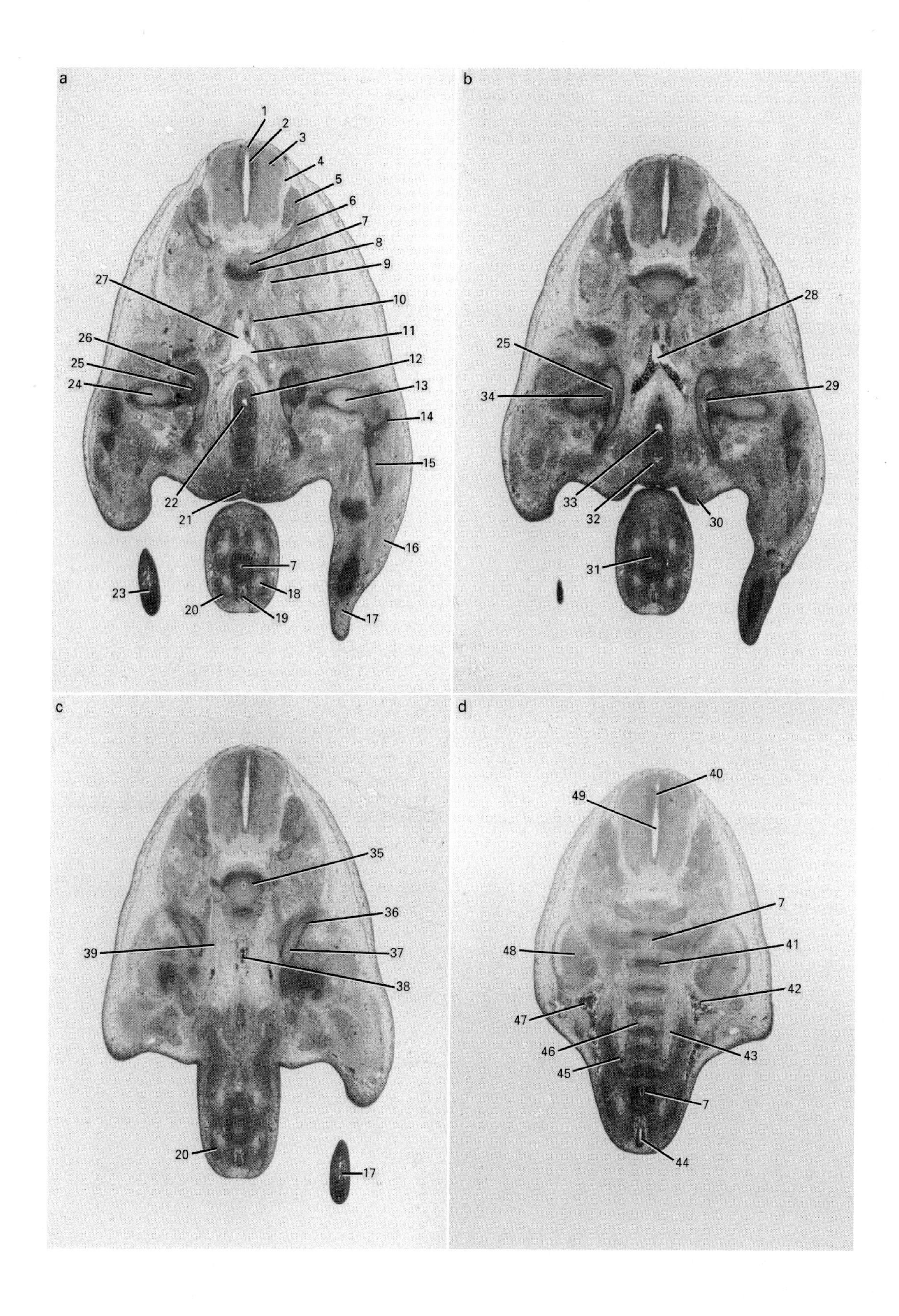
a
1
2
3
4
5
6
7
8
9
10
11
12
13
14
15
16
17
18
19
20
21
22
23
24
25
26
27
7
b
25
28
29
30
31
32
33
34
c
35
36
37
38
39
20
17
d
40
49
7
41
42
43
44
45
46
47
48
7

Plate 30l (13.5 days p.c.)

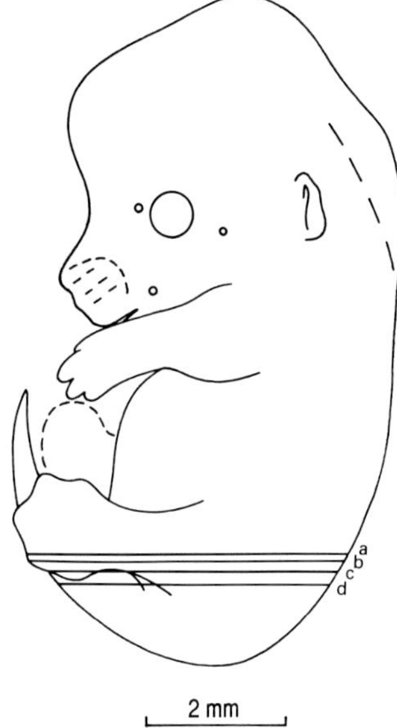

13.5 days p.c. Female embryo with about 60 pairs of somites. Crown–rump length about 9.1 mm (fixed). Transverse sections (Theiler Stage 22; rat, Witschi Stage 31; human, Carnegie Stages 20–23).

1. roof plate
2. ependymal layer
3. mantle layer
4. marginal layer
5. dorsal (posterior) root ganglion
6. cartilage primordium of neural arch
7. notochord
8. cartilage primordium of lumbar vertebral body (centrum)
9. lumbar plexus
10. descending (abdominal) aorta
11. left common iliac vein
12. wall of hindgut (rectum/upper part of anal canal)
13. cartilage primordium of shaft of femur
14. primordium of knee joint
15. cartilage primordium of tibia
16. left hindlimb
17. left "foot plate"
18. segmental nerve
19. filum terminale
20. proximal part of tail
21. definitive (phallic part) of urogenital sinus
22. lumen of hindgut
23. right "foot plate"
24. cartilage primordium of right femur
25. primordium of right hip joint
26. cartilage primordium of acetabular region of right hip bone (os innominatum)
27. postrenal segment of inferior vena cava
28. level of bifurcation of the inferior vena cava
29. primordium of the left hip joint
30. labial swelling
31. cartilage primordium of tail vertebral body
32. urogenital membrane
33. region just proximal to anal membrane
34. head of right femur
35. cartilage primordium of L4 vertebral body (centrum)
36. cartilage primordium of iliac crest
37. cartilage primordium of iliac bone
38. median sacral artery
39. components of lumbo-sacral plexus
40. region of localized neural luminal occlusion
41. cartilage primordium of upper sacral vertebral body
42. left femoral vein
43. sacral nerve trunk progressing distally towards the tail region
44. caudal extremity of normal neural tube morphology – the filum terminale extends distally from shortly after this level (from conus medullaris)
45. segmental artery
46. segmental interzone (future location of inter-vertebral disc)
47. right femoral vein
48. gluteal muscle mass
49. central canal

Plate 31a (13.5 days p.c.)

13.5 days p.c. Female embryo with about 60 pairs of somites. Crown–rump length about 9 mm (fixed). Sagittal sections (Theiler Stage 22; rat, Witschi Stage 31; human, Carnegie Stages 20–23).

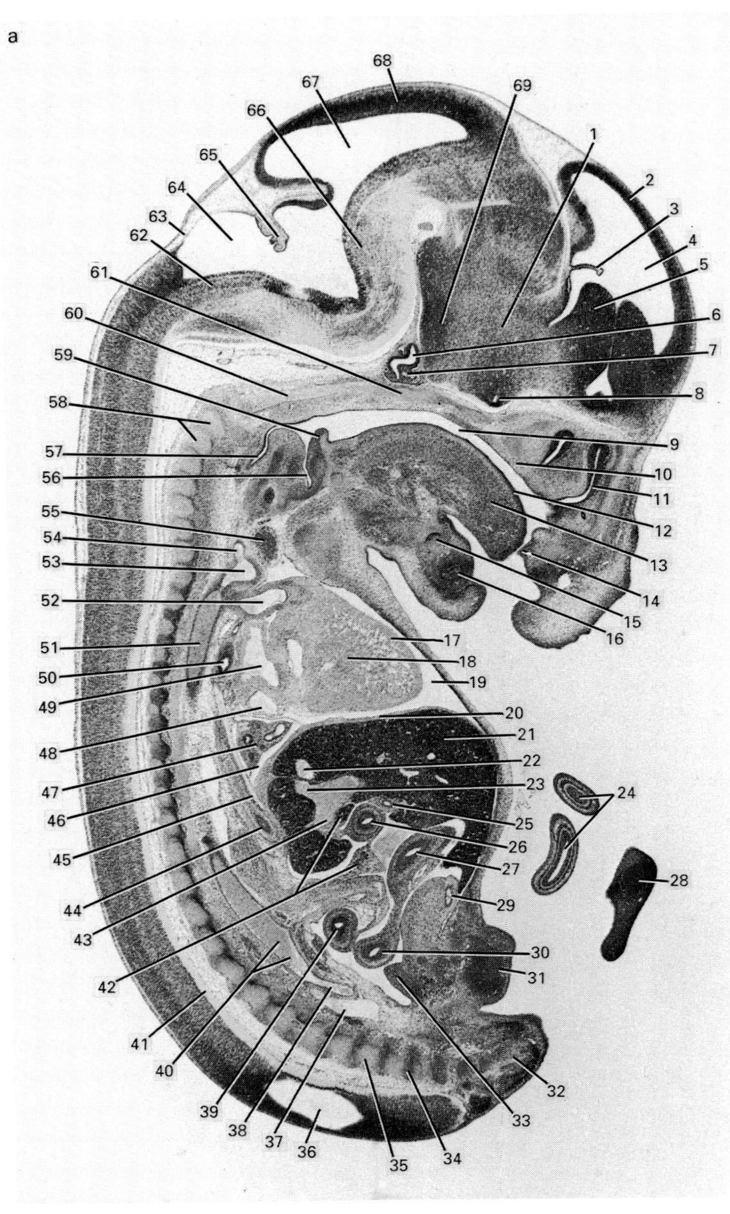

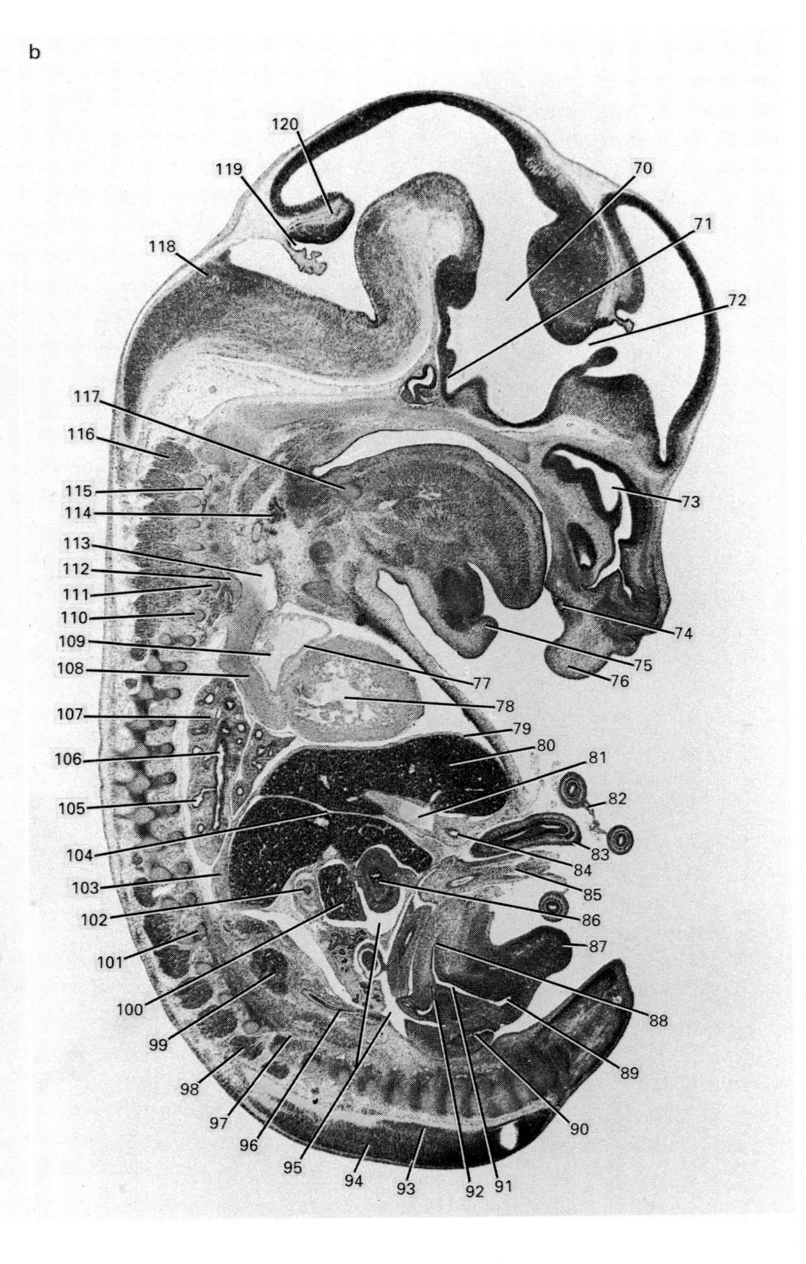

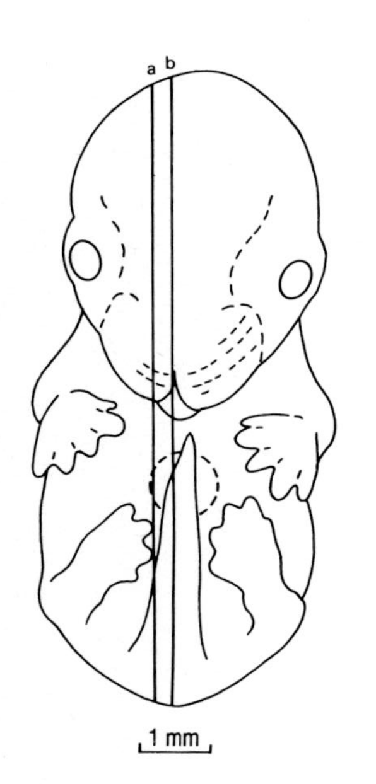

1. diencephalon (thalamus)
2. roof of neopallial cortex (future cerebral cortex)
3. choroid plexus extending into lateral ventricle
4. lateral ventricle
5. corpus striatum mediale (medial aspect of ganglionic eminence)
6. residual lumen of Rathke's pouch
7. early evidence of vascular differentiation which develops in close association with anterior wall of Rathke's pouch (future anterior pituitary)
8. optic recess of diencephalon
9. oropharynx
10. lower border of nasal septum
11. olfactory epithelium
12. dorsal surface of tongue
13. muscle mass of the tongue
14. rostral extremity of right palatal shelf of maxilla
15. Meckel's cartilage
16. primordium of right lower incisor tooth
17. wall of left ventricle
18. muscular part of interventricular septum
19. pericardial cavity
20. dome of diaphragm in region of central tendon
21. right lobe of liver
22. left hepatic vein draining towards the midline and then into the inferior vena cava
23. site of amalgamation of hepatic veins, ductus venosus and vitelline veins with inferior vena cava
24. loops of midgut within physiological umbilical hernia
25. cystic duct
26. first part of duodenum
27. lumen of distal part of midgut loop
28. "foot plate" of right hindlimb (pelvic limb)
29. right umbilical artery
30. intraperitoneal part of midgut
31. lateral border of genital tubercle
32. proximal part of tail
33. caudal part of right paramesonephric duct where it approaches the urogenital sinus
34. cartilage condensation being primordium of sacral vertebral body (centrum)
35. segmental interzone (future location of intervertebral disc)
36. central canal of spinal cord
37. right common iliac vein
38. right internal iliac artery
39. lumen of third part of duodenum
40. region of bifurcation of the aorta to give rise (in this location) to the right common iliac artery
41. marginal layer
42. pancreatic primordium
43. portal vein
44. caudal part of oesophagus where it is surrounded by the fibres of the right crus of the diaphragm
45. anterior vagal (X) trunk in close association with ventral wall of thoracic part of the oesophagus
46. extension of right pleural cavity around accessory lobe of right lung
47. accessory lobe of right lung
48. lumen of caudal part of right atrium just medial to site of entrance into it of inferior vena cava
49. lumen of right atrium
50. lumen of left main bronchus
51. descending (thoracic) aorta
52. outflow tract of right ventricle (pulmonary trunk leading to ductus arteriosus)
53. arch of aorta
54. origin of innominate artery from arch of aorta
55. right lobe of thymic rudiment
56. laryngeal aditus
57. entrance to the oesophagus
58. cartilage primordium of odontoid process (dens) of C2 vertebra and cartilage primordium of body of C2 vertebra
59. epiglottis
60. cartilage primordium of basioccipital bone (clivus)
61. cartilage primordium of basisphenoid bone
62. ventral part of medulla oblongata
63. roof of fourth ventricle
64. fourth ventricle
65. choroid plexus within central part of lumen of fourth ventricle
66. pons
67. mesencephalic vesicle
68. roof of midbrain
69. diencephalon (hypothalamus)
70. rostral part of third ventricle
71. infundibular recess of third ventricle
72. interventricular foramen
73. nasal cavity (right half)
74. primordium of right upper incisor tooth
75. lower lip
76. upper lip (right side)
77. wall of left atrium
78. lumen of left ventricle
79. left dome of the diaphragm
80. left lobe of the liver
81. ductus venosus
82. mesentery of midgut
83. proximal part of midgut loop passing into the physiological umbilical hernia
84. lumen of gallbladder
85. umbilical vein
86. gastro-duodenal junction in region of pyloric sphincter
87. tip of genital tubercle
88. urogenital sinus in region of proximal part of urachus
89. phallic part of urethra
90. anal canal
91. lumen of urogenital sinus (pelvic part)
92. site of entrance of ureter into vesical part of urogenital sinus
93. mantle layer (ventral grey horn)
94. mantle layer (dorsal grey horn)
95. peritoneal cavity
96. left ureter
97. segmental lumbar nerve progressing to form part of left lumbo-sacral plexus
98. left lumbar dorsal (posterior) root ganglion
99. medial part of left metanephros (definitive kidney)
100. caudate lobe of liver
101. cartilage primordium of head of left 13th rib
102. abdominal part of oesophagus
103. left crus of diaphragm
104. lobar interzone
105. branch of segmental bronchus within left lung
106. left segmental bronchus passing to lower part of left lung
107. left lung
108. left superior vena cava
109. lumen of left atrium
110. cartilage primordium of head of left second rib
111. cartilage primordium of head of left first rib
112. origin of left vertebral artery arising from the first part of the left subclavian artery
113. communication between left internal jugular vein and left jugular lymph sac
114. left lobe of thyroid gland
115. left vertebral artery
116. left dorsal (posterior) root ganglion of C2
117. cartilage primordium of body of hyoid bone
118. dorsal part of medulla oblongata
119. entrance to lateral recess of fourth ventricle
120. intraventricular portion of cerebellar primordium

PLATE
31b

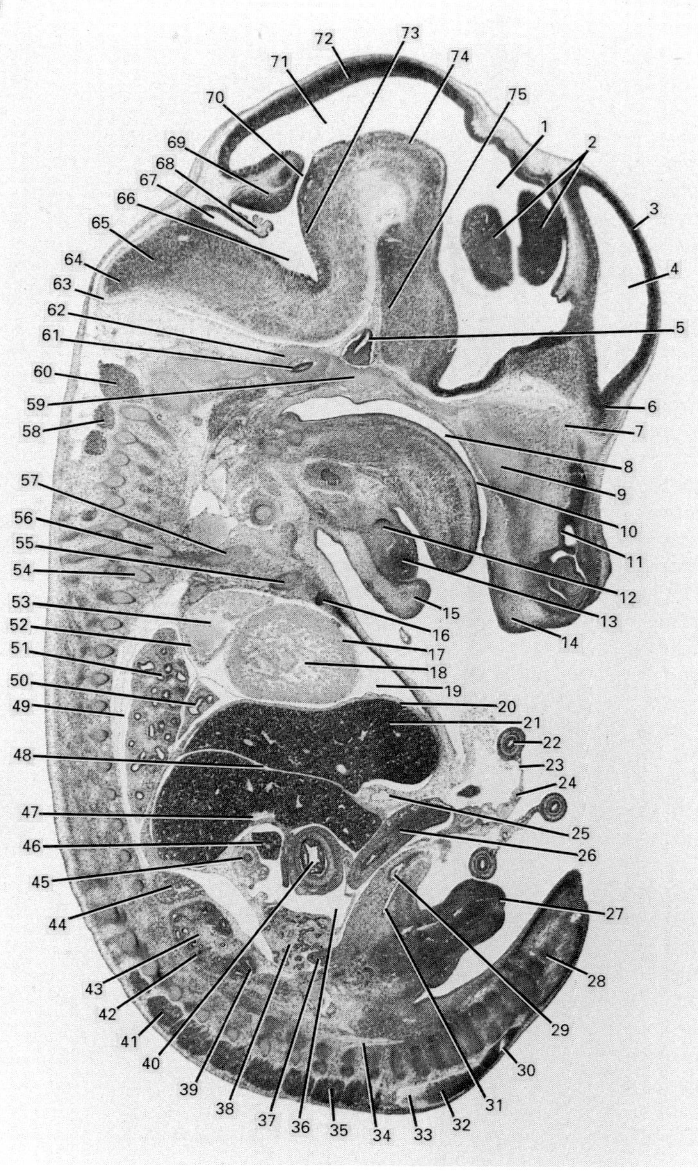

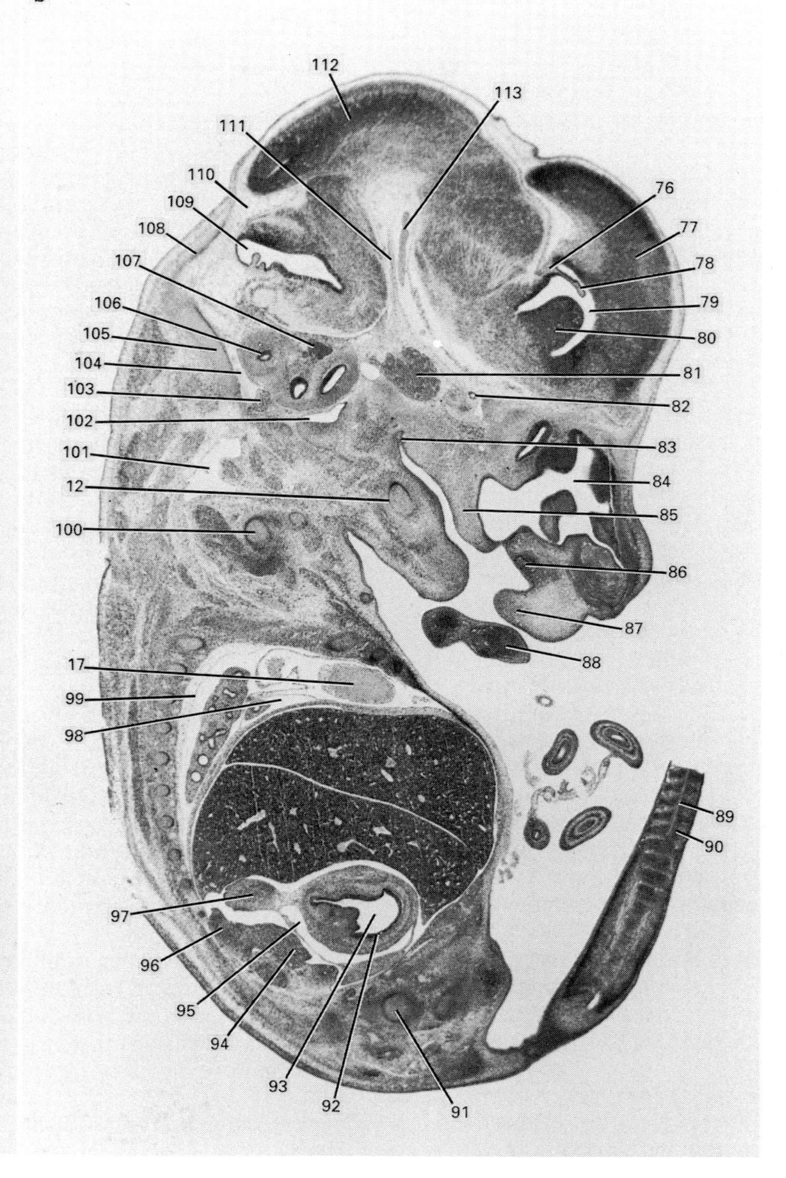

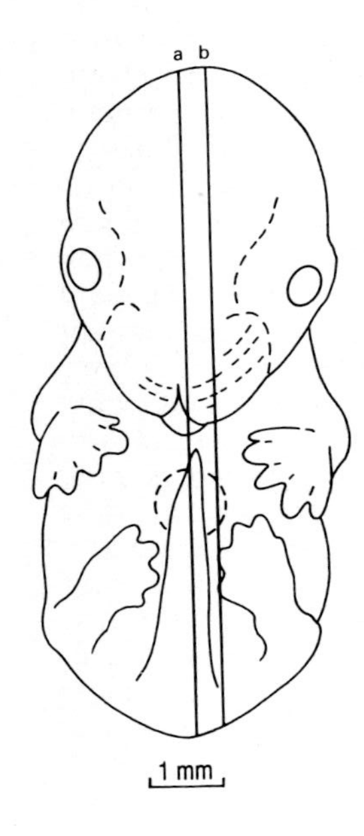

Plate 31b (13.5 days p.c.)

13.5 days p.c. Female embryo with about 60 pairs of somites. Crown–rump length about 9 mm (fixed). Sagittal sections (Theiler Stage 22; rat, Witschi Stage 31; human, Carnegie Stages 20–23).

1. rostral part of third ventricle
2. corpus striatum mediale and corpus striatum laterale (medial and lateral aspects of left ganglionic eminence)
3. roof of right neopallial cortex (future cerebral cortex)
4. right lateral ventricle
5. residual lumen of Rathke's pouch
6. medial margin of right olfactory lobe of brain
7. olfactory (I) nerves converging on olfactory lobe of brain
8. oropharynx
9. cartilage primordium of nasal septum
10. dorsal surface of tongue
11. olfactory epithelium
12. Meckel's cartilage
13. medial margin of tooth primordium of left lower incisor tooth
14. philtrum
15. lower lip
16. cartilage primordium of mid-shaft region of left third rib
17. wall of left ventricle
18. lumen of left ventricle
19. pericardial cavity
20. left dome of the diaphragm
21. left lobe of liver
22. loop of midgut within physiological umbilical hernia
23. mesentery of midgut
24. mesenteric vessels within mesentery of midgut
25. left umbilical vein passing dorsally to join the ductus venosus
26. proximal part of midgut loop passing into the physiological umbilical hernia
27. tip of genital tubercle
28. middle region of tail
29. left umbilical artery
30. central canal of spinal cord
31. urogenital sinus in region of urachus
32. mantle layer in proximal part of coccygeal region of spinal cord
33. marginal layer in proximal part of coccygeal region of spinal cord
34. left sacral nerve trunk progressing to form part of left lumbo-sacral plexus
35. left sacral dorsal (posterior) root ganglion
36. peritoneal cavity
37. caudal part of left paramesonephric duct
38. pancreatic primordium
39. rostral part of left ureter
40. lumen of pyloric antrum region of stomach
41. left lower lumbar dorsal (posterior) root ganglion
42. cortical region of left metanephros (definitive kidney)
43. medullary region of left metanephros (definitive kidney)
44. left adrenal primordium
45. abdominal part of oesophagus
46. caudate lobe of liver
47. portal vein
48. lobar interzone
49. left pleural cavity
50. accessory lobe of right lung
51. left lung
52. wall of left atrium
53. lumen of left atrium
54. cartilage primordium of neck of left second rib
55. cartilage primordium of mid-shaft region of left second rib
56. cartilage primordium of neck of left first rib
57. cartilage primordium of mid-shaft region of left first rib
58. left dorsal (posterior) root ganglion of C3
59. cartilage primordium of basisphenoid bone
60. left dorsal (posterior) root ganglion of C2
61. left cochlea
62. cartilage primordium of petrous part of left temporal bone
63. marginal layer in caudal region of medulla oblongata
64. mantle layer in caudal region of medulla oblongata
65. dorsal part of medulla oblongata
66. fourth ventricle
67. entrance to left lateral recess of fourth ventricle
68. choroid plexus within central part of lumen of fourth ventricle
69. intraventricular portion of cerebellar primordium
70. cerebral aqueduct
71. mesencephalic vesicle
72. roof of midbrain
73. pons
74. ventral part of midbrain
75. diencephalon (hypothalamus)
76. left choroid fissure
77. medial wall of left cerebral hemisphere
78. choroid plexus extending into left lateral ventricle
79. left lateral ventricle
80. left corpus striatum (ganglionic eminence) overlying the head of the caudate-putamen
81. left trigeminal (V) ganglion
82. left optic (I) nerve
83. entrance to left pharyngo-tympanic (Eustachian) tube
84. nasal cavity (left half)
85. palatal shelf of left maxilla
86. primordium of left upper incisor tooth
87. upper lip (left side)
88. "hand plate" of left forelimb (thoracic limb)
89. notochord
90. distal part of tail
91. cartilage primordium of head of left femur
92. mucosal lining of stomach
93. lumen of distal part of "body" of stomach
94. left ovary
95. lesser sac of peritoneal cavity (omental bursa)
96. degenerating left mesonephric tissue
97. medial wall of dorsal part of "body" of stomach
98. extension of right pleural cavity around accessory lobe of right lung
99. mesothelial cells (parietal pleura) lining left pleural cavity
100. cartilage primordium of proximal part of shaft of left humerus
101. left jugular lymph sac
102. left tubo-tympanic recess
103. left inferior ganglion of glossopharyngeal (IX) nerve
104. left internal jugular vein
105. cartilage primordium of exoccipital bone
106. posterior semicircular canal
107. vestibulocochlear (VIII) ganglion
108. cartilage primordium of squamous part of occipital bone
109. left lateral recess of fourth ventricle
110. loosely packed cephalic mesenchyme tissue in future subarachnoid space
111. left oculomotor (III) nerve
112. left lateral wall of midbrain
113. posterior communicating artery

Plate 31a,b), and separated from the ectodermal part of the cloaca by the urogenital membrane (*Plate 30l,b*). The urogenital sinus is now completely separated from the rectum by the earlier downgrowth of the urorectal septum, and this feature is most easily seen in sagittal sections through this region (*Plate 29a,b*). The primitive perineal "body" is formed where the urorectal septum makes contact with and separates the cloacal membrane into the urogenital membrane anteriorly and the anal membrane posteriorly.

Gonadal differentiation is also progressing apace, so that the histological differences previously noted between the ovary and testis are now becoming more marked. In the male, the testicular cords have given rise to the primitive seminiferous tubules (*Plate 68,h,j,k*). The latter consist of solid cords of tissue at this stage in which the germ cells are embedded, and the intervening interstitial tissue. Within the seminiferous cords, early events associated with spermatogenesis are taking place, and type A spermatogonia may be seen in division. While the mesonephros is regressing (*Plate 68,h,j*), the mesonephric duct persists, and will be utilized as the "drainage" system of the testis. In the female, the ovary still has few features of note (*Plate 68,i*). In both sexes, the gonads are attached to the posterior abdominal wall by a fairly narrow urogenital mesentery (termed the mesovarium in the female, and the mesorchium in the male). Both the ovaries and the testes are located in the upper part of the peritoneal cavity, and at this stage are anterolateral relations of the adrenals and the upper halves of the kidneys (*Plate 30j,c,d*; *Plate 30k,a,b*).

THE SKELETAL SYSTEM

Substantial changes have occurred between 12.5–13 and 13.5–14 days p.c., with regard to the degree of differentiation of the skeletal system. In the majority of sites where precartilage was present at the earlier time period, it is now possible to see that this has largely been replaced by cartilage. There are a few notable exceptions where precartilage is still in evidence, and other sites where, somewhat surprisingly, neither precartilage models nor mesenchymatous models are yet seen for some of the bony elements. For the sake of simplicity, it is probably easiest to briefly consider the various regions separately.

In the region of the base of the skull, all of the elements of the chondrocranium are now chondrified, so that, for example, it is possible to recognize the exoccipital bones (*Plate 30b,a,b*; *Plate 30c,a–c*), the supraoccipital bone (tectum posterius) (*Plate 30c,a*), the basioccipital bone (clivus) (*Plate 30c,c*; *Plate 31a,a*), the basisphenoid bone (*Plate 30c,c,d*; *Plate 31b,a*), as well as the petrous part of the temporal bones (the successor to the otic capsule) (*Plate 30b,a–d*; *Plate 30c,a–d*; *Plate 31b,b*). Additional cartilaginous elements seen in the anterior part of the facial region are the nasal capsule (*Plate 30c,d*; *Plate 30d,a–d*), and the nasal septum (vomer) (*Plate 30d,a–d*; Plate 30e,a–d; *Plate 30f,a,b*). Similarly, the following branchial arch cartilage derivatives are also seen at this stage, namely, Meckel's cartilage (of first arch origin) (*Plate 30c,b,c*; *Plate 30d,a–d*; *Plate 30e,a–d*), the various components of the hyoid bone (the lesser horns, which are of second arch origin, and the body and greater horns, which are of third arch origin) (*Plate 30d,c,d*), the three middle ear ossicles (the malleus and incus, which are of first arch origin, and the stapes which is of second arch origin) (*Plate 30c,b–d*). It is also just possible to discern at this stage the mesenchymatous condensations just lateral to Meckel's cartilage in which the mandible will soon differentiate (*Plate 30d,c,d*, unlabelled), and, but even more diffuse at this stage, the precursors of the other elements of the membranous viscerocranium and neurocranium.

Cartilaginous elements are now present in all but the most caudal of the vertebrae in the tail region, so that it is now possible to recognize their various constituent parts. Each possesses a precursor of the vertebral body (centrum) (*Plate 30e,a*), and a neural arch (in two separate units at this stage which extend posteriorly only as far as the anterior two-thirds of the lateral surface of the spinal cord) (*Plate 30d,a*). In the thoracic region, the costal elements are all chondrified, but it is not possible at this stage to distinguish between the ribs proper and the costal cartilages (*Plate 30f,b–d*), though this is just possible at about 14.5 days p.c. (*Plate 32g,b–d*). It is also possible to recognize the primordia of the articulations between the ribs and their associated vertebrae (*Plate 30g,b*; *Plate 31b,a*).

In the pectoral region, while the clavicle is said to form principally in membrane in man (Hamilton and Mossman, 1972), in the mouse it appears to have a pre-cartilage model which is only partly chondrified at 13.5–14 days p.c. (*Plate 30e,c*), and is seen to be fully chondrified by about 14.5 days p.c. (*Plate 32e,c*). However, no precartilage condensations are yet evident for the sternebrae, these being first evident at about 14.5 days p.c. (*Plate 32e,d*; *Plate 32f,a–c*). The elements of the larynx and the tracheal rings are also present in the form of pre-cartilage condensations at this stage (*Plate 30e,a*). The scapula and all of its component parts are also recognisable at this stage, and present in the form of a cartilage model (*Plate 30e,a–d*; *Plate 30f,a,b*). The long bones of the forelimb are also chondrified at this stage (humerus, *Plate 30f,a–d*; ulna and radius, *Plate 30f,d*; *Plate 30g,a,b*), while the more distal elements, such as the metacarpals and phalanges are still in the form of precartilage models (*Plate 30g,b–d*; *Plate 30h,a,b*). It is also possible to recognize at this stage the primordia of the shoulder joints (*Plate 30e,d*), and the elbow joints (*Plate 30g,b*).

Similar arrangements to those described above in relation to the pectoral girdle and forelimb, are evident with regard to the pelvic girdle and hindlimb. Thus while the pelvic bones (*Plate 30l,a–c*) and all of the long bones of the hindlimb (*Plate 30k,b,c*) are present in the form of cartilage models, the metatarsals and phalanges are still present in the form of precartilage models. The primordia of the hip and knee joints are also clearly evident at this stage (*Plate 30k,b*; *Plate 30l,a,b*).

Stage 23 Developmental age, 15 days p.c.
Plates 32a–k, Plates 33a, b, illustrate transverse and sagittal sections, respectively, through intact embryos, while Plates 34a, b, illustrate coronal sections through the head region of an embryo at a similar developmental stage.

EXTERNAL FEATURES
Embryos with over 60 pairs of somites (up to 65 pairs). Relatively few fundamental differences are observed between the external features of embryos at this stage and the previous stage of development, as all the events that are manifest from an external analysis of the embryo at this time evolve only gradually. Possibly the most obvious changes observed at this time are related to the appearance of the handplate and footplate. More subtle changes occur in the differentiation and general distribution of the hair follicles, though little change is seen in the external appearance of the vibrissae compared to that seen at 13.5–14 days p.c. In the cephalic region, the eyelid folds are becoming more prominent, while the external ear is seen to be folded slightly further forward than previously, and now covers just over half of the external acoustic meatus.

With regard to the differentiation of the hand plate, at 14.5–15 days p.c., it is possible to observe that any webbing that was previously present in the digital interzones at 13.5–14 days p.c. (*Plate 44,d,e*) has now disappeared (*Plate 44,g–i*). In the foot plate, which was clearly seen to be lagging behind the differentiation of the hand plate in this regard, some evidence of residual webbing is still apparent. This is particularly clearly seen in views of the dorsal surface of the foot plate (*Plate 45,a*), though digital differentiation in the foot plate has advanced significantly from the situation observed at 13.5–14 days p.c. (*Plate 44,d,e*), particularly with regard to the appearance of the plantar surface of this region (*Plate 45,b*; cf. *Plate 44,f*). The digits of both the forelimb and the hindlimb are fairly splayed out at this stage, though over the period of the next 1–2 days, the majority of the digits tend to become more parallel to each other. Thus by 16.5 days p.c., this modified arrangement of the digits is already evident in the forelimb (*Plate 45,g*), and a similar arrangement is seen with regard to the digits of the hindlimb by about 17.5 days p.c. (*Plate 45,n,o*).

At 14.5–15 days p.c., the digits of the hand plate are reasonably symmetrically arranged on either side of the middle digit (*Plate 44,h,i*). The situation with regard to the foot plate is slightly different, so that whereas at 13.5–14 days p.c. the arrangement of the digits of the foot plate was close to symmetrical (*Plate 44,f*), by about 14.5–15 days p.c., the first digit is seen to be slightly smaller than the fifth digit (*Plate 45,a,b*). However, over the period of the next few days, with the gradual medial migration of both the first and fifth digits, associated with the evolution of the definitive arrangement of the foot plate, the degree of asymmetry becomes less marked. The latter is evident from a view of the plantar surface of the foot plate at 17.5 days p.c. (*Plate 45,o*).

The earliest evidence of a generalized distribution of hair follicles is seen at about 13.5–14 days p.c., when they are principally seen to be confined to the pectoral, pelvic and trunk regions (*Plate 46,e,f*). By about 14.5–15 days p.c., the distribution of the hair follicles is seen to be more extensive, and they are now seen to cover much of the cephalic region (*Plate 46,g,h*). However, it is only by about 15.5–16 days p.c. that their full distribution is clearly evident, principally because there are few if any body creases present at this stage (*Plate 46,i,j*). Beyond about 16.5 days p.c., numerous skin creases are evident, particularly in the pectoral and trunk regions (*Plate 46,k–p*). There are, however, certain regions that are relatively devoid of hair follicles up to about 14.5–15 days p.c., for example, in the distal extremities of the limbs, so that in the forelimbs hair follicles are not normally seen more distally than the region overlying the wrist joint (*Plate 44,g*), and in the hindlimbs more distally than the region overlying the ankle (*Plate 44,g*). A similar arrangement with regard to the distribution of hair follicles persists until about 15.5 days p.c. (*Plate 44,j*). Similarly, few if any hair follicles extend beyond the proximal region of the tail, up until at least 17.5 days p.c. (*Plate 46,k–p*). The other region that initially (up until about 14.5–15 days p.c.) has relatively few hair follicles is the maxillary region immediately around the peripheral boundaries of the vibrissae-bearing areas of the upper lips (*Plate 46,g,h*), and even at 15.5 days p.c. (*Plate 46,i,j*), the hair follicles are less pronounced in this region than in the other follicle-bearing areas. Because the total number of hair follicles distributed over the trunk, cephalic region and proximal parts of the limbs is quite small, and though featured in the histological sections (e.g. *Plate 32i,a*, unlabelled), they are clearly not particularly prominent structures. This observation equally applies to the situation observed at later stages of gestation, so that while they are more in evidence at about 15.5–16 days p.c. (e.g. *Plate 35i, b*; *Plate 35k,b*), and at about 16.5–17 days p.c. (e.g. *Plate 37i,b*), where they are seen to be principally associated with the prominent skin folds, little evidence of differentiation of these structures is seen even at 17.5–18 days p.c. (e.g. *Plate 40i,c,d*).

Little difference is observed in the distribution and appearance of the vibrissae between this stage and the situation observed at the previous stage of development (*Plate 49,j,l*). However, the prominent tactile or sinus hair follicles characteristically found in various locations in relation to the orbital region, and just caudal to the angle of the mouth (*Plate 49,j,l*; *Plate 32b,d*; *Plate 32c,a*; *Plate 32d,a*; *Plate 32e,c*), show an increased degree of differentiation compared to the situation observed at 13.5–14 days p.c. (*Plate 30c,b*; *Plate 30d,b*; *Plate 30e,d*). It is also just possible to discern a degree of structural organization within them (e.g. *Plate 32b,d*), comparable to that seen in the vibrissae at this stage (*Plate 32d,c,d*).

The mammary gland primordia are first seen at about 12.5–13 days p.c. (*Plate 28c,c,e*, unlabelled; *Plate 28d,i*), and are more clearly seen at about 13.5

days p.c. (*Plate 30f,d*; *Plate 30j,c,d*). By 14.5–15 days p.c., they are seen to extend caudally from the upper thoracic to the inguinal region along the mammary ridges (*Plate 32f,b*; *Plate 32g,c*; *Plate 32h,a*; *Plate 32i,d*). In the adult, five pairs are seen, three are in the thoracic region, and two are abdomino-inguinal in location. The primordium of the mammary gland consists of an ectodermal epithelial thickening which penetrates into the subjacent mesenchyme which becomes increasingly condensed. Growth of the primordium occurs slowly at first into the underlying dermis. Subsequently, secondary sprouts grow out from the primordium into the surrounding tissues, and fat is gradually deposited around the developing gland. The mammary ridges are particularly prominent in embryos at about 17.5 days p.c. (*Plate 40f,a–d*; *Plate 40g,a–d*). In the postnatal period, both the initial epithelial downgrowths and the sprouts derived from them become canalized. The secondary outgrowths become the lactiferous ducts, while the tertiary sprouts form the alveoli and small ducts of the mammary gland. The latter subsequently open into an epidermal pit formed from the original epithelial downgrowth. A mesodermal proliferation in the region of the pit causes its elevation above the surrounding skin, to form the nipple.

In the cephalic region, the pinna of the ear increases in size, and at 14.5–15 days p.c. is seen to be folded forwards to a greater extent than previously, and covers just over half of the external acoustic meatus (*Plate 46,g,h*). While little change is apparent in the differentiation of the eyelids (from an analysis of scanning micrographs of this region) compared with that seen at the previous stage, an analysis of histological sections through this region reveals a considerable increase in their degree of differentiation, compared with the situation observed at 13.5–14 days p.c. (*Plate 52,i*; cf. *Plate 52,g*). The tip and anterior part of the tongue extend forwards and often protrude over the labial surface of the mandible and between the two components of the upper lip at this stage (*Plate 49,j,l*).

THE HEART AND VASCULAR SYSTEM

All of the components of the heart are readily recognized at this stage, and both the heart and the vascular system have achieved their definitive prenatal configuration. The upper (auricular) surfaces of the two atria extend on either side of the pulmonary trunk and the ascending thoracic aorta (*Plate 32f,b*). Their walls are relatively thin, and only moderately trabeculated (*Plate 32f,c,d*; *Plate 32g,a–c*; *Plate 33b,a,b*). The two atrial chambers of the heart are in continuity through the foramen ovale (*Plate 32g,b*, unlabelled). The septum primum, ostium secundum and septum secundum are all readily identified at this stage (*Plate 32g,b*). The major (venous) vessels that enter the right atrium are also readily identified, and are the right superior (cranial) vena cava (*Plate 32f,c,d*; *Plate 32g,d*), the inferior vena cava (*Plate 32g,d*; *Plate 32h,a–c*), and the left superior (cranial) vena cava (*Plate 32f,a–d*; *Plate 32g,a–d*), and each of their portals of entry into the right atrium is guarded by a venous valve, only some of which can be identified in the sections illustrated (e.g. *Plate 32g,c*). The pulmonary veins enter the left atrium, but as they are of extremely small luminal diameter at this stage, they are not readily seen on the sections illustrated.

The walls of the two ventricles are of approximately equal thickness (*Plate 32g,b*), being smooth on their outer (pericardial) surface and trabeculated on their inner aspect. The trabeculae carneae are particularly clearly seen (*Plate 32g,a–c*), but the papillary muscles and chordae tendineae are difficult to recognize at this stage. The volume of the cavity of the left ventricle is slightly greater than that of the right side. While the two ventricles externally are separated by the interventricular groove (*Plate 32g,a–d*), internally they are separated by the interventricular septum which principally consists of a thick-walled muscular component (*Plate 32g,a,b*), which is supplemented by a smaller membranous component, which was formed from a downgrowth of the bulbar (spiral) ridges.

The atrio-ventricular valves form partly by the proliferation of the connective tissue subjacent to the endocardium of the atrio-ventricular canals, and partly from the fused endocardial cushions. The leaflets of both of the atrio-ventricular valves are readily recognized, namely, those of the right side (the tricuspid valve) (*Plate 32g,b,c*), and those of the left side (the mitral valve) (*Plate 32g,b*). Similarly, the leaflets of the aortic and pulmonary valves are also readily recognized (aortic valve leaflets, *Plate 32f,d*; *Plate 33a,a*; pulmonary valve leaflets, *Plate 32f,c*). At the commencement of the aorta, just rostral to the aortic valve, the aortic sinuses are evident (*Plate 32f,d*), and it is from this region that the right and left coronary arteries will soon proliferate. The right and left pulmonary arteries pass directly caudally from the inferior aspect of the pulmonary trunk (*Plate 32f,d*), before passing laterally then posteriorly with the left and right main bronchi towards their respective lungs (*Plate 32g,a–d*). The innominate (brachiocephalic) artery is seen to emerge from the ascending part of the arch of the aorta (*Plate 32f,a*; *Plate 33a,a*), and subsequently gives off the right common carotid artery and the right subclavian artery (*Plate 33a,b*), while the pulmonary trunk flows principally into the concavity of the arch of the aorta via the ductus arteriosus (*Plate 32f,c*) which, at this stage, is a vessel of equal luminal diameter to that of the thoracic aorta.

Since the other components of the vasculature are not significantly different from the situation described during the previous two stages, the reader is advised to refer to these sections for a detailed analysis of the principal arterial and venous vessels and their location in the embryo. However, it should be noted that the posthepatic inferior vena cava (*Plate 32h,a–c*; *Plate 33a,a,b*) is seen to form principally from the ductus venosus (*Plate 32h,d*; *Plate 33a,a*), which largely transmits oxygenated blood from the placenta, to which is added a relatively small volume of deoxygenated blood from the portal vein (*Plate 32i,c*) which drains the gut tube and its accessory structures, via the hepatic veins. An additional contribution of deoxygenated blood is also made from the prehepatic inferior vena cava, which drains the lower limbs, kidneys and the other paired abdominal organs (*Plate 32j,d*; *Plate 32k,a,b*; *Plate 33a,a*).

THE PRIMITIVE GUT AND ITS DERIVATIVES

The relationship between the tongue and the palatal shelves has substantially changed during the period between 13.5–14 and 14.5–15 days p.c., with the gradual apposition and eventual fusion of the palatal shelves. In developmentally more advanced embryos at about 14–14.5 days p.c., the medial borders of the posterior halves of the palatal shelves have elevated to the horizontal position (*Plate 60,b,c*). By about 15 days p.c., the medial borders of the anterior halves of the palatal shelves have made contact with each other and subsequently fused (*Plate 60,d–g*). The line of fusion in the midline between the two palatal shelves is particularly clearly seen in histological sections through this region (*Plate 32d,a*). In the region just in front of the initial site of fusion, the primary palate has yet to fuse with the anterior borders of the palatal shelves (*Plate 33a,a,b*), and it does not do so until about 16 days p.c. (*Plate 60,l*), though the slit-like orifices of the incisive canals remain patent at this time. It is also possible to see that the prominent rugae, first evident at about 12.5–13 days p.c. (*Plate 59,f*), now run almost continuously across the full width of the palatal shelves (*Plate 60,g*). The latter become more obvious when the initial site of fusion gradually disappears (*Plate 60,j–l*).

The relationship between the palatal shelves and the nasal septum over this period is most clearly seen in coronal sections through this region. At about 14.5 days p.c., the most anterior part of the nasal septum is seen to have fused with the primary palate (*Plate 62,a*). Slightly more posteriorly, and extending over the anterior half of the developing palate, the two palatal shelves are seen to be in apposition, but the inferior border of the nasal septum is still some distance above their superior surface (*Plate 62,c–d*). It is also possible to observe from these sections, that the posterior border of the nasal septum does not extend beyond the region overlying the anterior part of the palatal shelves (*Plate 33a,b*; *Plate 62,d,e*). The vomeronasal organ (Jacobson's organ) is seen to be a particularly prominent structure at this stage of development, and occupies a considerable amount of the anterior part of the nasal septum on either side of the midline, and these organs are seen to be located just below the lower border of the cartilaginous primordium of the nasal septum (*Plate 62,a*).

Whereas at about 13.5 days p.c., the dorsum of the tongue has a rounded cross-sectional profile (*Plate 58,h,i*), this gradually changes, so that by about 13.5–14 days p.c., the surface of the tongue is seen to be flattened (*Plate 58,k*). By about 15 days p.c., the cross-sectional profile of the entire tongue is now concave, while the anterior part is seen to have a spatulate profile (*Plate 58,l*). From about 13.5–14 days p.c., it is possible to recognize large numbers of fungiform papillae which are principally located over the anterior half of the dorsal and lateral surfaces of the tongue (*Plate 58,k*), though by about 15 days p.c. these papillae are more prominent than previously, and somewhat elevated above the general surface of the tongue (*Plate 58,l*). The intermolar eminence (believed to be derived from the tuberculum impar) is also first seen at about 13.5 days p.c. (*Plate 58,h,i*), but becomes more prominent over the next few days (cf. appearance at about 16 days p.c., *Plate 58,o*).

One of the most interesting structures observed on the dorsal surface of the posterior one-third of the tongue is the median circumvallate papilla (there is only one vallate papilla present in the mouse, as in most rodent species, though there are three present in the Japanese dormouse and two in the porcupine and in the anteater). The circumvallate papilla is first evident as a midline elevation on the dorsum of the tongue at about 13–13.5 days p.c. (*Plate 30d,a,b*). Initially, it has a dense nerve plexus at its core, which is bilaterally innervated by large diameter nerve trunks from the glossopharyngeal (IX) nerves which pass forwards and medially towards the papilla (AhPin *et al.*, 1989). The circumvallate papilla has a large oval-shaped dome-like central region, with accessory papillae associated with the lateral wall of the furrow that surrounds the central region of the papilla (*Plate 32d,b*; *Plate 33a,a*; *Plate 58,m,n*).

More posteriorly, the hyoglossus muscle is seen to take origin principally from the greater horns of the hyoid bone, as well as from the lateral part of its body (*Plate 32d,c,d*). All of the components of the hyoid are recognizable at this stage, namely, the lesser and greater horns, and the body (*Plate 32d,c,d*) (the lesser horns are derived from the second branchial/pharyngeal arch cartilage, while the body and greater horns are derived from the third arch cartilage). Other extrinsic muscles of the tongue are also distinguishable at this stage, such as the styloglossus (*Plate 32d,c,d*), and the genioglossus (*Plate 32e,a–c*). The epiglottis is most clearly seen in sagittal sections through this region (*Plate 33a,a*), and when differentiated, will guard the entrance to the larynx.

The pharyngo-tympanic (Eustachian) tubes extend laterally and posteriorly from the pharynx, and open into the tubo-tympanic recesses which become closely associated with the middle ear ossicles, and differentiate into the cavity of the middle ear. The relationship between these various components of the first pharyngeal pouch is most easily recognized in coronal sections through the posterior part of the pharynx (*Plate 34b,a,b*). The two lobes of the thyroid gland are seen to be located one on either side of the laryngeal aditus at this stage (*Plate 32e,a,b*; *Plate 33a,b*), and are seen to consist of large numbers of buds. The parathyroid glands are usually located directly posterior to the lobes of the thyroid gland (*Plate 32e,a*) (only a single pair of parathyroid glands are present in the mouse, and these are derived from the dorsal part of the third pharyngeal pouch, unlike the situation in man, where two pairs of parathyroid glands are formed, one pair each from the dorsal parts of the third and fourth pharyngeal pouches, respectively). The other derivatives of the third pharyngeal pouches (from their ventral tubular part) are the two lobes of the thymus gland. These are quite substantial structures at this stage, being first clearly seen at about 12.5–13 days p.c. (*Plate 28c,a–c*), after they have already descended almost to their definitive position in the ventral part of the superior mediastinum. At about 15 days p.c. the two lobes of the thymus gland are clearly seen to be close anterior and superior relations of the vessels of the outflow tracts of the heart (*Plate 32e,c,d*; *Plate 32f,a*), and their location is particularly well seen in sagittal sections through this region (*Plate 33a,a,b*). The histological morphology of the thymus at this stage is fairly homogeneous, and suggests that each lobe has yet to differentiate into medullary and cortical regions. Even at about 17.5 days p.c., when the two lobes of the thymus are seen to be extremely large structures and are now located in the ventral part of the superior mediastinum, there is no indication of their regional differentiation (*Plate 40e,b–d*; *Plate 40f,a,b*). Their base is attached to the superior part of the pericardium, while their most rostral parts are closely associated with the anterior wall of the trachea (*Plate 40e,b,c*). The ultimobranchial body, the derivative of the ventral part of the fourth pharyngeal pouch, has by this stage amalgamated with the thyroid gland, its cells have disseminated within it, and are believed to give rise to the parafollicular cells or C-cells of the thyroid, which are responsible for the production of calcitonin (thyrocalcitonin).

The inlet of the oesophagus is initially flattened from side to side, and located directly posterior to the laryngeal aditus (*Plate 32d,b*), but while the oesophagus rapidly narrows to become a discrete thick-walled muscular tube, the lumen of the proximal part of which is characteristically also flattened from side to side, the trachea has a large patent and rounded luminal diameter (*Plate 32d,c,d*). More caudally, the oesophagus is seen to be located just to the left of the midline, while the trachea remains close to the midline for at least the upper part of its course (*Plate 32e,a–d*), before shifting slightly to the right of the midline (it then lies just anterior to the oesophagus) (*Plate 32e,d*; *Plate 32f,a–c*). At the level of the tracheal bifurcation, the oesophagus is closely related to the anterior (ventral) part of the upper thoracic vertebral bodies (*Plate 32g,a*). The trachea bifurcates in the mid-thoracic region, and gives rise to the right and left main bronchi (*Plate 32g,a,b*). In all parts of their course, the luminal diameter of the right main bronchus is somewhat larger than that of the left main bronchus (*Plate 32g,b–d*). The right main bronchus breaks up into lobar bronchi just before entering the various lobes of the right lung (*Plate 32h,a*), while the left main bronchus enters the left lung directly and without first branching (*Plate 32g,d*). The lungs at this stage have a substantially finer bronchial architecture than previously, with the differentiation of terminal bronchi and bronchioles (*Plate 66,i*).

The diaphragm is completely differentiated at this stage, and the crura are seen to be quite substantial structures. These are pierced close to the midline by the oesophagus, with the two vagal (X) trunks lying anterior (previously the left vagal (X) trunk) and on the right posterolateral side (previously the right vagal (X) trunk) (*Plate 32i,a*). Having pierced the diaphragm, the oesophagus is then directed towards the left side of the upper abdominal (peritoneal) cavity (*Plate 32i,b,c*), before entering the stomach at the gastro-oesophageal junction (*Plate 32i,c*). The descending aorta passes behind the crura to enter the abdominal region (*Plate 32i,d*), and passes caudally retroperitoneally just marginally to the left of the midline (*Plate 32j,a–d*). While the stomach remains one of the most voluminous of the intra-abdominal contents, its walls appear to be less clearly regionalized than previously (*Plate 32j,a–c*; cf. *Plate 28e,a–d*; *Plate 31b,b*). At the gastro-duodenal junction (in the region of the pyloric sphincter) the lumen narrows quite dramatically (*Plate 33a,a*), before opening out once more in the region of the first part of the duodenum (*Plate 33a,b*; *Plate 33b,a*). The histological appearance of the mucous membrane along the entire length of the duodenum reveals early evidence of differentiation, with the first indication of villi and intestinal glands (*Plate 33b,a*; cf. *Plate 30j,a,b*). It is also possible to recognize the site of entry of the common bile duct (region of the ampulla of Vater) into the posteromedial aspect of the second part of the duodenum at the duodenal papilla (*Plate 32i,b*).

The liver is by far the largest of the abdominal organs, and now has a fairly homogeneous histological appearance. Detailed analysis, however, reveals the presence of numerous blood-filled hepatic sinusoids and bile capillaries surrounded by liver parenchymal cells, as well as substantial evidence of haematopoietic activity. The liver encroaches on the undersurface of the diaphragm in the region of its central tendon, and this part of the liver has no peritoneal covering (the "bare" area) (*Plate 32h,a–c*). On its anterior aspect, the remnant of the ventral mesentery of the foregut (the falciform ligament) is more clearly seen than previously (*Plate 32h,b,c*; cf. *Plate 301,c*), in the lower border of which is found the ductus venosus (*Plate 32h,d*), which when obliterated after birth forms a fibrous cord termed the ligamentum venosum. The gallbladder is located in the ventral midline (*Plate 32h,d*), and both the common bile duct and the right hepatic duct are clearly seen in the sections illustrated (*Plate 32i,a,b*), and the former is seen to drain into the duodenal papilla (*Plate 32i,b*). The pancreas at this stage is a quite diffuse gland, and much of it is located posteromedial and inferior to the stomach in the left iliac fossa (*Plate 32j,a–d*; *Plate 32k,a*), and is here an anterior relation of the lower pole of the left kidney (*Plate 32k,a*).

The physiological umbilical hernia is still quite large (*Plate 32h,a–d*; *Plate 32i,a*), and will only disappear at about 16–16.5 days p.c., with the return of the midgut loop into the peritoneal cavity (*Plate 37i,c,d*). The mucosal lining of the midgut shows increased evidence of differentiation at this stage than previously with, like the duodenum, the first evidence of villi and intestinal glands (*Plate 32i,b–d*; *Plate 32j,a–d*; cf. *Plate 30j,b–d*). The rectum, by contrast, shows little evidence of differentiation, and for much of its length has a long dorsal mesentery (*Plate 32j,a–d*), and this is only slightly reduced in length in the lower lumbar and pelvic regions (*Plate 32k,a,b*). While the splenic primordium is first clearly evident at about 13–13.5 days p.c., as mesenchymal aggregates within the dorsal mesentery of the stomach (dorsal mesogastrium) (*Plate 32k,a,b*), this organ is now seen to be substantially larger in size than previously, with an elongated ribbon-like shape, and is located at this stage in close proximity to the left gonad (*Plate 32j,c,d*; *Plate 32k,a*).

THE NERVOUS SYSTEM

The two regions of the brain that show the most marked degree of differentiation between 13.5–14 and 14.5–15 days p.c., are the cerebellar primordium, and the olfactory lobes. Whereas previously the rhombic lip (i.e. the dorsal part of the alar lamina of the metencephalon) was the first site of differentiation of the intraventricular part of the cerebellar primordium (*Plate 30a,c*), this region has considerably increased in size, and begins to encroach on the lumen of the lateral recesses of the fourth ventricle (*Plate 32a,d*; *Plate 32b,a*; *Plate 33a,b*; *Plate 33b,a*). The extent of cerebellar differentiation at this stage is particularly clearly seen in coronal sections through the region of the developing hindbrain (*Plate 34b,b–e*). Sections through this region also reveal that choroid plexus not only differentiates from the roof of the fourth ventricle (*Plate 32a,d*; *Plate 32b,a*), as is first seen at about 12.5–13 days p.c. (*Plate 28a,d,e*), but is also now apparent in association with the lateral walls of the lateral recesses of the fourth ventricle (*Plate 32b,a,b*; *Plate 33b,a,b*; *Plate 34b,c,d*). It is also of interest to note at this stage, that early evidence of choroid plexus differentiation is seen in the pineal recess of the third ventricle (*Plate 32a,b–d*; *Plate 34a,c*), though the pineal gland (epiphysis) itself shows little evidence of differentiation (*Plate 32a,a*; *Plate 33a,b*). Moderate amounts of choroid plexus are also found extending into the lateral ventricles (*Plate 32a,c,d*; *Plate 32b,a*; *Plate 34a,c–e*).

While the initial changes that take place in the future olfactory region of the brain are observed as early as about 12.5–13 days p.c. (*Plate 28a,g,h*), with the first indication of the presence of olfactory lobes, they are not at this stage readily distinguished from the rest of the neopallial cortex, and the lumen of the lateral ventricle is seen to be in direct continuity with that of the future olfactory lobe (*Plate 29a,b*). By about 13.5–14 days p.c., the olfactory lobes are now relatively easily distinguished both in transversely and in sagittally sectioned material (*Plate 30c,d*; *Plate 30d,a*; *Plate 31a,b*, unlabelled; *Plate 31b,a*). However, by about 14.5–15 days p.c., the olfactory lobes are seen to be not only reasonably distinct entities, but are clearly seen to be located directly above the roof of the nasal cavity (*Plate 32c,a–c*; *Plate 33a,a*; *Plate 34a,a*), and receive their sensory input from the olfactory epithelium via the olfactory (I) nerves (*Plate 32c,b*; *Plate 34a,b*).

Within the eye, increasing evidence of differentiation of the lens is observed, with the almost complete obliteration of the cavity of the lens vesicle (*Plate 53,a*). The other feature of note is the first indication of stratification within the nuclear region of the neural retina, located just subjacent to the nerve fibre layer (where nerve fibres pass centrally towards the optic disc from the primitive ganglion cells) (*Plate 52,i*). As has been indicated previously, the nerve fibre layer is only poorly seen in haematoxylin and eosin stained material, and is best revealed in silver stained sections (cf. *Plate 52,e,f*). Similarly, as has also been mentioned previously in an earlier section of this stage, the eyelids show increased evidence of differentiation compared to that seen at previous stages (*Plate 32c,a–d*; *Plate 32d,a*; *Plate 34a,c–e*; cf. *Plate 30c,b–d*; *Plate 30d,a*). The general appearance of the optic chiasma has also changed compared to the situation observed previously, due to the central migration of the nerve fibres that originate in the ganglion cells of the neural retina (*Plate 32c,c*; cf. *Plate 30c,d*).

Only slight changes are evident in the morphological appearance of the pituitary gland between the situation observed at 13.5–14 and 14.5–15 days p.c., with an increase in the degree of differentiation of the pars anterior, and the first indication of differentiation in the region of the pars intermedia (*Plate 55,c,d*; cf. *Plate 55,b*). A change is also observed in the region of the pars nervosa, where its wall shows early evidence of thickening and differentiation. In addition, the entrance to the infundibular recess is now represented by a narrow slit, and the lumen of the infundibulum is considerably diminished in volume compared to the situation observed previously (*Plate 55,c,d*; *Plate 57,a–f*).

In the cephalic region, most of the cranial ganglia and nerves are recognizable at this stage, principally because of their size and location in relation to other known structures. Similarly, while all the spinal dorsal (posterior) root ganglia are clearly seen, it is much more difficult to recognize, for example, all but the largest of the peripheral nerves, and even nerve trunks (such as those of the brachial plexus) in haematoxylin and eosin stained material, and for any detailed study along these lines, appropriately (i.e. silver) stained material is essential. The latter observation equally applies to the analysis of the autonomic (i.e. sympathetic and parasympathetic) systems, in which only the major components are recognizable in the haematoxylin and eosin stained material illustrated here. Thus the appearance of the adrenal gland (*Plate 32i,d*; *Plate 32j,a,b*), for example, is transformed when silver-stained material is studied, as this clearly demonstrates the sympathetic (neural crest-derived) origin of its medullary tissue, and this is equally applicable to the appearance of the pre-aortic abdominal sympathetic paraganglia (of Zuckerkandl) (*Plate 32j,b,c*).

In the mid-thoracic region, the spinal cord shows increasing evidence of differentiation, so that the central canal is now considerably smaller than previously, and has a diamond-shaped profile (*Plate 63,e*; cf. *Plate 63,d*). The shape of the dorsal grey horns is also more prominent, and the volume of the marginal layer (white matter) increases steadily, particularly in the ventral half of the spinal cord. The spinal cord is seen to extend caudally to the proximal part of the tail (*Plate 32k,c,d*), to the region of the conus medullaris. From this site, the filum terminale is seen to extend distally towards the tip of the tail (*Plate 32h,d*, unlabelled; *Plate 32i,a–d*).

THE UROGENITAL SYSTEM

Both kidneys are seen to be very close inferior relations of the adrenal glands, and the left kidney, in this embryo, is slightly more rostrally located than the right kidney (*Plate 32j,a*). This situation is quite unusual, as in the mouse, as in the rat (Hebel and Stromberg, 1986), the right kidney is almost always more rostrally located than the left kidney (in man, the situation is different in that the upper pole of the right kidney in the adult is about 1.25 cm *lower* than that of the left kidney). A brief review of the situation observed in the embryos illustrated in the Atlas confirms that it is the right kidney that is normally more rostrally located than the left kidney: 12–12.5 days p.c. embryo, both kidneys are pelvic and at a similar level, since they have yet to ascend to their definitive position (*Plate 27d,f*). In the 12.5–13 days p.c. embryo, the right side is considerably more rostrally located than the left (*Plate 28e,e*). In the 13.5 days p.c. embryo, the right side is marginally more rostrally located than the left (*Plate 30j,d*). In the 15.5 days p.c. embryo, the right side is marginally more rostrally located than the left (*Plate 35k,a*). In the 16.5 days p.c. embryo, the right side is considerably more rostrally located than the left (*Plate 37j,b*). In the 17.5 days p.c. embryo, the right side is considerably more rostrally located than the left (*Plate 40j,d*).

The histological features of the kidney at 14.5–15 days p.c. are in marked contrast to those observed at 13.5–14 days p.c., in that the kidneys now have reasonably well-defined cortical and medullary regions. In the inner part of the cortical region, substantial numbers of primitive glomeruli are found (*Plate 72,f*), whereas in the more peripheral part of this region, just subjacent to the developing kidney capsule, poorly differentiated metanephric cap tissue is still in abundance (*Plate 72,f*). More centrally, the medullary region of the kidney principally contains undifferentiated mesenchyme tissue with a few interspersed collecting tubules. These drain into the minor and major calyces, and thence into the renal pelvis. From this location, the ureters pass caudally towards the vesical component of the urogenital sinus (the region which is destined to form the bladder). The wall of the latter is particularly thick and muscular (*Plate 32i,d*; *Plate 32j,a,b*), and is initially continuous with the allantois via the urachus (*Plate 32i,b*). The allantois undergoes early retrogressive changes, though the urachus passes upwards from the apex of the bladder as a narrow diameter canal, in association with the inner aspect of the anterior abdominal wall, towards the umbilicus (the urachus itself regresses after birth, and it is replaced by a fibrous cord, termed the median umbilical ligament, which is covered by a fold of peritoneum). In man, the lumen of the proximal part of the urachus is said to remain patent, and frequently communicates with the bladder near its apex.

The principal difference between the appearance of the gonads at 13.5–14 and 14.5–15 days p.c. relates to their increase in size. Little difference, however, is observed in the histological morphology of the ovary between these two stages (*Plate 68,i*; cf. *Plate 68,f*), though the width of the mesovarium is considerably less than previously. It is also worth noting that the luminal volume of the paramesonephric duct has increased substantially (*Plate 68,i*), and is destined to give rise (in the female) to the oviduct, uterus and (probably) the upper one-third of the vaginal canal, as well as the ovarian capsule (*Plate 71,d–i*). At this stage of development, the ovary is ovoid in shape, and already substantially smaller than a testis at a comparable stage of development. Despite the unusual disposition of the right kidney (being more caudally located than the left, see above) in this embryo, the right ovary is in fact more rostrally located than the left (*Plate 32j,c,d*). A comparable situation is observed in the 16.5 days p.c. embryo illustrated (*Plate 37k,b–d*; *Plate 37l,a,b*), and this appears to be the normal disposition in adult rodents, and, in addition, the left ovary is generally slightly nearer to the midline than the right ovary (Hebel and Stromberg, 1986). It is of interest to note here that at 12.5–13 days p.c., both ovaries are at a similar lower thoracic/upper lumbar vertebral level (*Plate 28e,c,d*), and they are entirely rostral to the upper poles of the two kidneys (*Plate 28e,e,f*). At 13.5 days p.c., both ovaries are still at a similar lower thoracic/upper lumbar vertebral level, and while their upper poles are substantially more rostrally located than those of the kidneys, their lower poles overlap and are seen to be anterolateral relations of the upper halves of the two kidneys. The latter disposition results from the "ascent" of the kidneys, rather than as a result of any degree of descent of the ovaries or differential growth of the caudal part of the embryo. The relationship between the left ovary and the paramesonephric duct and the degenerating mesonephros is clearly illustrated (*Plate 32k,a*), and is reasonably similar to the situation observed at 15.5 days p.c. (*Plate 71,a,b*), though at this stage, the ovary is somewhat more caudally located than at 14.5–15 days p.c.

Between 13.5–14 and 14.5–15 days p.c., the testes increase in size and show early evidence of differentiation of the seminiferous tubules (*Plate 68,j,k*; cf. *Plate 68,h*). The latter occupy most of the medullary region of the testis at this stage, and only relatively little space is occupied by the interstitial gland tissue (Leydig cells), one of whose functions is to secrete testosterone. The coelomic/peritoneal cellular layer that surrounds all but the mediastinum region of the testis gives rise to the tunica vaginalis, while the tissue just subjacent to this layer shows early evidence of differentiation to form the tunica albuginea. The latter eventually becomes the dense fibrous layer (or capsule) which surrounds the majority of the testis.

The testes are substantially larger than the ovaries at this stage of development, are ovoid in shape, and have already "descended" to a limited extent from their site of origin in the lower thoracic/upper lumbar region. The part played by the gubernaculum in the latter process has yet to be fully established. At this stage, the testes are anterolateral to the lower one-third of the kidneys (*Plate 70,b*). The mesorchium is also substantially narrower than previously (*Plate 68,j*; *Plate 68,g*). While the mesonephros shows increased evidence of degenerative changes, the mesonephric duct is differentiating to form the "drainage" system of the testis. Some of the mesonephros is believed to be incorporated into the region of the rete testis. At this stage, the two testes are at a similar vertebral level, whereas by 15.5 days p.c., the right testis is seen to be slightly more rostrally located than the left testis (*Plate 35k,b–d*; *Plate 35l,a*), and a similar situation is observed at 17.5 days p.c. (*Plate 40k,b,c*).

THE SKELETAL SYSTEM

Many centres of ossification are now present throughout the skeleton, though principally confined to some of the elements of the chondrocranium, viscerocranium, calvarium and mandible, as well as in the periosteum of some of the long bones. However, in developmentally more advanced embryos at this stage, centres of ossification may also be found in the elements of the pectoral and pelvic girdles, in the ribs and in the dorsal arches of the upper cervical vertebrae (*Plate 81,a–d*).

All of the elements of the chondrocranium are now present as cartilage models, and in some of these, the first evidence of ossification is seen. Possibly the largest primary centre of ossification is seen within the parachordal plate (the future basioccipital bone) (*Plate 32c,i*; *Plate 33a,a*; *Plate 81,d*), and ossification extends forwards to surround the rostral extremity of the notochord (*Plate 32c,c*). No ossification centre is yet seen in the cartilage primordium of the basisphenoid bone (*Plate 32c,d*). Although one of the first bones to chondrify, no ossification centre is yet evident within the cartilage primordium of the petrous part of the temporal bone (*Plate 32c,a–d*). Elsewhere in the skull, a number of membrane bones also show early evidence of ossification. This phenomenon is particularly well seen in the primordium of the maxillary bone in which several distinct centres of ossification are evident, such as in the lateral margins of the palatal shelves (*Plate 32d,a*), in the periorbital region both lateral and infero-lateral to the nasal capsule (*Plate 32c,a,b*; *Plate 33b,b*, unlabelled; *Plate 34a,b*, unlabelled), and on either side just deep to the first upper molar tooth primordium (*Plate 32d,c,d*). The mandible also shows substantial evidence of ossification around, but principally lateral to, Meckel's cartilage (*Plate 32d,b–d*; *Plate 32e,a–d*; *Plate 32f,a*; *Plate 81,a–d*). In the rest of the skull, the only other region where ossification is clearly evident at this stage, is within the outer table of the membranous primordium of the frontal and parietal bones (*Plate 32b,c,d*; *Plate 81,a–d*).

In the posterior part of the pharynx and in the neck region, the cartilaginous elements of the larynx and lower respiratory tract are now clearly seen, whereas at 13.5–14 days p.c., the laryngeal and tracheal elements were mostly in the form of precartilage, and consequently only poorly defined (e.g. *Plate 30d,d*; *Plate 30e,a–d*; *Plate 31a,a*). At this stage, the relationship between the various cartilaginous elements is particularly clearly seen in sagittal sections (*Plate 33a,a*), where it is possible to recognize the thyroid and cricoid cartilages, as well as the segmentally arranged tracheal "rings".

In the pectoral region, the scapula is seen to be ossified at this stage, (*Plate 32d,d*; *Plate 32e,a–d*; *Plate 81,a–d*). In the pelvic girdle, all of the bony elements are present in the form of cartilaginous primordia, though a centre of ossification is seen in the iliac bone at this stage (*Plate 32k,a–d*). By contrast, the clavicle which is at 13.5–14 days p.c. only partially chondrified (*Plate 30e,c*), now has extensive evidence of ossification, though this is confined to the periosteal region of its middle one-third (*Plate 32e,b*; *Plate 81,a–d*). Similarly, the components of the sternum, which at 13.5–14 days p.c. showed no evidence of chondrification, are now almost completely chondrified. Thus the primordium of the manubrium sterni is now clearly seen (*Plate 32c,d*; *Plate 32f,a,b*), as well as the cartilage primordia of the sternebrae, some of which have two precursor elements, one on either side of the ventral midline (*Plate 32f,c,d*; *Plate 32g,a–d*; *Plate 33a,a,b*; *Plate 33b,a,b*). The ribs also show early evidence of periosteal ossification, and this is particularly well seen at their angles, and in the proximal part of the shaft of the ribs (*Plate 32g,a–c*; *Plate 32h,a*; *Plate 81,a–d*). No evidence of ossification is yet seen in the vertebral bodies at this stage, though ossification centres are present in the dorsal arches of the cervical vertebrae (*Plate 81,c,d*).

In the upper limbs, a considerable degree of periosteal ossification is seen in the middle one-third of the humerus (*Plate 32f,a,b*), but the proximal and distal regions are as yet unossified. Within the ulna and radius, periosteal ossification is also confined to their middle one-thirds (*Plate 32f,d*; *Plate 32g,a–c*; *Plate 81,a–d*). A similar picture is seen in relation to the femur, tibia and the fibula, where their middle one-thirds each show early evidence of periosteal ossification (femur, *Plate 32j,a*; tibia and fibula, *Plate 32j,d*; *Plate 32k,a–c*; *Plate 81,c,d*). As yet, no centres of ossification are present in either the carpal, tarsal or phalangeal bones (*Plate 32g,a–d*; *Plate 32i,a–d*). The joint primordia are all well-defined at this stage, and

(*continued on page 248*)

PLATE 32a

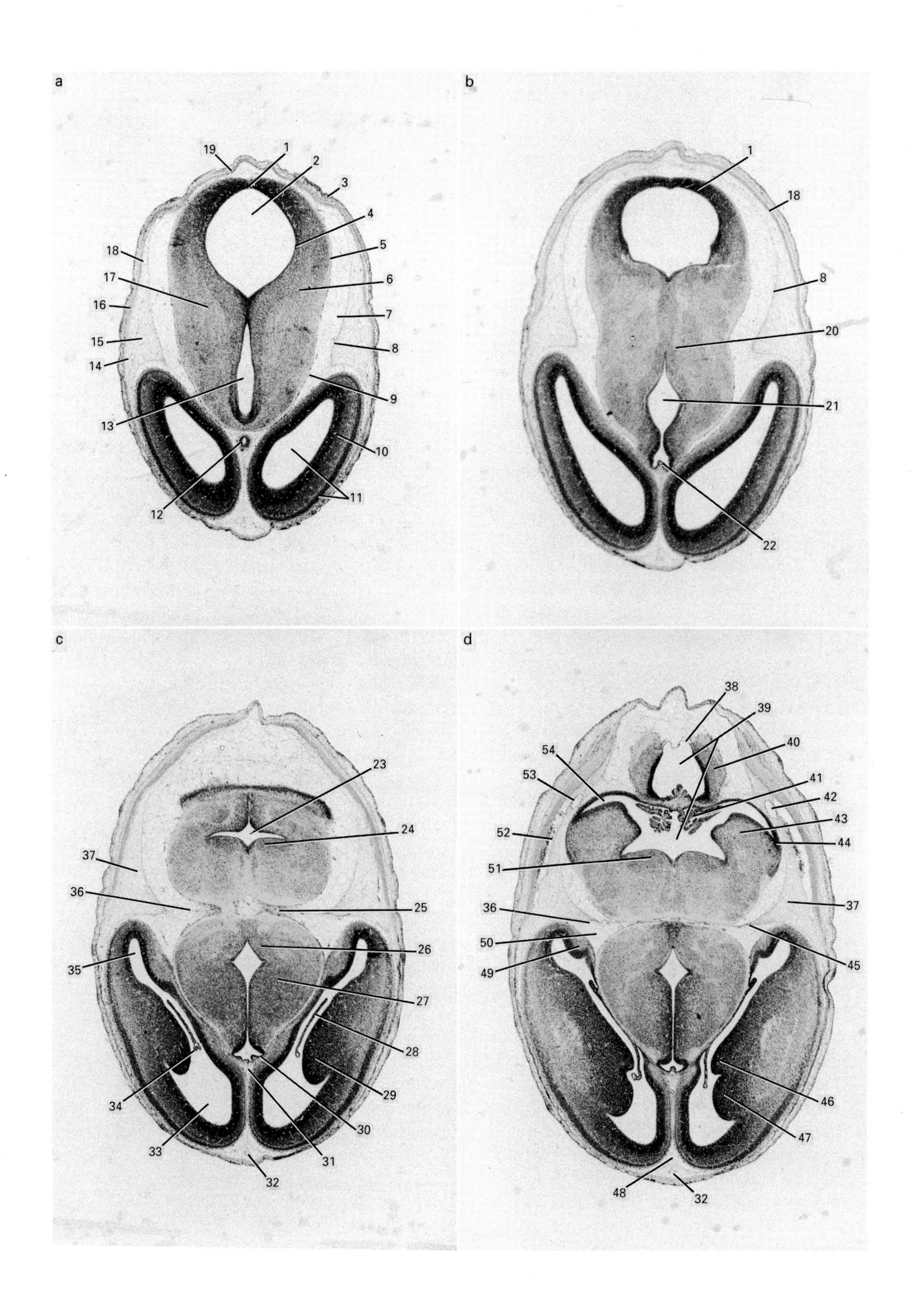
a
b
c
d
1
2
3
4
5
6
7
8
9
10
11
12
13
14
15
16
17
18
19
20
21
22
23
24
25
26
27
28
29
30
31
32
33
34
35
36
37
38
39
40
41
42
43
44
45
46
47
48
49
50
51
52
53
54

Plate 32a (14.5 days p.c.)

14.5 days p.c. Female embryo with over 60 pairs of somites. Crown–rump length about 10 mm (fixed). Transverse sections (Theiler Stages 22–23; rat, Witschi Stage 32).

1. roof of midbrain
2. mesencephalic vesicle (future aqueduct)
3. surface ectoderm (skin)
4. ependymal layer
5. marginal layer
6. mantle layer
7. loosely packed cephalic mesenchyme tissue in future subarachnoid space
8. mesenchymal condensation to form inner table of future dura mater
9. vascular elements of pia mater
10. wall of telencephalic vesicle
11. telencephalic vesicle (future lateral ventricle) and neopallial cortex (future cerebral cortex)
12. pineal primordium (epiphysis)
13. dorsal part of third ventricle
14. primitive venous plexus, precursors of dural venous sinuses
15. mesenchyme (future dura mater)
16. junction between upper border of chondrocranium (precursor of squamous part of occipital bone) and mesenchymal precursor of parietal bone
17. midbrain
18. cartilage primordium of squamous part of occipital bone
19. cartilage primordium of supraoccipital bone (occipital tectum, tectum posterius)
20. floor of the midbrain (tegmentum, in region of midbrain flexure)
21. third ventricle
22. early elements of choroid plexus associated with pineal recess of third ventricle
23. caudal extent of cerebral aqueduct
24. junction between pons and midbrain
25. rootlets of origin of oculomotor (III) nerve
26. diencephalon (hypothalamus)
27. diencephalon (thalamus)
28. choroid plexus arising from medial wall of lateral ventricle
29. corpus striatum (ganglionic eminence) overlying the head of the caudate–putamen
30. roof of the third ventricle
31. inferior sagittal dural venous sinus
32. superior sagittal dural venous sinus
33. anterior horn of lateral ventricle
34. rostral extremity of choroid plexus
35. posterior horn of lateral ventricle
36. oculomotor (III) nerve
37. trochlear (IV) nerve
38. roof of the fourth ventricle
39. fourth ventricle
40. alar plate of myelencephalon (medulla oblongata)
41. choroid plexus within fourth ventricle
42. rostral extremity of left endolymphatic sac
43. intraventricular portion of cerebellar primordium
44. rhombic lip (dorsal part of alar lamina of metencephalon – location of the lateral part of the cerebellar primordium)
45. mesenchymal condensation forming tentorium cerebelli
46. corpus striatum mediale (medial aspect of ganglionic eminence)
47. corpus striatum laterale (lateral aspect of ganglionic eminence)
48. mesenchymal condensation forming falx cerebri
49. tail of the caudate nucleus
50. rostral extension of posterior communicating artery
51. tegmentum of the pons (basal plate)
52. transverse dural venous sinus
53. rostral extremity of right endolymphatic sac
54. lateral recess of fourth ventricle

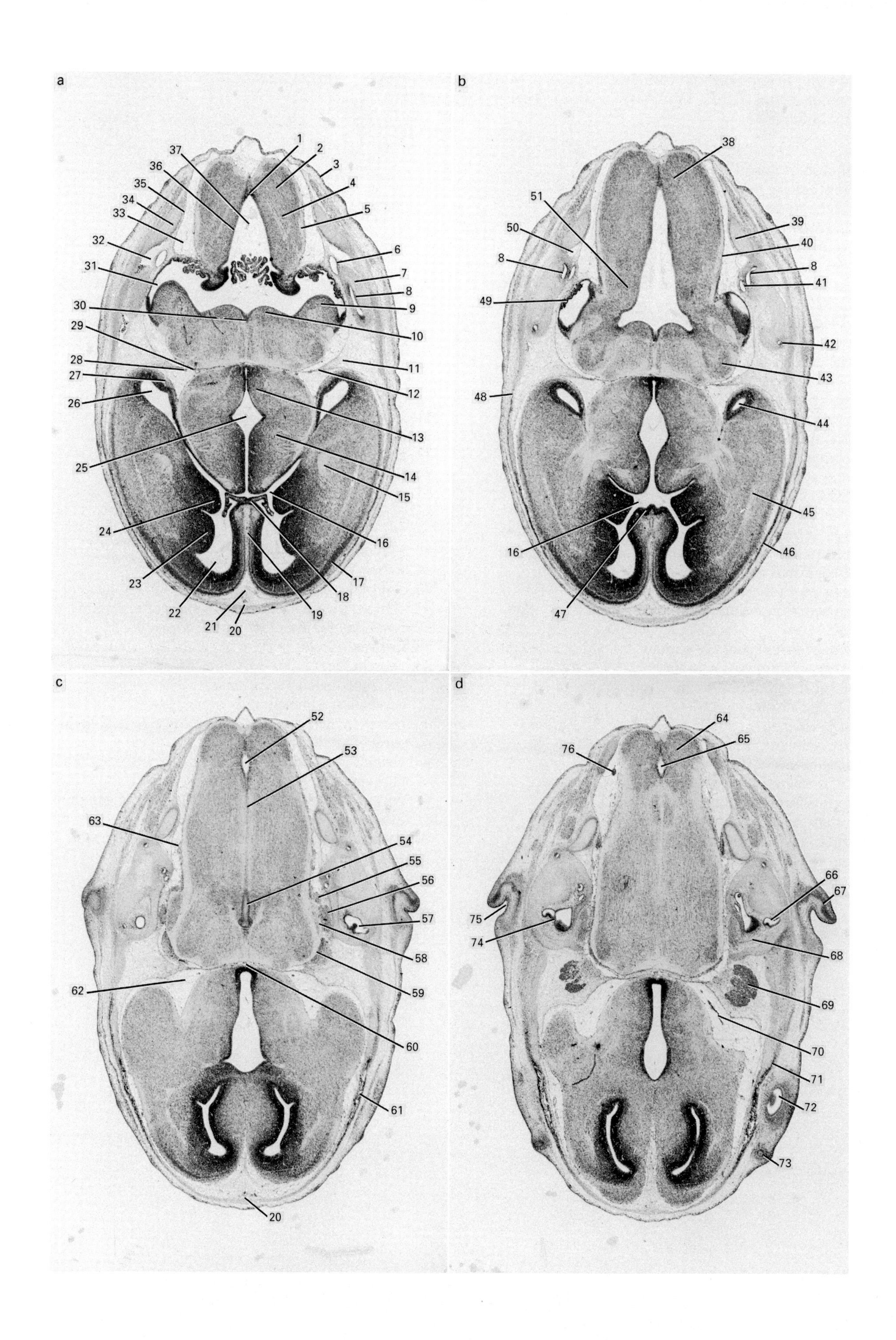
a
b
c
d
1
2
3
4
5
6
7
8
9
10
11
12
13
14
15
16
17
18
19
20
21
22
23
24
25
26
27
28
29
30
31
32
33
34
35
36
37
38
39
40
41
42
43
44
45
46
47
48
49
50
51
52
53
54
55
56
57
58
59
60
61
62
63
64
65
66
67
68
69
70
71
72
73
74
75
76

Plate 32b (14.5 days p.c.)

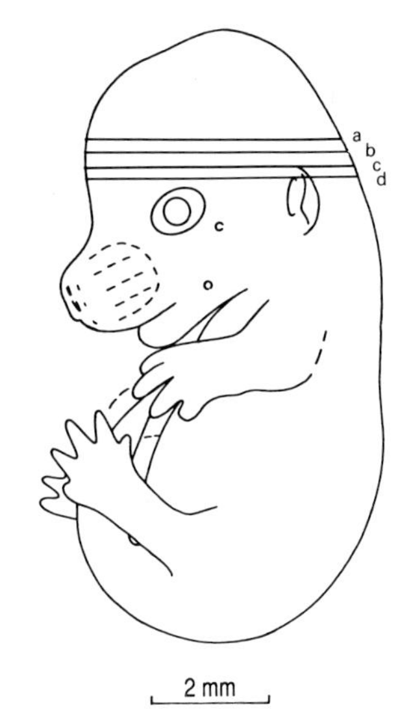

14.5 days p.c. Female embryo with over 60 pairs of somites. Crown–rump length about 10 mm (fixed). Transverse sections (Theiler Stages 22–23; rat, Witschi Stage 32).

1. roof plate of medulla oblongata
2. mantle layer
3. surface ectoderm (skin)
4. medulla oblongata (myelencephalon)
5. marginal layer
6. rostral extremity of endolymphatic sac
7. cartilage primordium of petrous part of temporal bone
8. posterior semicircular canal
9. intraventricular portion of cerebellar primordium
10. tegmentum of the pons (basal plate)
11. trochlear (IV) nerve
12. mesenchymal condensation forming tentorium cerebelli
13. diencephalon (hypothalamus)
14. diencephalon (thalamus)
15. fibres of internal capsule
16. interventricular foramen
17. neopallial cortex (future cerebral cortex)
18. choroidal fissure
19. inferior sagittal dural venous sinus
20. superior sagittal dural venous sinus
21. mesenchymal condensation forming falx cerebri
22. anterior horn of lateral ventricle
23. corpus striatum laterale (lateral aspect of ganglionic eminence)
24. corpus striatum mediale (medial aspect of ganglionic eminence)
25. third ventricle
26. posterior horn of lateral ventricle
27. tail of the caudate nucleus
28. oculomotor (III) nerve
29. rostral extension of posterior communicating artery
30. median sulcus
31. lateral recess of fourth ventricle
32. transverse dural venous sinus
33. loosely packed cephalic mesenchyme in future subarachnoid space
34. cartilage primordium of supraoccipital bone
35. mesenchyme (future dura mater)
36. ependymal layer
37. fourth ventricle
38. region of caudal medulla oblongata
39. cartilage primordium of exoccipital bone
40. vascular elements of pia mater
41. endolymphatic duct
42. superior semicircular canal
43. basal region of the pons
44. caudal extremity of posterior horn of lateral ventricle
45. internal capsule
46. mesenchymal precursor of frontal bone
47. lamina terminalis (roof of third ventricle)
48. mesenchymal precursor of parietal bone
49. choroid plexus in roof of fourth ventricle
50. superior petrosal dural venous sinus
51. region of rostral medulla oblongata
52. caudal region of fourth ventricle
53. medullary raphe (location of decussation of medial leminiscus)
54. floor of fourth ventricle
55. cochlear (VIII) ganglion
56. vestibular (VIII) ganglion
57. saccule
58. rootlets of vestibulocochlear (VIII) nerve
59. rootlets of trigeminal (V) nerve
60. basilar artery
61. first evidence of ossification within outer table of membranous primordium of frontal bone
62. rostral extremity of internal carotid artery
63. rootlets of glossopharyngeal (IX) nerve
64. rostral extremity of cervical region of spinal cord
65. rostral extremity of central canal
66. lateral semicircular canal
67. pinna of ear
68. nerve to ampulla of lateral semicircular canal
69. trigeminal (V) ganglion
70. middle cerebral artery
71. first evidence of ossification in outer table of membranous primordium of frontal bone in periorbital region
72. upper recess of conjunctival sac
73. primordium of prominent tactile or sinus hair follicle characteristically found in this location
74. origin of lateral semicircular canal from saccule
75. external auditory meatus
76. spinal accessory (XI) nerve

PLATE 32c

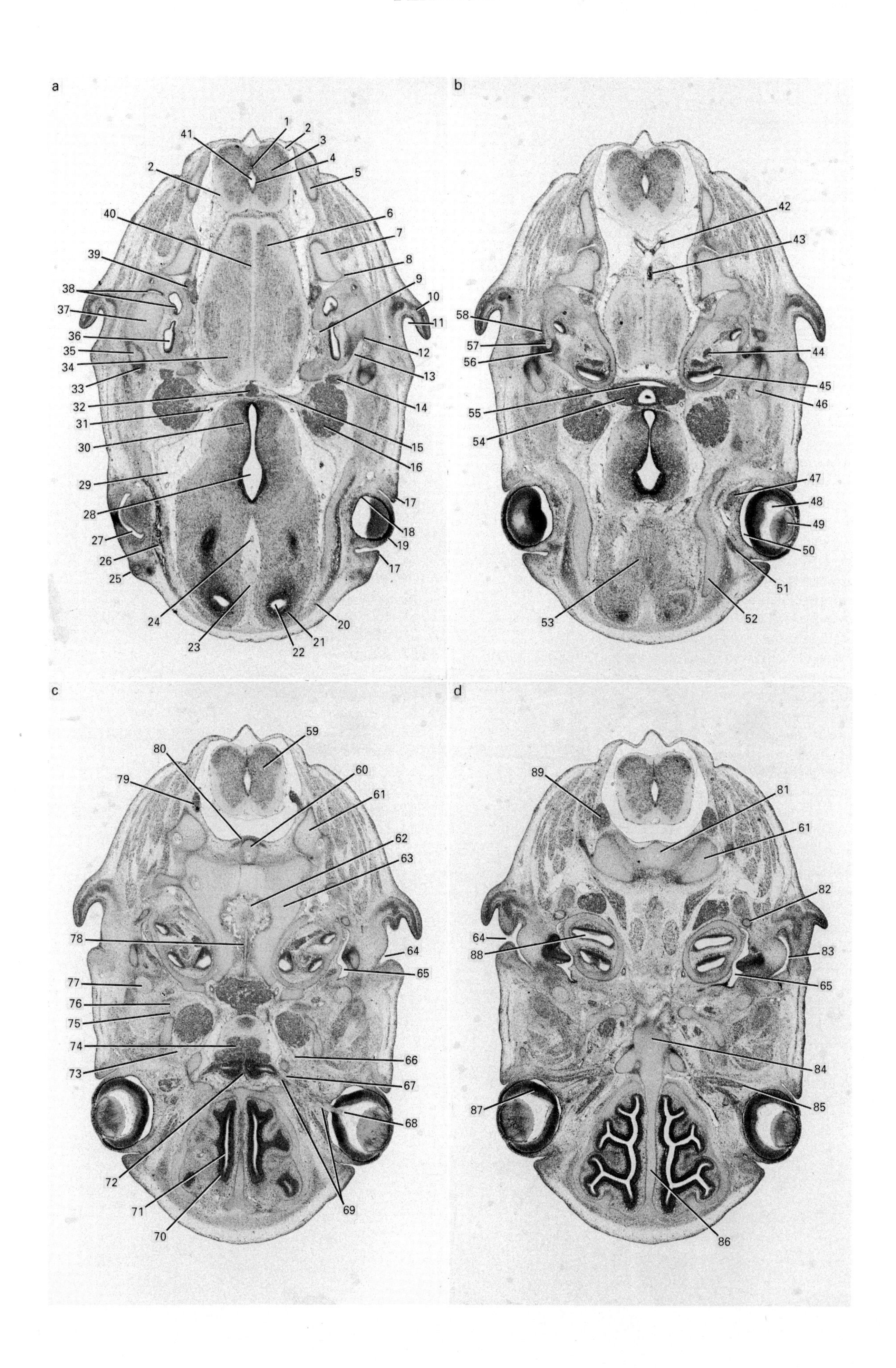
a
1
2
3
4
5
6
7
8
9
10
11
12
13
14
15
16
17
18
19
20
21
22
23
24
25
26
27
28
29
30
31
32
33
34
35
36
37
38
39
40
41
b
42
43
44
45
46
47
48
49
50
51
52
53
54
55
56
57
58
c
59
60
61
62
63
64
65
66
67
68
69
70
71
72
73
74
75
76
77
78
79
80
d
81
82
83
84
85
86
87
88
89

Plate 32c (14.5 days p.c.)

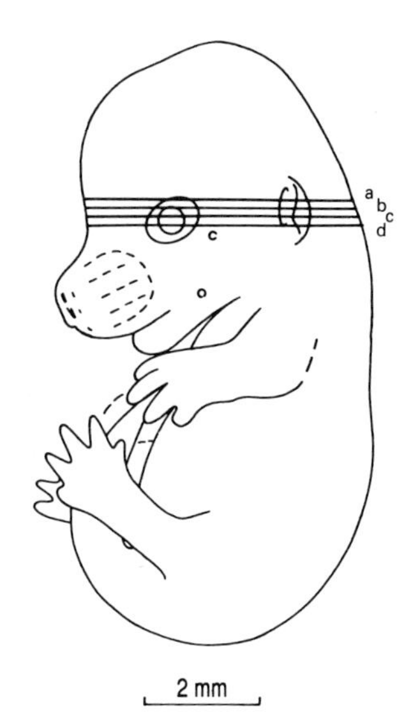

14.5 days p.c. Female embryo with over 60 pairs of somites. Crown–rump length about 10 mm (fixed). Transverse sections (Theiler Stages 22–23; rat, Witschi Stage 32).

1. roof plate
2. marginal layer
3. mantle layer
4. ependymal layer
5. cartilage primordium of neural arch of C1 vertebra (atlas)
6. medulla oblongata (myelencephalon)
7. cartilage primordium of exoccipital bone
8. jugular foramen
9. vestibulocochlear (VIII) nerve
10. pinna of ear
11. external auditory meatus
12. internal auditory (acoustic) meatus
13. facial (VII) nerve
14. facial (VII) ganglion
15. mesenchymal condensation forming tentorium cerebelli
16. trigeminal (V) ganglion
17. boundary of eyelid
18. pigment layer of retina
19. corneal epithelium
20. membranous primordium of frontal bone
21. neopallial cortex (future olfactory lobe)
22. lateral ventricle
23. mesenchymal condensation forming falx cerebri
24. median anterior cerebral artery (formed by amalgamation of right and left anterior cerebral arteries)
25. primordium of prominent tactile or sinus hair follicle characteristically found in this location
26. first evidence of ossification within outer table of membranous primordium of frontal bone
27. upper recess of conjunctival sac
28. third ventricle
29. loosely packed mesenchyme within future subarachnoid space
30. diencephalon (hypothalamus)
31. internal carotid artery
32. infundibulum (future pars nervosa) of pituitary
33. cartilage primordium of malleus
34. basal region of the pons
35. cartilage primordium of incus
36. crus commune of the labyrinth
37. cartilage primordium of petrous part of the temporal bone
38. posterior semicircular canal
39. superior ganglion of glossopharyngeal (IX) nerve
40. medullary raphe (location of decussation of the medial lemniscus)
41. central canal
42. left vertebral artery
43. basilar artery
44. vestibulocochlear (VIII) ganglion
45. superior part of cochlear duct
46. continuity between malleus and Meckel's cartilage
47. extrinsic ocular muscle (superior/dorsal rectus)
48. hyaloid cavity
49. lens
50. intra-retinal space
51. extrinsic ocular muscle (medial/nasal rectus)
52. superolateral part of nasal capsule
53. olfactory (I) nerves
54. site of active cellular and vascular proliferation in anterior wall of Rathke's pouch (pars anterior)
55. residual lumen of Rathke's pouch
56. cartilage primordium of stapes bone
57. stapedial artery
58. stapedius muscle
59. upper cervical region of spinal cord
60. cartilage primordium of dens (odontoid process of C2 vertebra)
61. cartilage primordium of lateral mass of C1 vertebra (atlas)
62. first evidence of ossification within parachordal plate
63. parachordal plate (future basioccipital bone)
64. entrance to first branchial cleft
65. tubo-tympanic recess
66. ophthalmic division of trigeminal (V) nerve
67. optic foramen
68. optic disc
69. optic (II) nerve
70. olfactory epithelium
71. nasal cavity
72. optic chiasma
73. superior orbital fissure
74. anterior part of floor of third ventricle
75. foramen rotundum
76. maxillary division of trigeminal (V) nerve
77. Meckel's cartilage
78. rostral extremity of notochord
79. dorsal (posterior) root ganglion of C1 (variable in size)
80. horizontal component of cruciate ligament (attached to the atlas) which helps to hold the odontoid process in position
81. cartilage primordium of body of C2 vertebra (axis)
82. Reichert's cartilage (region of future styloid process)
83. first branchial cleft
84. cartilage primordium of sphenoid bone
85. extrinsic ocular muscle (lateral/temporal rectus)
86. cartilage primordium of nasal septum
87. neural layer of retina
88. cochlea
89. dorsal (posterior) root ganglion of C2

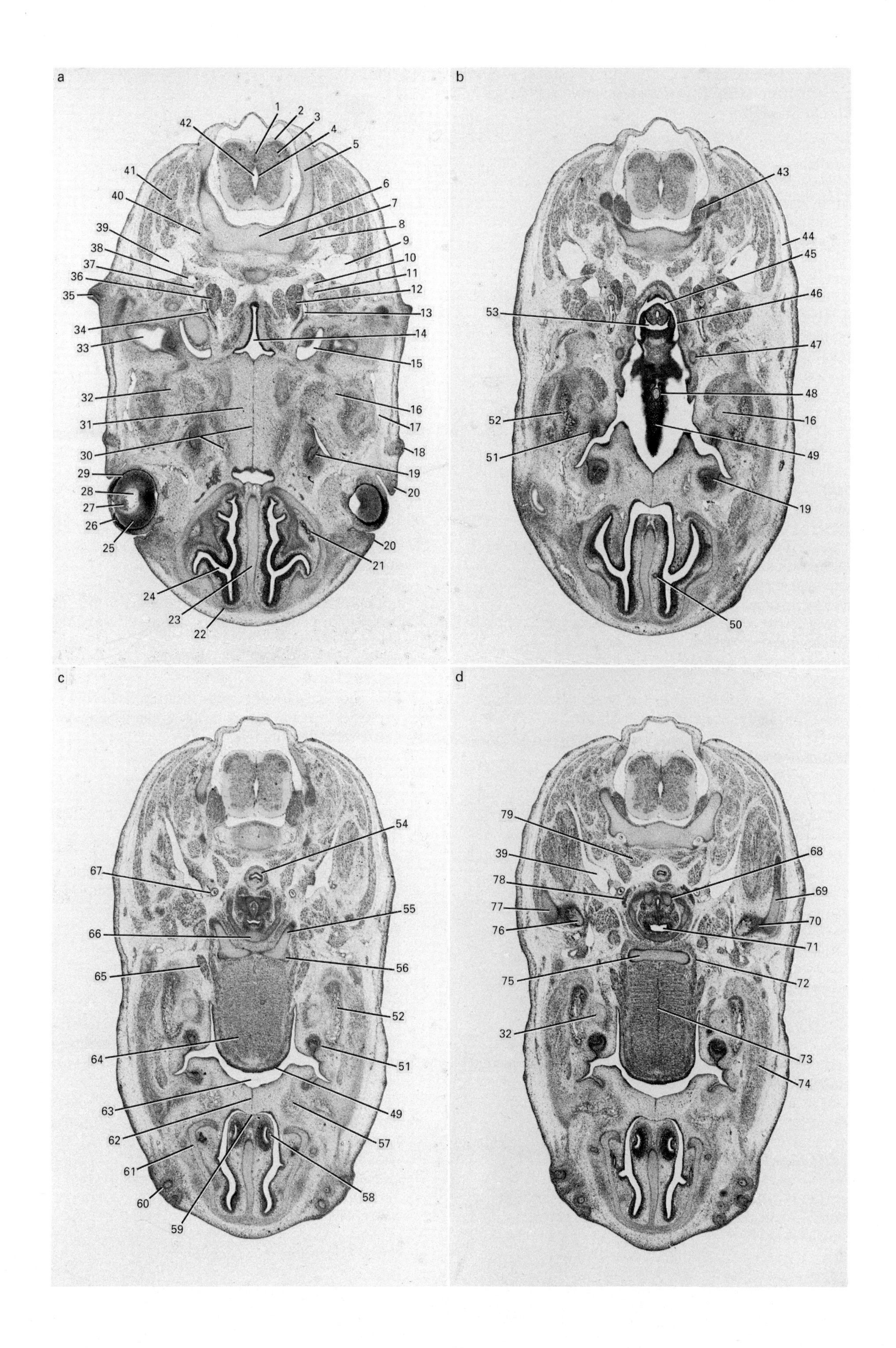
a
b
c
d
1
2
3
4
5
6
7
8
9
10
11
12
13
14
15
16
17
18
19
20
21
22
23
24
25
26
27
28
29
30
31
32
33
34
35
36
37
38
39
40
41
42
43
44
45
46
47
48
49
50
51
52
53
54
55
56
57
58
59
60
61
62
63
64
65
66
67
68
69
70
71
72
73
74
75
76
77
78
79

Plate 32d (14.5 days p.c.)

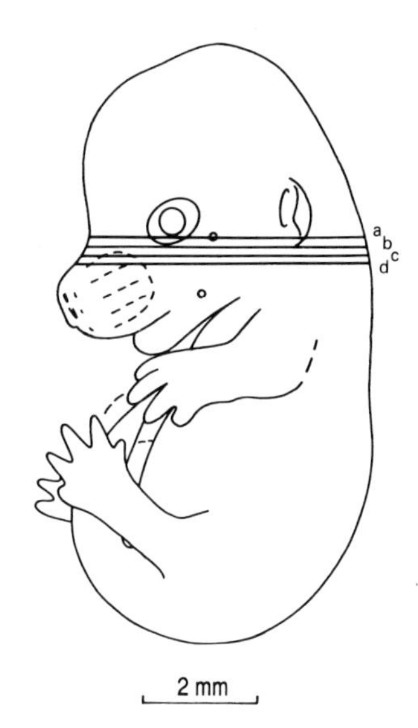

14.5 days p.c. Female embryo with over 60 pairs of somites. Crown–rump length about 10 mm (fixed). Transverse sections (Theiler Stages 22–23; rat, Witschi Stage 32).

1. roof plate
2. marginal layer
3. mantle layer
4. ependymal layer
5. cartilage primordium of neural arch of C2 vertebra (axis)
6. notochord (nucleus pulposus)
7. second cervical vertebral body (axis)
8. left vertebral artery
9. left jugular lymph sac
10. left internal jugular vein
11. left vagal (X) trunk
12. left superior cervical sympathetic ganglion
13. left internal carotid artery
14. entrance into nasopharynx
15. left pharyngo-tympanic tube (Eustachian tube)
16. left Meckel's cartilage
17. branch of primary head vein, which drains to external jugular vein
18. primordium of prominent tactile or sinus hair follicle characteristically found in this location
19. primordium of first upper molar tooth
20. margin of eyelid
21. tubules of serous nasal glands that course rostrally in the lateral wall of the middle meatus
22. nasal capsule
23. cartilage primordium of nasal septum
24. nasal cavity
25. neural layer of retina
26. corneal epithelium
27. lens
28. hyaloid cavity
29. pigment layer of retina
30. line of fusion (in the midline) of the two palatal shelves and (laterally) ossification centre in lateral part of right palatal shelf
31. horizontally directed palatal shelf of the maxilla
32. right Meckel's cartilage
33. right tubo-tympanic recess
34. right internal carotid artery
35. pinna of ear
36. right superior cervical sympathetic ganglion
37. right vagal (X) trunk
38. right internal jugular vein
39. right jugular lymph sac
40. right vertebral artery
41. long muscles of the neck: superficial layer – splenius; middle layer (lateral system) – iliocostalis cervicis, longissimus cervicis, capitis and atlantis; deep layer (medial system) – semispinalis capitis and complexus muscle
42. central canal
43. dorsal (posterior) root ganglion
44. superficial lymphatic vessel that drains into the jugular lymph sac
45. entrance to the oesophagus
46. piriform fossa
47. cartilage primordium of dorsal extremity of lesser horn of hyoid bone
48. median circumvallate papilla
49. dorsum of tongue covered with numerous filiform papillae
50. tubule of serous gland associated with nasal septum
51. primordium of lower molar tooth
52. first evidence of ossification within mandible (around, but principally lateral to, Meckel's cartilage)
53. entrance to larynx
54. oesophagus
55. cartilage primordium of greater horn of hyoid bone
56. cartilage primordium of proximal part of lesser horn of hyoid bone
57. early evidence of ossification within the maxilla
58. vomeronasal organ (Jacobson's organ)
59. site of fusion of the nasal septum in the midline with the dorsal surface of the palatal shelves
60. primordium of follicle of vibrissa
61. lateral part of nasal capsule
62. site of fusion in the midline of the two palatal shelves
63. oral cavity
64. intrinsic muscle of the tongue (transverse component)
65. extrinsic muscle of the tongue (styloglossus muscle)
66. primordium of thyroid cartilage (in continuity caudally with the primordium of the cricoid cartilage)
67. right common carotid artery
68. primordium of arytenoid cartilage
69. cartilage primordium of acromion of left scapula
70. primordium of left acromio-clavicular joint
71. laryngeal aditus
72. extrinsic muscle of the tongue (hyoglossus muscle)
73. precursor of median fibrous septum of tongue
74. masseter muscle
75. cartilage primordium of body of hyoid bone
76. early evidence of ossification within the clavicle (confined to its lateral half)
77. cartilage primordium of tip of acromion of right scapula
78. upper pole of right lobe of thyroid gland
79. prevertebral muscles of the neck (longus capitis, longus cervicis)

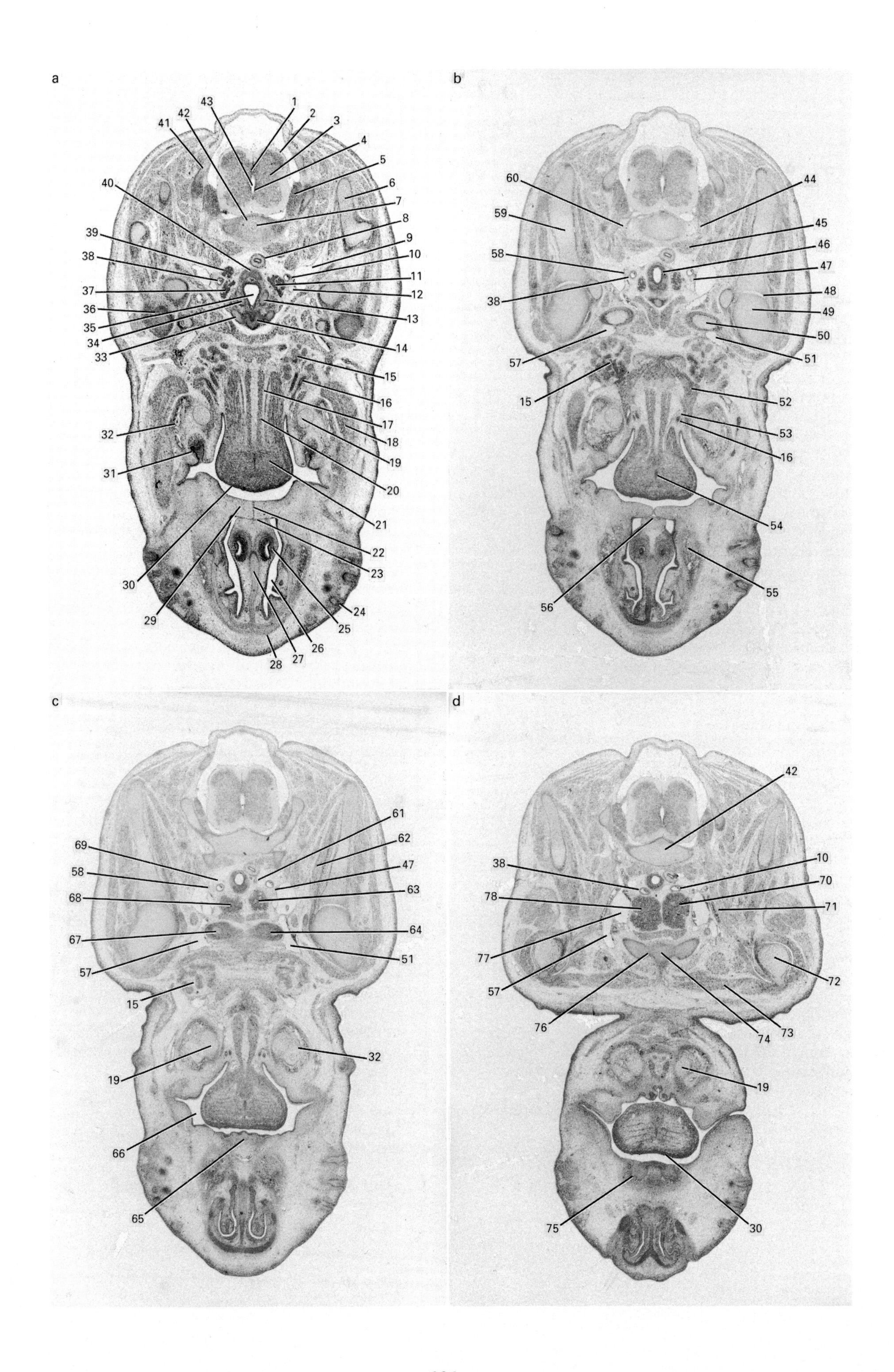
a
1
2
3
4
5
6
7
8
9
10
11
12
13
14
15
16
17
18
19
20
21
22
23
24
25
26
27
28
29
30
31
32
33
34
35
36
37
38
39
40
41
42
43
b
60
44
59
45
58
46
47
38
48
49
50
57
51
15
52
53
16
54
55
56
c
61
69
62
58
47
68
63
67
64
57
51
15
32
19
66
65
d
42
38
10
70
78
71
77
57
72
76
74
73
19
75
30

Plate 32e (14.5 days p.c.)

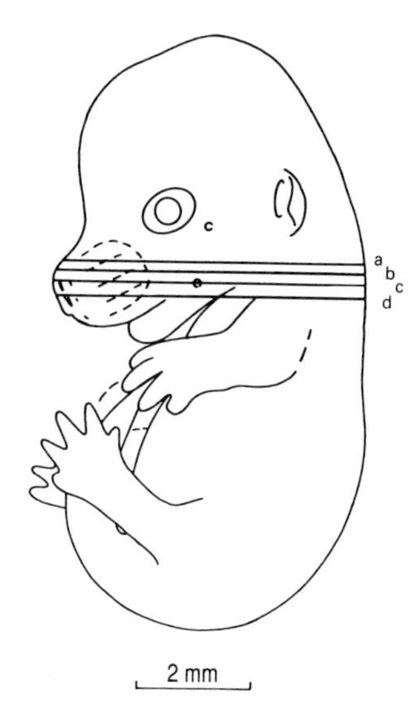

14.5 days p.c. Female embryo with over 60 pairs of somites. Crown–rump length about 10 mm (fixed). Transverse sections (Theiler Stages 22–23; rat, Witschi Stage 32).

1. roof plate
2. marginal layer
3. mantle layer
4. ependymal layer
5. dorsal (posterior) root ganglion
6. medial (posterior) border of left scapula
7. notochord (nucleus pulposus)
8. oesophagus
9. left jugular lymph sac
10. left common carotid artery
11. left thyroid lobe
12. left internal jugular vein
13. left arytenoid cartilage
14. cricoid cartilage (anterior arch)
15. submandibular gland
16. submandibular duct (medial) and sublingual duct (lateral)
17. hypoglossal (XII) nerve
18. masseter muscle
19. Meckel's cartilage
20. extrinsic muscle of the tongue (genioglossus)
21. intrinsic muscle of the tongue (transverse component)
22. site of fusion in the midline of the two palatal shelves
23. site of fusion of the nasal septum in the midline with the dorsal surface of the palatal shelves
24. primordium of follicle of vibrissa
25. vomeronasal organ (Jacobson's organ)
26. nasal cavity
27. cartilage primordium of nasal septum
28. nasal capsule
29. palatal shelf of maxilla (secondary palate)
30. dorsal surface of tongue
31. primordium of lower molar tooth
32. first evidence of ossification within mandible (around, but principally lateral to, Meckel's cartilage)
33. thyroid cartilage
34. right arytenoid cartilage
35. laryngeal aditus
36. rostral extremity of head of the right humerus
37. right thyroid lobe
38. right common carotid artery
39. right parathyroid gland (third pharyngeal pouch origin – only one pair is present, unlike the situation in man)
40. cricoid cartilage (posterior lamina)
41. cartilage primordium of neural arch
42. cartilage primordium of lower cervical vertebral body
43. central canal
44. left vertebral artery
45. prevertebral muscles of the neck (longus capitis, longus cervicis)
46. larynx
47. left vagal (X) trunk
48. left gleno-humeral joint
49. cartilage primordium of head of the left humerus
50. middle region of clavicle, with early evidence of ossification
51. left subclavian vein
52. extrinsic muscle of the tongue (hyoglossus)
53. lingual nerve hooking around submandibular duct
54. precursor of median fibrous septum of tongue
55. early evidence of ossification within the maxilla
56. site of apposition (but not yet fusion) of the two palatal shelves and the nasal septum
57. right subclavian vein
58. right vagal (X) trunk
59. primary ossification centre within cartilage primordium of the blade of the scapula with superficial layer of periosteal ossification
60. right vertebral artery
61. right recurrent laryngeal nerve
62. subscapularis muscle
63. rostral extremity of left thymic rudiment
64. medial one-third of left clavicle
65. primary palate
66. oral cavity
67. medial one-third of right clavicle
68. rostral extremity of right thymic rudiment
69. right recurrent laryngeal nerve
70. left thymic rudiment
71. components of left brachial plexus
72. cartilage primordium of upper shaft region of left humerus
73. pectoral muscle (pectoralis profundus)
74. cartilage primordium of manubrium sterni
75. tooth primordium of right upper incisor tooth
76. primordium of sterno-costal joint between first rib and manubrium
77. right jugular lymph sac
78. right thymic rudiment

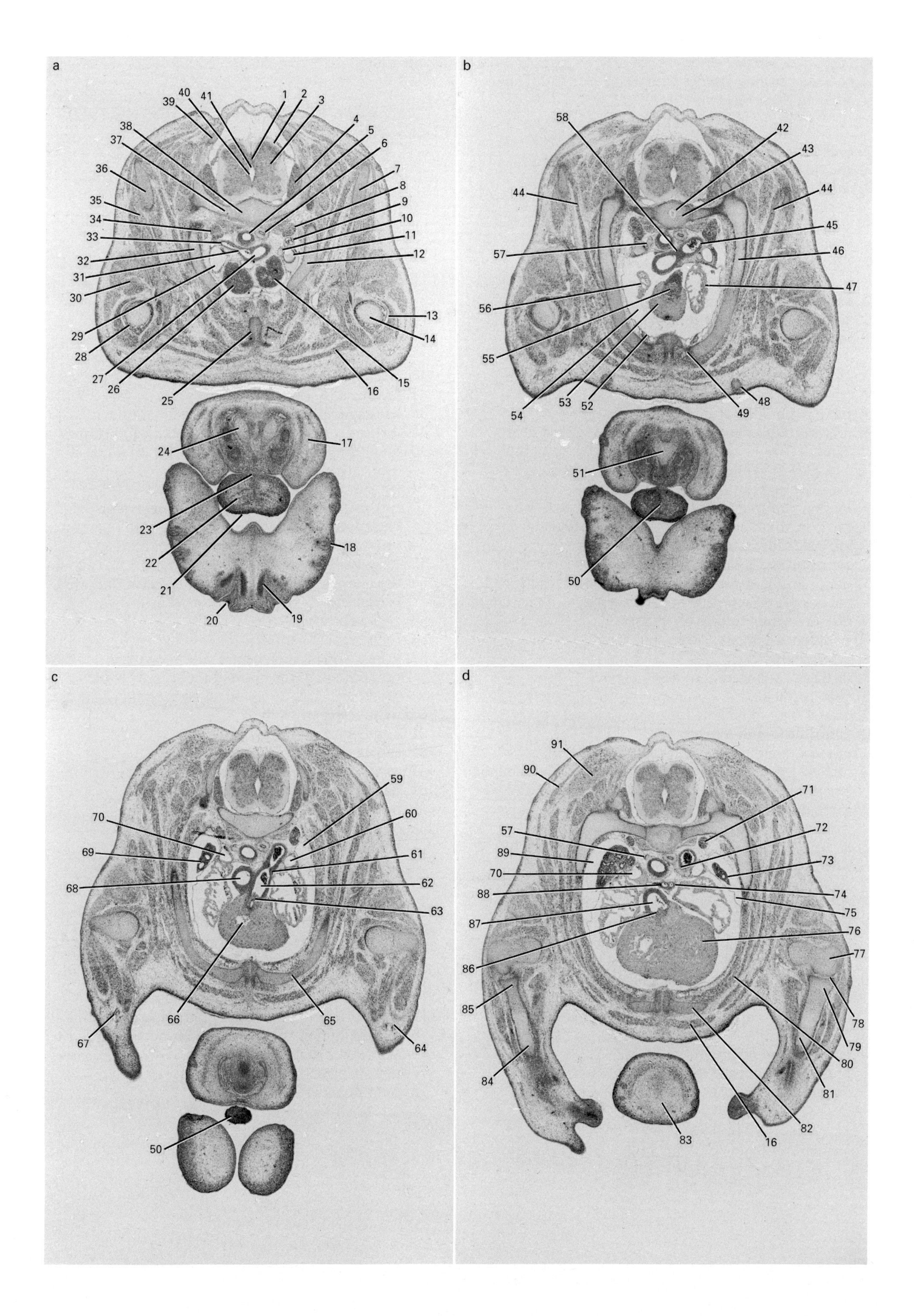

a
b
c
d
1
2
3
4
5
6
7
8
9
10
11
12
13
14
15
16
17
18
19
20
21
22
23
24
25
26
27
28
29
30
31
32
33
34
35
36
37
38
39
40
41
42
43
44
45
46
47
48
49
50
51
52
53
54
55
56
57
58
59
60
61
62
63
64
65
66
67
68
69
70
71
72
73
74
75
76
77
78
79
80
81
82
83
84
85
86
87
88
89
90
91

Plate 32f (14.5 days p.c.)

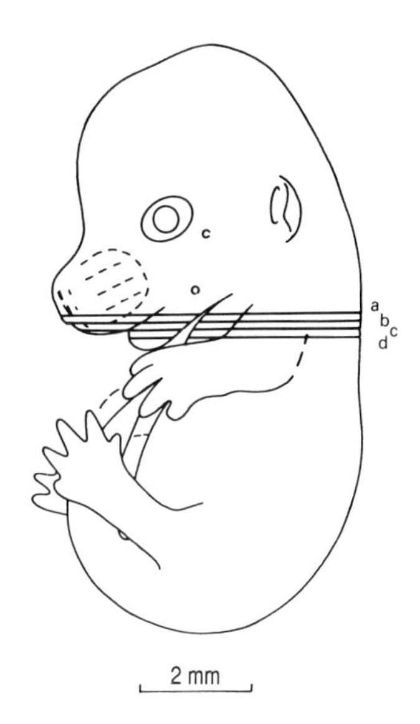

14.5 days p.c. Female embryo with over 60 pairs of somites. Crown–rump length about 10 mm (fixed). Transverse sections (Theiler Stages 22–23; rat, Witschi Stage 32).

1. roof plate
2. marginal layer
3. mantle layer
4. dorsal (posterior) root ganglion
5. trachea
6. oesophagus
7. medial (posterior) border of left scapula
8. left inferior cervical/first thoracic sympathetic ganglion (stellate ganglion)
9. left subclavian artery
10. left vagal (X) trunk
11. communication between left jugular lymph sac and left internal jugular vein (origin of left superior (cranial) vena cava)
12. ossification within cartilage primordium of shaft of left first rib
13. radial nerve
14. ossification within cartilage primordium of mid-shaft region of the left humerus
15. left lobe of thymic rudiment
16. pectoral muscle (pectoralis profundus)
17. lower jaw
18. primordium of follicle of vibrissa
19. rostral extremity of nasal cartilage
20. entrance into nasal cavity (external naris)
21. dorsal surface of tongue
22. intrinsic muscle of tongue (transverse component)
23. frenulum of the tongue
24. Meckel's cartilage
25. cartilage primordium of manubrium sterni
26. right lobe of thymic rudiment
27. rostral extremity of arch of the aorta
28. first evidence of ossification in periosteum of mid-shaft region of the humerus
29. communication between right jugular lymph sac and right internal jugular vein (origin of right superior (cranial) vena cava)
30. triceps brachii muscle
31. origin of the right subclavian artery from the brachiocephalic (innominate) artery
32. ossification within cartilage primordium of shaft of the right first rib
33. right subclavian artery
34. right inferior cervical/first thoracic sympathetic ganglion (stellate ganglion)
35. teres major muscle
36. medial (posterior) border of right scapula
37. location of articulation between cartilage primordium of head of the first rib and the body of the first thoracic vertebra
38. cartilage primordium of first thoracic vertebral body
39. cartilage primordium of neural arch
40. central canal
41. ependymal layer
42. notochord (nucleus pulposus)
43. cartilage primordium of second thoracic vertebral body
44. serratus anterior muscle
45. arch of the aorta just proximal to the site of entrance of the ductus arteriosus
46. ossification within cartilage primordium of the shaft of the right second rib
47. wall of the left atrium
48. mammary gland primordium
49. articulation between cartilage primordium of second rib and lower border of manubrium sterni
50. tip of the tongue
51. rostral extremity of Meckel's cartilages (site of fusion in ventral midline)
52. mesothelial lining of pericardial cavity (parietal pericardium)
53. wall of right ventricle
54. rostral extremity of pericardial cavity
55. origin of the pulmonary trunk
56. wall of the right atrium
57. right vagal (X) trunk
58. wall of the arch of the aorta
59. rostral extremity of left pleural cavity
60. left superior (cranial) vena cava
61. wall of ductus arteriosus
62. pulmonary trunk/origin of ductus arteriosus
63. leaflet of pulmonary valve
64. origin of left forelimb
65. cartilage primordium of anterior extremity of third rib
66. lumen of right ventricle
67. right forelimb
68. proximal part of ascending (thoracic) aorta
69. cranial lobe of right lung
70. right superior (cranial) vena cava
71. sympathetic ganglion
72. left recurrent laryngeal nerve
73. apex of left lung
74. left pulmonary artery
75. pleuro-pericardial membrane
76. wall of left ventricle
77. cartilage primordium of capitulum of the left humerus
78. primordium of proximal radio-humeral joint
79. cartilage primordium of head of the left radius
80. intercostal muscles (external and internal layers)
81. ossification within cartilage primordium of shaft of the left radius
82. primordium of costal cartilage
83. ventral extremity of the lower jaw
84. ossification within cartilage primordium of shaft of the right radius
85. cartilage primordium of head of the right radius
86. leaflet of aortic valve
87. origin of the thoracic aorta (region of aortic sinuses)
88. right pulmonary artery
89. right pleural cavity
90. trapezius muscle
91. latissimus dorsi muscle

PLATE 32g

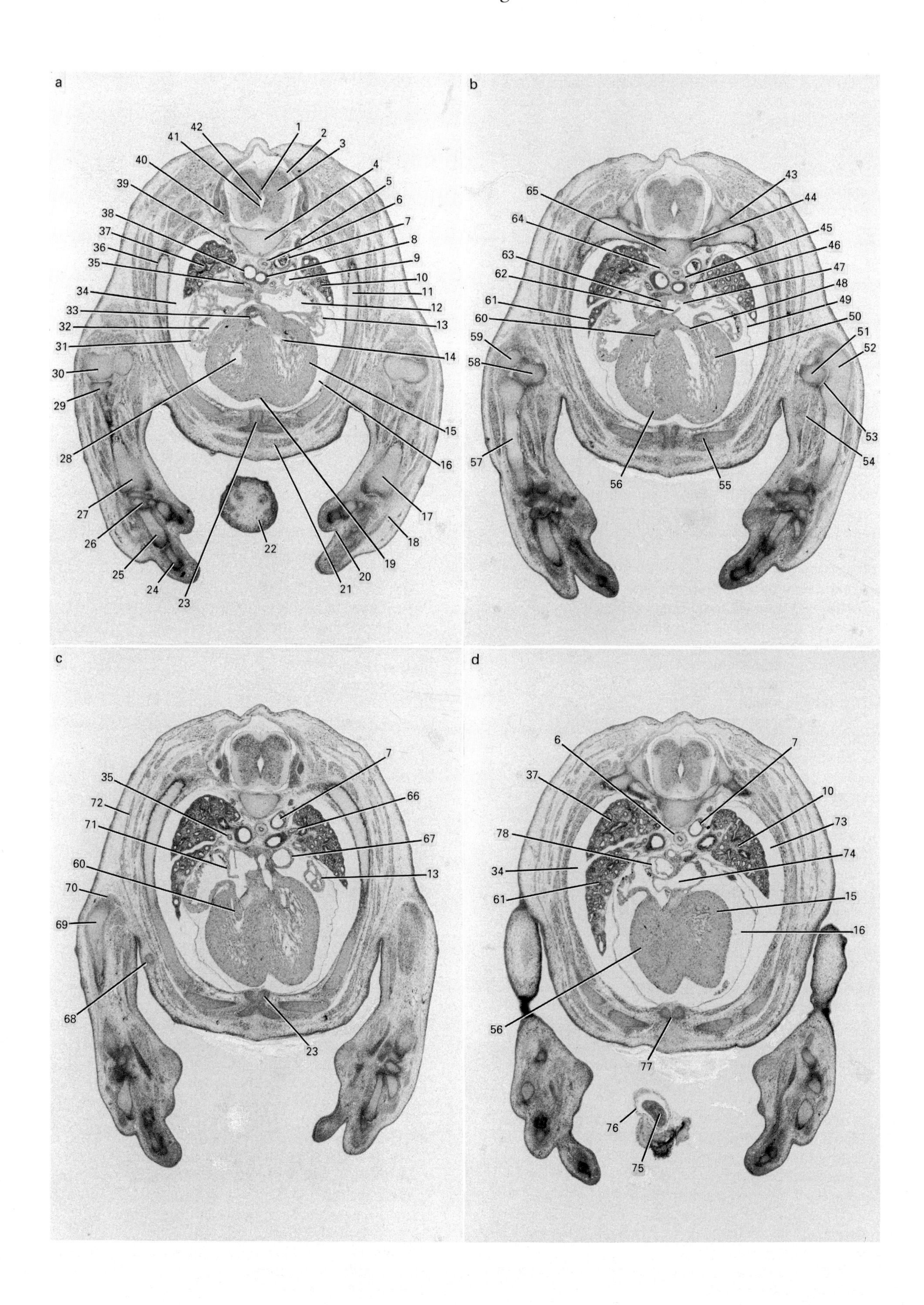
a
b
c
d
1
2
3
4
5
6
7
8
9
10
11
12
13
14
15
16
17
18
19
20
21
22
23
24
25
26
27
28
29
30
31
32
33
34
35
36
37
38
39
40
41
42
43
44
45
46
47
48
49
50
51
52
53
54
55
56
57
58
59
60
61
62
63
64
65
66
67
68
69
70
71
72
73
74
75
76
77
78

Plate 32g (14.5 days p.c.)

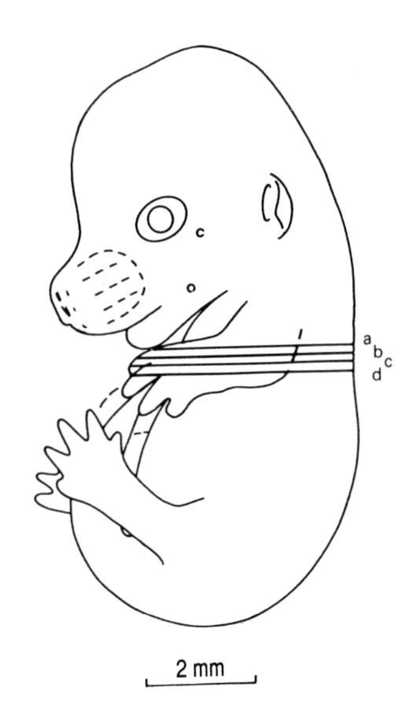

14.5 days p.c. Female embryo with over 60 pairs of somites. Crown–rump length about 10 mm (fixed). Transverse sections (Theiler Stages 22–23; rat, Witschi Stage 32).

1. roof plate
2. marginal layer
3. mantle layer
4. cartilage primordium of thoracic vertebral body
5. sympathetic ganglion
6. oesophagus
7. descending (thoracic) aorta
8. left main bronchus
9. left superior (cranial) vena cava
10. left lung
11. ossification within cartilage primordium of rib
12. left atrial chamber of heart
13. wall of left atrium (this chamber of the heart is substantially more rostral than the right atrium)
14. left ventricular chamber
15. trabeculated wall of left ventricle
16. pericardial cavity
17. distal part of shaft of left radius
18. left forelimb
19. interventricular groove
20. digital interzone
21. pectoral muscle (pectoralis profundus)
22. ventral extremity of the lower jaw
23. cartilage primordium of sternebral bone
24. cartilage primordium of phalangeal bone
25. cartilage primordium of metacarpal bone
26. cartilage primordium of carpal bones
27. cartilage primordium of distal part of shaft of right radius
28. right ventricular chamber
29. cartilage primordium of head of the radius
30. cartilage primordium of capitulum of humerus
31. wall of right atrium (this chamber of the heart is more caudal than the left atrium)
32. right atrial chamber of the heart
33. leaflet of aortic valve
34. right pleural cavity
35. right pulmonary artery
36. periosteum of rib primordium
37. cranial lobe of right lung
38. right main bronchus
39. right sympathetic chain
40. dorsal (posterior) root ganglion
41. ependymal layer
42. central canal
43. primordium of joint between tubercle of rib and neural arch of its own vertebra
44. primordium of joint between head of rib and its own vertebral body
45. left/anterior vagal (X) trunk
46. left pulmonary artery
47. ostium secundum
48. leaflet of left atrio-ventricular (mitral) valve
49. pleuro-pericardial membrane
50. trabeculae carneae
51. cartilage primordium of the trochlea of the left humerus
52. cartilage primordium of the proximal part of the left ulna
53. primordium of the left humero-ulnar joint
54. extensor muscles of the forelimb
55. primordium of costal cartilage
56. wall of right ventricle
57. early evidence of ossification within cartilage primordium of shaft of right ulna
58. cartilage primordium of the trochlea of the right humerus
59. cartilage primordium of the olecranon process of the right ulna
60. leaflet of right atrio-ventricular (tricuspid) valve
61. middle lobe of right lung
62. septum primum
63. septum secundum (dorsal component)
64. right vagal (X) trunk
65. notochord (nucleus pulposus)
66. left pulmonary artery entering root of the left lung
67. left superior (cranial) vena cava as it progresses towards midline
68. mammary gland primordium
69. distal part of shaft of right humerus
70. insertion of the tendon of the triceps muscle into the olecranon process of the right ulna
71. leaflet of venous valve at site of entrance of inferior vena cava into right atrium
72. cutaneous muscle of the trunk (panniculus carnosus)
73. left pleural cavity
74. opening of left superior (cranial) vena cava into cavity of right atrium
75. wall of midgut within physiological umbilical hernia
76. wall of physiological umbilical hernia
77. cartilage primordia of xiphoid process
78. opening of inferior vena cava into cavity of right atrium

PLATE 32h

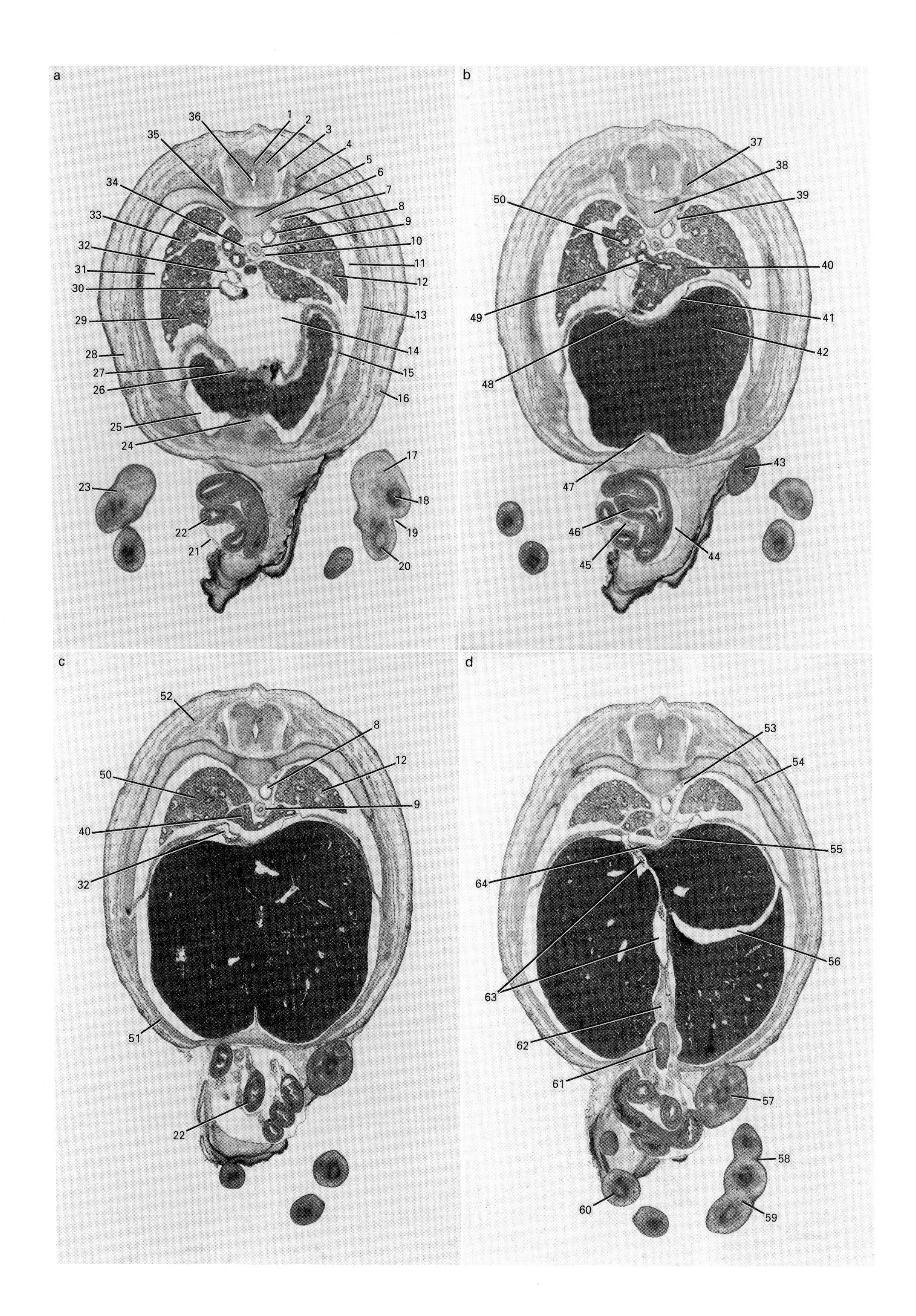

a
1
2
3
4
5
6
7
8
9
10
11
12
13
14
15
16
17
18
19
20
21
22
23
24
25
26
27
28
29
30
31
32
33
34
35
36
b
37
38
39
40
41
42
43
44
45
46
47
48
49
50
c
52
8
12
50
9
40
32
51
22
d
53
54
55
56
57
58
59
60
61
62
63
64

Plate 32h (14.5 days p.c.)

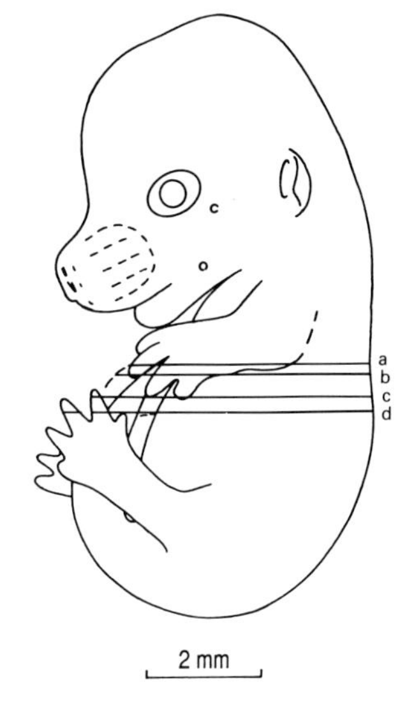

14.5 days p.c. Female embryo with over 60 pairs of somites. Crown–rump length about 10 mm (fixed). Transverse sections (Theiler Stages 22–23; rat, Witschi Stage 32).

1. roof plate
2. marginal layer
3. mantle layer
4. cartilage primordium of neural arch
5. notochord (nucleus pulposus)
6. ossification within cartilage primordium of rib
7. sympathetic trunk
8. descending (thoracic) aorta
9. oesophagus
10. left (becoming anterior) vagal (X) trunk
11. left pleural cavity
12. left lung
13. intercostal muscles (external and internal layers)
14. caudal extremity of pericardial cavity
15. peripheral margin (left dome) of the diaphragm
16. mammary gland primordium
17. left forelimb
18. cartilage primordium of phalangeal bone
19. digital interzone (left forelimb)
20. primordium of digit (left forelimb)
21. wall of physiological umbilical hernia
22. midgut loop within physiological umbilical hernia
23. right forelimb
24. rostral extremity of falciform ligament, continuous above with body wall (future sternal) component of ventral region of the diaphragm
25. sub-phrenic recess of peritoneal cavity
26. periphery of central tendon of the diaphragm
27. upper part of the right lobe of the liver
28. cutaneous muscle of trunk (panniculus carnosus)
29. middle lobe of right lung
30. caudal part of wall of right atrium
31. right pleural cavity
32. inferior vena cava
33. cranial lobe of right lung
34. right (becoming posterior) vagal (X) trunk
35. primordium of joint at location of rib articulation with its own vertebra
36. central canal
37. dorsal (posterior) root ganglion
38. cartilage primordium of thoracic vertebral body
39. hemiazygos vein
40. accessory lobe of right lung
41. left boundary of bare area of the liver
42. left lobe of the liver
43. caudal extremity of the tail
44. extension of abdominal cavity into the physiological umbilical hernia
45. mesentery of midgut
46. lumen of Meckel's diverticulum
47. falciform ligament
48. right boundary of bare area of the liver
49. lobar bronchus to accessory lobe of the right lung
50. caudal lobe of the right lung
51. rectus abdominis muscle
52. deep muscles of the back (multifidus, longissimus lumborum, longissimus thoracis)
53. sympathetic ganglion
54. periosteum of rib primordium
55. left crus of the diaphragm
56. lobar interzone
57. caudal region of the tail
58. digital interzone (left hindlimb)
59. left hindlimb
60. primordium of digit (right hindlimb)
61. proximal part of midgut loop
62. gallbladder bed
63. ductus venosus
64. right crus of the diaphragm

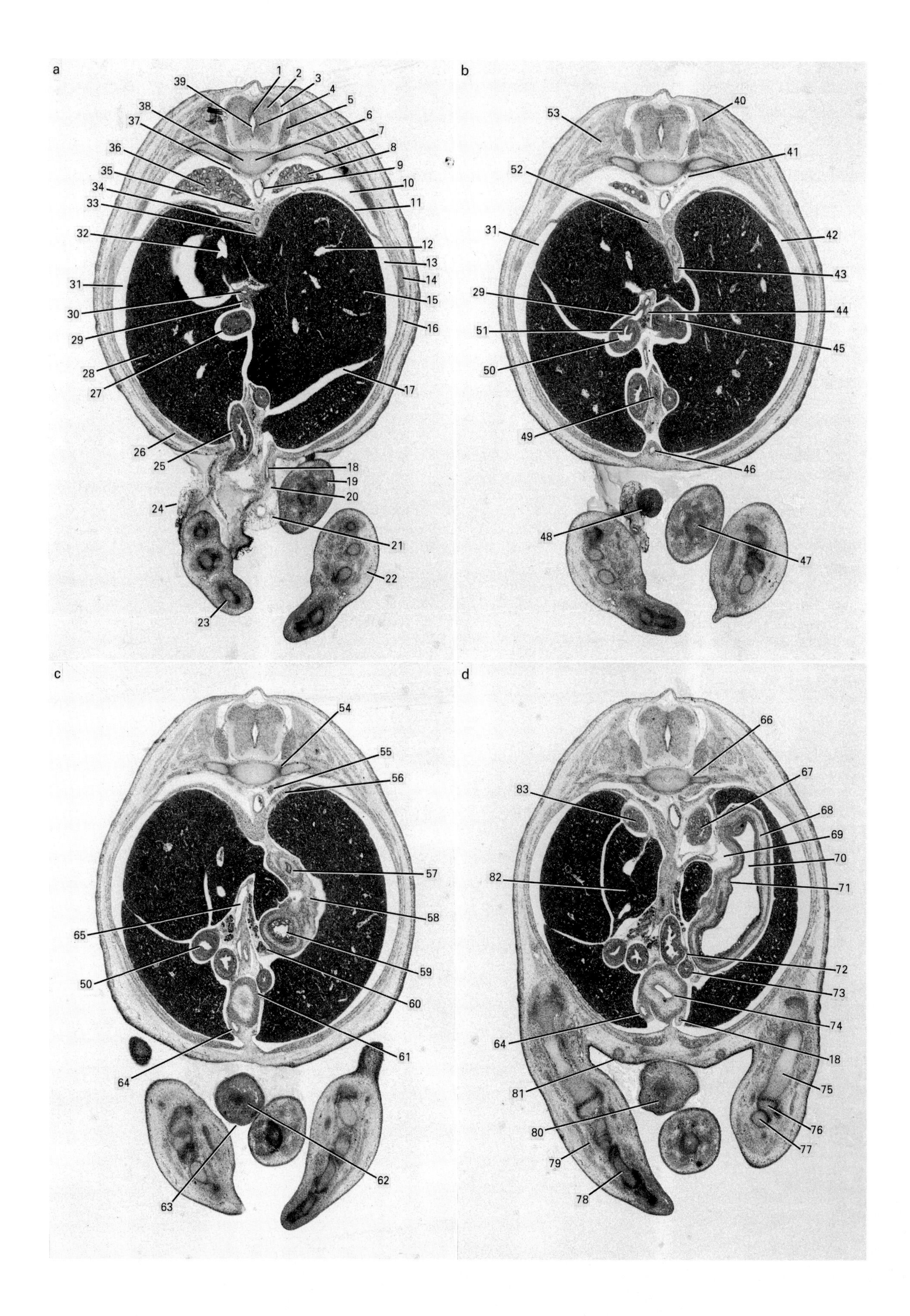
a
b
c
d

Plate 32i (14.5 days p.c.)

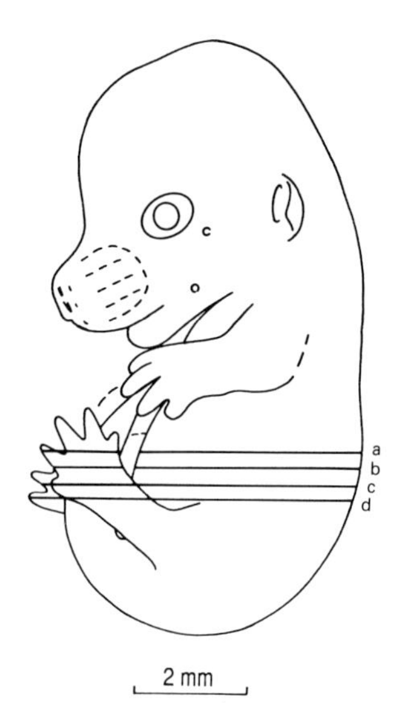

14.5 days p.c. Female embryo with over 60 pairs of somites. Crown–rump length about 10 mm (fixed). Transverse sections (Theiler Stages 22–23; rat, Witschi Stage 32).

1. roof plate
2. mantle layer
3. marginal layer
4. ependymal layer
5. dorsal (posterior) root ganglion
6. notochord (nucleus pulposus)
7. ossification within cartilage primordium of proximal part of shaft of rib
8. caudal extremity of left lung
9. descending (thoracic) aorta
10. left pleural cavity
11. left dome of the diaphragm
12. part of hepatic venous plexus
13. abdominal/peritoneal cavity
14. section through shaft of rib primordium
15. left lobe of liver
16. three layers of abdominal wall musculature (external oblique, internal oblique and transversus abdominis muscles)
17. interlobar space
18. left umbilical artery
19. tail
20. wall of physiological umbilical hernia
21. Wharton's jelly
22. left hindlimb
23. phalangeal primordium
24. amnion
25. proximal part of midgut loop
26. rectus abdominis muscle
27. wall of duodenum
28. right lobe of liver
29. common bile duct
30. right hepatic duct
31. right sub-phrenic space of abdominal/peritoneal cavity
32. origin of posthepatic inferior vena cava
33. anterior vagal (X) trunk
34. oesophagus
35. posterior vagal (X) trunk
36. caudal lobe of right lung
37. right sympathetic trunk
38. cartilage primordium of thoracic vertebral body
39. central canal
40. cartilage primordium of neural arch
41. hemiazygos vein
42. left sub-phrenic space of abdominal/peritoneal cavity
43. ventral mesentery of oesophagus
44. primordium of right lobe and body of the pancreas
45. pyloric region of stomach
46. urachus
47. filum terminale
48. dorsum of genital tubercle
49. vitelline vessels (superior mesenteric) within root of the mesentery of the midgut
50. lumen of the second (descending) part of the duodenum
51. site of entrance of the common bile duct into the duodenum at the duodenal papilla (ampulla of Vater); this opening into the duodenum is surrounded by a defined layer of circular muscle (the sphincter of Oddi)
52. right crus of the diaphragm
53. multifidus and longissimus lumborum muscles
54. rib articulation with lower thoracic vertebral body
55. sympathetic ganglion
56. left crus of the diaphragm
57. gastro-oesophageal junction
58. vagal (X) plexus ramifying on dorsal surface of stomach
59. lumen of pyloric antrum
60. primordium of left lobe of the pancreas
61. wall of urogenital sinus
62. genital tubercle
63. primitive urethral groove
64. right umbilical artery
65. portal vein
66. articulation of 13th rib primordium with thoracic vertebral body
67. upper pole of left adrenal primordium
68. anterior wall of stomach
69. lesser sac of omental bursa
70. lumen of body of stomach
71. posterior wall of stomach
72. dorsal mesentery of hindgut
73. hindgut
74. lumen of urogenital sinus (bladder)
75. early evidence of ossification within cartilage primordium of shaft of tibia
76. cartilage primordium of talus
77. cartilage primordium of calcaneum
78. cartilage primordium of metatarsal bone
79. right hindlimb
80. endoderm-lined phallic part of urethra
81. mammary gland primordium
82. caudate lobe of the liver
83. upper pole of right adrenal primordium

PLATE 32j

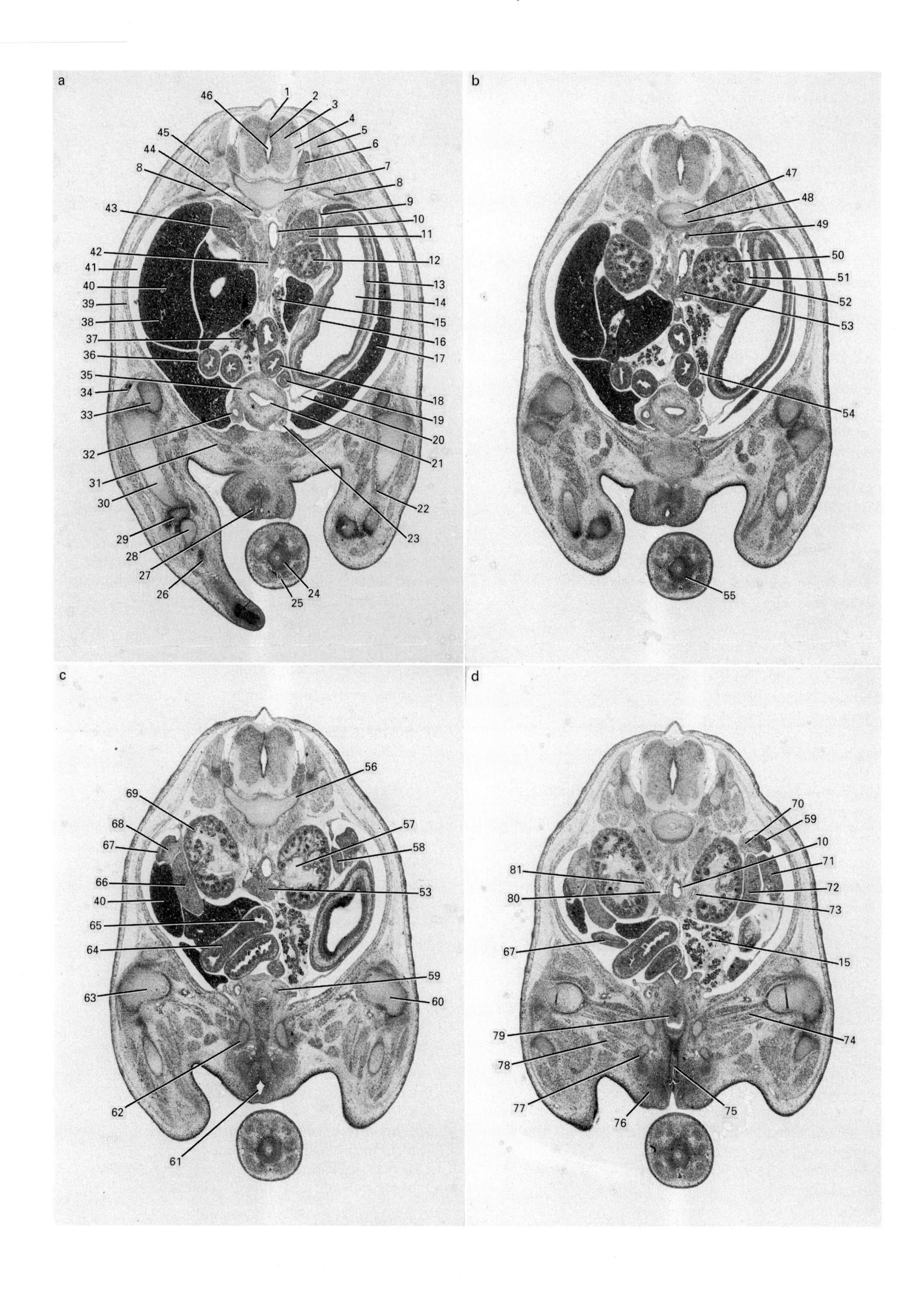

a
46
1
2
3
4
5
6
7
8
9
10
11
12
13
14
15
16
17
18
19
20
21
22
23
24
25
26
27
28
29
30
31
32
33
34
35
36
37
38
39
40
41
42
43
44
45
8
b
47
48
49
50
51
52
53
54
55
c
56
57
58
53
59
60
61
62
63
64
65
40
66
67
68
69
d
70
59
10
71
72
73
15
74
75
76
77
78
79
80
81
67

Plate 32j (14.5 days p.c.)

14.5 days p.c. Female embryo with over 60 pairs of somites. Crown–rump length about 10 mm (fixed). Transverse sections (Theiler Stages 22–23; rat, Witschi Stage 32).

1. roof plate
2. ependymal layer
3. mantle layer
4. marginal layer
5. cartilage primordium of neural arch
6. dorsal (posterior) root ganglion
7. cartilage primordium of thoracic vertebral body
8. ossification within cartilage primordium of proximal part of shaft of 13th rib
9. dorsal mesogastrium
10. descending (abdominal) aorta
11. left adrenal primordium
12. upper pole of left metanephros (definitive kidney)
13. anterior wall of stomach
14. lumen of body of stomach
15. primordium of left lobe of the pancreas
16. posterior wall of stomach
17. lower border of left lobe of liver
18. midgut
19. hindgut
20. ventral mesentery of the stomach (lesser omentum)
21. lumen of urogenital sinus (bladder)
22. periosteal ossification in cartilage primordium of left tibia
23. left umbilical artery
24. cartilage primordium of tail vertebral body
25. filum terminale
26. right "foot plate"
27. endoderm-lined phallic part of urethra
28. cartilage primordium of calcaneum
29. cartilage primordium of talus
30. ossification within cartilage primordium of shaft of tibia
31. rectus abdominis muscle
32. right umbilical artery
33. cartilage primordium of distal part of femur
34. ligamentum patellae
35. wall of urogenital sinus (bladder)
36. duodenum
37. primordium of right lobe and body of the pancreas
38. part of hepatic venous plexus
39. three layers of abdominal wall musculature (external oblique, internal oblique and transversus abdominis muscles)
40. right lobe of liver
41. abdominal/peritoneal cavity
42. hypogastric nerves
43. right adrenal primordium
44. sympathetic ganglion
45. multifidus and longissimus lumborum muscles
46. central canal
47. notochord (nucleus pulposus)
48. primordium of first lumbar vertebral body
49. left sympathetic trunk
50. metanephric tubule
51. superior recess of omental bursa (lesser sac)
52. metanephric glomerulus
53. pre-aortic abdominal sympathetic paraganglia (of Zuckerkandl)/para-aortic bodies
54. dorsal mesentery of hindgut
55. middle region of tail
56. cartilage primordium of first lumbar transverse process
57. medullary region of left kidney
58. upper pole of left ovary
59. left paramesonephric duct
60. cartilage primordium of mid-shaft region of left femur
61. definitive urogenital sinus (phallic part)
62. cartilage primordium of pubic bone
63. ossification within cartilage primordium of mid-shaft region of right femur
64. wall of third part of duodenum
65. lumen of jejunum
66. right ovary
67. right paramesonephric duct
68. mesonephric tissue (degenerating)
69. cortical region of right kidney
70. left mesonephric duct
71. splenic primordium
72. left ovary
73. upper region of left ureter
74. quadriceps (extensor) group of muscles
75. definitive urogenital sinus (pelvic part)
76. labial swelling
77. cartilage primordium of ischial tuberosity
78. hamstring (flexor) group of muscles
79. Müllerian tubercle
80. inferior vena cava
81. upper region of right ureter

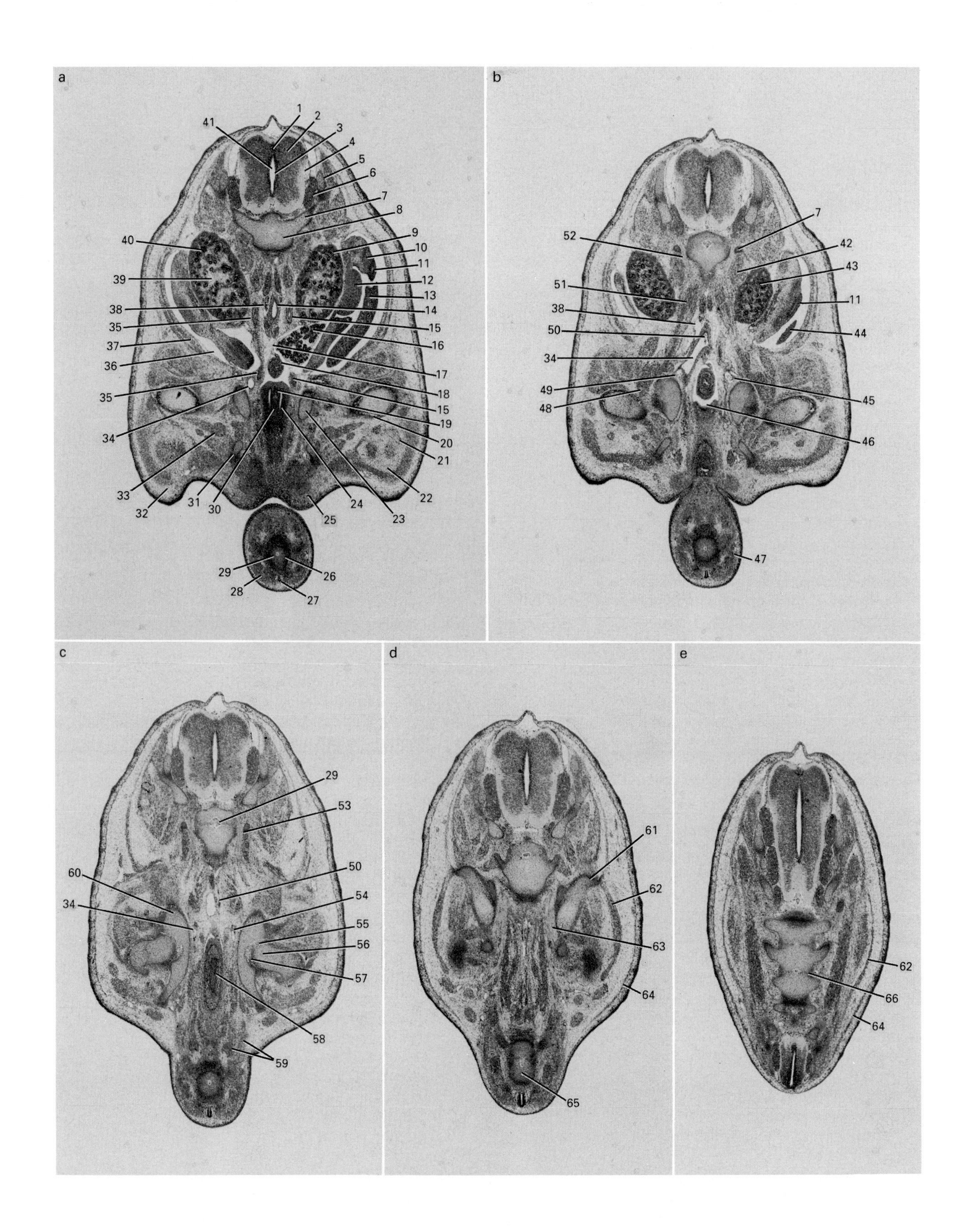
a
b
c
d
e
1
2
3
4
5
6
7
8
9
10
11
12
13
14
15
16
17
18
19
20
21
22
23
24
25
26
27
28
29
30
31
32
33
34
35
36
37
38
39
40
41
42
43
44
45
46
47
48
49
50
51
52
53
54
55
56
57
58
59
60
61
62
63
64
65
66

Plate 32k (14.5 days p.c.)

14.5 days p.c. Female embryo with over 60 pairs of somites. Crown–rump length about 10 mm (fixed). Transverse sections (Theiler Stages 22–23; rat, Witschi Stage 32).

1. roof plate
2. mantle layer
3. ependymal layer
4. marginal layer
5. cartilage primordium of neural arch
6. dorsal (posterior) root ganglion
7. cartilage primordium of lumbar transverse process
8. cartilage primordium of lumbar vertebral body
9. cortical region of left kidney
10. mesonephric tissue (degenerating)
11. left paramesonephric duct
12. left ovary
13. splenic primordium
14. abdominal aorta
15. left ureter
16. primordium of left lobe of the pancreas
17. dorsal mesentery of hindgut (rectum)
18. hindgut (rectum)
19. apposition of the two paramesonephric ducts
20. ossification within cartilage primordium of mid-shaft region of left femur with periosteal collar of bone
21. sciatic nerve
22. proximal part of left hindlimb
23. cartilage primordium of pubic bone
24. left mesonephric duct
25. labial swelling
26. cartilage primordium of tail vertebral body
27. proximal part of filum terminale
28. sacrococcygeus dorsalis lateralis muscle
29. notochord (nucleus pulposus)
30. Müllerian tubercle
31. cartilage primordium of ischial tuberosity
32. proximal part of right hindlimb
33. hamstring (flexor) group of muscles
34. right common iliac artery
35. right ureter
36. pelvic part of abdominal/peritoneal cavity
37. three layers of abdominal wall musculature (external oblique, internal oblique and transversus abdominis muscles)
38. inferior vena cava
39. medullary region of right kidney
40. metanephric glomerulus
41. central canal
42. quadratus lumborum muscle
43. lower pole of left kidney
44. lower border of splenic primordium
45. left common iliac vein
46. pelvic recess of peritoneal cavity (recto-uterine pouch of Douglas)
47. proximal region of tail
48. ilio-psoas muscle
49. right common iliac vein
50. bifurcation of aorta
51. psoas muscle
52. segmental lumbar nerve
53. components of lumbar plexus
54. left common iliac artery
55. primordium of hip joint
56. cartilage primordium of head of the femur
57. acetabular fossa
58. wall of hindgut (rectum)
59. sacrococcygeus ventralis and lateralis muscles
60. early evidence of ossification within cartilage primordium of iliac bone with superficial layer of periosteal bone
61. cartilage primordium of iliac crest
62. gluteus maximus (superficialis) muscle
63. lumbo-sacral trunk
64. cutaneous muscle of trunk (panniculus carnosus)
65. cartilage primordium of coccygeal vertebral body
66. segmental inter-zone, future location of intervertebral disc

PLATE
33a

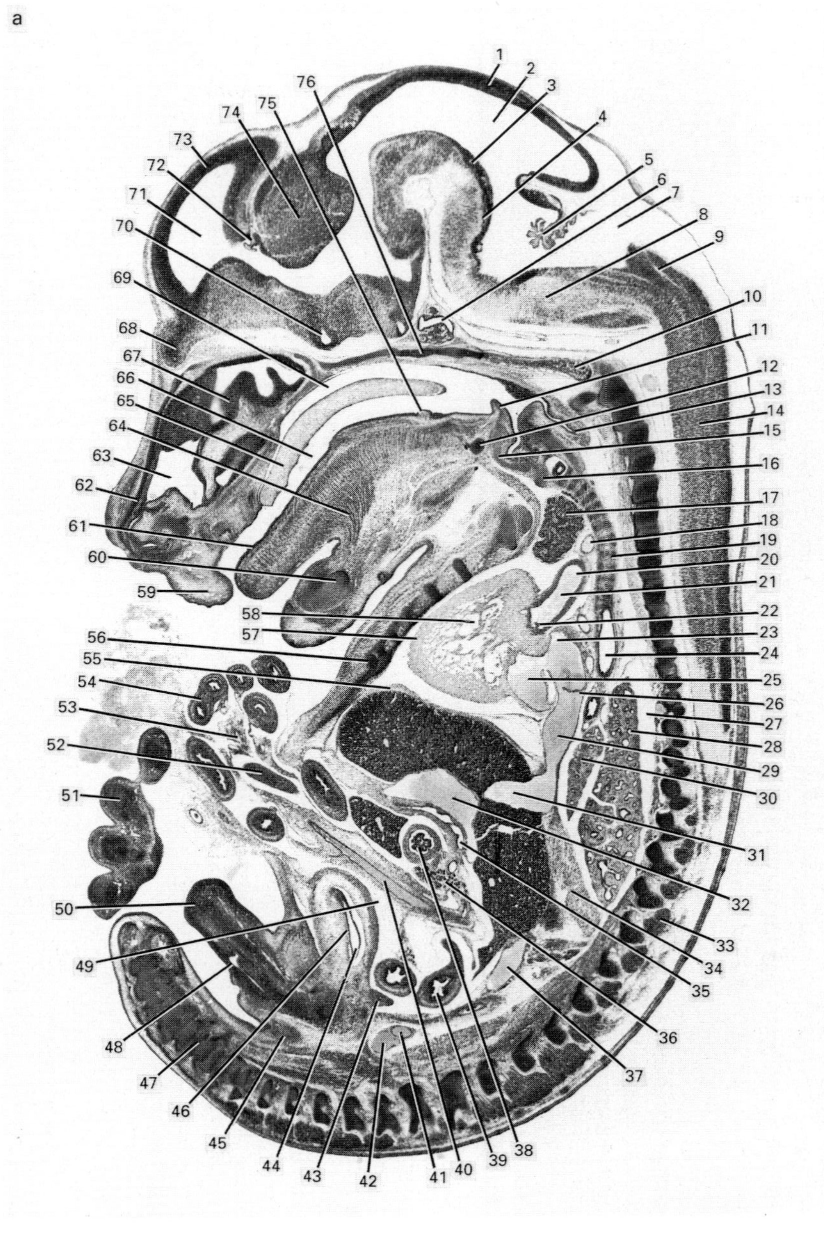
a
1
2
3
4
5
6
7
8
9
10
11
12
13
14
15
16
17
18
19
20
21
22
23
24
25
26
27
28
29
30
31
32
33
34
35
36
37
38
39
40
41
42
43
44
45
46
47
48
49
50
51
52
53
54
55
56
57
58
59
60
61
62
63
64
65
66
67
68
69
70
71
72
73
74
75
76

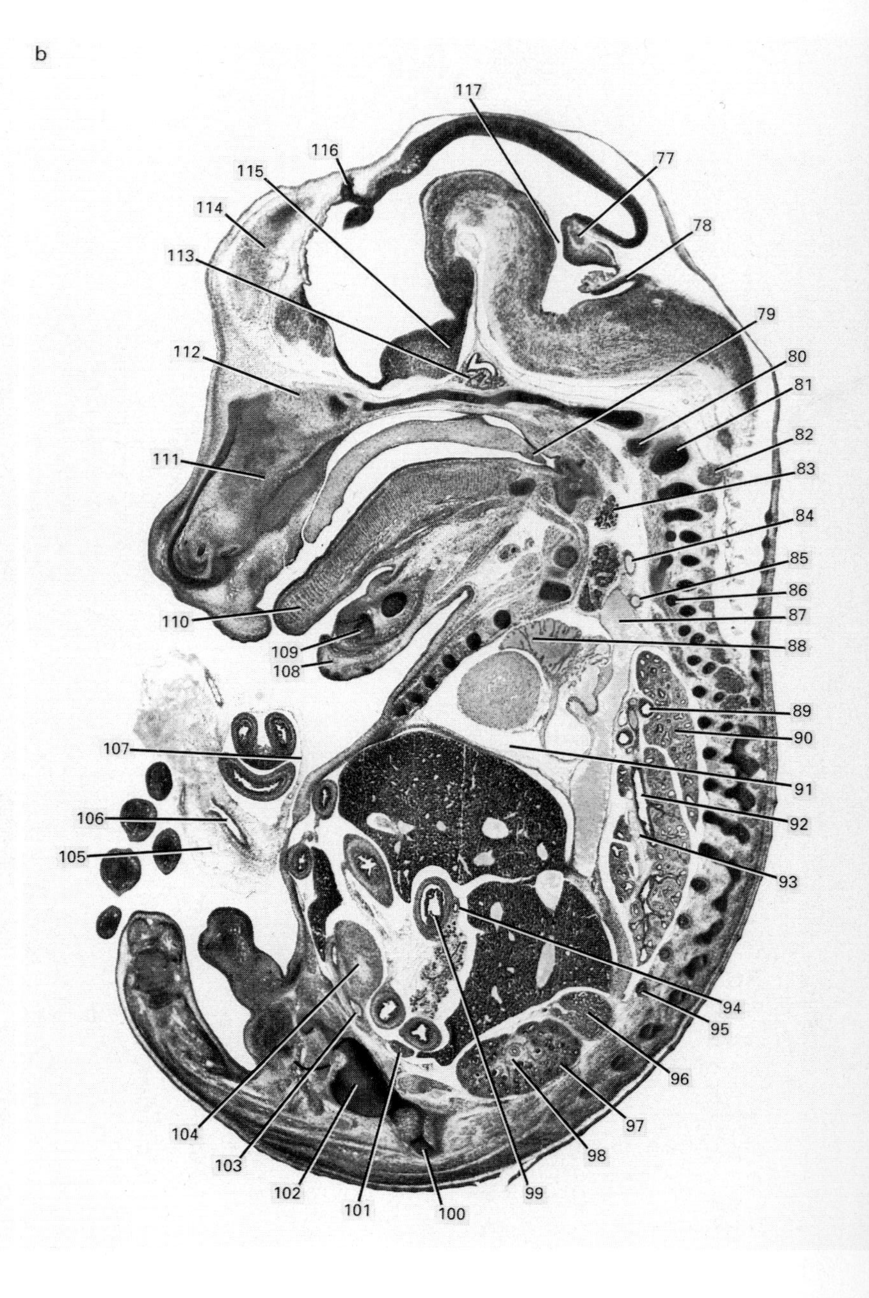
b
77
78
79
80
81
82
83
84
85
86
87
88
89
90
91
92
93
94
95
96
97
98
99
100
101
102
103
104
105
106
107
108
109
110
111
112
113
114
115
116
117

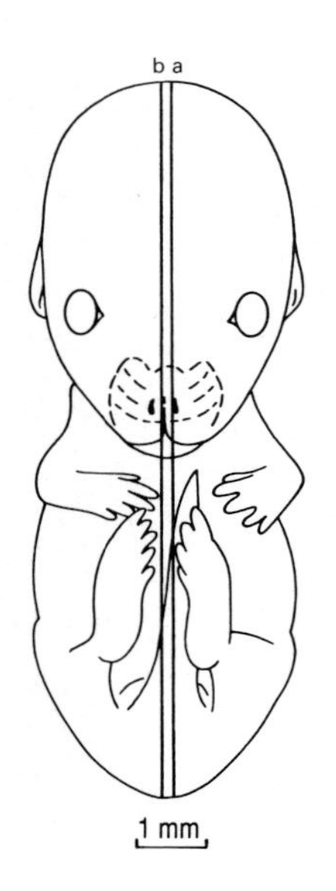
b a
1 mm

Plate 33a (14.5 days p.c.)

14.5 days p.c. Male embryo with over 60 pairs of somites. Crown–rump length 10.5 mm (fixed). Sagittal sections (Theiler Stages 22–23; rat, Witschi Stage 32).

1. roof of midbrain
2. mesencephalic vesicle
3. ventral part of midbrain
4. pons
5. choroid plexus within central part of lumen of fourth ventricle
6. residual lumen of Rathke's pouch
7. fourth ventricle
8. ventral part of medulla oblongata
9. dorsal part of medulla oblongata
10. early evidence of ossification within cartilage primordium of basioccipital bone (clivus)
11. epiglottis
12. cartilage primordium of hyoid bone
13. entrance to oesophagus
14. upper cervical region of spinal cord
15. thyroid cartilage
16. cricoid cartilage
17. right lobe of thymus gland
18. brachiocephalic (innominate) artery
19. trachea with cartilage "rings"
20. arch of the aorta
21. outflow tract of left ventricle (aortic trunk)
22. aortic valve
23. caudal part of right superior (cranial) vena cava
24. lumen of right main bronchus
25. lumen of right atrium
26. leaflet of venous valve at site of entrance of inferior vena cava into right atrium
27. right pleural cavity
28. apical part of caudal lobe of right lung
29. posthepatic part of inferior vena cava
30. accessory lobe of right lung
31. intra-hepatic part of inferior vena cava where it is joined by the ductus venosus
32. ductus venosus
33. lower thoracic dorsal (posterior) root ganglion
34. right crus of diaphragm
35. common bile duct
36. pancreatic primordium
37. prehepatic part of inferior vena cava
38. gastro-duodenal junction in region of pyloric sphincter
39. intraperitoneal part of midgut
40. superior mesenteric artery
41. right internal iliac artery
42. right common iliac vein
43. caudal part of right mesonephric duct (ductus deferens)
44. lumen of urogenital sinus (future bladder)
45. lateral wall of anal canal
46. endodermal lining of urogenital sinus
47. proximal part of tail
48. phallic part of urethra
49. peritoneal cavity
50. tip of genital tubercle (penis)
51. digit of left hindlimb
52. wall of distal part of midgut loop as it passes back into the peritoneal cavity from the physiological umbilical hernia
53. dorsal mesentery of midgut containing superior mesenteric vessels
54. loop of midgut within physiological umbilical hernia
55. right dome of diaphragm
56. costal cartilage
57. wall of right ventricle
58. lumen of right ventricle
59. upper lip (right side)
60. Meckel's cartilage
61. dorsal surface of tongue
62. cartilage primordium of nasal bone
63. nasal cavity (left half)
64. extrinsic muscle of the tongue (genioglossus)
65. cartilage primordium of palatal shelf of maxilla
66. oropharynx
67. olfactory epithelium
68. left olfactory lobe
69. nasopharynx
70. optic recess of diencephalon
71. left lateral ventricle
72. choroid plexus extending into lateral ventricle
73. roof of neopallial cortex (future cerebral cortex)
74. diencephalon (thalamus)
75. median circumvallate (vallate) papilla
76. cartilage primordium of basisphenoid bone
77. intraventricular portion of cerebellar primordium
78. lateral part of roof of fourth ventricle
79. soft palate
80. cartilage primordium of anterior arch of C1 vertebra (atlas)
81. cartilage primordium of body of C2 vertebra (axis)
82. right dorsal (posterior) root ganglion of C2
83. right lobe of thyroid gland
84. right common carotid artery
85. right subclavian artery
86. cartilage primordium of head of right first rib
87. right superior (cranial) vena cava
88. auricular region of right atrium
89. lobar bronchus to cranial lobe of right lung
90. cranial lobe of right lung
91. pericardial cavity
92. lobar bronchus to caudal lobe of right lung
93. branch of pulmonary vein draining basal part of caudal lobe of right lung
94. pancreatic duct
95. cartilage primordium of head of right 13th rib
96. right adrenal gland
97. cortical region of right metanephros (definitive kidney)
98. medullary region of right metanephros (definitive kidney)
99. lumen of first part of duodenum
100. cartilage primordium of right iliac crest
101. right mesonephric duct (ductus deferens) within right mesorchium
102. cartilage primordium of right innominate bone, just deep to hip joint
103. right superior vesical artery which continues as right umbilical artery
104. right lateral wall of urogenital sinus (future bladder)
105. Wharton's jelly
106. umbilical artery within proximal part of umbilical cord
107. wall of physiological umbilical hernia
108. lower lip
109. primordium of lower right incisor tooth
110. intrinsic muscle of tongue (vertical component)
111. cartilage primordium of nasal septum
112. olfactory (I) nerves converging on right olfactory lobe of brain
113. vascular differentiation which develops in close association with anterior wall of Rathke's pouch (future anterior pituitary)
114. medial wall of right cerebral hemisphere
115. diencephalon (hypothalamus)
116. lateral wall of pineal recess of third ventricle (wall of pineal primordium)
117. cerebral aqueduct

PLATE
33b

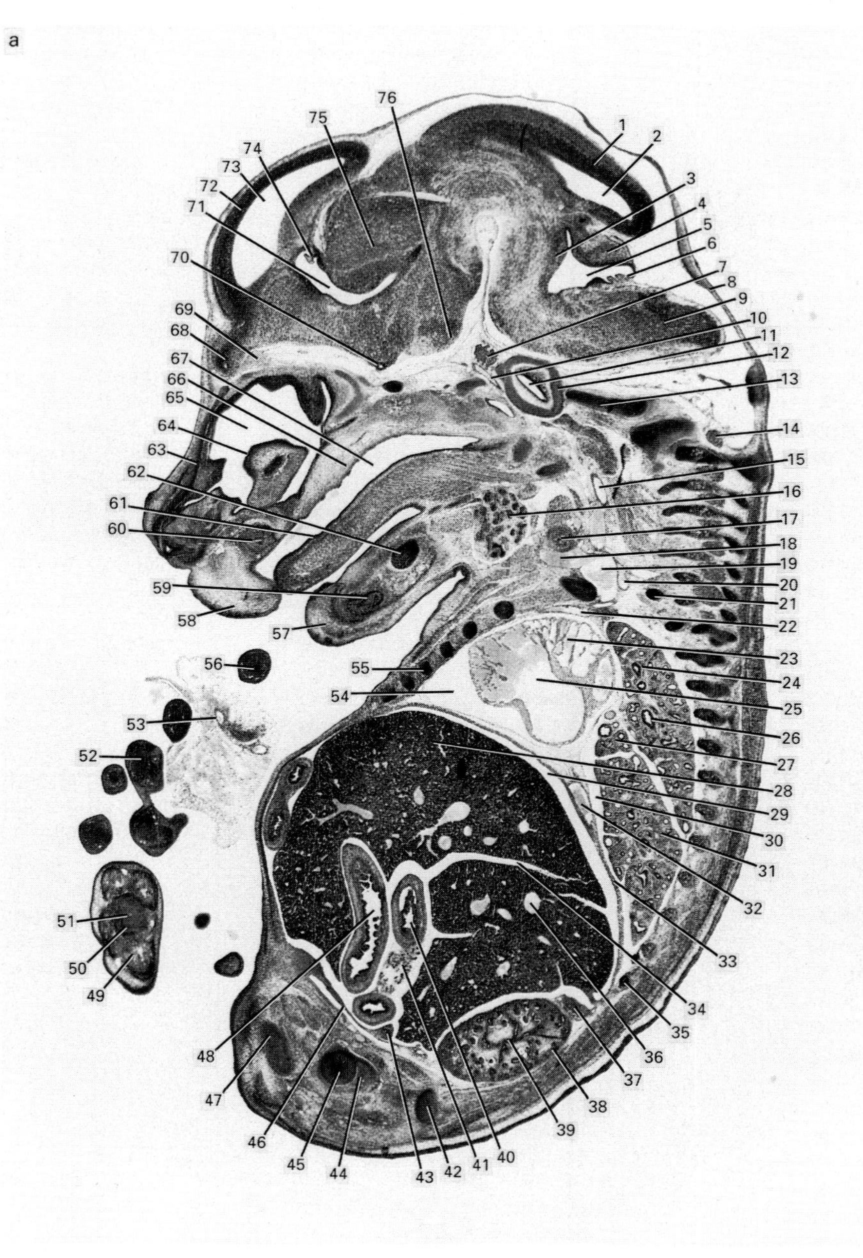
a
1
2
3
4
5
6
7
8
9
10
11
12
13
14
15
16
17
18
19
20
21
22
23
24
25
26
27
28
29
30
31
32
33
34
35
36
37
38
39
40
41
42
43
44
45
46
47
48
49
50
51
52
53
54
55
56
57
58
59
60
61
62
63
64
65
66
67
68
69
70
71
72
73
74
75
76

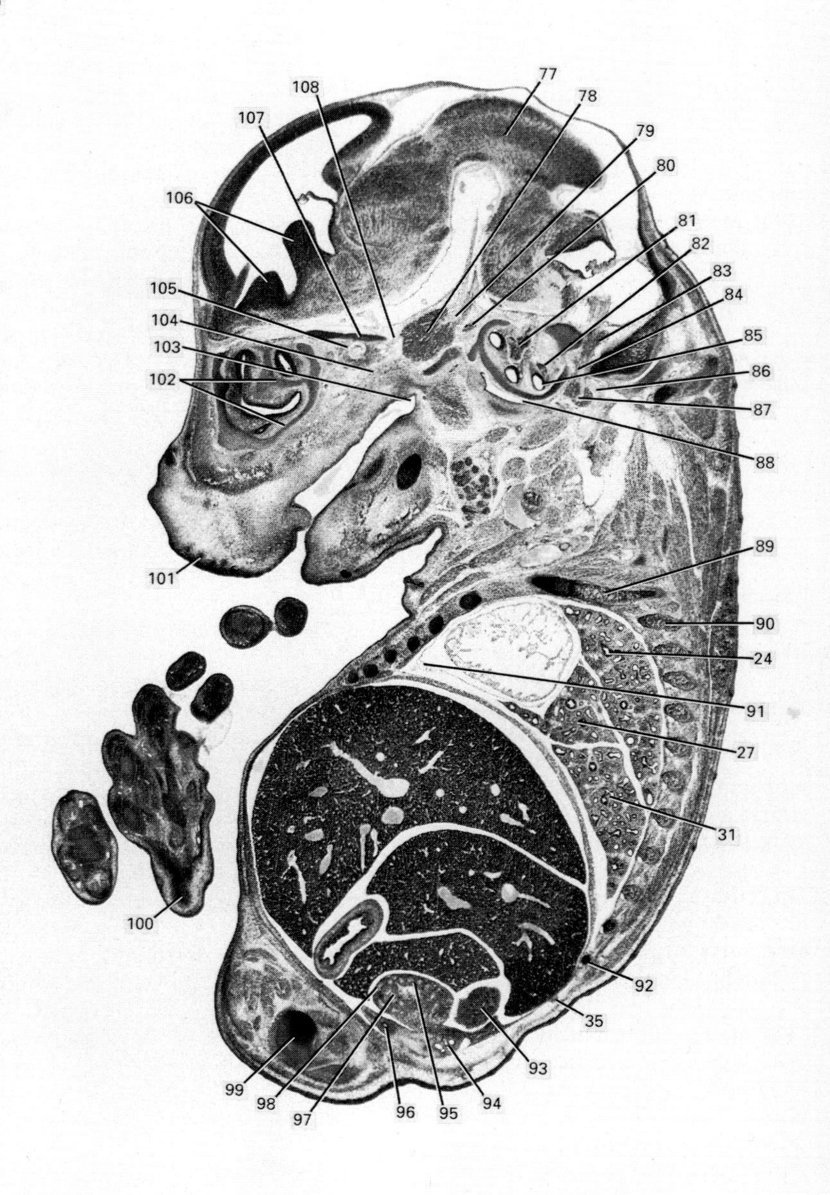
b
77
78
79
80
81
82
83
84
85
86
87
88
89
90
24
91
27
31
92
35
93
94
95
96
97
98
99
100
101
102
103
104
105
106
107
108

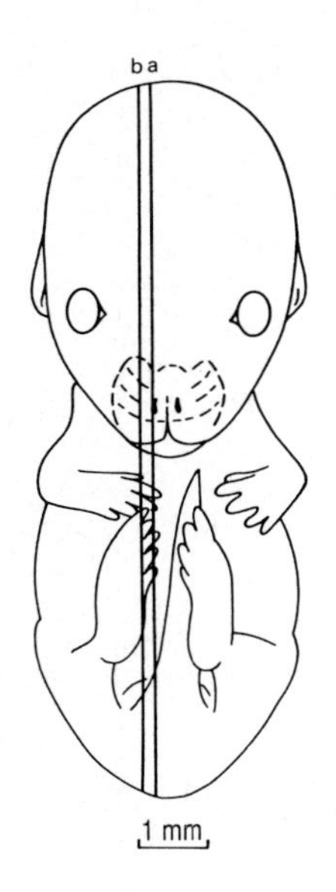
b a
1 mm

Plate 33b (14.5 days p.c.)

14.5 days p.c. Male embryo with over 60 pairs of somites. Crown–rump length 10.5 mm (fixed). Sagittal sections (Theiler Stages 22–23; rat, Witschi Stage 32).

1. roof of caudal part of midbrain
2. caudal part of mesencephalic vesicle
3. pons
4. intraventricular portion of cerebellar primordium
5. entrance to right lateral recess of fourth ventricle
6. choroid plexus within central part of lumen of fourth ventricle
7. lateral part of pituitary primordium
8. cartilage primordium of squamous part of occipital bone
9. medulla oblongata
10. right internal carotid artery
11. cochlea
12. cartilage primordium of petrous part of temporal bone
13. cartilage primordium of lateral part of basioccipital bone (clivus)
14. right upper cervical dorsal (posterior) root ganglion
15. right common carotid artery
16. submandibular gland
17. cartilage primordium of middle third of right clavicle
18. right subclavian vein
19. rostral part of right superior (cranial) vena cava
20. right subclavian artery
21. ossification within cartilage primordium of mid-shaft region of first rib
22. right internal thoracic (mammary) vein
23. auricular region of right atrium
24. apical region of cranial lobe of right lung
25. lumen of right atrium
26. segmental bronchus within cranial lobe of right lung
27. middle lobe of right lung
28. right lobe of liver
29. sub-phrenic recess of peritoneal cavity
30. right pleural cavity
31. caudal lobe of right lung
32. posthepatic inferior vena cava
33. caudal component of right dome of diaphragm
34. lobar interzone
35. cartilage primordium of mid-shaft region of 13th rib
36. part of right hepatic venous plexus
37. lateral margin (cortical region) of right adrenal gland
38. cortical region of right metanephros (definitive kidney)
39. medullary region of right metanephros (definitive kidney)
40. lumen of second part of duodenum
41. pancreatic primordium
42. cartilage primordium of rostral part of iliac bone
43. right mesonephric duct (ductus deferens) within right mesorchium
44. lateral margin of right hip joint
45. cartilage primordium of head of right femur
46. gubernaculum testis
47. cartilage primordium of ischial tuberosity
48. lumen of intraperitoneal part of midgut
49. middle region of tail
50. cartilage primordium of tail vertebral body
51. nucleus pulposus
52. digit of right hindlimb
53. umbilical vessels in proximal part of umbilical cord
54. pericardial cavity
55. cartilage primordium of ventral part of shaft of 5th rib
56. digit of right forelimb
57. lower lip
58. upper lip (right side)
59. primordium of lower right incisor tooth
60. primordium of upper right incisor tooth
61. dorsal surface of tongue
62. Meckel's cartilage
63. cartilage primordium of nasal bone
64. olfactory epithelium
65. nasal cavity (right half)
66. lateral margin of palatal shelf of right maxilla
67. oropharynx
68. olfactory lobe
69. olfactory (I) nerves converging on olfactory lobe of brain
70. lateral extension of optic recess of diencephalon (optic (II) nerve)
71. right interventricular foramen
72. roof of neopallial cortex
73. right lateral ventricle
74. choroid plexus extending into lateral ventricle
75. diencephalon (thalamus)
76. diencephalon (hypothalamus)
77. lateral wall of midbrain
78. right trigeminal (V) ganglion
79. rootlets of trigeminal (V) nerve
80. right facial (VII) ganglion
81. vestibular component of vestibulocochlear (VIII) ganglion
82. cochlear component of vestibulocochlear (VIII) ganglion
83. superior ganglion of glossopharyngeal (IX) nerve
84. right jugular foramen
85. origin of posterior semicircular canal
86. right internal jugular vein shortly after it has emerged from the jugular foramen
87. inferior ganglion of glossopharyngeal (IX) nerve
88. right tubo-tympanic recess
89. ossification within cartilage primordium of anterior two-thirds of right first rib
90. ossification within cartilage primordium of mid-shaft region of right second rib
91. mesothelial cells (parietal pericardium) lining pericardial cavity
92. cartilage primordium of mid-shaft region of right 12th rib
93. lateral margin (cortical region) of right metanephros (definitive kidney)
94. degenerating mesonephric tissue containing mesonephric tubules
95. seminiferous tubules within medullary region of right testis
96. proximal part of right mesonephric duct (ductus deferens)
97. medullary region of right testis
98. cortical region ("germinal" epithelium) of right testis
99. cartilage primordium of upper shaft region of right femur
100. heel region of right hindlimb
101. primordium of follicle of vibrissa
102. cartilage primordia of turbinate bones
103. entrance to right pharyngo-tympanic (Eustachian) tube
104. maxillary (V2) nerve
105. right optic (II) nerve surrounded by fibres of origin of extrinsic ocular muscles
106. corpus striatum mediale and corpus striatum laterale (medial and lateral aspects of right ganglionic eminence)
107. cartilage primordium of orbito-sphenoid bone
108. ophthalmic branch of trigeminal (V1) nerve

PLATE 34a

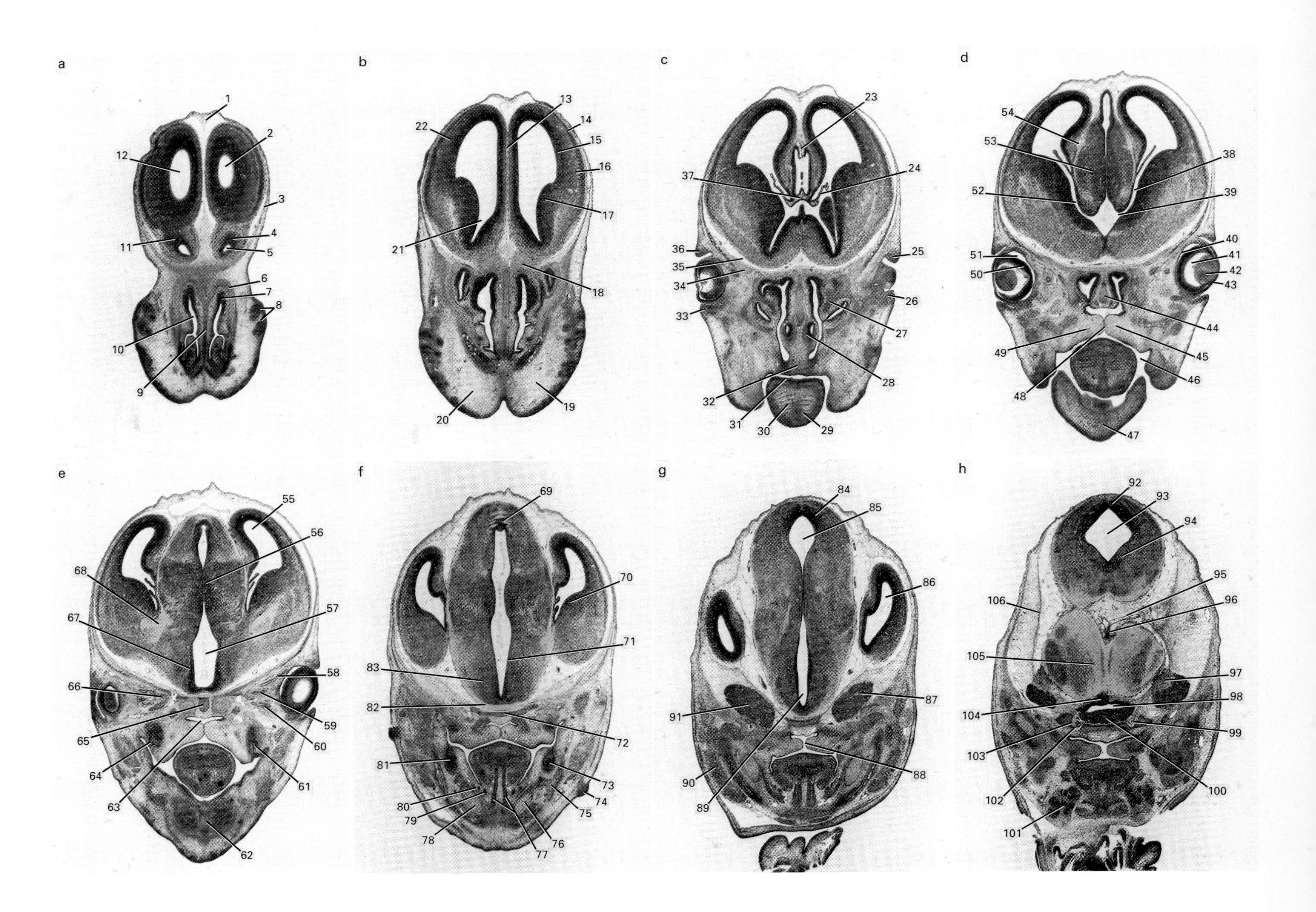
a
1
2
3
4
5
6
7
8
9
10
11
12
b
13
14
15
16
17
18
19
20
21
22
c
23
24
25
26
27
28
29
30
31
32
33
34
35
36
37
d
38
39
40
41
42
43
44
45
46
47
48
49
50
51
52
53
54
e
55
56
57
58
59
60
61
62
63
64
65
66
67
68
f
69
70
71
72
73
74
75
76
77
78
79
80
81
82
83
g
84
85
86
87
88
89
90
91
h
92
93
94
95
96
97
98
99
100
101
102
103
104
105
106

Plate 34a (14.5 days p.c.)

14.5 days p.c. Coronal sections through head (Theiler Stages 22–23; rat, Witschi Stage 32).

1. superior sagittal dural venous sinus
2. anterior extremity of left lateral ventricle
3. mesenchymal precursor of lateral part of frontal bone
4. left olfactory lobe (bulb)
5. extension of anterior horn of left lateral ventricle into olfactory lobe
6. nasal capsule
7. olfactory epithelium
8. primordia of follicles of vibrissae
9. cartilage primordium of nasal septum
10. nasal cavity
11. right olfactory lobe (bulb)
12. anterior extremity of right lateral ventricle
13. falx cerebri
14. neopallial cortex (future cerebral cortex)
15. ventricular zone of telencephalon
16. intermediate zone of wall of telencephalon
17. striatum (ganglionic eminence) overlying the head of the caudate nucleus
18. olfactory (I) nerves passing from olfactory epithelium towards olfactory lobe
19. left upper lip
20. right upper lip
21. inferior part of right lateral ventricle
22. wall of right telencephalic vesicle
23. choroid plexus in pineal recess of third ventricle
24. choroid plexus within left lateral ventricle and arising from its medial wall
25. left upper eyelid
26. left lower eyelid
27. cartilage primordium of turbinate bone
28. left vomeronasal organ (Jacobson's organ)
29. tongue (just subjacent to its tip)
30. intrinsic muscle of tongue (transverse component)
31. dorsum of tongue
32. primary palate
33. right lower eyelid
34. cartilage primordium of ala orbitalis (lesser wing of sphenoid bone)
35. loosely packed mesenchyme tissue in future subarachnoid space
36. right upper eyelid
37. anterior extremity of roof of third ventricle (region of lamina terminalis)
38. left interventricular foramen
39. entrance from third ventricle into left interventricular foramen
40. pigment layer of retina
41. hyaloid cavity
42. lens
43. corneal epithelium
44. posterior part of cartilage primordium of nasal septum
45. palatal shelf of left maxilla
46. oral cavity
47. anterior extremity of lower jaw
48. site of apposition in midline of the left and right palatal shelves
49. right palatine vessels
50. neural layer of retina
51. intra-retinal space
52. right interventricular foramen
53. diencephalon (thalamus)
54. right choroid fissure
55. superior part of left lateral ventricle
56. interthalamic adhesion (massa intermedia)
57. inferior part of third ventricle
58. extrinsic ocular muscle (left superior rectus)
59. left optic (II) nerve
60. extrinsic ocular muscle (left inferior rectus)
61. primordium of left upper molar tooth
62. site of fusion in ventral midline of left and right Meckel's cartilages
63. nasopharynx
64. primordium of right upper molar tooth
65. cartilage primordium of pre-sphenoid component of sphenoid bone
66. right optic (II) nerve
67. diencephalon (hypothalamus)
68. fibres of thalamic origin passing into caudate nucleus
69. diencephalon (epithalamus)
70. region overlying tail of caudate nucleus
71. ependymal layer of diencephalon (hypothalamus)
72. cartilage primordium of basisphenoid bone
73. primordium of left lower molar tooth
74. primordium of prominent tactile or sinus hair follicle characteristically found in this location
75. early evidence of ossification within left half of mandible
76. left Meckel's cartilage
77. extrinsic muscle of tongue (genioglossus)
78. right Meckel's cartilage
79. right submandibular duct
80. right sublingual duct
81. primordium of right lower molar tooth
82. marginal layer of diencephalon (hypothalamus)
83. mantle layer of diencephalon (hypothalamus)
84. junction between diencephalon and midbrain
85. rostral part of mesencephalic vesicle
86. posterior part of left lateral ventricle
87. left trigeminal (V) ganglion
88. site of communication between posterior parts of oro- and nasopharynx
89. infundibular recess in floor of third ventricle
90. right masseter muscle
91. right trigeminal (V) ganglion
92. roof of midbrain

93. mesencephalic vesicle
94. ventral part of midbrain
95. origin of left posterior cerebral and cerebellar arteries
96. basilar artery
97. rootlets of origin of left trigeminal (V) ganglion
98. residual lumen of Rathke's pouch
99. left internal carotid artery
100. site of active cellular and vascular proliferation in anterior wall of Rathke's pouch
101. right submandibular gland
102. right internal carotid artery
103. right carotid canal
104. infundibulum (future pars nervosa) of pituitary
105. posterior part of diencephalon (hypothalamus)
106. mesenchymal condensation to form inner table of future dura mater

PLATE
34b

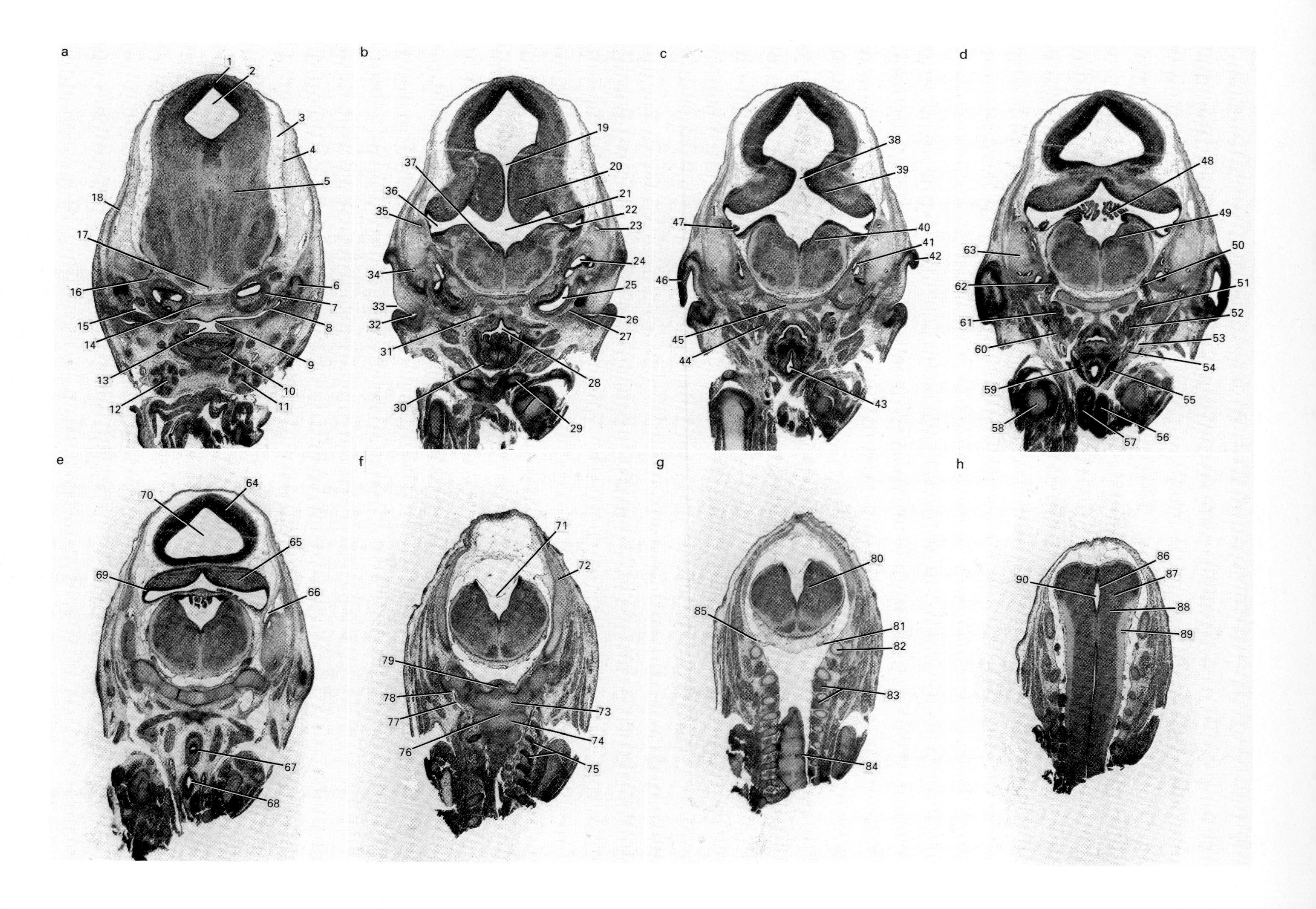
a
1 2 3 4 5 6 7 8 9 10 11 12 13 14 15 16 17 18
b
19 20 21 22 23 24 25 26 27 28 29 30 31 32 33 34 35 36 37
c
38 39 40 41 42 43 44 45 46 47
d
48 49 50 51 52 53 54 55 56 57 58 59 60 61 62 63
e
64 65 66 67 68 69 70
f
71 72 73 74 75 76 77 78 79
g
80 81 82 83 84 85
h
86 87 88 89 90

Plate 34b (14.5 days p.c.)

14.5 days p.c. Coronal sections through head (Theiler Stages 22–23; rat, Witschi Stage 32).

1. roof of midbrain
2. mesencephalic vesicle
3. loosely packed mesenchyme in future subarachnoid space
4. mesenchymal condensation to form inner table of future dura mater
5. junction between midbrain and pons
6. cartilage primordium of left malleus
7. cartilage primordium of petrous part of left temporal bone
8. left pharyngo-tympanic (Eustachian) tube
9. entrance from pharynx into left pharyngo-tympanic (Eustachian) tube
10. cartilage primordium of body of hyoid bone
11. left submandibular gland
12. right submandibular gland
13. site of communication between posterior parts of oro- and nasopharynx
14. cartilage primordium of basisphenoid bone
15. right tubo-tympanic recess
16. right facial (VII) nerve
17. basilar artery
18. mesenchymal precursor of right parietal bone
19. outflow from mesencephalic vesicle into cerebral aqueduct
20. rostral part of cerebellar primordium
21. central part of fourth ventricle
22. left lateral recess of fourth ventricle
23. left superior semicircular canal
24. left saccule
25. left cochlea
26. left external auditory (acoustic) meatus
27. left tubo-tympanic recess
28. entrance into oesophagus
29. cartilage primordium of medial one-third of left clavicle
30. laryngeal cartilage
31. prevertebral muscle (longus capitis)
32. right external auditory (acoustic) meatus
33. entrance into right external auditory (acoustic) meatus
34. right lateral semicircular canal
35. right superior semicircular canal
36. right lateral recess of fourth ventricle
37. pons
38. cerebral aqueduct
39. left part of cerebellar primordium
40. junction between pons and medulla oblongata
41. proximal part of left endolymphatic duct
42. pinna of left ear
43. entrance into trachea
44. right superior cervical sympathetic ganglion
45. ossification within cartilage primordium of basioccipital bone
46. pinna of right ear
47. choroid plexus arising from roof of right lateral recess of fourth ventricle and extending into its lumen
48. choroid plexus arising from roof of medial part of fourth ventricle and extending into its lumen
49. rostral extremity of medulla oblongata
50. superior ganglion of left vagus (X) nerve
51. inferior ganglion of left vagus (X) nerve
52. left superior cervical sympathetic ganglion
53. left internal jugular vein
54. left common carotid artery
55. left lobe of thyroid gland
56. left lobe of thymic rudiment
57. right lobe of thymic rudiment
58. cartilage primordium of head of right humerus
59. right lobe of thyroid gland
60. right vagus (X) nerve
61. inferior ganglion of right vagus (X) nerve
62. superior ganglion of right vagus (X) nerve
63. cartilage primordium of petrous part of right temporal bone
64. roof of caudal part of midbrain
65. caudal part of cerebellum
66. left endolymphatic duct
67. oesophagus
68. trachea
69. rhombic lip
70. caudal part (posterior extremity) of mesencephalic vesicle
71. roof of fourth ventricle
72. cartilage primordium of supraoccipital bone
73. cartilage primordium of body of C2 vertebra (axis)
74. cartilage primordium of body of C3 vertebra
75. segmental cervical spinal nerves
76. nucleus pulposus
77. right vertebral artery
78. right vertebral artery and associated veins in incompletely formed foramen transversarium
79. cartilage primordium of dens (odontoid process of C2 vertebra)
80. caudal part of medulla oblongata
81. left vertebral artery where it passes medially to unite with the right vertebral artery to form the basilar artery
82. ossification within cartilage primordium of neural arch of C1 vertebra (atlas)
83. upper cervical dorsal (posterior) root ganglia
84. location of future intervertebral disc
85. right vertebral artery where it passes medially to unite with the left vertebral artery to form the basilar artery
86. ependymal layer
87. upper cervical region of spinal cord
88. mantle layer
89. marginal layer
90. central canal

a b c d e f g h

2 mm

the shoulder joint (*Plate 32e,b*), elbow joint (*Plate 32f,d*), wrist joint (*Plate 32g,a*) hip joint (*Plate 32k,c*), knee joint (*Plate 32j,a,b*) and ankle joint (*Plate 32i,d*; *Plate 32j,a*) are all clearly seen.

Along the vertebral axis, all of the vertebral bodies are chondrified, and the primordia of the intervertebral joints (the "discs") are also easily recognized. In the central region of each disc is located the nucleus pulposus (a derivative of the notochord) (e.g. *Plate 32h,a,c*).

Stage 24 Developmental age, 16 days p.c.
Plates 35a–l; Plates 36a–e

EXTERNAL FEATURES

The two embryos that were isolated at about 15.5 days p.c. and used to illustrate this stage of development are both representative examples of embryos at the earlier end of the range expected at this stage, so that whereas certain developmentally more advanced embryos at this stage would almost certainly have had closed eyelids, in the two specimens illustrated, little evidence of eyelid closure is seen. By contrast, the scanning series illustrates all stages of eyelid closure, despite the fact that these embryos were all isolated at a similar stage p.c. Similarly, developmentally less advanced embryos at this stage would be expected to have a quite conspicuous physiological umbilical hernia, whereas in developmentally more advanced embryos in which the midgut loop has almost completely returned to the peritoneal cavity, the volume of the hernial sac may be quite small. An overall increase in the volume of the peritoneal cavity compared to that seen over the previous 1–2 days is also seen, which undoubtedly facilitates the return of the midgut loop. The appearance of the hair follicles, rather than any change in their distribution, is also substantially more marked than the situation observed at 14.5–15 days p.c. The other external feature of particular note at this stage is the decrease in the ventral curvature (i.e. flexion) of the postcranial vertebral axis, which was a feature of embryos at earlier stages of development. The overall appearance of the hand plate and foot plate also show increased evidence of differentiation.

One of the important events that occurs during this stage is the closure of the eyelids. As indicated above, in embryos isolated during the early part of this stage, the eyelids are still seen to be widely separated (*Plate 49,c*; *Plate 35b,b,c*; *Plate 35c,a,b*; *Plate 36c,b*), while in developmentally more advanced embryos, the eyelids may be almost completely closed (*Plate 49,g*), and by about 16.5–17 days p.c., the eyelids are invariably closed (*Plate 49,h*). The fact that even within a single litter a wide variation in the degree of eyelid closure is seen, suggests that the whole process probably takes no more than about 12–18 hours, and indeed in some embryos may even be of shorter duration. What is of particular interest is that the pattern of eyelid closure observed in the mouse seems to be quite dissimilar to the events described in the human embryo, where it is suggested that the free edges of the upper and lower eyelids grow rapidly towards each other until their margins make contact, and the eyelids eventually unite by epithelial fusion, while at the same time forming the conjunctival sac (Hamilton and Mossman (1972), though O'Rahilly and Müller (1987) suggest that in some human embryos, the medial and lateral borders of the eyelids may fuse first.

In the mouse, the scanning micrographs clearly reveal that the events associated with eyelid closure are quite dissimilar to the standard description outlined above. In the mouse, the initial event appears to be associated with an increase in cellular activity around the entire peripheral margin of the eyelid folds (*Plate 49,d*). The cellular activity appears to rapidly become more marked, and the circumference of the eyelid margins diminishes, as occurs with the tightening of a purse-string (*Plate 49,e–g*). The latter would seem to indicate that the whole process is mediated via the active contraction of cytoskeletal elements (possibly principally microfilaments) at the eyelid margins, which mediate the cell shape changes required to bring these events about (see Harris and McLeod (1982) for SEM analysis of eyelid growth and fusion in mice). Alternatively, the surface epithelial cells may migrate towards the gap to form a plug-like mass, so that at no time is there fusion of the subjacent mesenchyme. This mechanism allows the possibility that when eyelid separation eventually occurs, the epithelial "plug" is broken down and removed without associated damage to the eyelid mesenchyme. As occurs in man, the closure of the eyelids results in the formation of the conjunctival sac (and probably also facilitates the differentiation of the cornea) (*Plate 53,b,c*). Unlike the situation in man, where the eyelids normally reopen during the seventh month of gestation, in the mouse, this event does not occur until about 12–14 days after birth.

Two other external features are worthy of note at this stage. Between 14.5–15 and 15.5–16 days p.c., the appearance of the external ear changes, so that whereas previously the pinna was seen to be folded forwards and covered about one-half of the external acoustic meatus (*Plate 46,g,h*; *Plate 32c,a–c*), it is now seen to cover approximately the posterior two-thirds of this orifice (*Plate 46,i,j*; *Plate 35b,d*; *Plate 35c,a*). In relation to the vibrissae, whereas at 14.5–15 days p.c., cross-sections of these structures reveal little more than what appears to be concentric rings in which the hair root, bulb and dermal papilla are seen to be surrounded by the outer root sheath (e.g. *Plate 32d,c,d*), at 15.5–16 days p.c., most of the sections through the vibrissae show more detailed evidence of their differentiation. For example, it is now possible to recognize at their base the prominent and darkly-staining hair bulb, which is surrounded by the dermal papilla. More distally along the shaft of the vibrissa, the elements of the inner and outer root sheath are themselves surrounded by the dermal sheath (*Plate 35d,d*; *Plate 35e,a,b*, unlabelled). While the tip of the follicle of the vibrissa clearly protrudes well beyond the mound at the base of the vibrissa (*Plate 49,m*), it does not erupt through this epidermal evagination until about 16.5 days p.c. (*Plate 49,o*).

Various changes are also evident in the trunk region, one of the most important of which is related to the volume of the liver *vis-à-vis* that of the peritoneal cavity during this period. During the latter part of this stage of development, the midgut loop is seen to be returning to the peritoneal cavity, and this event is associated with a corresponding diminution in the volume of the physiological umbilical hernia. Various factors are believed to play critical roles in facilitating this manoeuvre, one of which is the overall increase in the volume of the peritoneal cavity that gradually occurs over the previous day or so. While the liver remains the largest of the intra-abdominal organs, its increase in volume does not keep pace with the increase in the volume of the peritoneal cavity that occurs over this period of time. The latter is believed to be largely due to the rapid increase that occurs in the growth of the caudal half of the embryo at this time. When the physiological umbilical hernia is first apparent, at about 10.5–11 days p.c. (*Plate 24c,b–g*), the peritoneal/coelomic cavity is quite small, the gut tube relatively undifferentiated, and the liver (though also relatively undifferentiated) nevertheless still occupies the majority of the available space. During the period from about 12.5 to 13 days p.c., as development proceeds, the liver and to a lesser extent the stomach, and increasingly the pancreas, occupy virtually all of the available space (*Plate 29b,a,b*). A similar situation is observed at about 13.5–14.5 days p.c., though by this time the liver appears to occupy an even greater proportion of the peritoneal cavity than previously (13.5 days p.c., *Plate 31a,a,b*; *Plate 31b,a,b*; 14.5 days p.c., *Plate 33a,a,b*; *Plate 33b,a,b*). By about 15.5 days p.c., while the caudal half of the embryo enlarges and the volume of the peritoneal cavity increases, the volume of the liver does not increase to the same extent (*Plate 36b,a,b*; *Plate 36c,a,b*). Thus the increase in available space, which is principally located in the anterior and inferior parts of the peritoneal cavity, consequently substantially facilitates the return of the midgut loop. The differential growth in volume between that of the peritoneal cavity and that of the liver is even more marked at about 16.5 days p.c., when it is possible to see that approximately the lower half of the peritoneal cavity is now occupied by the midgut and, but only to a lesser extent, by the hindgut. At about 16.5 days p.c., the liver is now seen to occupy about two-thirds of the volume of the right side of the peritoneal cavity (*Plate 38a,a*), and only about one-half to one-third of the available space in the midline (*Plate 38b,a,b*), though to the far left of the midline, the left lobe of the liver still occupies about one-half of the available space (*Plate 38c,b*).

The relative diminution in volume of certain retroperitoneal structures over the latter part of this period, is also believed to play a key role in facilitating the return of the midgut loop to the peritoneal cavity. Thus the kidneys and gonads, which undoubtedly increase in their absolute volume over this period, do so more slowly than that of the peritoneal cavity. The lungs also tend to "ascend" into the thoracic region proper over the latter part of this period, and consequently tend to release even more space not only to facilitate the return, but also the progressive elongation and differentiation of the midgut in particular, and also the differentiation of the other intra-abdominal contents.

A marked difference is also observed in the shape of the back of embryos between about 14.5–15 and 15.5–16 days p.c. Whereas previously the embryo has a characteristically curved (i.e. flexed) postcranial vertebral axis (*Plate 46,g,h*), during this and subsequent stages, the degree of flexion diminishes, and the back tends to become increasingly straightened (*Plate 46,l,m*). The latter is in fact associated with an increase in the lordotic curvature (ie deflexion) of the vertebral column, which is particularly well seen in the lower thoracic/upper lumbar region, in more advanced stages of development (*Plate 46,k–p*).

The hand plate and foot plate also show increased evidence of differentiation, so that, for example, the individual digits are more clearly defined than previously. The latter is particularly well seen in relation to the foot plate, where the residual webbing still present at 14.5–15 days p.c. (*Plate 44,g*; *Plate 45,a,b*) has now virtually completely disappeared (*Plate 44,j*; *Plate 45,e,f*). In the hindlimbs, the first evidence of nail primordia, associated with the dorsal surface of the distal phalanges is also seen (*Plate 45,e*). With regard to the overall shape of the foot plate, it is apparent that while previously the digits were seen to be splayed out, at 15.5–16 days p.c., the middle three digits are largely parallel to each other, while the first and fifth digits are only marginally less splayed out than previously (*Plate 45,e*). In the hand plate, while the first suggestion of nail primordia was seen at about 14.5–15 days p.c. (*Plate 44,h*), these are now more clearly seen in the lateral four digits (*Plate 45,c*), and become increasingly more clearly evident over the next few days (*Plate 45,l,n*). The lateral four digits are also becoming increasingly more finger-like, elongated and parallel than previously, while the first digit (the pollex) tends over the next few days to become proportionately progressively smaller than the other digits (*Plate 45,c,d*; *Plate 45,g,h*). The prominent pads characteristically seen on the plantar surface of the foot plate, and palmar surface of the hand plate, are also more clearly evident at this stage than previously (*Plate 45,d,f*), though they are more readily seen in embryos at more advanced stages of development (*Plate 45,m,o*).

THE HEART AND VASCULAR SYSTEM

The most significant change that takes place in relation to the heart between 14.5–15 and 15.5–16 days p.c. is an alteration in its axis, and this will be discussed after the volumes of the various chambers of the heart, and the appearance of the heart valves are briefly considered. The increase in the volume of the right atrial chamber compared to that of the left, was alluded to previously, but is now more marked (*Plate 35f,b–d*; *Plate 35g,a–d*; cf. *Plate 32f,b–d*; *Plate 32g,a–d*). The volume of the left ventricle continues to be slightly greater than that of the right ventricle (*Plate 35g,b–d*). The leaflets of the pulmonary and aortic valves are considerably more differentiated than previously (pulmonary valve, *Plate 35f,d*; cf. *Plate 32f,c*; aortic valve, *Plate 35g,a*; *Plate 36c,a*, unlabelled; cf. *Plate 33a,a*). This equally applies to the atrio-ventricular valves (*Plate 35g,c*; *Plate 36c,a,b*; cf. *Plate 33a,a*, unlabelled).

Of greater interest is the fact that the right atrium is now considerably more caudally located than the left atrium, whereas at 14.5–15 days p.c., the difference in the levels of these two chambers of the heart was less obvious. Similarly, the interventricular groove is now only evident towards the apical region of the heart, and the groove is clearly directed anterolaterally towards the left (*Plate 35h,a,b*), whereas at 14.5–15 days p.c. the interventricular groove was seen to run along almost the entire length of the interventricular region, and was clearly directed towards the median plane (*Plate 32g,a–d*). These features reflect a significant change in the orientation of the heart that occurs over this relatively brief period, so that the axis of the heart appears to have changed from a principally anteroposterior direction, to a more oblique orientation, with the main axis of the heart being from the upper right to the lower left part of the thoracic cavity. The axis of the heart is clearly seen to be in the direction of the muscular part of the interventricular septum (*Plate 35g,b–d*; *Plate 35h,a*). This modification to the principal axis of the heart persists, and is clearly seen in embryos at 16.5 days p.c. (*Plate 37f,c,d*; *Plate 37g,a–d*; *Plate 37h,a,b*), and at subsequent stages of development (e.g. 17.5 days p.c., *Plate 40g,a–d*).

With the eventual return of the midgut loop into the peritoneal cavity, the umbilical cord is then seen to contain only the umbilical vessels surrounded by Wharton's jelly. Elsewhere, the disposition of the major components of the vasculature of the embryo is similar to that outlined in the previous three stages of development. Both the arterial and venous systems remain essentially unmodified until shortly after birth, with the constriction of various vessels and their eventual replacement by fibrous tissue to form such structures as the ligamentum arteriosum, the ligamentum teres hepatis, etc. Often the umbilical arteries have different luminal diameters (see observations in Stage 21), and the smaller of these vessels (usually that of the left side) may atrophy, so that in those cases only a single umbilical artery would be present in the umbilical cord. The latter situation commonly occurs in the rat, where only a single umbilical artery and vein are usually present in the umbilical cord after about 17 days p.c. (Hebel and Stromberg, 1986).

THE PRIMITIVE GUT AND ITS DERIVATIVES

During the period between about 15.5 and 16 days p.c., the final stages of palatal shelf closure occur, so that the nasopharynx is now completely separated from the oropharynx (*Plate 60,h–l*). Over this period, the deep groove (naso-palatine canal) which previously separated the anterior borders of the palatal shelves of the maxillae from the primary palate (the intermaxillary segment) are finally obliterated, leaving only the slit-like orifice of the incisive canal (*Plate 60,l*). On the oropharyngeal surface of the palate, a series of prominent rugae are evident, which were first observed at about 13.5–14 days p.c., well before the medial borders of the palatal shelves first showed evidence of elevation (*Plate 59,j*). Beyond the most posterior part of the palate, the entrance to the nasopharynx is clearly seen (*Plate 60,h*).

The nasal septum at this stage fuses in the midline with the upper surface of the anterior half of the palatal shelves and with the primary palate. Much of this region of the nasal septum is seen to be occupied by the paired vomeronasal organs (Jacobson's organs), which are located within the soft tissues on either side and just inferior to the cartilage primordium of the nasal septum (*Plate 62,h,i*). Slightly more posteriorly, in the region opposite the first upper molar tooth, the nasal septum fails to fuse with the upper surface of the palate, and in this location the two halves of the nasopharynx are seen to be in continuity (*Plate 62,j*). Further posteriorly, the lateral borders of the lower part of the soft tissues of the nasal septum fuse with the soft tissues overlying the maxillary bones just superomedial and posterior to the origin of the second upper molar teeth (*Plate 62,k,l*). At this stage of development, minimal evidence of precartilage condensations are seen within the anterior half of the palate, though condensations of precartilage/cartilage are present within its posterior half (*Plate 62,j–l*). The presence of ossification centres had previously been noted within the maxillary bones at the lateral borders of the palatal shelves at about 14.5–15 days p.c. (*Plate 32d,a*, unlabelled), and these regions show increased evidence of differentiation (*Plate 35c,c*).

In the nasopharynx, the conchae are well delineated (e.g. *Plate 62,j*), and the majority of its upper part is lined by specialized olfactory epithelium (*Plate 35c,a–c*; *Plate 36a,a,b*). Glandular tissue is found in association with the lateral walls of the middle meatus (*Plate 62,j*), as well as in the soft tissues on either side of the upper part of the nasal septum (*Plate 62,h,i*), and the volume of this tissue is considerably more marked than previously (cf. *Plate 32d,a*; *Plate 32e,a*). Tubules of similar glandular tissue are also found within the soft palate (*Plate 35c,d*). The pharyngo-tympanic (Eustachian) tubes pass rostrally and laterally towards the region of the middle ear, where they open into the tubo-tympanic recesses (*Plate 35c,a–c*), and are now seen to be in particularly close association with the middle ear ossicles (*Plate 35c,a,b*).

Between 14.5–15 and 15.5–16 days p.c., the tongue elongates, increases in volume and continues to differentiate. A superficial examination of its surface features reveals that the intermolar eminence becomes more prominent than previously, as do the numerous fungiform papillae associated principally with its dorsal surface (*Plate 58,m,o*). The median circumvallate papilla, first clearly seen at about 14.5 days p.c. (*Plate 32d,b*; *Plate 33a,a*; *Plate 58,l*), is now much more prominent (*Plate 58,m,n*), and its features are clearly seen. More posteriorly, the epiglottis has increased in size compared to the situation observed previously (cf. *Plate 31a,a*; *Plate 33a,a*), and now guards the entrance to the larynx (*Plate 58,o*), and is also seen to be an anterior relation of the arytenoid swellings (*Plate 58,o*). At the histological level, the various components which make up the intrinsic musculature of the tongue are easily seen, as are the individual pairs of extrinsic muscles (*Plate 35d,b–d*; *Plate 35e,a,b*). These sections also reveal the pivotal roles played by the hyoid bone and mandible, in particular, as sites of attachment of some of these extrinsic tongue muscles.

Whereas at 14.5–15 days p.c., it was quite difficult to recognize all of the components of the tooth primordia (e.g. *Plate 32d,c,d*; *Plate 32e,a*), by 15.5–16 days p.c., it is possible to see the stellate reticulum (formed by the downgrowth of the oral epithelium), which partially surrounds the dental papilla. The inner and outer boundaries of the stellate reticulum are formed by the inner and outer enamel epithelia. The tooth primordium is also seen to be surrounded on its deep surface by centres of ossification within either the maxillae (in the case of the upper incisor and molar teeth) or by centres of ossification within the mandible (in the case of the lower incisor and molar teeth) (molars, *Plate 35c,d*; *Plate 35d,a–d*; incisors, *Plate 35e,a,b*; *Plate 35f,a,b*).

In the neck region, three paired glands are present, namely the thyroid, parathyroid and thymus glands. At this stage, the upper poles of the thyroid gland are in close proximity to the posterior parts of the thyroid cartilage, and more caudally to the posterior parts of the cricoid cartilage (*Plate 35d,c,d*; *Plate 36b,b*). The thyroid gland extends further caudally, where it is seen to become closely related to the rostral part of the trachea (*Plate 35e,a*). Only a very narrow isthmus is present which passes in front of the trachea, and connects the lower parts of the two lobes. The isthmus at this stage usually lies in front of the first or second tracheal "ring" (*Plate 35e,b*). At more advanced stages of development, the isthmus tends to "descend" to a more caudal level (e.g. at 17.5 days p.c., it is seen to be located at the level of the third and forth tracheal "rings", *Plate 41*). The thyroid gland is also seen to be a more substantial structure than previously (cf. *Plate 32e,a,b*; *Plate 33a,b*).

The relationship between the parathyroid and thyroid glands is not illustrated in these sections, but is clearly seen in the previous stage (*Plate 32e,a*), where the parathyroid glands are seen to be posterior relations of the upper one-third to one-half of the thyroid gland. In more advanced stages of development, the relationship changes somewhat, so that by 16.5 days p.c., for example, the parathyroid glands are more laterally located than previously (*Plate 39b,d*).

The two thymic primordia in the mouse remain as two distinct entities, and are remarkably large structures at this stage of development. Their upper poles are seen to be inferior relations of the lower poles of the thyroid gland. They now extend caudally from the lower part of the neck well into the superior part of the anterior mediastinum, where their lower poles are seen to have descended (by about 12.5–13 days p.c.) to make contact with the superior part of the pericardial cavity (*Plate 28c,a–c*). By about 14.5–15 days p.c., they are seen to have descended even more caudally (*Plate 32f,a*; *Plate 33a,a*), to become anterior relations of the principal branches of the arch of the aorta. A similar relationship is seen to be the case at about 15.5–16 days p.c. (*Plate 35f,a,b*; *Plate 36c,a*).

The posterior part of the oropharynx is continuous caudally with the inlets to the lower respiratory tract (the larynx, trachea, the main bronchi and the more peripheral components of the respiratory tree), and the oesophageal region of the foregut. As indicated previously, at this stage of development, the epiglottis is a quite prominent structure, which guards the entrance to the larynx (*Plate 35d,a,b*; *Plate 36b,b*; *Plate 58,o*). In the laryngeal region, it is now possible to identify the thyroid cartilage and more caudally the cricoid cartilage (*Plate 35d,c,d*), as well as the arytenoid cartilages more centrally at the inlet to the larynx (*Plate 35d,b–d*). More caudally, the individual tracheal "rings" of cartilage are easily identified both in transverse sections, where their U-shaped structure can be appreciated (*Plate 35e,b–d*; *Plate 35f,a–c*), but particularly in median (sagittal) sections (cf. *Plate 33a,a*). In the sagittally sectioned series used to illustrate this stage of development, the interpretation of the head and neck sections is complicated because this region of the embryo is directed away from the median plane.

Of the various salivary glands, the submandibular and sublingual glands are quite substantial structures at this stage, though they are not readily separable (*Plate 35d,d*; *Plate 35e,a–d*). These were first recognized at about 14.5 days p.c. (*Plate 32e,a–c*), and have differentiated to a considerable extent since that time. The parotid gland, by contrast, is a quite distinct but relatively small structure which is first recognized at this stage of development (*Plate 35d,a*). The sublingual and submandibular ducts are easily recognized (*Plate 35d,d*; *Plate 35e,a–c*), and their paths can be traced as they proceed anteriorly from these glands towards the sublingual caruncles where they both open separately, on its rostral aspect, into the floor of the mouth just proximal to the distal one-third of the tongue (*Plate 35e,d*). The sublingual duct opens below the lateral tip, and the submandibular duct below the medial tip of the caruncle.

A comparison between the histological appearance of the lungs in embryos at about 14.5–15 days and 15.5–16 days p.c. is interesting, in that it clearly indicates that a considerable increase has occurred in the degree of differentiation of the lungs over this period. There appears to be a diminution in the number of tertiary bronchi present, and an increase in the number of bronchioles per unit volume of lung tissue, which is particularly evident in the lower magnification sections (*Plate 66,i*; cf. *Plate 66,k*), as well as in the overviews of the thoracic region provided in the intermittent serial sections used to illustrate these stages of development (e.g. *Plate 35h,a*; cf. *Plate 32h,a*). The lungs therefore appear to be more homogeneous histologically than previously, and less sponge-like. The volume of the thoracic region occupied by the lungs over this period, however, does not appear to increase to any extent, and the level of the bifurcation of the trachea at these stages is also quite similar to that observed previously.

The disposition and size of the stomach is quite similar to that seen at the previous stage. Whereas previously the thickness of the anterior and posterior walls was fairly similar (*Plate 33j,a,b*), the thickness of the posterior wall tends to be slightly greater than that of the anterior wall at this stage, largely due to the differentiation of its mucosal lining and submucosa (*Plate 35j,d*; *Plate 35k,a,b*). The splenic primordium continues to enlarge, and extends forwards in close association with the inferolateral surface of the stomach (*Plate 35k,b–d*; *Plate 35l,a,b*). In this embryo, the tail of the spleen is seen to make contact with the lateral surface of the left testis (*Plate 35k,d*). The pyloric region of the stomach leads directly into the first part of the duodenum (*Plate 35i,d*), and within the lumen here, and within that of the rest of the duodenum, it is possible to see that mucosal differentiation is considerably more advanced than previously (*Plate 35i,d*; *Plate 35j,a–d*; cf. *Plate 32i,b–d*), with the presence of large numbers of well-formed villi, and early evidence of intestinal glands within the submucosa (e.g. *Plate 35i,d*).

While the upper half of the peritoneal cavity is largely occupied by the liver, and to a lesser extent by the stomach (*Plate 35h,d*; *Plate 35i,a–d*), as previously (*Plate 32h,b–d*), it is clear from an analysis of the transverse sections of this embryo that more space is available in the anterior and inferior parts of the peritoneal cavity than previously, that could accommodate the midgut when it eventually returns into the abdominal cavity (*Plate 35j,c,d*; *Plate 35k,a–c*; cf. *Plate 32i,b–d*; *Plate 32j,a–d*). Apart from the loops of midgut, and to a much smaller degree those of the hindgut, the only other substantial viscera in the lower abdominal cavity are the pancreas, the gonads (testes in this embryo, though the ovaries in embryos of similar developmental age are significantly more rostrally located, see *Plate 71,a*; cf. location of testes, see *Plate 70,c*), and the bladder, while the largest of the retroperitoneal structures in this general region are the kidneys (*Plate 35k,a–d*). The physiological umbilical hernia is seen to be quite large in the case of the transversely sectioned embryo illustrated, and contains a considerable amount of midgut (*Plate 35h,c,d*; *Plate 35i,a–c*), whereas in the sagittally sectioned embryo illustrated, only a relatively small volume of the midgut remains in the physiological umbilical hernia (*Plate 36b,a*), the majority having previously returned to the peritoneal cavity.

A more detailed histological analysis of the liver reveals a considerable increase in the amount of haematopoietic activity in this organ at this time, compared to previously. In addition, the definitive arrangement of the lobes are now achieved. In the pancreas, the exocrine (glandular) component shows a considerable degree of branching, and some of the ducts are seen to have distinct lumina. Early evidence of islet (of Langerhans) formation is seen, though the characteristic beta-cells are not recognized until about 17–18 days p.c.

THE NERVOUS SYSTEM

Between 14.5–15 and 15.5–16 days p.c., both the eye and the pituitary gland display increased evidence of differentiation. The changes that occur in the eye principally reflect the fact that over this period eyelid closure takes place (for observations on the external events associated with this process, see earlier section). The closure of the eyelids is associated with the formation of the conjunctival sac (*Plate 53,b*). The enclosure of the anterior surface of the eye probably facilitates the further differentiation of the rudimentary cornea, which is first clearly recognizable as a distinct entity at about 14–14.5 days p.c. (*Plate 52,i*). At 15.5–16 days p.c., the cornea is fairly homogeneous in structure, with an inner limiting layer of flattened cells (the beginning of Descemet's endothelium), and an outer limiting layer of endothelial cells. Between these two layers is a region of stellate-shaped mesodermal cells which are destined to become the ground substance of the substantia propria. At later stages of development, the cornea has a stratified appearance, the inner one-third to one-half of the intermediate zone having a more condensed appearance than the outer part (e.g. *Plate 53,f,g*; *Plate 53,h,i*). At this, and at subsequent stages of development, it is possible to recognize the line of fusion of the upper and lower eyelids (e.g. *Plate 53,d*). While only minimal evidence of an anterior chamber is seen at 14.5–15 days p.c. (*Plate 53,a*), a considerable increase in the volume of this region of the eye is seen by about 15.5–16 days p.c. (*Plate 53,b,c*), though this region expands to an even greater extent at about 16.5–17 days p.c. (*Plate 53,d*), when it delineates the outer aspect of the iris and its medial margin (i.e. the boundary of the pupil).

The anterior part (or capsule) of the lens is showing evidence of thinning, while the lens fibres, particularly towards the central region of the lens, are now very elongated and almost parallel to each other. The distinction between these central fibres and those at the equatorial region of the lens is more clearly seen than previously (*Plate 53,c*; cf. *Plate 53,a*), and the nuclei of the central lens fibres are tending to stain more irregularly than previously. The neural retina also shows increased evidence of differentiation, but it is only at later stages of development that the stratification characteristically associated with this region of the eye is clearly seen (e.g. at 16.5 days p.c., *Plate 53,e*; at 17.5 days p.c., *Plate 53,f*; at 18.5 days p.c., *Plate 53,h*). The histological evidence of neural retinal differentiation is also more clearly seen in silver stained than in haematoxylin and eosin stained material. Overviews of the optic region, reveal that the extrinsic ocular muscles are substantially more differentiated than previously (*Plate 53,b*; cf. *Plate 52,i*).

The pituitary gland also shows increased evidence of differentiation, and between 14.5–15 and 15.5–16 days p.c., substantial changes are evident in its histological morphology. The latter is most evident from an analysis of median sagittal sections through the pituitary gland. While at 14.5–15 days p.c., the entrance to the infundibular recess was widely open from the third ventricle (*Plate 55,c*), this channel becomes progressively narrower (*Plate 55,d–f*). More interestingly, the infundibular recess itself becomes progressively obliterated to form the definitive pars nervosa of the pituitary gland (*Plate 55,e,f*). The tissue comprising the pars anterior also becomes more consolidated than previously (*Plate 55,f*; cf. *Plate 55,c*). The examination of contemporaneous transverse sections through the region of the pituitary stalk is essential to appreciate the close relationship that exists between the stalk and the pars tuberalis (derived from the rostral extremity of the lateral part of the pars anterior), which tends to surround it (*Plate 35b,d*; *Plate 57,g*).

The first evidence of the pineal primordium (epiphysis) is seen at about 14.5 days p.c. (*Plate 32a,a*; *Plate 33a,b*), as a hollow outgrowth from the epithalamic region of the diencephalon, and possesses a lumen which is in continuity with the dorsal part of the third ventricle. Little evidence of differentiation is seen, however, until about 15.5–16 days p.c., when a thickening of its dorsal and lateral walls is first evident (*Plate 35a,a*; *Plate 36a,a*). In addition, choroid plexus material, which is associated with the pineal recess of the third ventricle, appears to extend into the proximal part of its lumen. There is also a slight increase in the volume of choroid plexus material associated with the lateral recesses of the fourth ventricle (*Plate 35a,d*; cf. *Plate 32b,a*), and a similar minor increase in the amount of choroid plexus material associated with the lateral ventricles is also evident (*Plate 35a,b–d*; cf. *Plate 32a,c,d*).

A considerable degree of differentiation is apparent in the inner ear apparatus between the situation observed at this stage, and that seen during the previous stage of development. Thus the semicircular canals are seen to be more well defined than previously, and have a narrower lumen (*Plate 35b,b*; cf. *Plate 32b,a,b*), and this is equally applicable to the cochlear apparatus (*Plate 35b,d*; *Plate 35c,a,b*; cf. *Plate 32c,b–d*). The middle ear ossicles, which were previously quite difficult to distinguish (*Plate 32c,a–d*), are now seen to be more distinct entities, though the malleus is still seen to be in continuity with the dorsal part of Meckel's cartilage (*Plate 35b,b,c*; cf. *Plate 32c,b*).

In the cervical region, it is possible to recognize the superior (*Plate 35c,c,d*), middle (*Plate 35f,a*) and cervico-thoracic (stellate) (*Plate 35f,b*) sympathetic ganglia, as well as some of the ventral primary rami of the cervical nerves that are progressing distally to form components of the brachial plexus (e.g. *Plate 35e,c*; and similarly components of the lumbo-sacral plexus, *Plate 35l,b,c*), but any detailed analysis of even the larger components of the peripheral nervous system, including nerve trunks, is more easily studied in silver-stained material. In the thoracic region, nevertheless, it is possible to recognize the sympathetic trunks and ganglia in haematoxylin and eosin-stained material (e.g. *Plate 35g,a*), and in the abdomen, the hypogastric plexus of nerves (*Plate 35j,d*) is also easily seen. The pre-aortic abdominal sympathetic paraganglia (of Zuckerkandl), and the extremely large para-aortic bodies (consisting of chromaffin tissue), first recognized at about 13.5 days p.c., when they were seen to be rather diffuse masses of material associated with the anterior and anterolateral surfaces of the abdominal aorta (*Plate 30k,b*), are now more well defined than previously (*Plate 35k,a–c*; cf. *Plate 32j,b–d*). Despite their enormous size, the physiological role of the para-aortic bodies has yet to be fully determined.

The appearance of a typical section through the spinal cord in the mid-thoracic region is quite similar to that seen at 14.5–15 days p.c., except that the luminal diameter of the central canal has diminished considerably (*Plate 63,f*; cf. *Plate 63,e*). The spinal cord extends caudally to the base of the tail, where the conus medullaris is located (*Plate 35l,c,d*), and more distally the filum terminale (e.g. *Plate 35j,a*) extends towards the tip of the tail.

The rostral extremity of the adrenal glands at this stage is approximately at the level of the 13th rib (*Plate 35j,d*; *Plate 36b,b*; *Plate 36c,a*), while their lower poles are closely associated with the upper poles of the kidneys (*Plate 35k,a,b*). At this stage, the adrenals appear to have a reasonably homogeneous histological morphology, the majority of the cells being mesodermal in origin, and destined to form the cortical region of the gland. A few cord-like cellular columns are also present within the medial side of the glands. The latter consist of neural crest-derived sympathetic tissue, and are destined to form the adrenal medulla.

THE UROGENITAL SYSTEM

The histological appearance of the kidney at 15.5–16 days p.c. represents a stage with intermediate features between those seen at 14.5–15 days p.c. and those seen at 16.5–17 days p.c. This is particularly evident in relation to the division of the kidney into an outer cortical region and an inner medullary region. At 14.5–15 days p.c., the future medullary region still contains substantial numbers of primitive glomeruli as well as large numbers of thick-walled collecting tubules (*Plate 72,e,f*), whereas at 15.5–16 days p.c., the glomeruli show increasing evidence of differentiation, and are tending to be concentrated in the outer one-third of the kidney, the region destined to form the cortex. Similarly, at this time, the collecting ducts tend to have thinner walls than previously, and these are now largely radially arranged, and drain into a dilated renal pelvis (*Plate 72,g,h*). By contrast, at 16.5–17 days p.c., the glomeruli are now exclusively seen to be located in the outer one-third of the kidney, and the collecting ducts now drain into a substantially enlarged renal pelvis (*Plate 72,i,j*). Throughout these stages, but more particularly in the earlier stages, an outer undifferentiated rim of metanephric cap tissue is seen to be located just subjacent to the capsule of the kidney. At 17.5–18 days p.c., major and minor calyces drain into the renal pelvis, and the cortical region is seen to contain, in addition to glomeruli, numerous proximal and distal convoluted tubules (*Plate 72,k,l*).

Whereas at 14.5–15 days p.c., the testis was ovoid in shape (*Plate 70,b*), at 15.5–16 days p.c. it is seen to have a more rounded profile (*Plate 70,c,d*). The location of the testis is similar to that observed at the previous stage of development, though over this and during the next stage of development, the testes continue to descend towards the brim of the pelvis (cf. at 16.5 days p.c., *Plate 70,e*). On a close inspection of the scanning micrographs, it is possible to observe the close relationship that exists between the testis and the epididymis. The latter lies posterolateral and slightly inferior to the testis, and is subdivided into the rostrally located caput (head), the intermediate part or corpus (body), and inferiorly is located the cauda (tail) (*Plate 70,d*). The epididymis drains into the ductus deferens, which in turn drains via the ejaculatory duct into the prostatic region of the urethra (*Plate 35k,c,d*).

The appendix testis is a quite substantial structure at this stage, and is seen to be attached to the rostral pole of the testis (*Plate 70,d*), and is believed to be of paramesonephric duct origin. The appendix epididymis (not seen here) represents the blind cranial end of the mesonephric duct (for additional discussion about this structure, see text associated with *Plates 68–71*), and with other aberrant gonadal duct tissue usually completely regresses. At the histological level, little difference is observed between the features seen at this and at the previous stage of development (*Plate 68,l,m*; cf. *Plate 68,j,k*). The volume of the testis occupied by mesenchyme tissue at this stage varies considerably (e.g. *Plate 35k,b–d*; *Plate 68,l,m*), whereas at later stages, the volume of this tissue tends to decrease, and is largely replaced by interstitial tissue (which occupies much of the space between the seminiferous tubules) (*Plate 69,e*; *Plate 69,n*). The mesenchymatous capsule (tunica albuginea) that surrounds the testis also increases in thickness over this period (*Plate 68,m*; cf. *Plate 69,d*).

The ovaries are substantially smaller than the testes at this stage of development (*Plate 71,a,b*; *Plate 70,c,d*), are more rostrally located, and partially surrounded, particularly on their superior, inferior and posterior surfaces, by the rostral part of the paramesonephric duct (*Plate 71,a,b*). The ovaries are also more ovoid in shape than the testes, and though they gradually increase in size, their increase in volume fails to keep pace with the overall increase in the volume of the peritoneal cavity and that of the other intra-abdominal viscera. A notable feature at this stage is the presence of the suspensory ligament, which appears to be attached to the upper (rostral) pole of the paramesonephric duct.

With the gradual enlargement of the kidneys that occurs over the next few days, the ovaries tend to become obscured by the lower poles of the kidneys, and are relegated to the paracolic gutters (*Plate 71,h*). Over the period of the next few days, the ovaries gradually become surrounded by paramesonephric duct tissue, so that by about 19.5 days p.c., the ovary is no longer visible, since it is completely embraced by this tissue, which is destined to form the ovarian capsule (*Plate 71,i*). At this stage, it is possible to recognize the entrance to the paramesonephric duct from the peritoneal cavity (the future fimbriated os) (*Plate 71,g*), though at later stages, the entire oviduct as well as the ovary is enclosed within the ovarian capsule. Similarly, there appears to be little if any periovarian fat seen at this stage of development which is characteristically present in the adult.

At the histological level, the ovary has a "spotty" appearance, being swollen by the presence of ovigerous cords (egg clusters), and the characteristic whorl-like stromal cells (*Plate 69,a*). Numerous oogonia are seen to be in mitosis, and the primary oocytes are mostly in leptotene (one stage of prophase of the first meiotic division) (*Plate 69b,b*). The germinal (coelomic) epithelium of the ovary, which is continuous with the peritoneal mesothelium, forms a monolayer of cells around this organ, there being no equivalent to the fibrous capsule (tunica albuginea) of the testis (*Plate 69,b*; cf. *Plate 68,o*). While the paramesonephric ducts form the two uterine horns and (probably) the upper part of the vagina, as well as the ovarian capsule in the female, the first evidence of differentiation to form the oviduct occurs during the late prenatal period. The morphological appearance of the external genitalia in the male and female is remarkably similar at this stage, and cannot be used as an unequivocal means of distinguishing between the sexes (male, *Plate 35j,d*; *Plate 35k,a–c*; cf. female, *Plate 37j,d*; *Plate 37k,a–c*).

THE SKELETAL SYSTEM

Since the two embryos used to illustrate the features of this stage of development are both examples of embryos at the less advanced end of the range, relatively few significant differences are observed in the number of new ossification centres present than might have been observed in developmentally more advanced embryos. This observation equally applies to the extent of ossification seen in already existing centres observed at 14.5–15 days p.c. However, some additional cartilage primordia are evident, and those already seen at 14.5–15 days p.c. are generally substantially more clearly defined than previously. In the chondrocranium, for example, all of the centres of ossification previously noted have increased in size, such as those present in the exoccipital bones, which showed early evidence of periosteal ossification (*Plate 35c,a,b*). A moderate degree of periosteal ossification is now seen in the exoccipital bones (*Plate 35b,c,d*), and in the basisphenoid (*Plate 35c,b*), as well as in the membranous bones of the cranial vault. Thus ossification is now seen in the temporal, parietal and frontal bones (*Plate 35b,a–c*), and in all of the components of the viscerocranium, as well as in the mandible (*Plate 35c,a–d*; *Plate 35d,a–d*; *Plate 36a,a,b*; *Plate 36b,a,b*; *Plate 81,e,f*). New centres of ossification have also appeared in the nasal bones (*Plate 35b,b–d*).

More caudally, the ossification centres within the cartilage primordia of the neural arch of the C1 vertebra (atlas) are more extensive than previously, and are extremely useful landmarks in this region (*Plate 35c,d*). The earliest evidence of periosteal ossification is also seen in relation to the proximal parts of the neural arches of each of the vertebrae as far distally as the mid-lumbar region (but this is initially more obvious in the more rostral vertebral elements) (*Plate 81,e,f*).

(*continued on page 290*)

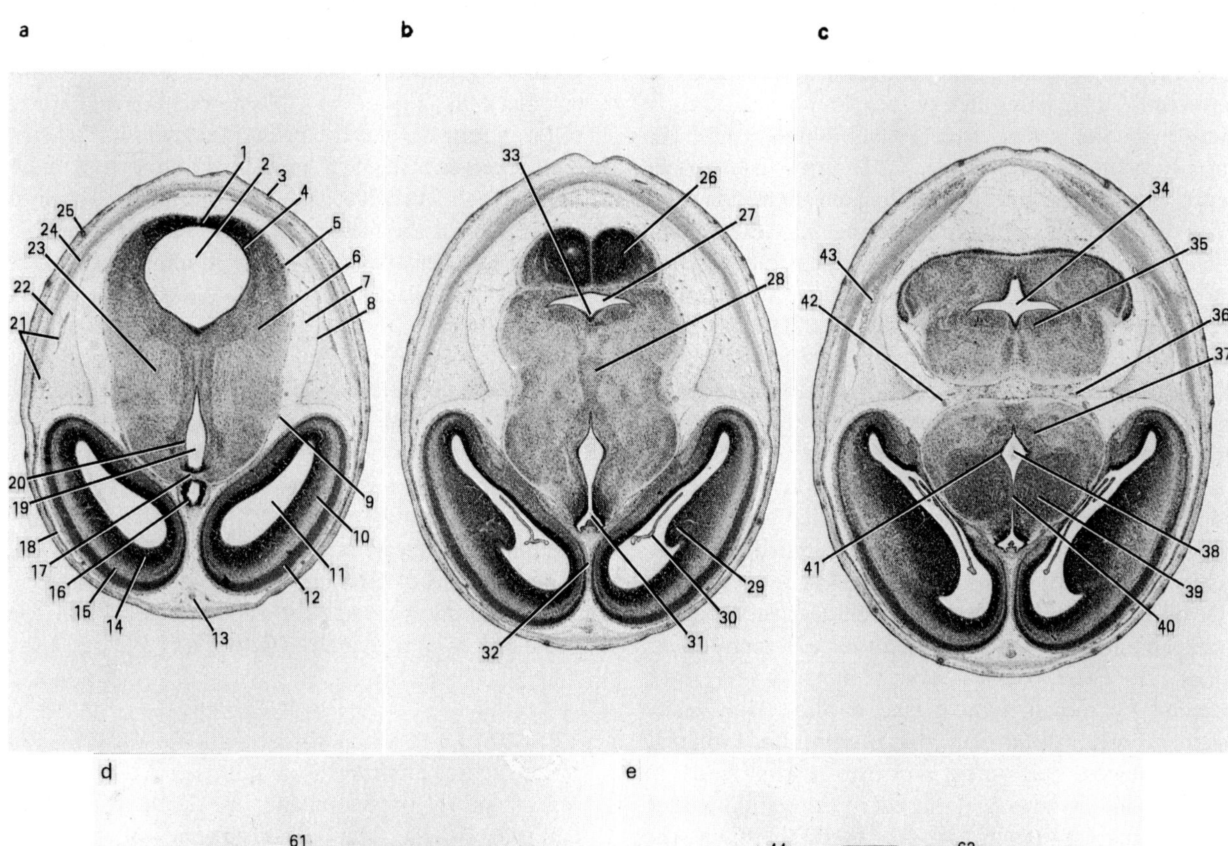
a
b
c
1
2
3
4
5
6
7
8
9
10
11
12
13
14
15
16
17
18
19
20
21
22
23
24
25
26
27
28
29
30
31
32
33
34
35
36
37
38
39
40
41
42
43

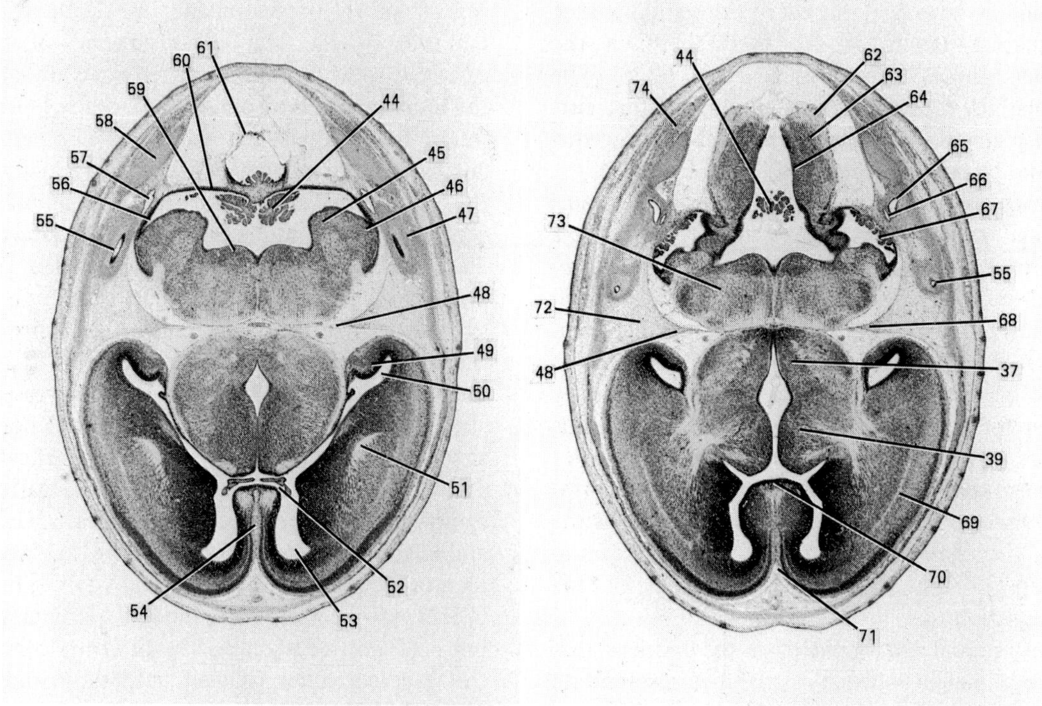
d
e
44
45
46
47
48
49
50
51
52
53
54
55
56
57
58
59
60
61
62
63
64
65
66
67
68
37
39
69
70
71
72
73
74

Plate 35a (15.5 days p.c.)

15.5 days p.c. Male embryo. Crown–rump length 12.2 mm (fixed). Transverse sections (Theiler Stage 24; rat, Witschi Stage 33).

1. roof of dorsal part of midbrain
2. mesencephalic vesicle
3. surface ectoderm (skin)
4. ependymal layer of mesencephalic vesicle
5. marginal layer
6. mantle layer
7. loosely packed cephalic mesenchyme in future subarachnoid space
8. mesenchymal condensation to form inner table of future dura mater
9. vascular elements of pia mater
10. wall of telencephalic vesicle
11. telencephalic vesicle (future lateral ventricle)
12. neopallial cortex (future cerebral cortex) composed of inner broad nuclear layer (cortical plate) with narrow outer relatively anuclear layer (marginal zone)
13. anterior extension of superior sagittal dural venous sinus
14. ventricular zone
15. intermediate zone
16. pineal primordium which arises as a hollow outgrowth of the epithalamus
17. epithalamic region
18. mesenchymal precursor of parietal bone
19. dorsal part of third ventricle
20. ependymal layer
21. primitive venous plexus, precursors of dural venous sinuses in this location
22. anterior margin of cartilage primordium of squamous part of occipital (supraoccipital) bone
23. midbrain
24. cartilage primordium of squamous part of occipital (supraoccipital) bone
25. primordium of hair follicle
26. rostral extension of primitive cerebellum
27. caudal extension of mesencephalic vesicle
28. floor of midbrain (tegmentum, in region of midbrain flexure)
29. striatum (ganglionic eminence) overlying the head of the caudate nucleus
30. choroid plexus within lateral ventricle and arising from its medial wall
31. pineal recess of third ventricle
32. primitive venous plexus, precursors of inferior sagittal dural venous sinus
33. median sulcus
34. cerebral aqueduct
35. junctional region between midbrain and pons
36. rootlets of origin of oculomotor (III) nerve
37. diencephalon (hypothalamus)
38. third ventricle
39. diencephalon (thalamus)
40. interthalamic adhesion (massa intermedia)
41. hypothalamic sulcus
42. rostral extension of posterior communicating artery
43. transverse dural venous sinus
44. choroid plexus in central region of fourth ventricle
45. intraventricular portion of cerebellar primordium
46. rhombic lip (dorsal part of alar lamina of metencephalon – location of the lateral part of the cerebellar primordium)
47. cartilage primordium of petrous part of temporal bone
48. oculomotor (III) nerve
49. region overlying tail of caudate nucleus
50. caudal extremity of posterior horn of lateral ventricle
51. fibres of internal capsule
52. interventricular foramen
53. anterior portion of superior horn of lateral ventricle
54. inferior sagittal dural venous sinus
55. superior semicircular canal
56. lateral extremity of fourth ventricle
57. rostral extremity of endolymphatic sac
58. cartilage primordium of exoccipital bone
59. tegmentum of the pons (basal plate)
60. roof of the fourth ventricle
61. ependymal diverticulum of fourth ventricle
62. marginal layer of medulla oblongata
63. mantle layer of medulla oblongata
64. ependymal layer of medulla oblongata
65. posterior semicircular canal
66. endolymphatic duct
67. choroid plexus within lateral extremity of fourth ventricle
68. mesenchymal condensation forming tentorium cerebelli
69. internal capsule
70. roof of third ventricle
71. mesenchymal condensation forming falx cerebri
72. trochlear (IV) nerve
73. pons
74. superficial and deep groups of occipital muscles – principally extensors of the head

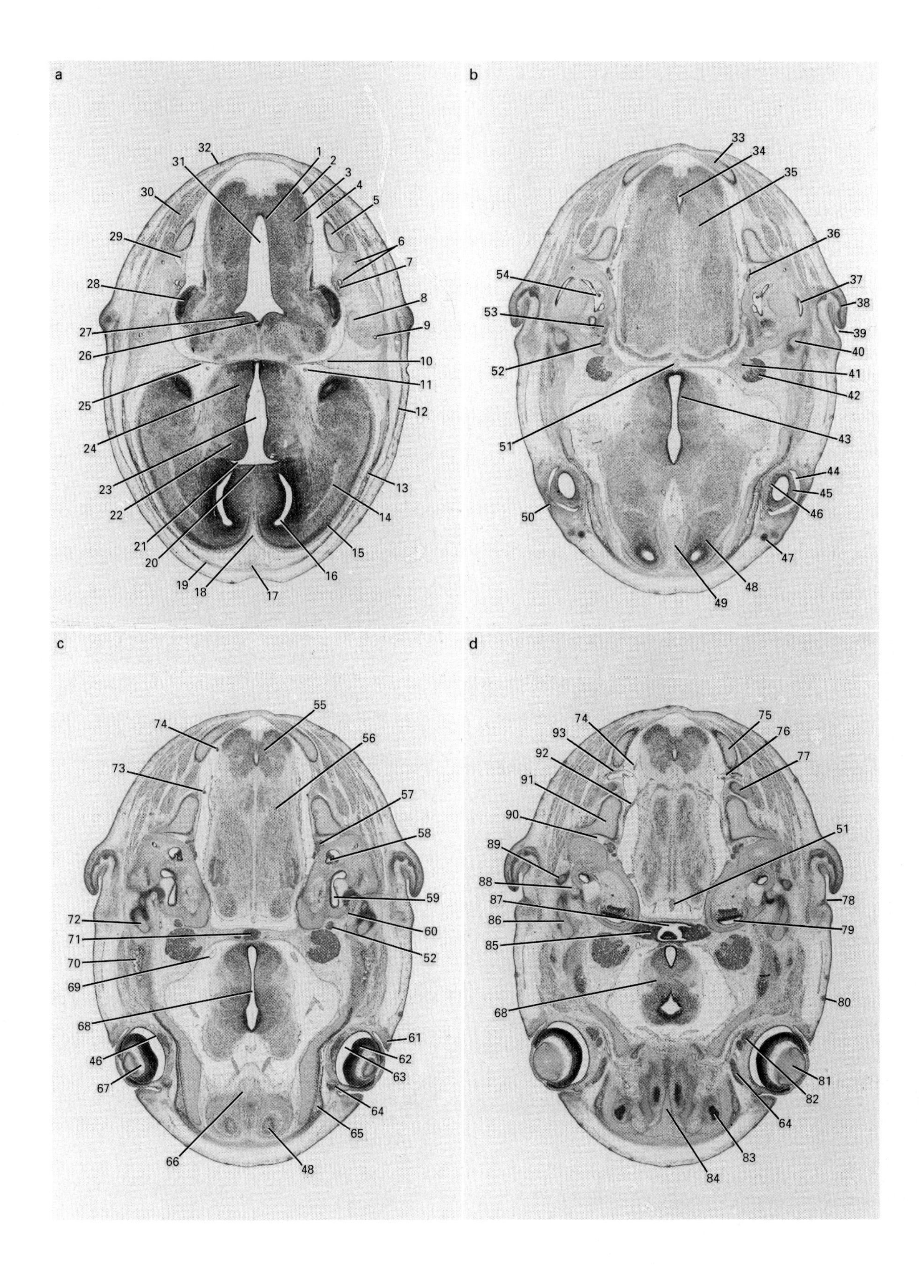
a
b
c
d
1
2
3
4
5
6
7
8
9
10
11
12
13
14
15
16
17
18
19
20
21
22
23
24
25
26
27
28
29
30
31
32
33
34
35
36
37
38
39
40
41
42
43
44
45
46
47
48
49
50
51
52
53
54
55
56
57
58
59
60
61
62
63
64
65
66
67
68
69
70
71
72
73
74
75
76
77
78
79
80
81
82
83
84
85
86
87
88
89
90
91
92
93

Plate 35b (15.5 days p.c.)

15.5 days p.c. Male embryo. Crown–rump length 12.2 mm (fixed). Transverse sections (Theiler Stage 24; rat, Witschi Stage 33).

1. ependymal layer of medulla oblongata
2. marginal layer of medulla oblongata
3. mantle layer of medulla oblongata
4. loosely packed cephalic mesenchyme in future subarachnoid space
5. periosteal ossification associated with cartilage primordium of exoccipital bone
6. posterior semicircular canal
7. endolymphatic duct
8. cartilage primordium of petrous part of temporal bone
9. superior semicircular canal
10. mesenchymal condensation forming tentorium cerebelli
11. rostral extension of posterior communicating artery
12. temporalis muscle
13. ossification within outer table of membranous primordium of frontal bone
14. internal capsule
15. neopallial cortex in wall of telencephalic vesicle
16. anterior portion of superior horn of lateral ventricle
17. superior sagittal dural venous sinus
18. inferior sagittal dural venous sinus
19. membranous primordium of frontal bone
20. lamina terminalis (roof of third ventricle)
21. interventricular foramen
22. diencephalon (thalamus)
23. third ventricle
24. diencephalon (hypothalamus)
25. oculomotor (III) nerve
26. median sulcus
27. junctional region between pons and medulla oblongata
28. lateral part of cerebellar primordium
29. sigmoid dural venous sinus
30. superficial and deep groups of occipital muscles – principally extensors of the head
31. caudal part of fourth ventricle
32. surface ectoderm (skin)
33. ossification within cartilage primordium of dorsal part of neural arch of C1 vertebra (atlas)
34. junctional region between caudal extremity of fourth ventricle and rostral extremity of central canal
35. caudal region of medulla oblongata
36. rootlets of glossopharyngeal (IX) nerve
37. lateral semicircular canal
38. pinna of ear
39. external auditory meatus
40. cartilage primordium of incus
41. rootlets of trigeminal (V) nerve
42. trigeminal (V) ganglion
43. ependymal lining of third ventricle
44. upper recess of conjunctival sac
45. corneal epithelium
46. pigment layer of retina
47. primordium of prominent tactile or sinus hair follicle characteristically found in this location
48. olfactory lobe of brain
49. rostral extremity of nasal capsule
50. upper part of eyelid
51. basilar artery
52. facial (VII) ganglion
53. vestibulocochlear (VIII) ganglion
54. origin of lateral semicircular canal from saccule
55. rostral extremity of cervical region of spinal cord
56. medulla oblongata
57. glossopharyngeal (IX) ganglion
58. origin of posterior semicircular canal
59. saccule/utricle
60. facial (VII) nerve
61. margin of eyelid
62. neural layer of retina
63. intra-retinal space
64. extrinsic ocular muscle (medial/nasal rectus)
65. ossification in outer table of cartilage primordium of superolateral part of nasal capsule
66. olfactory (I) nerves
67. hyaloid cavity
68. region of floor of third ventricle
69. internal carotid artery
70. ossification in zygomatic process of temporal bone
71. infundibulum (future pars nervosa) of pituitary
72. cartilage primordium of malleus
73. cranial accessory (XI) nerve
74. spinal accessory (XI) nerve
75. ossification within cartilage primordium of ventral part of neural arch of C1 vertebra (atlas)
76. vertebral artery
77. cartilage primordium of lateral mass of C1 vertebra (atlas)
78. entrance to first branchial cleft
79. lumen of cochlea
80. primordium of hair follicle
81. lens
82. extrinsic ocular muscle (superior rectus)
83. olfactory epithelium
84. cartilage primordium of nasal septum
85. site of active cellular and vascular proliferation in anterior wall of Rathke's pouch (inferior part of future pars tuberalis)
86. continuity between malleus and Meckel's cartilage
87. residual lumen of Rathke's pouch
88. stapedial artery
89. cartilage primordium of stapes
90. site of continuity between sigmoid dural venous sinus and internal jugular vein
91. inferior part of exoccipital bone
92. rootlet of vagus (X) nerve
93. first cervical (C1) nerve just proximal to its division into anterior and posterior primary divisions

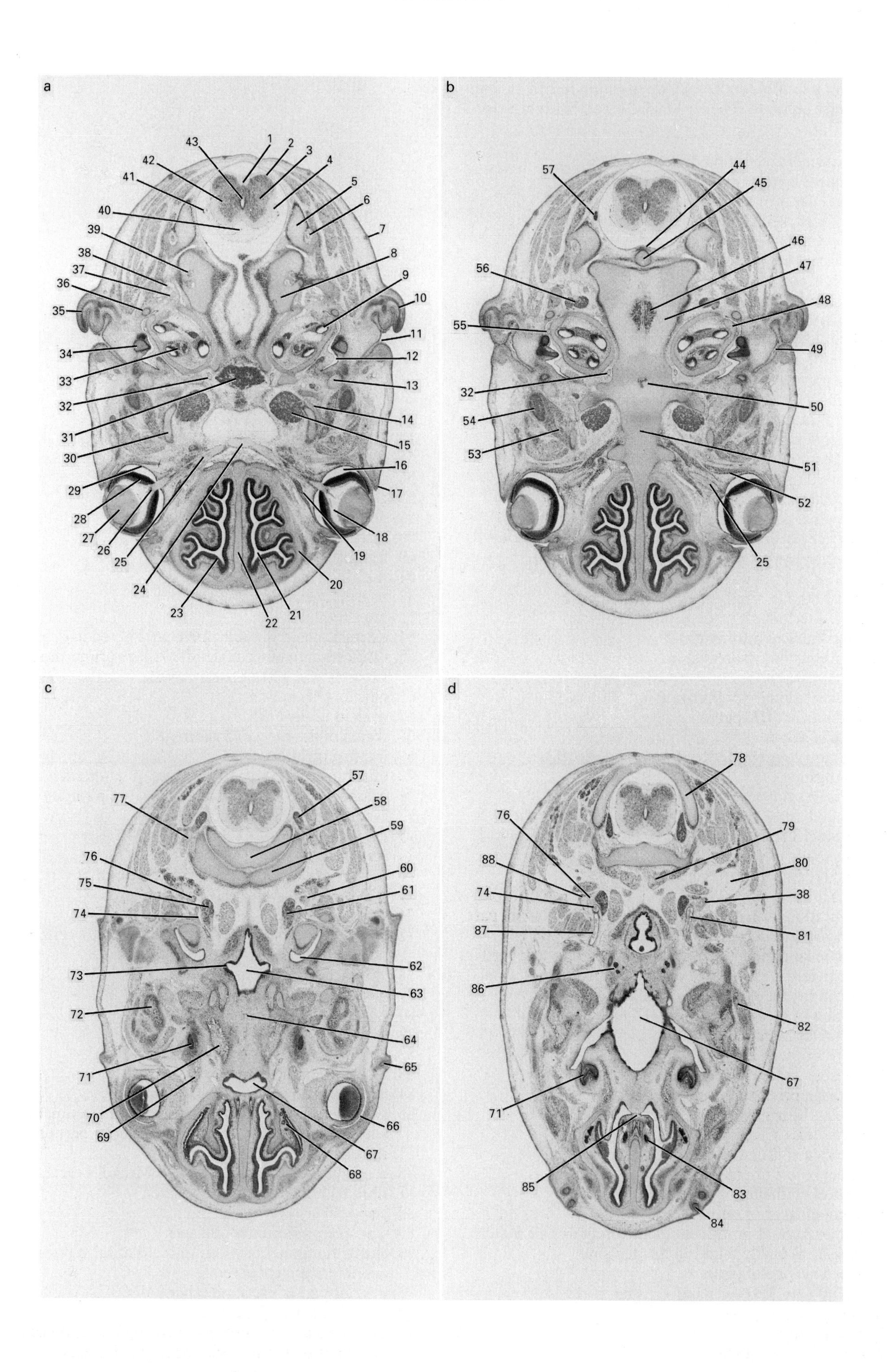
a
b
c
d

Plate 35c (15.5 days p.c.)

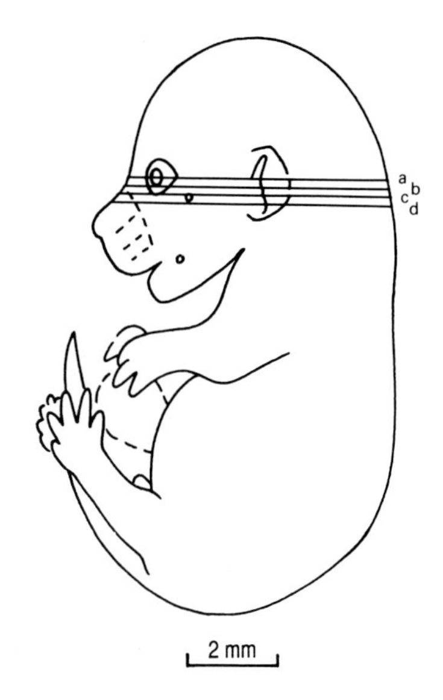

15.5 days p.c. Male embryo. Crown–rump length 12.2 mm (fixed). Transverse sections (Theiler Stage 24; rat, Witschi Stage 33).

1. roof plate
2. marginal layer
3. mantle layer
4. loosely packed mesenchyme tissue in future subarachnoid space
5. lateral mass of C1 vertebra (atlas)
6. vertebral artery
7. primordium of hair follicle
8. lateral mass of basioccipital bone
9. cochlea
10. pinna of ear
11. entrance to first branchial cleft
12. tubo-tympanic recess
13. Meckel's cartilage
14. maxillary division of trigeminal (V) nerve
15. trigeminal (V) ganglion
16. intra-retinal space (artefactually increased due to shrinkage)
17. margin of eyelid
18. hyaloid cavity
19. extrinsic ocular muscle (medial/nasal rectus)
20. nasal capsule
21. nasal cavity
22. cartilage primordium of nasal septum
23. olfactory epithelium
24. optic chiasma
25. optic (II) nerve
26. optic disc
27. lens
28. neural layer of retina
29. pigment layer of retina
30. cartilage primordium of orbito-sphenoid bone
31. site of active cellular and vascular proliferation in anterior wall of Rathke's pouch
32. internal carotid artery
33. cochlear (VIII) ganglion
34. cartilage primordium of stapes
35. condensation of precartilage tissue within pinna
36. Reichert's cartilage (region of future styloid process)
37. glossopharyngeal (IX) nerve
38. internal jugular vein
39. hypoglossal (XII) nerve passing through hypoglossal (anterior condylar) canal
40. anterior spinal artery
41. vascular elements of pia mater
42. upper cervical region of spinal cord
43. central canal
44. horizontal component of cruciate ligament (attached to the atlas) which helps to hold the odontoid process in position
45. cartilage primordium of dens (odontoid process of C2 vertebra)
46. ossification within parachordal plate
47. early ossification within cartilage primordium of parachordal plate (future basioccipital bone) with extensive degree of periosteal ossification
48. cartilage primordium of petrous part of temporal bone
49. first branchial cleft
50. floor of pituitary fossa (sella turcica) with early evidence of ossification
51. cartilage primordium of sphenoid bone
52. extrinsic ocular muscle (lateral/temporal rectus)
53. medial pterygoid muscle
54. cartilage primordium of ramus of mandible
55. stapedial artery
56. glossopharyngeal (IX) ganglia
57. dorsal (posterior) root ganglion of C2
58. cartilage primordium of centrum of C2 vertebra (axis)
59. cartilage primordium of anterior arch of C1 vertebra (atlas) – a hypochordal bow
60. left vagal (X) trunk
61. left superior cervical sympathetic ganglion
62. left pharyngo-tympanic tube (Eustachian tube)
63. entrance into nasopharynx
64. cartilage primordium of hard palate (fused palatal shelves of right and left maxillary bones)
65. primordium of prominent tactile or sinus hair follicle characteristically found in this location
66. lower recess of conjunctival sac
67. oral cavity (oropharynx)
68. tubules of serous glands that course rostrally in the lateral wall of the middle meatus
69. ophthalmic division of trigeminal (V) nerve
70. ossification within maxilla at lateral boundary of palatal shelf
71. primordium of first upper molar tooth
72. ossification in proximal part of body of the mandible around but principally lateral to Meckel's cartilage
73. entrance to right pharyngo-tympanic tube
74. internal carotid artery
75. right superior cervical sympathetic ganglion
76. right vagal (X) trunk
77. second cervical (C2) nerve
78. ossification within cartilage primordium of neural arch of C2 vertebra (axis)
79. prevertebral muscles of the neck (longus capitis, longus cervicis)
80. left jugular lymph sac
81. left facial artery as it branches from the external carotid artery
82. masseter muscle
83. vomeronasal organ (Jacobson's organ)
84. primordium of follicle of vibrissa
85. site of fusion of the nasal septum in the midline with the dorsal surface of the palatal shelves
86. developing glandular ducts within soft palate
87. right facial artery
88. carotid sheath

PLATE 35d

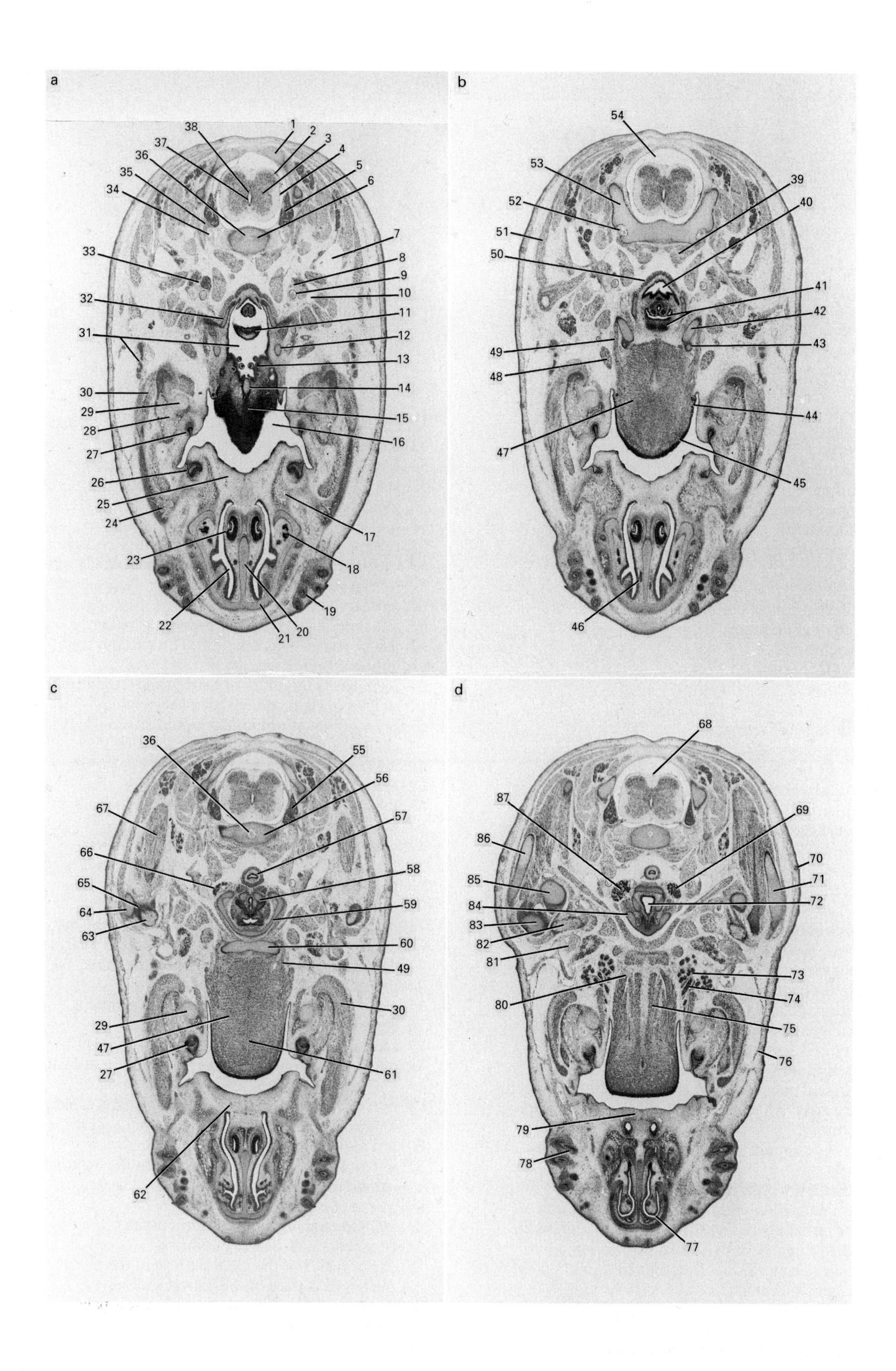
a
1
2
3
4
5
6
7
8
9
10
11
12
13
14
15
16
17
18
19
20
21
22
23
24
25
26
27
28
29
30
31
32
33
34
35
36
37
38
b
39
40
41
42
43
44
45
46
47
48
49
50
51
52
53
54
c
36
55
56
57
58
59
60
49
30
61
62
27
47
29
63
64
65
66
67
d
68
69
70
71
72
73
74
75
76
77
78
79
80
81
82
83
84
85
86
87

Plate 35d (15.5 days p.c.)

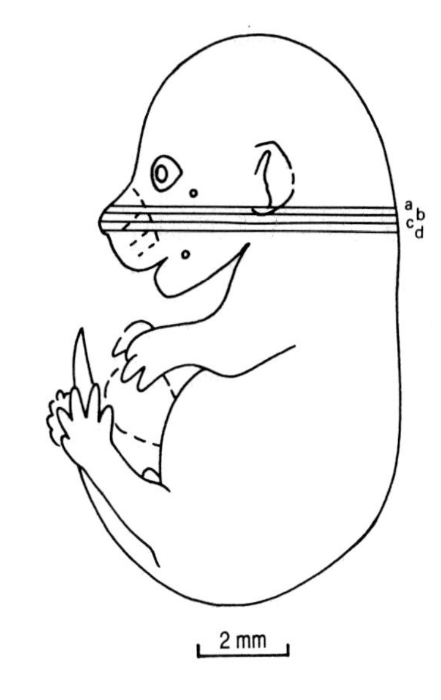

15.5 days p.c. Male embryo. Crown–rump length 12.2 mm (fixed). Transverse sections (Theiler Stage 24; rat, Witschi Stage 33).

1. ossification within cartilage primordium of neural arch of C2 vertebra (axis)
2. marginal layer
3. mantle layer
4. vascular elements of pia mater
5. dorsal (posterior) root ganglion of C3
6. cartilage primordium of centrum of C3 vertebra
7. left jugular lymph sac
8. left vagal (X) trunk
9. left internal carotid artery
10. left internal jugular vein
11. epiglottis
12. cartilage primordium of dorsal extremity of lesser horn of hyoid bone
13. developing glandular ducts on dorsal surface of root of the tongue
14. median circumvallate (vallate) papilla
15. heavy concentration of filiform papillae on dorsal surface of tongue
16. oral cavity (oropharynx)
17. ossification within maxilla at lateral boundary of palatal shelf
18. tubules of serous glands that course rostrally in the lateral wall of the middle meatus
19. primordium of follicle of vibrissa
20. cartilage primordium of nasal septum
21. nasal capsule
22. nasal cavity
23. vomeronasal organ (Jacobson's organ)
24. ossification within cartilage primordium of zygomatic bone
25. cartilage primordium of palatal shelf of maxilla
26. primordium of first upper molar tooth
27. primordium of lower molar tooth
28. ossification in body of the mandible around but principally lateral to Meckel's cartilage
29. Meckel's cartilage
30. masseter muscle
31. depression between the epiglottis and root of the tongue (vallecula) and laterally primordium of parotid gland
32. cricoid cartilage
33. carotid sheath
34. vertebral artery
35. third cervical (C3) nerve
36. nucleus pulposus
37. ependymal layer
38. central canal
39. prevertebral muscles of the neck (longus capitis, longus cervicis)
40. entrance to the oesophagus
41. entrance to the larynx and (anteriorly) the base of the epiglottis
42. cartilage primordium of greater horn of hyoid bone
43. cartilage primordium of lesser horn of hyoid bone
44. fungiform papilla on lateral border of the tongue
45. dorsal surface of tongue
46. tubules of serous glands associated with nasal septum
47. intrinsic muscle of the tongue (transverse component)
48. extrinsic muscle of the tongue (styloglossus)
49. extrinsic muscle of the tongue (hyoglossus)
50. inferior constrictor muscle
51. trapezius muscle
52. vertebral artery and associated veins in incompletely formed foramen transversarium
53. pedicle of C3 vertebra
54. spinal canal
55. dorsal (posterior) root ganglion of C4
56. cartilage primordium of centrum of C4 vertebra
57. oesophagus
58. left arytenoid cartilage
59. primordium of thyroid cartilage (in continuity caudally with the primordium of the cricoid cartilage)
60. cartilage primordium of body of hyoid bone
61. precursor of median fibrous septum of tongue
62. greater palatine artery
63. ossification within the lateral part of the right clavicle
64. cartilage primordium of acromion of right scapula
65. primordium of acromio-clavicular joint
66. upper pole of right thyroid lobe
67. supraspinatus muscle
68. posterior spinal vein
69. left lobe of thyroid gland
70. primordium of hair follicle
71. cartilage primordium of spine of left scapula
72. laryngeal aditus
73. sublingual/submandibular gland
74. submandibular duct
75. extrinsic muscle of the tongue (genioglossus)
76. cutaneous muscle of the lower face and neck – platysma (panniculus carnosus); this muscle extends cranially to become the m. cutaneus faciei et labiorum
77. entrance into nasal cavity (external naris)
78. first evidence of definitive vibrissa within its follicle
79. region of primary palate
80. hypoglossal (XII) nerve
81. external jugular vein
82. ossification within lateral one-third of right clavicle
83. rostral extremity of cartilage primordium of the right humerus
84. primordium of cricoid cartilage
85. cartilage primordium of coracoid process of right scapula
86. cartilage primordium of spine of the right scapula
87. right lobe of thyroid gland

PLATE 35e

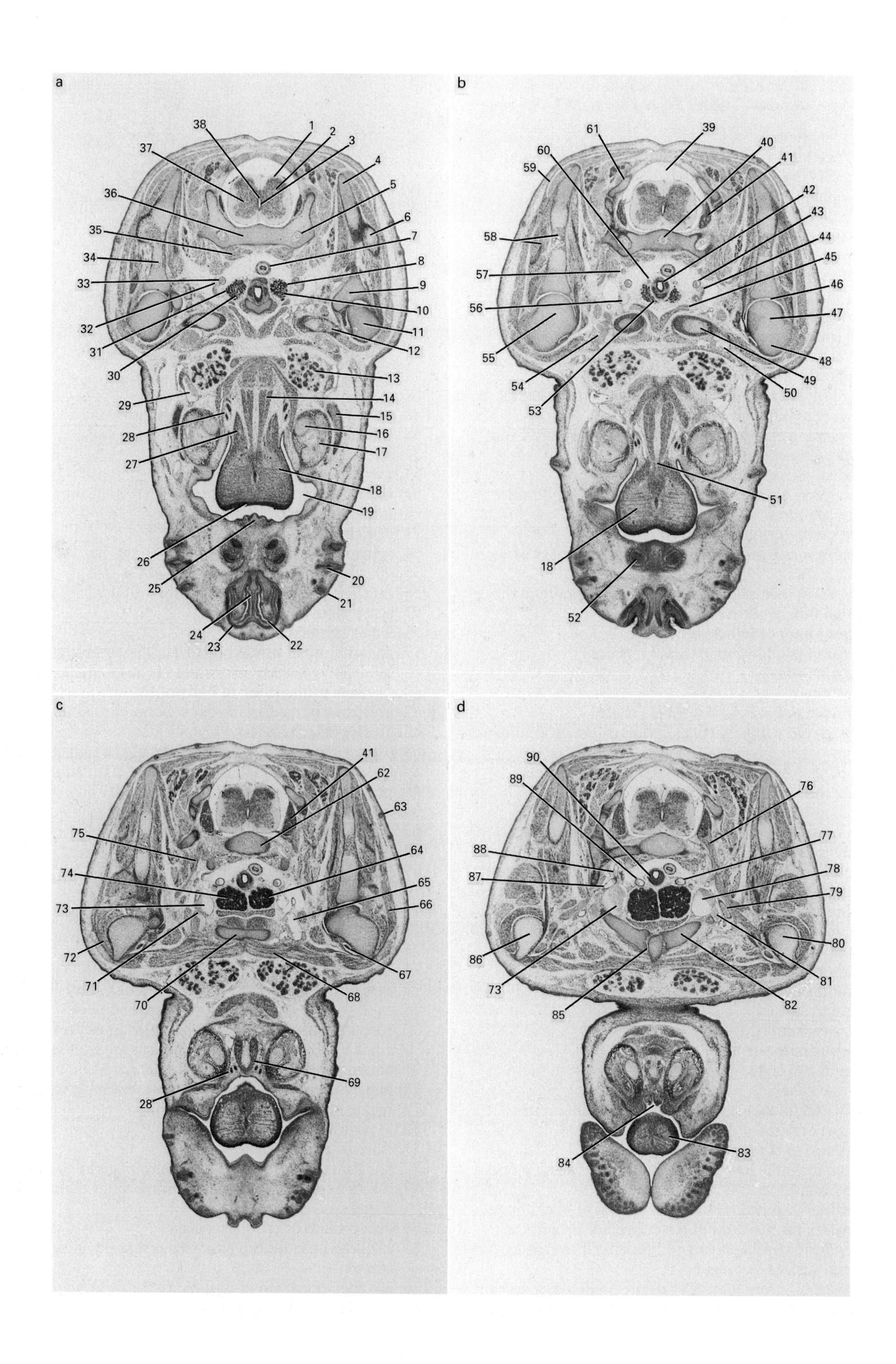
a
38 1 2 3 37 4 36 5 6 35 7 34 8 33 9 10 32 11 31 12 30 13 29 14 15 28 16 27 17 18 19 26 25 20 21 24 23 22
b
61 39 60 40 59 41 42 43 58 44 45 57 46 56 47 55 48 54 49 50 53 51 18 52
c
41 62 63 75 64 74 65 73 66 72 67 71 68 70 69 28
d
90 89 76 88 77 87 78 79 86 80 73 81 85 82 84 83

Plate 35e (15.5 days p.c.)

15.5 days p.c. Male embryo. Crown–rump length 12.2 mm (fixed). Transverse sections (Theiler Stage 24; rat, Witschi Stage 33).

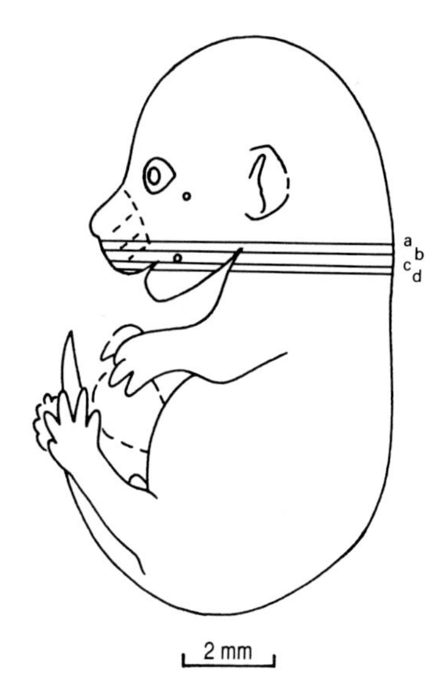

1. marginal layer
2. mantle layer
3. ependymal layer
4. cartilage primordium of posterior (medial) border of left scapula
5. vertebral artery and vein within foramen transversarium
6. ossification at base of the spine of the left scapula
7. oesophagus
8. left parathyroid gland
9. cartilage primordium of coracoid process of left scapula
10. left lobe of thyroid gland
11. rostral extremity of cartilage primordium of head of the left humerus
12. ossification within middle one-third of left clavicle
13. sublingual/submandibular gland
14. extrinsic muscle of the tongue (genioglossus)
15. masseter muscle
16. Meckel's cartilage
17. ossification in body of the mandible around but principally lateral to Meckel's cartilage
18. intrinsic muscle of the tongue (transverse component)
19. oral cavity (buccopharynx)
20. first evidence of definitive vibrissa within its follicle
21. site of eruption of vibrissa
22. entrance into nasal cavity (external naris)
23. nasal capsule
24. cartilage primordium of nasal septum
25. region of the primary palate
26. dorsal surface of the tongue
27. extrinsic muscles of the tongue (styloglossus/hyoglossus)
28. submandibular duct (medial) and sublingual duct (lateral)
29. external jugular vein
30. right lobe of thyroid gland
31. right parathyroid gland
32. right vagal (X) trunk
33. right common carotid artery
34. ossification within cartilage primordium of blade of right scapula
35. prevertebral muscles of the neck (longus capitis, longus cervicis)
36. cartilage primordium of C5 vertebra
37. mid-cervical region of the spinal cord
38. central canal
39. spinal canal
40. nucleus pulposus
41. dorsal (posterior) root ganglion of C6
42. trachea
43. left common carotid artery
44. left vagal (X) trunk
45. sternohyoid muscle
46. left gleno-humeral (shoulder) joint
47. cartilage primordium of head of left humerus
48. cartilage primordium of proximal part of shaft of left humerus
49. cartilage primordium of medial one-third of left clavicle
50. left axillary vein
51. frenulum of the tongue
52. primordium of right upper incisor tooth
53. isthmus of thyroid gland
54. confluence of the right axillary and external jugular veins to form the subclavian vein
55. cartilage primordium of head of the right humerus
56. internal jugular vein
57. sympathetic trunk
58. infraspinatus muscle and ossification within blade of right scapula
59. trapezius muscle
60. right recurrent laryngeal nerve
61. periosteal ossification associated with cartilage primordium of neural arch of C6 vertebra
62. cartilage primordium of centrum of C6 vertebra
63. primordium of hair follicle
64. left lobe of thymic rudiment
65. left subclavian vein
66. left deltoid muscle
67. tendon of the long head of the biceps brachii muscle
68. pectoral muscle (pectoralis superficialis)
69. tendon of insertion of genioglossus muscle
70. cartilage primordium of manubrium sterni
71. confluence of the right subclavian vein with the right superior (cranial) vena cava
72. right deltoid muscle
73. right superior (cranial) vena cava
74. right lobe of thymic rudiment
75. ventral primary ramus of C6 progressing to form a component of the right brachial plexus
76. ventral primary ramus of C7 progressing to form a component of the left brachial plexus
77. proximal part of left common carotid artery
78. left superior (cranial) vena cava
79. left subclavian artery where it passes over the first rib
80. cartilage primordium of upper shaft region of left humerus
81. confluence of the left subclavian vein with the left superior (cranial) vena cava
82. cartilage primordium of anterior part of the first rib
83. tip of the tongue
84. sublingual caruncles where submandibular and sublingual ducts open into floor of mouth
85. primordium of first chondro-sternal joint
86. cartilage primordium of upper shaft region of right humerus
87. right subclavian artery
88. origin of the right vertebral artery
89. proximal part of right common carotid artery
90. tracheal "ring" of hyaline cartilage

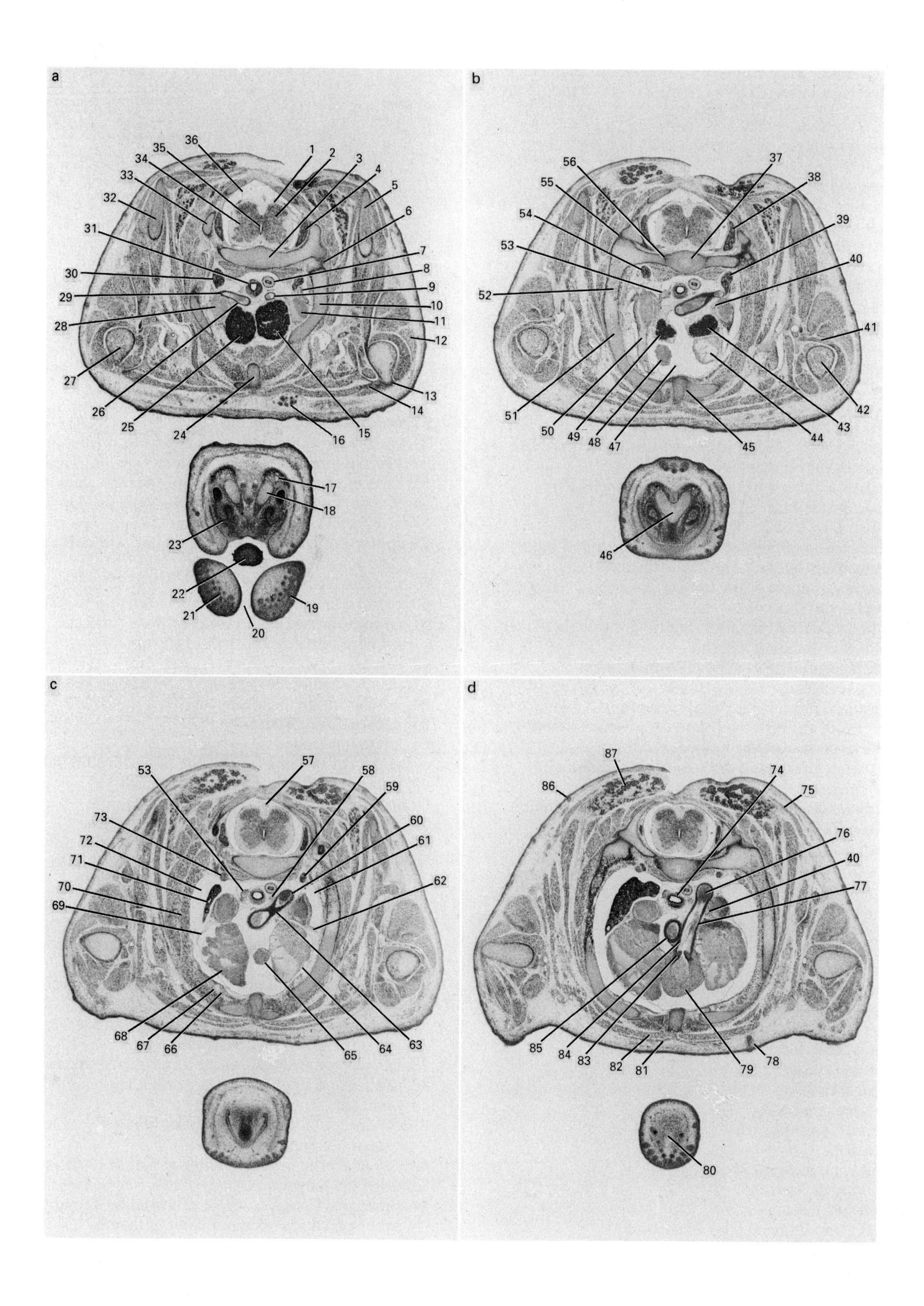
a
b
c
d
1
2
3
4
5
6
7
8
9
10
11
12
13
14
15
16
17
18
19
20
21
22
23
24
25
26
27
28
29
30
31
32
33
34
35
36
37
38
39
40
41
42
43
44
45
46
47
48
49
50
51
52
53
54
55
56
57
58
59
60
61
62
63
64
65
66
67
68
69
70
71
72
73
74
75
76
77
78
79
80
81
82
83
84
85
86
87

Plate 35f (15.5 days p.c.)

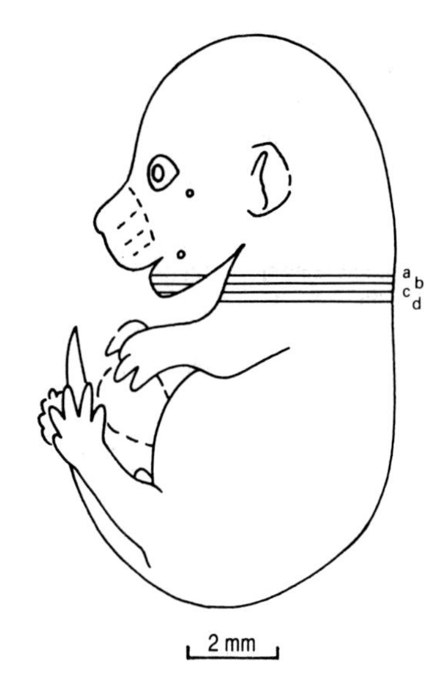

15.5 days p.c. Male embryo. Crown–rump length 12.2 mm (fixed). Transverse sections (Theiler Stage 24; rat, Witschi Stage 33).

1. marginal layer
2. mantle layer
3. cartilage primordium of T1 vertebra
4. dorsal (posterior) root ganglion of T1
5. cartilage primordium of medial (posterior) border of left scapula
6. primordium of first costo-vertebral joint
7. oesophagus
8. left subclavian artery
9. left common carotid artery
10. ossification within cartilage primordium of left first rib
11. left superior (cranial) vena cava
12. deltoid muscle
13. cartilage primordium of lower border of deltoid tuberosity with periosteal collar of bone
14. tendon of insertion of pectoralis profundus muscle
15. left lobe of thymic rudiment
16. lower pole of submandibular gland
17. ossification within incisive region of the mandible around but principally lateral to Meckel's cartilage
18. Meckel's cartilage
19. primordium of follicle of vibrissa
20. philtrum
21. right upper lip
22. tip of the tongue
23. tooth primordium of right lower incisor tooth
24. cartilage primordium of first sternebra
25. right lobe of thymic rudiment
26. brachiocephalic (innominate) artery which gives rise to right common carotid artery and right subclavian artery
27. ossification within cartilage primordium of upper shaft region of right humerus
28. right superior (cranial) vena cava
29. right subclavian artery
30. middle cervical sympathetic ganglion
31. trachea
32. cartilage primordium of medial (posterior) border of right scapula
33. ossification within cartilage primordium of neural arch of T1 vertebra
34. upper thoracic region of the spinal cord
35. central canal
36. spinal canal
37. nucleus pulposus
38. dorsal (posterior) root ganglion of T2
39. left cervicothoracic (stellate) sympathetic ganglion (formed by the fusion of the caudal cervical and first two thoracic ganglia) which extends between the heads of the second and third ribs
40. left vagal (X) trunk
41. axillary (circumflex) nerve
42. ossification within mid-shaft region of left humerus
43. lower pole of left lobe of thymic rudiment
44. rostral extremity of left atrium
45. primordium of second sterno-costal joint
46. site of fusion in the ventral midline of the left and right Meckel's cartilages
47. rostral extremity of pericardial cavity
48. rostral extremity of right atrium
49. lower pole of right lobe of thymic rudiment
50. right internal thoracic (mammary) vein where it joins the right superior (cranial) vena cava
51. cartilage primordium of right second rib with periosteal collar of bone
52. ossification within cartilage primordium of rib
53. right vagal (X) trunk
54. right cervicothoracic (stellate) sympathetic ganglion
55. primordium of articulation between cartilage primordium of tubercle of second rib and neural arch of its own vertebra
56. primordium of articulation between cartilage primordium of head of the second rib and the body of the second thoracic vertebra.
57. posterior spinal vein
58. left recurrent laryngeal nerve
59. left sympathetic trunk
60. arch of the aorta just proximal to the site of entrance of the ductus arteriosus
61. left pleural cavity
62. left pleuro-pericardial membrane
63. aortic wall in the concavity of the arch of the aorta
64. wall of left atrium
65. wall of pulmonary trunk/proximal part of the ductus arteriosus
66. internal thoracic (mammary) artery
67. internal thoracic (mammary) vein
68. wall of right atrium
69. right pleuro-pericardial membrane
70. intercostal muscles (external and internal layers)
71. cranial lobe of right lung
72. right pleural cavity
73. azygos vein
74. trachea just proximal to its bifurcation (at the carina) to form the right and left main bronchi
75. trapezius muscle
76. arch of the aorta at site of entrance of the ductus arteriosus
77. wall of the ductus arteriosus
78. mammary gland primordium
79. wall of right ventricle
80. ventral extremity of the lower jaw
81. pectoralis superficialis
82. pectoralis profundus
83. leaflet of pulmonary valve
84. wall of pulmonary trunk
85. wall of proximal part of arch of the aorta
86. primordium of hair follicle
87. deposits of brown (multilocular) fat

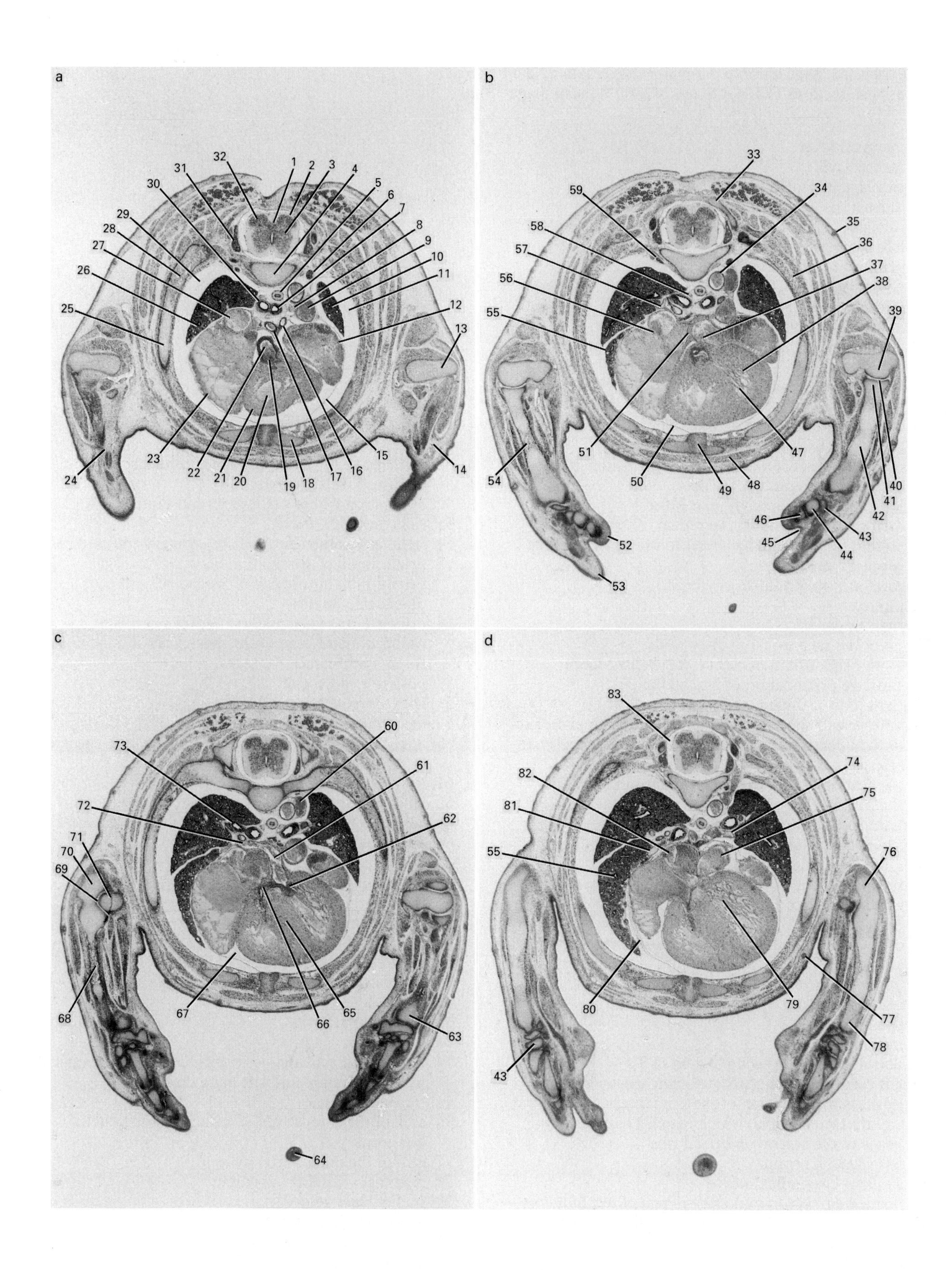
a
b
c
d
1
2
3
4
5
6
7
8
9
10
11
12
13
14
15
16
17
18
19
20
21
22
23
24
25
26
27
28
29
30
31
32
33
34
35
36
37
38
39
40
41
42
43
44
45
46
47
48
49
50
51
52
53
54
55
56
57
58
59
60
61
62
63
64
65
66
67
68
69
70
71
72
73
74
75
76
77
78
79
80
81
82
83

Plate 35g (15.5 days p.c.)

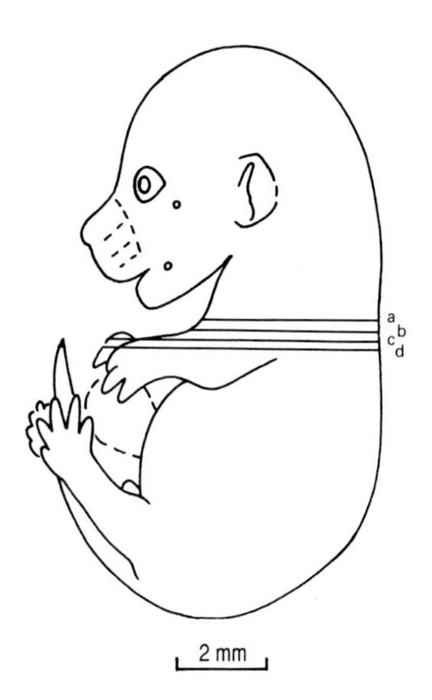

15.5 days p.c. Male embryo. Crown–rump length 12.2 mm (fixed). Transverse sections (Theiler Stage 24; rat, Witschi Stage 33).

1. roof plate
2. marginal layer
3. mantle layer
4. cartilage primordium of mid-thoracic vertebral body
5. oesophagus
6. thoracic sympathetic ganglion
7. left main bronchus
8. left vagal (X) trunk
9. left superior (cranial) vena cava
10. apex of left lung
11. left pleural cavity
12. wall of left atrium
13. distal part of cartilage primordium of shaft of left humerus
14. origin of left forelimb
15. left pleuro-pericardial membrane
16. left pulmonary artery
17. right pulmonary artery
18. primordium of costal cartilage
19. leaflet of aortic valve
20. lumen of right ventricle
21. wall of right ventricle
22. origin of the thoracic aorta
23. wall of right atrium
24. origin of right forelimb
25. cartilage primordium of right third rib with periosteal collar of bone
26. cranial lobe of right lung
27. right superior (cranial) vena cava
28. ossification within rib primordium
29. right pleural cavity
30. right main bronchus
31. dorsal (posterior) root ganglion
32. upper thoracic region of the spinal cord
33. ossification within cartilage primordium of neural arch
34. postductal part of thoracic aorta
35. primordium of hair follicle
36. intercostal muscles (external and internal layers)
37. septum primum
38. trabeculae carneae within wall of left ventricle
39. cartilage primordium of capitulum of the left humerus
40. primordium of radio-humeral joint
41. cartilage primordium of head of the left radius
42. ossification within cartilage primordium of shaft of the left radius
43. cartilage primordium of carpal bone
44. cartilage primordium of first metacarpal bone
45. digital interzone
46. cartilage primordium of first phalangeal bone
47. lumen of left ventricle
48. pectoral muscle (pectoralis profundus)
49. cartilage primordium of sternebra
50. right pleuro-pericardial membrane
51. rostral border of septum secundum
52. first digit
53. second digit
54. ossification in mid-shaft region of cartilage primordium of right radius
55. middle lobe of right lung
56. venous valve at site of entrance of right superior (cranial) vena cava into right atrium
57. hilus (root) of the right lung
58. right vagal (X) trunk
59. right sympathetic trunk
60. hemiazygos vein
61. posterior wall of left atrium at site of entrance of pulmonary veins
62. left atrio-ventricular (mitral) valve
63. distal part of cartilage primordium of shaft of left radius
64. tip of the tail
65. muscular part of interventricular septum
66. right atrio-ventricular (tricuspid) valve
67. pericardial cavity
68. ossification within mid-shaft region of cartilage primordium of right ulna
69. primordium of right humero-ulnar joint
70. cartilage primordium of proximal part of right ulna
71. cartilage primordium of the trochlea of the right humerus
72. pulmonary vein draining the cranial lobe of the right lung
73. lobar bronchus to cranial lobe of the right lung
74. left pulmonary vein
75. left superior (cranial) vena cava at its site of entrance into the right atrium
76. cartilage primordium of proximal part of left ulna
77. mammary gland primordium
78. cartilage primordium of distal part of left ulna
79. chordae tendineae
80. primordium of fibrous pericardium
81. pulmonary vein draining the middle lobe of the right lung
82. site of entrance of inferior vena cava into right atrium
83. spinal canal

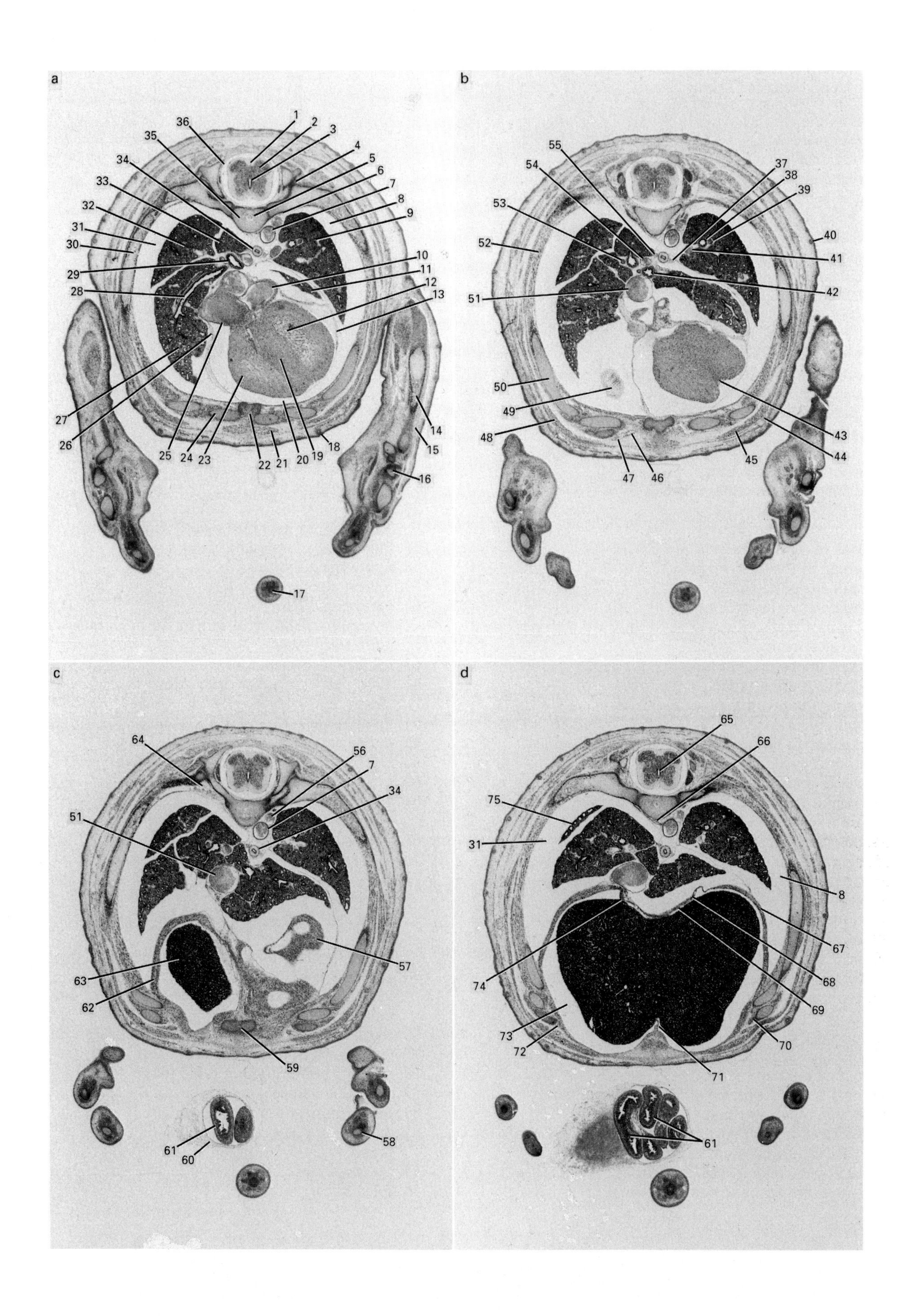
a
b
c
d
1
2
3
4
5
6
7
8
9
10
11
12
13
14
15
16
17
18
19
20
21
22
23
24
25
26
27
28
29
30
31
32
33
34
35
36
37
38
39
40
41
42
43
44
45
46
47
48
49
50
51
52
53
54
55
56
57
58
59
60
61
62
63
64
65
66
67
68
69
70
71
72
73
74
75

Plate 35h (15.5 days p.c.)

15.5 days p.c. Male embryo. Crown–rump length 12.2 mm (fixed). Transverse sections (Theiler Stage 24; rat, Witschi Stage 33).

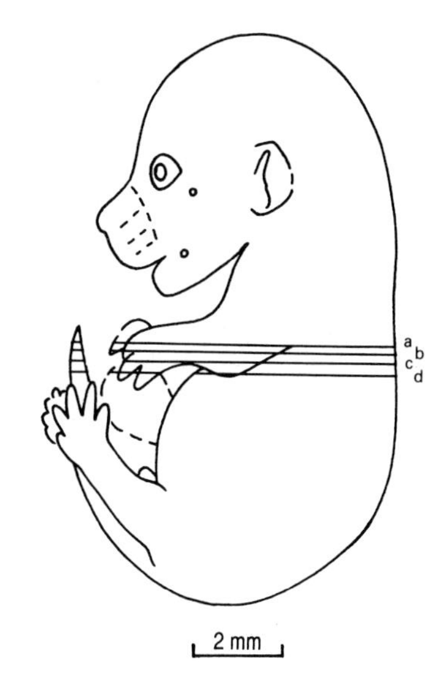

1. marginal layer
2. mantle layer
3. ependymal layer
4. primordium of joint between tubercle of rib and neural arch of its own vertebra
5. nucleus pulposus
6. thoracic sympathetic trunk
7. thoracic aorta
8. left pleural cavity
9. left lung
10. left superior (cranial) vena cava at its site of entrance into the right atrium
11. accessory lobe of right lung
12. chordae tendineae
13. left pleuro-pericardial membrane
14. ossification within cartilage primordium of shaft of left ulna
15. left forelimb
16. cartilage primordium of carpal bone
17. tip of the tail
18. mammary gland primordium
19. muscular part of interventricular septum
20. pericardial cavity
21. pectoral muscle (pectoralis profundus)
22. cartilage primordium of sternebra
23. wall of right ventricle
24. primordium of costal cartilage
25. wall of right atrium
26. primordium of fibrous pericardium
27. middle lobe of right lung
28. lobar bronchus to middle lobe of right lung
29. right main bronchus
30. ossification within cartilage primordium of rib
31. right pleural cavity
32. cranial lobe of the right lung
33. caudal lobe of the right lung
34. oesophagus
35. centrum of thoracic vertebra
36. spinal canal
37. left vagal (X) trunk
38. left main bronchus
39. left pulmonary artery
40. primordium of hair follicle
41. left pulmonary vein
42. lobar bronchus to accessory lobe of right lung
43. interventricular groove
44. superior epigastric artery and vein
45. left rectus abdominis muscle
46. internal thoracic (mammary) artery
47. internal thoracic (mammary) vein
48. right rectus abdominis muscle
49. pleural component of right pleuro-peritoneal membrane
50. cartilage primordium of rib with periosteal collar of bone
51. posthepatic inferior vena cava
52. intercostal muscles (external and internal layers)
53. branch from right pulmonary artery to caudal lobe of the right lung
54. right pulmonary vein
55. right vagal (X) trunk
56. hemiazygos vein
57. pleural component of left pleuro-peritoneal membrane
58. cartilage primordium of phalangeal bone
59. cartilage primordium of xiphoid process
60. wall of physiological umbilical hernia
61. midgut loop within physiological umbilical hernia
62. right dome of the diaphragm
63. apex of right lobe of the liver
64. intercostal vein
65. central canal
66. intercostal artery
67. left dome of the diaphragm
68. left boundary of the bare area of the liver
69. central tendon of the diaphragm overlying bare area of the liver
70. left costal margin
71. falciform ligament
72. right costal margin
73. sub-phrenic recess of peritoneal cavity
74. right boundary of the bare area of the liver
75. caudal margin of cranial lobe of the right lung

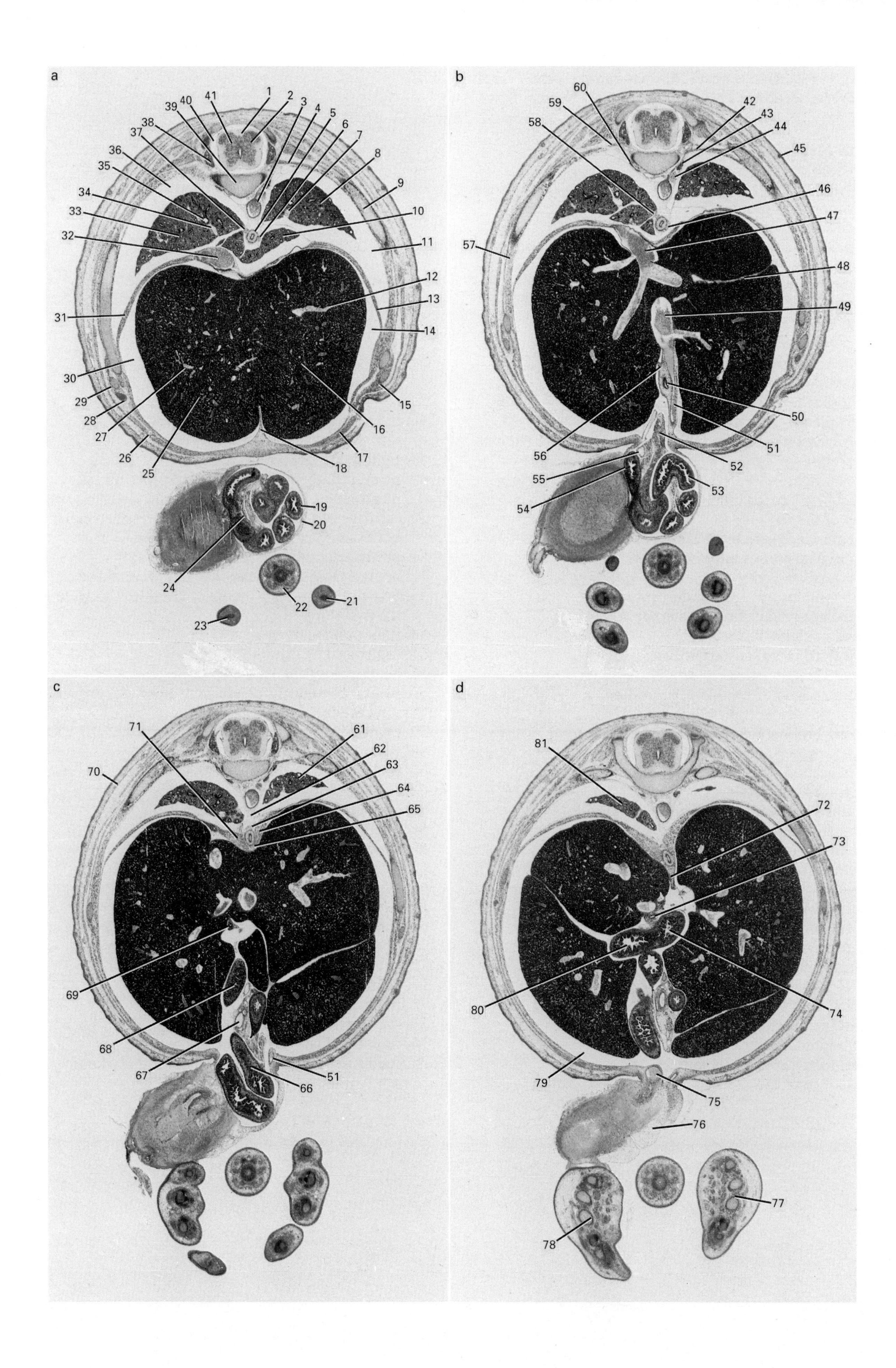
a
b
c
d
1
2
3
4
5
6
7
8
9
10
11
12
13
14
15
16
17
18
19
20
21
22
23
24
25
26
27
28
29
30
31
32
33
34
35
36
37
38
39
40
41
42
43
44
45
46
47
48
49
50
51
52
53
54
55
56
57
58
59
60
61
62
63
64
65
66
67
68
69
70
71
72
73
74
75
76
77
78
79
80
81

Plate 35i (15.5 days p.c.)

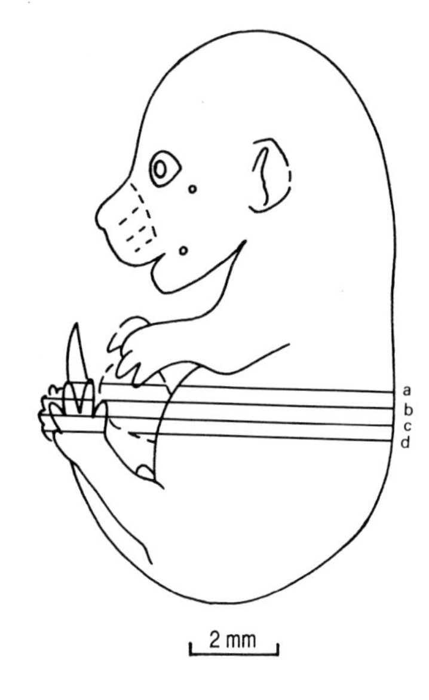

15.5 days p.c. Male embryo. Crown–rump length 12.2 mm (fixed). Transverse sections (Theiler Stage 24; rat, Witschi Stage 33).

1. marginal layer
2. mantle layer
3. thoracic aorta
4. intercostal nerve
5. oesophagus
6. left vagal (X) trunk
7. branch of left pulmonary vein
8. left lung
9. ossification within cartilage primordium of rib with periosteal collar of bone
10. accessory lobe of right lung
11. left pleural cavity
12. part of hepatic venous plexus
13. left pleuro-peritoneal membrane (peripheral margin of diaphragm)
14. left sub-phrenic recess of peritoneal cavity
15. left costal margin
16. left lobe of liver
17. left rectus abdominis muscle
18. falciform ligament
19. loop of midgut within physiological umbilical hernia
20. wall of physiological umbilical hernia
21. cartilage primordium of phalangeal bone (digit of left hindlimb)
22. caudal region of tail
23. digit of right hindlimb
24. superior mesenteric vessels within midgut mesentery
25. right lobe of liver
26. superior epigastric vessels
27. hepatic venous sinusoid
28. right costal margin
29. primordium of costal cartilage
30. right sub-phrenic recess of peritoneal cavity
31. right pleuro-peritoneal membrane (peripheral margin of diaphragm)
32. proximal part of posthepatic inferior vena cava
33. caudal lobe of right lung
34. branch of the right pulmonary vein
35. lobar bronchus to caudal lobe of the right lung
36. right pleural cavity
37. right vagal (X) trunk
38. dorsal (posterior) root ganglion
39. cartilage primordium of centrum of thoracic vertebra
40. spinal canal
41. spinal cord in lower thoracic region
42. intercostal artery
43. left sympathetic trunk
44. hemiazygos vein
45. primordium of hair follicle
46. left (becoming anterior) vagal (X) trunk
47. origin of posthepatic inferior vena cava formed primarily from ductus venosus, left umbilical vein and right vitelline venous system
48. lobar interzone
49. ductus venosus
50. lumen of gallbladder
51. left umbilical vein passing dorsally to join the ductus venosus
52. mesenteric vessels within the dorsal mesentery of the midgut
53. lumen of midgut loop
54. proximal part of midgut loop
55. right umbilical vein
56. gallbladder bed
57. intercostal muscles (external and internal layers)
58. right (becoming posterior) vagal (X) trunk
59. intercostal nerve
60. right sympathetic trunk
61. caudal extremity of left lung
62. dorsal meso-oesophagus
63. posterior vagal (X) trunk
64. left crus of the diaphragm
65. anterior vagal (X) trunk
66. distal part of midgut loop
67. superior mesenteric vessels
68. jejunum
69. right hepatic duct
70. cutaneous muscle of trunk (panniculus carnosus)
71. right crus of the diaphragm
72. ventral mesentery of the oesophagus
73. common bile duct
74. pyloric region of the stomach
75. right umbilical artery
76. Wharton's jelly
77. left hindlimb
78. right hindlimb
79. peritoneal cavity
80. lumen of first part of the duodenum
81. caudal extremity of caudal lobe of right lung

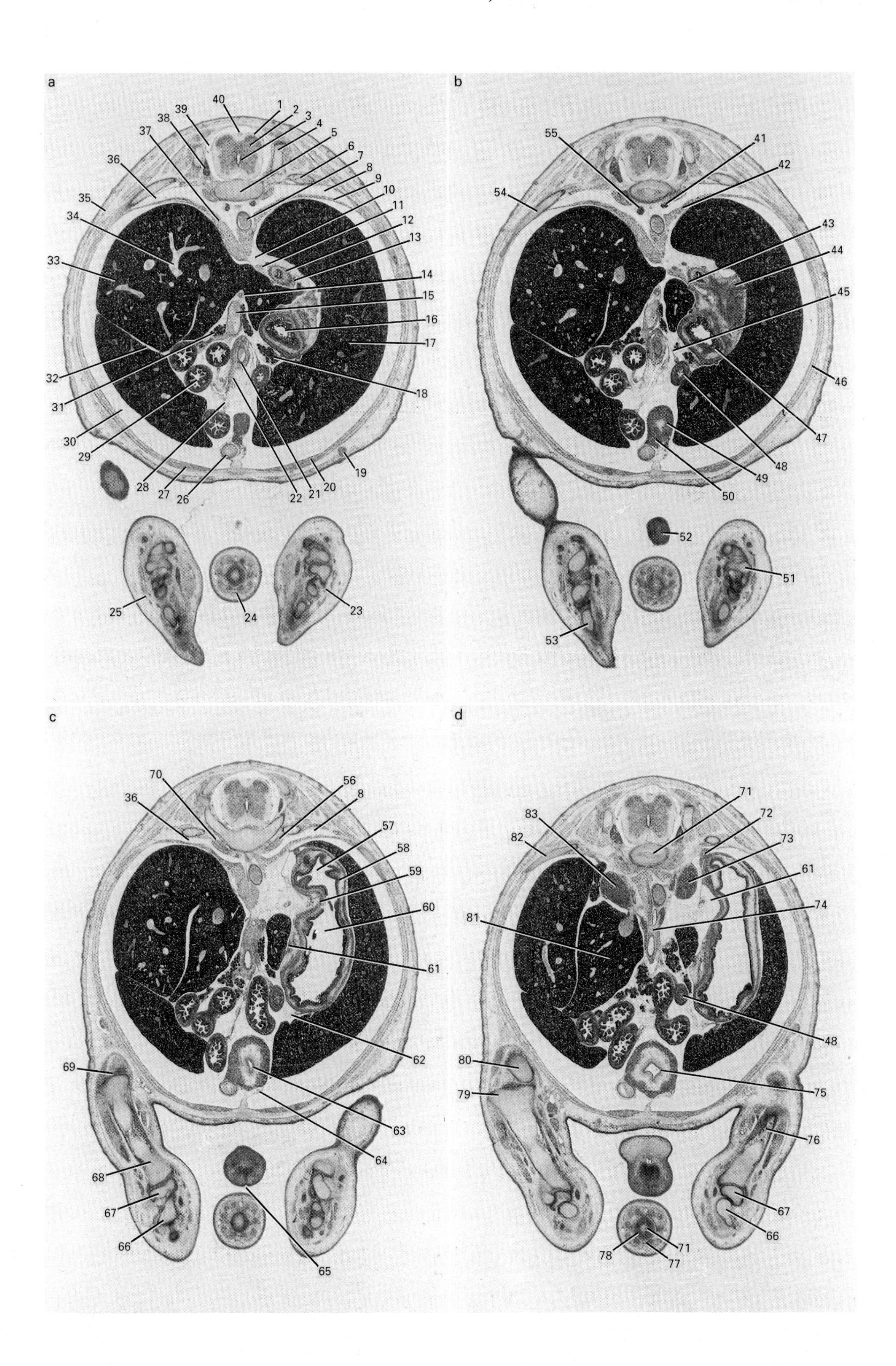
a
b
c
d
1
2
3
4
5
6
7
8
9
10
11
12
13
14
15
16
17
18
19
20
21
22
23
24
25
26
27
28
29
30
31
32
33
34
35
36
37
38
39
40
41
42
43
44
45
46
47
48
49
50
51
52
53
54
55
56
57
58
59
60
61
62
63
64
65
66
67
68
69
70
71
72
73
74
75
76
77
78
79
80
81
82
83

Plate 35j (15.5 days p.c.)

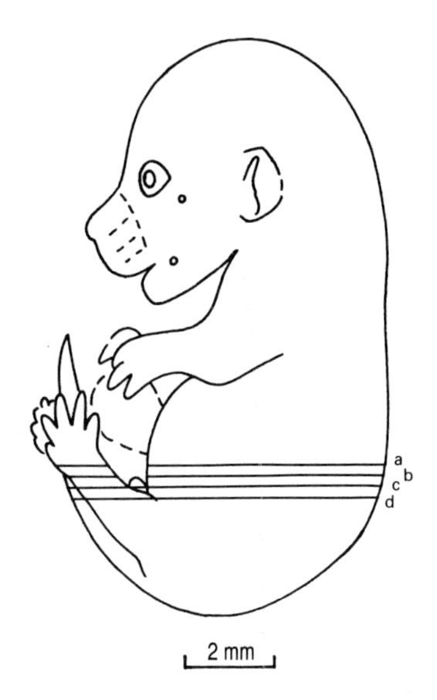

15.5 days p.c. Male embryo. Crown–rump length 12.2 mm (fixed). Transverse sections (Theiler Stage 24; rat, Witschi Stage 33).

1. marginal layer
2. mantle layer
3. ependymal layer
4. cartilage primordium of neural arch
5. cartilage primordium of caudal thoracic vertebra
6. ossification within cartilage primordium of rib with periosteal collar of bone
7. caudal part of thoracic aorta
8. caudal extremity of left pleural cavity
9. left pleuro-peritoneal membrane (peripheral margin of diaphragm)
10. dorsal mesentery of the oesophagus
11. posterior vagal (X) trunk
12. lower end of the oesophagus
13. anterior vagal (X) trunk
14. common bile duct
15. portal vein
16. pyloric antrum of the stomach
17. left lobe of the liver
18. primordium of left lobe of the pancreas
19. mammary gland primordium
20. left rectus abdominis muscle
21. superior mesenteric artery
22. superior mesenteric vein
23. left hindlimb
24. filum terminale
25. right hindlimb
26. right umbilical artery
27. right rectus abdominis muscle
28. dorsal mesentery of midgut
29. lumen of midgut
30. peritoneal cavity
31. primordium of right lobe of the pancreas
32. lobar interzone
33. right lobe of the liver
34. hepatic vein
35. cutaneous muscle of the trunk (panniculus carnosus)
36. caudal extremity of right pleural cavity
37. right crus of the diaphragm
38. dorsal (posterior) root ganglion
39. spinal canal
40. posterior spinal vein
41. left sympathetic trunk
42. left crus of the diaphragm
43. branches of posterior vagal (X) trunk (from vagal plexus) ramifying on the dorsal surface of stomach
44. rostral surface in the region of the "body" of the stomach
45. dorsal mesentery of the hindgut
46. three layers of abdominal wall musculature (external oblique, internal oblique and transversus abdominis muscles)
47. mucosal lining of the pyloric antrum of the stomach
48. hindgut
49. wall of the apical region of the urogenital sinus (future bladder)
50. base of urachus
51. cartilage primordium of tarsal bone
52. dorsum of genital tubercle
53. cartilage primordium of phalangeal bone
54. ossification within cartilage primordium of 12th rib with periosteal collar of bone
55. right sympathetic trunk
56. left most caudal thoracic sympathetic ganglion
57. lumen of fundus region of the stomach
58. anterior wall of stomach
59. posterior wall of stomach
60. lumen of the "body" of the stomach
61. lesser sac of peritoneal cavity (omental bursa)
62. caudal part of dorsal mesogastrium
63. lumen of the apical region of the urogenital sinus (future bladder)
64. left umbilical artery
65. urethral groove on ventral aspect of phallus (penis)
66. cartilage primordium of calcaneum
67. cartilage primordium of talus
68. ossification within cartilage primordium of shaft of right tibia
69. primordium of right knee joint
70. right most caudal thoracic sympathetic ganglion
71. nucleus pulposus
72. dorsal mesogastrium
73. upper pole of left adrenal primordium (future cortical component)
74. hypogastric plexus of nerves
75. endodermal lining of urogenital sinus
76. ossification within cartilage primordium of shaft of the left tibia
77. filum terminale
78. cartilage primordium of tail vertebral body
79. cartilage primordium of head of the right tibia
80. cartilage primordium of distal part of right femur
81. caudate lobe of the liver
82. ossification within cartilage primordium of 13th rib
83. upper pole of right adrenal primordium (cortical component)

PLATE 35k

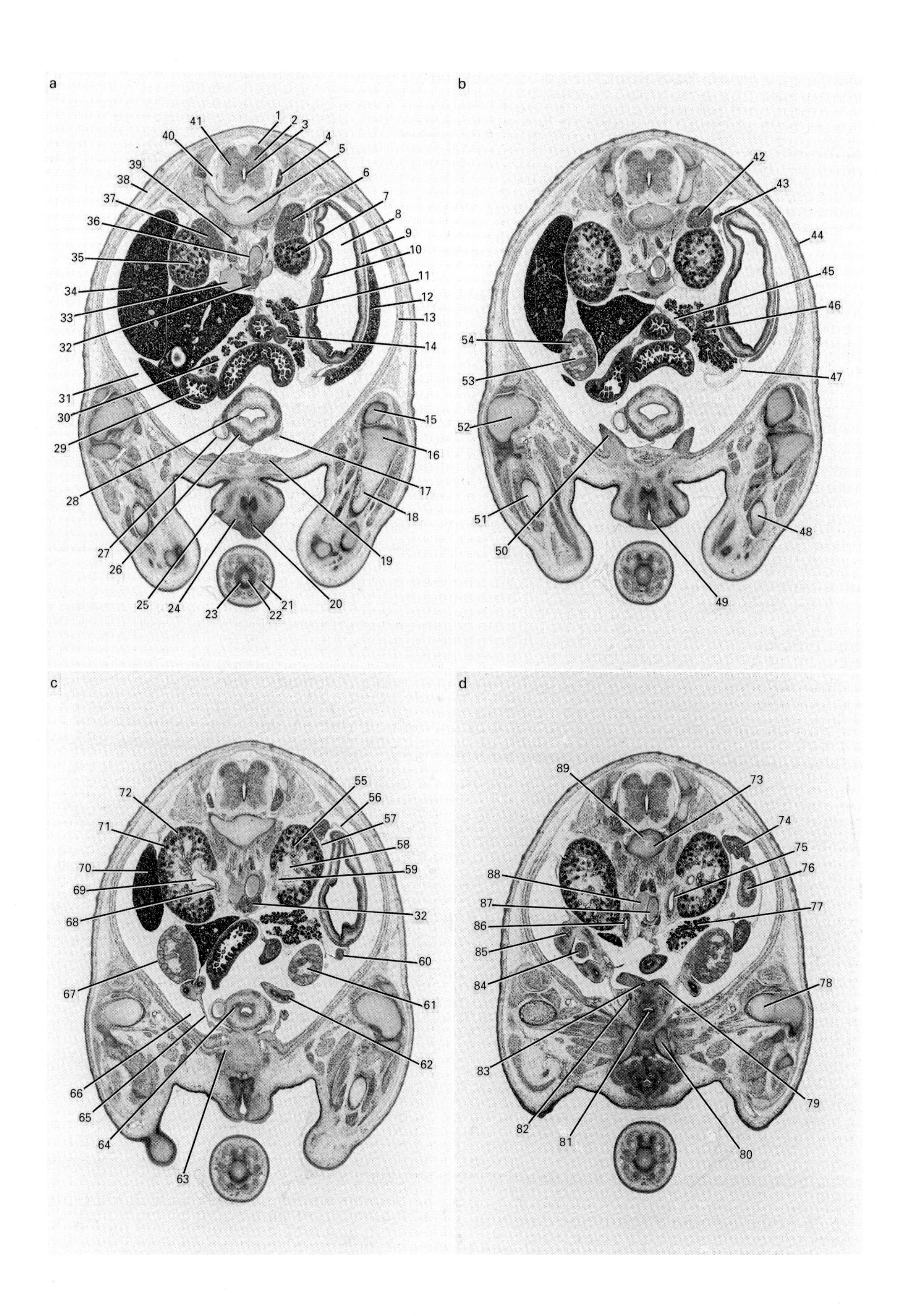
a
1
2
3
4
5
6
7
8
9
10
11
12
13
14
15
16
17
18
19
20
21
22
23
24
25
26
27
28
29
30
31
32
33
34
35
36
37
38
39
40
41
b
42
43
44
45
46
47
48
49
50
51
52
53
54
c
55
56
57
58
59
32
60
61
62
63
64
65
66
67
68
69
70
71
72
d
73
74
75
76
77
78
79
80
81
82
83
84
85
86
87
88
89

Plate 35k (15.5 days p.c.)

15.5 days p.c. Male embryo. Crown–rump length 12.2 mm (fixed). Transverse sections (Theiler Stage 24; rat, Witschi Stage 33).

1. marginal layer
2. mantle layer
3. ependymal layer
4. dorsal (posterior) root ganglion of L1
5. cartilage primordium of L1 vertebra
6. left adrenal primordium (future cortical component)
7. upper pole of left kidney
8. lumen of the "body" of the stomach
9. anterior wall of the stomach
10. posterior wall of the stomach
11. primordium of the left lobe of the pancreas
12. caudal extremity of the left lobe of the liver
13. three layers of abdominal wall musculature (external oblique, internal oblique and transversus abdominis muscles)
14. hindgut
15. cartilage primordium of distal part of left femur
16. cartilage primordium of head of the left tibia
17. left umbilical artery
18. ossification within cartilage primordium of shaft of left tibia
19. insertion of rectus abdominis muscle
20. rostral component of penile/phallic urethra (yet to canalize in this location)
21. tail
22. nucleus pulposus
23. cartilage primordium of tail vertebral body
24. glans penis
25. prepuce
26. wall of urogenital sinus (future bladder)
27. right umbilical artery
28. endodermal lining of urogenital sinus
29. third part of duodenum
30. primordium of right lobe of the pancreas
31. peritoneal cavity
32. pre-aortic abdominal sympathetic paraganglia (of Zuckerkandl)/para-aortic bodies
33. prehepatic part of inferior vena cava
34. right lobe of the liver
35. upper pole of right kidney
36. abdominal aorta
37. lower pole of right adrenal primordium (cortical component)
38. cutaneous muscle of the trunk (panniculus carnosus)
39. right upper lumbar sympathetic ganglion
40. spinal canal
41. upper lumbar region of spinal cord
42. lower pole of left adrenal primordium (cortical component)
43. upper pole of splenic primordium
44. primordium of hair follicle
45. dorsal mesentery of the hindgut
46. primordium of body of the pancreas
47. dorsal mesogastrium
48. early ossification within cartilage primordium of left fibula
49. penile/phallic part of urethra (canalized in this location)
50. caudal extremity of mesorchium
51. early ossification within cartilage primordium of right fibula
52. cartilage primordium of distal part of right femur
53. medullary region of right testis
54. seminiferous tubule (not yet canalized)
55. collecting tubules
56. gastro-splenic ligament
57. lieno-renal ligament
58. medullary region of left kidney
59. branch of left renal vein
60. "tail" of splenic primordium in ventral extension of dorsal mesogastrium
61. left testis
62. mesonephric duct (ductus deferens)
63. primordium of pubic tubercle
64. site of entrance of right ureter into lumen of the urogenital sinus (future ureteric orifice at upper/outer angle of the trigone of the bladder)
65. gubernaculum testis
66. pelvic part of peritoneal cavity
67. cortical region of right testis
68. rostral extremity of right ureter
69. pelvis of right kidney
70. caudal extremity of right lobe of the liver
71. glomerular apparatus
72. cortical region of right kidney
73. cartilage primordium of mid-lumbar vertebral body
74. splenic primordium
75. left ureter
76. caudal wall of "body" of the stomach
77. tail region of pancreatic primordium
78. ossification within cartilage primordium of shaft of left femur with periosteal collar of bone
79. left ejaculatory duct
80. cartilage primordium of left pubic bone
81. prostatic region of urethra
82. caudal extremity of right ureter
83. right ejaculatory duct
84. caudal pole of right testis
85. degenerating mesonephric tubules within remnant of mesonephros (Wolffian body)
86. right ureter
87. postrenal part of abdominal aorta
88. inferior vena cava
89. nucleus pulposus

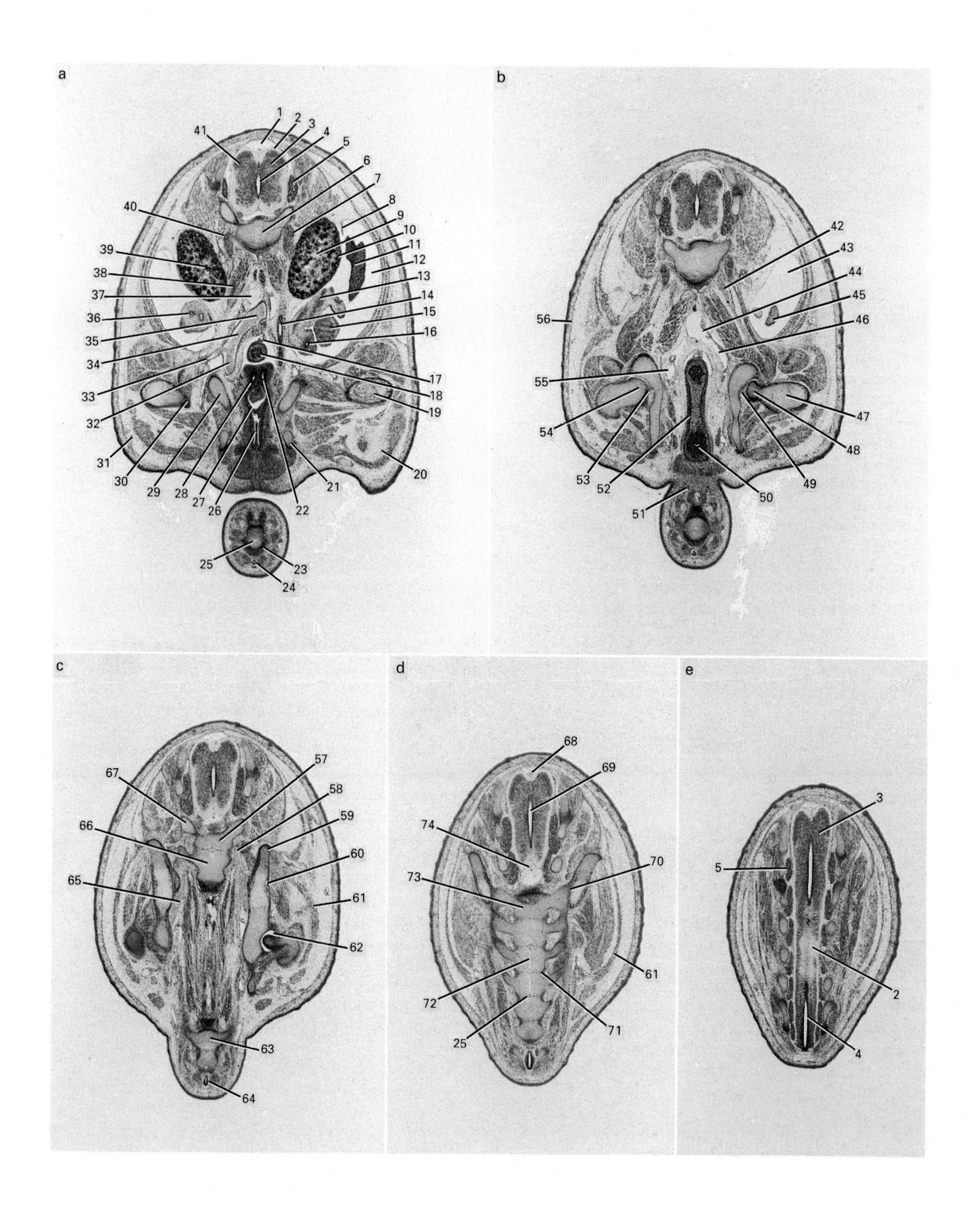
a
b
c
d
e
1
2
3
4
5
6
7
8
9
10
11
12
13
14
15
16
17
18
19
20
21
22
23
24
25
26
27
28
29
30
31
32
33
34
35
36
37
38
39
40
41
42
43
44
45
46
47
48
49
50
51
52
53
54
55
56
57
58
59
60
61
62
63
64
65
66
67
68
69
70
71
72
73
74

Plate 351 (15.5 days p.c.)

15.5 days p.c. Male embryo. Crown–rump length 12.2 mm (fixed). Transverse sections (Theiler Stage 24; rat, Witschi Stage 33).

1. spinal canal
2. marginal layer
3. mantle layer
4. ependymal layer
5. dorsal (posterior) root ganglion
6. cartilage primordium of mid-lumbar vertebral body
7. left sympathetic trunk
8. lieno-renal ligament
9. medullary region of left kidney
10. cortical region of left kidney
11. splenic primordium
12. lower abdominal region of peritoneal cavity
13. degenerating mesonephric tubules
14. left ureter
15. seminiferous tubules within medullary region of left testis
16. left mesonephric duct (ductus deferens)
17. dorsal mesentery of hindgut
18. hindgut
19. ossification within mid-shaft region of cartilage primordium of left femur
20. left hindlimb
21. cartilage primordium of ischial bone
22. left ejaculatory duct
23. cartilage primordium of tail vertebral body
24. filum terminale
25. nucleus pulposus
26. pelvic part of urethra
27. prostatic region of urethra
28. right ejaculatory duct
29. cartilage primordium of right pubic bone
30. insertion of ilio-psoas tendon
31. right hindlimb
32. right internal iliac artery
33. right external iliac/femoral artery
34. right common iliac artery
35. bifurcation of the abdominal aorta to give rise to the right and left common iliac arteries
36. right mesonephric duct (ductus deferens)
37. inferior vena cava
38. psoas major muscle
39. lower pole of right kidney
40. right sympathetic trunk
41. spinal cord in mid-lumbar region
42. lumbar nerve trunk (component of lumbar plexus)
43. pelvic region of peritoneal cavity
44. bifurcation of inferior vena cava at site of union of the right and left common iliac veins
45. tip of splenic primordium
46. left common iliac vein
47. ossification within cartilage primordium of shaft of left femur with periosteal collar of bone
48. cartilage primordium of head of left femur
49. primordium of hip joint
50. anal canal
51. base of the tail
52. wall of rectum/anal canal
53. primordium of acetabular fossa
54. cartilage primordium of head of the right femur
55. right common iliac vein
56. cutaneous muscle of lower trunk region (panniculus carnosus)
57. cartilage primordium of L5 vertebral body
58. lumbar nerve trunk (component of left lumbo-sacral plexus)
59. cartilage primordium of iliac crest
60. periosteal ossification associated with cartilage primordium of iliac bone
61. gluteus maximus (superficialis) muscle
62. synovial cavity of hip joint
63. cartilage primordium of coccygeal vertebra
64. caudal extremity of spinal lumen
65. lumbar nerve trunk (component of right lumbo-sacral plexus)
66. cartilage primordium of L6 vertebra
67. cartilage primordium of transverse process of L5 vertebra
68. posterior spinal vein
69. central canal
70. primordium of sacro-iliac joint
71. intervertebral disc between L3 and L4 vertebrae
72. remnant of notochord running through cartilage primordium of vertebral body
73. cartilage primordium of S1 vertebral body
74. anterior spinal artery

PLATE
36a

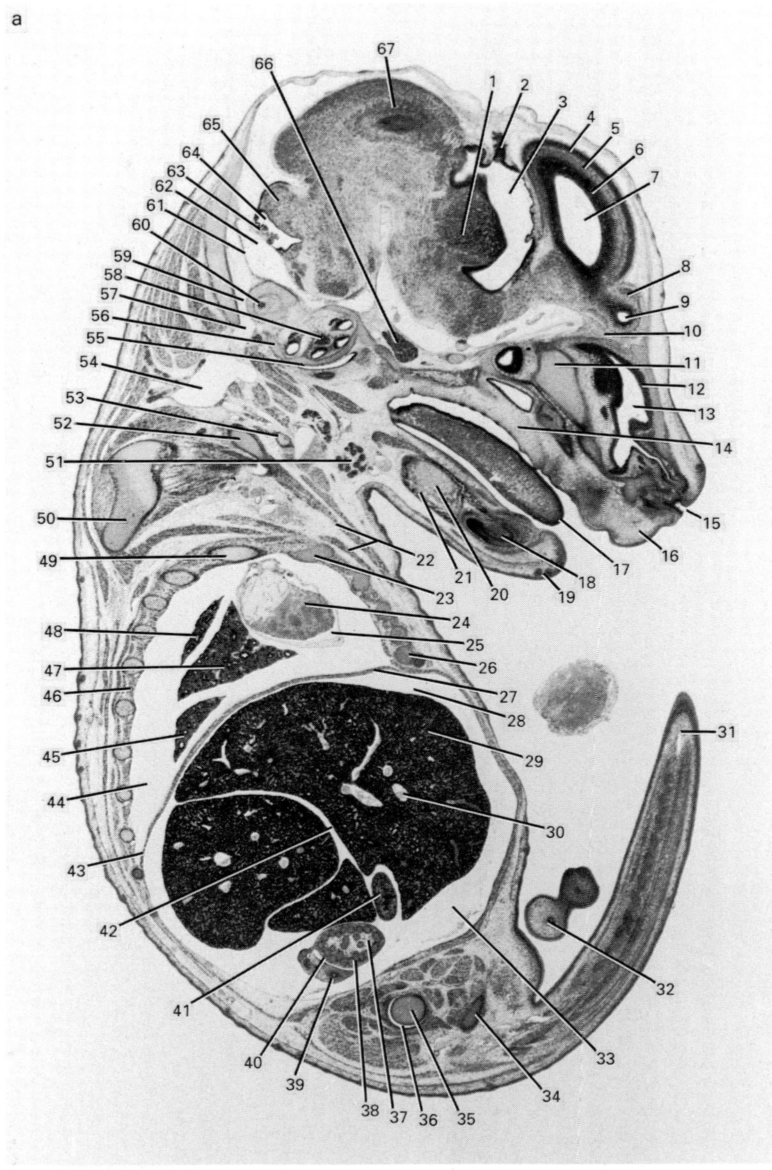
a
1 2 3 4 5 6 7 8 9 10 11 12 13 14 15 16 17 18 19 20 21 22 23 24 25 26 27 28 29 30 31 32 33 34
35 36 37 38 39 40 41 42 43 44 45 46 47 48 49 50 51 52 53 54 55 56 57 58 59 60 61 62 63 64 65 66 67

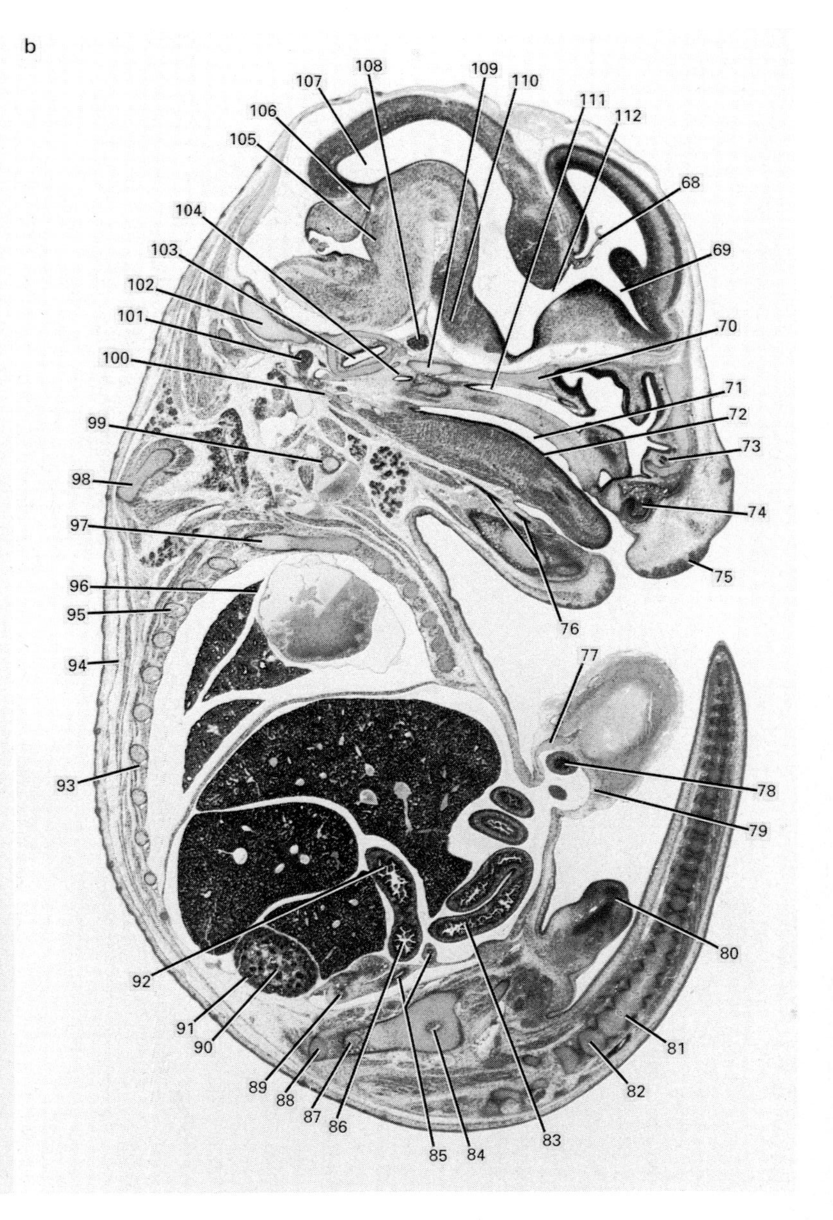
b
68 69 70 71 72 73 74 75 76 77 78 79 80 81 82 83 84 85 86 87 88 89 90 91 92
93 94 95 96 97 98 99 100 101 102 103 104 105 106 107 108 109 110 111 112

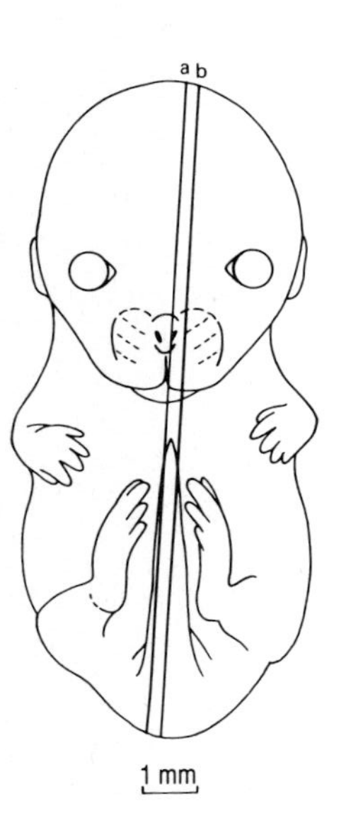
a b
1 mm

Plate 36a (15.5 days p.c.)

15.5 days p.c. Male embryo. Crown–rump length 12 mm (fixed). Sagittal sections (Theiler Stage 24; rat, Witschi Stage 33).

1. diencephalon (thalamus)
2. left lateral wall of pineal gland
3. rostral part of third ventricle
4. neopallial cortex (future cerebral cortex) composed of inner broad nuclear layer (cortical plate) with narrow outer relatively anuclear layer (marginal zone)
5. intermediate zone
6. ventricular zone
7. left lateral ventricle
8. left olfactory lobe
9. extension of anterior horn of left lateral ventricle within left olfactory lobe
10. olfactory (I) nerves converging on olfactory lobe of brain
11. dorsal part of cartilage primordium of nasal septum
12. olfactory epithelium
13. nasal cavity (left half)
14. cartilage primordium of palatal shelf of right maxilla
15. entrance into nasal cavity (external naris)
16. upper lip (left side)
17. tip of tongue
18. primordium of right lower incisor tooth
19. primordia of follicles of vibrissae associated with lower lip
20. right Meckel's cartilage
21. ossification within right half of mandible
22. pectoralis superficialis and pectoralis profundus muscles
23. right third costal cartilage
24. lumen of right atrium
25. pericardial cavity
26. right seventh costal cartilage
27. right dome of diaphragm
28. sub-phrenic recess of peritoneal cavity
29. right lobe of liver
30. part of right hepatic venous plexus
31. distal part of tail
32. right lateral border of genital tubercle (penis)
33. peritoneal cavity
34. cartilage primordium of ischial tuberosity
35. cartilage primordium of head of right femur
36. cavity of right hip joint
37. medullary region of right testis
38. seminiferous tubules within medullary region of right testis
39. proximal part of right mesonephric duct (ductus deferens)
40. cortical region of right testis
41. right lateral wall of middle of second part of duodenum
42. lobar interzone
43. costal origin of dorsal part of diaphragm
44. right pleural cavity
45. caudal lobe of right lung
46. intercostal muscles (external and internal layers)
47. middle lobe of right lung
48. cranial lobe of right lung
49. cartilage primordium of shaft of right third rib with periosteal collar of bone
50. ossification within cartilage primordium of blade of right scapula
51. right submandibular gland
52. cartilage primordium of tip of coracoid process of right scapula
53. ossification within lateral one-third of right clavicle
54. right jugular lymph sac
55. right tubo-tympanic recess
56. cartilage primordium of petrous part of right temporal bone
57. proximal part of right internal jugular vein shortly after its emergence from the jugular foramen
58. right cochlear (VIII) ganglion
59. right jugular foramen
60. lateral wall of right posterior semicircular canal
61. roof of ependymal diverticulum of fourth ventricle
62. ependymal diverticulum of fourth ventricle
63. choroid plexus arising from roof of right lateral recess of fourth ventricle
64. right lateral recess of fourth ventricle
65. cerebellar primordium
66. right trigeminal (V) ganglion
67. right lateral wall of midbrain
68. choroid plexus within anterior horn of left lateral ventricle and arising from its medial wall
69. communication between left lateral ventricle and cavity within left olfactory lobe
70. cartilage primordium of orbito-sphenoid bone
71. oropharynx
72. dorsal surface of tongue
73. tubules of serous glands associated with lateral wall of nasal cavity
74. primordium of left upper incisor tooth
75. primordia of follicles of vibrissae associated with upper lip
76. right submandibular and sublingual ducts
77. umbilical vein in proximal part of umbilical cord
78. loop of midgut within physiological umbilical hernia
79. wall of physiological umbilical hernia
80. glans penis
81. nucleus pulposus in central region of future intervertebral disc
82. cartilage primordium of coccygeal vertebral body
83. lumen of intraperitoneal part of midgut
84. central part of acetabular fossa
85. right mesonephric duct (ductus deferens) within right mesorchium
86. lumen of distal part of second part of duodenum
87. ossification within cartilage primordium of right iliac bone
88. cartilage primordium of right iliac crest
89. degenerating mesonephric tissue containing mesonephric tubules
90. medullary region of right kidney
91. cortical region of right kidney
92. lumen of proximal part of second part of duodenum
93. intercostal neurovascular bundle (intercostal vein, artery and nerve)
94. cutaneous muscle of thorax and trunk regions (panniculus carnosus)
95. ossification within cartilage primordium of mid-shaft region of right fifth rib
96. wall of right atrium
97. ossification within cartilage primordium of shaft of right second rib
98. medial border and inferior angle of right scapula
99. cartilage primordium of middle one-third of right clavicle with periosteal collar of bone
100. right glossopharyngeal (IX) nerve passing towards inferior glossopharyngeal (IX) ganglion, conveying taste and general sensibility from posterior one-third of the tongue (including the sulcus terminalis and vallate papilla)
101. right inferior glossopharyngeal (IX) ganglion
102. ossification within cartilage primordium of exoccipital bone
103. cochlea
104. left pharyngo-tympanic (Eustachian) tube
105. pons
106. cerebral aqueduct
107. caudal part of mesencephalic vesicle
108. right lateral part of pituitary primordium
109. cartilage primordium of basisphenoid bone
110. diencephalon (hypothalamus)
111. nasopharynx
112. left interventricular foramen

Plate 36b (15.5 days p.c.)

15.5 days p.c. Male embryo. Crown–rump length 12 mm (fixed). Sagittal sections (Theiler Stage 24; rat, Witschi Stage 33).

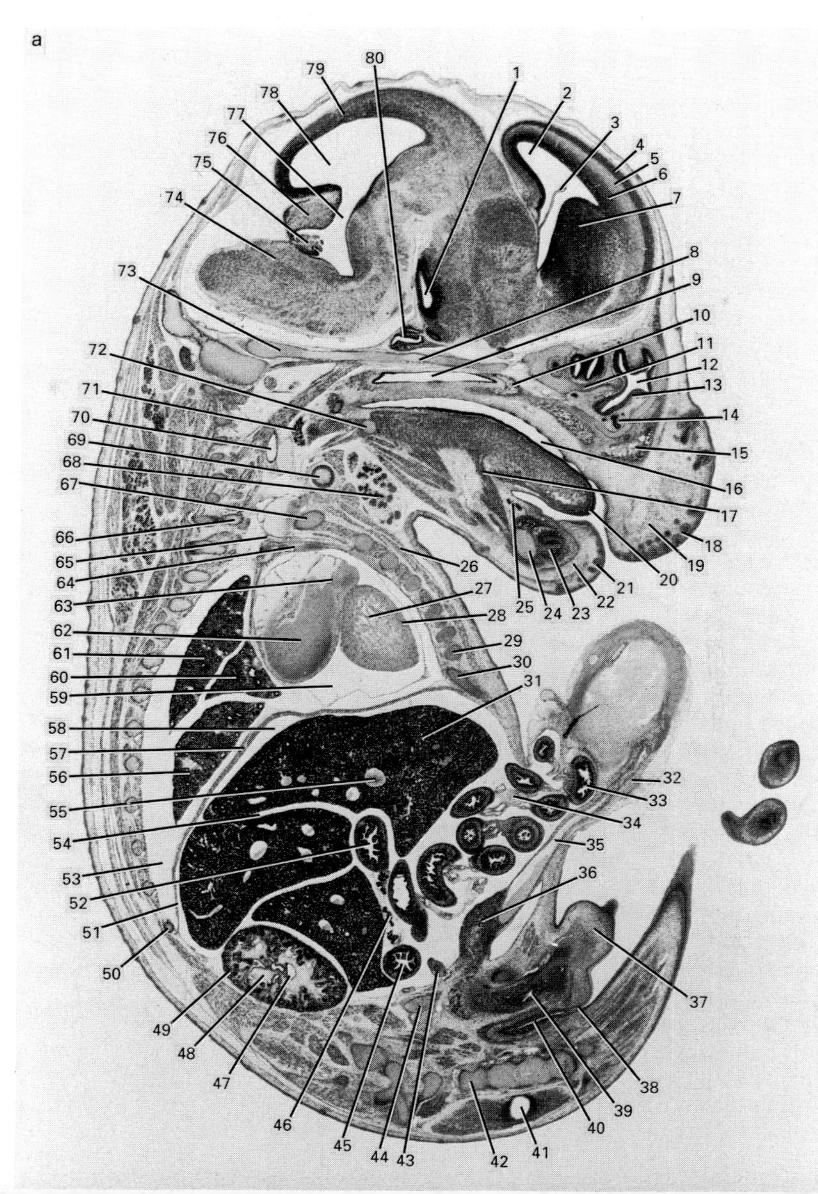

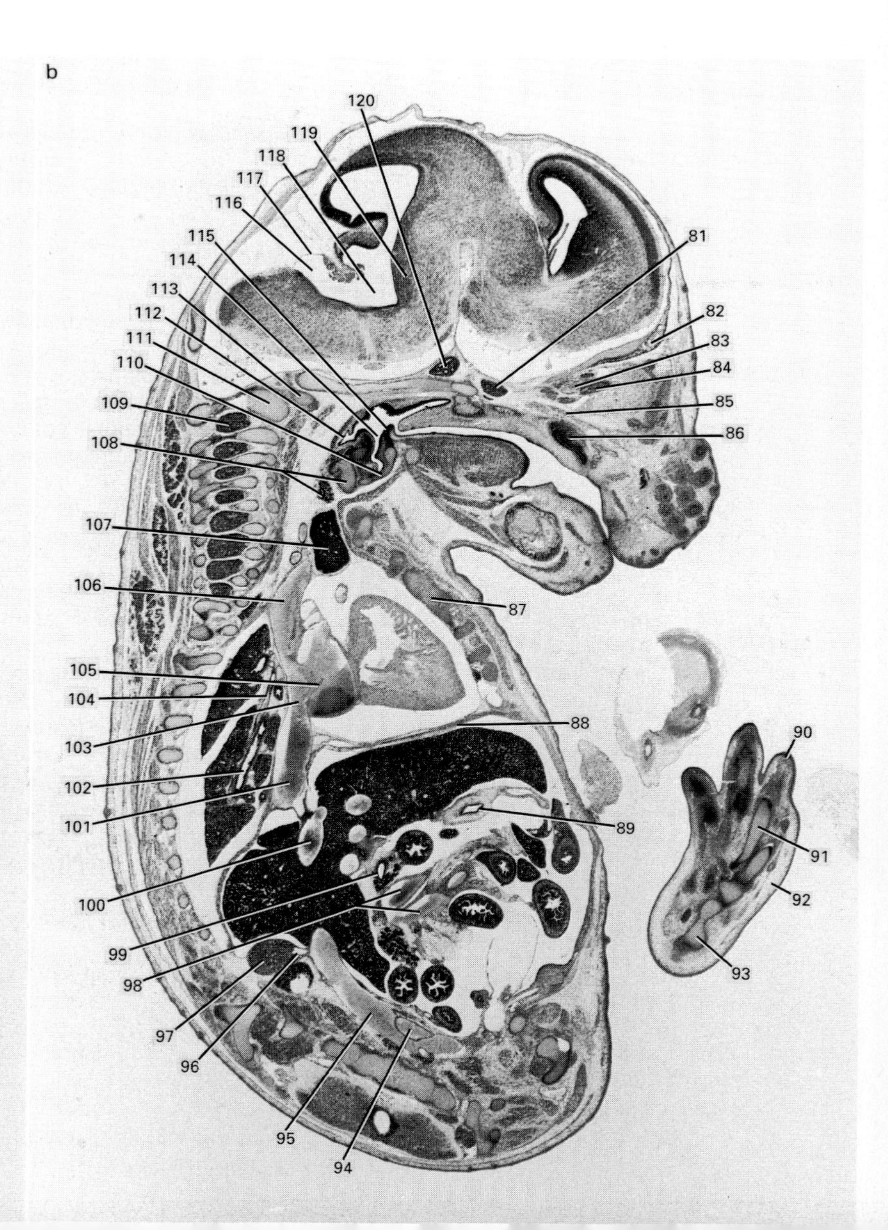

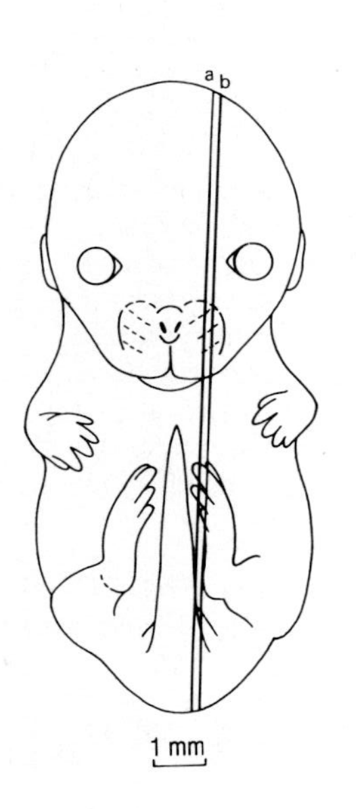

1. infundibular recess of third ventricle
2. left lateral ventricle
3. choroid plexus within anterior horn of left lateral ventricle and arising from its medial wall
4. neopallial cortex (future cerebral cortex) composed of inner broad nuclear layer (cortical plate) with narrow outer relatively anuclear layer (marginal zone)
5. intermediate zone
6. ventricular zone
7. striatum (ganglionic eminence) overlying the head of the caudate nucleus
8. cartilage primordium of basisphenoid bone
9. nasopharynx
10. ossification within palatal shelf of left maxilla
11. cartilage primordia of turbinate bones
12. nasal cavity (left half)
13. olfactory epithelium
14. tubules of serous glands associated with lateral wall of nasal cavity
15. ossification within left maxilla
16. oropharynx
17. extrinsic muscle of tongue (genioglossus)
18. primordia of follicles of vibrissae associated with upper lip
19. upper lip (left side)
20. tip of tongue
21. primordia of follicles of vibrissae associated with lower lip
22. lower lip
23. primordium of left lower incisor tooth
24. left Meckel's cartilage
25. left submandibular and sublingual ducts
26. pectoralis superficialis and pectoralis profundus muscles
27. lumen of right ventricle
28. wall of right ventricle
29. seventh right costal cartilage
30. cartilage primordium of xiphoid process (xiphisternum)
31. right lobe of liver
32. wall of physiological umbilical hernia
33. loop of midgut within physiological umbilical hernia
34. dorsal mesentery of midgut containing mesenteric vessels
35. right umbilical artery
36. right lateral wall of bladder
37. tip of genital tubercle (penis)
38. anal canal
39. proximal part of penile urethra
40. distal part of rectum
41. central canal in sacro-coccygeal region of spinal cord
42. cartilage primordium of sacral vertebral body
43. distal part of right mesonephric duct (ductus deferens)
44. right internal iliac artery
45. lumen of third part of duodenum
46. pancreatic primordium
47. pelvis of right kidney
48. medullary region of right kidney
49. cortical region of right kidney
50. cartilage primordium of dorsal part of shaft region of right 12th rib with periosteal collar of bone
51. costal origin of dorsal part of diaphragm
52. lumen of first part of duodenum
53. right pleural cavity
54. lobar interzone
55. part of right hepatic venous plexus
56. caudal lobe of right lung
57. dorsal part of right dome of diaphragm
58. sub-phrenic recess of peritoneal cavity
59. pericardial cavity
60. middle lobe of right lung
61. cranial lobe of right lung
62. lumen of right atrium
63. lumen of auricular part of right atrium
64. right internal thoracic (mammary) vein
65. right internal thoracic (mammary) artery
66. cartilage primordium of tubercle of right first rib
67. right first costal cartilage
68. right submandibular gland
69. cartilage primordium of medial one-third of right clavicle with periosteal collar of bone
70. right common carotid artery
71. right lobe of thyroid gland
72. cartilage primordium of body of hyoid bone
73. ossification within cartilage primordium of basioccipital bone (clivus)
74. medulla oblongata
75. choroid plexus arising from roof of right lateral recess of fourth ventricle
76. medial part of right half of cerebellar primordium
77. cerebral aqueduct
78. caudal part of mesencephalic vesicle
79. roof of midbrain
80. residual lumen of Rathke's pouch within pituitary primordium
81. medial part of left trigeminal (V) ganglion
82. ossification within medial part of roof of left orbit (orbital plate of left frontal bone)
83. left optic (I) nerve
84. extrinsic ocular muscle (medial/nasal rectus)
85. branch of maxillary division of right trigeminal (V) nerve passing from plexuses around follicles of vibrissae to trigeminal (V) ganglion
86. primordium of left upper molar tooth
87. cartilage primordium of lower part of manubrium sterni
88. ventral part of central region of dome of diaphragm
89. lumen of gall bladder
90. third digit of left hindlimb
91. cartilage primordium of third metatarsal bone
92. lateral border of left hindlimb
93. cartilage primordium of tarsal bone
94. left common iliac artery
95. inferior vena cava
96. adrenal vein where it drains into the inferior vena cava
97. right adrenal gland
98. dilated right superior mesenteric veins within dorsal mesentery of midgut
99. pancreatic duct
100. intrahepatic part of inferior vena cava
101. proximal part of posthepatic inferior vena cava
102. lobar bronchus to caudal lobe of right lung
103. inferior vena cava where it enters the right atrium
104. branch of right pulmonary artery passing to caudal lobe of right lung
105. leaflet of venous valve at site of entrance of inferior vena cava into right atrium
106. right superior (cranial) vena cava
107. right lobe of thymus gland
108. cricoid cartilage with left lobe of thyroid gland in close proximity to it
109. right second cervical dorsal (posterior) root ganglion
110. thyroid cartilage
111. cartilage primordium of body of C2 vertebra (axis)
112. entrance to oesophagus
113. cartilage primordium of anterior arch of C1 vertebra (atlas)
114. entrance to larynx (laryngeal aditus)
115. epiglottis
116. ependymal diverticulum of fourth ventricle
117. roof of ependymal diverticulum of fourth ventricle
118. fourth ventricle
119. pons
120. site of active cellular and vascular proliferation in anterior wall of Rathke's pouch (future pars anterior, anterior lobe of pituitary)

Plate 36c (15.5 days p.c.)

15.5 days p.c. Male embryo. Crown–rump length 12 mm (fixed). Sagittal sections (Theiler Stage 24; rat, Witschi Stage 33).

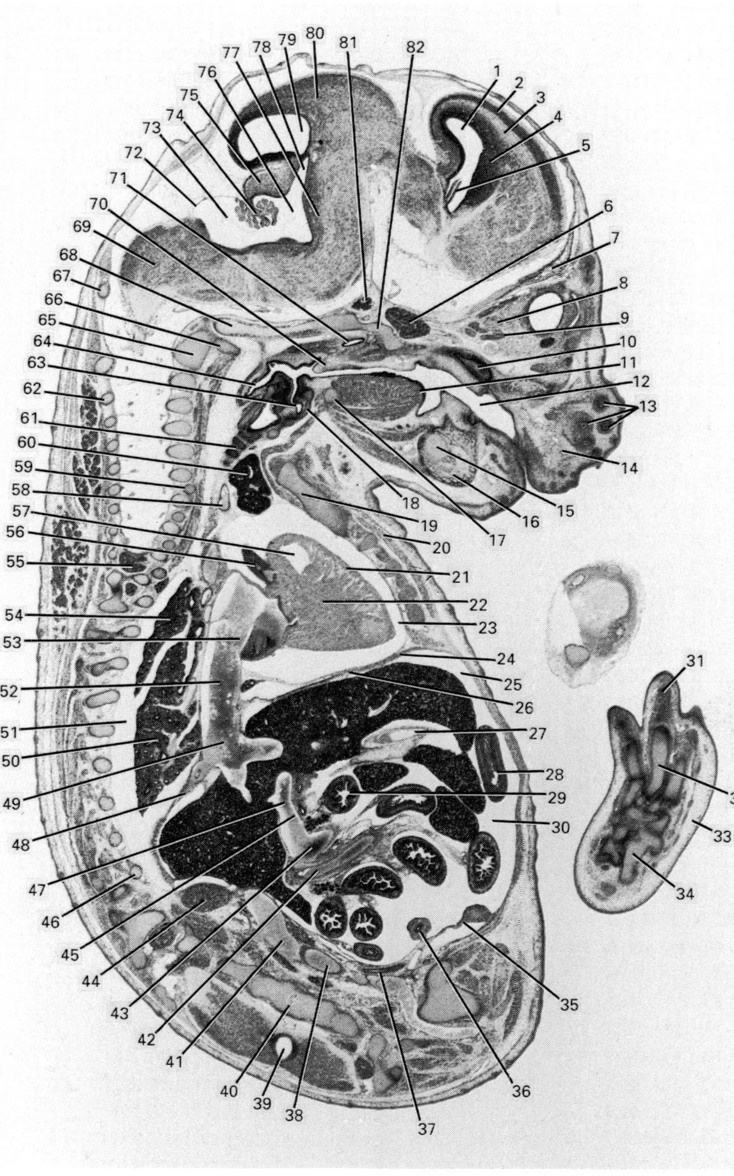

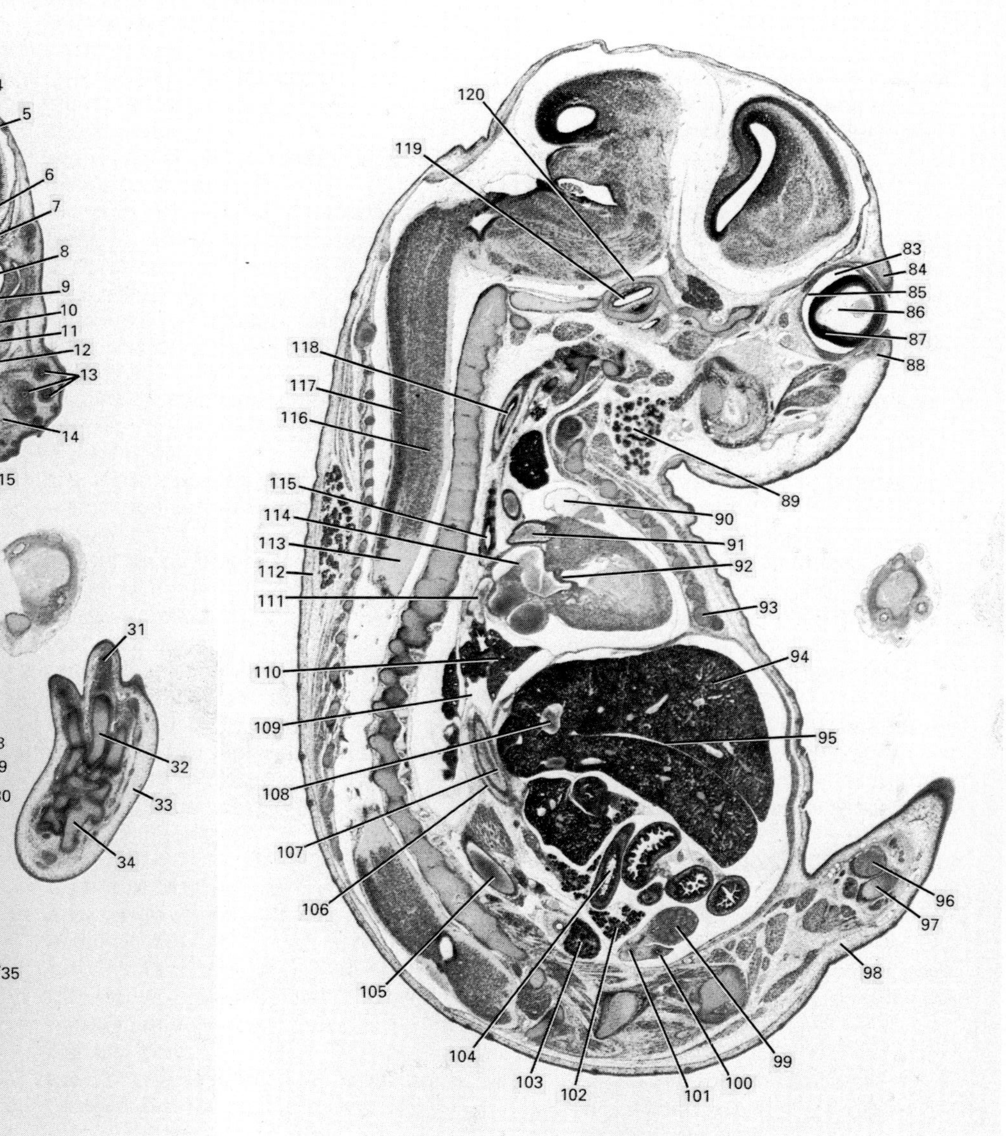

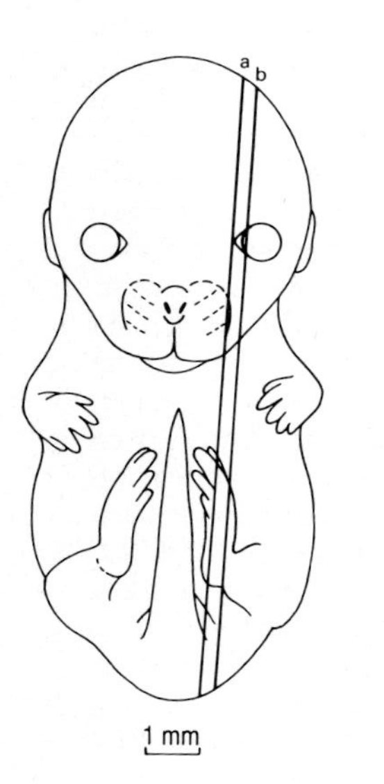

1. left lateral ventricle
2. neopallial cortex (future cerebral cortex) composed of inner broad nuclear layer (cortical plate) with narrow outer relatively anuclear layer (marginal zone)
3. intermediate zone
4. ventricular zone
5. choroid plexus within anterior horn of left lateral ventricle and arising from its medial wall
6. left trigeminal (V) ganglion
7. ossification within medial part of roof of left orbit (orbital plate of left frontal bone)
8. left optic (II) nerve
9. extrinsic ocular muscle (inferior rectus)
10. primordium of left upper molar tooth
11. left lateral border of tongue
12. oropharynx
13. primordia of follicles of vibrissae associated with upper lip (left side)
14. upper lip (left side)
15. left Meckel's cartilage
16. ossification within left half of mandible
17. cartilage primordium of body of hyoid bone
18. thyroid cartilage
19. cartilage primordium of manubrium sterni
20. pectoralis superficialis and pectoralis profundus muscles
21. wall of right ventricle
22. muscular part of interventricular septum
23. pericardial cavity
24. sternal fibres of origin of diaphragm
25. sub-phrenic recess of peritoneal cavity
26. central tendon of diaphragm
27. ductus venosus
28. lumen of intraperitoneal part of midgut
29. lumen of first part of duodenum
30. peritoneal cavity
31. second digit of left hindlimb (pelvic limb)
32. cartilage primordium of left second metatarsal bone
33. lateral border of hindlimb
34. cartilage primordium of left tarsal bone
35. attachment of left gubernaculum testis
36. left mesonephric duct (ductus deferens)
37. left ureter
38. left common iliac artery
39. central canal of spinal cord
40. nucleus pulposus
41. inferior vena cava
42. superior mesenteric artery within dorsal mesentery of midgut
43. superior mesenteric vein where it drains into the hepatic portal vein
44. left adrenal gland
45. common bile duct
46. cartilage primordium of dorsal part of shaft region of left 13th rib
47. hepatic portal vein
48. dorsolateral part of diaphragm close to where it is pierced by the inferior vena cava
49. proximal part of posthepatic inferior vena cava
50. caudal lobe of right lung
51. right pleural cavity
52. posthepatic inferior vena cava
53. inferior vena cava where it enters the right atrium
54. cranial lobe of right lung
55. lower right thoracic dorsal (posterior) root ganglion
56. origin of thoracic aorta (aortic trunk)
57. lumen of right ventricle
58. brachiocephalic (innominate) artery
59. cartilage primordium of neck region of right first rib with periosteal collar of bone
60. right lobe of thymus gland
61. trachea
62. cartilage primordium of neural arch of C3 vertebra with periosteal collar of bone
63. entrance to larynx (laryngeal aditus)
64. entrance to oesophagus
65. cartilage primordium of odontoid process (dens) and body of C2 vertebra (axis)
66. cartilage primordium of anterior arch of C1 vertebra (atlas)
67. cartilage primordium of neural arch of C1 vertebra (atlas) with periosteal collar of bone
68. ossification within cartilage primordium of basioccipital bone (clivus)
69. medulla oblongata
70. left posterolateral border of palatal shelf of left maxilla
71. left pharyngo-tympanic (Eustachian) tube
72. roof of ependymal diverticulum of fourth ventricle
73. ependymal diverticulum of fourth ventricle
74. choroid plexus arising from roof of fourth ventricle
75. cerebellar primordium
76. fourth ventricle
77. pons
78. cerebral aqueduct
79. caudal part of mesencephalic vesicle
80. roof of midbrain
81. left lateral part of pituitary primordium
82. cartilage primordium of basisphenoid bone
83. intra-retinal space
84. upper eyelid
85. pigment layer of retina with subjacent sclera
86. hyaloid cavity
87. inner (neural) layer of retina
88. lower eyelid
89. left submandibular gland
90. auricular part of left atrium
91. outflow tract of right ventricle (pulmonary trunk)
92. posterior cusp of left atrio-ventricular (mitral) valve
93. seventh left costal cartilage
94. left lobe of liver
95. lobar interzone
96. ossification within cartilage primordium of mid-shaft region of left tibia
97. ossification within cartilage primordium of mid-shaft region of left fibula
98. proximal part of left hindlimb
99. seminiferous tubules within medullary region of left testis
100. proximal part of left mesonephric duct (ductus deferens)
101. degenerating mesonephric tissue containing mesonephric tubules
102. pancreatic primordium
103. medial margin (cortical region) of left kidney
104. lumen of duodenal-jejunal junction
105. abdominal aorta
106. posterior vagal (X) trunk which is closely associated with the posteromedial wall of oesophagus where it emerges from the right crus of the diaphragm
107. anterior vagal (X) trunk
108. part of left hepatic venous plexus
109. extension of right pleural cavity around accessory lobe of right lung
110. accessory lobe of right lung
111. right pulmonary artery with branch to accessory lobe of right lung
112. cutaneous muscle of thorax and trunk regions (panniculus carnosus)
113. marginal layer of spinal cord
114. lumen of left atrium
115. lumen of caudal part of trachea
116. mantle layer (ventral grey horn) in mid-cervical region of spinal cord
117. mantle layer (dorsal grey horn) in mid-cervical region of spinal cord
118. lumen of oesophagus
119. left cochlea
120. cartilage primordium of petrous part of left temporal bone

Plate 36d (15.5 days p.c.)

15.5 days p.c. Male embryo. Crown–rump length 12 mm (fixed). Sagittal sections (Theiler Stage 24; rat, Witschi Stage 33).

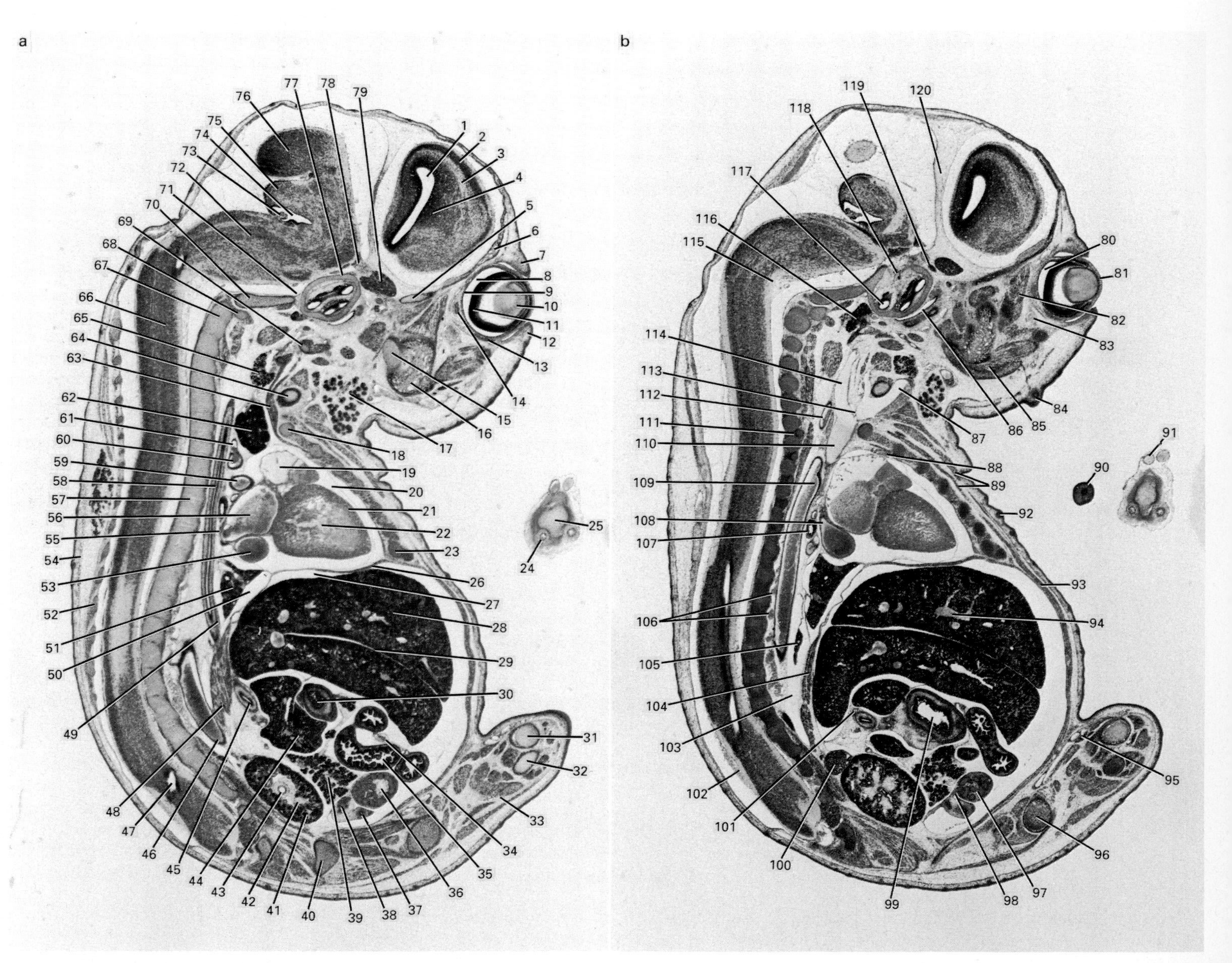

1. left lateral ventricle
2. neopallial cortex (future cerebral cortex) composed of inner broad nuclear layer (cortical plate) with narrow outer relatively anuclear layer (marginal zone)
3. intermediate zone
4. ventricular zone
5. cartilage primordium of sphenoid bone
6. ossification within medial part of roof of left orbit (orbital plate of left frontal bone)
7. upper eyelid of left eye
8. intra-retinal space
9. pigment layer of retina with subjacent sclera
10. lens
11. inner (neural) layer of retina
12. lower eyelid
13. extrinsic ocular muscle (inferior rectus)
14. extrinsic ocular muscle (lateral rectus)
15. left Meckel's cartilage
16. ossification within left half of mandible
17. left submandibular gland
18. left first costal cartilage
19. lumen of auricular part of left atrium
20. pericardial cavity
21. wall of left ventricle
22. lumen of left ventricle
23. seventh left costal cartilage
24. umbilical artery within loop of umbilical cord
25. Wharton's jelly
26. left dome of diaphragm
27. sub-phrenic recess of peritoneal cavity
28. left lobe of liver
29. lobar interzone
30. gastro-duodenal junction in region of pyloric sphincter
31. ossification within cartilage primordium of mid-shaft region of left tibia
32. ossification within cartilage primordium of mid-shaft region of left fibula
33. proximal part of left hindlimb
34. dorsal mesentery of midgut containing mesenteric vessels
35. lumen of intraperitoneal part of midgut
36. medullary region of left testis
37. proximal part of left mesonephric duct (ductus deferens)
38. degenerating mesonephric tissue containing mesonephric tubules
39. pancreatic primordium
40. cartilage primordium of iliac bone
41. cortical region of left kidney
42. medullary region of left kidney
43. proximal part of left ureter at junction with renal pelvis
44. caudate lobe of liver
45. intra-abdominal (sub-diaphragmatic) part of oesophagus
46. thoracic (descending) aorta where it passes through the aortic aperture of the diaphragm
47. central canal in mid-lumbar region of spinal cord
48. fibres of right crus of the diaphragm forming margins of the aortic aperture
49. anterior vagal (X) trunk
50. extension of right pleural cavity around accessory lobe of right lung
51. accessory lobe of right lung
52. cartilage primordium of neural arch of lower thoracic vertebra with periosteal collar of bone
53. lumen of left superior (cranial) vena cava as it passes medially towards the right atrium
54. cutaneous muscle of thorax and trunk regions (panniculus carnosus)
55. dorsal wall of left atrium
56. lumen of left atrium
57. nucleus pulposus in central region of future intervertebral disc
58. lumen of oesophagus
59. continuation of pulmonary trunk to form the ductus arteriosus
60. brachiocephalic (innominate) artery where it gives off the right common carotid artery
61. proximal part of right common carotid artery
62. left lobe of thymus gland
63. sternothyroid muscle (posteriorly) which arises from the posterior surface of the manubrium sterni with sternohyoid muscle (anteriorly)
64. cartilage primordium of middle one-third of left clavicle with periosteal collar of bone
65. left lobe of thyroid gland
66. upper cervical region of spinal cord
67. cartilage primordium of odontoid process (dens) and body of C2 vertebra (axis)
68. cartilage primordium of anterior arch of C1 vertebra (atlas)
69. cartilage primordium of left greater horn of hyoid bone
70. ossification within cartilage primordium of basioccipital bone (clivus)
71. lateral part of left jugular foramen
72. medulla oblongata
73. choroid plexus within left lateral recess of fourth ventricle
74. left lateral recess of fourth ventricle
75. cerebellar primordium
76. left lateral wall of midbrain
77. cartilage primordium of petrous part of left temporal bone
78. rootlets of left trigeminal (V) nerve
79. left trigeminal (V) ganglion
80. hyaloid cavity
81. conjunctival epithelium (continuous with corneal epithelium)
82. anterior fibres of left temporalis muscle
83. left inferior orbital vein
84. primordium of follicle of vibrissa
85. superficial and deep portions of left masseter muscle
86. left tubo-tympanic recess
87. left subclavian vein
88. left internal thoracic (mammary) vessels
89. pectoralis superficialis and pectoralis profundus muscles
90. first digit left forelimb
91. umbilical vein within loop of umbilical cord
92. mammary gland primordium
93. left rectus abdominis muscle
94. part of left hepatic venous plexus
95. left femoral vessels
96. ossification within cartilage primordium of mid-shaft region of left femur
97. seminiferous tubules within medullary region of left testis
98. proximal part of left mesorchium
99. lumen of proximal part of pyloric antrum region of the stomach lined by glandular epithelium
100. left adrenal gland
101. posterior vagal (X) trunk
102. posterior spinal vein
103. caudal part of left pleural cavity
104. left crus of the diaphragm
105. medial margin of caudal part of left lung
106. left intercostal veins which drain into the inferior hemiazygos vein
107. left main bronchus
108. site where descending part of left superior (cranial) vena cava passes medially and transversely towards the right atrium
109. site where ductus arteriosus enters the distal part of the arch of the aorta
110. left superior (cranial) vena cava
111. cartilage primordium of head of left first rib
112. left common carotid artery
113. small venous valves at termination of left subclavian vein
114. left internal jugular vein
115. cervical flexure
116. left inferior vagal (X) ganglion
117. cochlea
118. rootlet of left vestibulocochlear (VIII) nerve
119. medial part of left facial (VII) ganglion
120. mesenchymal condensation forming tentorium cerebelli

PLATE
36e

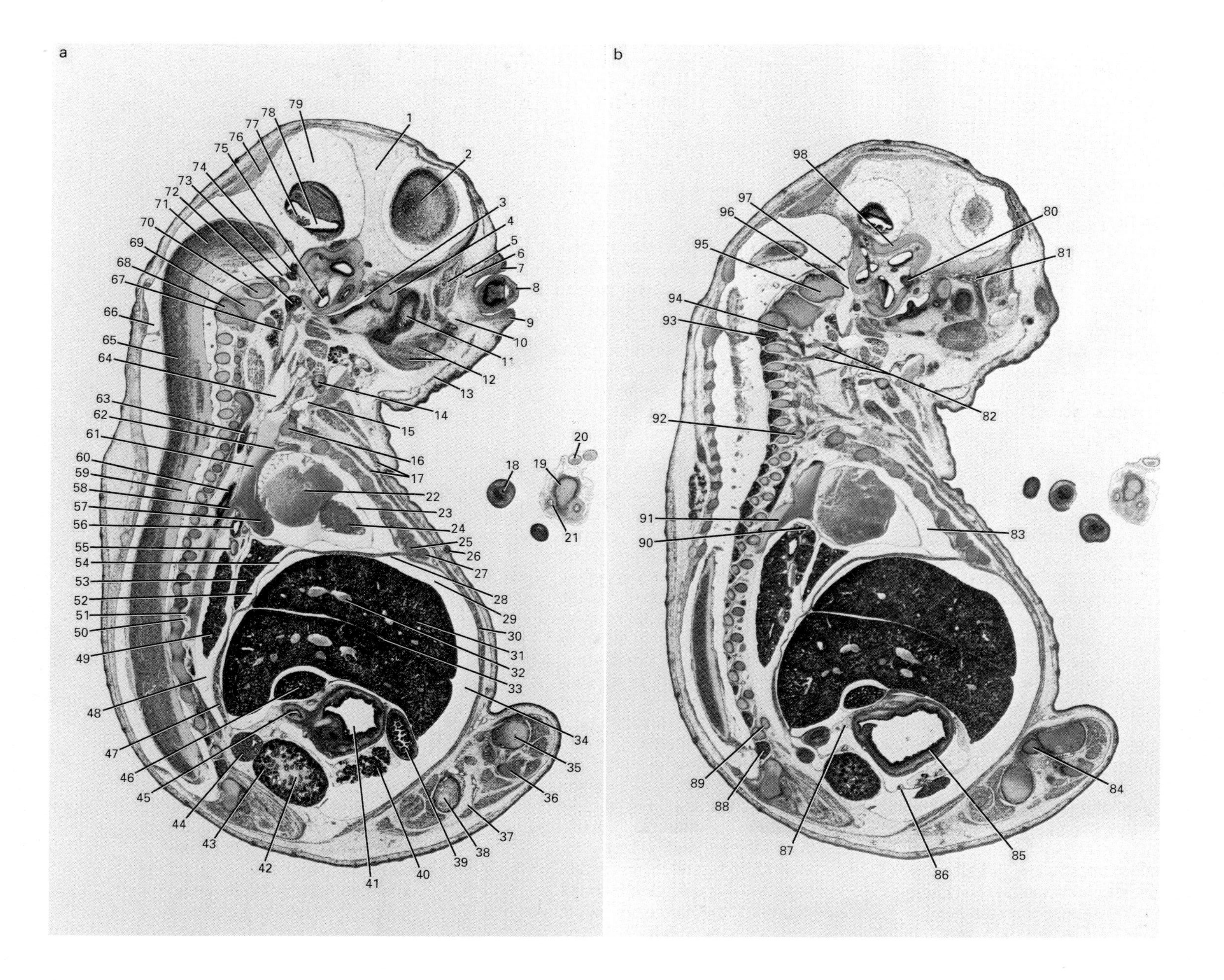

a
b
1
2
3
4
5
6
7
8
9
10
11
12
13
14
15
16
17
18
19
20
21
22
23
24
25
26
27
28
29
30
31
32
33
34
35
36
37
38
39
40
41
42
43
44
45
46
47
48
49
50
51
52
53
54
55
56
57
58
59
60
61
62
63
64
65
66
67
68
69
70
71
72
73
74
75
76
77
78
79
80
81
82
83
84
85
86
87
88
89
90
91
92
93
94
95
96
97
98

Plate 36e (15.5 days p.c.)

15.5 days p.c. Male embryo. Crown–rump length 12 mm (fixed). Sagittal sections (Theiler Stage 24; rat, Witschi Stage 33).

1. mesenchymal condensation forming left lateral origin of tentorium cerebelli
2. left lateral wall of left cerebral hemisphere
3. left tubo-tympanic recess
4. left Meckel's cartilage
5. anterior fibres of left temporalis muscle
6. left superior orbital vein
7. upper eyelid of left eye
8. conjunctival epithelium (continuous with corneal epithelium)
9. lower eyelid
10. left inferior orbital vein
11. ossification within ramus of left mandible
12. superficial and deep portions of left masseter muscle
13. platysma muscle (panniculus carnosus)
14. ossification within lateral one-third of left clavicle
15. left subclavian vein
16. cartilage primordium of mid-shaft region of left first rib with periosteal collar of bone
17. pectoralis superficialis and pectoralis profundus muscles
18. first digit left forelimb (thoracic limb)
19. Wharton's jelly
20. umbilical vein within loop of umbilical cord
21. umbilical artery within loop of umbilical cord
22. lumen of left atrium
23. pericardial cavity
24. wall of left ventricle in region of apex of heart
25. seventh left costal cartilage
26. mammary gland primordium
27. eighth left costal cartilage
28. left dome of diaphragm
29. sub-phrenic recess of peritoneal cavity
30. left rectus abdominis muscle
31. left lobe of liver
32. part of left hepatic venous plexus
33. lobar interzone
34. peritoneal cavity
35. ossification within cartilage primordium of shaft region of left tibia
36. ossification within cartilage primordium of shaft region of left fibula
37. proximal part of left hindlimb
38. ossification within cartilage primordium of upper shaft region of left femur
39. lumen of midgut
40. pancreatic primordium
41. lumen of "body" of the stomach
42. medullary region of left kidney
43. cortical region of left kidney
44. left adrenal gland
45. lumen of caudal part of oesophagus just rostral to gastro-oesophageal junction
46. caudate lobe of liver
47. costal fibres of origin of diaphragm
48. caudal part of left pleural cavity
49. caudal part of left lung
50. left intercostal artery
51. left intercostal vein where it drains into the inferior hemiazygos vein
52. extension of right pleural cavity around accessory lobe of right lung
53. accessory lobe of right lung
54. central tendon of diaphragm
55. left pulmonary vein
56. left main bronchus
57. caudal part of left superior (cranial) vena cava where it passes medially and transversely towards the right atrium
58. inferior hemiazygos vein where it joins the left superior (cranial) vena cava
59. marginal layer of spinal cord
60. left lateral wall of thoracic (descending) aorta
61. lumen of left superior (cranial) vena cava
62. left internal thoracic (mammary) vein where it joins the left superior (cranial) vena cava
63. left cervicothoracic (stellate) sympathetic ganglion
64. left jugular lymph sac
65. mantle layer of spinal cord
66. spinal canal
67. left vagal (X) trunk
68. cartilage primordium of body of C2 vertebra (axis)
69. cartilage primordium of anterior arch of C1 vertebra (atlas)
70. ossification within lateral part of cartilage primordium of basioccipital bone (clivus)
71. medulla oblongata
72. inferior ganglion of left vagus (X) nerve
73. left jugular foramen
74. superior ganglion of left vagus (X) nerve
75. cochlea
76. primordium of interparietal bone
77. choroid plexus within lateral extremity of fourth ventricle
78. lateral extremity of fourth ventricle
79. loosely packed cephalic mesenchyme in future subarachnoid space
80. cartilage primordium of left stapes
81. ossification within lateral part of left maxilla
82. anterior primary rami of upper cervical nerves
83. ventral extension of left pleural cavity
84. cartilage primordium of left patella
85. gastric mucosa lining the "body" of the stomach (area of fundic glands)
86. dorsal mesentery of midgut with associated mesenteric vessels
87. dorsal mesogastrium
88. upper lumbar dorsal (posterior) root ganglion
89. cartilage primordium of tubercle of left 13th rib
90. branches of left pulmonary artery passing into left lung
91. inferior hemiazygos vein
92. cartilage primordium of tubercle of left first rib
93. upper cervical dorsal (posterior) root ganglion
94. rostral part of left vertebral artery
95. cartilage primordium of exoccipital bone with periosteal ossification
96. rostral part of left internal jugular vein within the left jugular foramen
97. left sigmoid dural venous sinus where it drains into the left internal jugular vein
98. cartilage primordium of petrous part of left temporal bone

a b

1 mm

While the clavicle previously showed early evidence of ossification (*Plate 32e,b,c*), this process is now seen to involve a more extensive region of this bone, but ossification appears to be particularly marked in its middle and lateral one-third (*Plate 35d,d*; *Plate 35e,a,b*). Elsewhere in the pectoral region, much of the scapula now shows extensive evidence of ossification (*Plate 35e,a–c*; *Plate 36a,a*). In the region of the sternum, each of the cartilage primordia of the sternebrae are more clearly defined than previously (*Plate 35g,a–d*; cf. *Plate 32f,b*), but particularly the xiphoid process which is now seen to be a quite substantial structure (*Plate 35h,c*; cf. *Plate 32g,d*). Similarly, while the ribs previously showed early evidence of periosteal ossification, this process is now seen to be more extensive, and involves approximately the posterior one-third to one-half of the shaft region of the ribs (e.g. *Plate 35f,b–d*).

In relation to the long bones, in the upper part of the shaft region of the humerus, an extensive area of ossification is now evident (*Plate 35f,a,b*), and more extensive regions of ossification are present than observed previously in the mid-shaft region of both the ulna and radius (*Plate 35g,b–d*; *Plate 81,e,f*). In the pelvic girdle, periosteal ossification is apparent in the upper part of the cartilage primordium of the iliac bone (*Plate 35l,c,d*; *Plate 36a,b*). Larger centres of ossification are also seen in the mid-shaft region of the femur, tibia and fibula (*Plate 35j,c,d*; *Plate 35k,a*; *Plate 81,e,f*). No ossification is yet seen in either the carpal or metacarpal (or tarsal or metatarsal) bones, nor in the phalanges at this stage.

Stage 25 Developmental age, 17 days p.c.
Plates 37a–l, Plates 38a–c illustrate transverse and sagittal sections, respectively, through intact embryos, while Plates 39a–d illustrate coronal sections through the head region of an embryo at a similar developmental stage

EXTERNAL FEATURES

The most obvious difference observed between embryos at 15.5–16 and 16.5–17 days p.c. relates to the appearance of the skin. Whereas previously, apart from the presence of large numbers of hair follicle primordia in the head, trunk and proximal part of the limbs, the skin was smooth and closely followed the contours of these regions (*Plate 46,i,j*), at 16.5–17 days p.c., the skin, particularly in the neck, trunk and limbs distally as far as the carpal and tarsal regions, is now seen to contain substantial numbers of wrinkles. The latter are mostly parallel to each other, and run circumferentially around the trunk, and in a less obvious pattern in the other areas where wrinkles are present. At this stage, the facial region and the distal parts of the limbs are almost completely devoid of wrinkles (*Plate 46,k,l*). As a direct consequence of an increase in the thickness of the skin, and the presence of the prominent wrinkles, the superficial veins that were clearly seen through the almost transparent skin, and were a feature of unfixed embryos from about 12 days p.c., are mostly no longer visible.

In the facial region, the most obvious difference between the appearance observed at this and the previous stage of development is seen in the vibrissae-bearing areas of the upper lips (*Plate 39a,a*), where the vibrissae are now seen to have erupted through the epidermal evaginations, leaving a characteristic mound of cellular debris at their site of eruption (*Plate 49,n,o*). The prominent tactile or sinus hair follicles have yet to erupt (*Plate 49,n*). The other characteristic feature of this stage of development is the fused eyelids (*Plate 49,h*). The upper and lower eyelids fuse towards the end of the previous stage of development, so that during that stage, all degrees of eyelid closure were encountered (*Plate 49,c–g*). The external appearance of the fused eyelids is unusual, because there appears to be an extensive ridge of cellular excrescences and/or extruded epithelial cells along the outer surface of the line of fusion of the upper and lower eyelids. It must be assumed that at least some of these cells are the rounded clumps seen previously at or close to the eyelid margins during the closure process (*Plate 49,d–f*). The latter are sloughed off by the time of birth. While these cells are not readily seen in conventional histological sections (though they are readily seen in sections of plastic-embedded material), the line of fusion between the upper and lower eyelids is usually easily recognized, being most readily appreciated in coronal sections through the head region (*Plate 39a,e*; *Plate 53d*). Since the lower eyelid grows to a greater extent than the upper eyelid, the final line of fusion is represented by a line approximately 10–15° from the horizontal (being higher anteriorly, and lower posteriorly), at the junction overlying the upper one-third and the lower two-thirds of the subjacent cornea (*Plate 53,b*). However, the actual location of the line of fusion of the eyelids on any particular section, depends on the orientation and level of the section (e.g. *Plate 39a,f*; *Plate 53,d*). It is also evident that the pinna of the ear now extends even further forwards than previously (*Plate 46,l*; cf. *Plate 46,j*), and covers all but the anterior one-third to one-quarter of the external auditory meatus.

In the trunk region, the umbilical hernia, a feature of the umbilical region since about 12–12.5 days p.c. (*Plate 27c,a–e*), has now completely disappeared with the return of the midgut loop into the peritoneal cavity (*Plate 37i,c*; *Plate 38a,b*; *Plate 38b,a*; cf. *Plate 35h,d*; *Plate 35i,a–c*; *Plate 36b,a*).

An analysis of the limbs reveals that substantial changes are evident in their degree of differentiation between the situation observed at 15.5–16 and 16.5–17 days p.c. In relation to the handplate region of the forelimb, the discrepancy between the length of the first digit (the pollex) compared to the other four digits is considerably more marked than previously (*Plate 45,g,h*; cf. *Plate 45,c,d*). Similarly, whereas previously all of the digits tended to be fairly splayed out, at this stage, the medial three or four digits are now seen to be almost parallel to each other. As a consequence of the latter, the proximal part of the digital interzones on either side of the middle digit are slightly less well defined than previously. The well-defined pads characteristically seen on the palmar surface of the metacarpal region of the handplate are, however, significantly more prominent than previously (*Plate 45,h*; cf. *Plate 45,d*).

In the footplate region of the hindlimb, the changes that have occurred over this period are quite similar to those seen in relation to the comparable region of the forelimb, namely that the middle three digits tend now to be almost parallel to each other, while the first and fifth digits are still slightly splayed out, whereas previously all of the digits were markedly splayed out (*Plate 45,i–k*; cf. *Plate 45,e,f*). By contrast to the situation observed in relation to the forelimb, while the middle three digits are of approximately equal length, the first and fifth digits, though slightly shorter than the others, are still of similar length (*Plate 45,k*). The pads associated with the plantar surface of the footplate region of the hindlimbs are more prominent than previously (*Plate 45,j,k*; cf. *Plate 45,f*), but are less prominent than those associated with the palmar surface of the handplate region of the forelimbs. In both the forelimb and the hindlimb, the nail (claw) primordia are more obvious than previously (forelimb, *Plate 45,g*; cf. *Plate 45,c*; hindlimb, *Plate 45,i*; cf. *Plate 45,e*).

THE HEART AND VASCULAR SYSTEM

The orientation of the heart is essentially identical to that described in the previous stage, though even less is evident of the interventricular groove, apart from a shallow depression just proximal to the apex of the heart (*Plate 37g,c*, unlabelled), compared to the situation observed previously, though even at 15.5–16 days p.c. (*Plate 35h,a,b*), this is a relatively insignificant feature of the anterior surface of the heart (cf. 14.5–15 days p.c., *Plate 32g,a–d*). The walls of the ventricles, as well as the muscular part of the interventricular septum, are more consolidated than previously, and less of their lumina appear to be occupied by trabeculae carneae (*Plate 37f,c*; *Plate 37g,a,b*; cf. *Plate 35g,b–d*; *Plate 35h,a*). While the muscular (trabeculated) walls of the atria are still extremely thin (compared to those of the ventricles), the volumes of these chambers of the heart are considerable, being far greater at this stage than those of the ventricles (*Plate 38a,a,b*; *Plate 38b,a*). However, this situation is not markedly different from the position observed at the previous stage of development (*Plate 36a,a,b*; *Plate 36b,a,b*).

Detailed analysis of the proximal part of the aortic trunk at about 15–15.5 days p.c. reveals the first indication of the coronary vessels. At 16.5–17 days p.c., however, these vessels are seen to have a considerable luminal diameter, and they are easily traced throughout their course. The left coronary artery arises from the left posterior aortic sinus (*Plate 37f,b*), and in this embryo gives off the large anterior interventricular branch which passes distally towards the apex of the heart. However, this vessel may sometimes arise from the right coronary artery, shortly after its origin from the anterior aortic sinus. All of the other principal features of the heart are as indicated previously. It should be noted, however, that at this stage of development, no clearly defined visceral pericardium (or epicardium) is seen, though the myocardium, in places, appears to be covered by a monolayer of squamous (mesothelial) cells, which probably represents their precursor elements. The pleuro-pericardial membrane (later to form the fibrous pericardium) is extremely thin at this stage (*Plate 37f,a–d*), and consists only of three layers of cells, namely a central core of mesenchyme, with a covering on one side of a single layer of (parietal) pericardial cells, and on the other side by a monolayer of parietal pleura.

By tracing the right and left common carotid arteries rostrally, it is now possible to see at their bifurcation (which gives origin to the internal and external carotid arteries), a localized thickened region of their wall, which constitutes the carotid bodies. The latter may either be located between the origins of the internal and external carotid arteries, or be posterior to the bifurcation (as in this embryo). The dilated region of the common carotid artery at its bifurcation constitutes the carotid sinus (the carotid bodies and sinuses are chemo- and baroreceptors, respectively) (*Plate 39b,a*). These regions are innervated by the glossopharyngeal (IX) nerve, and are of third branchial arch origin.

While the internal carotid arteries give off no extracranial branches, the external carotid arteries give off a number of named branches (such as the lingual, facial, maxillary, superficial temporal etc.), most of which can be traced at this stage. This exercise is facilitated if the circulation is injected with either coloured latex or Indian ink (which should contain carbon particles). Intracardiac injection of these agents can be successfully performed from the early somite stage onwards, and the specimens can then either be embedded and sectioned, or preferably cleared with methyl salicylate, or a range of other clearing agents, in order to view the entire circulation, or specific parts of it, *in situ*, in intact cleared embryos (Moffat, 1959).

THE PRIMITIVE GUT AND ITS DERIVATIVES

No significant differences are observed in either the appearance of the tongue or in its relationship to the palate, from the situation observed at the previous stage of development. The relationship between the nasopharynx and the oropharynx at this stage is, however, slightly different than observed previously, principally because of the elongation of the face that occurs during this time, and this is best appreciated from an analysis of coronal and sagittal sections through this region. These sections clearly indicate the increased degree of differentiation seen over this relatively short period in the region of the nasopharynx, in particular. This is evident when coronal sections through the anterior part of this region in embryos at 16.5–17 days p.c. are compared with similar sections through this region in embryos at about 14.5 days p.c. (*Plate 39a,a–e*; cf. *Plate 34a,a–d*), at a stage prior to palatal shelf fusion, when both the oropharynx and nasopharynx are in extensive continuity.

The increased degree of differentiation of the conchae, and the vomeronasal organs (Jacobson's organs) are immediately apparent. More particularly, at 16.5–17 days p.c., it is possible for the first time to clearly see the close relationship that exists between the olfactory lobes of the brain and the structures associated with the roof of the nasopharynx (in the region of the nasal cavity) (*Plate 39a,d,e*). The numerous branches of the olfactory (I) nerves pass upwards from the olfactory epithelium towards the inferior surface of the olfactory cortex, through the cartilage primordia of the cribriform plate of the ethmoid bones (previously part of the primitive nasal capsule) (*Plate 39a,e*).

The dorsal surface of the anterior one-third to one-half of the tongue is clearly seen to have a concave/spatulate profile (*Plate 39a,b*), while the posterior half is rounded and convex (*Plate 39a,d,e*). The most posterior region (at the base of the tongue) has a reasonably flat profile (*Plate 39a,f*; *Plate 39b,a*), in the middle of which is located the single circumvallate papilla (*Plate 39b,a*). Analyses of median sagittal sections through this region are also instructive, in that they serve to emphasize that the posterior half of the nasopharynx contains no olfactory epithelium, and is merely a passageway which connects the nasal cavity (anteriorly) with the oropharynx (posteriorly) (*Plate 38b,b*; *Plate 38c,a*). The exact location of the median circumvallate papilla is also readily appreciated from the analysis of such sections (*Plate 38b,b*).

Sagittal sections through the posterior part of the oropharynx also allow the relationship between the epiglottis and the entrance to the larynx to be appreciated (*Plate 38c,a*), as well as the direct continuity that exists between the upper and lower respiratory tracts. In appropriate transverse (and sagittal) sections, it is possible to see the site of attachment of the base of the epiglottis to the inner aspect of the middle part of the thyroid cartilage (*Plate 37c,b*). What is less easy to appreciate from an analysis of these sections, is the relationship that exists between the posterior part of the oropharynx and the cavity of the middle ear. This relationship is most easily appreciated from an analysis of transverse sections through this region, as the tubular connection between these two regions (via the pharyngo-tympanic (or Eustachian) tube) runs almost directly posterior (but slightly upwards) to open into the tubo-tympanic recess (*Plate 37b,c,d*). Only after birth does the cavity of the middle ear communicate with the air cells contained within the mastoid process of the temporal bone.

Analysis of transverse sections through the common entrance from the oropharynx into the oesophagus and trachea allows the relationship between the thyroid, cricoid and arytenoid cartilages to be appreciated, as well as their relationship to the body and greater horns (cornua) of the hyoid bone (*Plate 37c,b–d*; *Plate 37d,a*). These views clearly complement the sagittal sections through this region (*Plate 38b,b*; *Plate 38c,a*), as well as the coronal sections, which are rather more difficult to interpret in isolation (principally because of the obliquity of the structures being sectioned) (*Plate 39b,c,d*).

The parotid gland, though the smallest of the salivary glands, is seen to be substantially more consolidated than previously (*Plate 37b,c*; cf. *Plate 35d,a–c*), while the sublingual and submandibular glands (which are still not readily separable) also show an increased degree of differentiation (e.g. *Plate 37d,a,b*; cf. *Plate 35e,a–d*). The thyroid, parathyroid and thymus glands are the other important glandular structures in this region. The thyroid gland has marginally increased in size compared to the situation observed previously (*Plate 37c,d*; *Plate 37d,a,b*; cf. *Plate 35d,c,d*; *Plate 35e,a,b*), and the two lobes are closely associated with the posterolateral surfaces of the lower half of the thyroid cartilage at this stage. The narrow isthmus of the thyroid gland passes across the midline directly in front of the second or third tracheal "ring" (*Plate 38c,a*), though previously it was located at the level of the lower border of the cricoid cartilage/first tracheal "ring" (*Plate 35e,b*).

The parathyroid glands tend to be embedded in the posterolateral part of the lobes of the thyroid gland at this stage, and this relationship is particularly well illustrated in the coronal section through this region (*Plate 39b,d*). At earlier stages, the parathyroid glands tended to be quite small when compared to the volume of the thyroid, and they gradually became apposed to, and then often incorporated into the posterolateral part of the thyroid gland (*Plate 35e,a*; cf. *Plate 32,a*). The thymus gland, consisting of two closely apposed lobes, is by far the largest "solid" structure in this region (*Plate 37d,d*; *Plate 37e,a,b*), and extends caudally from the level opposite the lower poles of the thyroid gland into the superior part of the anterior mediastinum. The thymus is almost completely retrosternal, and is a directly superior relation of the walls of the right and left atria (*Plate 38b,a,b*; *Plate 38c,a,b*), with only parietal pericardium and a small volume of cervical mesenchyme intervening.

With regard to the lower respiratory tract, it is interesting to note that between 15.5–16 and 16.5–17 days p.c., the overall volume of the thoracic region occupied by the pleural cavities increases, and this is consistent with the general growth of the embryo over this period, but the actual volume occupied by lung tissue does not appear to increase to the same degree (e.g. *Plate 37g,a–d*; cf. *Plate 35g,a–d*). As a consequence of this, space becomes available within the thorax into which the lungs can eventually expand. As marked difference in the volume of the lungs is seen by about 17.5–18 days p.c., when the alveoli in this embryo are seen to have expanded for the first time (e.g. *Plate 40g,a–f*). Clearly at this time, *in utero*, the lungs are filled with amniotic fluid and bronchial secretions, but once these are expelled, the alveoli can fill with air immediately thereafter, and the lungs can expand to fill all the available space within the thorax. The possibility must be considered that the histological appearance of the lungs noted here (with regard to the 17.5–18 days p.c. embryo) may, in a sense, be artefactual, in that this embryo may have gasped for air in the period between its isolation and fixation, and that the dilated alveoli may in fact be a reflection of such a scenario.

At the histological level, the lungs at 16.5–17 days p.c. appear to be fairly homogeneous in consistency, and apart from a few secondary and tertiary bronchi/bronchioles could almost be mistaken for solid organs. However, towards their periphery, alveolar ducts are first apparent which are lined by cuboidal cells, though at this time no evidence of alveoli are seen. The difference in the appearance of the lungs between 16.5–17 days and the 17.5–18 days p.c. embryo, is therefore all the more dramatic (*Plate 66,m,n*; cf. *Plate 66,o,p*), particularly at the relatively low magnifications used to illustrate these stages of development (for possible explanation, however, see above).

While the appearance of both the liver and the stomach are in most regards quite similar to the situation observed at the previous stage of development, a considerable change is apparent in the gross and histological appearance of the small intestine. Possibly the most interesting difference relates to the increase in the length of this region of the gut tube that has occurred between 15.5–16 and 16.5–17 days p.c. The latter is evident from a cursory inspection of the contents of the lower half of the peritoneal cavity in both the transversely and sagittally sectioned embryos used to illustrate this stage of development (*Plate 37j,a–d*; *Plate 37k,a,b*; *Plate 38a,b*; *Plate 38b,a,b*; *Plate 38c,a,b*; cf. *Plate 35j,a–d*; *Plate 35k,a–c*; *Plate 36b,a,b*; *Plate 36c,a,b*). This substantial increase in length is also associated with an increase in its luminal diameter, as well as in the complexity of the villous lining of this region of the gut (e.g. *Plate 37j,d*; cf. *Plate 35j,d*). Throughout this period, the appearance of the hindgut region remains virtually identical to that seen previously.

The volume of the pancreas also increases to a considerable degree during this period, and is now principally located on the left side of the peritoneal cavity, its tail region being directed towards the anterolateral surface of the lower half of the left kidney (*Plate 37k,a–d*). The spleen also continues to increase in volume, but maintains its characteristic ribbon-like form, being closely applied to the posterolateral surface of the stomach and left kidney (*Plate 37j,c,d*; *Plate 37k,a*). The spleen is also seen to be a very vascular organ at this stage, and its vascular tree is seen to contain numerous lymphocytes.

THE NERVOUS SYSTEM

Many components of the nervous system show increased evidence of differentiation. This particularly applies to the eye, where, apart from the fact that eyelid fusion occurred during the latter part of the previous stage of development, other significant changes also occur in its detailed morphology between 15.5–16 and 16.5–17 days p.c. Possibly the most obvious difference relates to the increase in the volume of the conjunctival sac formed (in its most complete state) once the eyelids have fused (*Plate 53,d*). Initially, before this event occurs, the conjunctival sacs are largely confined to the territories immediately beneath the upper and lower eyelids (though they also extend medially and laterally beneath the inner and outer canthus of the eye) (*Plate 52,i*). However, with the differentiation and growth of the eyelids, the volume of the conjunctival sacs increases, so that they eventually consist of the space enclosed between the inner aspect of the fused eyelids anteriorly, and the cornea posteriorly, while still extending superiorly and inferiorly marginally beyond the territory of the bulb of the eye.

It is unclear exactly why the eyelids should close at this time, then reopen at a later stage. The timing of this event varies considerably between species. In man, the eyelids fuse at about 8 weeks p.c., at about the time of the transition from the embryonic to the fetal period, and reopen during about the 7th month of pregnancy. The eyes are also open in newborn calves and guinea-pigs, but closed in newborn dogs, cats and mice, and the state of eyelid closure at birth appears to depend on the stage of development at which the species is born, since eyelid closure is one of the last morphogenetic events of embryogenesis. It is most likely that the conjunctival sac affords protection for the cornea during its differentiation, and provides a microenvironment in which the necessary processes involved can take place in isolation from the influence of the potentially toxic substances present in the amniotic fluid.

The iris also begins to differentiate, and progressively thins out to form the boundary of the pupil. The ciliary body equally shows evidence of differentiation over this period, though the morphological features of this structure are not easily seen until later stages of development. The pigment layer of the retina is seen to extend to the peripheral margin of the iris, and then extends slightly further onto the inner aspect of the iris to make contact with the peripheral margin of the optic cup (the pars iridis retinae). The neural retina also shows increased evidence of stratification, so that it is now seen to consist of an outer nuclear (neuroblastic) layer, which is destined to form the horizontal cells and the nuclei of the photoreceptor cells, a transient intermediate anuclear layer (of Chievitz), and an inner nuclear (neuroblastic) layer which is destined to give rise principally to the ganglion cells (*Plate 53,e*). However, the histological morphology of all of these cells is still at a primitive stage of differentiation, even at the time of birth.

The lens now shows the first evidence of the formation of the embryonic/fetal "nucleus", though this is more clearly seen at later stages of development (*Plate 53,i*). On careful inspection of the medial margin of the iris, it is possible to recognize the vascular capsule of the lens (the pupillary membrane) that extends across the anterior surface of the lens, the central part of which (the part not supported by the pars iridis retinae) breaks down to form the pupil. The blood vessels that supply the dorsal part of the capsule are derived from the hyaloid artery, while those that supply the ventral part are derived from the anterior ciliary arteries. These vessels largely atrophy before the pupillary membrane breaks down.

If the optic (II) nerves are traced back posteriorly through the optic foramina, the nerve fibres are seen to converge on the optic chiasma (*Plate 37b,b*; *Plate 39b,a,b*). Thence the optic pathway progresses posteriorly via the optic tract to the lateral geniculate body, then via the geniculocalcarine tract to the visual part of the occipital cortex.

The histological morphology of the pituitary gland is quite similar to the situation described at the latter part of the previous stage of development (*Plate 55,g,h*; c.f. *Plate 55,e,f*), though at the earlier part of the previous stage, the lumen of the infundibulum is still present (*Plate 55,d*). The volume of the lumen of Rathke's pouch has also diminished slightly, and continues to do so throughout this stage, so that by 17.5 days p.c., the lumen is now only present as a narrow cleft (*Plate 55,i*; *Plate 57,m–q*).

The other glandular derivative of the diencephalon (which forms from the caudal part of its roof plate), namely the pineal gland (epiphysis), also shows increased evidence of differentiation. It is now directed posteriorly and rostrally (*Plate 38c,a*), and has a relatively thin wall at this stage. The proximal part of its lumen is in direct continuity with the pineal (epithalamic) recess of the third ventricle (*Plate 39c,d*). At the entrance to the epithalamic recess, choroid plexus partially obstructs the lumen (*Plate 39c,a–c*). A substantial venous plexus surrounds the superior (rostral) part of the body of the pineal gland (*Plate 39d,a*), and the plexus is in direct communication with the superior sagittal dural venous sinus as well as various anastomotic veins (*Plate 39d,a,b*). Over the period of the next 1–2 days, the lumen of the pineal gland becomes increasingly obliterated by the differentiation of its walls (*Plate 41*).

Despite the fact that all components of the brain are recognizable in transverse sections through the cephalic region at this stage of development, most individuals who work exclusively on the brain are only likely to be familiar with coronal sections through this region. For completeness, all of the major components of the brain seen in the transverse sections are appropriately labelled, but an additional series of intermittent serial coronal sections is provided for those with a specific interest in the differentiation of the fine and gross morphology of the brain and its closely allied structures.

In the coronal sections, working posteriorly, the first of the brain structures to be encountered are the olfactory lobes (bulbs) (*Plate 39a,d,e*). These are located directly above the posterior part of the nasal cavities, and will in due course receive sensory information from the olfactory epithelium via numerous branches of the olfactory (I) nerves which pass upwards through the cartilage primordia of the cribriform plates of the ethmoid bones (*Plate 39a,e*) (for further details, see earlier section). These lobes of the brain contain an extension of the anterior horn of the lateral ventricles. The relationship between the olfactory lobes and the cerebral hemispheres is clearly illustrated in the sagittal sections through this region (e.g. *Plate 38b,b*; *Plate 38c,a*). The two cerebral hemispheres (previously termed telencephalic vesicles) are essentially outgrowths of the primitive forebrain, and are the regions in the fully differentiated brain where the higher cortical centres are localized. The cortex of the brain is in the process of becoming differentiated, and is clearly seen at this stage to be stratified into a number of well-defined layers, one of which is the neopallial cortex which is itself composed of an inner broad nuclear layer (cortical plate), with a narrow outer relatively anuclear layer (marginal zone). Subjacent to the neopallial cortex is the intermediate zone, and this is seen to surround the ventricular zone (e.g. *Plate 38c,a*; *Plate 39b,a,b*).

Within the cerebral hemispheres are located the lateral ventricles, and these are subdivided into anterior and posterior horns, and the former are further divided into superior and inferior parts (*Plate 39b,a–d*; *Plate 39c,a–d*; *Plate 39d,a*). The lateral ventricles are seen to contain choroid plexus which arises from their medial walls (*Plate 39c,a,b*). The lateral ventricles communicate with the third ventricle (of the diencephalon) via the interventricular foramina (of Monro) (*Plate 39b,c,d*), and the third ventricle in turn communicates with the fourth ventricle (*Plate 39d,d*) via the mesencephalic vesicle (the future cerebral aqueduct – of Sylvius) (*Plate 39d,a–c*). The latter relationship is clearly seen in the paramedian sagittal sections through this region (*Plate 38c,a*). The third ventricle is also seen to extend inferiorly to form its infundibular recess, and superiorly to form its pineal recess, while at earlier stages it additionally expanded laterally to form the primitive optic vesicles. The latter then differentiate to form major components of the eye, though the optic nerves are now clearly seen to be solid structures, due to the central migration of nerve fibres originating in the ganglionic cell layer of the neural retina.

Various components of the diencephalon bulge into the third ventricle and distort its lumen, the largest of which are the left and right components of the thalamus. The latter may be directly in contact with each other across the midline at the so-called interthalamic adhesion (massa intermedia) (*Plate 39b,d*). More inferiorly is located the hypothalamus (*Plate 39b,d*), from which arises the neural component of the pituitary (*Plate 38b,b*; *Plate 38c,a*; *Plate 39b,d*). More superiorly is located the epithalamus, from which arises the pineal gland (body) (*Plate 39c,c,d*) (for details of pituitary and pineal morphology, and for detailed structure of the eye, see earlier sections). One particularly clearly defined structure seen at this stage, and in this location, is the anterior commissure (*Plate 39b,b*), which consists of a large bundle of association fibres which cross the midline in the lamina terminalis.

The midbrain (mesencephalon) connects the primitive forebrain (prosencephalon) with the hindbrain. The primitive hindbrain (rhombencephalon) is subdivided into a more rostral metencephalon and a more caudally located myelencephalon. The former gives rise to the pons and cerebellum, while the latter gives rise to the medulla oblongata. Because of the various brain flexures, it is easier to appreciate the relationship between the various components of the primitive hindbrain from an initial analysis of sagittal sections through this region (*Plate 38c,a*). These clearly demonstrate that the floor of the rostral part of the hindbrain is formed by the pons, whereas the roof of this region is formed by the cerebellum. Reference now to the coronal sections through this region reveals that the cerebellum extends posteriorly from the rhombic lips (dorsal part of the alar laminae of the metencephalon) that are located on either side of the lateral recesses of the fourth ventricle (*Plate 39d,c*), and fuse across the dorsal midline to form the cerebellar plate (*Plate 39d,d*). Possibly the most impressive feature of the fourth ventricle is the choroid plexus, which is particularly florid at this stage of development, and arises both from its roof and from the walls of its lateral recesses (*Plate 39d,b,c*). The roof of the fourth ventricle is (initially) characteristically extremely thin, but becomes progressively covered and reduced in size due to the caudal expansion of the cerebellum (*Plate 38c,b*).

While transverse sections through the brain may in some respects be more difficult to interpret, such sections clearly complement the coronal sections discussed above, and indeed can greatly facilitate the understanding of certain regions of the brain. The latter particularly applies to the junctional region between the midbrain and the hindbrain, and to the understanding of the gross morphology of the hindbrain. The caudal part of the ventricular system, for example, is easily followed, with the passage of the

(*continued on page 334*)

PLATE 37a

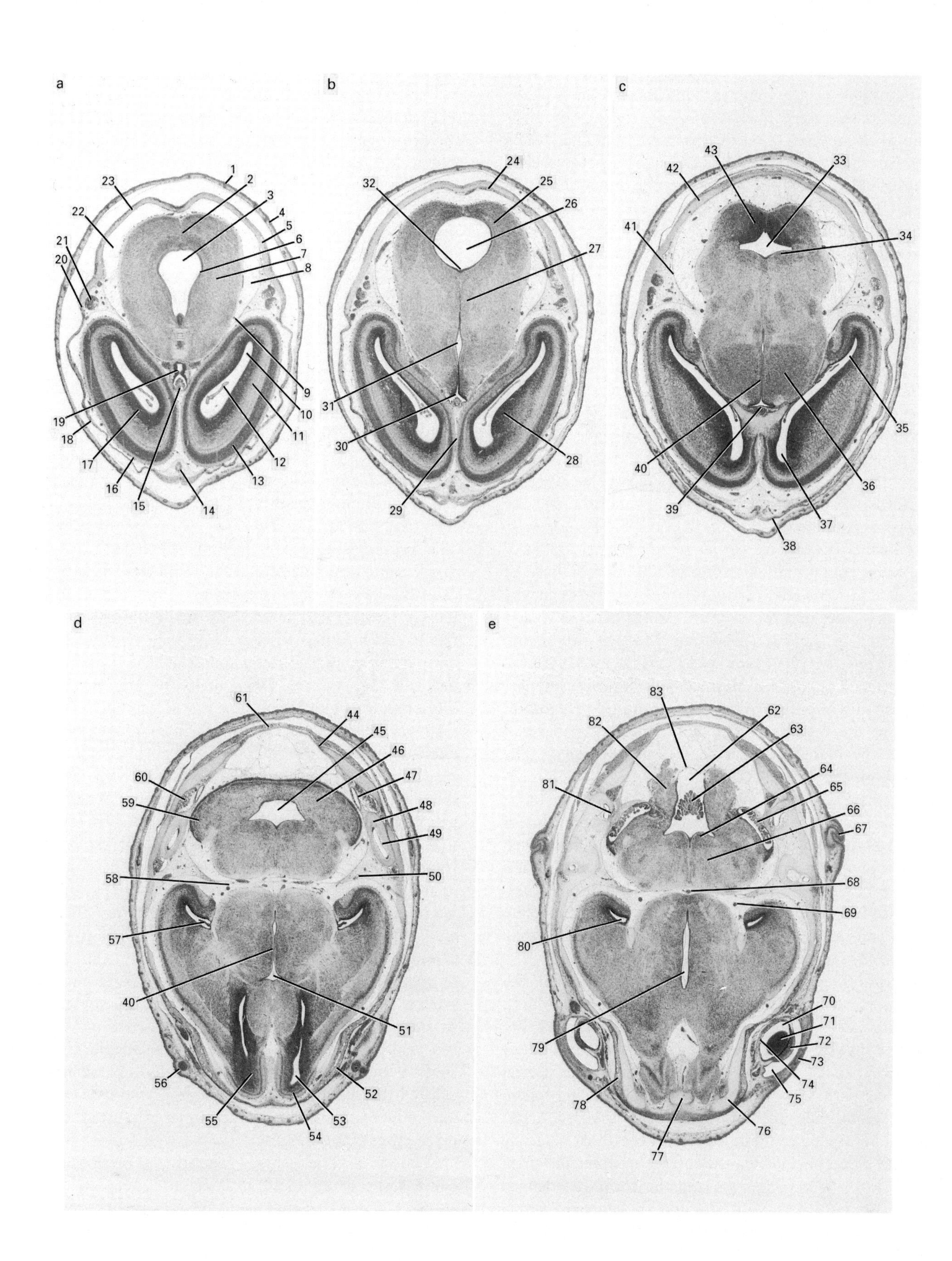

a
b
c
d
e
1
2
3
4
5
6
7
8
9
10
11
12
13
14
15
16
17
18
19
20
21
22
23
24
25
26
27
28
29
30
31
32
33
34
35
36
37
38
39
40
41
42
43
44
45
46
47
48
49
50
51
52
53
54
55
56
57
58
59
60
61
62
63
64
65
66
67
68
69
70
71
72
73
74
75
76
77
78
79
80
81
82
83

Plate 37a (16.5 days p.c.)

16.5 days p.c. Female embryo. Crown–rump length 14.2 mm (fixed). Transverse sections (Theiler Stage 25; rat, Witschi Stage 34).

1. primordium of hair follicle in skin
2. roof of dorsal part of midbrain
3. mesencephalic vesicle
4. skin (and subjacent epicranial aponeurosis (galea aponeurotica))
5. cartilage primordium of squamous part of occipital (supraoccipital) bone
6. ventricular zone of mesencephalic vesicle
7. mantle layer of midbrain
8. loosely packed cephalic mesenchyme in future subarachnoid space
9. vascular elements of pia mater around midbrain
10. telencephalic vesicle (future lateral ventricle)
11. intermediate zone of wall of telencephalon
12. choroid plexus within lateral ventricle and arising from its medial wall
13. neopallial cortex (future cerebral cortex) composed of inner broad nuclear layer (cortical plate) with narrow outer relatively anuclear layer (marginal zone)
14. anterior extension of superior sagittal dural venous sinus
15. outgrowth of choroid plexus from roof of third ventricle which appears to be invaginating into pineal primordium (epiphysis)
16. vascular elements of pia mater around telencephalon
17. ventricular zone of telencephalon
18. early evidence of ossification within mesenchymal (membranous) precursor of parietal bone
19. pineal primordium (epiphysis) which develops as an outgrowth from the diencephalon (epithalamic region)
20. anterior margin of cartilage primordium of occipital (supraoccipital) bone
21. anterior margin of transverse dural venous sinus
22. loose connective tissue subjacent to epicranial aponeurosis (galea aponeurotica)
23. upper margin of cartilage primordium of occipital (supraoccipital) bone
24. dorsal part of transverse dural venous sinus
25. junctional zone between caudal part of midbrain and rostral part of primitive cerebellum
26. caudal part of mesencephalic vesicle
27. floor of midbrain (tegmentum, in region of midbrain flexure)
28. striatum (ganglionic eminence) overlying the head of the caudate nucleus
29. falx cerebri
30. pineal recess of third ventricle
31. dorsal part of third ventricle
32. median sulcus
33. cerebral aqueduct
34. junctional region between midbrain and pons
35. caudal extremity of posterior horn of lateral ventricle
36. diencephalon (thalamus)
37. anterior portion of superior horn of lateral ventricle
38. membrane-covered anterior fontanelle
39. roof of third ventricle
40. interthalamic adhesion (massa intermedia)
41. mesenchymal condensation forming tentorium cerebelli
42. cartilage primordium of occipital (supraoccipital) bone
43. rostral extension of primitive cerebellum
44. posterior margin of exoccipital bone
45. rostral part of fourth ventricle
46. cerebellar plate (region of cerebellum formed by medial growth and eventual apposition of intraventricular portions of cerebellar primordia)
47. rostral extension of endolymphatic sac
48. cartilage primordium of petrous part of temporal bone
49. superior semicircular canal
50. trochlear (IV) nerve
51. rostral extension of third ventricle just caudal to the interventricular foramina
52. ossification in outer table of primordium of frontal bone
53. extension of anterior horn of lateral ventricle into olfactory lobe
54. olfactory cortex
55. olfactory lobe
56. primordium of prominent tactile or sinus hair follicle characteristically found in this location
57. choroid plexus within base of posterior horn of lateral ventricle
58. oculomotor (III) nerve
59. lateral part of cerebellar primordium
60. rostral part of sigmoid dural venous sinus
61. membrane-covered fontanelle (region bounded by membranous primordium of supraoccipital bone above, occipital arch below, and laterally by the cartilage primordia of the two exoccipital bones)
62. ependymal diverticulum of fourth ventricle
63. choroid plexus arising from roof of fourth ventricle
64. tegmentum of pons (basal plate)
65. choroid plexus arising from roof of lateral recess of fourth ventricle
66. pons
67. rostral extremity of pinna of ear
68. basilar artery
69. rostral extremity of posterior communicating artery
70. intra-retinal space
71. neural layer of retina (grazing section)
72. corneal epithelium (grazing section)
73. upper eyelid (fused to lower eyelid)
74. pigment layer of retina
75. conjunctival sac
76. nasal capsule
77. cartilage primordium of nasal septum
78. ossification within outer table of orbital part of frontal/maxillary bone
79. region of floor of third ventricle
80. base of posterior horn of lateral ventricle
81. endolymphatic duct
82. rostral part of medulla oblongata
83. roof of ependymal diverticulum of fourth ventricle

Plate 37b (16.5 days p.c.)

16.5 days p.c. Female embryo. Crown–rump length 14.2 mm (fixed). Transverse sections (Theiler Stage 25; rat, Witschi Stage 34).

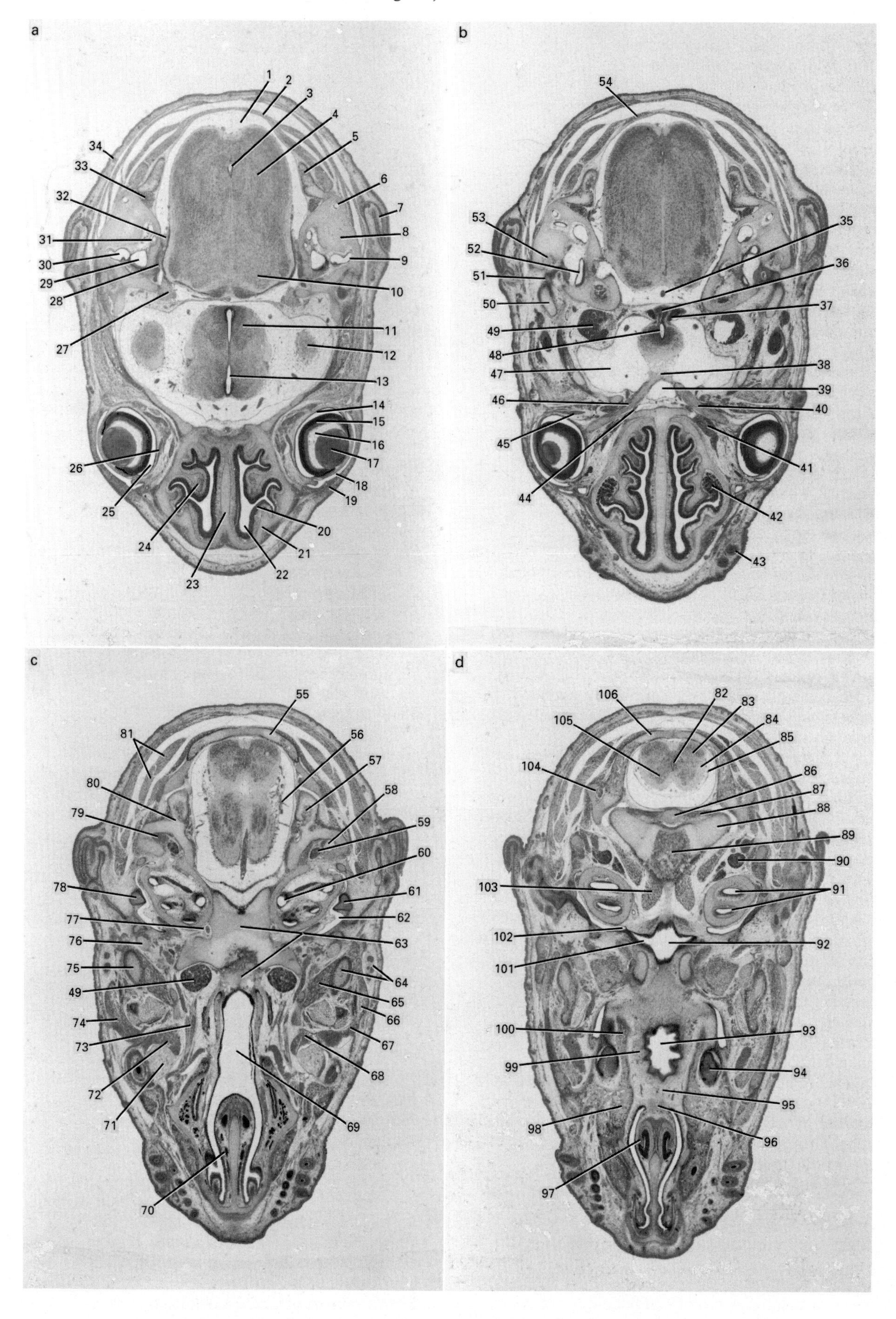

1. loosely-packed cephalic mesenchyme in future subarachnoid space
2. membrane-covered posterior fontanelle
3. caudal extremity of fourth ventricle
4. mantle layer of medulla oblongata
5. ossification within cartilage primordium of exoccipital bone
6. posterior semicircular canal
7. pinna of ear
8. cartilage primordium of petrous part of temporal bone
9. origin of left lateral semicircular canal
10. junctional zone between pons and medulla oblongata
11. floor of diencephalon (hypothalamus)
12. grazing section through wall of telencephalic vesicle
13. region of floor (i.e. ventral part) of third ventricle
14. pigment layer of retina
15. neural layer of retina
16. hyaloid cavity
17. lens
18. conjunctival sac
19. anterior margin of fused eyelids
20. olfactory epithelium
21. nasal capsule
22. nasal cavity
23. cartilage primordium of nasal septum
24. cartilage primordium of turbinate bone
25. intra-retinal space
26. extrinsic ocular muscle (medial/ nasal rectus)
27. rootlets of trigeminal (V) nerve
28. vestibulocochlear (VIII) ganglion
29. utricle (utriculus)
30. ampulla of right lateral semicircular canal
31. proximal part of endolymphatic duct
32. rootlets of vestibulocochlear (VIII) nerve
33. sigmoid dural venous sinus
34. primordium of hair follicle in skin
35. basilar artery
36. infundibulum (future pars nervosa) of pituitary
37. pars tuberalis of pituitary
38. optic chiasma
39. cisterna chiasmatica (of subarachnoid space)
40. optic nerve (II) passing through the optic foramen
41. ophthalmic vein
42. tubules of serous glands that course rostrally in the lateral wall of the middle meatus
43. primordium of follicle of vibrissa
44. right ophthalmic artery
45. extrinsic ocular muscle (lateral/ temporal rectus)
46. optic foramen
47. basal (interpeduncular) cistern (cisterna interpeduncularis, of subarachnoid space)
48. infundibular recess of third ventricle
49. trigeminal (V) ganglion
50. cartilage primordium of malleus
51. cartilage primordium of incus
52. caudal part of utricle just below site of entrance of endolymphatic duct
53. facial (VII) nerve
54. rectus capitis muscle
55. ossification within cartilage primordium of occipital arch (tectum posterior, supraoccipital bone)
56. rootlets of spinal and cranial parts of accessory (XI) nerve
57. ossification within basioccipital bone
58. origin of internal jugular vein
59. glossopharyngeal (IX) ganglion
60. proximal part of cochlear duct
61. cartilage primordium of left stapes
62. left tubo-tympanic recess
63. cartilage primordium of basiocciput and (anteriorly) ossification within cartilage primordium of basisphenoid bone
64. cartilage primordium of ramus of left half of mandible with periosteal ossification and (laterally) the parotid gland
65. medial pterygoid muscle
66. left facial vein
67. left inferior orbital vein where it drains into the left facial vein
68. left inferior orbital vein where it drains into the left cavernous sinus
69. nasopharynx
70. tubules of serous glands associated with nasal septum
71. orbital fat pad
72. right ophthalmic vein
73. ophthalmic division of trigeminal (V) nerve
74. right facial vein
75. cartilage primordium of ramus of right half of mandible with periosteal ossification
76. Meckel's cartilage
77. right internal carotid artery within carotid canal
78. cartilage primordium of right stapes
79. jugular foramen
80. site of fusion between cartilage primordium of basioccipital bone and cartilage primordium of posterior part of temporal bone
81. splenius and rhomboideus capitis muscles
82. central canal of upper cervical region of spinal cord
83. region of dorsal horn (mantle layer) of spinal cord

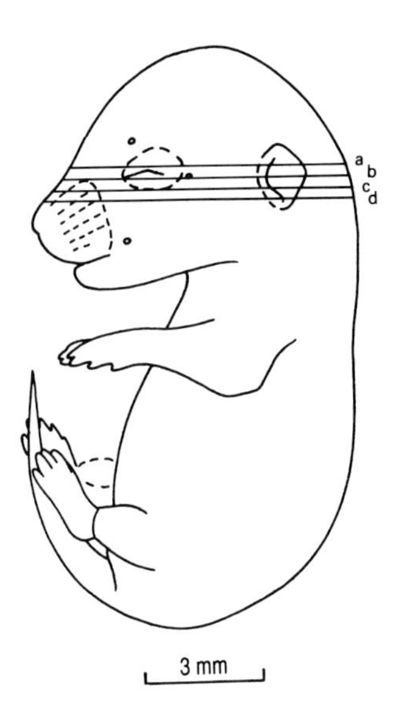

84. white matter (marginal layer)
85. rootlet of spinal part of accessory (XI) nerve
86. cartilage primordium of rostral part of dens (odontoid process of C2 vertebra (axis)) with early evidence of periosteal ossification
87. membrana tectoria (accessory ligament of atlanto-occipital joint)
88. cartilage primordium of lateral part of basioccipital bone
89. ossification within central part of basioccipital bone (clivus)
90. vagal (X) ganglion
91. cochlea
92. oropharynx
93. oral cavity
94. primordium of left upper molar tooth
95. ventral (anterior) part of secondary palate
96. site of fusion between primary palate and inferior border of nasal septum
97. vomeronasal organ (Jacobson's organ)
98. ossification within maxilla
99. palatal shelf of maxilla (secondary palate)
100. ossification within maxilla at lateral boundary of palatal shelf
101. opening of pharyngo-tympanic (Eustachian) tube into oropharynx
102. opening of pharyngo-tympanic (Eustachian) tube into tubo-tympanic recess
103. prevertebral muscles of the neck (longus capitis, longus cervicis)
104. right vertebral artery within foramen transversarium (incomplete at this stage) of C1 vertebra (atlas)
105. ventral horn (mantle layer)
106. ossification within cartilage primordium of neural arch of C1 vertebra (atlas)

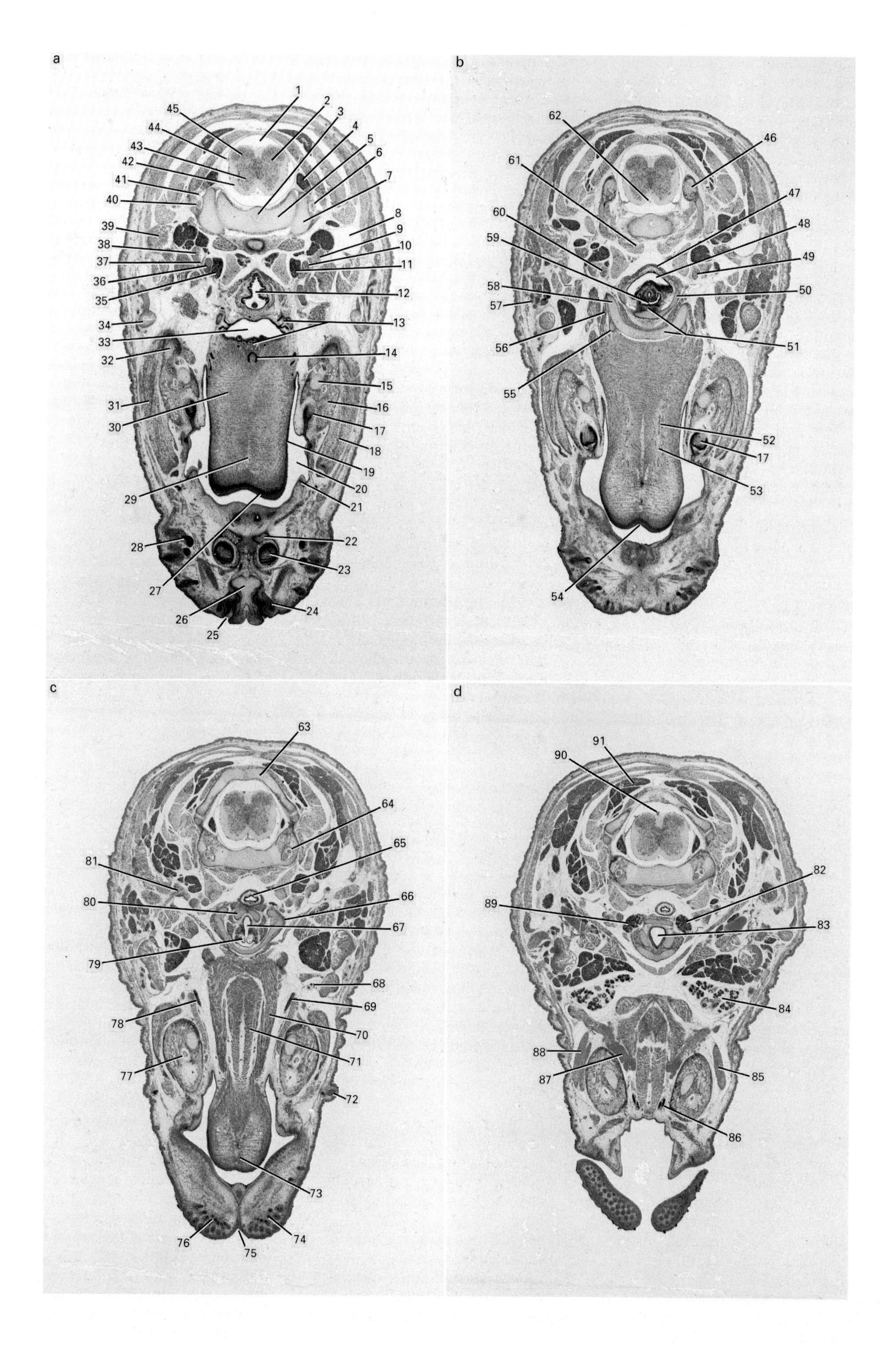
a
b
c
d
1
2
3
4
5
6
7
8
9
10
11
12
13
14
15
16
17
18
19
20
21
22
23
24
25
26
27
28
29
30
31
32
33
34
35
36
37
38
39
40
41
42
43
44
45
46
47
48
49
50
51
52
53
54
55
56
57
58
59
60
61
62
63
64
65
66
67
68
69
70
71
72
73
74
75
76
77
78
79
80
81
82
83
84
85
86
87
88
89
90
91

Plate 37c (16.5 days p.c.)

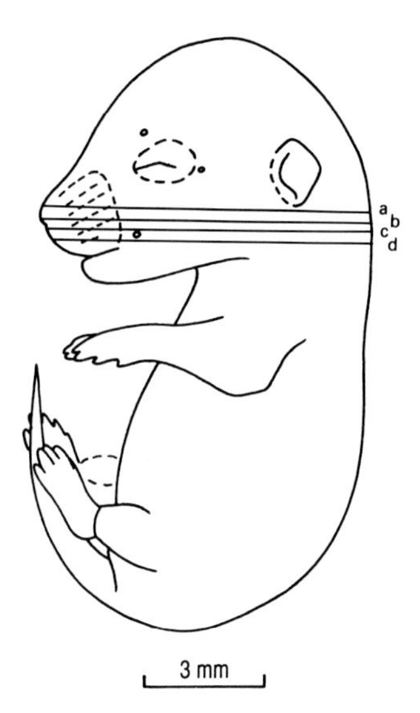

16.5 days p.c. Female embryo. Crown–rump length 14.2 mm (fixed). Transverse sections (Theiler Stage 25; rat, Witschi Stage 34).

1. spinal canal
2. upper cervical region of spinal cord
3. nucleus pulposus
4. dorsal (posterior) root ganglion of C2
5. cartilage primordium of centrum of C2 vertebra (axis)
6. left vertebral artery
7. cartilage primordium of lateral mass of C1 vertebra (atlas)
8. left external jugular vein
9. left internal jugular vein
10. left internal carotid artery
11. left superior cervical sympathetic ganglion
12. common entrance from pharynx into oesophagus and trachea
13. developing glandular ducts opening into pharynx and onto dorsal surface of root of the tongue
14. median circumvallate (vallate) papilla
15. Meckel's cartilage
16. ossification in body of the mandible around but principally lateral to Meckel's cartilage
17. primordium of left lower molar tooth
18. left masseter muscle
19. lateral margin of tongue
20. oral cavity
21. anterior margin of palatal shelf of maxilla
22. ossification within inferomedial part of maxilla
23. primordium of left upper incisor tooth
24. nasal cartilage
25. entrance to external naris
26. most ventral part of cartilage primordium of nasal septum
27. dorsal surface of the tongue
28. follicle of vibrissa
29. precursor of median fibrous septum of tongue
30. intrinsic muscle of the tongue (transverse component)
31. right masseter muscle
32. cartilage primordium of ramus of right half of mandible with periosteal ossification
33. oropharynx
34. right linguofacial vein
35. right superior cervical sympathetic ganglion
36. right internal carotid artery
37. right vagal (X) trunk
38. right internal jugular vein
39. right external jugular vein
40. right vertebral artery
41. right vertebral vein
42. marginal layer of spinal cord
43. mantle layer (ventral horn)
44. origin of dorsal spinal root
45. mantle layer (dorsal horn)
46. ossification within cartilage primordium of pedicle of cervical vertebra
47. inferior constrictor muscle
48. entrance into oesophagus
49. distal part of left common carotid artery
50. lateral part of thyroid cartilage
51. cartilage primordium of body of hyoid bone and (posteriorly) site of attachment of base of epiglottis to inner aspect of thyroid cartilage
52. branch of lingual artery
53. lingual nerve
54. dorsal median sulcus of tongue
55. cartilage primordium of proximal part of inferior horn of hyoid bone
56. origin of lateral part of hyoglossus muscle
57. parotid gland
58. cartilage primordium of greater horn of hyoid bone
59. entrance into larynx
60. distal part of right common carotid artery
61. prevertebral muscles of the neck (longus capitis, longus cervicis)
62. anterior spinal artery
63. ossification within cartilage primordium of neural arch
64. foramen transversarium (completely formed at this level)
65. oesophagus
66. superior tubercle of thyroid cartilage
67. laryngeal aditus
68. upper pole of sublingual gland
69. left sublingual duct
70. extrinsic muscle of the tongue (hyoglossus)
71. extrinsic muscle of the tongue (genioglossus)
72. primordium of prominent hair follicle characteristically found in this location
73. tip of the tongue
74. left upper lip containing numerous follicles of vibrissae
75. philtrum
76. right upper lip
77. inferior alveolar canal containing inferior alveolar vessels and nerve
78. right sublingual duct
79. vocal process of arytenoid cartilage
80. cricoid cartilage
81. mixed (segmental) spinal nerve
82. left lobe of thyroid gland
83. proximal part of trachea
84. left sublingual/submandibular salivary gland
85. left facial vein
86. submandibular duct (medial) and sublingual duct (lateral)
87. deep lingual vein which joins the sublingual vein to enter either the facial, internal jugular or lingual veins
88. right facial vein
89. right lobe of thyroid gland
90. posterior spinal vein
91. deposits of brown (multilocular) fat (found principally between the scapulae, in the ventral part of the neck and in the axillary region)

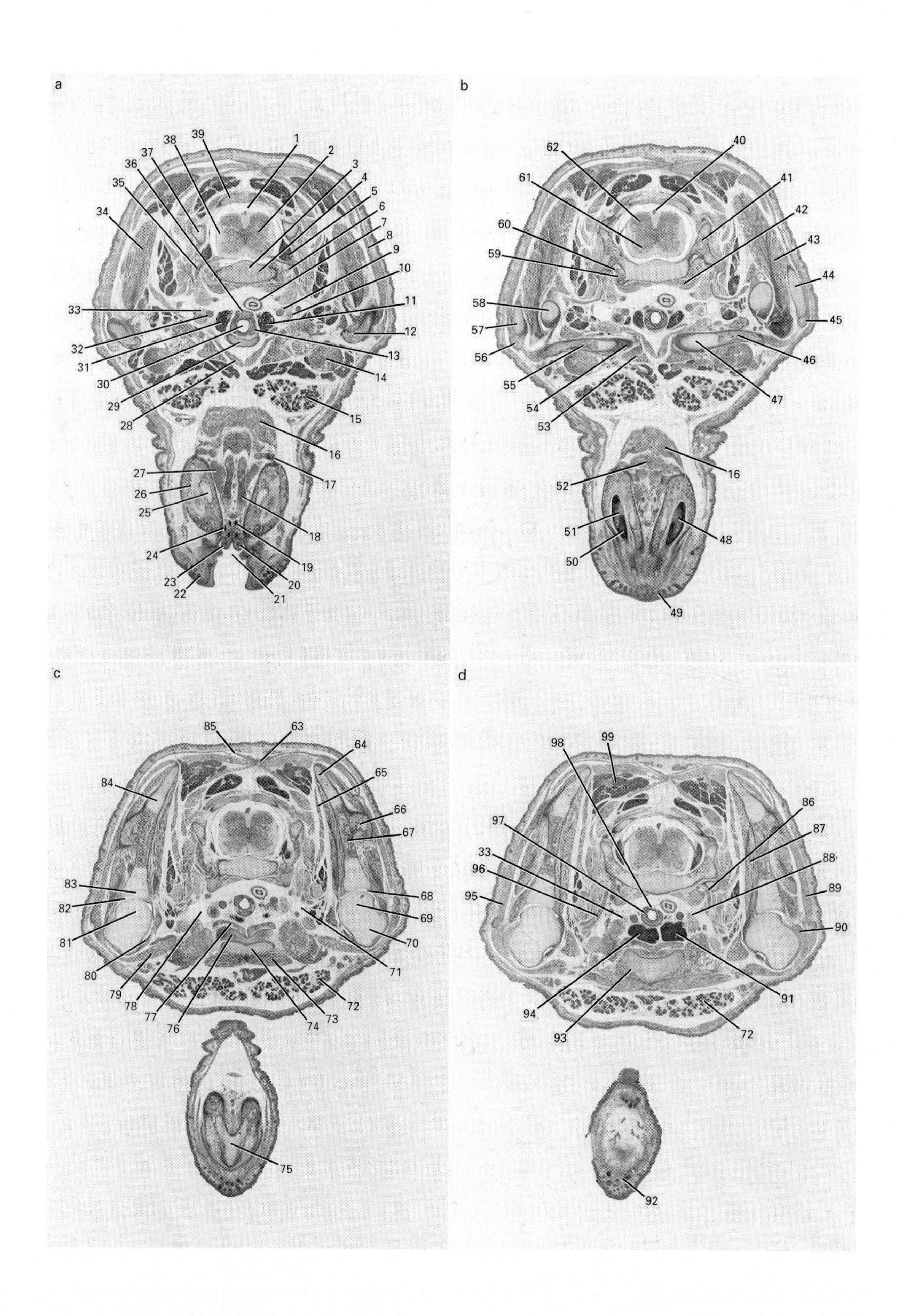
a
b
c
d
1
2
3
4
5
6
7
8
9
10
11
12
13
14
15
16
17
18
19
20
21
22
23
24
25
26
27
28
29
30
31
32
33
34
35
36
37
38
39
40
41
42
43
44
45
46
47
48
49
50
51
52
53
54
55
56
57
58
59
60
61
62
63
64
65
66
67
68
69
70
71
72
73
74
75
76
77
78
79
80
81
82
83
84
85
86
87
88
89
90
91
92
93
94
95
96
97
98
99

Plate 37d (16.5 days p.c.)

16.5 days p.c. Female embryo. Crown–rump length 14.2 mm (fixed). Transverse sections (Theiler Stage 25; rat, Witschi Stage 34).

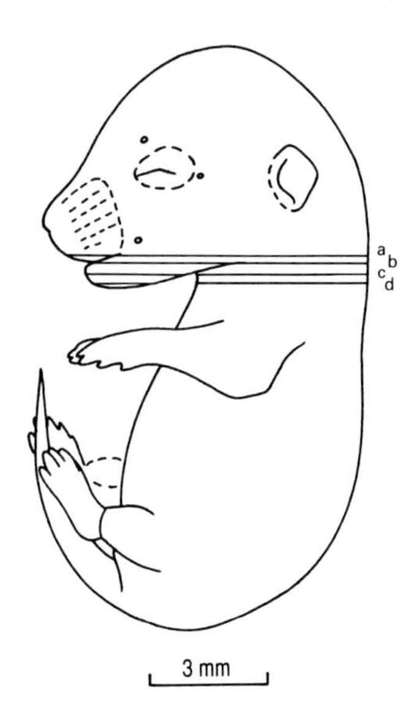

1. spinal canal
2. mid-cervical region of spinal cord
3. nucleus pulposus
4. cartilage primordium of centrum of cervical vertebra
5. dorsal (posterior) root ganglion
6. left vertebral artery
7. oesophagus
8. trapezius muscle
9. left common carotid artery
10. left internal jugular vein
11. left lobe of thyroid gland
12. ossification within lateral one-third of left clavicle
13. first tracheal "ring" (incomplete) of cartilage
14. left external jugular vein
15. sublingual/submandibular salivary gland
16. anterior belly of digastric muscle
17. origin of mylohyoid muscle
18. combined origin of genioglossus and geniohyoid muscles
19. left submandibular duct
20. opening of left submandibular and sublingual ducts into left sublingual caruncle
21. median cleft between right and left halves of lower lip
22. right half of lower lip with large number of hair follicles
23. right sublingual caruncle
24. right sublingual duct
25. Meckel's cartilage
26. ossification in body of the mandible around but principally lateral to Meckel's cartilage
27. right deep lingual vein
28. sternohyoid muscle
29. anterior border of cricoid cartilage
30. lumen of proximal (rostral) part of trachea
31. right lobe of thyroid gland
32. right common carotid artery
33. right vagal (X) trunk
34. right supraspinatus muscle
35. mixed (segmental) spinal nerve
36. posterior border of cricoid cartilage
37. cartilage primordium of pedicle of cervical vertebra
38. marginal zone of spinal cord
39. ossification within cartilage primordium of neural arch
40. posterior spinal vein
41. ossification within cartilage primordium of pedicle of cervical vertebra
42. prevertebral muscles of the neck (longus capitis, longus cervicis)
43. left supraspinatus muscle
44. ossification within cartilage primordium of anterior part of spine of left scapula
45. left cephalic vein
46. ossification within middle one-third of left clavicle
47. cartilage primordium of medial one-third of left clavicle with periosteal collar of bone
48. primordium of left lower incisor tooth
49. heavy concentration of hair follicles in lower lip (left side)
50. enamel epithelium of right lower incisor tooth primordium
51. dental papilla (pulp) of right lower incisor tooth primordium
52. mylohyoid muscle
53. lower part of belly of right sternomastoid muscle
54. right sterno-clavicular joint
55. ossification within middle one-third of right clavicle
56. right cephalic vein
57. ossification within cartilage primordium of anterior part of spine of right scapula
58. cartilage primordium of proximal part of acromion of right scapula with periosteal collar of bone
59. right vertebral vein
60. right vertebral artery
61. mantle layer (ventral horn)
62. mantle layer (dorsal horn)
63. ligamentum nuchae
64. cartilage primordium of posterior (medial) border of left scapula
65. left serratus anterior muscle
66. ossification at base of the spine of the left scapula
67. ossification within cartilage primordium of blade of left scapula
68. left gleno-humeral (shoulder) joint
69. cartilage primordium of head of the left humerus
70. cartilage primordium of proximal part of shaft of the left humerus
71. left jugular lymph sac
72. submandibular gland
73. sternal fibres of origin of left pectoralis superficialis muscle
74. cartilage primordium of rostral part of manubrium sterni
75. site of fusion in the ventral midline of the left and right Meckel's cartilages
76. sternal origin of right sternomastoid muscle
77. sternal origin of right pretracheal "strap" muscles (sternohyoid, sternothyroid)
78. right jugular lymph sac
79. right pectoralis profundus muscle
80. tendon of the long head of the right biceps brachii muscle
81. cartilage primordium of head of the right humerus
82. right gleno-humeral (shoulder) joint
83. glenoid region of cartilage primordium of right scapula
84. ossification within cartilage primordium of blade of right scapula
85. medial attachment of right trapezius muscle
86. ventral primary ramus of lower cervical nerve progressing to form a component of the left brachial plexus
87. left subscapularis muscle
88. left vagal (X) trunk
89. left deltoid muscle
90. upper fibres of insertion of left triceps brachii muscle
91. left lobe of thymus
92. tip of lower jaw
93. site of articulation between cartilage primordium of right first rib and manubrium sterni
94. right lobe of thymus
95. right deltoid muscle
96. ventral primary rami of lower cervical nerves progressing to form components of the right brachial plexus
97. incomplete "ring" of tracheal cartilage occupying anterior two-thirds of circumference of the trachea
98. posterior one-third of trachea completed by fibrous and elastic tissue, and smooth muscle (trachealis muscle)
99. deposits of brown (multilocular) fat

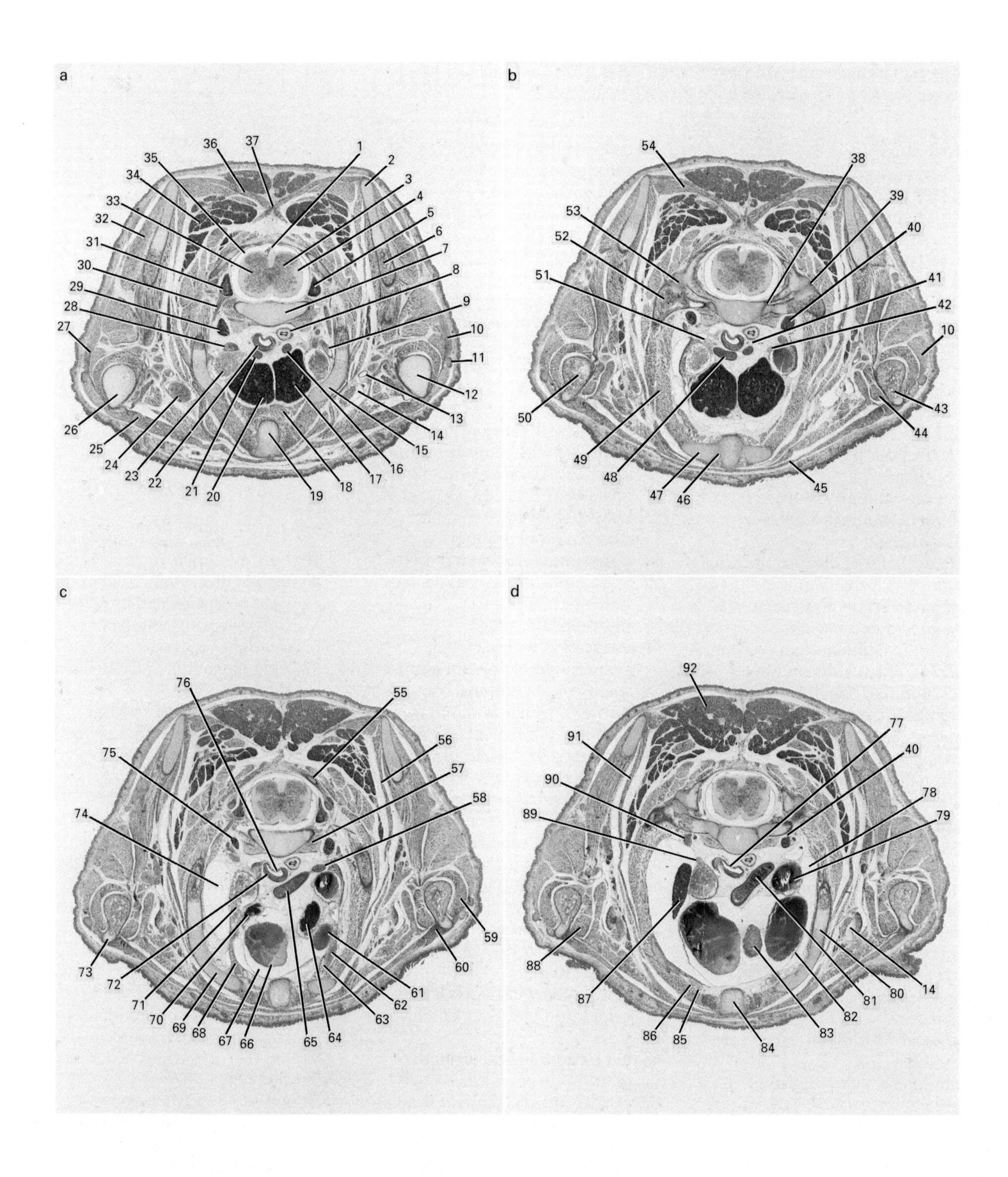
a
b
c
d

Plate 37e (16.5 days p.c.)

16.5 days p.c. Female embryo. Crown–rump length 14.2 mm (fixed). Transverse sections (Theiler Stage 25; rat, Witschi Stage 34).

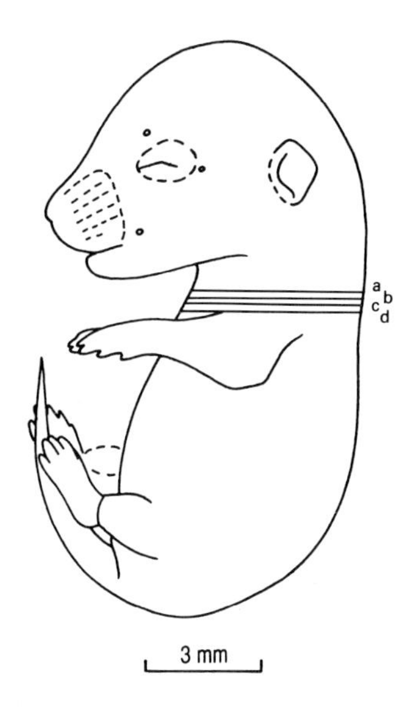

1. posterior spinal vein
2. cartilage primordium of posterior (medial) border of left scapula
3. mantle layer
4. marginal layer
5. left dorsal (posterior) root ganglion of T1
6. ossification within cartilage primordium of blade of left scapula
7. cartilage primordium of T1 vertebral body
8. oesophagus
9. left subclavian artery
10. left deltoid muscle
11. left cephalic vein
12. cartilage primordium of upper shaft region of left humerus with periosteal collar of bone
13. proximal part of axillary artery
14. left axillary vein
15. ossification within cartilage primordium of left first rib
16. left common carotid artery
17. left lobe of thymus
18. caudal attachment of sternomastoid and sternohyoid muscles
19. cartilage primordium of first sternebra
20. right lobe of thymus
21. right common carotid artery
22. trachea
23. right superior (cranial) vena cava
24. right axillary vein
25. pectoralis profundus muscle
26. cartilage primordium of upper shaft region of right humerus with periosteal collar of bone
27. right deltoid muscle
28. right subclavian artery
29. right middle cervical sympathetic ganglion
30. ventral primary ramus of T1 progressing to form a component of the right brachial plexus
31. right dorsal (posterior) root ganglion of T1
32. infraspinatus muscle
33. spinal cord in upper thoracic region
34. longissimus thoracis and cervicis, and iliocostalis thoracis and cervicis muscles
35. spinal canal
36. cervical/thoracic components of trapezius muscle
37. ligamentum nuchae
38. articulation between cartilage primordium of head of the second rib and the body of the second thoracic vertebra
39. articulation between cartilage primordium of tubercle of the second rib and transverse process/neural arch of the same vertebra
40. left cervicothoracic (stellate) sympathetic ganglion
41. left recurrent laryngeal nerve
42. left vagal (X) trunk
43. ossification within cartilage primordium of upper border of left deltoid tuberosity
44. tendon of the long head of the biceps brachii muscle
45. pectoralis superficialis muscle (becomes continuous with pectoralis profundus muscle)
46. second costo-sternal joint
47. primordium of second costal cartilage
48. brachiocephalic (innominate) artery which gives rise to the right subclavian artery and right common carotid artery
49. intercostal muscles (external and internal layers)
50. ossification within mid-shaft region of right humerus
51. right vagal (X) trunk at level of origin of the right recurrent laryngeal nerve
52. ossification within cartilage primordium of neck of second rib
53. cartilage primordium of transverse process of T2 vertebra
54. semispinalis dorsi and cervicis muscles
55. ossification within cartilage primordium of neural arch
56. left serratus anterior muscle
57. thoracic part of longus colli muscle
58. origin of the left subclavian artery from the arch of the aorta
59. triceps brachii muscle
60. common tendon of insertion of pectoral muscles (pectoralis profundus and superficialis)
61. apical region of wall of left atrium
62. left internal thoracic (mammary) vein
63. left internal thoracic (mammary) artery
64. lower pole of left lobe of the thymus
65. convex surface of arch of the aorta
66. wall of right atrium
67. pericardial cavity
68. right pleuro-pericardial membrane
69. ossification within cartilage primordium of right second rib
70. lower pole of right lobe of the thymus
71. right internal thoracic (mammary) vein where it joins the right superior (cranial) vena cava
72. cartilage ring of trachea
73. ossification within cartilage primordium of right deltoid tuberosity
74. right pleural cavity
75. right sympathetic trunk
76. trachealis muscle
77. level of the bifurcation of the trachea
78. left pleural cavity
79. left superior (cranial) vena cava
80. ossification within cartilage primordium of left second rib
81. lumen of arch of the aorta
82. left pleuro-pericardial membrane
83. wall of outflow tract of right ventricle in region of origin of the pulmonary trunk
84. cartilage primordium of second sternebra
85. right internal thoracic (mammary) artery
86. right internal thoracic (mammary) vein
87. cranial lobe of right lung
88. biceps brachii muscle
89. azygos vein where it joins the right superior (cranial) vena cava
90. azygos vein
91. right serratus anterior muscle
92. deposits of brown (multilocular) fat

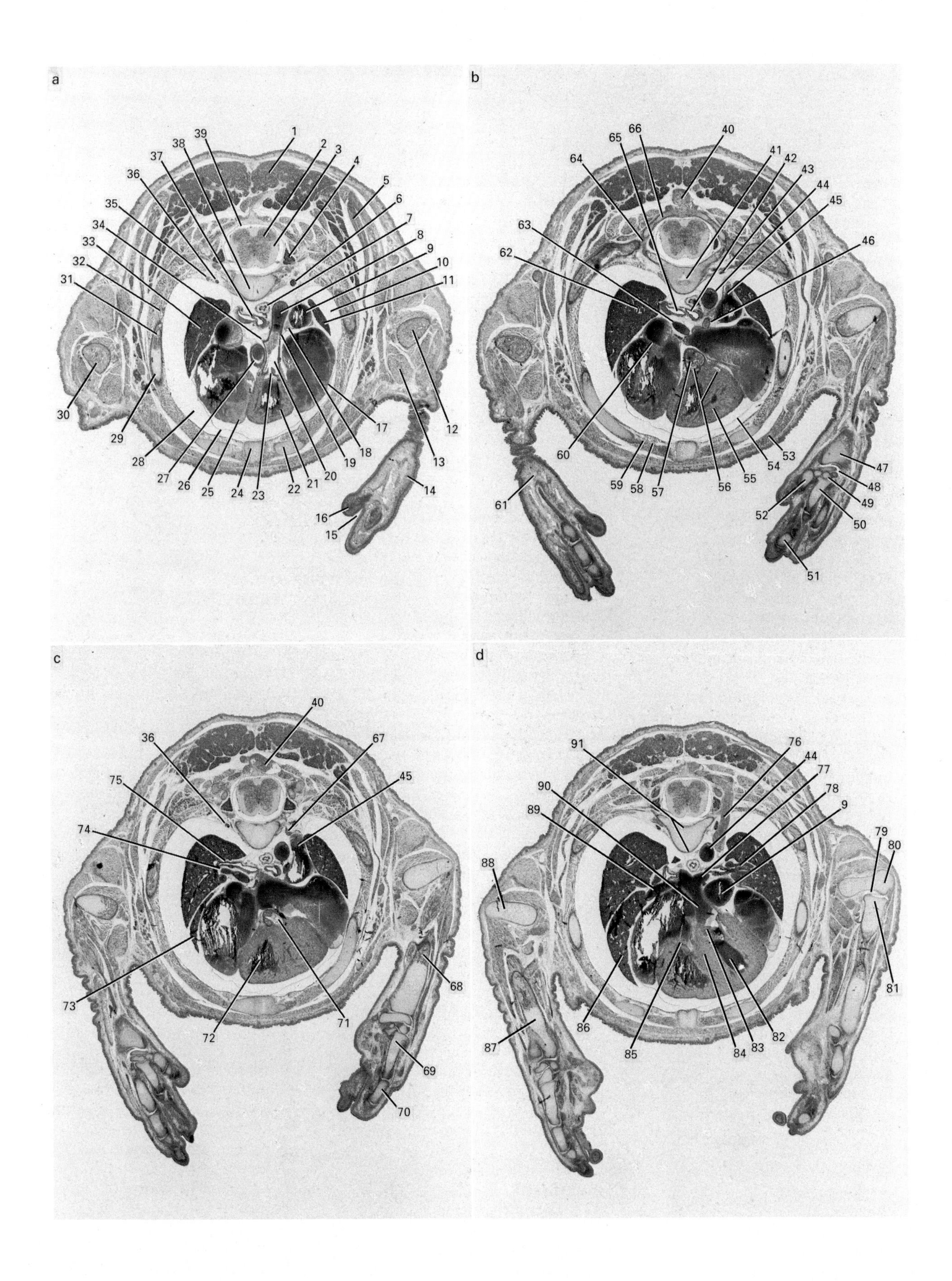
a
b
c
d
1
2
3
4
5
6
7
8
9
10
11
12
13
14
15
16
17
18
19
20
21
22
23
24
25
26
27
28
29
30
31
32
33
34
35
36
37
38
39
40
41
42
43
44
45
46
47
48
49
50
51
52
53
54
55
56
57
58
59
60
61
62
63
64
65
66
67
68
69
70
71
72
73
74
75
76
77
78
79
80
81
82
83
84
85
86
87
88
89
90
91

Plate 37f (16.5 days p.c.)

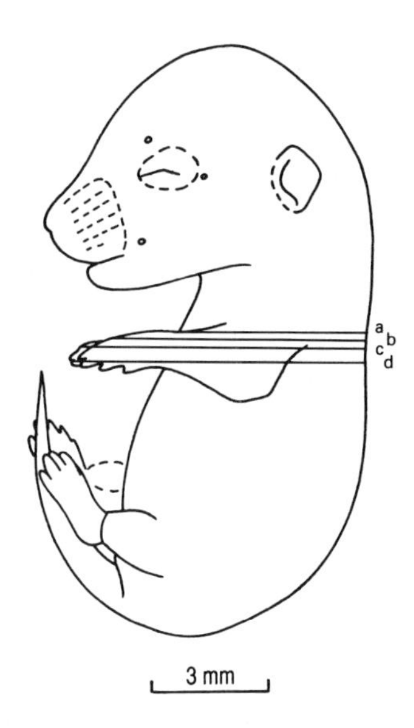

16.5 days p.c. Female embryo. Crown–rump length 14.2 mm (fixed). Transverse sections (Theiler Stage 25; rat, Witschi Stage 34).

1. deposits of brown (multilocular) fat
2. mantle layer
3. marginal layer
4. dorsal (posterior) root ganglion
5. serratus anterior muscle
6. third thoracic sympathetic ganglion
7. oesophagus
8. arch of the aorta at site of entrance of the ductus arteriosus
9. left superior (cranial) vena cava
10. left pleural cavity
11. apex of left lung
12. ossification within cartilage primordium of lower shaft region of left humerus
13. biceps brachii muscle
14. left forelimb
15. digital interzone
16. first digit
17. wall of left atrium
18. left vagal (X) trunk
19. wall of the ductus arteriosus
20. origin of the pulmonary trunk
21. wall of outflow tract of right ventricle
22. primordium of third costal cartilage
23. leaflet of pulmonary valve
24. cartilage primordium of third sternebra
25. wall of right atrium
26. right pleuro-pericardial membrane
27. proximal part of thoracic aorta
28. right pleural cavity
29. cartilage primordium of third rib with periosteal collar of bone
30. ossification within cartilage primordium of lower shaft region of right humerus
31. ossification within cartilage primordium of right third rib
32. intercostal muscles (external and internal layers)
33. cranial lobe of right lung
34. origin of the right pulmonary artery
35. azygos vein
36. right sympathetic trunk
37. bifurcation of the trachea
38. cartilage primordium of T3 vertebral body
39. spinal canal
40. part of extra-vertebral venous plexus
41. nucleus pulposus
42. left main bronchus
43. left sympathetic trunk
44. arch of the aorta just distal to site of entrance of the ductus arteriosus
45. left accessory hemiazygos vein where it joins the left superior (cranial) vena cava
46. left pulmonary artery
47. cartilage primordium of distal part of shaft of left radius
48. left radio-carpal (wrist) joint
49. cartilage primordium of carpal bone
50. cartilage primordium of second metacarpal bone
51. cartilage primordium of phalangeal bone
52. cartilage primordium of first metacarpal bone
53. pectoralis profundus muscle
54. left pleuro-pericardial membrane
55. wall of ventricle overlying the muscular part of the interventricular septum
56. leaflet of aortic valve and left coronary artery which arises from the left posterior aortic sinus
57. origin of the thoracic aorta
58. right internal thoracic (mammary) artery
59. right internal thoracic (mammary) vein
60. venous valve at site of entrancc of right superior (cranial) vena cava into right atrium
61. right forelimb
62. lobar bronchus within cranial lobe of right lung
63. right pulmonary artery
64. articulation between cartilage primordium of tubercle of third rib and neural arch of its own vertebra
65. left main bronchus
66. ossification within cartilage primordium of neural arch of T3 vertebra
67. small venous communication between hemiazygos vein and extra-vertebral venous plexus
68. ossification within cartilage primordium of mid-shaft region of left radius
69. cartilage primordium of left third metacarpal bone
70. cartilage primordium of proximal phalangeal bone
71. outlet of left ventricle just proximal to the aortic valve
72. lumen of right ventricle
73. apex of middle lobe of right lung
74. lobar branch of right pulmonary artery where it enters the cranial lobe of the right lung
75. lobar bronchus where it enters the cranial lobe of the right lung
76. large vascular communication between extra-vertebral venous plexus and inferior hemiazygos vein
77. left pulmonary vein
78. left main bronchus and left pulmonary artery where they enter the hilus (root) of the left lung
79. radio-humeral joint
80. cartilage primordium of capitulum of the left humerus
81. cartilage primordium of head of the left radius
82. lumen of left ventricle
83. leaflet of left atrio-ventricular (mitral) valve
84. muscular component of interventricular septum
85. leaflet of right atrio-ventricular (tricuspid) valve
86. middle lobe of right lung
87. ossification within cartilage primordium of mid-shaft region of right radius
88. cartilage primordium of distal part of the shaft of the right humerus
89. right pulmonary vein
90. right lateral margin of site of entrance of the pulmonary venous blood into the left atrium
91. thin rim of periosteal ossification around mid-thoracic vertebral body

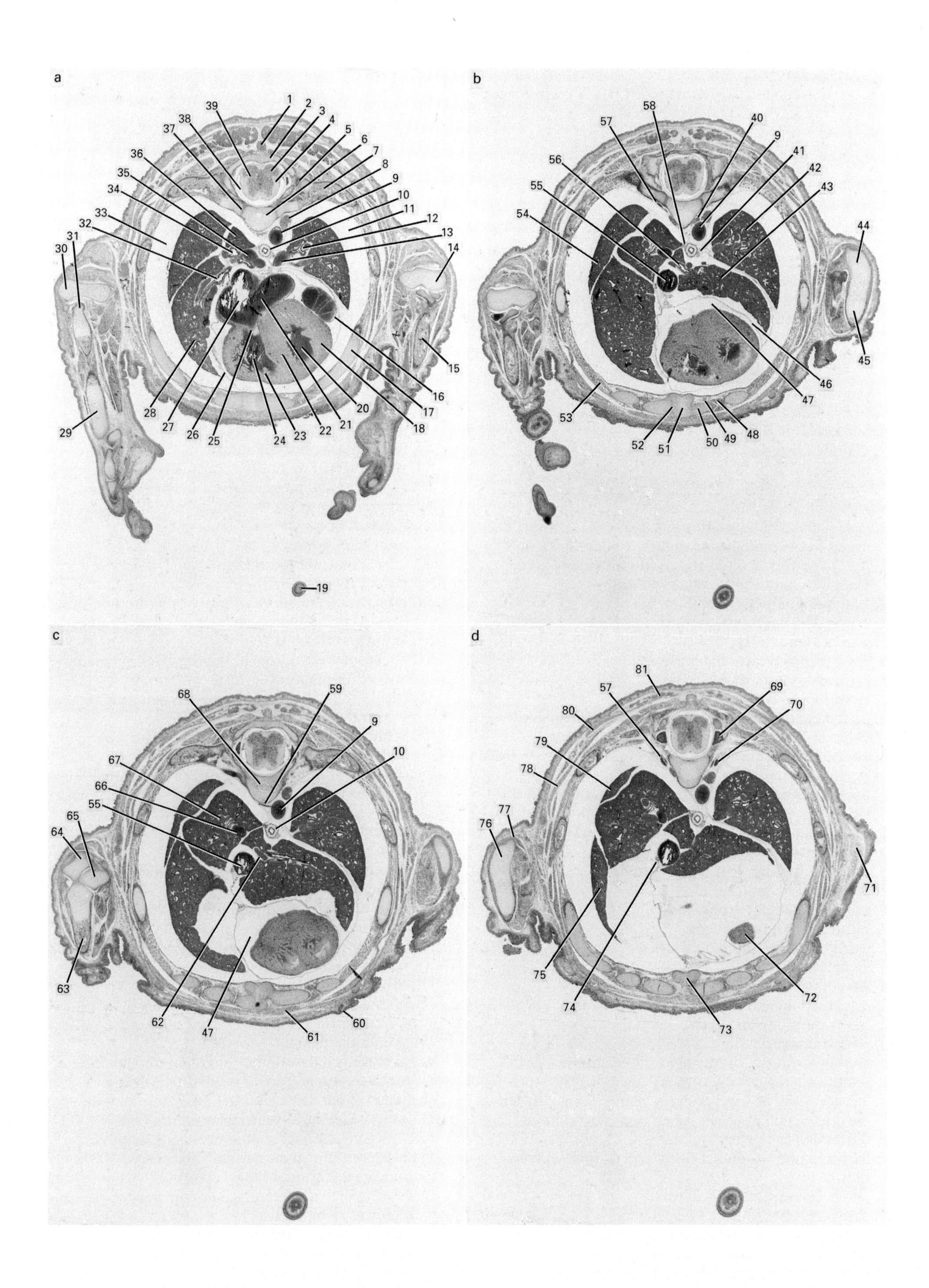
a
b
c
d

Plate 37g (16.5 days p.c.)

16.5 days p.c. Female embryo. Crown–rump length 14.2 mm (fixed). Transverse sections (Theiler Stage 25; rat, Witschi Stage 34).

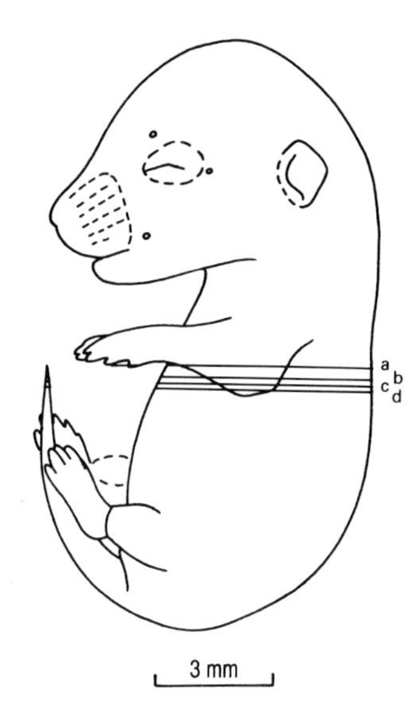

1. part of extra-vertebral venous plexus
2. cartilage primordium of neural arch of T4 vertebra
3. mantle layer of dorsal horn of grey matter
4. marginal layer
5. articulation between cartilage primordium of tubercle of fourth rib and neural arch of its own vertebra
6. cartilage primordium of T4 vertebral body with narrow zone of periosteal ossification
7. ossification within cartilage primordium of fourth rib
8. hemiazygos vein
9. thoracic aorta
10. oesophagus
11. left pleural cavity
12. left main bronchus where it enters the hilus (root) of the left lung
13. left pulmonary vein
14. cartilage primordium of distal part of shaft of left humerus
15. ossification within cartilage primordium of proximal part of shaft of left ulna
16. wall of left atrium
17. ossification within cartilage primordium of shaft of rib
18. wall of right ventricle
19. tip of the tail
20. lumen of the left ventricle
21. left superior (cranial) vena cava at its site of entrance into the right atrium
22. muscular part of interventricular septum
23. wall of right ventricle
24. lumen of right ventricle
25. leaflet of right atrio-ventricular (tricuspid) valve
26. right pleuro-pericardial membrane
27. lumen of right atrium
28. middle lobe of right lung
29. cartilage primordium of distal part of shaft of right ulna
30. cartilage primordium of distal part of the shaft of the right humerus
31. cartilage primordium of the head of the right radius
32. lobar bronchus where it enters the middle lobe of the right lung
33. right pleural cavity
34. right main bronchus
35. right pulmonary vein
36. caudal lobe of the right lung
37. articulation between cartilage primordium of head of the fourth rib and body of the fourth thoracic vertebra
38. spinal canal
39. spinal cord in mid-thoracic region
40. vascular communication between azygos and inferior hemiazygos veins which passes across the ventral surface of the vertebral body
41. left vagal (X) trunk
42. left lung
43. accessory lobe of the right lung
44. cartilage primordium of proximal part of left ulna
45. ossification within cartilage primordium of mid-shaft region of left ulna
46. left pleuro-pericardial membrane
47. pericardial cavity
48. internal thoracic (mammary) vein
49. internal thoracic (mammary) artery
50. primordium of costal cartilage
51. cartilage primordium of sternebra
52. costo-sternal joint
53. intercostal muscles (external and internal layers)
54. cranial lobe of right lung
55. inferior vena cava
56. lobar bronchus where it enters the accessory lobe of the right lung
57. right sympathetic trunk
58. right vagal (X) trunk
59. segmental arterial branch from descending thoracic aorta
60. mammary gland primordium
61. upper fibres of left rectus abdominis muscle
62. branch of right pulmonary artery to the accessory lobe of the right lung
63. ossification within cartilage primordium of mid-shaft region of the right ulna
64. cartilage primordium of olecranon process of right ulna
65. cartilage primordium of the trochlea of the right humerus
66. branch of right pulmonary vein draining the caudal lobe of the right lung
67. branch of right pulmonary artery to the caudal lobe of the right lung
68. nucleus pulposus
69. dorsal (posterior) root ganglion
70. left sympathetic trunk
71. caudal border of left forelimb
72. caudal surface of wall of left ventricle
73. cartilage primordium of xiphoid process
74. right phrenic nerve
75. caudal border of middle lobe of the right lung
76. cartilage primordium of proximal part of shaft of right ulna
77. insertion of the triceps brachii muscle into the olecranon process of the right ulna
78. serratus dorsalis muscle
79. caudal border of the cranial lobe of the right lung
80. latissimus dorsi muscle
81. deposits of brown (multilocular) fat

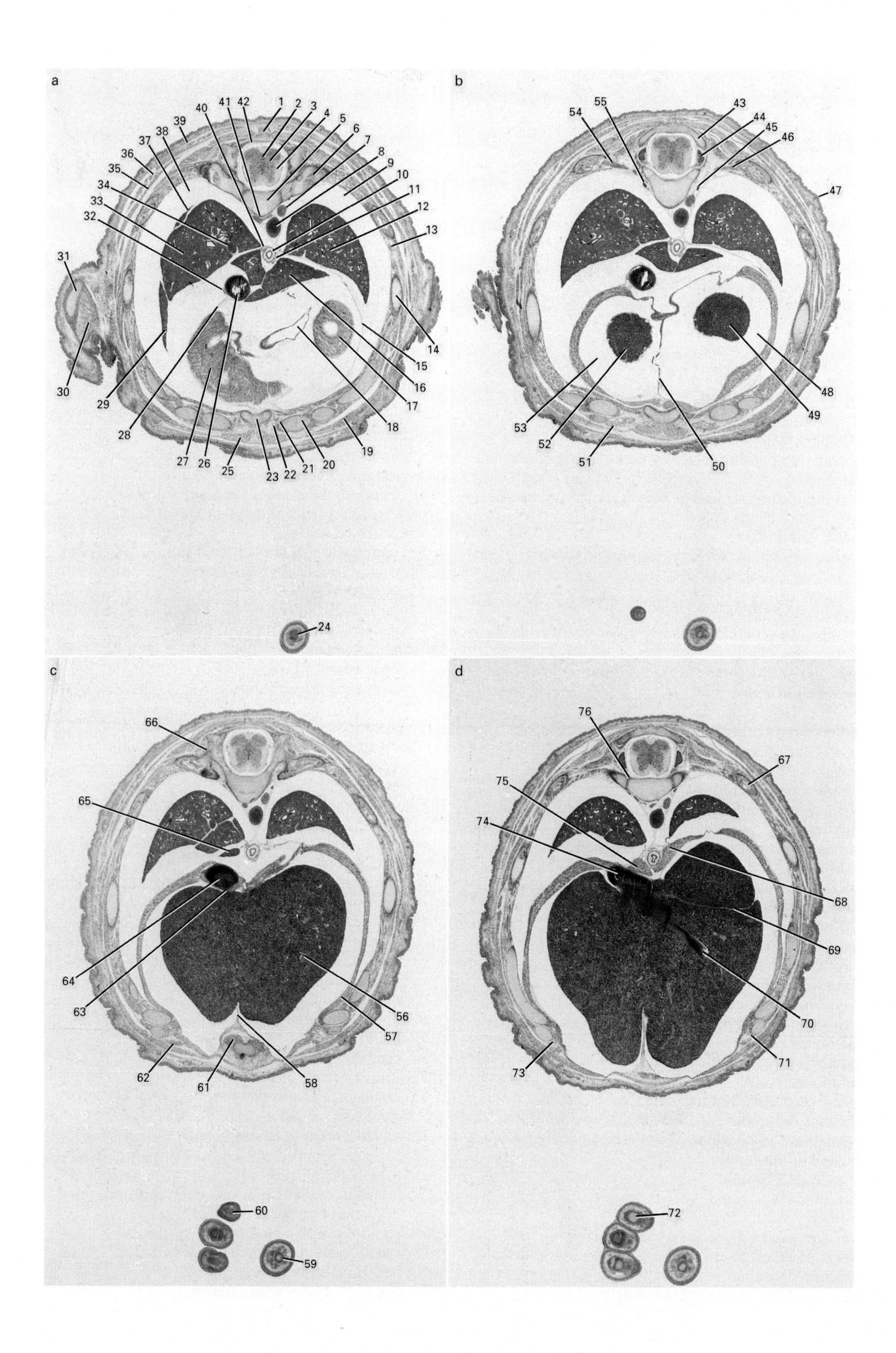
a
b
c
d
1
2
3
4
5
6
7
8
9
10
11
12
13
14
15
16
17
18
19
20
21
22
23
24
25
26
27
28
29
30
31
32
33
34
35
36
37
38
39
40
41
42
43
44
45
46
47
48
49
50
51
52
53
54
55
56
57
58
59
60
61
62
63
64
65
66
67
68
69
70
71
72
73
74
75
76

Plate 37h (16.5 days p.c.)

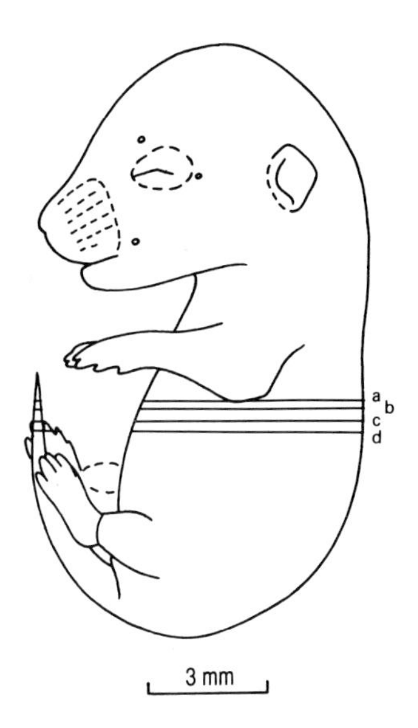

16.5 days p.c. Female embryo. Crown–rump length 14.2 mm (fixed).
Transverse sections (Theiler Stage 25; rat, Witschi Stage 34).

1. part of extra-vertebral venous plexus
2. spinal cord in mid-thoracic region
3. mantle layer of dorsal horn of grey matter
4. marginal layer
5. cartilage primordium of vertebral body with narrow zone of periosteal ossification
6. ossification within cartilage primordium of neck region of rib
7. hemiazygos vein
8. descending part of thoracic aorta
9. left pleural cavity
10. oesophagus
11. left (becoming anterior) vagal (X) trunk
12. left lung
13. intercostal muscles (external and internal layers)
14. cartilage primordium of mid-shaft region of rib with periosteal collar of bone
15. left pleuro-pericardial membrane
16. accessory lobe of right lung
17. left dome of the diaphragm
18. peritoneal cavity
19. external and internal oblique abdominal muscles
20. primordium of costal cartilage
21. internal thoracic (mammary) vein
22. internal thoracic (mammary) artery
23. cartilage primordium of xiphoid process with thin rim of periosteal ossification
24. caudal region of the tail
25. rectus abdominis muscle
26. inferior vena cava
27. right dome of the diaphragm
28. right pleuro-pericardial membrane
29. caudal border of middle lobe of the right lung
30. caudal border of right forelimb
31. cartilage primordium of proximal part of shaft of right ulna
32. right phrenic nerve
33. caudal lobe of the right lung
34. branch of right pulmonary vein draining the caudal lobe of the right lung
35. serratus dorsalis muscle
36. latissimus dorsi muscle
37. caudal border of the cranial lobe of the right lung
38. right pleural cavity
39. cutaneous muscle of trunk (panniculus carnosus)
40. right (becoming posterior) vagal (X) trunk
41. segmental thoracic vein draining to the inferior hemiazygos vein
42. spinal canal
43. ossification within cartilage primordium of neural arch
44. dorsal (posterior) root ganglion
45. left sympathetic trunk
46. an intercostal vein which drains to the inferior hemiazygos vein
47. primordium of hair follicle
48. left sub-phrenic recess of peritoneal cavity
49. apex of left lobe of the liver
50. rostral extension of falciform ligament
51. right musculo-phrenic vein
52. apex of right lobe of the liver
53. right sub-phrenic recess of peritoneal cavity
54. a right intercostal vein
55. right sympathetic trunk
56. hepatic venous sinusoid
57. costal origin of the diaphragm
58. falciform ligament
59. cartilage primordium of tail vertebral body
60. second digit from right hindlimb
61. xiphisternal origin of diaphragm
62. musculo-phrenic artery
63. right boundary of the bare area of the liver
64. origin of posthepatic component of inferior vena cava
65. caudal border of accessory lobe of the right lung
66. joint between tubercle of rib and neural arch of its own vertebra
67. ossification within cartilage primordium of mid-shaft region of rib
68. left crus of the diaphragm
69. lobar interzone
70. component of hepatic venous plexus
71. left costal margin
72. cartilage primordium of phalangeal bone
73. right costal margin
74. inferior vena cava formed principally from ductus venosus, left umbilical vein and right vitelline venous system
75. right crus of the diaphragm
76. joint between cartilage primordium of head of rib and body of its own vertebra

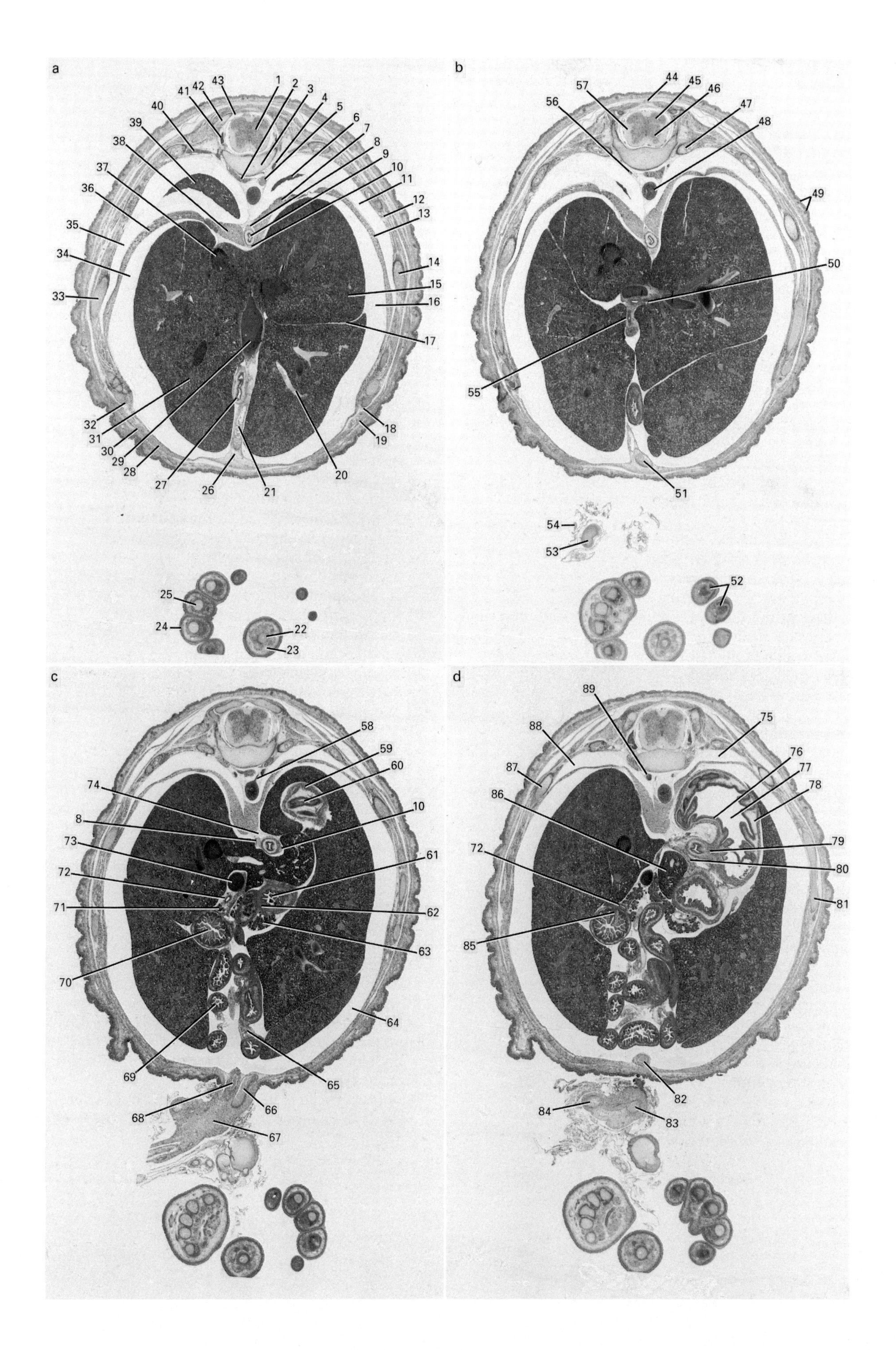
a
b
c
d

Plate 37i (16.5 days p.c.)

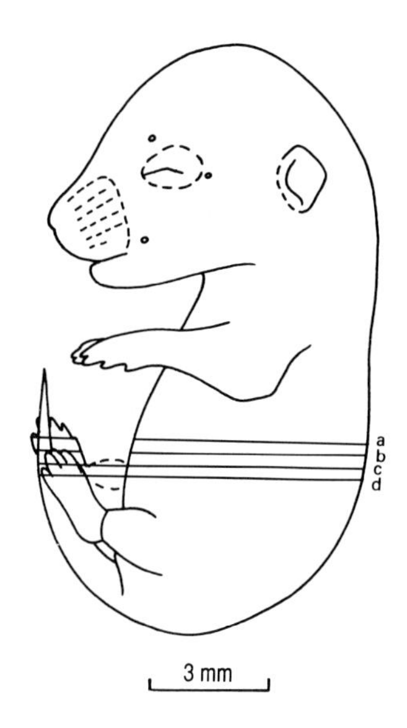

16.5 days p.c. Female embryo. Crown–rump length 14.2 mm (fixed).
Transverse sections (Theiler Stage 25; rat, Witschi Stage 34).

1. spinal cord in lower thoracic region
2. right intercostal vein draining across midline into hemiazygos vein
3. cartilage primordium of vertebral body with narrow zone of periosteal ossification
4. left sympathetic trunk
5. inferior hemiazygos vein
6. caudal border of left lung
7. left crus of the diaphragm
8. right (posterior) vagal (X) trunk
9. oesophagus
10. left (anterior) vagal (X) trunk
11. left pleural cavity
12. intercostal muscles (external and internal layers)
13. left dome of the diaphragm
14. ossification within cartilage primordium of mid-shaft region of left rib
15. left lobe of liver
16. left sub-phrenic recess of peritoneal cavity
17. lobar interzone
18. left costal margin
19. left musculo-phrenic vessels
20. component of hepatic venous plexus
21. left umbilical vein passing dorsally to join the ductus venosus
22. cartilage primordium of tail vertebral body
23. caudal region of tail
24. digit of right hindlimb
25. cartilage primordium of phalangeal bone
26. falciform ligament
27. lumen of gall bladder
28. right rectus abdominis muscle
29. ductus venosus
30. right superior epigastric vessels
31. right hepatic vein
32. right costal margin
33. cartilage primordium of mid-shaft region of right rib
34. right sub-phrenic recess of peritoneal cavity
35. right pleural cavity
36. right dome of the diaphragm
37. inferior vena cava (intra-hepatic at this level) formed principally from ductus venosus, left umbilical vein and right vitelline venous system
38. right crus of the diaphragm
39. caudal border of caudal lobe of the right lung
40. a right intercostal artery
41. dorsal (posterior) root ganglion
42. ossification within cartilage primordium of neural arch
43. spinal canal
44. part of extra-vertebral venous plexus
45. mantle layer (alar plate – dorsal grey horn)
46. mantle layer (basal plate – ventral grey horn)
47. ossification within cartilage primordium of head of rib
48. descending part of thoracic aorta
49. primordia of hair follicles
50. left hepatic duct
51. left umbilical vein within falciform ligament
52. digits of left hindlimb
53. umbilical vessel within umbilical cord
54. amnion
55. cystic duct
56. right sympathetic trunk
57. marginal layer in region of lateral white column
58. segmental artery
59. muscular wall of fundus region of stomach
60. gastric mucosa in fundus region of stomach
61. lumen of pyloric region of stomach
62. location of pyloric sphincter
63. left lobe of the pancreas
64. peritoneal cavity
65. dorsal mesentery of midgut containing vitelline vessels
66. left umbilical vein as it leaves the umbilical cord
67. Wharton's jelly
68. umbilical artery as it enters the umbilical cord
69. lumen of midgut loop of bowel
70. lumen of proximal part of second part of the duodenum
71. right lobe of the pancreas
72. common bile duct
73. portal vein
74. dorsal mesentery of oesophagus
75. caudal extremity of left pleural cavity
76. posterior wall of stomach
77. lumen of the "body" of the stomach
78. anterior wall of stomach
79. gastro-oesophageal junction
80. arterial branch from coeliac trunk to posterior wall of "body" of the stomach
81. cartilage primordium of shaft of left 13th rib with periosteal collar of bone
82. left (median in this location) umbilical artery
83. left umbilical vein within the umbilical cord
84. umbilical artery within the umbilical cord
85. site of entrance of common bile duct into posteromedial aspect of lumen of second part of the duodenum (ampulla of Vater)
86. caudate lobe of liver
87. ossification within cartilage primordium of mid-shaft region of right 12th rib
88. caudal extremity of right pleural cavity
89. right caudal thoracic sympathetic ganglion

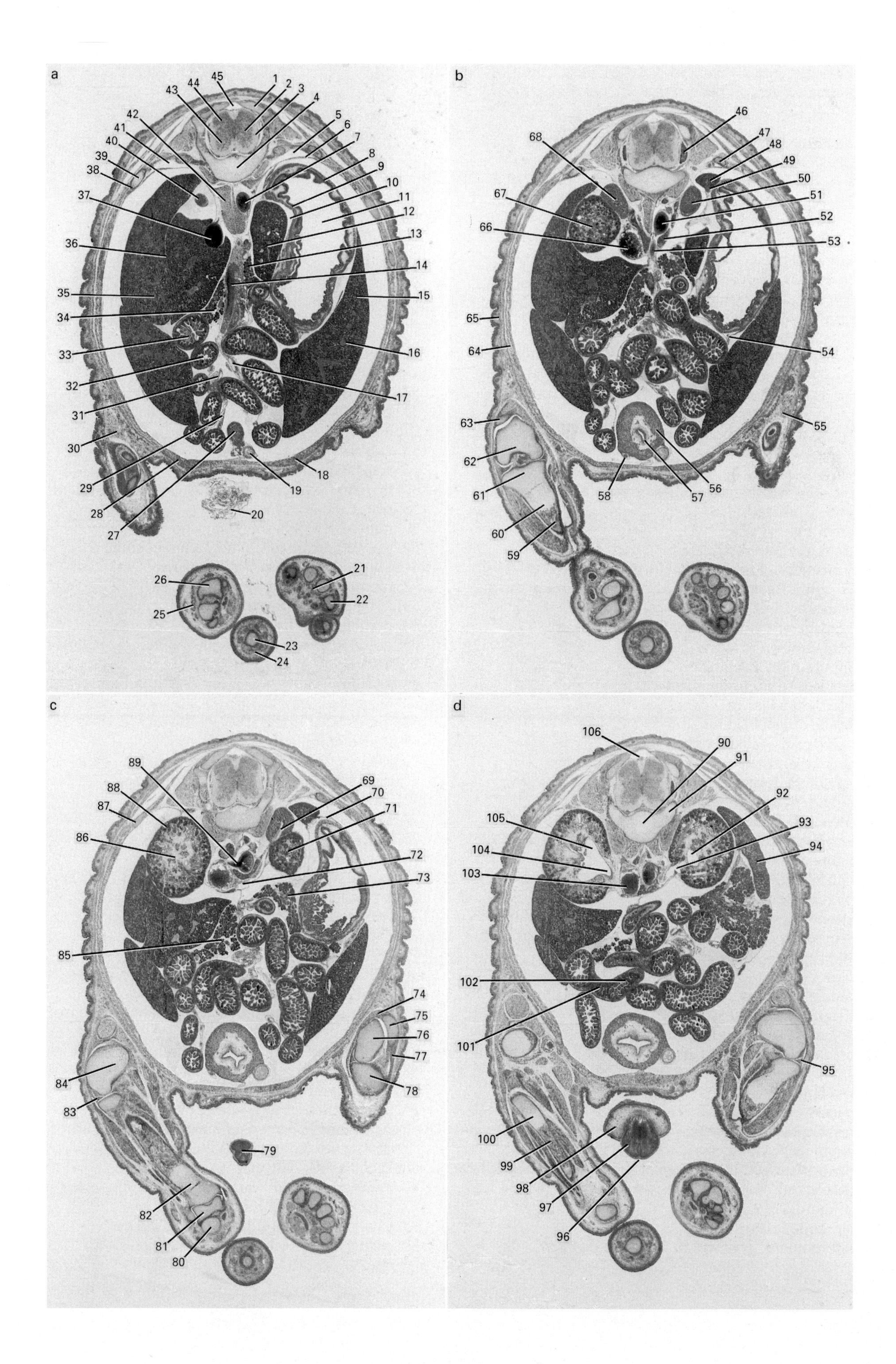
a
b
c
d
1
2
3
4
5
6
7
8
9
10
11
12
13
14
15
16
17
18
19
20
21
22
23
24
25
26
27
28
29
30
31
32
33
34
35
36
37
38
39
40
41
42
43
44
45
46
47
48
49
50
51
52
53
54
55
56
57
58
59
60
61
62
63
64
65
66
67
68
69
70
71
72
73
74
75
76
77
78
79
80
81
82
83
84
85
86
87
88
89
90
91
92
93
94
95
96
97
98
99
100
101
102
103
104
105
106

Plate 37j (16.5 days p.c.)

16.5 days p.c. Female embryo. Crown–rump length 14.2 mm (fixed). Transverse sections (Theiler Stage 25; rat, Witschi Stage 34).

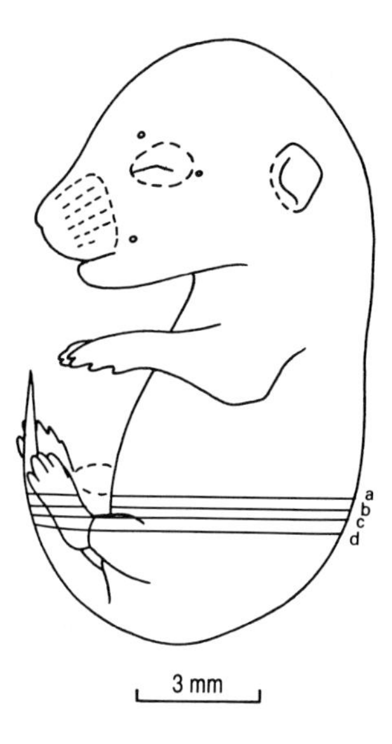

1. ossification within cartilage primordium of neural arch
2. spinal cord in lower thoracic region
3. marginal layer in region of lateral white column
4. cartilage primordium of 13th thoracic vertebral body
5. caudal extremity of left pleural cavity
6. left pleuro-peritoneal membrane (dorsal component of the diaphragm)
7. descending most caudal part of thoracic aorta
8. anterior wall of the stomach
9. posterior wall of the stomach
10. lesser sac of peritoneal cavity (omental bursa)
11. lumen of the "body" of the stomach
12. caudate lobe of the liver
13. left lobe of the pancreas
14. superior mesenteric vein passing dorsally to form the portal vein
15. left lobe of the liver
16. component of hepatic venous plexus
17. branch of superior mesenteric vein within the dorsal mesentery of the midgut
18. left rectus abdominis muscle
19. left umbilical artery
20. amnion surrounding umbilical cord
21. footplate region of left hindlimb
22. cartilage primordium of metatarsal bone
23. cartilage primordium of tail vertebral body
24. caudal region of tail
25. ankle region of right hindlimb
26. cartilage primordium of tarsal bone
27. rostral extremity of wall of bladder/region of urachus
28. right rectus abdominis muscle
29. dorsal mesentery of midgut
30. proximal part of right hindlimb
31. branch of superior mesenteric artery within dorsal mesentery of midgut
32. lumen of midgut
33. distal region of second part of the duodenum
34. right lobe of the pancreas
35. right lobe of the liver
36. lobar interzone
37. inferior vena cava in region of bare area of the liver
38. right costal margin
39. cartilage primordium of right 12th rib with periosteal collar of bone
40. upper pole of right adrenal gland
41. caudal extremity of right pleural cavity
42. right crus of the diaphragm
43. mantle layer (basal plate – ventral grey horn)
44. mantle layer (alar plate – dorsal grey horn)
45. spinal canal
46. dorsal (posterior) root ganglion
47. ossification within cartilage primordium of neck of 13th rib
48. upper pole of spleen
49. splenic artery where it enters the hilum of the spleen
50. upper pole of left adrenal gland
51. rostral part of abdominal aorta
52. pre-aortic abdominal sympathetic paraganglia (of Zuckerkandl)
53. superior mesenteric artery
54. ventral mesentery of the stomach (ventral mesogastrium – lesser omentum)
55. proximal part of the left hindlimb
56. wall of urogenital sinus (future bladder)
57. endodermal lining of urogenital sinus
58. right umbilical artery
59. ossification within cartilage primordium of mid-shaft region of right tibia
60. cartilage primordium of proximal part of right tibia with periosteal collar of bone
61. cartilage primordium of head of the right tibia
62. cartilage primordium of inter-condylar region of distal part of right femur
63. cartilage primordium of right patella
64. three layers of abdominal wall musculature (external oblique, internal oblique and transversus abdominis muscles)
65. cutaneous muscle of the trunk (panniculus carnosus)
66. prehepatic part of inferior vena cava
67. upper pole of right kidney
68. right adrenal gland
69. left adrenal vein
70. dorsal mesogastrium
71. upper pole of left kidney
72. left renal vein
73. tail of the pancreas
74. quadriceps femoris tendon
75. cartilage primordium of left patella
76. cartilage primordium of distal part of the left femur
77. ligamentum patellae
78. cartilage primordium of head of the left tibia
79. dorsum of phallus (clitoris)
80. cartilage primordium of right calcaneum
81. cartilage primordium of right talus
82. cartilage primordium of distal part of shaft of right tibia
83. right knee joint
84. cartilage primordium of distal part of shaft of right femur
85. body of the pancreas
86. medullary region of right kidney
87. cartilage primordium of tip of 13th rib
88. cortical region of right kidney
89. origin of right renal artery from the abdominal aorta
90. cartilage primordium of L1 vertebral body
91. cartilage primordium of L1 transverse process
92. pelvis of left kidney
93. left renal vein
94. body of spleen
95. left knee joint
96. urethral groove on ventral aspect of phallus (clitoris)
97. glans clitoridis
98. prepuce
99. ossification within mid-shaft region of right fibula
100. cartilage primordium of upper (proximal) shaft region of right fibula
101. third part of duodenum
102. proximal part of jejunum
103. inferior vena cava
104. proximal part of right ureter
105. pelvis of right kidney
106. posterior spinal vein

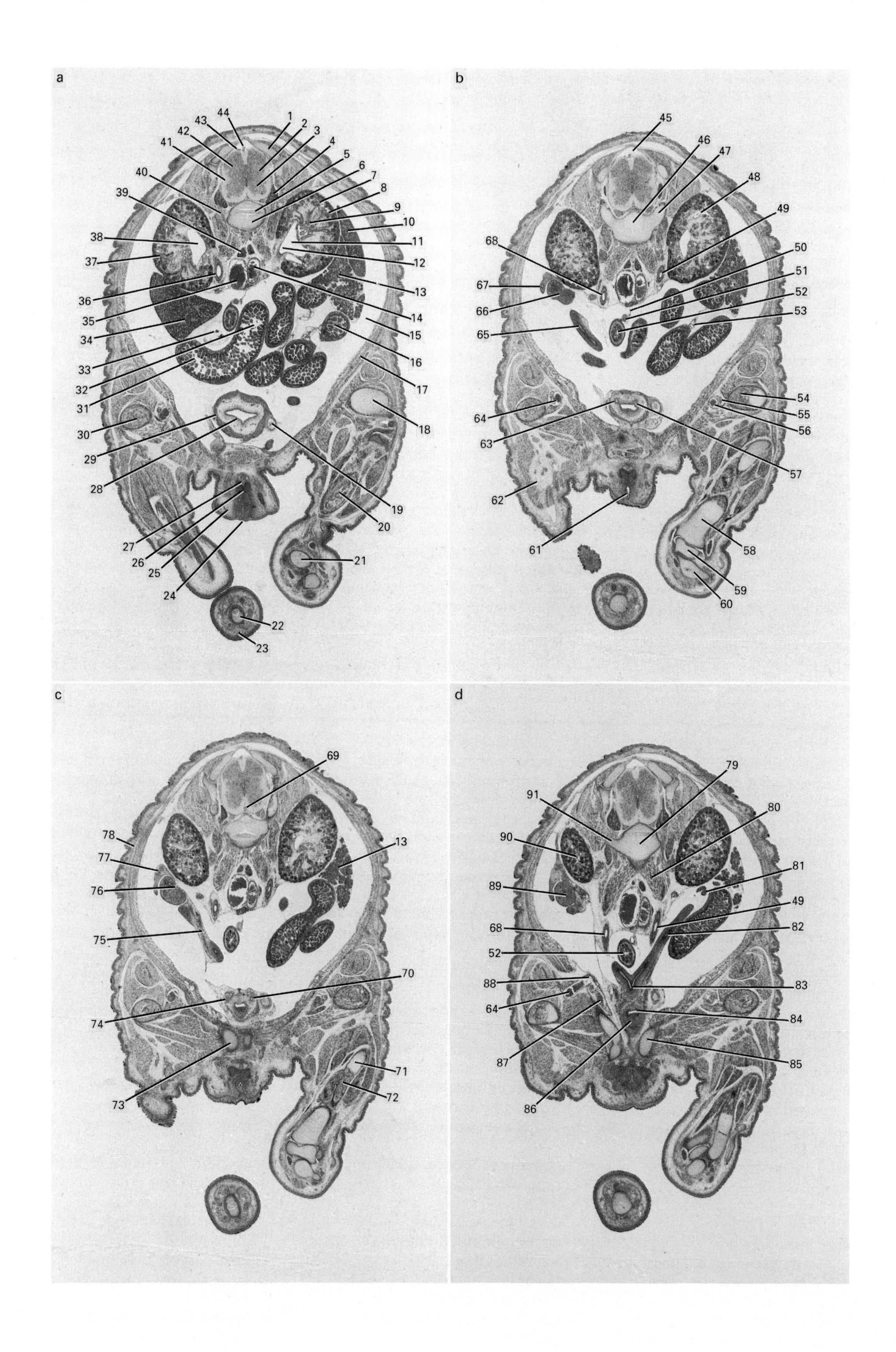
a
b
c
d
1
2
3
4
5
6
7
8
9
10
11
12
13
14
15
16
17
18
19
20
21
22
23
24
25
26
27
28
29
30
31
32
33
34
35
36
37
38
39
40
41
42
43
44
45
46
47
48
49
50
51
52
53
54
55
56
57
58
59
60
61
62
63
64
65
66
67
68
69
70
71
72
73
74
75
76
77
78
79
80
81
82
83
84
85
86
87
88
89
90
91

Plate 37k (16.5 days p.c.)

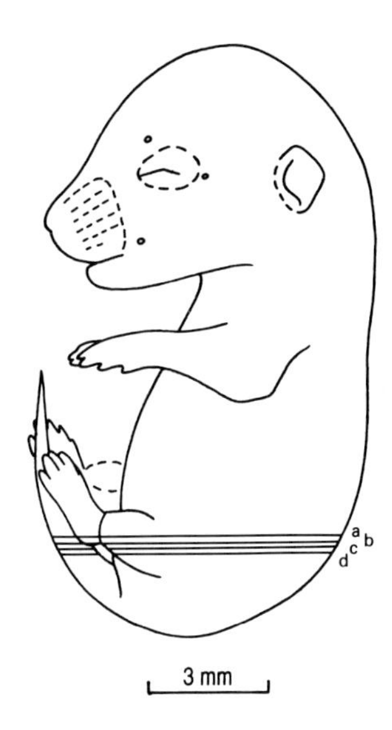

16.5 days p.c. Female embryo. Crown–rump length 14.2 mm (fixed). Transverse sections (Theiler Stage 25; rat, Witschi Stage 34).

1. ossification within cartilage primordium of neural arch
2. spinal cord in upper lumbar region
3. mantle layer (basal plate – ventral grey horn)
4. dorsal (posterior) root ganglion
5. nucleus pulposus
6. cartilage primordium of L1 vertebral body
7. cortical region of left kidney
8. glomerulus
9. collecting tubules
10. spleen
11. pelvis of left kidney
12. proximal part of left ureter
13. tail of pancreas
14. abdominal aorta
15. peritoneal cavity
16. lumen of midgut
17. three layers of abdominal wall musculature (external oblique, internal oblique and transversus abdominis muscles)
18. cartilage primordium of distal part of the left femur
19. left umbilical artery
20. ossification within cartilage primordium of mid-shaft region of left tibia
21. cartilage primordium of metatarsal bone
22. cartilage primordium of tail vertebral body
23. caudal region of tail
24. urethral groove on ventral aspect of phallus (clitoris)
25. prepuce
26. crus of clitoris
27. deep dorsal vein of clitoris
28. endodermal lining of urogenital sinus (future bladder)
29. wall of urogenital sinus
30. ossification within cartilage primordium of mid-shaft region of right femur
31. third part of duodenum
32. proximal part of jejunum
33. caudal part of uncinate process of pancreas
34. caudal part of right lobe of liver
35. inferior vena cava
36. proximal part of right urethra
37. cortical region of right kidney
38. pelvic region of right kidney
39. right upper lumbar sympathetic ganglion
40. segmental lumbar nerve
41. marginal layer in region of lateral white column
42. mantle layer (alar plate – dorsal grey horn)
43. posterior spinal artery (lateral)
44. posterior spinal vein (median)
45. spinal canal
46. cartilage primordium of L2 vertebral body
47. cartilage primordium of L2 transverse process
48. branch of renal vein
49. left ureter
50. dorsal mesentery of hindgut
51. branches of inferior mesenteric vessels within dorsal mesentery of hindgut
52. lumen of hindgut (rectum)
53. branch of superior mesenteric vein within dorsal mesentery of midgut
54. ossification within cartilage primordium of mid-shaft region of left femur
55. left femoral artery
56. left femoral vein
57. site of entrance of left ureter into lumen of the urogenital sinus (future ureteric orifice at upper/outer angle of the trigone of the bladder)
58. cartilage primordium of distal part of left tibia
59. cartilage primordium of left talus
60. cartilage primordium of left calcaneum
61. phallic part of urethra
62. caudal border of right hindlimb
63. site of entrance of right ureter into lumen of the urogenital sinus
64. right femoral vein
65. lumen of right paramesonephric duct (future right uterine horn)
66. upper pole of right ovary
67. right paramesonephric duct (future oviduct)
68. right ureter
69. anterior spinal artery
70. distal part of left ureter as it passes through the wall of the urogenital sinus
71. cartilage primordium of proximal part of shaft of left fibula
72. ossification within cartilage primordium of mid-shaft region of left fibula
73. cartilage primordium of right pubic bone
74. distal part of right ureter as it passes through the wall of the urogenital sinus
75. mesometrium
76. rete ovarii
77. degenerating mesonephric tubules
78. cutaneous muscle of trunk (panniculus carnosus)
79. cartilage primordium of L3 vertebral body
80. psoas major muscle
81. upper pole of left ovary
82. lumen of left paramesonephric duct (future left uterine horn)
83. site of apposition in the midline of the two paramesonephric ducts
84. lumen of vesical part of urethra
85. cartilage primordium of left pubic bone
86. neck of the bladder
87. vein draining to the external iliac vein from the region of the urethra and pelvic floor
88. external iliac vein
89. right ovary
90. lower pole of right kidney
91. cartilage primordium of L3 transverse process

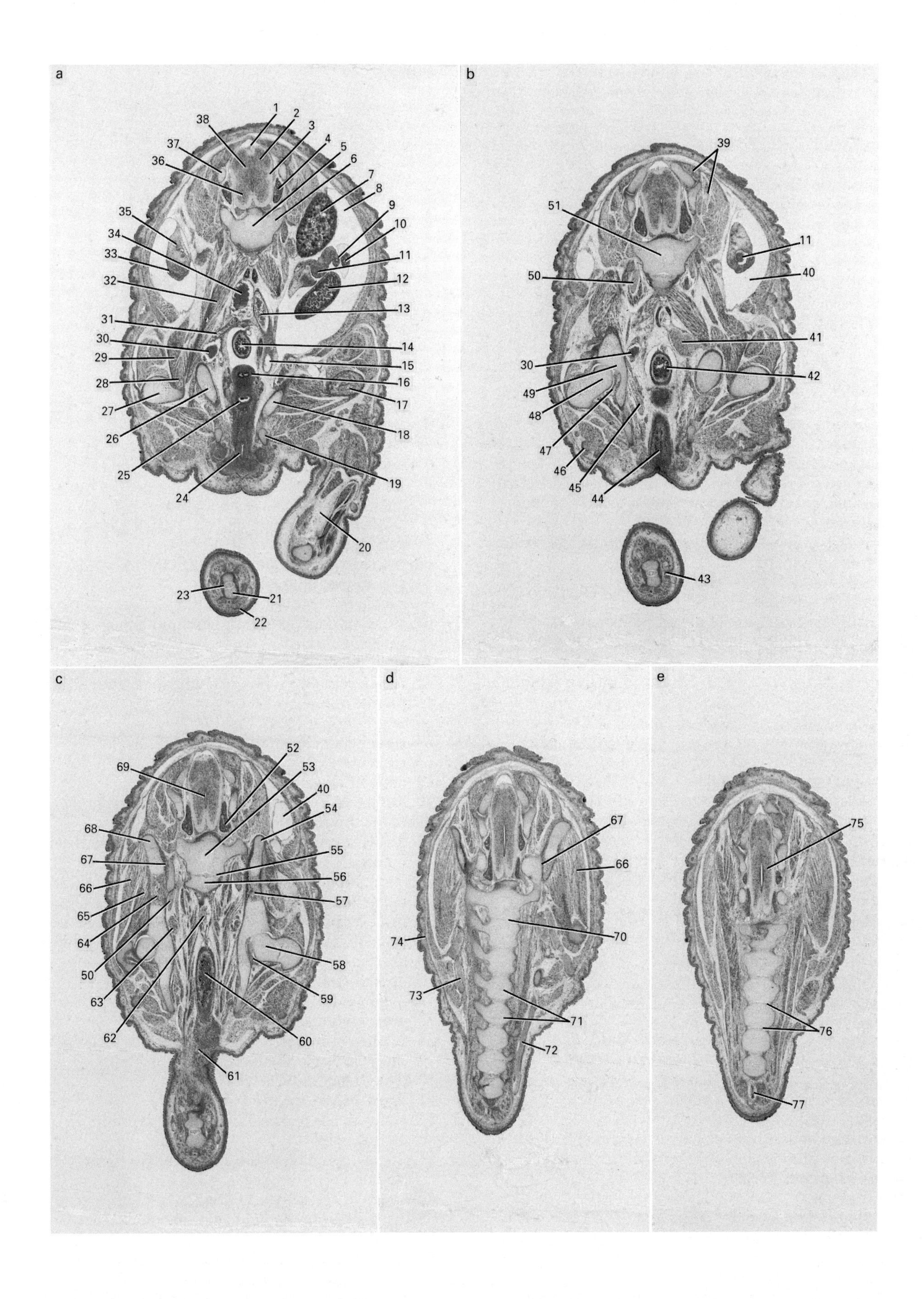
a
b
c
d
e
1
2
3
4
5
6
7
8
9
10
11
12
13
14
15
16
17
18
19
20
21
22
23
24
25
26
27
28
29
30
31
32
33
34
35
36
37
38
39
40
41
42
43
44
45
46
47
48
49
50
51
52
53
54
55
56
57
58
59
60
61
62
63
64
65
66
67
68
69
70
71
72
73
74
75
76
77

Plate 371 (16.5 days p.c.)

16.5 days p.c. Female embryo. Crown–rump length 14.2 mm (fixed). Transverse sections (Theiler Stage 25; rat, Witschi Stage 34).

1. posterior spinal vein
2. mantle layer (alar plate – dorsal grey horn)
3. marginal layer in region of lateral white column
4. dorsal (posterior) root ganglion
5. cartilage primordium of L4 vertebral body
6. cartilage primordium of L4 transverse process
7. lower pole of left kidney
8. peritoneal cavity
9. left paramesonephric duct (future oviduct)
10. tip of the tail of the pancreas
11. lower pole of left ovary
12. loop of midgut
13. location of bifurcation of the aorta to form the right and left common iliac arteries
14. hindgut (rectum)
15. left common iliac artery
16. caudal extremity of paramesonephric ducts (future cervical region)
17. ossification within cartilage primordium of mid-shaft region of left femur
18. cartilage primordium of pubic component of left innominate bone
19. cartilage primordium of ischial component of innominate bone
20. left hindlimb
21. cartilage primordium of tail vertebral body
22. caudal region of tail
23. nucleus pulposus
24. pubic/pelvic part of urethra
25. neck of the bladder/vesical part of the urethra
26. cartilage primordium of acetabular region of right innominate bone
27. cartilage primordium of proximal part of the shaft of the right femur
28. tendon of insertion of ilio-psoas muscle
29. tendon of iliacus muscle
30. right common iliac vein
31. right common iliac artery
32. psoas major muscle
33. right paramesonephric duct (future oviduct)
34. inferior vena cava
35. degenerating mesonephric tubules
36. mantle layer (basal plate – ventral grey horn)
37. cartilage primordium of ventral component of neural arch
38. spinal cord in lower lumbar region
39. multifidus and longissimus muscles which blend with the sacrococcygeus dorsalis medialis muscle
40. caudal extremity of peritoneal cavity
41. left common iliac vein
42. caudal part of the rectum
43. middle region of the tail
44. ano-rectal junction just proximal to proctodaeum (anal pit)
45. levator ani muscle
46. gluteal vessels
47. right hip joint
48. cartilage primordium of head of the right femur
49. cartilage primordium of acetabular fossa
50. lumbar nerve trunk (component of right lumbo-sacral plexus)
51. cartilage primordium of L5 vertebral body
52. lower lumbar dorsal (posterior) root ganglion
53. cartilage primordium of L6 vertebral body
54. cartilage primordium of left iliac crest
55. site of future intervertebral disc between L6 and S1 vertebral body
56. cartilage primordium of S1 vertebral body
57. ossification within cartilage primordium of iliac component of left innominate bone
58. cartilage primordium of head of the left femur
59. left hip joint
60. endodermal lining of caudal part of rectum
61. base of the tail
62. median sacral artery
63. internal iliac artery
64. ossification within cartilage primordium of iliac component of right innominate bone
65. gluteus medius muscle
66. gluteus maximus (superficialis) muscle
67. sacro-iliac joint
68. cartilage primordium of right iliac crest
69. ependymal layer surrounding central canal of spinal cord
70. site of future intervertebral disc between S1 and S2 vertebral bodies
71. cartilage primordia of coccygeal vertebrae
72. left lateral caudal artery
73. right lateral caudal artery
74. cutaneous muscle of lower trunk region (panniculus carnosus)
75. spinal cord in lumbo-sacral region
76. site of future intervertebral discs between coccygeal vertebral bodies
77. caudal region of spinal cord (conus medullaris) in proximal part of the tail

Plate 38a (16.5 days p.c.)

16.5 days p.c. Male embryo. Crown–rump length 15.5 mm (fixed). Sagittal sections (Theiler Stage 25; rat, Witschi Stage 34).

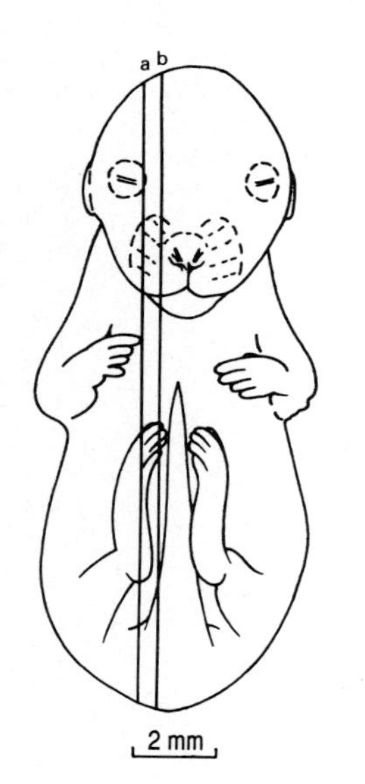

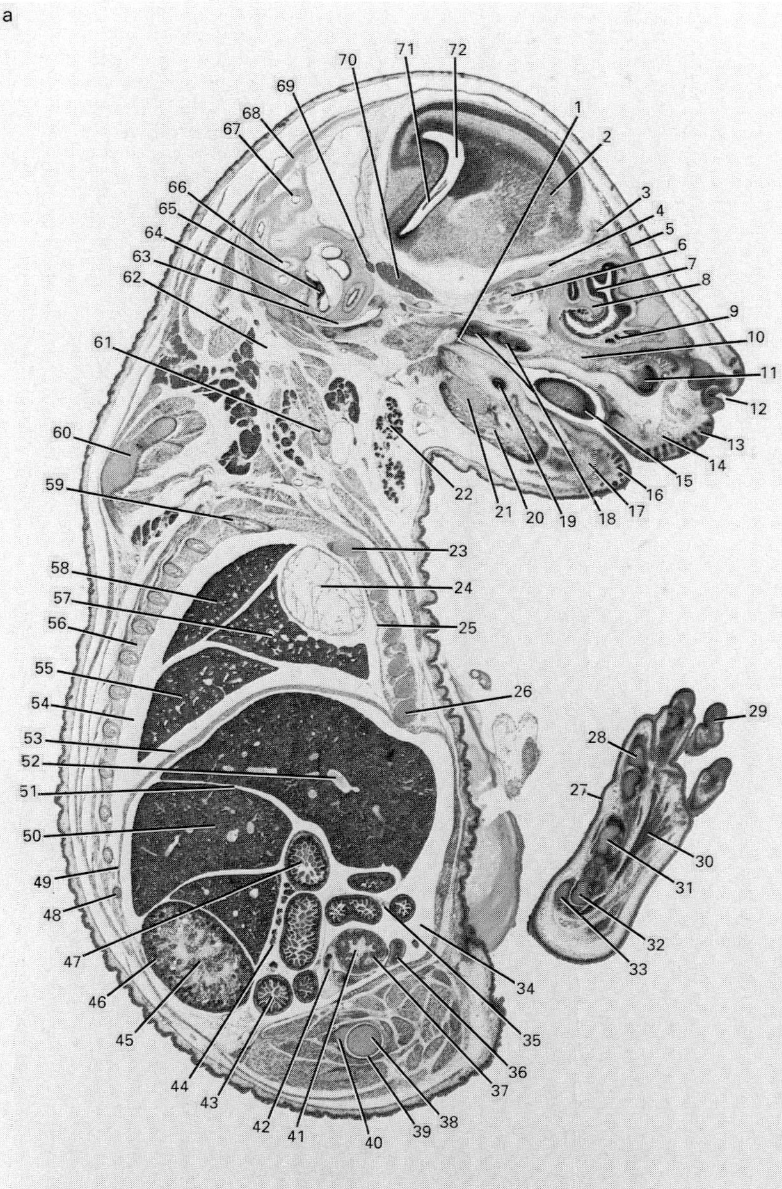

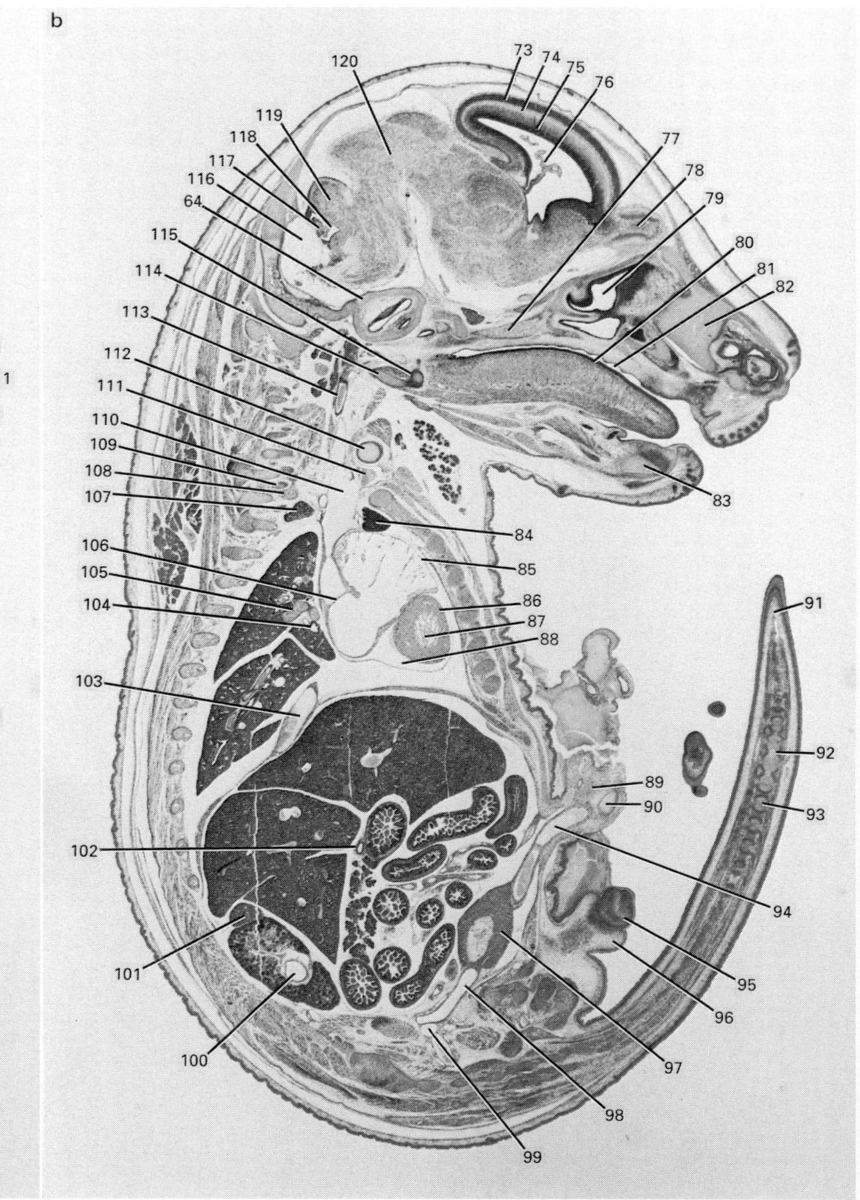

1. entrance to right pharyngo-tympanic (Eustachian) tube
2. lateral border of right cerebral hemisphere
3. lateral border of right olfactory lobe
4. cartilage primordium of right orbito-sphenoid bone
5. ossification within cartilage primordium of nasal bone
6. right optic (II) nerve surrounded by fibres of origin of extrinsic ocular muscles
7. olfactory epithelium
8. cartilage primordia of turbinate bones
9. tubules of serous glands associated with nasal septum
10. ossification within anterolateral part of palatal shelf of right maxilla
11. primordium of upper right incisor tooth
12. entrance into nasal cavity (external naris)
13. primordia of follicles of vibrissae associated with upper lip
14. upper lip (right side)
15. lateral border of tip of tongue
16. primordia of follicles of vibrissae associated with lower lip
17. lower lip
18. primordia of right upper molar teeth
19. primordium of first lower molar tooth
20. ossification within right half of mandible around but principally lateral to Meckel's cartilage
21. Meckel's cartilage
22. right submandibular gland
23. second right costal cartilage
24. lumen of right atrium
25. mesothelial cells (parietal pleura) lining pleural cavity
26. sixth right costal cartilage
27. medial (plantar) aspect of right hindlimb
28. cartilage primordium of phalangeal bone
29. digit of right hindlimb
30. tendon of flexor digitorum profundus muscle
31. cartilage primordium of metatarsal bone
32. cartilage primordium of right talus bone
33. cartilage primordium of distal part of right tibia
34. peritoneal cavity
35. dorsal mesentery of midgut with associated mesenteric vessels
36. right mesonephric duct (ductus deferens) within right mesorchium
37. seminiferous tubules within medullary region of right testis
38. cartilage primordium of head of right femur
39. cavity of right hip joint
40. cartilage primordium of right innominate bone at margin of acetabular fossa
41. medullary region of right testis
42. degenerating mesonephric tissue containing mesonephric tubules
43. lumen of third part of duodenum
44. pancreas
45. medullary region of right kidney
46. cortical region of right kidney
47. lumen of first part of duodenum
48. cartilage primordium of dorsal part of shaft of right 13th rib
49. costal origin of dorsal part of diaphragm
50. right lobe of liver
51. lobar interzone
52. part of right hepatic venous plexus
53. caudal component of right dome of diaphragm
54. right pleural cavity
55. caudal lobe of right lung
56. intercostal muscles (external and internal layers)
57. middle lobe of right lung
58. cranial lobe of right lung
59. ossification within mid-shaft region of cartilage primordium of right second rib
60. cartilage primordium of medial border of right scapula
61. ossification within middle one-third of cartilage primordium of right clavicle
62. rostral part of right internal jugular vein shortly after it has emerged from the jugular foramen
63. right tubo-tympanic recess
64. cartilage primordium of petrous part of right temporal bone
65. cochlea
66. right posterior semicircular canal
67. rostral part of right endolymphatic duct
68. cartilage primordium of occipital (supraoccipital) bone
69. facial (VII) ganglion
70. trigeminal (V) ganglion
71. choroid plexus within caudal extremity of posterior horn of right lateral ventricle
72. posterior horn of right lateral ventricle
73. neopallial cortex (future cerebral cortex)
74. intermediate zone
75. ventricular zone
76. choroid plexus within anterior horn of right lateral ventricle
77. ossification within posterolateral part of palatal shelf of right maxilla
78. right olfactory lobe
79. nasal cavity (right half)
80. dorsal surface of tongue
81. oropharynx
82. cartilage primordium of nasal septum
83. ventral extremity of right Meckel's cartilage
84. right lobe of thymus gland
85. wall of right atrium
86. wall of right ventricle
87. lumen of right ventricle
88. pericardial cavity
89. Wharton's jelly
90. umbilical vein within proximal part of umbilical cord
91. distal part of tail
92. nucleus pulposus
93. cartilage primordium of tail vertebral body
94. right umbilical artery where it enters the umbilical cord
95. glans penis
96. prepuce
97. right lateral wall of bladder
98. right internal iliac artery where it proceeds to form the right superior vesical artery
99. right internal iliac vein
100. pelvic region of right kidney
101. right adrenal gland
102. pancreatic duct
103. posthepatic part of inferior vena cava
104. lobar bronchus within middle lobe of right lung
105. branch of right pulmonary vein
106. leaflet of venous valve at site of entrance of right superior vena cava into right atrium
107. right cervicothoracic (stellate) sympathetic ganglion
108. ossification within cartilage primordium of tubercle of first rib
109. right dorsal (posterior) root ganglion of C8
110. right superior (cranial) vena cava just below level it receives the right subclavian vein
111. right subclavian vein
112. cartilage primordium of medial one-third of right clavicle with periosteal collar of bone
113. right internal carotid artery
114. cartilage primordium of right greater horn of hyoid bone
115. cartilage primordium of right lesser horn of hyoid bone
116. ependymal diverticulum of fourth ventricle
117. choroid plexus arising from roof of right lateral recess of fourth ventricle
118. right lateral recess of fourth ventricle
119. cerebellar primordium
120. right lateral wall of midbrain

Plate 38b (16.5 days p.c.)

16.5 days p.c. Male embryo. Crown–rump length 15.5 mm (fixed). Sagittal sections (Theiler Stage 25; rat, Witschi Stage 34).

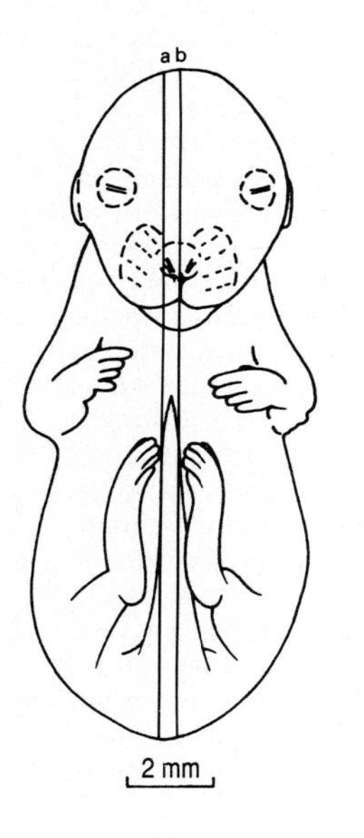

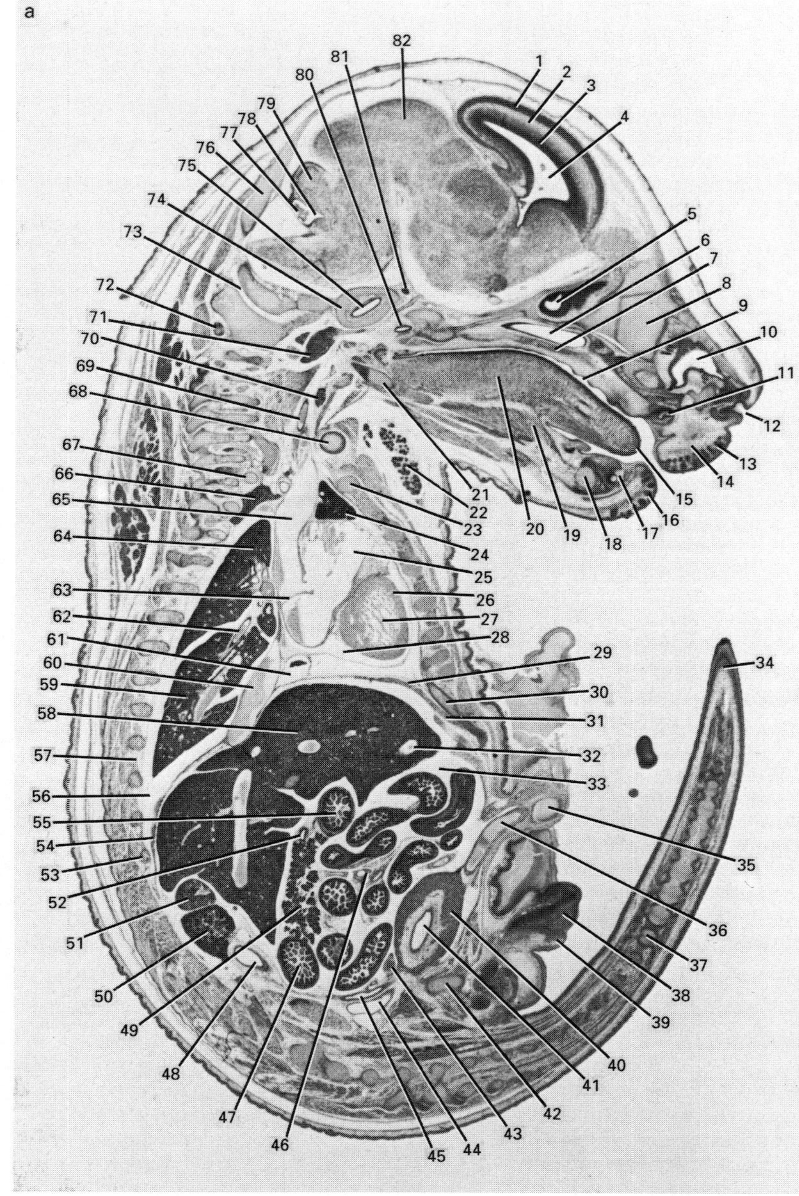

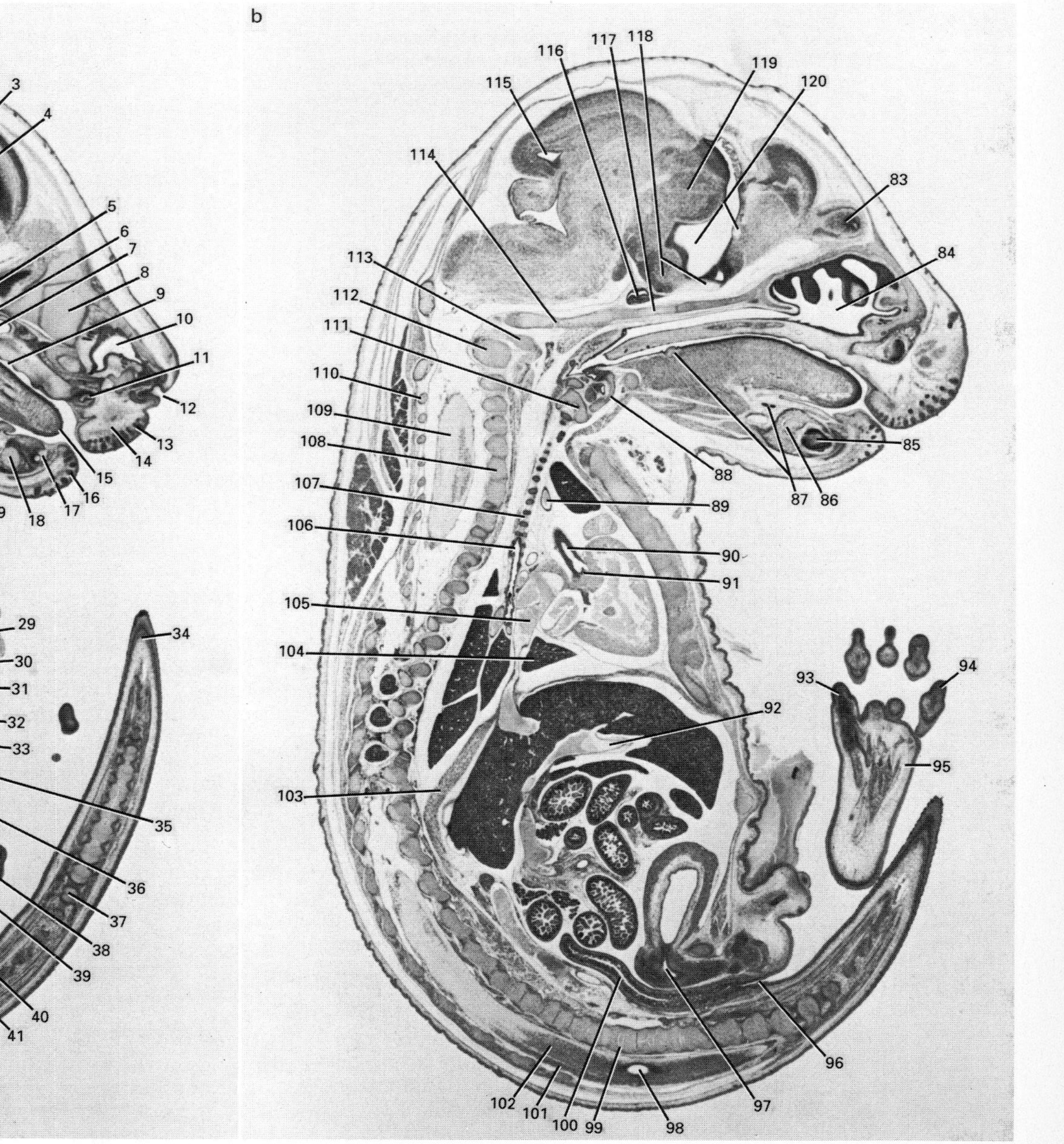

1. neopallial cortex (future cerebral cortex)
2. intermediate zone
3. ventricular zone
4. right lateral ventricle
5. dorsal part of nasal cavity
6. nasopharynx
7. ossification within palatal shelf of right maxilla
8. cartilage primordium of nasal septum
9. oropharynx
10. ventral part of nasal cavity (left half)
11. primordium of upper left incisor tooth
12. entrance into nasal cavity (external naris)
13. primordia of follicles of vibrissae associated with upper lip
14. upper lip (right side)
15. tip of tongue
16. primordia of follicles of vibrissae associated with lower lip
17. primordium of lower left incisor tooth
18. Meckel's cartilage
19. extrinsic muscle of tongue (genioglossus)
20. intrinsic muscles of tongue (superior and inferior longitudinal, transverse and vertical components)
21. cartilage primordium of lateral part of body of hyoid bone
22. submandibular gland
23. right first costal cartilage
24. right lobe of thymus gland
25. lumen of right atrium
26. wall of right ventricle
27. lumen of right ventricle
28. pericardial cavity
29. right dome of diaphragm
30. cartilage primordium of xiphoid process (xiphisternum)
31. sternal fibres of origin of diaphragm
32. part of right hepatic venous plexus
33. peritoneal cavity
34. distal part of tail
35. umbilical vein within proximal part of umbilical cord
36. right umbilical artery where it enters the umbilical cord
37. cartilage primordium of coccygeal vertebral body
38. glans penis
39. ventral part of prepuce
40. wall of bladder
41. lumen of bladder
42. cartilage primordium of right pubic bone
43. caudal part of right mesonephric duct (ductus deferens) where it approaches the region of the neck of the bladder
44. right internal iliac artery
45. distal part of right ureter
46. mesenteric vessels within dorsal mesentery of midgut
47. lumen of third part of duodenum
48. caudal part of pelvis of right kidney where it forms the rostral part of the ureter
49. pancreas
50. medial margin of right kidney
51. right adrenal gland
52. pancreatic duct
53. ossification in mid-shaft region of cartilage primordium of right 13th rib
54. costal origin of dorsolateral part of diaphragm
55. lumen of first part of duodenum
56. right pleural cavity
57. intercostal muscles (external and internal layers)
58. right lobe of liver
59. caudal lobe of right lung
60. posthepatic part of inferior vena cava
61. extension of right pleural cavity around accessory lobe of right lung
62. branch of right pulmonary artery passing to caudal lobe of right lung
63. leaflet of venous valve at site of entrance of right superior (cranial) vena cava into right atrium
64. apical part of cranial lobe of right lung
65. right superior (cranial) vena cava
66. right cervicothoracic (stellate) sympathetic ganglion
67. ossification within mid-shaft region of cartilage primordium of right first rib
68. cartilage primordium of medial one-third of right clavicle with periosteal collar of bone
69. right internal carotid artery
70. right lobe of thyroid gland
71. right superior cervical sympathetic ganglion
72. right dorsal (posterior) root ganglion of C2
73. atlanto-occipital joint
74. cartilage primordium of petrous part of right temporal bone
75. cochlea
76. cartilage primordium of occipital (supraoccipital) bone with early evidence of periosteal ossification
77. choroid plexus arising from roof of right lateral recess of fourth ventricle
78. right lateral recess of fourth ventricle
79. cerebellar primordium
80. left pharyngo-tympanic (Eustachian) tube
81. lateral extremity of pituitary
82. right lateral wall of midbrain
83. medial wall of left olfactory lobe
84. nasal cavity (left half)
85. primordium of left lower incisor tooth
86. ossification within left half of mandible
87. submandibular and sublingual ducts and median circumvallate papilla
88. thyroid cartilage
89. brachiocephalic (innominate) artery
90. origin of thoracic aorta (aortic trunk)
91. leaflet of aortic valve
92. ductus venosus
93. first digit of left hindlimb
94. fifth digit of left hindlimb
95. plantar surface (lateral border) of left hindlimb
96. anal canal
97. prostatic region of urethra
98. central canal
99. nucleus pulposus in central region of future intervertebral disc
100. lumen of rectum
101. mantle layer (dorsal grey horn)
102. mantle layer (ventral grey horn)
103. right crus of diaphragm
104. accessory lobe of right lung
105. inferior vena cava where it enters the right atrium
106. level of bifurcation of the trachea
107. trachea
108. cartilage primordium of lateral part of body of C7 vertebra
109. marginal zone in mid-cervical region of spinal cord
110. cartilage primordium of neural arch with periosteal ossification
111. cricoid cartilage
112. cartilage primordium of odontoid process (dens) and body of C2 vertebra (axis) with early evidence of periosteal ossification
113. ossification within cartilage primordium of anterior arch of C1 vertebra (atlas)
114. ossification within cartilage primordium of basioccipital bone (clivus)
115. caudal part of mesencephalic vesicle
116. pituitary
117. ossification within cartilage primordium of basisphenoid bone
118. diencephalon (hypothalamus) and (anteriorly) the optic chiasma
119. diencephalon (thalamus)
120. rostral part of third ventricle and (anteriorly) the anterior commissure

Plate 38c (16.5 days p.c.)

16.5 days p.c. Male embryo. Crown–rump length 15.5 mm (fixed). Sagittal sections (Theiler Stage 25; rat, Witschi Stage 34).

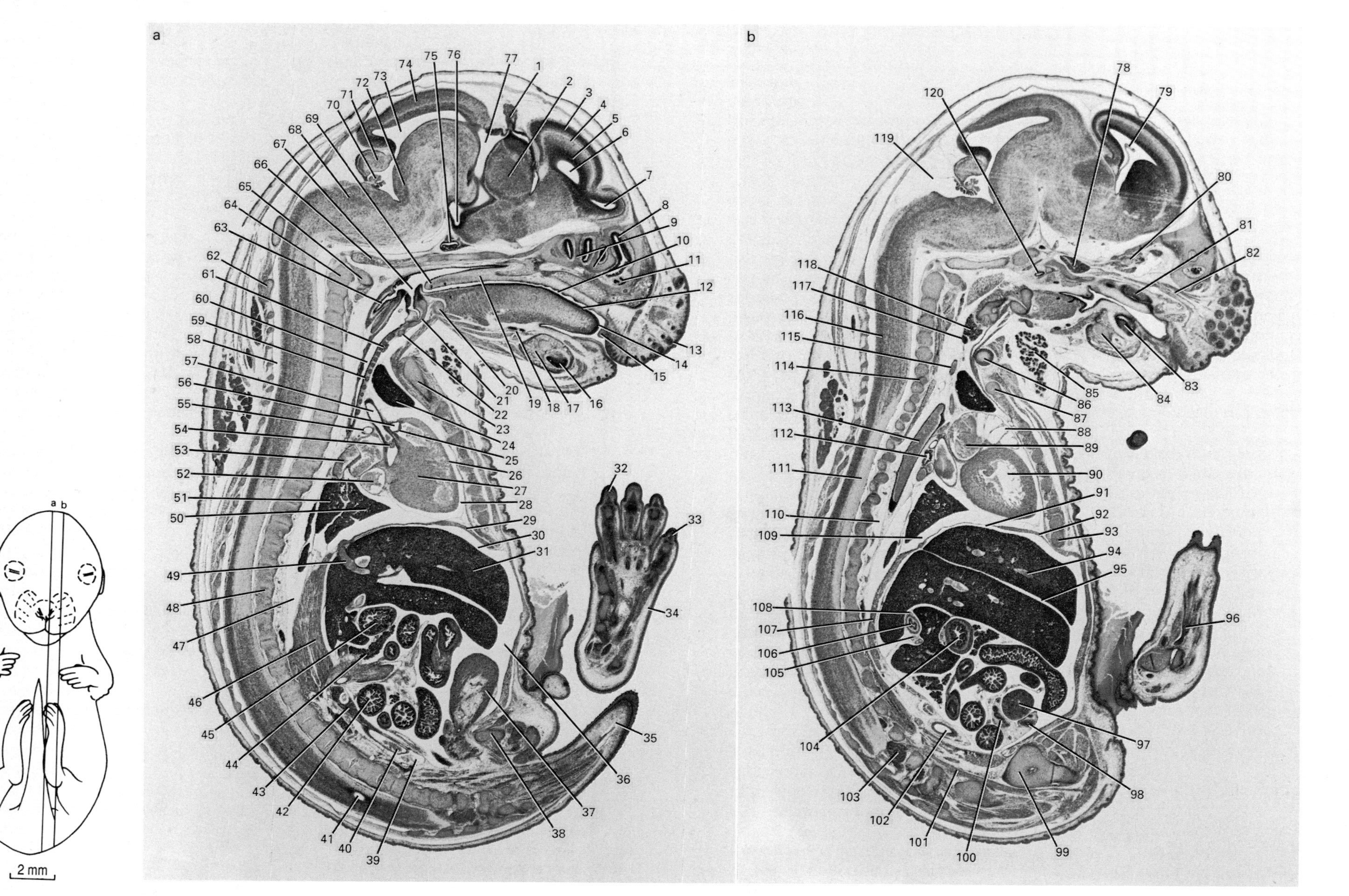

1. left lateral wall of pineal gland
2. diencephalon (thalamus)
3. neopallial cortex composed of inner broad nuclear layer (cortical plate) with narrow outer relatively anuclear layer (marginal zone)
4. intermediate zone
5. ventricular zone
6. left lateral ventricle
7. extension of anterior horn of left lateral ventricle into left olfactory lobe
8. olfactory epithelium
9. cartilage primordia of turbinate bones
10. oropharynx
11. tubules of serous glands associated with nasal septum
12. dorsum of tongue
13. primordia of follicles of vibrissae associated with upper lip
14. tip of tongue
15. elevation which forms boundary between oral mucosa and surface ectoderm
16. primordium of left lower incisor tooth
17. ossification within left half of mandible
18. submandibular and sublingual ducts
19. palatal shelf of left maxilla
20. cartilage primordium of body of hyoid bone
21. thyroid cartilage
22. cricoid cartilage
23. cartilage primordium of manubrium sterni
24. right lobe of thymus gland
25. leaflet of pulmonary valve
26. wall of right ventricle
27. muscular part of interventricular septum
28. pericardial cavity
29. right dome of diaphragm
30. sub-phrenic recess of peritoneal cavity
31. right lobe of liver
32. second digit of left hindlimb
33. fifth digit of left hindlimb
34. lateral border of left hindlimb
35. middle region of the tail
36. peritoneal cavity
37. left lateral wall of bladder
38. cartilage primordium of left pubic bone
39. site of amalgamation of the right and left common iliac veins to form the inferior vena cava
40. proximal part of left common iliac artery at level of bifurcation of the abdominal aorta
41. central canal
42. third part of duodenum
43. pre-aortic abdominal sympathetic ganglion (coeliac ganglion)
44. pancreas
45. first part of duodenum
46. right crus of diaphragm
47. right pleural cavity
48. ossification within cartilage primordium of lower thoracic vertebral body
49. left hepatic vein where it joins the inferior vena cava
50. accessory lobe of right lung
51. caudal lobe of right lung
52. lumen of right atrium
53. inferior vena cava where it enters the right atrium
54. leaflet of aortic valve
55. right pulmonary artery
56. proximal part of pulmonary trunk
57. aortic trunk/proximal part of arch of aorta
58. spinal canal
59. cartilage ring of trachea
60. lumen of trachea
61. isthmus of thyroid gland
62. ossification within cartilage primordium of neural arch of C2 vertebra (axis)
63. rostral part of oesophagus
64. cartilage primordium of odontoid process (dens) of C2 vertebra with early evidence of periosteal ossification
65. ossification within cartilage primordium of anterior arch of C1 vertebra (atlas)
66. ossification within cartilage primordium of basioccipital bone (clivus)
67. entrance into larynx (laryngeal aditus)
68. rostral part of medulla oblongata
69. soft palate
70. choroid plexus within central part of lumen of fourth ventricle
71. cerebellar primordium
72. pons
73. caudal part of mesencephalic vesicle
74. roof of midbrain
75. pituitary
76. infundibular recess of third ventricle
77. rostral part of third ventricle
78. left trigeminal (V) ganglion
79. choroid plexus within lateral ventricle
80. left optic (II) nerve surrounded by fibres of origin of left extrinsic ocular muscles
81. primordia of left upper molar teeth
82. branches of maxillary division of trigeminal (V) nerve passing from plexuses around follicles of vibrissae to trigeminal (V) ganglion
83. primordium of left lower molar tooth
84. Meckel's cartilage
85. submandibular gland
86. cartilage primordium of medial one-third of left clavicle with periosteal collar of bone
87. left first costal cartilage
88. auricular part of left atrium
89. lumen of left atrium
90. lumen of left ventricle
91. central tendon of diaphragm
92. sternal fibres of origin of diaphragm
93. left seventh costal cartilage
94. branch of left hepatic venous plexus within left lobe of liver
95. lobar interzone
96. tendon of left extensor digitorum longus muscle
97. seminiferous tubules within medullary region of left testis
98. left mesonephric duct (ductus deferens) within left mesorchium
99. cartilage primordium of left innominate bone just deep to hip joint
100. degenerating mesonephric tissue containing mesonephric tubules
101. lumbar nerve trunk progressing caudally to form a component of left lumbo-sacral plexus
102. caudal part of pelvis of left kidney where it forms the rostral part of the ureter
103. lower lumbar dorsal (posterior) root ganglion
104. gastro-duodenal junction in region just distal to pyloric sphincter
105. anterior vagal (X) trunk
106. lumen of caudal part of oesophagus
107. medial fibres of right crus of diaphragm
108. posterior vagal (X) trunk
109. extension of right pleural cavity around accessory lobe of right lung
110. inferior hemiazygos vein
111. marginal layer of spinal cord in mid-thoracic region
112. left main bronchus
113. thoracic (descending) aorta
114. ossification within cartilage primordium of head of first rib
115. left internal carotid artery
116. mantle layer (dorsal grey horn) in mid-cervical region of spinal cord
117. left parathyroid gland
118. left lobe of thyroid gland
119. roof of ependymal diverticulum of fourth ventricle
120. left trigeminal (V) ganglion

Plate 39a

16.5 days p.c. Coronal sections through head (Theiler Stage 25; rat, Witschi Stage 34)

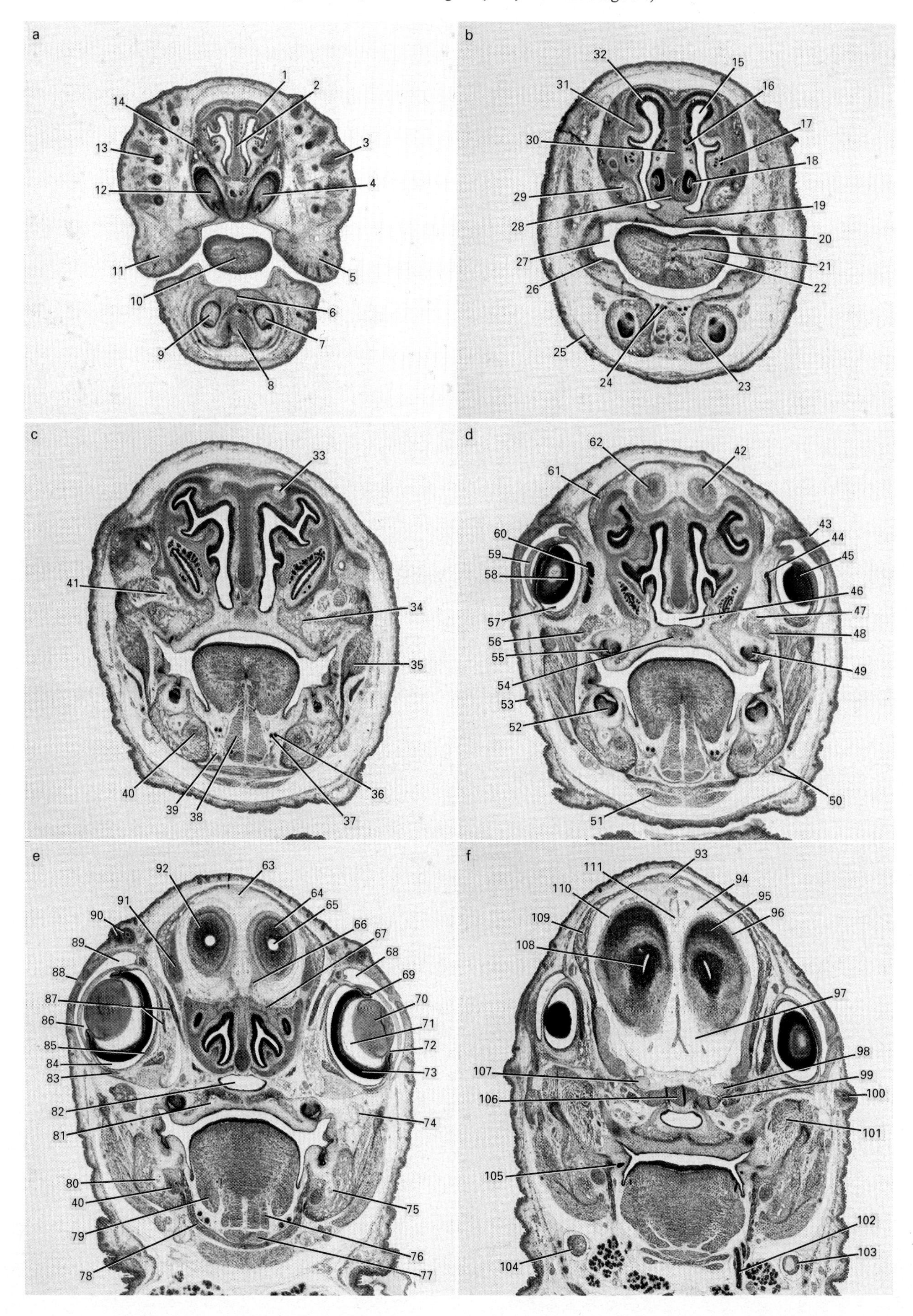

1. nasal capsule
2. anterior extremity of cartilage primordium of nasal septum
3. longitudinal section of follicle of vibrissa
4. primordium of left upper incisor tooth
5. left upper lip
6. line of fusion between the right and left halves of the mandible (site of future symphysis menti)
7. primordium of left lower incisor tooth
8. site of fusion of the right and left Meckel's cartilages
9. primordium of right lower incisor tooth
10. tip of tongue
11. right upper lip
12. primordium of right upper incisor tooth
13. transverse section of follicle of vibrissa
14. ossification in membranous primordium of intermaxillary bone
15. nasal cavity
16. tubules of serous glands associated with nasal septum
17. tubules of serous glands located in the lateral wall of the middle meatus
18. vomeronasal organ (Jacobson's organ)
19. anterior extremity of secondary palate (palatal shelf of maxilla)
20. dorsal surface of tongue
21. intrinsic muscle of the tongue (transverse component)
22. intrinsic muscle of the tongue (vertical component)
23. ossification in body of left half of mandible
24. ventral extremity of right submandibular (medial) and sublingual (lateral) ducts just proximal to their openings on the medial and lateral parts of the tip of the right sublingual caruncle
25. cutaneous muscle of the lower face and neck – platysma (panniculus carnosus); this muscle extends cranially to become the m. cutaneus faciei et labiorum
26. mucous membrane of oral cavity
27. oral cavity
28. ossification in inferomedial part of nasal septum
29. ossification in right maxilla
30. cartilage primordium of nasal septum
31. cartilage primordium of turbinate bone
32. olfactory epithelium
33. branches of left olfactory (I) nerve passing centrally from olfactory epithelium towards olfactory bulb
34. ossification in left maxilla
35. anterior fibres of left masseter muscle
36. left sublingual duct
37. left submandibular duct
38. right genioglossus muscle
39. right lingual vein
40. right Meckel's cartilage
41. right infraorbital nerve bundle
42. anterior extremity of left olfactory bulb
43. line of fusion between left upper and lower eyelids
44. left Harderian gland
45. neural layer of left retina (grazing section)
46. communication between right and left halves of nasopharynx beneath inferior part of nasal septum
47. left infraorbital nerve bundle
48. left inferior orbital vein
49. primordium of left upper molar tooth
50. facial artery and vein
51. anterior belly of digastric muscle
52. primordium of right lower molar tooth
53. hair follicle in skin
54. ossification in ventral/anterior extremity of secondary palate (two centres of ossification, one on either side of the midline
55. primordium of right upper molar tooth
56. right inferior orbital vein
57. intra-retinal space (grazing section)
58. neural layer of right retina (grazing section)
59. cornea of right eye
60. right Harderian gland
61. ossification in membranous primordium of right nasal bone
62. anterior extremity of right olfactory bulb
63. superior sagittal dural venous sinus
64. wall of left olfactory bulb
65. extension of anterior horn of left lateral ventricle into olfactory bulb
66. left olfactory (I) nerve
67. cartilage primordium of cribriform plate of left ethmoid bone
68. left conjunctival sac
69. superior margin of left iris (boundary of pupil)
70. lens
71. hyaloid cavity
72. inferior margin of left iris
73. neural layer of retina
74. part of pterygoid plexus of veins
75. left inferior alveolar vessels and nerve
76. left genioglossus muscle
77. geniohyoid muscle
78. mylohyoid muscle
79. hyoglossus muscle
80. right inferior alveolar vessels and nerve
81. greater palatine vessels and nerve
82. nasopharynx
83. extrinsic ocular muscle (right inferior rectus)
84. intra-retinal space
85. pigment layer of retina with subjacent sclera
86. anterior chamber of right eye
87. extrinsic ocular muscles (right superior oblique and right superior rectus)
88. line of fusion between right upper and lower eyelids
89. right conjunctival sac
90. prominent tactile or sinus hair follicles characteristically found in this location
91. cartilage primordium of orbito-sphenoid bone
92. wall of right olfactory bulb
93. venous plexus within membranous primordium of frontal bone
94. subarachnoid space
95. inner broad nuclear layer (cortical plate) of neopallial cortex
96. outer narrow relatively anuclear layer (marginal zone) of left neopallial cortex
97. basal (interpeduncular) cistern (cisterna interpeduncularis, of subarachnoid space)
98. left optic (II) nerve
99. optic foramen
100. prominent tactile or sinus hair follicle characteristically found in this location
101. left pterygoid muscles
102. left submandibular duct as it emerges from the submandibular gland
103. capsule of left carotid body
104. right internal carotid artery
105. developing glandular duct opening into lateral part of oral cavity
106. cartilage primordium of presphenoid bone
107. right optic (II) nerve
108. anterior portion of superior horn of right lateral ventricle
109. ossification in membranous primordium of right frontal bone
110. vascular elements of pia mater on surface of neopallial cortex
111. anterior extension of falx cerebri

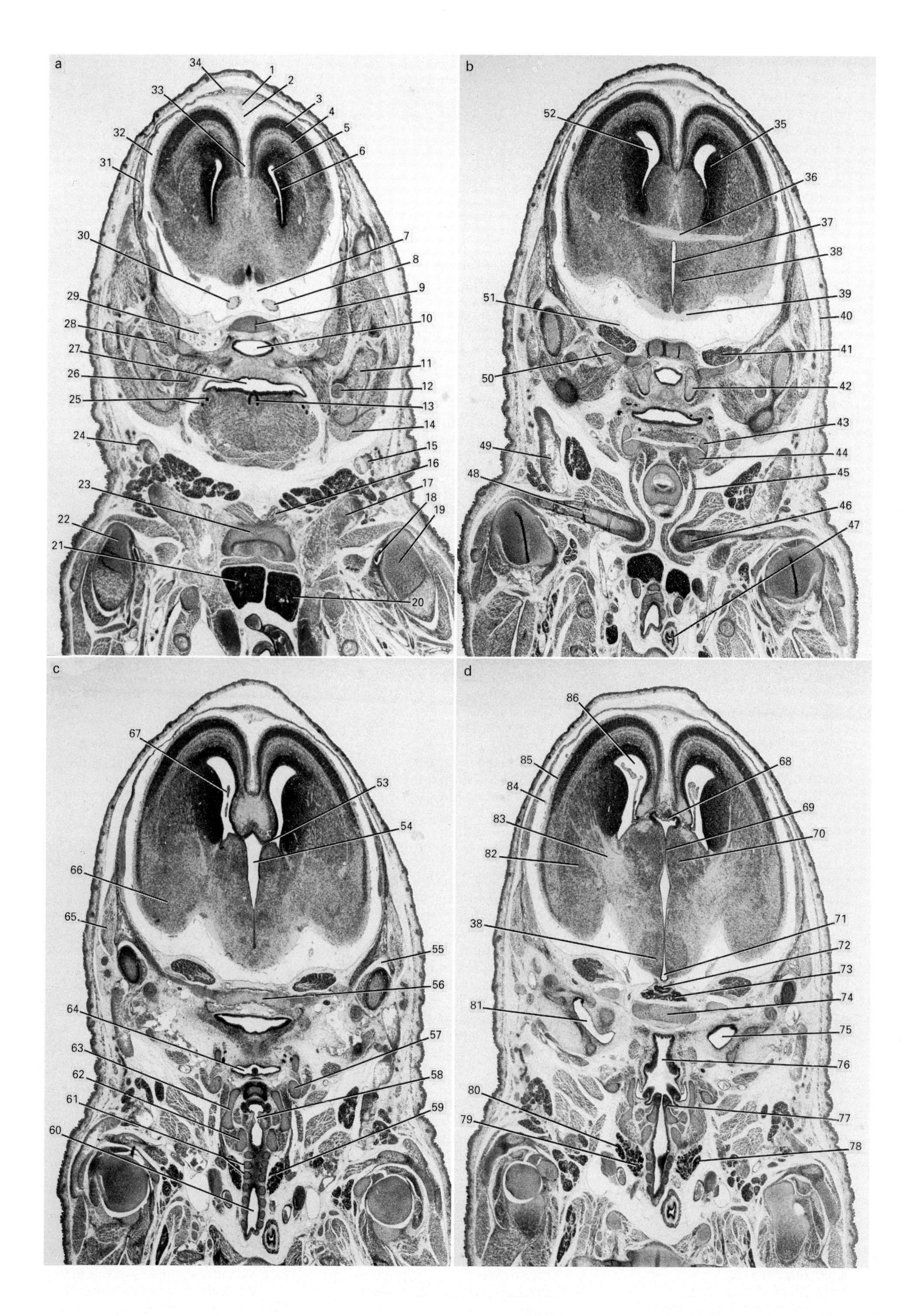
a
b
c
d
1
2
3
4
5
6
7
8
9
10
11
12
13
14
15
16
17
18
19
20
21
22
23
24
25
26
27
28
29
30
31
32
33
34
35
36
37
38
39
40
41
42
43
44
45
46
47
48
49
50
51
52
53
54
55
56
57
58
59
60
61
62
63
64
65
66
67
68
69
70
71
72
73
74
75
76
77
78
79
80
81
82
83
84
85
86

Plate 39b (16.5 days p.c.)

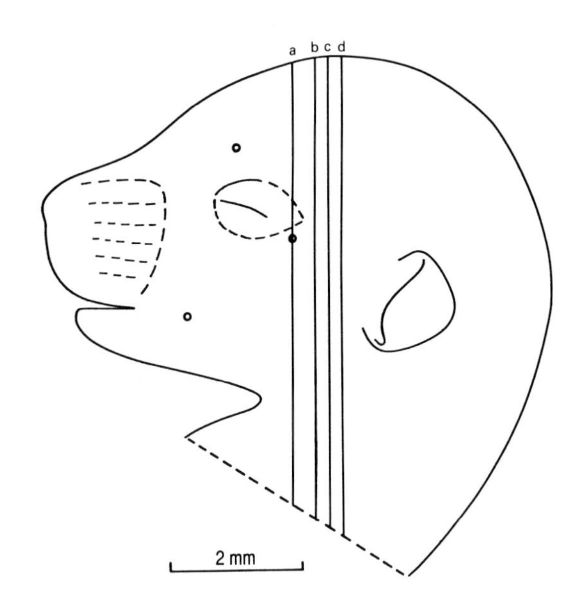

16.5 days p.c. Coronal sections through head (Theiler Stage 25; rat, Witschi Stage 34)

1. superior sagittal dural venous sinus
2. falx cerebri
3. inner broad nuclear layer (cortical plate) of neopallial cortex
4. intermediate zone
5. anterior extremity of superior part of anterior horn of left lateral ventricle
6. ependymal layer
7. basal (interpeduncular) cistern (cisterna interpeduncularis, of subarachnoid space)
8. left optic (II) nerve
9. cartilage primordium of presphenoid bone
10. nasopharynx
11. ossification within ramus of the mandible
12. left Meckel's cartilage
13. median circumvallate (vallate) papilla
14. inferior fibres of left masseter muscle
15. left carotid body
16. lower fibres of insertion of left sternomastoid muscle
17. left external jugular vein
18. tendon of the long head of the left biceps brachii muscle
19. cartilage primordium of upper part of shaft of left humerus
20. left lobe of thymus
21. right lobe of thymus
22. cartilage primordium of upper part of shaft of right humerus
23. cartilage primordium of manubrium sterni
24. right carotid body
25. developing glandular ducts opening onto posterior and lateral surface of the tongue
26. right medial pterygoid muscle
27. oropharynx
28. lateral pterygoid muscle
29. branches of maxillary division of right trigeminal (V) nerve
30. right optic (II) nerve
31. ossification in membranous primordium of right frontal bone
32. subarachnoid space
33. inferior sagittal dural venous sinus
34. mesenchyme in region of future anterior fontanelle
35. striatum (ganglionic eminence) overlying the head of the caudate nucleus
36. anterior commissure
37. anterior extremity of third ventricle
38. diencephalon (hypothalamus)
39. optic chiasma
40. cutaneous muscle of the lower face and neck – platysma (panniculus carnosus); this muscle in this location is termed the m. cutaneus faciei et labiorum
41. trigeminal (V) ganglion in Meckel's cave
42. medial pterygoid plate
43. cartilage primordium of lesser horn of hyoid bone
44. cartilage primordium of lateral part of body and greater horn of hyoid bone
45. sternohyoid muscle
46. ossification within cartilage primordium of medial one-third of left clavicle
47. oesophagus
48. ossification in middle one-third of cartilage primordium of right clavicle
49. right external jugular vein
50. cartilage primordium of apex of petrous part of right temporal bone
51. fibrous floor of middle cranial fossa
52. anterior horn of right lateral ventricle
53. region just anterior to left interventricular foramen
54. third ventricle
55. left temporo-mandibular joint
56. ossification within cartilage primordium of basioccipital bone
57. cartilage primordium of greater horn of hyoid bone
58. laryngeal aditus
59. left lobe of thyroid gland
60. lumen of trachea
61. tracheal rings
62. cricoid cartilage
63. thyroid cartilage
64. entrance from pharynx into the oesophagus
65. right temporalis muscle
66. inferior part of right telencephalic hemisphere (future temporal lobe)
67. choroid plexus within right lateral ventricle
68. choroid plexus in pineal recess of third ventricle
69. interthalamic adhesion (massa intermedia)
70. diencephalon (thalamus)
71. infundibular recess of third ventricle
72. infundibulum (future pars nervosa) of pituitary
73. pars anterior of pituitary
74. ossification within cartilage primordium of hypophyseal fossa (sella turcica) of sphenoid bone
75. left pharyngo-tympanic (Eustachian) tube
76. entrance from pharynx into oesophagus (posteriorly) and trachea (anteriorly)
77. left arytenoid cartilage
78. left parathyroid gland
79. right lobe of thyroid gland
80. right parathyroid gland
81. right pharyngo-tympanic (Eustachian) tube
82. lenticular nucleus
83. right internal capsule
84. vascular elements of pia mater on surface of neopallial cortex
85. outer narrow relatively anuclear layer (marginal zone) of left neopallial cortex
86. superior part of body of right lateral ventricle

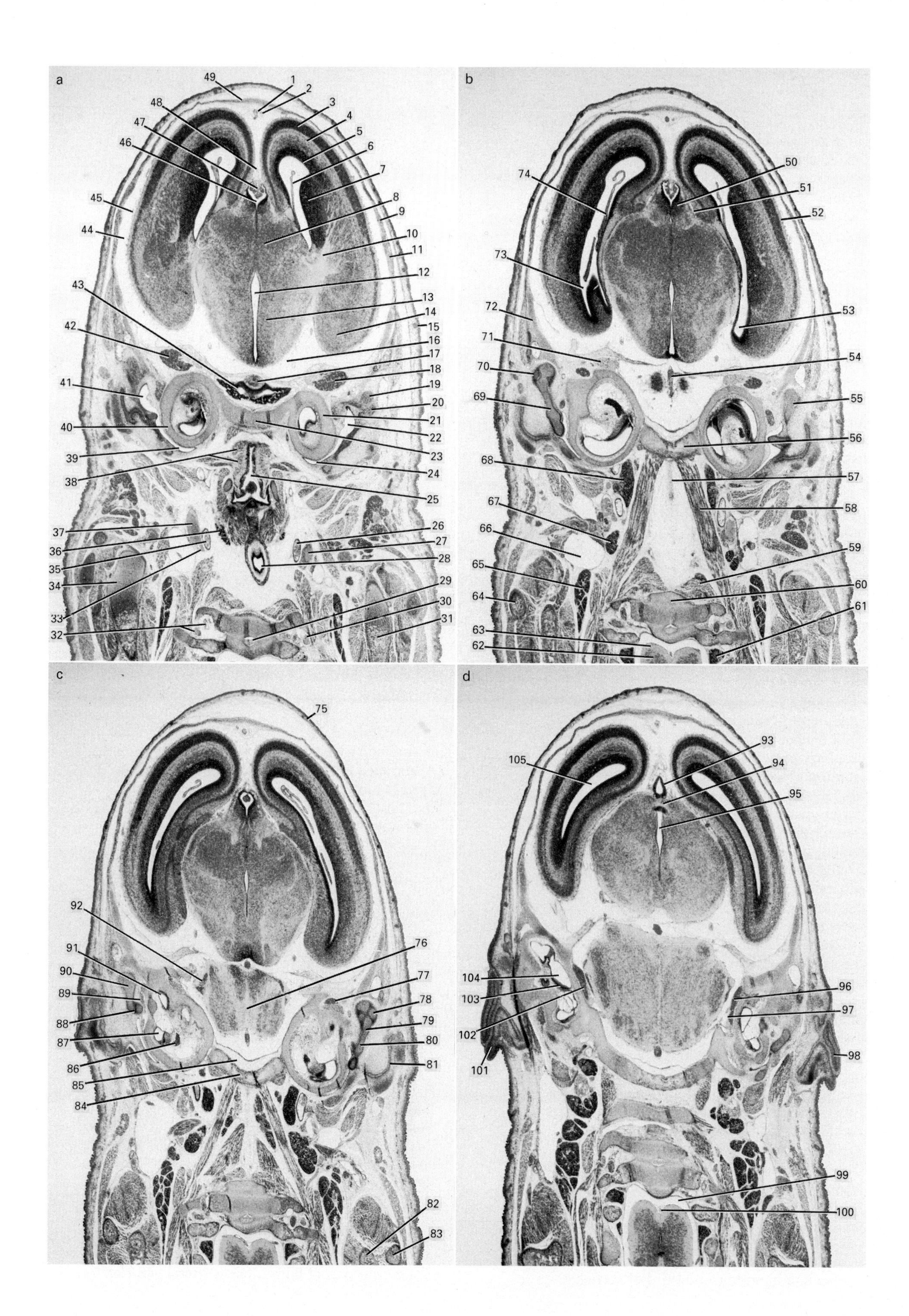
a
1
2
3
4
5
6
7
8
9
10
11
12
13
14
15
16
17
18
19
20
21
22
23
24
25
26
27
28
29
30
31
32
33
34
35
36
37
38
39
40
41
42
43
44
45
46
47
48
49
b
50
51
52
53
54
55
56
57
58
59
60
61
62
63
64
65
66
67
68
69
70
71
72
73
74
c
75
76
77
78
79
80
81
82
83
84
85
86
87
88
89
90
91
92
d
93
94
95
96
97
98
99
100
101
102
103
104
105

Plate 39c (16.5 days p.c.)

16.5 days p.c. Coronal sections through head (Theiler Stage 25; rat, Witschi Stage 34)

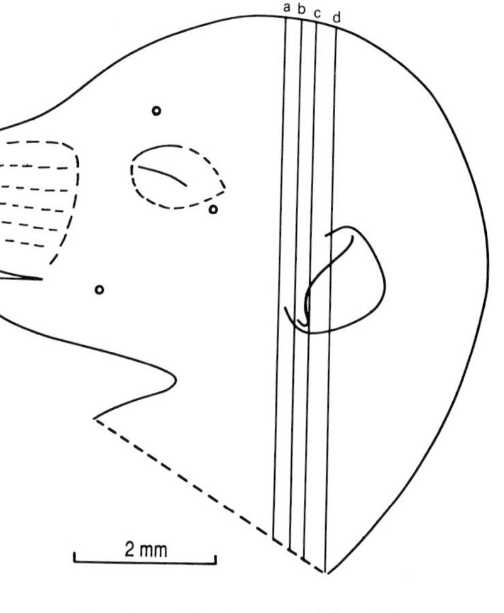

1. superior sagittal dural venous sinus
2. falx cerebri
3. inner broad nuclear zone (cortical plate) of neopallial cortex
4. intermediate zone
5. ependymal zone
6. choroid plexus within superior part of body of left lateral ventricle
7. striatum (ganglionic eminence) overlying the head of the caudate nucleus
8. diencephalon (thalamus)
9. ossification in outer table of cartilage primordium of squamous part of left temporal bone
10. left internal capsule
11. cartilage primordium of squamous part of left temporal bone
12. third ventricle
13. diencephalon (hypothalamus)
14. inferior part of right telencephalic hemisphere (future temporal lobe)
15. primordium of hair follicle
16. basal (interpeduncular) cistern (cisterna interpeduncularis, of subarachnoid space)
17. infundibulum (future pars nervosa) of pituitary
18. left trigeminal (V) ganglion
19. left Meckel's cartilage
20. left facial (VII) nerve
21. cartilage primordium of petrous part of left temporal bone
22. left tubo-tympanic recess
23. ossification within cartilage primordium of hypophyseal fossa (sella turcica) of sphenoid bone
24. left pharyngo-tympanic (Eustachian) tube
25. posterior part of pharynx
26. left common carotid artery
27. left internal jugular vein
28. oesophagus
29. nucleus pulposus
30. segmental cervical spinal nerve
31. ossification in cartilage primordium of shaft of left humerus
32. right vertebral artery and vein in foramen transversarium of cervical vertebra
33. right vagal (X) trunk
34. cartilage primordium of proximal part of right humerus
35. right common carotid artery
36. right lobe of thyroid gland
37. right internal jugular vein
38. fibres of superior and middle constrictor muscle
39. right pharyngo-tympanic (Eustachian) tube
40. cartilage primordium of petrous part of right temporal bone
41. right tubo-tympanic recess
42. right trigeminal (V) ganglion
43. residual lumen of Rathke's pouch
44. subarachnoid space
45. vascular elements of pia-arachnoid
46. pineal recess of third ventricle
47. choroid plexus within pineal recess of third ventricle
48. inferior sagittal dural venous sinus
49. mesenchyme in region of anterior fontanelle
50. diencephalon (epithalamus)
51. left choroid fissure
52. outer narrow relatively anuclear layer (marginal zone) of neopallial cortex
53. inferior part of posterior horn of left lateral ventricle
54. basilar artery
55. continuity between left Meckel's cartilage and cartilage primordium of the incus
56. ossification within cartilage primordium of anterior part of basioccipital bone (clivus)
57. pharyngeal raphe
58. prevertebral muscle (longus capitis)
59. prevertebral muscle of the neck (longus cervicis)
60. cartilage primordium of cervical vertebral body
61. dorsal (posterior) root ganglion
62. ventral horn of spinal cord
63. marginal zone (layer) forming white matter of cervical spinal cord
64. periosteum covering tip of spine of the left scapula
65. right cervicothoracic (stellate) sympathetic ganglion
66. right jugular lymph sac
67. right middle cervical sympathetic ganglion
68. right superior cervical sympathetic ganglion
69. cartilage primordium of right incus
70. articulation between cartilage primordia of right malleus and incus and Meckel's cartilage
71. rootlets of right trigeminal (V) nerve
72. cartilage primordium of squamous part of right temporal bone
73. inferior part of posterior horn of right lateral ventricle
74. choroid plexus within right lateral ventricle
75. skin (and subjacent epicranial aponeurosis (galea aponeurotica))
76. floor of pons
77. facial (VII) ganglion
78. cartilage primordium of head (caput) of the left malleus
79. cartilage primordium of body (corpus) of the left incus
80. cartilage primordium of long limb (crus longum) of left incus
81. external acoustic meatus (not yet canalized)
82. ossification in cartilage primordium of upper part of blade of left scapula
83. ossification in cartilage primordium of spine of left scapula
84. ossification in central part of cartilage primordium of basioccipital bone (clivus)
85. fibrous floor of middle cranial fossa
86. cochlear (VIII) ganglion (spiral ganglion)
87. lateral semicircular canal
88. articulation between cartilage primordia of right incus and stapes
89. right stapedial artery
90. cartilage primordium of body (corpus) of the right incus
91. superior semicircular canal
92. rootlets of origin of right facial (VII) nerve
93. wall of pineal stalk (peduncle)
94. rostral part of roof of midbrain
95. rostral part of mesencephalic vesicle
96. left vestibular (VIII) ganglion
97. left internal acoustic meatus
98. pinna of left ear
99. spinal canal
100. anterior spinal artery
101. pinna of right ear
102. right vestibular (VIII) ganglion
103. spiral organ of Corti
104. cochlear duct
105. posterior horn of right lateral ventricle

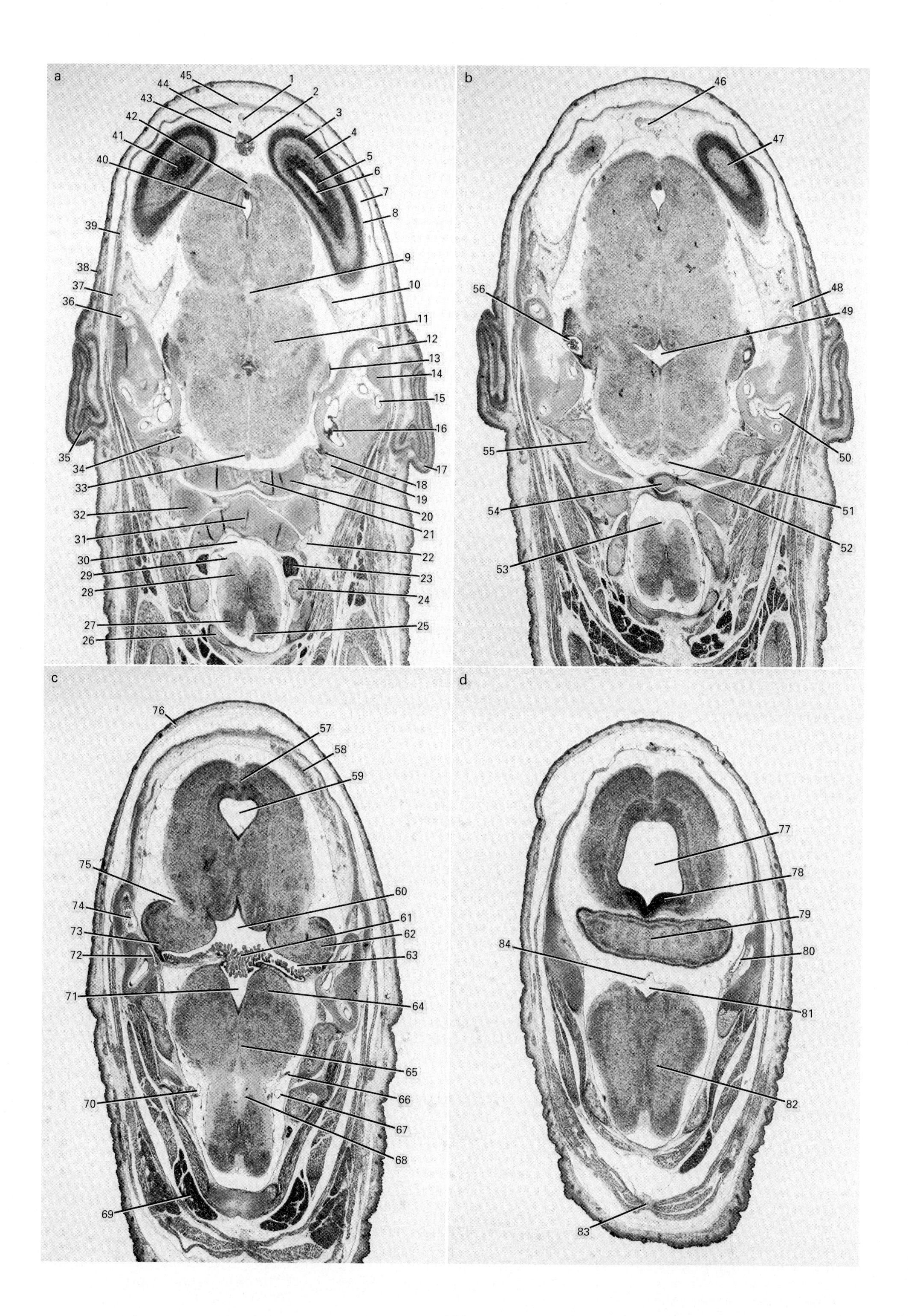
a
b
c
d
1
2
3
4
5
6
7
8
9
10
11
12
13
14
15
16
17
18
19
20
21
22
23
24
25
26
27
28
29
30
31
32
33
34
35
36
37
38
39
40
41
42
43
44
45
46
47
48
49
50
51
52
53
54
55
56
57
58
59
60
61
62
63
64
65
66
67
68
69
70
71
72
73
74
75
76
77
78
79
80
81
82
83
84

Plate 39d (16.5 days p.c.)

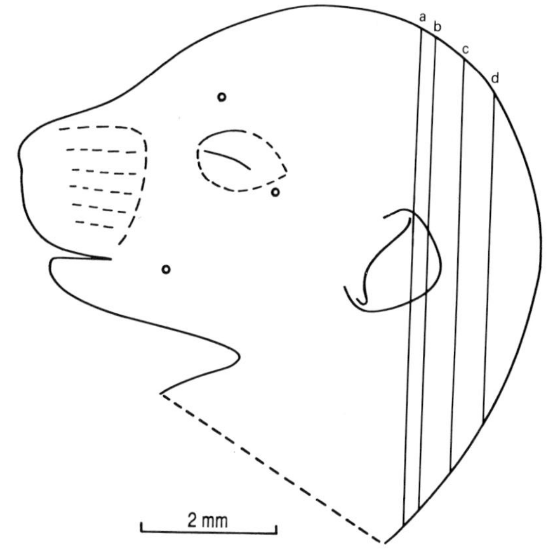

16.5 days p.c. Coronal sections through head (Theiler Stage 25; rat, Witschi Stage 34)

1. superior sagittal dural venous sinus
2. pineal body
3. inner broad nuclear zone (cortical plate) of neopallial cortex
4. intermediate zone
5. ependymal zone
6. superior part of posterior horn of left lateral ventricle
7. subarachnoid space
8. vascular elements of pia-arachnoid
9. junction between pons and midbrain
10. left half of tentorium cerebelli
11. pons
12. superior semicircular canal
13. fibres of origin of left vestibular (VIII) ganglion
14. cartilage primordium of petrous part of left temporal bone
15. lateral semicircular canal
16. spiral organ of Corti
17. pinna of left ear
18. glossopharyngeal (IX) ganglion
19. origin of left internal jugular vein in jugular foramen
20. ossification within cartilage primordium of lateral part of basioccipital bone (clivus)
21. ossification in central part of cartilage primordium of basioccipital bone (clivus)
22. left vertebral artery
23. upper cervical dorsal (posterior) root ganglion
24. ossification in cartilage primordium of lamina of cervical vertebra
25. posterior spinal vein
26. ossification within cartilage primordium of right half of neural arch of cervical vertebra
27. dorsal horn of cervical spinal cord
28. ventral horn of cervical spinal cord
29. white matter of cervical spinal cord
30. spinal canal
31. cartilage primordium of body of C2 vertebra (axis)
32. cartilage primordium of lateral mass of C1 vertebra (atlas)
33. basilar artery
34. right jugular foramen
35. pinna of right ear
36. right superior semicircular canal
37. cartilage primordium of squamous part of right temporal bone
38. primordium of hair follicle
39. ossification in membranous primordium of right parietal bone
40. rostral part of mesencephalic vesicle (future cerebral aqueduct)
41. posterior part of wall of right neopallial cortex (right cerebral hemisphere)
42. roof of rostral part of midbrain (tectum)
43. part of substantial venous anastomosis around posterior part of pineal body
44. falx cerebri
45. mesenchymal condensation in interparietal region in location of future sagittal suture
46. large anastomotic vein which communicates with the superior sagittal dural venous sinus and the right inferior anastomotic vein
47. posterior part of wall of left neopallial cortex (left cerebral hemisphere)
48. origin of left sigmoid dural venous sinus
49. floor of fourth ventricle
50. cochlea
51. site of anastomosis between left and right vertebral arteries to form the basilar artery
52. transverse ligament of atlas
53. anterior spinal artery
54. cartilage primordium of rostral part of dens (odontoid process of C2 vertebra (axis)) with periosteal ossification
55. ossification in cartilage primordium of right exoccipital bone
56. choroid plexus in right lateral recess of fourth ventricle
57. caudal part of roof of midbrain
58. large anastomotic vein which communicates with the superior sagittal dural venous sinus and the left inferior anastomotic vein
59. caudal part of mesencephalic vesicle (future cerebral aqueduct)
60. rostral part of fourth ventricle
61. choroid plexus arising from roof of fourth ventricle
62. rostral extension of primitive cerebellum
63. left lateral recess of fourth ventricle
64. rostral part of medulla oblongata
65. floor of medulla oblongata
66. rootlets of spinal part of accessory (XI) nerve
67. left vertebral artery (intracranial part)
68. junction between caudal part of medulla oblongata and rostral part of cervical spinal cord
69. deposits of brown (multilocular) fat (found principally between the scapulae, in the ventral part of the neck and in the axillary region)
70. right vertebral artery (intracranial part)
71. caudal part of fourth ventricle
72. right endolymphatic duct
73. rhombic lip (dorsal part of alar lamina of metencephalon – location of the lateral part of the cerebellar primordium)
74. sigmoid dural venous sinus
75. cerebellar artery
76. skin (and subjacent epicranial aponeurosis (galea aponeurotica))
77. junction between caudal part of mesencephalic vesicle and rostral part of fourth ventricle
78. tegmentum of pons (basal plate)
79. cerebellar plate
80. left endolymphatic duct
81. caudal extremity of fourth ventricle
82. medulla oblongata
83. ligamentum nuchae
84. roof of ependymal diverticulum of fourth ventricle

cerebral aqueduct into the rostral part of the fourth ventricle (*Plate 37a,a–e*), and the location and gross morphology of the primitive cerebellum is easily understood. Similarly, the location of the choroid plexus that arises in various parts of the fourth ventricle is also clearly demonstrated in these sections (*Plate 37a,e*), as well as the extremely thin roof of the ependymal diverticulum of the fourth ventricle, which at this stage is virtually the only part of the roof of the fourth ventricle that still displays this feature (*Plate 37a,e*).

The other structures that are perhaps more easily recognized in transverse than in coronal sections through the head are the various cranial nerve ganglia. The latter are principally recognized because of their size and characteristic cellular contents (the ganglion cells), but also because of their location in relation to other intra-cranial (and extra-cranial) structures. The trigeminal (V) is by far the largest of the cranial nerve ganglia, and is easily recognized whatever the plane of section (*Plate 37b,b,c*; *Plate 39b,b–d*; *Plate 39c,a*). The facial (VII) and vestibular/acoustic (VIII) ganglia are also easily recognized because of their proximity to the medial/posteromedial aspect of the cartilage primordia of the petrous part of the temporal bone (*Plate 37b,a*; *Plate 39c,c,d*; *Plate 39d,a*), Similarly, the glossopharyngeal (IX) and vagal (X) ganglia are recognized because of their relationship to the jugular foramen (*Plate 37b,c,d*). While the major branches of the trigeminal (V) nerve are readily followed in low magnification sections, and are labelled in the appropriate sections, the smaller cranial nerves invariably require to be studied in higher magnification sections than those illustrated here, and preferably in silver-stained material. While much can be seen in haematoxylin and eosin stained sections, such as, for example, the branches of the maxillary division of the trigeminal (V) nerve which pass from the plexuses around the follicles of the vibrissae to the trigeminal (V) ganglion (*Plate 38c,b*), for a more detailed study, silver-stained material is essential.

The transition from the medulla oblongata to the rostral part of the cervical spinal cord is seen in both the coronal (*Plate 39d,c*) and transversely sectioned embryos (*Plate 37b,b–d*), as is its relationship to the basilar artery. In this regard, the rootlets of both the spinal and cranial parts of the accessory (XI) nerves are also clearly seen (*Plate 37b,c*). Because the features of the spinal cord differ according to their location along the spinal axis, and in order to emphasize this point, sections through the upper cervical, mid-thoracic and mid-lumbar regions of the spinal cord at this stage of development are illustrated (*Plate 63,g–i*, respectively). The difference in their cross-sectional areas are particularly marked, the increased size of the sections through the cervical and lumbar regions of the spinal cord being related to their proximity to the brachial and lumbosacral plexuses, involved principally in the innervation of the fore- and hindlimbs, respectively. It is of particular note that at this stage of development the lumen of the central canal is almost completely obliterated throughout its length. Furthermore, it is of interest that the spinal cord extends caudally to the mid/lower coccygeal region (*Plate 37l,e*), and that the level of the conus medullaris is slightly less caudally located than the situation described previously (*Plate 35l,c–e*).

The adrenal glands are particularly well vascularized at this stage (*Plate 37j,b,c*; *Plate 38a,b*; *Plate 38b,a*). The cortical tissue tends to consist principally of large cells with pale staining cytoplasm arranged in columns around the primitive cortical sinusoids, while the medullary cells, which are located in the hilum and towards the central region of the gland, tend to be of smaller dimensions, with darker staining nuclei, and are aggregated together in rounded masses. The adrenal gland at this stage is also seen to be surrounded by a primitive capsule.

THE UROGENITAL SYSTEM

The kidneys are seen to be more rounded, and are more clearly divided into a peripherally located cortical region and a centrally located medullary region than previously (*Plate 72,i*; cf. *Plate 72,g*). At the histological level, the majority of the medullary region consists of undifferentiated mesenchyme tissue with radially arranged large-diameter collecting tubules. The glomeruli now tend to be confined to the periphery of the cortical region, and just subjacent to these are located the proximal and distal convoluted tubules (*Plate 72,j*), which are first clearly seen at this stage of development. Towards thc periphery of the kidney, and just subjacent to its fibrous capsule, undifferentiated metanephric cap tissue is still seen to be present.

Little difference is observed in the external appearance of the testes and their associated ducts between 15.5–16 and 16.5–17 days p.c., though the testes have descended over this period to a slightly more caudal position than previously (*Plate 70,c,d*; cf. *Plate 70,e,f*). The testes are partially surrounded, but particularly on their posterior aspects by the various components of the epididymis, at the lower pole of which the ductus deferens emerges. From the lower pole of the testis, the gubernaculum testis may be seen as it descends towards the internal inguinal ring (see *Plate 40k,d*). In sagittal sections, it is possible to see that the testis has almost completed its descent, and its lower pole is closely associated with the fascia overlying the pelvic wall musculature (*Plate 38a,a*). At the histological level, the primordia of the seminiferous tubules are seen to be solid structures, and are principally located towards the outer one-half or slightly more of the testis. The remainder of the medullary region of the testis consists of mesenchyme tissue (*Plate 69,c,d*). Between the seminiferous tubules, early evidence of interstitial tissue is seen (*Plate 69,a*). The tunica albuginea (formed by the condensation of superficial mesenchyme tissue, with a covering layer of peritoneal mesothelial cells) is substantially thicker than previously (*Plate 69,e*; cf. *Plate 68,o*).

The ovaries are markedly different in their appearance at 16.5–17 days p.c. compared with the situation observed at 15.5–16 days p.c., principally because their growth in size over this period does not appear to keep pace with that of the surrounding structures (*Plate 71,d*; cf. *Plate 71,a*). Whereas the overall length of the ovary was about one-half of the length of the kidney at 15.5–16 days p.c., it is now seen to be closer to 20% of the length of the kidney. The ovaries are still ovoid in shape, and located at about the same level as previously, namely at about the level of the lower poles of the kidneys (*Plate 37k,b–d*; *Plate 37l,a,b*; *Plate 71,c–e*). The ovaries, over this period, have become increasingly surrounded by the paramesonephric duct tissue that is destined to form the capsule of the ovary (see *Plate 71,i*). The suspensory ligament of the ovary is still seen to be a prominent feature at this stage of development (*Plate 71,e*). The two paramesonephric ducts (the future uterine horns) are directed caudally and towards the midline, and meet just behind the neck of the bladder. During their course, they pass in front of the two ureters (*Plate 81k,d*). Where they meet in the midline, the two paramesonephric ducts become closely apposed, but their lumina are separated by their two intervening walls. Within the ovary at this stage, many of the oogonia have progressed to form primary oocytes, and are seen to have entered prophase of the first meiotic division.

THE SKELETAL SYSTEM

Analysis of alizarin-stained "cleared" embryos provides the most useful guide to the degree of ossification present within the embryo at this stage of development. However, it is also still advantageous to analyse serially sectioned material to assess the degree of ossification present within individual components of the skeletal system. It should be appreciated, however, that all of the centres of ossification noted at the previous stage of development have increased in size, and in some cases quite substantially so. Possibly the most impressive changes may be observed with regard to some of the bones of the skull, so that an increased degree of ossification is now evident in the maxillary, basioccipital and basisphenoid bones, as well as in the mandible (*Plate 37b,b,c*; *Plate 37c,a–d*).

A marked change is also observed with regard to the ossification present within the vertebral axis. Whereas extensive areas of ossification are observed in the pedicles and laminae as well in the proximal one-half to two-thirds of nearly all of the neural arches (*Plate 37b,d*; *Plate 37d,b*) (the so-called "lateral masses" of these vertebrae), only a thin rim of periosteal ossification has recently appeared in the mid-thoracic and upper lumbar vertebral bodies. In the cervical (with the exception of C1), upper thoracic and lower lumbar regions, as well as in the more caudal parts of the vertebral axis, no evidence of ossification in relation to the bodies of the vertebrae is yet seen (*Plate 81,g,h*). The pattern of ossification described here would seem to be remarkably similar to that described for the human embryo/fetus, where the centres of ossification in the vertebral bodies are first seen in the lower thoracic region (at the ninth to tenth week p.c.), and then develop at successively higher and lower levels, so that the ossification centre in the second cervical vertebral body appears by the end of the fourth month (Warwick and Williams, 1973).

In the scapulae, clavicles and iliac bones, which previously showed moderate degrees of ossification, more extensive areas of ossification are now seen (*Plate 37d,c,d*; *Plate 37e,a*), though the most peripheral parts of these cartilage primordia have yet to be involved in this process. An ossification centre may also be seen in the cartilage primordium of the ischium (*Plate 81,h*), and a smaller centre may be present in the middle of the pubic bone at this time.

With regard to the long bones, all now also show increased levels of ossification within their shafts (e.g. *Plate 37e,b–e*; *Plate 37j,c,d*), though the distal parts of these bones show no evidence of this process (*Plate 81,g,h*). In the more distal parts of the limbs, a limited degree of periosteal ossification may be evident in the metacarpal and metatarsal bones in the most advanced embryos at this stage, though the carpal and tarsal bones as well as the phalanges show, as yet, either no evidence or only minimal evidence of ossification. The only other bones in which extensive areas of ossification are seen at this stage are the ribs (e.g. *Plate 37f,a–c*), and in the sternebrae a thin rim of periosteal ossification may also be seen in the most advanced embryos at this stage of development (*Plate 37h,a,b*).

Stage 26 Developmental age, 18 days p.c.
Plates 40a–l, Plate 41

EXTERNAL FEATURES

Most of the external features of embryos at this stage of development are similar to those seen in newborn mice, and are not markedly different from the features displayed by embryos at the previous stage of development. Possibly the most obvious change is a particularly gradual one, and this relates to the difference in the proportion of the cranial compared with the postcranial axis observed at this stage, when compared with the situation observed at previous stages of development. If the crown–rump length is taken as the unit of measurement, and then the proportion of this unit accounted for by the cranial region (taken for simplicity as the region rostral to the upper part of the shoulder/where the lower border of the mandible encroaches on the soft tissues of the neck), is plotted against the developmental age of the embryo (in days p.c.), this provides a reasonable indication of the gradual (linear) change that occurs in the bodily proportions of the embryo over the period from 11.5 to 18.5 days p.c. (see Fig. 11). While the figures obtained are very approximate, they do show a distinct trend, namely that with increasing developmental age, a diminishing proportion of the crown–rump length is accounted for by the cranial region (the values used to plot this graph were obtained from an analysis of the embryos illustrated in *Plate 46*).

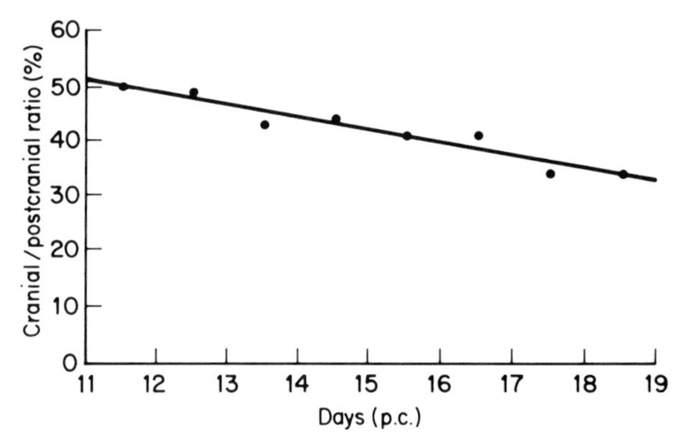

Figure 11 The relationship between cranial and postcranial dimensions between 11.5 and 18.5 days p.c.

While the trunk region of the embryo gradually increases in length, that of the tail appears not to keep pace with the rest of the vertebral axis. Whereas up to about 12.5 days p.c. the tail is seen to be extremely long compared to the rest of the vertebral axis, its proportionate length gradually diminishes, so that at about 18.5 days p.c., the tail is proportionately shorter than it has been at any previous stage of development. Over this period, the tail also becomes proportionately wider, and its tip less pointed than previously. The surface texture of the skin is similar to that observed at the previous stage of development, and the wrinkles are now seen to cover all but the most distal parts of the limbs, the tail, and the facial region.

With the passage of each successive day of gestation, an increasing proportion of the external acoustic meatus becomes covered by the pinna of the ear, so that at this time, relatively little of its lumen is now visible. Elsewhere in the head region, in the vibrissae-bearing areas of the upper lips, the bases of the vibrissae are more prominent, and their length and diameter is somewhat greater than previously. The appearance of the eyelids and the rest of the facial region is similar to that described previously.

Little difference is observed in the appearance of the limbs between the situation observed at 16.5–17 and 17.5–18 days p.c. The dorsal surface of the hand plate region of the forelimb is similar to that observed previously (*Plate 45,l*; cf. *Plate 45,g*), and this equally applies to the palmar surface of the hand plate (*Plate 45,m*; cf. *Plate 45,h*). In the case of the foot plate region of the hindlimb, while the appearance of the dorsal surface is similar at these two stages of development (*Plate 45,n*; cf. *Plate 45,i*), that of the plantar surface is somewhat different, in that the pads characteristically seen in this location are more prominent than the situation observed previously (*Plate 45,o*; cf. *Plate 45,j,k*). In both the fore- and hindlimbs, the appearance of the claw primordia in the region principally overlying the dorsum and tip of the distal phalanges is also quite similar.

THE HEART AND VASCULAR SYSTEM

The heart has reached its definitive prenatal state, and shows only a few changes in its gross morphology from the situation observed at 16.5–17 days p.c., though the coronary vessels are now considerably more easily seen and their courses more easily traced than previously (e.g. *Plate 40f,d*). The difference in thickness between the walls of the atria and ventricles is essentially the same as indicated previously, as is the substantial difference in the volumes of the atria compared to those of the ventricles (*Plate 40f,b–d*; *Plate 40g,a–d*). In certain parts of the wall of the ventricles, the thickness appears to be substantially less than that observed previously, and this is particularly noticeable in the region subjacent to and just to the right of the interventricular groove (*Plate 40g,b–d*; cf. *Plate 37f,c,d*; *Plate 37g,a,b*). The ventricular wall is now substantially less consolidated than previously, and this may be accounted for by the significant increase in the degree of trabeculation observed at this time (e.g. *Plate 40g,c*). This change may be a reflection of the differentiation of the ventricular muscle that occurs at this time, in preparation for the increased functional load that it will have to sustain after birth.

No obvious differences are observed between the components of the vascular system seen at 16.5–17 and 17.5–18 days p.c., though some of the major vessels are particularly well seen at this stage of development. This particularly applies to the larger components of the venous system, such as the external jugular veins and the subclavian veins (*Plate 40d,a–d*; *Plate 40e,a,b*), and the right and left superior (cranial) venae cavae, all of which are seen to be vessels with very considerable luminal diameters (*Plate 40e,c,d*; *Plate 40f,a–d*). Both of the superior (cranial) venae cavae enter the right atrium, the left superior (cranial) vena cava passing across the midline posterior to the heart to do so (*Plate 40g,a,b*). The posthepatic inferior vena cava is a vessel of about the same luminal diameter as each of the superior (cranial) venae cavae, and its pathway rostrally towards the right atrium can also very easily be traced from its origin in the liver to its termination in the right atrium (*Plate 40g,a–e*; *Plate 40h,a,b*).

Within the liver, the ductus venosus is seen to be another vessel with a large luminal diameter, being formed principally from the left umbilical vein (*Plate 40h,c,d*; *Plate 40i,a–d*). By contrast, the proximal (pre- and early postductal) component of the thoracic aorta has a substantial luminal diameter, though this is smaller than that of any of the major veins indicated above. However, as the aorta passes caudally from the thorax, its lumen appears to rapidly diminish, so that by the time it passes behind the crura of the diaphragm into the abdomen, its lumen already appears to be substantially reduced (*Plate 40i,a–c*), though it becomes dilated once more in its immediate prerenal course (*Plate 40i,d*; *Plate 40j,a–c*), before narrowing again in its post-renal course (*Plate 40k,a–d*; *Plate 40l,a*). If the diameters of the common iliac veins and arteries are compared, those of the veins are seen to be significantly greater than those of the arteries (*Plate 40l,b,c*). Within the umbilical cord, the diameters of the umbilical arteries are seen to be considerably smaller than those of the veins, though somewhat surprisingly, the left umbilical artery has a much greater luminal diameter than that of the right umbilical artery (*Plate 40j,b–d*; *Plate 40k,a–d*). In general terms, it appears that the volume of the venous system (including that of the atria) is significantly greater than that of the arterial side of the circulation, at this stage of development.

THE PRIMITIVE GUT AND ITS DERIVATIVES

As with the heart and vascular system, little difference is observed between the appearance of the gut and its derivatives between 16.5–17 and 17.5–18 days p.c. Both the tongue and the palate are essentially fully differentiated, and the relationship between the oropharynx and the nasopharynx is in its definitive state. It is interesting to note, however, that, as observed previously, the posterior/inferior border of the nasal septum subjacent to the location of the vomeronasal organs (Jacobson's organs), represents the only site where the two nasal cavities are in continuity (*Plate 40b,a*; cf. *Plate 39a,d*). Further posteriorly, the nasopharynx passes backwards to the site where the oropharynx and nasopharynx are in continuity (*Plate 40c,c*; cf. *Plate 39b,d*). Slightly anterior to this location, the two pharyngo-tympanic (Eustachian) tubes pass laterally and slightly posteriorly towards the tubo-tympanic recess, to allow continuity between the oropharynx and the cavity of the middle ear (*Plate 40b,d*; *Plate 40c,a*). The latter is gradually increasing in volume, and will in due course provide the space in which the middle ear ossicles can function as mechanical transducers between the tympanic membrane and the cochlear apparatus of the inner ear.

Two openings are located close to the beginning of the cochlea, namely the fenestra cochleae, which communicates with the tympanic cavity, and is closed by the secondary tympanic membrane, and the fenestra vestibuli which is occupied by the base of the stapes. The latter is attached at the borders of its oval baseplate by the annular ligament (a third minute orifice is also present, termed the cochlear canaliculus, which transmits a small vein which joins the inferior petrosal sinus, and also allows communication between the subarachnoid space and the scala tympani).

Whereas at the previous stage of development, it was possible to clearly recognize a single upper molar tooth (the first upper molar tooth), there being only a suggestion of a second molar tooth more posteriorly (*Plate 37b,d*), these tooth primordia being located in the maxilla just lateral to the lateral margin of the palatal shelf, by 17.5–18 days p.c., both a first and a second upper molar tooth are now recognized (*Plate 40b,d*). The degree of differentiation of the second upper molar tooth, in developmental terms, is initially up to about 3–4 days behind that of the first upper molar tooth, since the first upper molar tooth is first clearly recognized at about 12.5–13 days p.c. (*Plate 28b,e,f*). However, despite the fact that over the period between 16.5–17 and 17.5–18 days p.c., the second upper molar tooth differentiates rapidly, it is still significantly less well differentiated than the first upper molar tooth at this stage of development. In the upper jaw, there is no suggestion at this stage of a third molar tooth, and in the lower jaw there is no suggestion of either a second or a third molar tooth at this stage. The teeth which show the most advanced degree of differentiation at this stage are the incisors (upper incisors, *Plate 40b,a–c*; lower incisors, *Plate 40c,b–d*).

In the neck region, the isthmus of the thyroid gland is at about the same level as previously, and crosses the midline at the level of the third tracheal "ring". At the histological level, it is now possible to recognize large numbers of small colloid-filled follicles within the thyroid gland. The arterial blood supply to the gland comes principally from several branches of the superior thyroid artery, and a smaller branch from the inferior thyroid artery, all of which enter the gland from its posteromedial aspect (*Plate 40d,c*). In this embryo, the left parathyroid gland is embedded within the middle part of the left lobe of the thyroid gland, whereas the right parathyroid gland is found in an aberrant location, namely within the carotid sheath some distance lateral to the right lobe of the thyroid gland, and at the level of the isthmus (*Plate 40d,d*). Both parathyroid glands are seen to have a greater degree of vascularity than previously. The U-shaped tracheal "rings" are now fully differentiated, and it is possible to recognize the trachealis muscle which runs along the posterior part of the trachea to complete the "rings" (*Plate 40d,d*).

The thymus gland has markedly increased in size between the situation observed at 16.5–17 and 17.5–18 days p.c. The gland is now seen to have a thin fibrous capsule, and has descended further into the superior part of the anterior mediastinum than previously (*Plate 40e,b–d*; *Plate 40f,a,b*; cf. *Plate 37d,d*; *Plate 37e,a–c*). This gland is now extremely large, highly vascularized and reasonably homogeneously darkly staining due to the presence of enormous numbers of lymphocytes. Each half of the thymus gland does not appear to be divided into separate lobules, but dispersed within the lymphoid tissue are moderately large aggregations of cells with a pale-staining nucleus, surrounded by a rim of pale-staining cytoplasm. The appearance of the latter suggests that this tissue may be the precursor of the medullary core of the gland. The characteristic interlobar septa, related to the vascular architecture, seen in the adult gland, has yet to appear. The posterior surface of the thymus is indented by the right and left superior (cranial) venae cavae, anteriorly by the posterior surface of the manubrium and upper sternebrae (*Plate 40e,c,d*), and inferiorly by the upper surface of the walls of the right and left atria (*Plate 40f,b*). The medial surface of the apical region of the cranial lobe of the right lung also impinges on its right inferolateral surface (*Plate 40f,b*).

The relationship between the oesophagus and the trachea is similar to that seen at the previous stage of development. As the oesophagus and trachea descend in the neck towards the thoracic inlet, their relationship to each other gradually changes. The trachea is seen to remain a midline structure along its entire length (to the level of its bifurcation) (*Plate 40d,b–d*; *Plate 40e,a–d*; *Plate 40f,a–c*), while the oesophagus from shortly after its origin becomes deviated towards the left of the midline, and a posterolateral relation of the trachea (*Plate 40d,b–d*). Further caudally, at the level of the clavicles, the oesophagus is now seen to be a lateral relation of the trachea, and the latter is an anterior relation of the lower cervical and upper thoracic vertebral bodies. As the oesophagus descends through the thorax, it tends towards the midline once more (*Plate 40g,a–e*; *Plate 40h,a–c*), and only deviates towards the left side once more after it has passed through the diaphragm into the abdomen, to enter the stomach at the gastro-oesophageal junction (*Plate 40i,d*).

Beyond about the mid-thoracic level, the oesophagus is seen to be closely associated with the anteromedial/medial wall of the descending thoracic aorta (*Plate 40f,c,d*), and then becomes a slightly more distant anterior relation of this vessel (*Plate 40g,d–f*; *Plate 40h,a–d*), until the oesophagus passes towards the left to enter the stomach. The aorta, however, continues caudally just marginally to the left of the midline (e.g. *Plate 40j,a,b*).

If the trachea is now followed caudally, it bifurcates just below the level of the arch of the aorta, to give rise to the left and right main bronchi (*Plate 40f,c,d*). At about the same level, the pulmonary trunk gives off the right and left pulmonary arteries from its posterior aspect (*Plate 40f,c*). The latter are directed laterally, and follow the main bronchi into the hilar region of the lungs. A single left main bronchus is present, consistent with the presence of a single (undivided) left lung, while the right main bronchus divides soon after it enters the right lung into lobar bronchi, which then pass to the cranial, middle, caudal and accessory lobes of the right lung (*Plate 40f,d*; *Plate 40g,a–c*).

As noted in the description of the detailed histological morphology of the lungs in the previous stage of development, while differentiation of the lungs undoubtedly occurs during the period between 16.5–17 and 17.5–18 days p.c., the appearance of the lung fields in this embryo probably does not represent the embryonic situation seen *in utero*. It is likely that the alveolar expansion seen here is due to the embryo taking several gasps of air between the period of its isolation from within the uterus and its fixation. The histological features of the lungs in this embryo (*Plate 66,o,p*), therefore, more closely resemble the appearance of the lungs in the *very early* newborn period, as full alveolar expansion has yet to occur. In this embryo, consistent with its alveolar immaturity, the cells that line the alveolar sacs are cuboidal, whereas in developmentally more differentiated lung tissue, these cells would be expected to have a squamous morphology, in order to facilitate gaseous exchange across the gas/liquid interface.

It is possible to trace the two vagal (X) trunks from their origin at the vagal (X) ganglia in the region of the jugular foramen (*Plate 40c,b*), through the neck (e.g. *Plate 40d,a,b*; *Plate 40e,a*), where they are close medial relations of the right and left superior (cranial) venae cavae. The right vagal (X) trunk is seen to pass in front of the right subclavian artery, where it gives off a recurrent (laryngeal) branch which ascends rostrally after it hooks under the right subclavian artery. On the left side, the vagal (X) trunk descends further caudally, then passes in front of the arch of the aorta before giving off its recurrent laryngeal branch. The latter passes initially medially under the concavity of the arch of the aorta, just distal to the site of entrance into the aorta of the ductus arteriosus (*Plate 40f,b*), before ascending (rostrally) to supply all of the muscles of the larynx, excepting the cricothyroid, as well as being sensory to the mucous membrane of the larynx below the level of the vocal folds.

Only in the mid-thoracic region do the two vagal (X) trunks become closely associated with the oesophagus (as the right and left vagal (X) trunks), and this relationship is retained until the lower thoracic region, where the right vagal (X) trunk becomes more posteriorly related, and the left trunk more anteriorly related than previously (*Plate 40h,b*). The two vagal (X) trunks are first seen to be directly anterior and posterior relations of the oesophagus where it passes through the diaphragm (*Plate 40h,d*). This relationship is retained until the gastro-oesophageal junction is reached, when the posterior vagal (X) trunk is seen to be located in the dorsal meso-oesophagus (*Plate 40i,d*), while the anterior vagal (X) trunk is still closely adherent to the anterior wall of the oesophagus.

The stomach is an extremely dilated region of the foregut, and is clearly seen to be divided into two (functionally) distinct parts, principally by virtue of the difference in their mucosal lining. Thus, there is a cutaneous (non-glandular) portion (which includes the region of the fundus), and a glandular portion which constitutes most of the "body" of the stomach. The line of transition from the glandular to the non-glandular portion is particularly obvious at this stage (*Plate 40j,a–d*). The first evidence of the "compound" stomach that occurs in this species is seen at about 12 days p.c.

The spleen, which develops from aggregations of mesenchyme tissue within the dorsal mesogastrium, continues to differentiate, and maintains its characteristic ribbon-like form (*Plate 40j,c,d*; *Plate 40k,a–c*). The two components of the dorsal mesogastrium within which the spleen is suspended, namely the gastro-splenic and lieno-renal "ligaments", are clearly seen at this stage (*Plate 40j,c*). The spleen is seen to be covered by a single layer of peritoneal mesothelial cells which, with the subjacent mesenchyme tissue, will form the capsule of the spleen. In the adult, the capsule also contains a small amount of smooth muscle.

At the histological level, the spleen is seen to be packed with small lymphocytes, these cells being particularly concentrated at the periphery of the spleen, and in the proximity of its large blood vessels. The latter probably represent the primordia of the white pulp (the perivascular lymphatic aggregates), while the paler less densely packed intervening regions probably represent the primordia of the red pulp, which predominantly contains venous sinusoids and intervening cellular cords. A substantial amount of haematopoietic activity occurs in the spleen at this time. The medullary region of the spleen at this stage is also seen to contain the earliest suggestions of lobules. A series of large arteries and their corresponding veins enter and leave the spleen at its hilum, passing along the lieno-renal "ligament" to do so. The distal part of the body and the tail of the pancreas make contact with the hilar region of the spleen and much of its anteromedial surface (*Plate 40k,b–d*).

Beyond indicating that the liver is by far the largest of the intra-abdominal organs from very shortly after its first appearance on or about 11.5–12 days p.c., little mention has been made during the previous stages of development regarding either the functional activity of this organ, or of its detailed cellular architecture (though any detailed description of its fine structure is beyond the scope of this Atlas). From an early stage, the liver takes over from the yolk sac as the principal source of red blood cell production, with the formation of the definitive red cell lineage. It is believed that the haematopoietic stem cells that colonize the liver migrate from the wall of the yolk sac to establish new colonies which provide foci of haematopoietic activity in the liver. Haematopoiesis then rapidly becomes the principal activity of this organ throughout the majority of gestation, though at this advanced stage of embryonic development, the spleen and subsequently the bone marrow take over this critical role.

The venous system of the liver is derived from a variety of sources, but in principle the venous blood that is derived from the portal system (*Plate 40i,c*) (previously predominantly from the right vitelline vein) differentiates to form the hepatic venous sinusoids. The oxygenated blood that arrives via the left umbilical vein (*Plate 40h,c,d*; *Plate 40i,a–d*) largely bypasses the parenchyma of the liver (within the ductus venosus) (*Plate 40h,d*), and passes directly into the posthepatic inferior vena cava (*Plate 40h,c*). The oxygenated blood from the placenta, mixed with a smaller volume of deoxygenated blood from both the portal system and the prehepatic inferior vena cava, is directed to the right atrium (*Plate 40g,c*) for distribution principally to the head and neck region and to the upper limbs. The detoxifying activity of the liver is principally a postnatal function.

The biliary system is well differentiated at this stage, and the gall bladder is located just to the right of the midline in the lower border of the falciform ligament (*Plate 40h,b–d*). Both the cystic duct (*Plate 40h,d*; *Plate 40i,a*) and the common bile duct can be readily traced at this stage (*Plate 40i,b,c*). These drain into the lumen of the duodenum via the duodenal papilla (*Plate 40j,a*), and are usually joined along their course by a number of pancreatic ducts (*Plate 40i,d*), which open into it at various levels, or occasionally some of these may drain directly into the duodenum.

The pancreas is a particularly large organ at this stage of development. At the histological level, the pancreas is seen to consist of branched glandular tissue divided by a delicate meshwork of connective tissue into small lobules. The secretory cells (i.e. the cells that constitute the exocrine component of the pancreas) have a dark uniformly staining basophilic cytoplasm, with a peripherally (basally) located relatively small nucleus. Interspersed between the lobules of the gland are relatively large diameter components of the duct system, which eventually drain via the main pancreatic duct(s) (which are lined by cuboidal cells) into the second part of the duodenum at the duodenal papilla (*Plate 40i,d*; *Plate 40j,a*). Interspersed apparently randomly throughout the gland are large clusters of pancreatic islet tissue (of Langerhans). Special staining techniques, however, are required to distinguish between the various cell types that are present within the islet complex at this stage. The islet tissue is often found in association with capillary beds. In mature islets, the A-cells tend to be located towards the periphery of the islets, while the B-cells tend to be located towards the centre of the islets. As indicated previously, not infrequently several main pancreatic ducts form instead of a single duct (which normally drains into the duodenal papilla), which then drain individually into the lumen of the duodenum.

The gut tube constitutes the other principal structure that is located within the peritoneal cavity, and very substantial changes are observed in the detailed morphology of its mucosal lining between 16.5–17 and 17.5–18 days p.c. The increased degree of differentiation observed is also reflected in the changes that occur in the lining of the glandular region of the stomach over this period (*Plate 40i,d*; *Plate 40j,a–d*). The only region of the gut tube where the degree of differentiation observed is less marked than elsewhere, is in the distal part of the large intestine and in the hindgut. In this region, the thickness of the wall and submucosa is generally considerably greater than elsewhere along the gut tube, and in this region, the crypts are quite shallow. However, the most impressive feature of this region is the characteristic presence of large numbers of goblet cells. The lumen of the rectum also contains a characteristic proteinaceous precipitate that is not seen elsewhere along the gut tube at this stage of development. A considerable volume of fairly homogeneous eosinophilic material is, however, found in the lumen of the proximal part of the large intestine, and this is seen to contain small irregular sized masses of darkly staining material which are probably bile pigments, being breakdown products of haemoglobin of erythroblastic origin.

In the duodenum, the lumen is wide, and the mucous membrane characteristically present in the form of circularly or spirally arranged folds whose surface is covered with filiform projections (the intestinal villi) (*Plate 40i,b–d*; *Plate 40j,a–d*). Elsewhere, the gut is clearly regionalized, so that in the large intestine, for example, the mucosa that projects into the lumen appears to be thrown into relatively few folds, with no obvious villi on their surface. The latter region has features that are transitional between those seen in the proximal part of the midgut, and those observed in the rectal component of the hindgut. The mucosa of this region also contains goblet cells, and this region of the gut therefore constitutes the primordium of the proximal to middle part of the large intestine.

THE NERVOUS SYSTEM

An overview of the major components of the brain and nervous system was provided in relation to the previous stage of development, where it was possible to analyse the morphological features observed in both transversely and coronally sectioned material, supplemented with a number of representative sagittal sections. In the most advanced embryos, such as those used to illustrate this stage of development, because of the increasing complexity of the brain, while the analysis of transverse and coronal sections can be extremely instructive, sagittal sections, other then those in the median plane, are of relatively little value. For this reason, and because little substantial changes are evident between the situation observed at 16.5–17 and 17.5–18 days p.c., only a single sagittal section has been used to illustrate the general features observed in such sections of embryos at this stage of development (*Plate 41*).

In gross terms, however, the parts of the brain and associated structures that display an increased degree of differentiation compared with those observed at the previous stage of development, are the olfactory lobes, the pineal and pituitary glands, and the eye. Similarly, transverse sections through the spinal cord at various defined levels clearly demonstrate that the disposition of the grey and white matter, for example, is beginning to reflect the quite different functional activities associated with these different spinal levels.

At the histological level, it is clear that substantial changes are also occurring within the substance of the brain. However, beyond indicating in the broadest of terms the most obvious changes that are observed in the various regions of the brain, it is clear that the detailed analysis of this region is beyond the scope of this Atlas. Moreover, as indicated elsewhere in this Atlas, the use of haematoxylin and eosin is only of limited value in the analysis of the embryonic brain and nervous tissue. Furthermore, the majority of the structures recognized in the neonatal and adult brain, in particular, are only poorly differentiated during even the later part of the embryonic period. Most of the nuclei and other features would not, in any case, be seen in the relatively low magnification sections of this region provided here. This is clearly an area where a specialized atlas is required, and an attempt to fill this deficiency with regard to the rat embryo has recently been published (Paxinos *et al.*, 1991).

With regard to the olfactory lobe of the brain, its lumen (which is an extension from the anterior horn of the lateral ventricle) is clearly seen to be diminished in volume compared with the situation observed previously (*Plate 40a,b*; cf. *Plate 37a,d*). This is largely due to the increased degree of cellular proliferation that occurs within the ventricular zone and the immediately subjacent areas of the olfactory bulb. This latter observation equally applies to the corresponding innermost layers of the cerebral hemispheres, so that the net result of the increased degree of cellular proliferation that occurs in these locations is a decrease in the volume of the various regions of the lateral ventricles, compared with the situation observed previously. The latter is particularly clearly seen in relation to the superior and posterior horns of the lateral ventricles (*Plate 40a,b–e*; cf. *Plate 37a,a–e*), and in the case of the third ventricle, where the lumen is now little more than a narrow slit which separates the various components of the diencephalon (*Plate 40a,b–e*; *Plate 40b,a,b*; cf. *Plate 37a,b–e*; *Plate 37b,a*). In the majority of embryos, the lumen of the middle/upper part of the third ventricle is already largely obliterated due to the interthalamic adhesion, which is first seen in some embryos at about 14.5 days p.c. (*Plate 34a,d,e*), but is more clearly seen in embryos at more advanced stages of development (e.g. *Plate 35a,c,d*). The vascularity of the meninges, particularly associated with the dorsal, medial and lateral surfaces of the olfactory bulbs, is also substantially increased compared to the situation observed previously. This observation regarding the meningeal vascularity equally applies to the very substantially increased volume of the dural venous sinuses seen at this stage (*Plate 40a,a–d*; cf. *Plate 37a,a–d*).

The pineal gland (epiphysis) has increased in volume, and its lumen is now obliterated as a result of the proliferation of the cells of its walls. The body of the gland and its anteriorly directed stalk (peduncle) is now located immediately above the roof of the midbrain (tectum), and is surrounded by a considerable venous plexus (*Plate 40a,a*). The latter principally drains into the great cerebral vein (of Galen), and thence into the straight sinus, which is situated in the midline of the tentorium cerebelli. The pineal recess, which is located in the dorsal part (i.e. in the roof) of the third ventricle, is seen to be largely obliterated, due to the florid growth of the choroid plexus in this location. The latter extends ventrally, associated with the inner aspect of the lamina terminalis, almost down to the level of the dorsal part of the anterior commissure (*Plate 40a,b*; cf. *Plate 38b,b*; *Plate 39b,d*).

The pineal gland has a thin capsule which is surrounded by the pia mater. At the histological level, the pineal gland is seen to be extremely vascular, and contains cords of darkly staining cells (the pinealocytes) which appear in the form of rosettes on the sections, and these are separated by connective tissue. Some of the cords surround blood vessels. The pinealocytes are of uniform appearance at this stage, but will differentiate within the first 2–3 weeks after birth into populations of darkly and lightly staining cells.

The pituitary gland is of considerable size at this stage of development, and the pars anterior region, in particular, has a substantially greater volume than that observed previously, due to the increased cellular and vascular proliferation that occurs in this region (*Plate 55,i*; cf. *Plate 55,h*). Of particular note, however, is the fact that the residual lumen of Rathke's pouch has also diminished in volume due to cellular proliferation in both the pars anterior as well as in the pars intermedia (*Plate 55,i*; cf. *Plate 55,h*). The degree of cellular proliferation in the region of the pars anterior is particularly clearly seen in the transverse sections through this region (*Plate 57,q,r*), as is the relationship of the pars tuberalis to the pituitary stalk (*Plate 57,m,n*). At the histological level, the pars anterior is seen to be particularly richly vascularized, and its cells are arranged in columns to form a meshwork surrounding the blood vessels. While a few of the cells in this location have a granular cytoplasm, the majority are seen to be uniformly darkly stained. Within the pars nervosa, moderate numbers of primitive pituicytes are located. By contrast, the pituitary stalk contains relatively few cells, but large numbers of nerve fibres are present in this location. These originate in some of the hypothalamic nuclei and are directed towards the pars nervosa.

Various components of the eye are seen to have differentiated over the period between 16.5–17 and 17.5–18 days p.c. This is particularly evident with regard to the appearance of the neural part of the retina, which shows increased evidence of stratification (*Plate 53,f*; cf. *Plate 53,e*). In sections through the lens, the lens fibres are seen to have increased in length, whereas the cells that comprise the anterior part of the lens (or capsule) are still seen to be cuboidal in shape, and show minimal difference in their volume or shape compared with the situation observed previously. The nuclei of the lens fibres located towards the centre of the lens (in the region of the embryonic/fetal "nucleus") are seen to be almost translucent at this stage (*Plate 53,g*), and this situation is even more clearly seen in developmentally more advanced embryos (e.g. *Plate 55,i*). The posterior lens "suture", which is clearly seen at 16.5 days p.c., diminishes in length with the increase in the volume of the embryonic/fetal "nucleus". The medial margin of the primitive iris becomes increasingly thinner as it differentiates (*Plate 53,h,i*; cf. *Plate 53,e*), and the ciliary body also shows increased evidence of differentiation. The cornea has a distinct inner and outer cellular layer, as previously described, but the intermediate layer is substantially thicker than previously. A substantial condensation of mesenchyme tissue is now seen to be located just subjacent to the pigment layer of the retina, to form the sclera of the eye. In appropriate sections, the primordia of the tear ducts are also clearly seen.

An analysis of representative transverse sections through the spinal cord at various defined levels, such as through the upper cervical region (*Plate 63,j*), the mid-thoracic region (*Plate 63,k*), and the mid-lumbar region (*Plate 63,l*) is informative. The most obvious difference between the appearance of the spinal cord at these levels relates to their significantly different cross-sectional areas and their overall shapes. Thus in both the upper cervical and mid-thoracic regions, the cord is ovoid in shape, with the ratio of the side-to-side diameter to the dorsal–ventral diameter being about 1.8:1. The cross-sectional area is, however, substantially greater in the cervical than in the thoracic region. The former section is at a level just above the cervical enlargement, where most of the corresponding spinal nerves are involved in the formation of the brachial plexus for the innervation of the upper limbs. By contrast, in the mid-lumbar region, these two diameters of the cord are fairly similar. This section is through the cord just above the lumbosacral enlargement, where the corresponding nerves constitute most of the lumbosacral plexuses for the innervation of the lower limbs. At each of these levels, the luminal diameter of the central canal is extremely small, but nevertheless similar to that seen at 16.5 days p.c. (*Plate 63,g–i*). The overall cross-sectional appearance of the cord at these levels at this stage of development is transitional between that seen at 18.5–19 days p.c., and that seen at earlier stages of development (e.g. *Plate 63,f*).

The general configuration of the grey matter, and the ratio of the white:grey matter at each of these levels in the spinal cord is beginning to reflect the different functional activities associated with these various levels. The proportion of white:grey matter increases as one passes from a caudal to a more rostral level. In the mid-thoracic region, the first indication of the lateral grey horns is seen at this stage (*Plate 63,k*), which are associated with sympathetic efferent neurons. At all levels of the cord, the peripheral parts of the ventral grey horns already contain substantial numbers of large cells, being the cell bodies of alpha motor neurons (lower motor neurons) involved in the innervation of muscle. These cells are particularly conspicuous in the cervical and lumbosacral enlargements. By contrast, the nuclei of the cells in the dorsal grey horns of the cord tend to be more rounded and much smaller in volume than those in the ventral grey horns, and are of relatively uniform appearance, at this stage of development. These cells also have a relatively small cytoplasmic volume.

THE UROGENITAL SYSTEM

The kidneys at 17.5–18 days p.c. are seen to be at an advanced stage of their differentiation. This is particularly clearly seen at the histological level, where a marked difference is seen in the appearance of the cortical region compared with the features observed at 16.5–17 days p.c. (*Plate 72,k,l*; cf. *Plate 72,i,j*). In the cortical region, in addition to the presence of more differentiated glomeruli, larger numbers of proximal and distal convoluted tubules are also recognized. The latter were only relatively uncommonly encountered at the previous stage of development. Similarly, the medullary region is now seen to contain large numbers of ascending and descending components of the loops of Henle, as well as other components of the collecting duct system. In more gross terms, whereas previously the renal pelvis tended to be a relatively simple structure, this region of the kidney is now seen to be composed of both minor and major calyces. On transverse sections through the lower abdominal and pelvic regions, it is relatively easy to trace the path of descent of the ureters to where they open into the lumen of the bladder at the lateral angles of the trigone (*Plate 40l,a*).

The bladder is a particularly prominent structure at this stage (*Plate 40j,d*; *Plate 40k,a–d*; *Plate 41*). Its wall is seen to be thick and muscular. The latter constitutes the detrusor muscle, and this consists of three layers of non-striated muscle fibres, an external and an internal layer of longitudinal fibres, and a middle layer of circularly arranged fibres. The mucous membrane is of the transitional variety, and the loose nature of the submucosa allows the mucosa to be thrown into folds when the bladder is empty.

In the male, the majority of the components of the reproductive duct system are recognizable at this stage of development (*Plate 70g,h*), though like the histological features of the testes (see below), they are only moderately differentiated. A histological analysis of the testes reveals that the seminiferous tubules, which occupy the majority of the medullary region of the gonad, appear in most locations to be solid structures. On closer inspection, however, it is clear that the walls of the seminiferous tubules are lined by fairly uniformly-staining relatively small rounded cells, many of which represent the primordia of the sustentacular (Sertoli) cells. Towards the centre of the tubules are located the spermatogenic cells, and these tend to have a considerably larger volume than those of the Sertoli cells. Much of the region between the spermatogenic cells appears to be occupied by a granular precipitate. The latter in fact fills the lumen of the tubules, and gives the false impression that these are still solid structures at this stage of development. The relatively little interstitial space located between the seminiferous tubules is occupied by a wide variety of cell types, some of which are the precursors of the hormone-secreting Leydig cells. The tunica albuginea of the testis is also more condensed than previously (*Plate 70,l* unlabelled; cf. *Plate 70,e*).

In the region of the rete testis are a considerable number of tubules, which make contact with the epididymis via the efferent ductules (*Plate 69,l*). The tail of the epididymis, in turn, drains into the ductus deferens (*Plate 40k,b,c*). The latter has an outer (external) areolar layer and a thick muscular wall composed of an outer longitudinal and an inner circular layer of smooth muscle. Within the central region of the ductus, the narrow lumen is surrounded by a single layer of darkly staining cuboidal/columnar epithelial cells. The ductus deferens drains into the prostatic region of the urethra, via the ejaculatory duct (*Plate 41*), and these enter the urethra on either side of the protrusion produced by the median lobe of the prostate. The prostate gland is a reasonably large structure at this stage, and surrounds the outlet region (neck) of the bladder (*Plate 41*). The bulbo-urethral glands are at an early stage in their differentiation, and open into the bulbar part of the urethra (*Plate 41*).

The seminal vesicles are quite substantial structures at this stage, and are easily seen on scanning electron micrographs of this region (*Plate 70,g*). They are somewhat larger at 18.5 days than at 17.5 days p.c. (*Plate 70,i*), and become increasingly enlarged during the first few weeks after birth. The appendix testis (*Plate 70,h*) and gubernaculum (*Plate 40k,d*; *Plate 70,l*) are also clearly seen at this stage of development. Apart from the appendix testis and the prostatic utricle (which are of paramesonephric duct origin), all of the components of the male reproductive duct system are derived from the mesonephric duct.

The ovaries, by contrast to the testes which are now located within the pelvis (*Plate 70,g*), are found at the level of the lower pole of the kidneys (*Plate 71,f*), and are partially hidden by them. At this stage of development, the ovarian capsule is still incomplete and covers all of the posterior surface and slightly more than the upper one-half of the anterior surface of the ovary (*Plate 71,g*). However, by about 18.5 days p.c. (*Plate 71,i*), the capsule almost completely surrounds the ovary. The two uterine horns are rather more elongated than previously (*Plate 71,f*; cf. *Plate 71,d*), and their most caudal part, at their point of union in the midline, is posterior to the trigone region of the bladder. At the histological level, the ovary still has a fine-grained "spotty" appearance (*Plate 69,h,i*), but it is possible to see that some of the oocytes in the cortical region are surrounded by follicle cells to form primary follicles. These progressively increase in number over the next few days, so that by the early postnatal period, the ovaries contain large numbers of primary follicles. These are at first principally located towards the periphery of the ovary, but will later be

(*continued on page 370*)

PLATE 40a

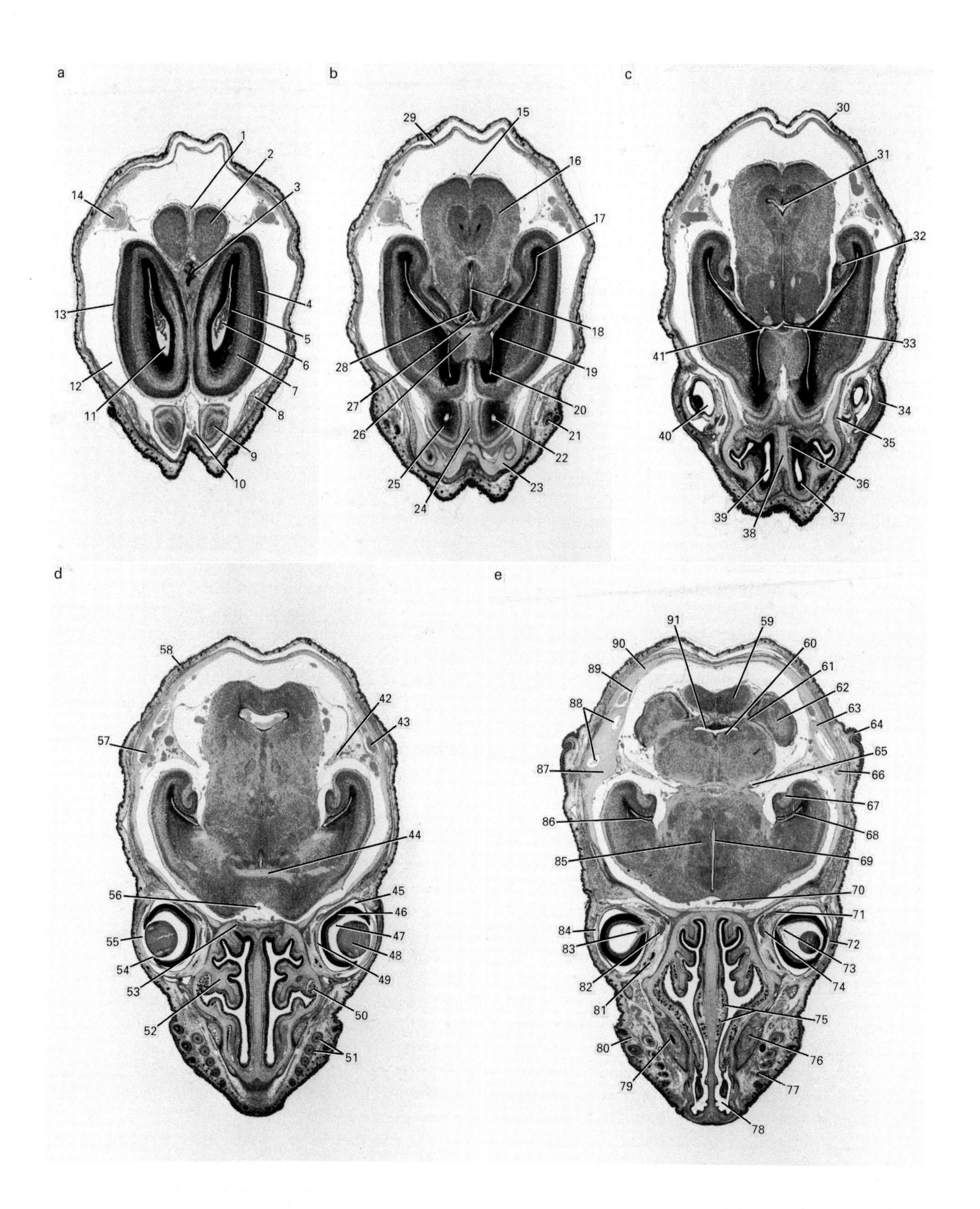
a
b
c
d
e
1
2
3
4
5
6
7
8
9
10
11
12
13
14
15
16
17
18
19
20
21
22
23
24
25
26
27
28
29
30
31
32
33
34
35
36
37
38
39
40
41
42
43
44
45
46
47
48
49
50
51
52
53
54
55
56
57
58
59
60
61
62
63
64
65
66
67
68
69
70
71
72
73
74
75
76
77
78
79
80
81
82
83
84
85
86
87
88
89
90
91

Plate 40a (17.5 days p.c.)

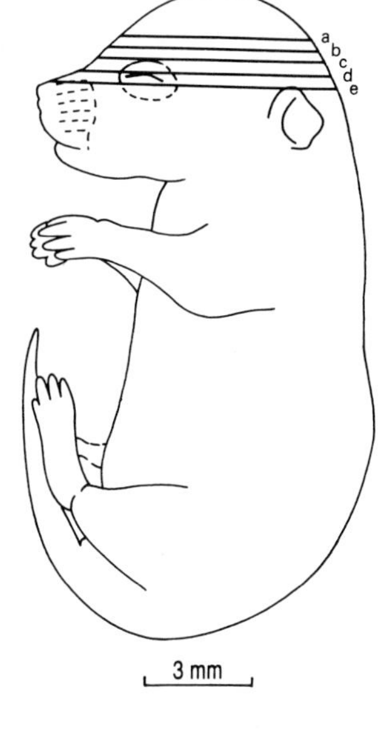

17.5 days p.c. Male embryo. Crown–rump length 17.8 mm (fixed).
Transverse sections (Theiler Stage 26; rat, Witschi Stage 35)

1. proximal part of inferior sagittal dural venous sinus into which the pineal venous plexus drains
2. dorsal part of midbrain
3. substantial venous plexus around dorsal part of pineal gland
4. neopallial cortex (future cerebral cortex) composed of inner broad nuclear layer (cortical plate) with narrow outer relatively anuclear layer (marginal zone)
5. ventricular zone of lateral ventricle
6. choroid plexus within lateral ventricle and arising from its medial wall
7. intermediate zone of wall of lateral ventricle (white matter)
8. ossification in outer table of primordium of frontal bone
9. dorsal part of left olfactory bulb (grazing section)
10. anterior extension of superior sagittal dural venous sinus
11. lateral ventricle
12. subarachnoid space
13. vascular elements of pia mater
14. superior anastomotic vein linking the right transverse and superior sagittal dural venous sinuses
15. inferior sagittal dural venous sinus
16. midbrain
17. caudal extremity of posterior horn of lateral ventricle
18. dorsal part of third ventricle
19. striatum (ganglionic eminence) overlying the head of the caudate nucleus
20. anterior portion of superior horn of lateral ventricle
21. primordium of prominent tactile or sinus hair follicle characteristically found in this location
22. extension of anterior horn of lateral ventricle into olfactory bulb
23. nasal capsule
24. anterior extension of falx cerebri
25. right olfactory bulb
26. superior part of anterior commissure
27. outgrowth of choroid plexus from roof of third ventricle
28. pineal recess of third ventricle
29. ossification in outer table of upper part of cartilage primordium of supraoccipital bone
30. skin (and subjacent epicranial aponeurosis (galea aponeurotica))
31. cerebral aqueduct
32. hippocampus (hippocampal formation)
33. roof of third ventricle
34. upper eyelid (fused at this stage to lower eyelid)
35. ossification in orbital part of frontal/maxillary bone
36. ossification in outer table of cartilage primordium of nasal septum
37. olfactory epithelium
38. cartilage primordium of nasal septum
39. nasal cavity
40. intra-retinal space (grazing section)
41. interventricular foramen
42. tentorium cerebelli
43. left transverse dural venous sinus
44. inferior part of anterior commissure (fibres cross the midline in the lamina terminalis)
45. intra-retinal space
46. neural layer of retina
47. hyaloid cavity
48. lens
49. pigment layer of retina
50. serous glands in the lateral wall of the middle meatus
51. follicles of vibrissae
52. cartilage primordium of turbinate bone
53. site of fusion between inferior part of nasal capsule and inferolateral margin of nasal septum
54. medial margin of pupil of right eye
55. anterior chamber of eye
56. basilar artery
57. right transverse dural venous sinus
58. primordium of hair follicle
59. dorsal part of cerebellum (vermis)
60. pons
61. posterior cerebral and superior cerebellar arteries
62. lateral part of cerebellum
63. cartilage primordium of petrous part of left temporal bone
64. pinna of ear
65. rootlets of origin of left oculomotor (III) nerve
66. left sigmoid dural venous sinus
67. dentate gyrus
68. caudal (inferior) extremity of posterior horn of left lateral ventricle
69. third ventricle
70. anteromedial group of arteries arising from anterior cerebral and anterior communicating arteries
71. extrinsic ocular muscle (left lateral/temporal rectus)
72. cornea
73. hyaloid vessels
74. extrinsic ocular muscle (left medial/nasal rectus)
75. serous glands associated with nasal septum
76. ossification within left maxilla
77. left upper lip
78. anterior part of nasal cavity close to external naris
79. tooth primordium of right upper molar tooth
80. right upper lip
81. tubules of Harderian gland
82. extrinsic ocular muscle (right medial/nasal rectus)
83. optic disc
84. conjunctival sac
85. diencephalon (thalamus)
86. choroid plexus within base of posterior horn of lateral ventricle
87. cartilage primordium of petrous part of right temporal bone
88. superior semicircular canal
89. rostral extremity of right endolymphatic duct
90. cartilage primordium of peripheral part of supraoccipital bone
91. rostral part of fourth ventricle

Plate 40b (17.5 days p.c.)

17.5 days p.c. Male embryo. Crown–rump length 17.8 mm (fixed).
Transverse sections. (Theiler Stage 26; rat, Witschi Stage 35)

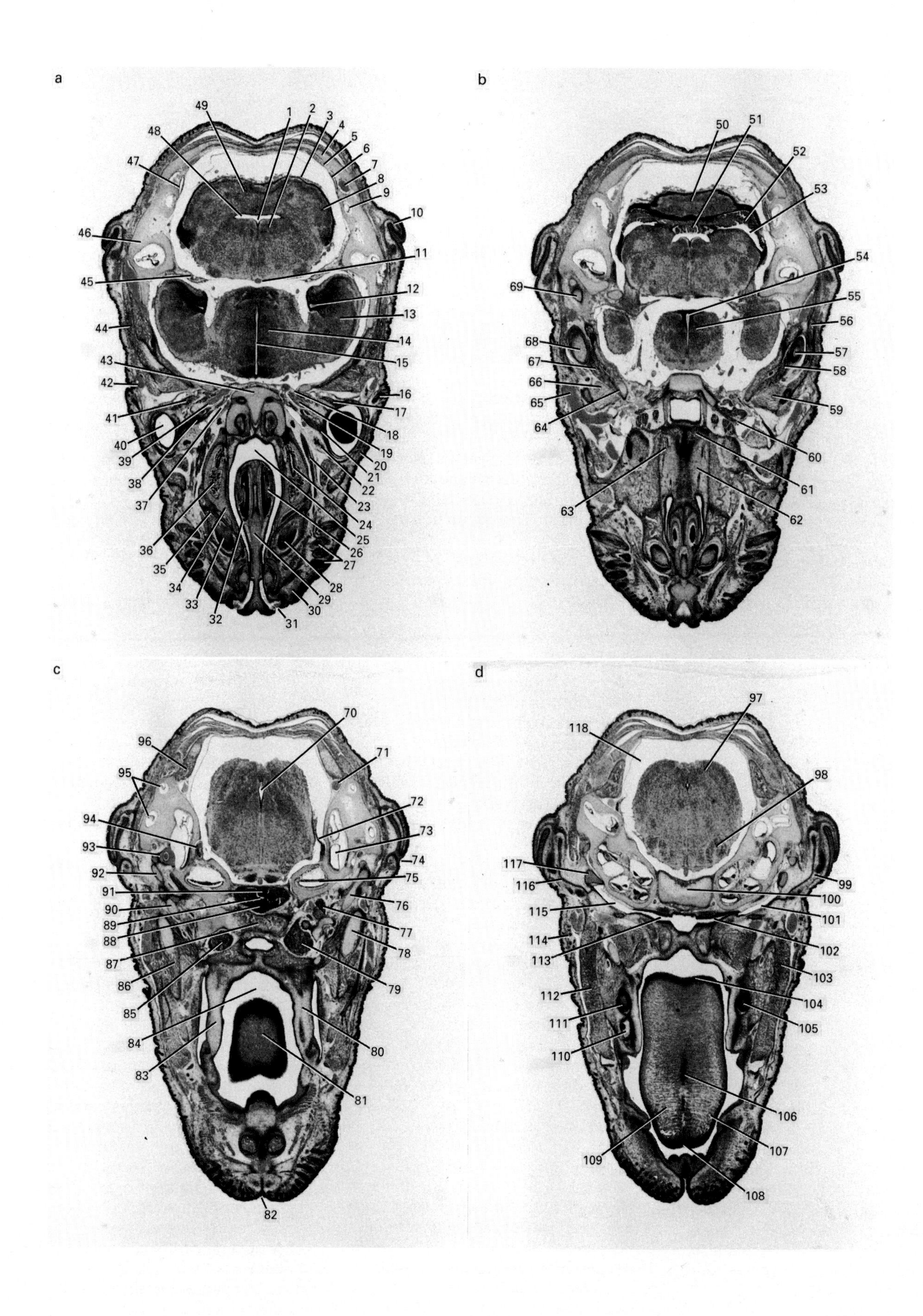

1. median sulcus
2. pons
3. vascular elements of pia mater
4. ossification within cartilage primordium of supraoccipital bone
5. left transverse dural venous sinus
6. rostral extremity of left endolymphatic duct
7. rostral part of left sigmoid dural venous sinus
8. cerebellar plate
9. posterior semicircular canal
10. pinna of ear
11. basilar artery
12. caudal extremity of posterior horn of lateral ventricle
13. inferior part of cerebral hemisphere (temporal lobe)
14. diencephalon (thalamus)
15. third ventricle
16. primordium of prominent tactile or sinus hair follicle characteristically found in this location
17. extrinsic ocular muscle (left lateral/temporal rectus)
18. left optic canal
19. left optic (II) nerve
20. lower eyelid
21. extrinsic ocular muscle (left medial/nasal rectus)
22. tubules of Harderian gland
23. branch of maxillary division of left trigeminal (V) nerve passing from plexuses around follicles of vibrissae to trigeminal (V) ganglion
24. olfactory epithelium
25. nasopharynx
26. vomeronasal organ
27. follicles of vibrissae
28. primordium of left upper incisor tooth
29. cartilage primordium of nasal septum
30. left upper lip
31. entrance into nasal cavity from external naris
32. tubules of serous glands associated with nasal septum
33. primordium of right upper incisor tooth
34. nasal capsule
35. ossification within maxilla
36. tubules of serous glands associated with lateral wall of the middle meatus
37. right inferior orbital (ophthalmic) vein where it drains into the right cavernous sinus
38. extrinsic ocular muscle (right medial/nasal rectus)
39. right optic (II) nerve
40. inferior part of intra-retinal space
41. extrinsic ocular muscle (right lateral/temporal rectus)
42. right superficial temporal vein
43. posterior border of cartilage primordium of nasal septum
44. right temporalis muscle
45. tentorium cerebelli
46. cartilage primordium of petrous part of right temporal bone
47. rostral extremity of right endolymphatic duct
48. rostral part of fourth ventricle
49. dorsal part of cerebellum
50. median part of cerebellum
51. choroid plexus arising from roof and projecting into central part of lumen of fourth ventricle
52. choroid plexus projecting into lumen of left lateral recess of fourth ventricle
53. left lateral recess of fourth ventricle
54. entrance to infundibular recess of third ventricle
55. diencephalon (hypothalamus)
56. left temporalis muscle
57. cartilage primordium of left mandibular condyle
58. left lateral pterygoid muscle
59. left medial pterygoid muscle
60. extrinsic ocular muscle (left inferior rectus)
61. ossification within posterior part of hard palate
62. palatal shelf of left maxilla
63. right greater palatine vessels and nerve
64. attachment of right lateral pterygoid muscle to infratemporal surface of skull
65. right inferior orbital (ophthalmic) vein where it drains into the right facial vein
66. right medial pterygoid muscle
67. right lateral pterygoid muscle
68. cartilage primordium of right mandibular condyle
69. cartilage primordium of head of right malleus
70. caudal part of fourth ventricle
71. caudal part of left sigmoid dural venous sinus where it enters the jugular foramen
72. rootlet of left vestibulocochlear (VIII) nerve
73. saccule
74. entrance to left external auditory (acoustic) meatus (not yet canalized)
75. cochlea
76. continuity between left malleus and Meckel's cartilage
77. left facial (VII) ganglion
78. ossification within ramus of left mandible
79. left trigeminal (V) ganglion
80. left primitive gum in which upper molar teeth differentiate
81. dorsum of tongue
82. philtrum
83. right primitive gum
84. oropharynx
85. right trigeminal (V) ganglion
86. ossification within ramus of right mandible
87. ossification within cartilage primordium of basisphenoid bone

88. pars anterior of pituitary
89. residual lumen of Rathke's pouch
90. right Meckel's cartilage
91. pars intermedia of pituitary
92. upper part of cartilage primordium of body of right incus
93. cartilage primordium of right stapes
94. right vestibulocochlear (VIII) ganglion
95. lateral semicircular canal
96. ossification within cartilage primordium of exoccipital bone
97. dorsal part of medulla oblongata
98. left inferior olivary nucleus
99. left external auditory (acoustic) meatus (not yet canalized)
100. cartilage primordium of basisphenoid bone
101. left tubo-tympanic recess
102. left pharyngo-tympanic (Eustachian) tube
103. left masseter muscle
104. dorsum of tongue
105. primordium of left second upper molar tooth
106. median fibrous septum of tongue
107. lingual veins
108. tip of tongue
109. intrinsic muscle of tongue (transverse component)
110. primordium of right first upper molar tooth
111. primordium of right second upper molar tooth
112. right masseter muscle
113. right pharyngo-tympanic (Eustachian) tube
114. right parotid gland
115. right tubo-tympanic recess
116. right external auditory (acoustic) meatus (not yet canalized)
117. cartilage primordium and upper part of crus longum (long limb) of right incus
118. loosely packed cephalic mesenchyme in future subarachnoid space

Plate 40c (17.5 days p.c.)

17.5 days p.c. Male embryo. Crown–rump length 17.8 mm (fixed).
Transverse sections (Theiler Stage 26; rat, Witschi Stage 35)

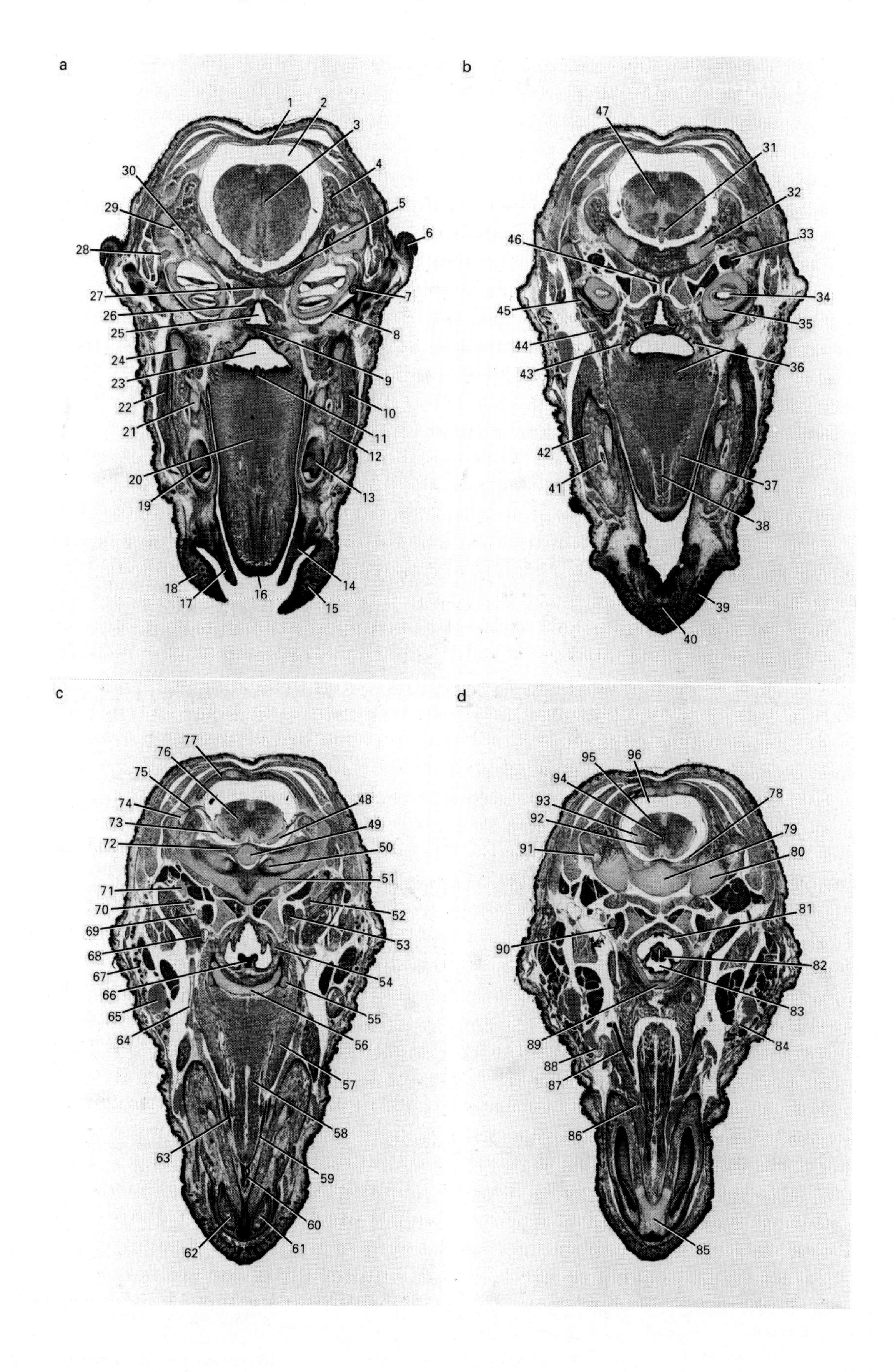

1. membrane covered fontanelle (region bounded by membranous primordium of supraoccipital bone above, occipital arch below, and laterally by the cartilage primordia of the two exoccipital bones
2. loosely packed cephalic mesenchyme in future subarachnoid space
3. caudal part of medulla oblongata
4. ossification within cartilage primordium of left exoccipital bone
5. ossification within cartilage primordium of rostral part of basioccipital bone (clivus)
6. inferior part of pinna of left ear
7. cartilage primordium of head of left malleus
8. left tubo-tympanic recess
9. posterior part of soft palate
10. left masseter muscle
11. median circumvallate (vallate) papilla
12. ossification within left half of mandible
13. primordium of left molar tooth
14. left upper lateral margin of lower lip
15. left upper lip containing numerous follicles of vibrissae
16. tip of tongue
17. right upper lateral margin of lower lip
18. right upper lip containing numerous follicles of vibrissae
19. primordium of right molar tooth
20. intrinsic muscle of tongue (transverse component)
21. right Meckel's cartilage
22. right masseter muscle
23. posterior part of oropharynx
24. ossification within ramus of right half of mandible
25. posterior part of nasopharynx
26. superior and middle constrictor muscles
27. pharyngeal tubercle
28. cartilage primordium of right stapes
29. caudal part of right sigmoid dural venous sinus where it enters the jugular foramen
30. right superior glossopharyngeal (IX) ganglion
31. basilar artery
32. ossification within cartilage primordium of lateral part of basioccipital bone (clivus)
33. left inferior vagal (X) ganglion
34. cochlea
35. cartilage primordium of petrous part of left temporal bone
36. developing glandular ducts opening into pharynx and onto dorsal surface of root of tongue
37. left lingual vessels and nerve
38. median fibrous septum of tongue
39. lower lip containing numerous follicles of vibrissae
40. tip of lower jaw
41. mandibular canal containing inferior alveolar vessels and nerve
42. ossification within body of right half of mandible
43. cartilage primordium of tip of right greater horn of hyoid bone
44. styloglossus muscle
45. cartilage primordium of styloid process of right temporal bone
46. midline thickening of pharyngo-basilar fascia forming pharyngeal ligament
47. junctional zone between caudal part of medulla oblongata and rostral part of cervical spinal cord
48. left vertebral artery as it passes medially to fuse with the right vertebral artery to form the basilar artery
49. ossification within cartilage primordium of rostral part of dens (odontoid process of C2 vertebra (axis))
50. ossification within medial part of cartilage primordium of anterior arch of C1 vertebra (atlas)
51. ossification within cartilage primordium of caudal part of basioccipital bone (clivus)
52. rostral extremity of left vagal (X) trunk
53. left superior cervical sympathetic ganglion
54. left superior cornu of thyroid cartilage
55. cartilage primordium of proximal part of left inferior horn of hyoid bone
56. cartilage primordium of body of hyoid bone
57. mylohyoid muscle
58. genioglossus muscle
59. left submandibular duct
60. opening of submandibular and sublingual ducts at sublingual caruncles
61. primordium of left lower incisor tooth
62. primordium of right lower incisor tooth
63. right submandibular duct (medial) and right sublingual duct (lateral)
64. right inferior alveolar vessels
65. right external jugular vein
66. epiglottis
67. right parotid gland
68. right external carotid artery
69. right internal carotid artery
70. rostral extremity of right vagal (X) trunk
71. right internal jugular vein
72. horizontal band of cruciate ligament
73. right vertebral artery as it passes medially to fuse with the left vertebral artery to form the basilar artery
74. right vertebral artery where it passes around the lateral mass of C1 vertebra (atlas)
75. anterior primary division of C1 nerve
76. upper cervical region of spinal cord
77. ossification within cartilage primordium of neural arch of C1 vertebra (atlas)
78. caudal intracranial occipital veins that drain into the vertebral plexus of veins
79. cartilage primordium of centrum of C2 vertebra (axis)
80. ossification within cartilage primordium of lateral mass of C1 vertebra (atlas)
81. entrance into oesophagus
82. vocal process of left arytenoid cartilage
83. laryngeal aditus
84. left facial vein where it enters the left external jugular vein
85. site of fusion of the left and right Meckel's cartilages
86. right deep lingual vein which joins the sublingual vein which then enters the facial vein
87. right submandibular duct
88. right facial vein
89. thyroid cartilage
90. location of division of right common carotid artery into internal and external carotid arteries
91. right vertebral vessels within foramen transversarium of C1 vertebra (atlas)
92. mantle layer (ventral grey horn)
93. marginal layer
94. central canal
95. mantle layer (dorsal grey horn)
96. spinal canal

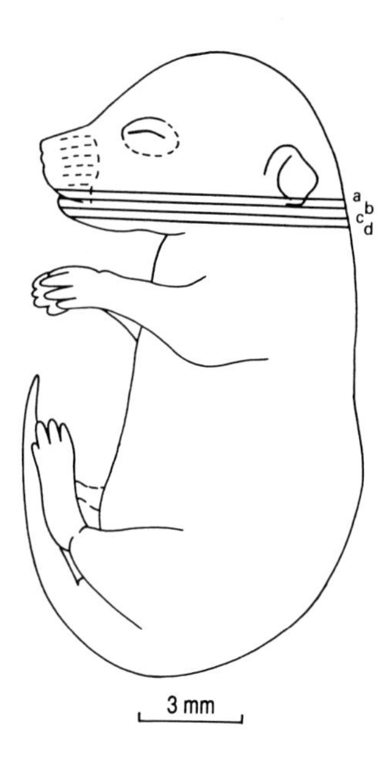

PLATE 40d

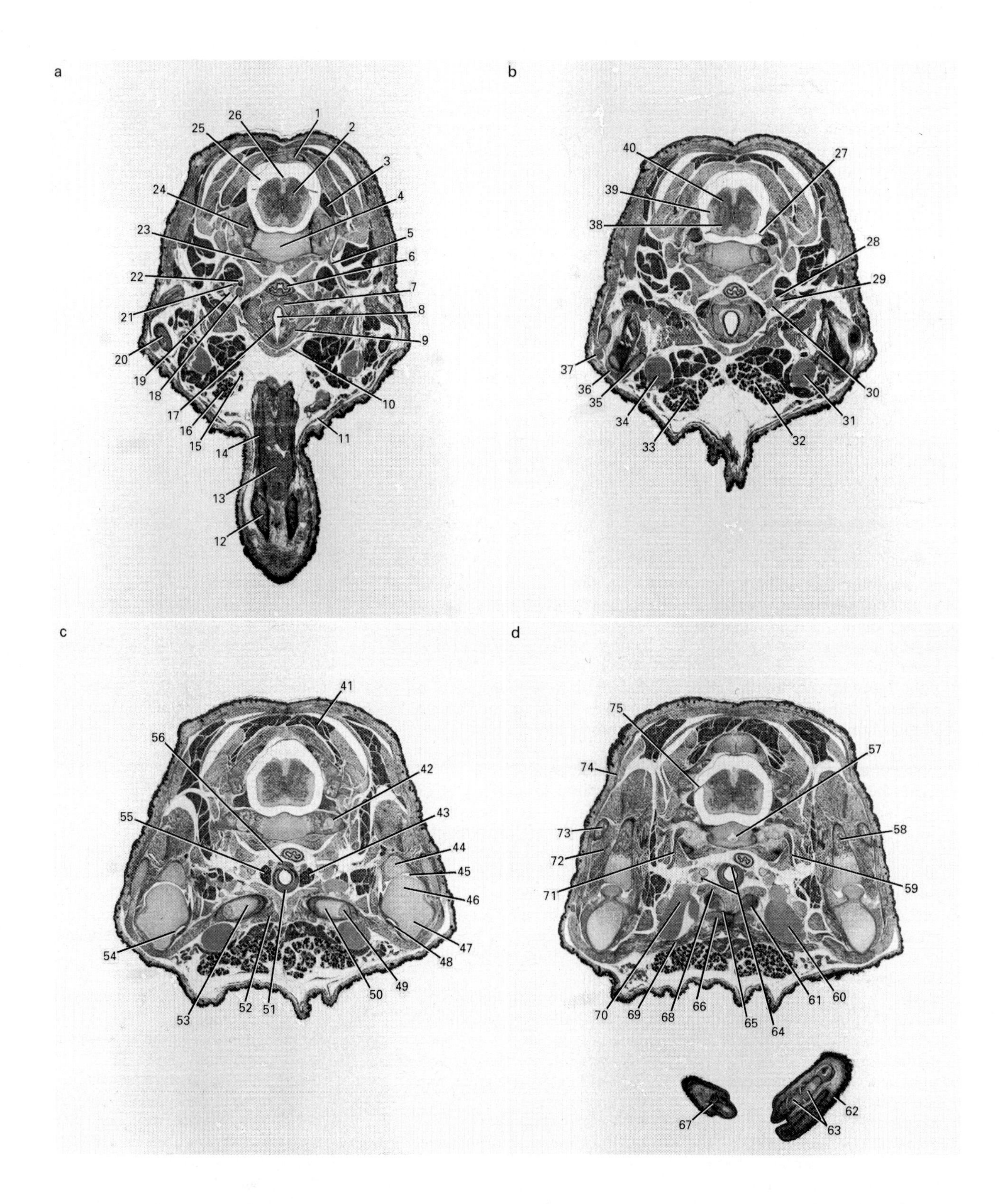
a
b
c
d
1
2
3
4
5
6
7
8
9
10
11
12
13
14
15
16
17
18
19
20
21
22
23
24
25
26
27
28
29
30
31
32
33
34
35
36
37
38
39
40
41
42
43
44
45
46
47
48
49
50
51
52
53
54
55
56
57
58
59
60
61
62
63
64
65
66
67
68
69
70
71
72
73
74
75

Plate 40d (17.5 days p.c.)

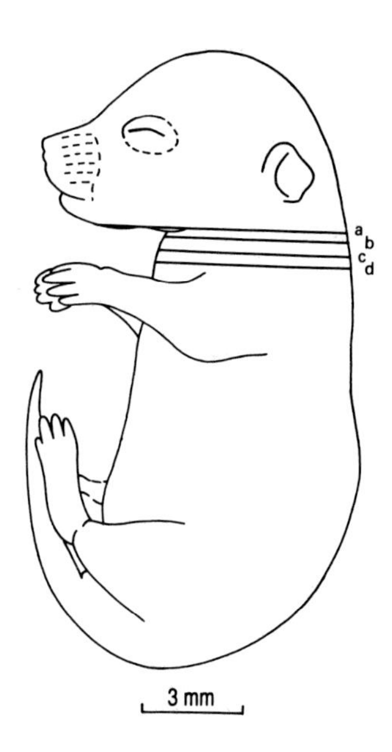

17.5 days p.c. Male embryo. Crown–rump length 17.8 mm (fixed). Transverse sections (Theiler Stage 26; rat, Witschi Stage 35)

1. ossification within cartilage primordium of neural arch of C3 vertebra
2. spinal cord in upper cervical region
3. dorsal (posterior) root ganglion
4. cartilage primordium of centrum of third cervical vertebra
5. C3 spinal nerve
6. oesophagus
7. lamina of cricoid cartilage
8. rostral extremity of lumen of trachea
9. thyroid cartilage
10. pretracheal "strap" muscles (sternohyoid (medial) and thyrohyoid (lateral))
11. platysma muscle (panniculus carnosus)
12. ossification within ventral extremity of right half of mandible
13. intrinsic muscle of tongue (transverse component)
14. genioglossus muscle
15. right submandibular duct
16. upper pole of right submandibular gland
17. inferior border of right vocal fold
18. right common carotid artery
19. right internal jugular vein
20. ossification within cartilage primordium of tip of acromion of right scapula
21. right vagal (X) trunk
22. caudal part of right superior cervical sympathetic ganglion
23. prevertebral muscles of neck (longus capitis, longus cervicis)
24. ossification within cartilage primordium of pedicle of third cervical vertebra
25. spinal canal
26. posterior spinal vein
27. efferent nerve fibres of ventral root
28. left vagal (X) trunk
29. left internal jugular vein
30. left internal carotid artery
31. left external jugular vein just rostral to where it drains into the left subclavian vein
32. left submandibular gland
33. right submandibular gland
34. right external jugular vein just rostral to where it drains into the right subclavian vein
35. ossification within lateral one-third of right clavicle
36. right deltoid muscle
37. right cephalic vein
38. mantle layer (ventral grey horn)
39. marginal layer
40. mantle layer (dorsal grey horn)
41. deposits of brown (multilocular) fat
42. foramen transversarium of cervical vertebra
43. left lobe of thyroid gland
44. glenoid region of cartilage primordium of left scapula
45. left shoulder joint
46. cartilage primordium of head of left humerus
47. cartilage primordium of proximal part of shaft of the left humerus
48. tendon of the long head of the biceps brachii muscle
49. ossification within lateral part of cartilage primordium of middle one-third of left clavicle
50. cartilage primordium of middle one-third of left clavicle
51. tracheal "ring" (incomplete) of cartilage
52. right sternomastoid muscle
53. cartilage primordium of middle one-third of right clavicle
54. intertubercular (bicipital) groove
55. right lobe of thyroid gland
56. inferior border of cricoid cartilage
57. nucleus pulposus
58. ossification within cartilage primordium of blade of left scapula
59. lower cervical nerve progressing to form a component of the left brachial plexus
60. left subclavian vein where it is joined by the left external jugular vein
61. medial margin of left clavicle
62. preaxial border of left forelimb
63. cartilage primordia of phalangeal bones
64. trachealis muscle and (laterally) lower pole of right parathyroid gland in aberrant location
65. right sterno-clavicular joint
66. right supero-lateral border of manubrium sterni
67. first digit of right forelimb
68. sternal origin of right pretracheal "strap" muscles (sternohyoid, sternothyroid)
69. sternal fibres of origin of pectoralis superficialis
70. right subclavian vein where it passes beneath the right clavicle and over the first rib
71. right lower cervical spinal nerve progressing to form a component of the right brachial plexus
72. supraspinatus muscle
73. ossification within cartilage primordium of spine of right scapula
74. cutaneous muscle of trunk (panniculus carnosus)
75. afferent nerve fibres of dorsal root

Plate 40e (17.5 days p.c.)

17.5 days p.c. Male embryo. Crown–rump length 17.8 mm (fixed).
Transverse sections (Theiler Stage 26; rat, Witschi Stage 35)

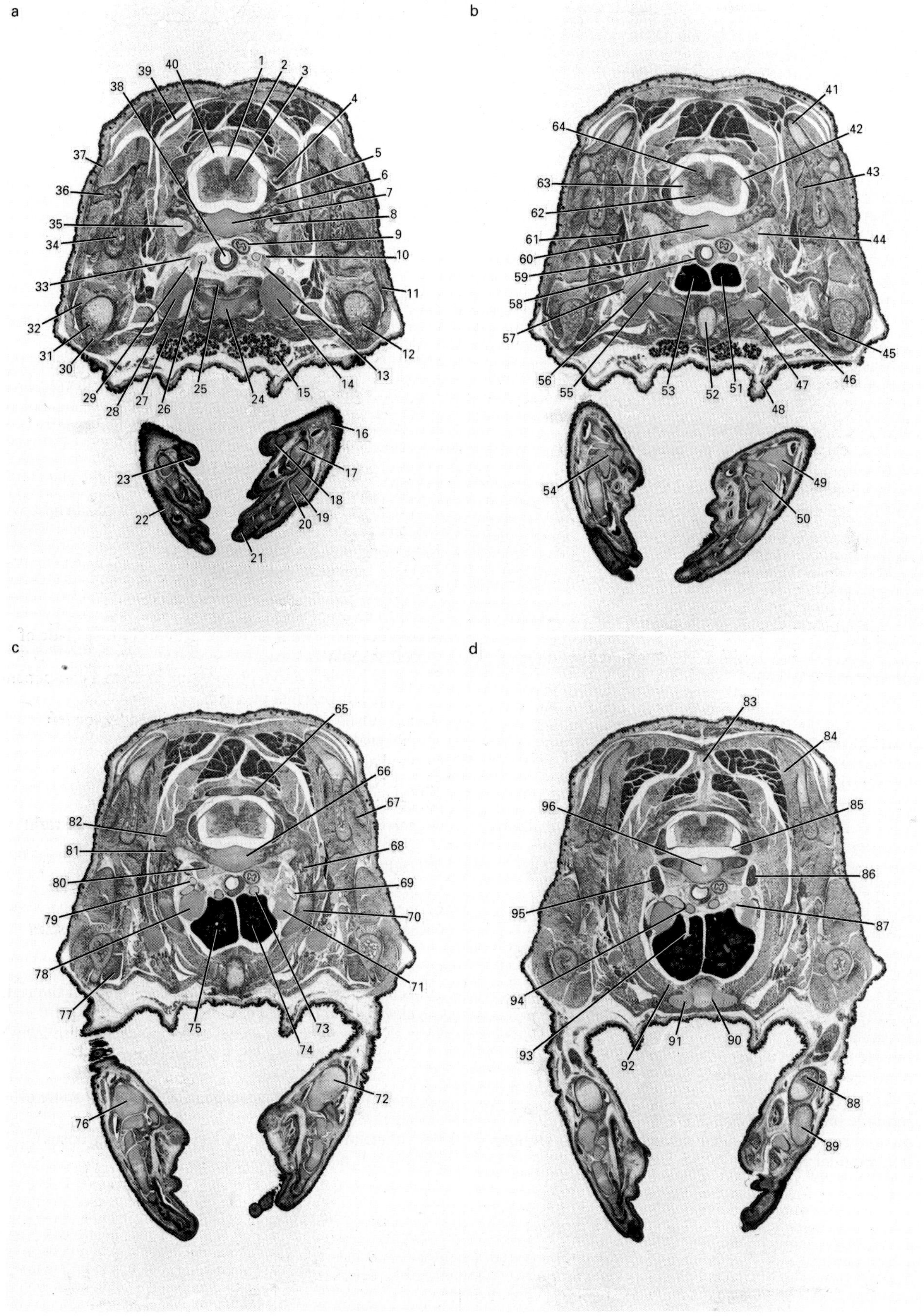

a

b

c

d

1. posterior spinal vein
2. deposits of brown (multilocular) fat
3. spinal cord in lower cervical region
4. ossification within cartilage primordium of lamina of eighth cervical vertebra
5. dorsal (posterior) root ganglion
6. ossification within cartilage primordium of pedicle of eighth cervical vertebra
7. cartilage primordium of centrum of eighth cervical vertebra
8. left vertebral artery within foramen transversarium (at this level, the lateral border of the foramen is completed by a band of fibrous tissue)
9. oesophagus
10. left vagal (X) trunk
11. left deltoid muscle
12. ossification within cartilage primordium of upper shaft region of left humerus
13. left internal jugular vein
14. left subclavian vein where it passes beneath the left clavicle and over the first rib
15. left submandibular gland
16. left forelimb
17. ossification within cartilage primordium of left second metatarsal bone
18. cartilage primordia of first and second phalangeal bones of left first digit
19. cartilage primordium of left third metacarpal bone
20. third left metacarpo-phalangeal joint
21. cartilage primordium of terminal phalanx of left third digit
22. dorsal surface of right forelimb
23. right first digit
24. cartilage primordium of upper part of manubrium sterni
25. origin of sternothyroid and sternohyoid muscles from posterior surface of manubrium sterni
26. right common carotid artery
27. sternal fibres of origin of right pectoralis superficialis muscle
28. right subclavian vein where it passes beneath the right clavicle and over the first rib
29. right internal jugular vein
30. ossification within cartilage primordium of deltoid tuberosity of right humerus
31. ossification within cartilage primordium of ventral aspect of upper shaft region of right humerus
32. right triceps brachii muscle
33. right vagal (X) trunk
34. ossification within cartilage primordium of blade of right scapula
35. right vertebral artery almost surrounded by right vertebral venous plexus
36. ossification within cartilage primordium of spine of right scapula
37. cutaneous muscle of trunk (panniculus carnosus)
38. lumen of trachea
39. cervical/thoracic components of right half of trapezius muscle
40. spinal canal
41. cartilage primordium of posterior (medial or vertebral) border of left scapula
42. afferent nerve fibres of dorsal root
43. left subscapularis muscle
44. left vertebral artery
45. left biceps brachii muscle
46. venous valves within the left superior (cranial) vena cava where it is formed by the fusion of the left subclavian and left internal jugular veins
47. left first costal cartilage
48. mammary elevation (ridge)
49. cartilage primordium of distal part of shaft of left radius
50. cartilage primordia of left carpal bones
51. upper pole of left lobe of thymus gland
52. ossification within cartilage primordium of caudal part of manubrium sterni
53. upper pole of right lobe of thymus gland
54. cartilage primordia of right carpal bones
55. rostral part of right superior (cranial) vena cava
56. venous valves where right subclavian vein enters the right superior (cranial) vena cava
57. right subclavian artery
58. tracheal cartilage
59. C8 spinal nerve
60. cartilage primordium of upper half of centrum of T1 vertebra
61. C7 spinal nerve
62. mantle layer (ventral grey horn)
63. marginal layer of upper thoracic region of spinal cord
64. mantle layer (dorsal grey horn)
65. ossification within cartilage primordium of neural arch of T1 vertebra
66. cartilage primordium of lower half of centrum of T1 vertebra
67. left infraspinatus muscle
68. ossification within cartilage primordium of dorsal part of shaft of left first rib
69. left subclavian artery
70. ossification within cartilage primordium of mid-shaft region of left first rib
71. proximal part of left superior (cranial) vena cava
72. cartilage primordium of distal one-third of shaft of left radius
73. left common carotid artery
74. left lobe of thymus gland

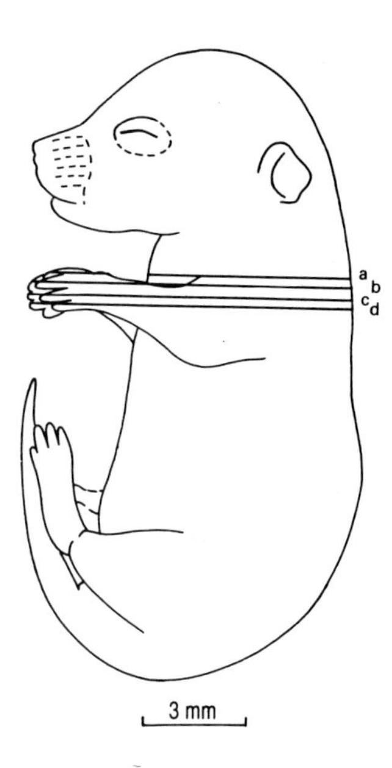

75. right lobe of thymus gland
76. cartilage primordium of distal part of shaft of right radius
77. right biceps brachii muscle
78. right superior (cranial) vena cava
79. proximal part of right vertebral artery
80. prevertebral muscles of the neck (longus capitis, longus cervicis)
81. ossification within cartilage primordium of tubercle of right first rib
82. ossification within cartilage primordium of right transverse process of T1 vertebra
83. ligamentum nuchae
84. left serratus anterior muscle
85. efferent nerve fibres of ventral root
86. left cervicothoracic (stellate) sympathetic ganglion
87. left vagal (X) trunk which in this location is closely apposed to the medial wall of the left superior (cranial) vena cava
88. ossification within cartilage primordium of mid-shaft region of left radius
89. cartilage primordium of distal one-third of shaft of left ulna
90. articulation between second left costal cartilage and cartilage primordium of caudal part of body of manubrium sterni
91. right second costal cartilage
92. right internal thoracic (mammary) vessels
93. thymic vein draining posterior part of right lobe of thymus gland
94. site where right subclavian artery and right common carotid artery amalgamate to form the brachiocephalic (innominate) artery
95. right cervicothoracic (stellate) sympathetic ganglion
96. rostral part of nucleus pulposus at site of intervertebral disc between T1 and T2 vertebral bodies

Plate 40f (17.5 days p.c.)

17.5 days p.c. Male embryo. Crown–rump length 17.8 mm (fixed).
Transverse section (Theiler Stage 26; rat, Witschi Stage 35)

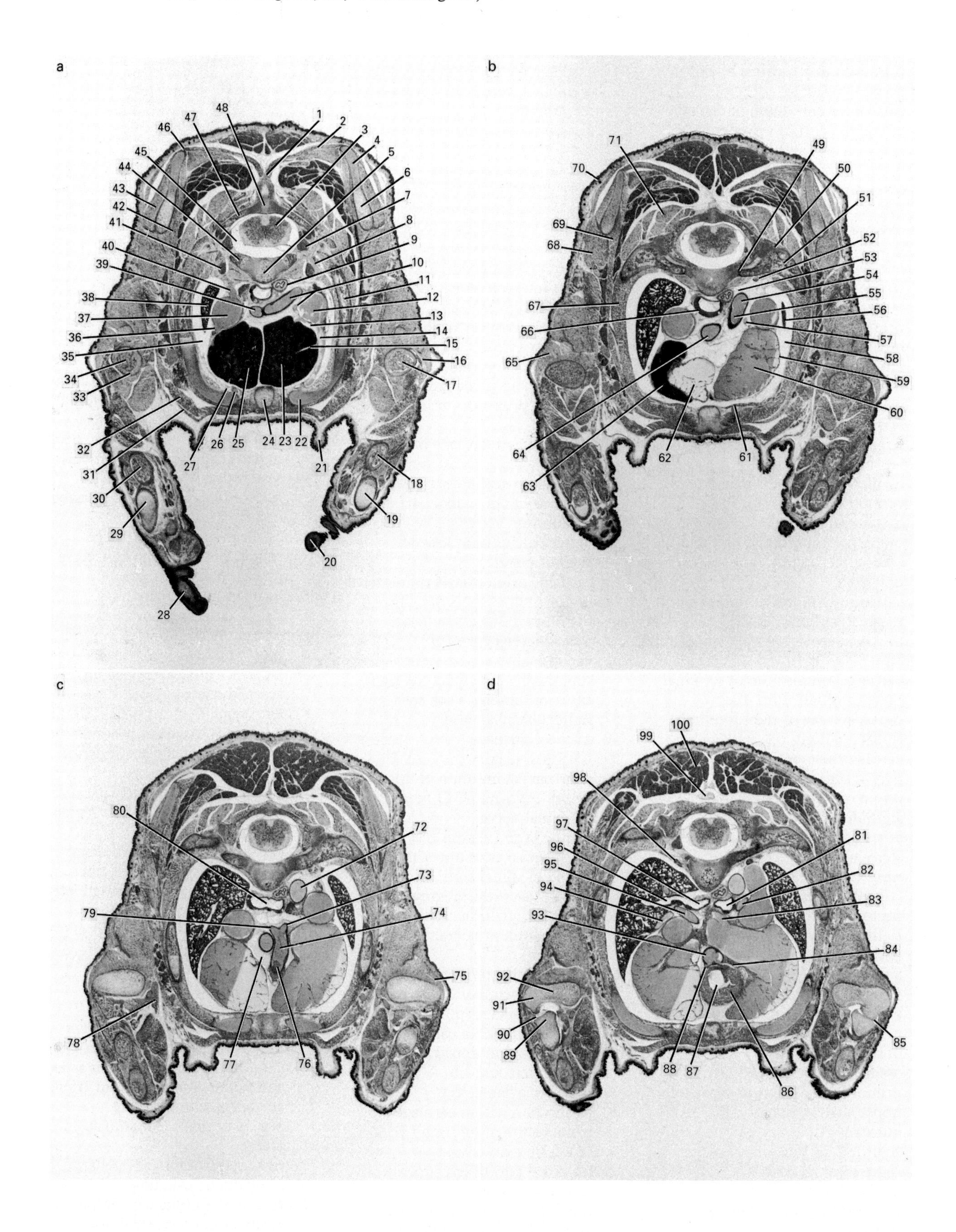

1. attachment of thoracic part of trapezius muscle to spinous process of upper thoracic vertebra
2. trapezius muscle
3. spinal cord in upper thoracic region
4. ossification within cartilage primordium of posterior (medial) border of left scapula
5. ossification within central part of cartilage primordium of upper thoracic vertebral body
6. ossification within cartilage primordium of blade of left scapula
7. left dorsal (posterior) root ganglion
8. upper thoracic (intercostal) nerve
9. oesophagus
10. lumen of arch of the aorta and (anteriorly) the left vagal (X) trunk
11. ossification within cartilage primordium of left third rib
12. left superior (cranial) vena cava
13. large posterior thymic vein which drains into the left internal thoracic (mammary) vein
14. left internal thoracic (mammary) vein just proximal to where it joins the left superior (cranial) vena cava
15. small lateral thymic vein
16. posterior circumflex humeral vessels and axillary (circumflex) nerve which principally supplies the deltoid muscle
17. ossification within cartilage primordium of mid-shaft region of left humerus
18. ossification within cartilage primordium of mid-shaft region of left radius
19. ossification within cartilage primordium of mid-shaft region of left ulna
20. postaxial border of left forelimb
21. mammary elevation
22. third costal cartilage
23. left lobe of the thymus
24. ossification within cartilage primordium of sternebra
25. right lobe of thymus
26. right internal thoracic (mammary) artery
27. right internal thoracic (mammary) vein
28. postaxial border of right forelimb
29. ossification within cartilage primordium of mid-shaft region of right ulna
30. ossification within cartilage primordium of mid-shaft region of right radius
31. pectoralis superficialis muscle
32. pectoralis profundus muscle
33. ossification within cortical region of cartilage primordium of mid-shaft region of right humerus
34. ossification within medullary region of mid-shaft region of right humerus
35. right pleural cavity
36. entrance into right superior (cranial) vena cava of right internal thoracic (mammary) vein
37. right superior (cranial) vena cava
38. cranial lobe of right lung
39. brachiocephalic (innominate) artery
40. azygos vein where it joins the right superior (cranial) vena cava
41. right cervicothoracic (stellate) sympathetic ganglion
42. subscapularis muscle
43. infraspinatus muscle
44. ossification within cartilage primordium of lateral part of vertebral body
45. spinal canal
46. insertion of right trapezius muscle into posterior (medial) border of right scapula
47. ossification within cartilage primordium of neural arch
48. cartilage primordium of spinous process of upper cervical vertebra
49. articulation between head of rib and body of corresponding thoracic vertebra
50. articulation between tubercle of rib and neural arch of corresponding thoracic vertebra
51. ossification within cartilage primordium of tubercle of rib
52. left sympathetic trunk
53. thoracic duct
54. arch of aorta at site of entrance of ductus arteriosus
55. distal part of ductus arteriosus
56. apex of left lung
57. left vagal (X) trunk
58. left pleural cavity
59. wall of left atrium
60. lumen of left atrium
61. left transversus thoracis muscle
62. lumen of right atrium
63. lower pole of right lobe of thymus
64. proximal (ascending) part of thoracic aorta
65. triceps brachii muscle
66. cartilage ring of trachea
67. intercostal muscle (external and internal layers)
68. right deltoid muscle
69. right serratus anterior muscle
70. hair follicle
71. semispinalis dorsi and cervicis muscles
72. arch of the aorta (descending thoracic part) just distal to site of entrance of the ductus arteriosus
73. origin of left pulmonary artery

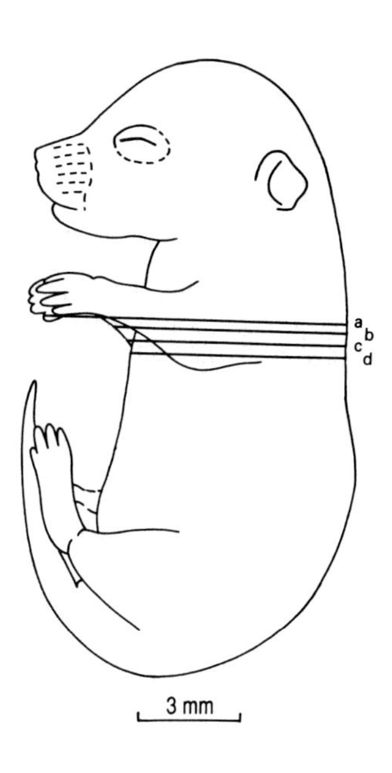

74. lumen of pulmonary trunk
75. tendon of insertion of left triceps brachii muscle into olecranon process of left ulna
76. leaflet of pulmonary valve
77. pericardial cavity
78. brachial artery
79. origin of right pulmonary artery
80. bifurcation of the trachea
81. left main bronchus
82. left pulmonary artery
83. site of entrance of right pulmonary vein into left atrium
84. left coronary artery which arises from the left posterior aortic sinus
85. radio-humeral component of left elbow joint
86. wall of right ventricle
87. lumen of right ventricle
88. origin of right coronary artery which arises from the anterior aortic sinus
89. cartilage primordium of head of the right radius
90. radio-humeral component of right elbow joint
91. cartilage primordium of capitulum of the right humerus
92. cartilage primordium of distal part of shaft of the right humerus with periosteal collar of bone
93. posterior aortic sinus
94. branch of right pulmonary vein which drains the cranial lobe of the right lung
95. right pulmonary artery
96. lobar bronchus to cranial lobe of the right lung
97. right main bronchus
98. right sympathetic trunk
99. part of extra-vertebral venous plexus
100. deposits of brown (multilocular) fat

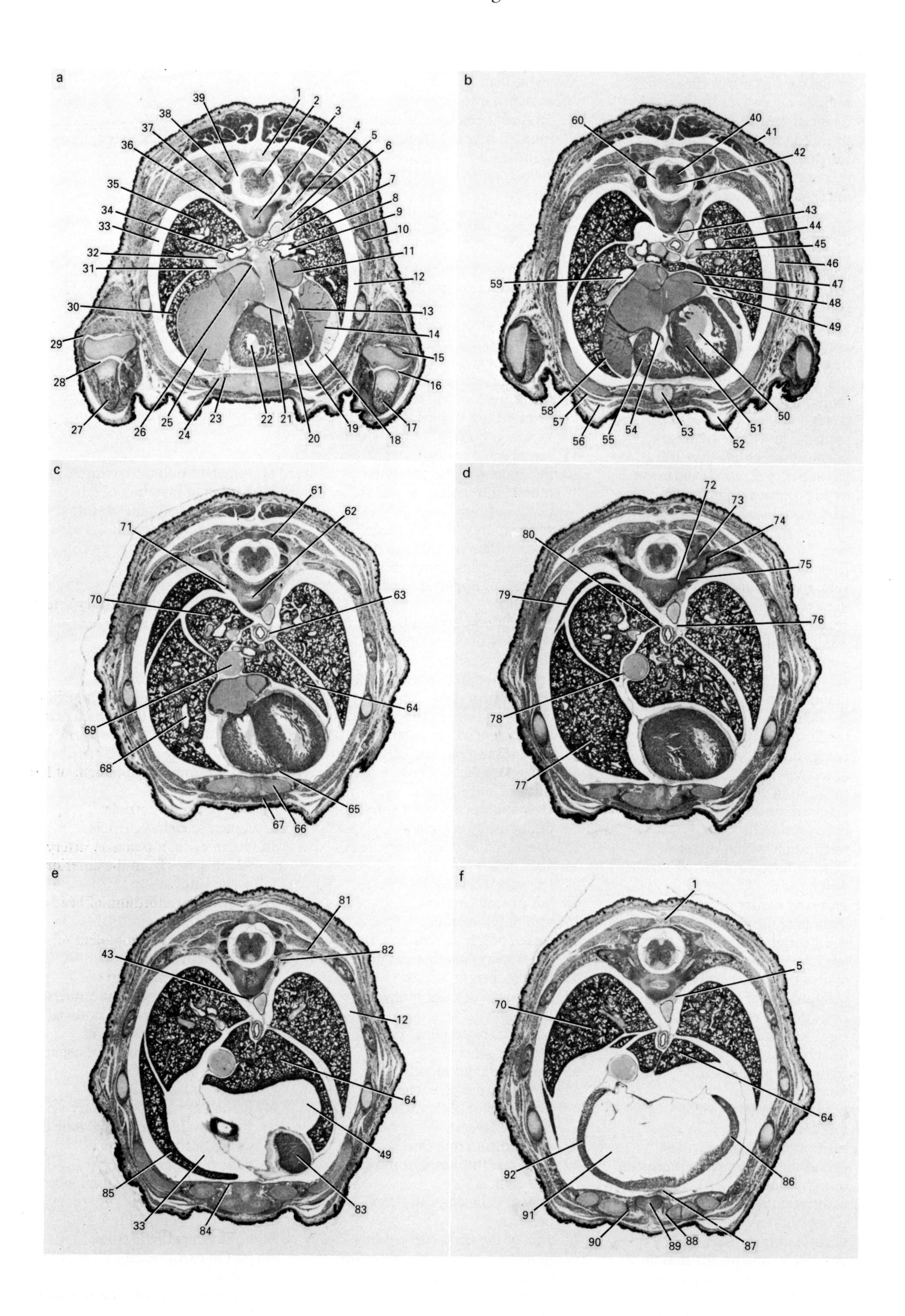
a
b
c
d
e
f
1
2
3
4
5
6
7
8
9
10
11
12
13
14
15
16
17
18
19
20
21
22
23
24
25
26
27
28
29
30
31
32
33
34
35
36
37
38
39
40
41
42
43
44
45
46
47
48
49
50
51
52
53
54
55
56
57
58
59
60
61
62
63
64
65
66
67
68
69
70
71
72
73
74
75
76
77
78
79
80
81
82
83
84
85
86
87
88
89
90
91
92

Plate 40g (17.5 days p.c.)

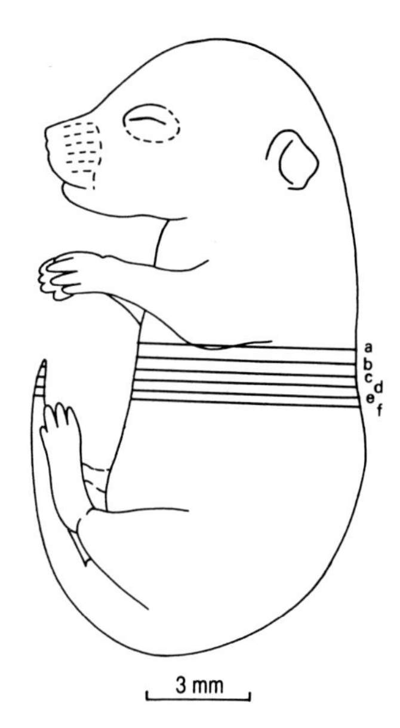

17.5 days p.c. Male embryo. Crown–rump length 17.8 mm (fixed). Transverse sections (Theiler Stage 26; rat, Witschi Stage 35)

1. part of extra-vertebral venous plexus
2. spinal cord in mid-thoracic region
3. ossification within central part of cartilage primordium of mid-thoracic vertebral body
4. left thoracic sympathetic ganglion
5. hemiazygos vein
6. thoracic (descending) aorta
7. oesophagus
8. left lung
9. left main bronchus
10. ossification within cartilage primordium of mid-shaft region of a left rib
11. left superior (cranial) vena cava
12. left pleural cavity
13. wall of left ventricle
14. lumen of auricular component of left atrium
15. cartilage primordium of olecranon process of left ulna
16. humero-ulnar component of left elbow joint
17. ossification within cartilage primordium of distal part of shaft of left ulna
18. wall of auricular component of left atrium
19. pericardial cavity
20. site of entrance of left pulmonary vein into left atrium
21. leaflet of left atrio-ventricular (mitral) valve
22. lumen of right ventricle
23. right internal thoracic (mammary) artery
24. right internal thoracic (mammary) vein
25. lumen of right atrium
26. site of entrance of right pulmonary vein into left atrium
27. ossification within cartilage primordium of distal part of shaft of right ulna
28. humero-ulnar component of right elbow joint
29. tendon of insertion of right triceps brachii muscle into olecranon process of right ulna
30. apex of middle lobe of right lung
31. right pulmonary vein
32. branch of right pulmonary artery
33. right pleural cavity
34. right main bronchus
35. cranial lobe of right lung
36. right thoracic sympathetic ganglion
37. dorsal (posterior) root ganglion
38. ossification within ventral part of neural arch of thoracic vertebra where it articulates with tubercle of rib
39. spinal canal
40. mantle layer (dorsal grey horn)
41. hair follicle
42. mantle layer (ventral grey horn)
43. thoracic duct
44. branch of left pulmonary artery
45. left pulmonary vein
46. intercostal muscles (external and internal layers)
47. apex of accessory lobe of right lung
48. site of entrance of left superior (cranial) vena cava into right atrium
49. extension of right pleural cavity around accessory lobe of right lung
50. lumen of left ventricle
51. muscular component of interventricular septum
52. mammary elevation (ridge)
53. ossification within cartilage primordium of sternebra
54. leaflet of right atrio-ventricular (tricuspid) valve
55. wall of right ventricle
56. pectoralis superficialis muscle
57. pectoralis profundus muscle
58. wall of auricular component of right atrium
59. lobar bronchus to middle lobe of right lung
60. marginal layer
61. ossification within cartilage primordium of neural arch
62. nucleus pulposus
63. left vagal (X) trunk
64. accessory lobe of right lung
65. interventricular groove
66. costal cartilage
67. upper fibres of left rectus abdominis muscle
68. segmental bronchus within middle lobe of right lung
69. inferior vena cava
70. caudal lobe of right lung
71. right sympathetic trunk
72. articulation between head of rib and body of corresponding thoracic vertebra
73. articulation between tubercle of rib and neural arch of corresponding thoracic vertebra
74. ossification within cartilage primordium of tubercle of rib
75. ossification within cartilage primordium of head of rib
76. dorsal meso-oesophagus
77. middle lobe of right lung
78. right phrenic nerve
79. caudal border of cranial lobe of the right lung
80. right vagal (X) trunk
81. intercostal nerve
82. intercostal artery
83. caudal surface of wall of left ventricle
84. transversus thoracis muscle
85. caudal border of middle lobe of the right lung
86. left dome of the diaphragm
87. xiphisternal origin of the diaphragm
88. left costal margin
89. ossification within cartilage primordium of xiphoid process (xiphisternum)
90. right costal margin
91. sub-phrenic recess of peritoneal cavity
92. right dome of the diaphragm

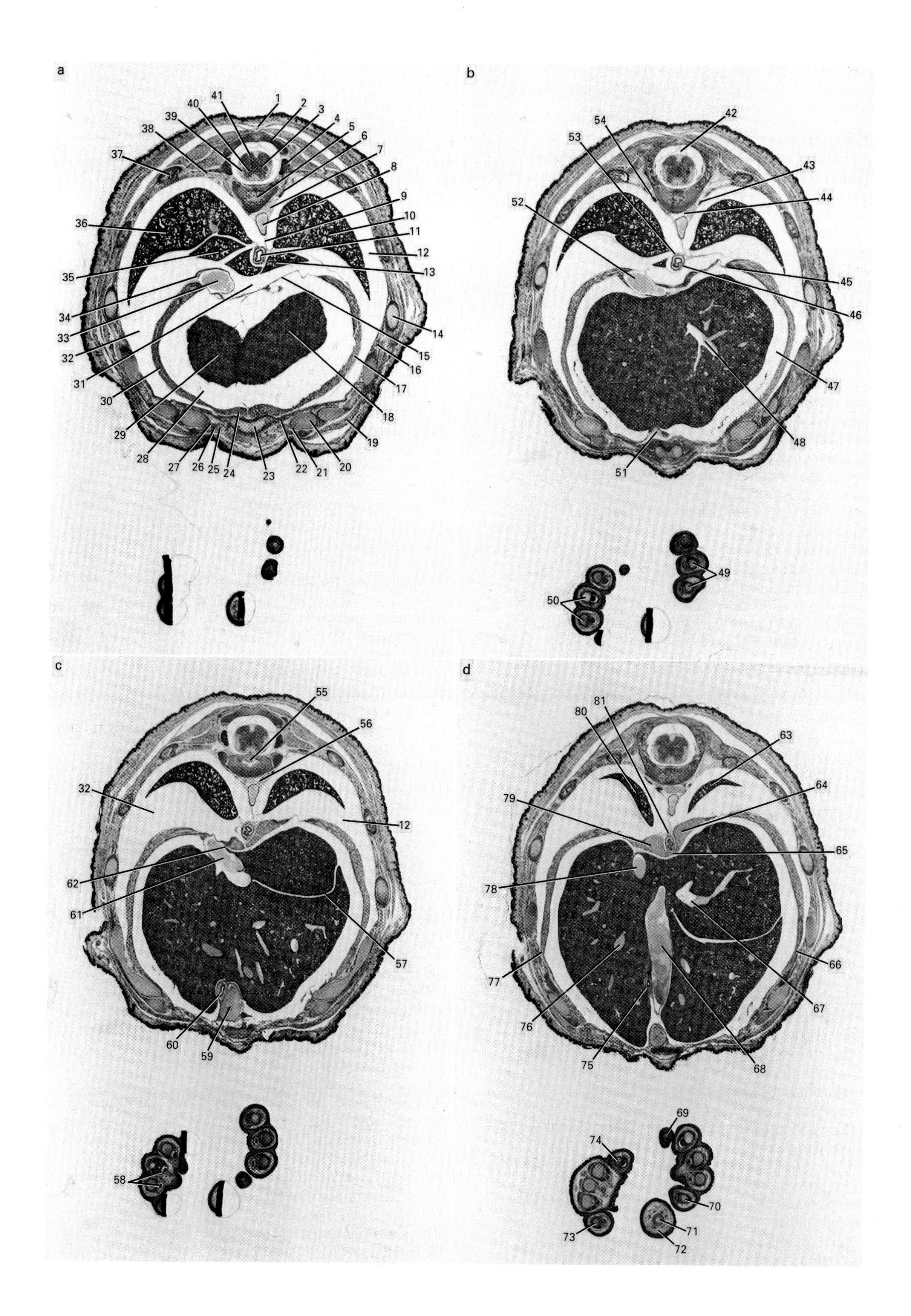
a
b
c
d
1
2
3
4
5
6
7
8
9
10
11
12
13
14
15
16
17
18
19
20
21
22
23
24
25
26
27
28
29
30
31
32
33
34
35
36
37
38
39
40
41
42
43
44
45
46
47
48
49
50
51
52
53
54
55
56
57
58
59
60
61
62
63
64
65
66
67
68
69
70
71
72
73
74
75
76
77
78
79
80
81

Plate 40h (17.5 days p.c.)

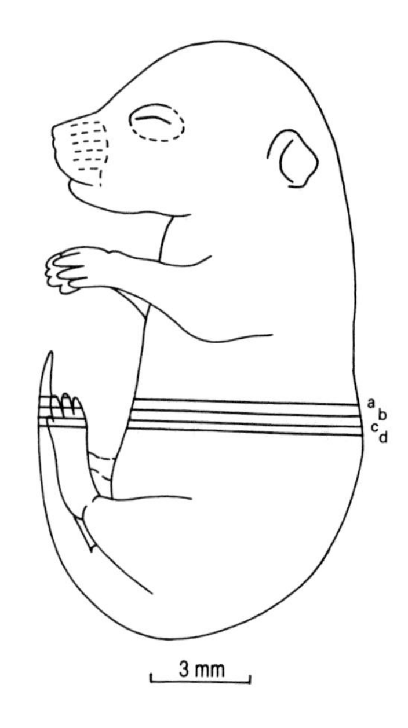

17.5 days p.c. Male embryo. Crown–rump length 17.8 mm (fixed).
Transverse sections (Theiler Stage 26; rat, Witschi Stage 35)

1. part of extra-vertebral venous plexus
2. ossification within cartilage primordium of neural arch
3. spinal cord in lower thoracic region
4. dorsal (posterior) root ganglion
5. ossification within central part of cartilage primordium of lower thoracic vertebral body
6. left thoracic sympathetic ganglion
7. hemiazygos vein
8. thoracic (descending) aorta
9. dorsal meso-oesophagus
10. left lung
11. oesophagus
12. left pleural cavity
13. accessory lobe of right lung
14. cartilage primordium of caudal asternal rib with periosteal collar of bone
15. central tendon of the diaphragm
16. intercostal muscles (external and internal layers)
17. left dome of the diaphragm (muscle fibres of costal origin)
18. apex of the left lobe of the liver
19. rectus abdominis muscle
20. costal cartilage
21. left costal margin
22. transversus thoracis muscle
23. ossification within cartilage primordium of xiphoid process (xiphisternum)
24. xiphisternal origin of the diaphragm
25. right internal thoracic (mammary) artery
26. right internal thoracic (mammary) vein
27. right costal margin
28. sub-phrenic recess of peritoneal cavity
29. apex of right lobe of the liver
30. right dome of the diaphragm (muscle fibres of costal origin)
31. extension of right pleural cavity around accessory lobe of right lung
32. right pleural cavity
33. inferior vena cava
34. right phrenic nerve
35. segment of caudal lobe of right lung
36. principal part of caudal lobe of right lung
37. ossification within cartilage primordium in proximal part of shaft of rib
38. intercostal vein
39. marginal layer
40. mantle layer (dorsal grey horn)
41. mantle layer (ventral grey horn)
42. spinal canal
43. left sympathetic trunk
44. thoracic duct
45. left boundary of bare area of the liver
46. left (almost anterior) vagal (X) trunk
47. peritoneal cavity
48. part of hepatic venous plexus
49. cartilage primordia of phalangeal bones of left hindlimb
50. cartilage primordia of phalangeal bones of right hindlimb
51. falciform ligament
52. proximal part of posthepatic inferior vena cava
53. right (almost posterior) vagal (X) trunk
54. right sympathetic trunk
55. nucleus pulposus
56. origin of intercostal artery from thoracic (descending) aorta
57. lobar interzone
58. long flexor digital tendons
59. left umbilical vein passing dorsally to join the ductus venosus
60. lumen of the gall bladder
61. inferior vena cava (partly intra-hepatic at this level) formed principally from ductus venosus, left umbilical vein and right vitelline venous system
62. right boundary of bare area of the liver
63. caudal border of left lung
64. left crus of the diaphragm
65. anterior vagal (X) trunk
66. left costal origin of the diaphragm
67. left hepatic vein (draining upper part of the left lobe of the liver)
68. ductus venosus
69. first digit left hindlimb
70. fifth digit left hindlimb
71. cartilage primordium of tail vertebral body
72. distal part of tail
73. fifth digit right hindlimb
74. first digit right hindlimb
75. distal part of cystic duct
76. part of right hepatic venous plexus which drains the upper part of the right lobe of the liver
77. right costal origin of the diaphragm
78. inferior vena cava (intra-hepatic at this level) before it receives hepatic venous blood
79. right crus of the diaphragm
80. caudal border of caudal lobe of right lung
81. posterior vagal (X) trunk

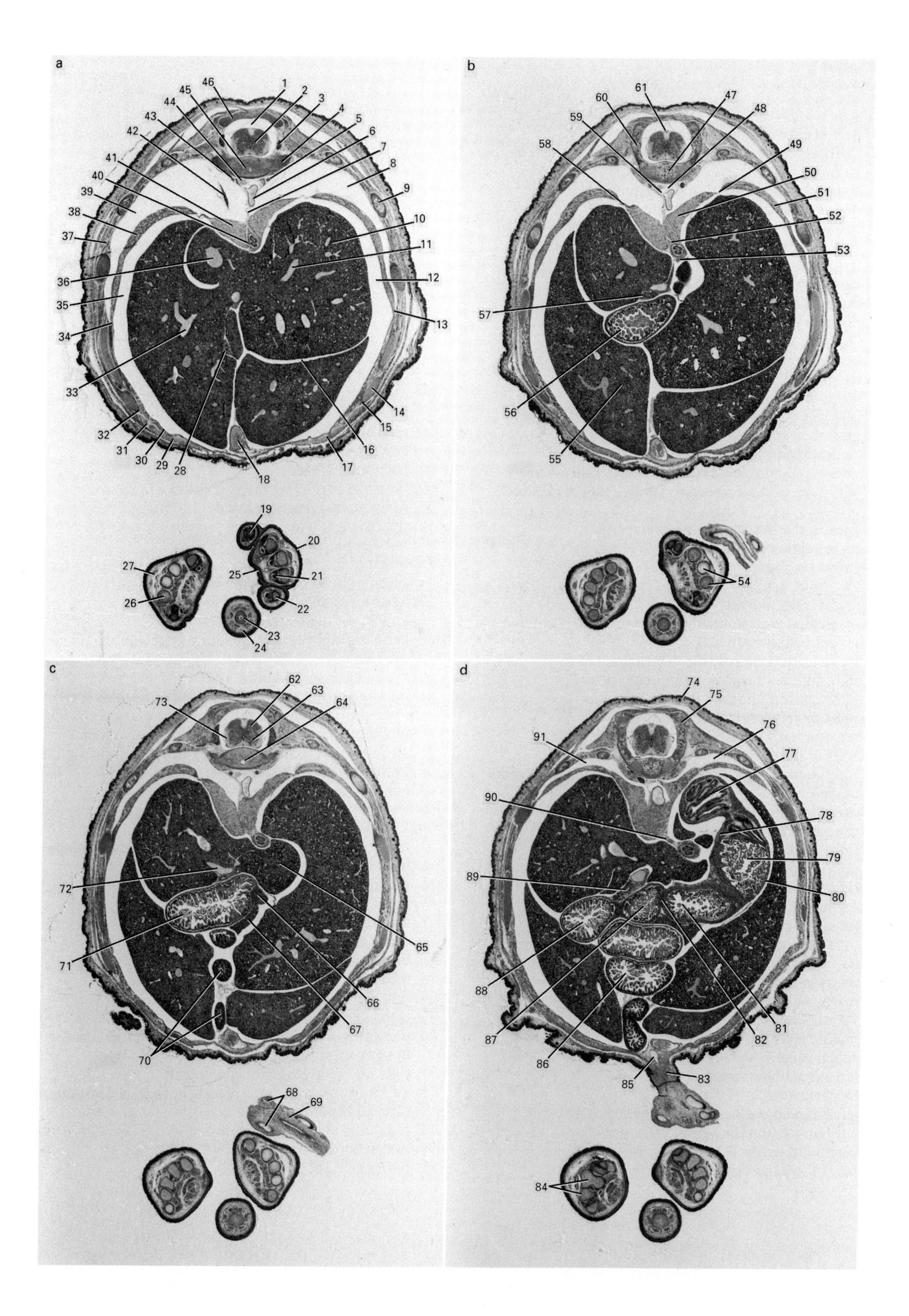
a
1
2
3
4
5
6
7
8
9
10
11
12
13
14
15
16
17
18
19
20
21
22
23
24
25
26
27
28
29
30
31
32
33
34
35
36
37
38
39
40
41
42
43
44
45
46
b
47
48
49
50
51
52
53
54
55
56
57
58
59
60
61
c
62
63
64
65
66
67
68
69
70
71
72
73
d
74
75
76
77
78
79
80
81
82
83
84
85
86
87
88
89
90
91

Plate 40i (17.5 days p.c.)

17.5 days p.c. Male embryo. Crown–rump length 17.8 mm (fixed). Transverse sections (Theiler Stage 26; rat, Witschi Stage 35)

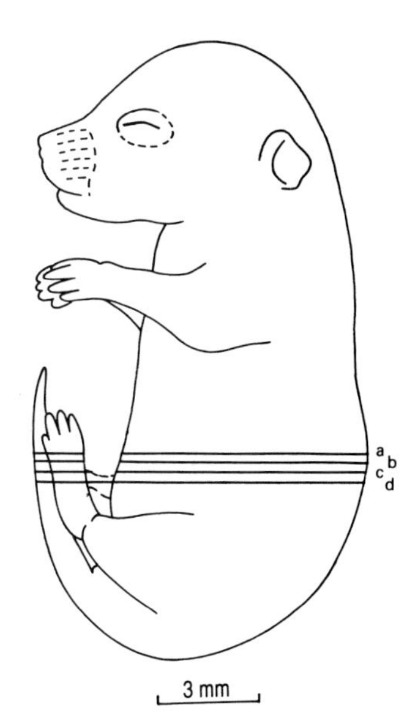

1. spinal canal
2. spinal cord in lower thoracic region
3. dorsal (posterior) root ganglion
4. ossification within cartilage primordium of head of rib
5. left sympathetic trunk
6. thoracic (descending) aorta
7. dorsal meso-oesophagus
8. left pleural cavity
9. ossification within cartilage primordium of mid-shaft region of a left rib
10. left lobe of the liver
11. branch of left hepatic vein (draining upper part of the left lobe of the liver)
12. left sub-phrenic recess of peritoneal cavity
13. left costal origin of the diaphragm
14. costal cartilage
15. left costal margin
16. lobar interzone
17. left rectus abdominis muscle
18. left umbilical vein which passes dorsally to join the ductus venosus
19. first digit of left hindlimb
20. dorsal surface of left hindlimb
21. long flexor digital tendons
22. fifth digit of left hindlimb
23. cartilage primordium of tail vertebral body
24. distal part of tail
25. plantar surface of left hindlimb
26. fourth metatarsal bone of right hindlimb
27. right hindlimb
28. distal part of cystic duct
29. right rectus abdominis muscle
30. transversus abdominis muscle
31. right superior epigastric vessels
32. right costal margin
33. branch of right hepatic vein (draining upper part of the right lobe of the liver)
34. right costal origin of the diaphragm
35. right sub-phrenic recess of peritoneal cavity
36. inferior vena cava (intra-hepatic at this level) before it receives hepatic venous blood
37. intercostal muscles (external and internal layers)
38. right dome of the diaphragm (muscle fibres of costal origin)
39. right pleural cavity
40. oesophagus
41. right crus of the diaphragm
42. caudal border of caudal lobe of right lung
43. right sympathetic trunk
44. azygos vein
45. cartilage primordium of lower thoracic vertebral body with periosteal ossification
46. cartilage primordium of dorsal part of neural arch
47. ossification within central part of cartilage primordium of lower thoracic vertebral body
48. thoracic duct
49. left boundary of bare area of the liver
50. left crus of the diaphragm
51. left dome of the diaphragm (muscle fibres of costal origin)
52. posterior vagal (X) trunk
53. anterior vagal (X) trunk
54. metatarsal bones of left hindlimb
55. right lobe of the liver
56. villous projections into lumen of first part of the duodenum
57. common bile duct
58. right boundary of bare area of the liver
59. proximal part of a right intercostal artery
60. ossification within ventral part of cartilage primordium of neural arch
61. posterior spinal vein
62. mantle layer (dorsal grey horn)
63. mantle layer (ventral grey horn)
64. nucleus pulposus
65. caudate lobe of the liver
66. wall overlying pyloric sphincter
67. pancreas
68. umbilical vessels within umbilical cord
69. Wharton's jelly
70. loops of small intestine
71. proximal part of second part of the duodenum
72. portal vein
73. marginal layer
74. hair follicle
75. ossification within dorsolateral part of cartilage primordium of neural arch
76. caudal extremity of left pleural cavity
77. lumen of fundus (proventricular) region of the stomach
78. posterior wall in region of the "body" of the stomach
79. lumen of the "body" (glandular region) of the stomach
80. anterior wall in region of the "body" of the stomach
81. lumen in pyloric antrum region of the stomach
82. pyloric sphincter at gastro-duodenal junction
83. proximal part of umbilical cord
84. metatarsal bones of right hindlimb
85. umbilical artery as it enters the umbilical cord
86. lumen of small intestine
87. lumen of first part of the duodenum
88. lumen in middle of the second (descending) part of the duodenum
89. pancreatic duct
90. posterior vagal (X) trunk in dorsal meso-oesophagus
91. caudal extremity of right pleural cavity

Plate 40j (17.5 days p.c.)

17.5 days p.c. Male embryo. Crown–rump length 17.8 mm (fixed).
Transverse sections (Theiler Stage 26; rat, Witschi Stage 35)

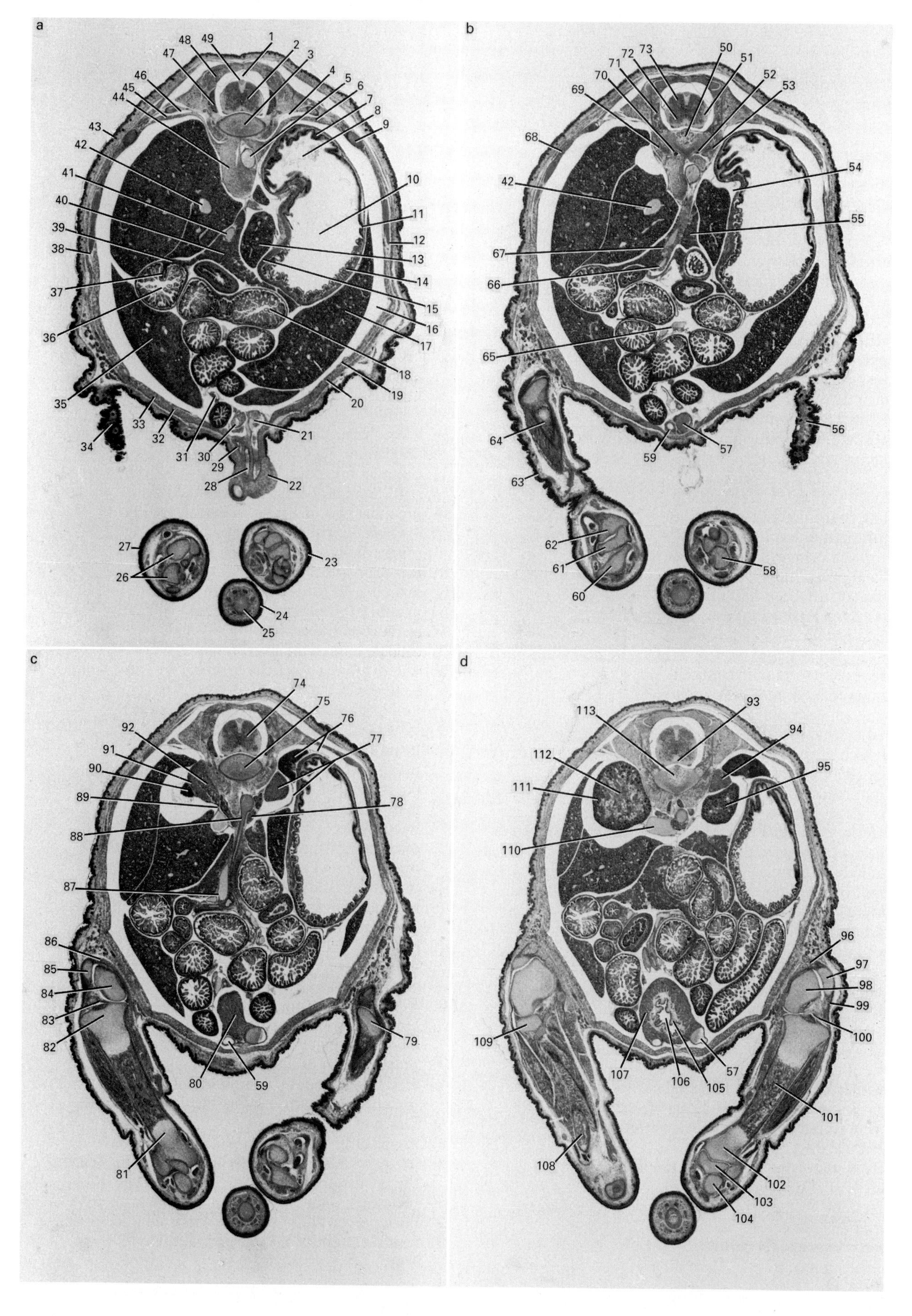

1. spinal canal
2. spinal cord in lower thoracic region
3. cartilage primordium of lower thoracic vertebral body with periosteal ossification
4. caudal extremity of left pleural cavity
5. thoracic (descending) aorta
6. diaphragmatic muscle fibres originating from left medial and lateral arcuate ligaments
7. gastric mucosa lining the fundus region of the stomach
8. lumen of fundus (proventricular) region of the stomach (cutaneous non-glandular area)
9. 12th costal cartilage
10. lumen of the "body" (glandular region) of the stomach
11. line of transition from cutaneous (non-glandular) to glandular mucous membrane
12. left costal margin (anterior border of 13th rib)
13. caudate lobe of the liver
14. gastric mucosa lining the "body" of the stomach (area of fundic glands)
15. posterior wall of the stomach
16. peritoneal cavity
17. left lobe of the liver
18. lumen of small intestine
19. transversus abdominis muscle
20. rectus abdominis muscle
21. left umbilical vein
22. Wharton's jelly within proximal part of umbilical cord
23. dorsal surface of left hindlimb
24. distal part of tail
25. cartilage primordium of tail vertebral body
26. cartilage primordia of tarsal bones of right hindlimb
27. dorsal surface of right hindlimb
28. left umbilical artery in proximal part of umbilical cord
29. right umbilical artery in proximal part of umbilical cord
30. vitelline vein
31. dorsal mesentery of midgut
32. right superior epigastric vessels
33. right rectus abdominis muscle
34. rostral (pre-axial) border of right hindlimb
35. right lobe of the liver
36. lumen in middle of the second (descending) part of the duodenum
37. duodenal papilla (site of entrance of pancreatic duct into second part of the duodenum)
38. right costal margin
39. pancreatic duct
40. body of pancreas
41. portal vein
42. inferior vena cava (intra-hepatic at this level) before it receives hepatic venous blood
43. hair follicle
44. right crus of the diaphragm
45. diaphragmatic muscle fibres originating from right medial and lateral arcuate ligaments
46. caudal extremity of right pleural cavity
47. dorsal (posterior) root ganglion
48. ossification within dorsolateral part of cartilage primordium of neural arch
49. posterior spinal vein
50. ossification within central part of cartilage primordium of lower thoracic vertebral body
51. marginal layer
52. left sympathetic trunk
53. left psoas major muscle
54. lesser sac of peritoneal cavity (superior recess of omental bursa)
55. proximal part of tail of pancreas
56. rostral (pre-axial) border of left hindlimb
57. left umbilical artery
58. cartilage primordium of tarsal bone
59. right umbilical artery
60. cartilage primordium of right calcaneum
61. cartilage primordium of right talus
62. cartilage primordium of distal part of right tibia
63. outer (lateral) border of right hindlimb
64. ossification within mid-shaft region of right tibia
65. superior mesenteric vessels within dorsal mesentery of midgut
66. superior mesenteric artery
67. upper pancreaticoportal vein where it drains into the portal vein
68. intercostal muscles (external and internal layers)
69. right psoas major muscle
70. right sympathetic trunk
71. ossification within ventral part of cartilage primordium of neural arch
72. mantle layer (ventral grey horn)
73. mantle layer (dorsal grey horn)
74. spinal cord in upper lumbar region
75. cartilage primordium of upper lumbar vertebral body with peritoneal ossification
76. upper pole of spleen and (laterally) the gastro-splenic "ligament"
77. upper pole of left adrenal gland and (laterally) the lieno-renal "ligament"
78. pre-aortic abdominal sympathetic ganglion (coeliac ganglion)
79. cartilage primordium of head of left tibia
80. rostral part of wall of bladder (region of urachus)
81. cartilage primordium of distal part of shaft of right tibia
82. cartilage primordium of head of right tibia
83. right knee joint
84. cartilage primordium of distal part of right femur

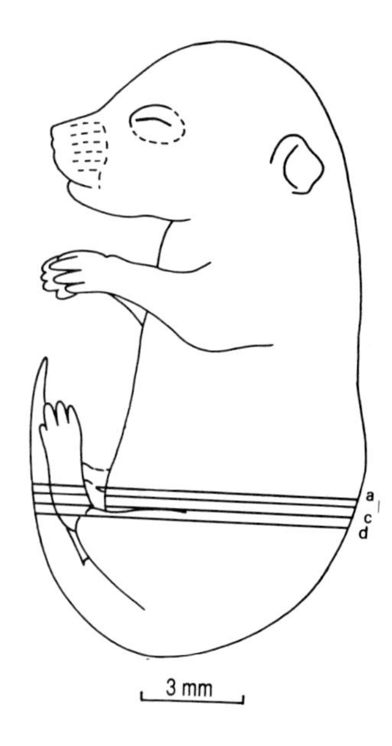

85. cartilage primordium of right patella
86. lower fibres of right quadriceps femoris muscle
87. superior mesenteric vein where it drains into the portal vein
88. coeliac artery where it takes origin from the abdominal aorta
89. suprarenal vein where it drains into the inferior vena cava
90. upper pole of right kidney
91. cortex of right adrenal gland
92. medullary region of right adrenal gland
93. anterior spinal artery
94. lower pole of left adrenal gland
95. upper pole of left kidney
96. lower fibres of left quadriceps femoris muscle
97. cartilage primordium of left patella
98. cartilage primordium of distal part of left femur
99. patellar ligament (tendon of insertion of quadriceps femoris muscle)
100. left knee joint
101. ossification within mid-shaft region of cartilage primordium of left tibia
102. cartilage primordium of distal part of left tibia
103. cartilage primordium of left talus
104. cartilage primordium of left calcaneum
105. mucosal lining of bladder
106. lumen of bladder
107. wall of bladder
108. ossification within distal part of shaft of right fibula
109. lateral border of head of right tibia
110. inferior vena cava (just distal to origin of right and left renal veins)
111. cortex of right kidney
112. medulla of right kidney
113. ossification within central part of cartilage primordium of upper lumbar vertebral body

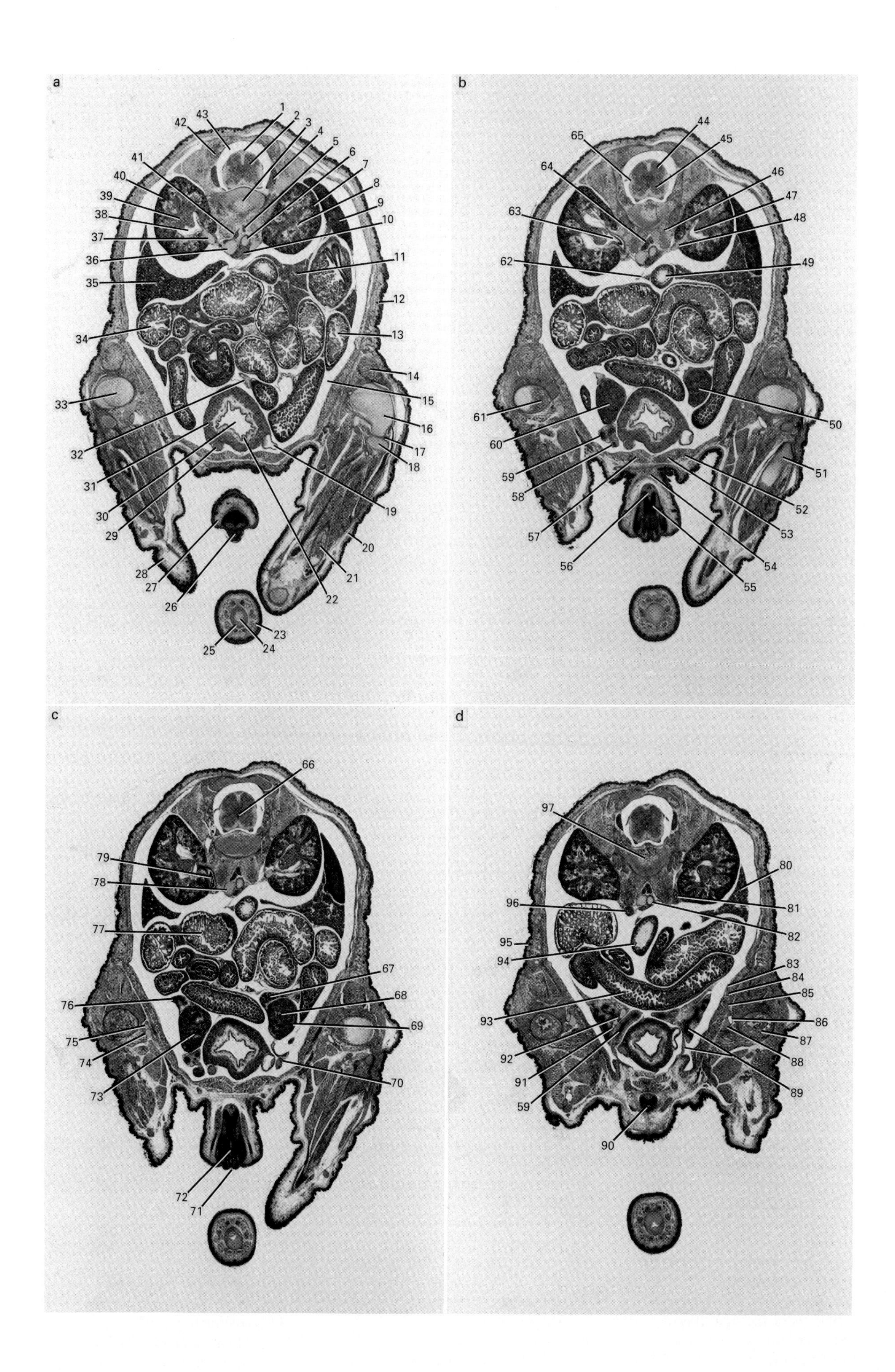
a
b
c
d
1
2
3
4
5
6
7
8
9
10
11
12
13
14
15
16
17
18
19
20
21
22
23
24
25
26
27
28
29
30
31
32
33
34
35
36
37
38
39
40
41
42
43
44
45
46
47
48
49
50
51
52
53
54
55
56
57
58
59
60
61
62
63
64
65
66
67
68
69
70
71
72
73
74
75
76
77
78
79
80
81
82
83
84
85
86
87
88
89
90
91
92
93
94
95
96
97

Plate 40k (17.5 days p.c.)

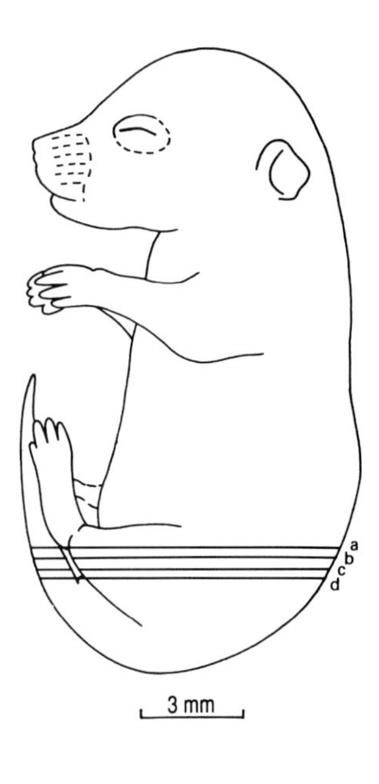

17.5 days p.c. Male embryo. Crown–rump length 17.8 mm (fixed). Transverse sections (Theiler Stage 26; rat, Witschi Stage 35)

1. posterior spinal vein
2. spinal cord in upper lumbar region
3. ossification within central part of cartilage primordium of upper lumbar vertebral body
4. dorsal (posterior) root ganglion
5. left sympathetic trunk
6. abdominal aorta just distal to origin of right and left renal arteries
7. cortical region of left kidney
8. medullary region of left kidney
9. spleen
10. pre-aortic abdominal sympathetic ganglion (coeliac ganglion)
11. tail of pancreas
12. hair follicle
13. lumen of small intestine
14. left quadriceps femoris muscle
15. peritoneal cavity
16. cartilage primordium of distal part of left femur
17. left knee joint
18. lateral border of head of left tibia
19. left umbilical artery
20. outer (lateral) border of left hindlimb
21. ossification within distal part of shaft of left fibula
22. mucosal lining of bladder
23. distal part of tail
24. nucleus pulposus
25. cartilage primordium of tail vertebral body
26. glans penis
27. prepuce
28. outer (lateral) border of right hindlimb
29. right umbilical artery
30. lumen of bladder
31. wall of bladder
32. superior mesenteric vessels within dorsal mesentery of midgut
33. cartilage primordium of distal part of right femur
34. lumen of second part of duodenum
35. caudal part of right lobe of the liver
36. inferior vena cava
37. right renal vein
38. lumen of minor calyx of right kidney
39. medullary region of right kidney
40. cortical region of right kidney
41. right sympathetic trunk
42. ossification within dorsolateral part of cartilage primordium of neural arch
43. spinal canal
44. mantle layer (dorsal grey horn)
45. mantle layer (ventral grey horn)
46. left psoas major muscle
47. lumen of minor calyx of left kidney
48. branch of left renal vein which drains the upper pole of the left kidney
49. hindgut (rectum)
50. upper pole of left testis
51. cartilage primordium of proximal part of shaft of left fibula
52. left inferior epigastric vessels
53. left rectus abdominis muscle
54. proximal part of penis
55. crus penis
56. deep dorsal vein of penis
57. right rectus abdominis muscle
58. right gubernaculum testis
59. right ductus deferens
60. upper pole of right testis
61. ossification within cartilage primordium of mid-shaft region of right femur
62. dorsal mesentery of hindgut
63. pelvis of right kidney
64. site of association in midline between pair of lumbar sympathetic ganglia
65. marginal layer
66. ependymal layer lining central canal
67. mesonephric tubules within degenerating left mesonephros becoming modified to form the vasa efferentia of the testis and epididymis
68. medullary region of left testis
69. cortical region of left testis
70. left gubernaculum testis
71. urethral groove at tip of penis
72. penile urethra
73. seminiferous tubules within medullary region of right testis
74. right femoral vein
75. right femoral artery
76. mesonephric tubules within degenerating right mesonephros
77. lumen of third part of the duodenum
78. postrenal part of inferior vena cava
79. proximal part of right ureter just distal to renal pelvis
80. distal part of tail of pancreas
81. proximal part of left ureter
82. postrenal part of abdominal aorta
83. transversus abdominis muscle
84. internal oblique muscle
85. external oblique muscle
86. left femoral artery
87. ossification within cartilage primordium of mid-shaft region of left femur
88. left femoral vein
89. left ductus deferens and (inferiorly) the gubernaculum testis
90. bulbar part of urethra
91. right testicular artery
92. pampiniform plexus of testicular veins
93. mucous membrane of small intestine with characteristic circularly or spirally arranged folds whose surface is covered with filiform projections (the intestinal villi)
94. mucous membrane of large intestine (rectum) with large number of goblet cells characteristically present
95. cutaneous muscle of the trunk (panniculus carnosus)
96. proximal part of right ureter
97. ossification within central part of cartilage primordium of mid-lumbar vertebral body

Plate 40l (17.5 days p.c.)

17.5 days p.c. Male embryo. Crown–rump length 17.8 mm (fixed).
Transverse sections (Theiler Stage 26; rat, Witschi Stage 35)

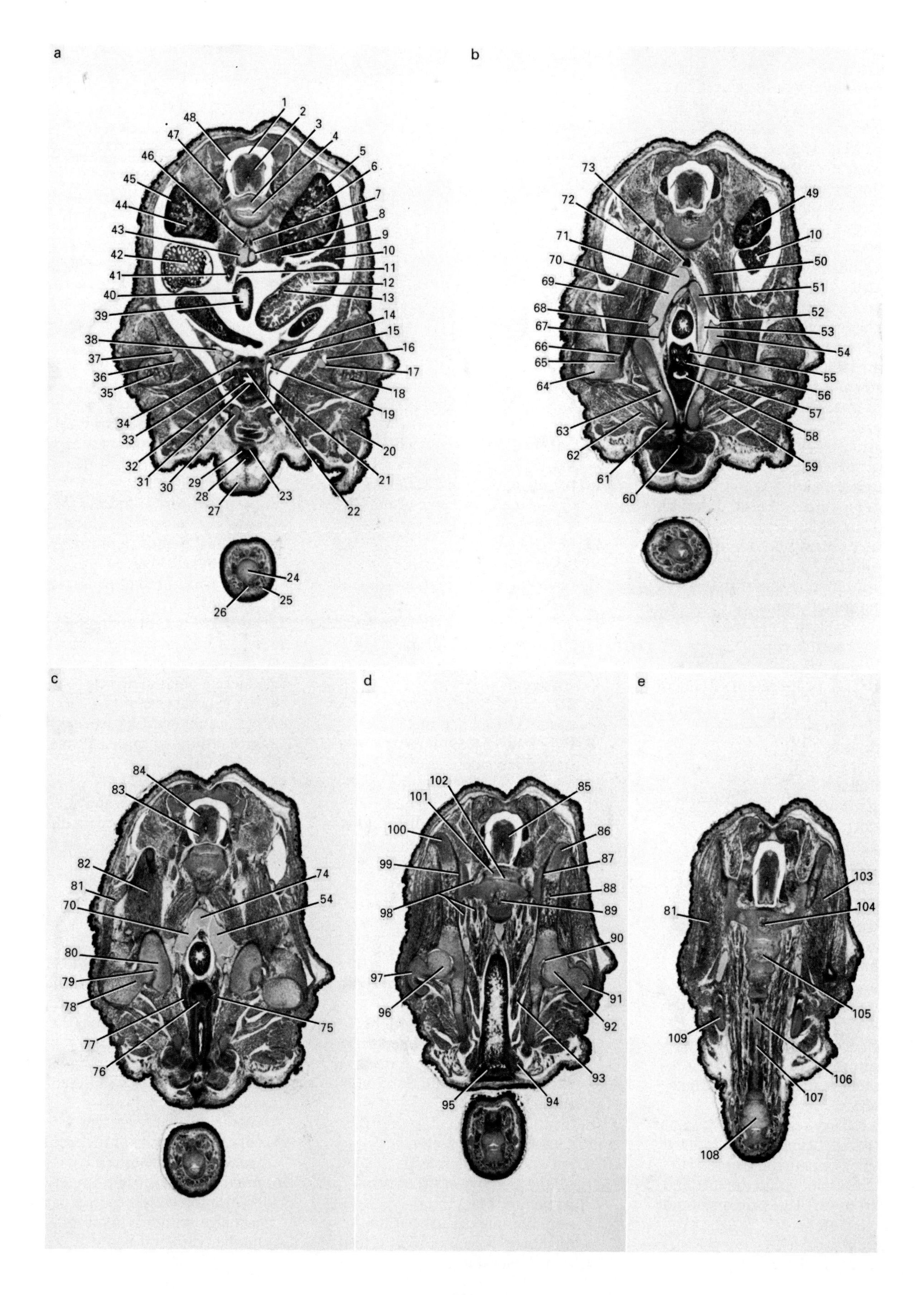

1. posterior spinal vein
2. spinal cord in mid-lumbar region
3. cartilage primordium of mid-lumbar vertebral body with periosteal ossification
4. nucleus pulposus
5. cortical region of left kidney
6. medullary region of left kidney
7. left sympathetic trunk
8. postrenal part of abdominal aorta
9. left ureter
10. distal part of tail of pancreas (islets of Langerhans are readily recognizable at this stage)
11. dorsal mesentery of hindgut
12. lumen of small intestine
13. mucous membrane of small intestine with characteristic circularly or spirally arranged folds whose surface is covered with filiform projections (the intestinal villi)
14. left testicular artery
15. pampiniform plexus of testicular veins
16. left femoral artery
17. left femoral vein
18. ossification within cartilage primordium of mid-shaft region of left femur
19. left superior vesical artery which continues as left umbilical artery
20. proximal part of left hindlimb
21. left umbilical artery
22. plexus of veins in region of neck of the bladder
23. deep dorsal vein of penis
24. cartilage primordium of tail vertebral body
25. middle region of tail
26. filum terminale
27. loose connective tissue in future scrotal region
28. bulbar part of urethra
29. crus penis
30. right internal pudendal vein where it joins the right internal iliac vein
31. mucosal lining of bladder
32. lumen of neck of the bladder
33. right ureter within tissues of wall of neck of the bladder before it passes rostrally (from this location) to open at the posterolateral angle of the trigone
34. right ductus deferens
35. ossification within cartilage primordium of mid-shaft region of right femur
36. right femoral vein
37. right femoral artery
38. mesorchium (a peritoneal fold attached to the lower pole of the testis)
39. lumen of hindgut (rectum)
40. mucous membrane of large intestine (rectum) with large number of goblet cells characteristically present
41. right ureter
42. lumen at proximal part of third part of duodenum
43. inferior vena cava
44. medullary region at lower pole of right kidney
45. cortical region of right kidney
46. right sympathetic trunk
47. dorsal (posterior) root ganglion
48. spinal canal
49. lower pole of left kidney
50. left psoas major muscle
51. left common iliac artery
52. left external iliac artery
53. left internal iliac vein
54. left common iliac vein
55. duct of left seminal vesicle
56. left ductus deferens
57. left obturator externus muscle
58. prostatic region of the urethra
59. left adductor magnus muscle
60. proximal part of bulbar urethra surrounded by bulbo-urethral gland
61. cartilage primordium of inferior ramus of pubic bone
62. right adductor magnus muscle
63. ossification within cartilage primordium of superior ramus of pubic bone
64. cartilage primordium of proximal part of shaft of right femur
65. cartilage primordium of lesser trochanter of right femur
66. tendon of insertion of right iliopsoas muscle
67. right internal iliac artery
68. right external iliac artery
69. right iliacus muscle
70. right common iliac vein
71. inferior vena cava where it is formed by the amalgamation of the right and left common iliac veins
72. right psoas major muscle
73. caudal pair of lumbar sympathetic ganglia
74. site of fusion of the right and left common iliac veins
75. left vesical and inferior hypogastric plexuses containing both sympathetic and parasympathetic fibres
76. prostate
77. right vesical and inferior hypogastric plexuses
78. cartilage primordium of neck of right femur
79. right hip joint
80. cartilage primordium of acetabular region of right iliac bone
81. right gluteus maximus muscle
82. right gluteus medius muscle
83. mantle layer (ventral grey horn)
84. mantle layer (dorsal grey horn)
85. spinal cord in lumbo-sacral region
86. cartilage primordium of left iliac crest
87. left sacro-iliac joint

88. ossification within cartilage primordium of the left ilium
89. ossification within cartilage primordium of S1 vertebral body
90. left hip joint
91. cartilage primordium of greater trochanter of left femur
92. cartilage primordium of head of left femur
93. left component of levator ani muscle
94. external anal sphincter
95. ano-rectal junction
96. cartilage primordium of head of right femur
97. cartilage primordium of greater trochanter of right femur
98. nerve trunks progressing to form components of right lumbo-sacral plexus
99. right sacro-iliac joint
100. cartilage primordium of right iliac crest
101. ossification within cartilage primordium of L6 vertebral body
102. site of future intervertebral disc betwen L6 and S1 vertebral bodies
103. left gluteus maximus muscle
104. ossification within cartilage primordium of S2 vertebral body
105. ossification within cartilage primordium of S3 vertebral body
106. anterior sacral vein
107. anterior sacral artery
108. cartilage primordium of coccygeal vertebral body in proximal part of tail
109. cartilage primordium of ischial tuberosity

Plate 41 (17.5 days p.c.)

17.5 days p.c. Male embryo. Crown–rump length 19 mm (fixed). Sagittal section (Theiler Stage 26; rat, Witschi Stage 35).

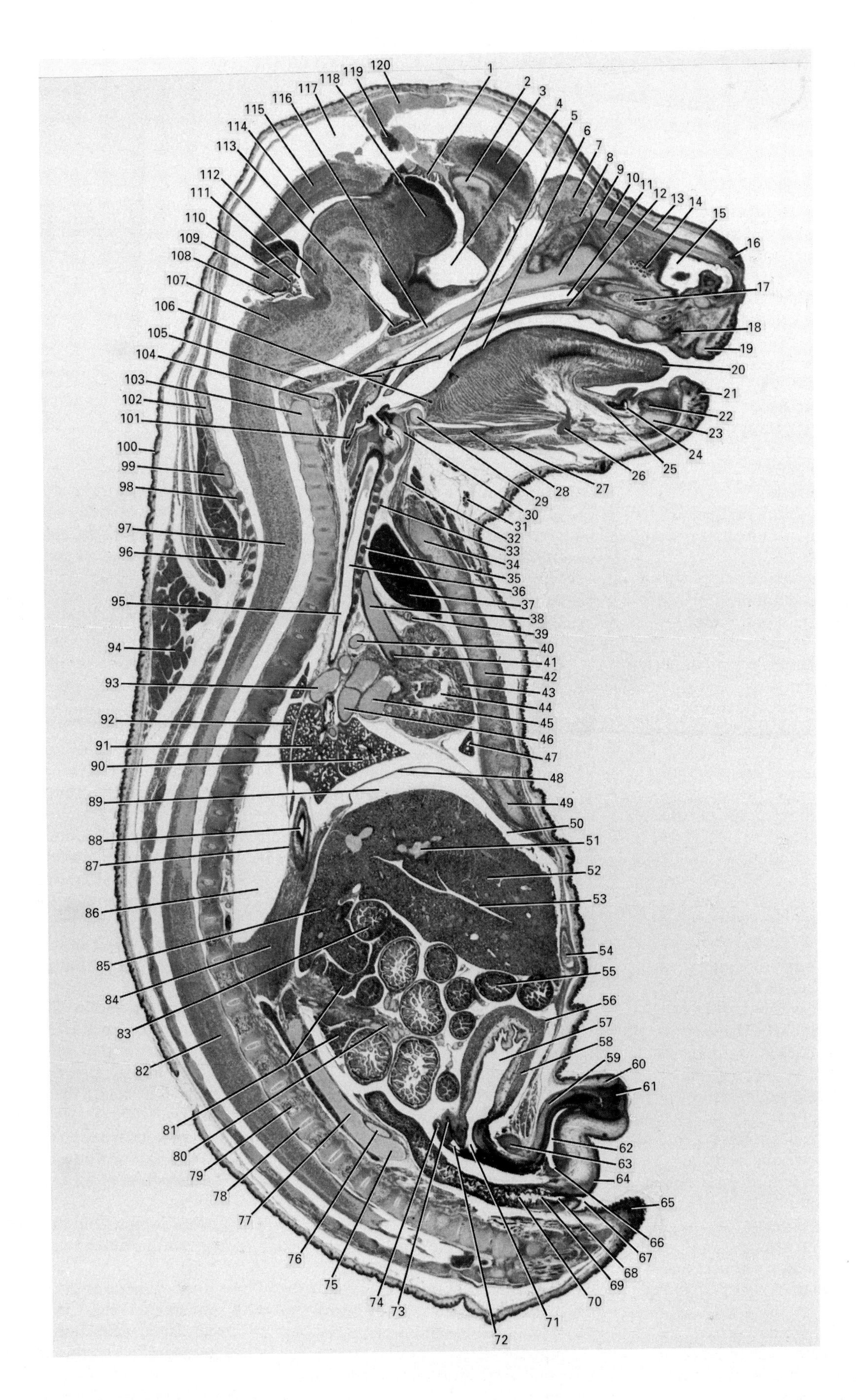

1. choroid plexus within lateral ventricle and arising from its medial wall
2. fibres of principally thalamic origin passing into the caudate nucleus
3. striatum (ganglionic eminence) overlying the head of the caudate nucleus
4. posterior horn of lateral ventricle
5. oropharynx
6. dorsum of tongue
7. olfactory lobe
8. cartilage primordium of cribriform plate of the ethmoid bone
9. cartilage primordium of nasal septum
10. olfactory (I) nerve fibres
11. nasopharynx
12. ossification within rostral part of secondary palate (palatal shelf of maxilla)
13. ossification within outer table of cartilage primordium of nasal bone
14. tubules of serous glands associated with wall of nasal septum
15. entrance into nasopharynx from external naris
16. vibrissa (upper lip)
17. ossification within primary palate (inter-maxillary segment)
18. incisor tooth (upper jaw)
19. upper lip
20. tip of tongue
21. vibrissa (lower lip)
22. periosteal ossification associated with upper part of body of mandible
23. ossification within cartilage primordium of body of mandible
24. site of common opening of submandibular and sublingual ducts
25. genioglossus muscle
26. lingual artery
27. mylohyoid muscle
28. hyoglossus muscle
29. cartilage primordium of body of hyoid bone
30. thyroid cartilage
31. submaxillary gland
32. sternohyoid muscle
33. isthmus of thyroid gland
34. ossification within cartilage primordium of manubrium sterni
35. tracheal ring
36. lumen of trachea
37. right lobe of thymus gland
38. ascending part of arch of the aorta
39. wall of left atrium
40. pulmonary trunk
41. leaflet of aortic valve
42. ossification within sternebra
43. wall of right ventricle of heart
44. lumen of right ventricle of heart
45. lumen of right atrium of heart
46. site of entrance of inferior vena cava into right atrium of the heart
47. ventral extremity of middle lobe of right lung
48. central tendon of diaphragm (region of dome of the diaphragm)
49. cartilage primordium of lower part of xiphisternum (xiphoid cartilage)
50. peritoneal cavity
51. right hepatic vein
52. right lobe of the liver
53. lobar interzone
54. umbilical vein
55. lumen of midgut (small intestine)
56. base of urachus
57. lumen of bladder
58. anterior wall of bladder
59. deep dorsal vein of penis
60. prepuce
61. glans penis
62. phallic (penile) part of urethra
63. ossification within cartilage primordium of pubic bone
64. skin of scrotum
65. proximal part of tail
66. bulbar recess
67. anal pit
68. lumen of anal canal
69. region of pectinate line (junctional zone between ectodermally derived anal pit, and endodermally derived embryonic hindgut)
70. lumen of rectum
71. prostatic part of urethra
72. ejaculatory duct
73. ductus deferens where it approaches the ejaculatory duct
74. lumen of proximal part of seminal vesicle
75. right common iliac vein
76. right common iliac artery
77. inferior vena cava
78. nucleus pulposus in central part of lumbar intervertebral disc
79. ossification within cartilage primordium of lumbar vertebral body
80. vitelline vessels (branches of superior mesenteric artery and vein) within dorsal mesentery of the midgut
81. pancreas
82. wall of spinal cord in mid-lumbar region
83. lumen of first part of duodenum
84. muscle fibres going to form the right dome of the diaphragm
85. right hepatic duct
86. right pleural cavity
87. posterior vagal (X) trunk
88. lumen of the distal part of the oesophagus
89. right sub-phrenic recess of peritoneal cavity
90. middle lobe of right lung
91. caudal lobe of right lung
92. cranial lobe of right lung

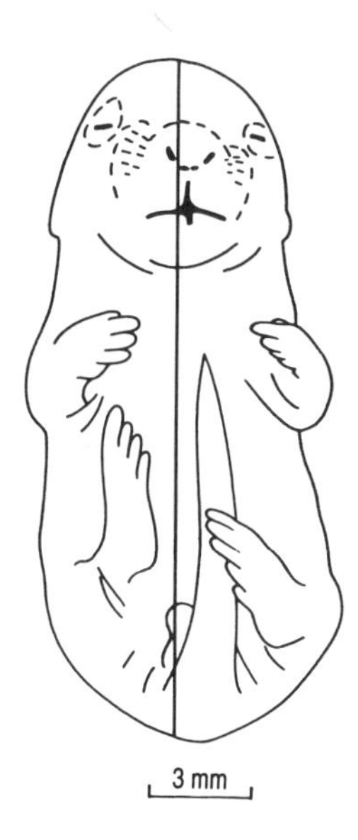

93. right pulmonary vein
94. deposits of brown (multilocular) fat
95. level of the bifurcation of the trachea
96. dorsal spinous process (tubercle) of thoracic vertebra
97. wall of spinal cord in mid-cervical region
98. neural arch of C3 vertebra
99. neural arch of C2 vertebra (axis)
100. hair follicle
101. entrance into oesophagus
102. neural arch of C1 vertebra (atlas)
103. cartilage primordium of odontoid process (dens) of C2 vertebra with periosteal ossification
104. ossification within cartilage primordium of anterior arch of C1 vertebra (atlas)
105. ossification within cartilage primordium of basioccipital bone (clivus)
106. ducts of mucous palatine glands located within soft palate, pharyngeal region, and dorsal surface of root of tongue
107. medulla oblongata
108. roof of fourth ventricle
109. cerebellum
110. choroid plexus within central part of lumen of fourth ventricle
111. rostral part of fourth ventricle
112. pons
113. pituitary gland
114. caudal part of mesencephalic vesicle (cerebral aqueduct)
115. roof of midbrain
116. ossification within cartilage primordium of basisphenoid bone
117. subarachnoid space
118. diencephalon (thalamus)
119. epiphysis (pineal gland)
120. superior sagittal dural venous sinus

found throughout the gonads, as this virtually all consists of cortical-derived tissue.

THE SKELETAL SYSTEM

Few substantial changes are observed in the skeletal system between 16.5–17 and 17.5–18 days p.c. In the skull, the ossification centres for the parietal, interparietal and frontal bones have all increased in size, as have those for the various facial bones. The ossification within the mandible (*Plate 40b,d*; *Plate 40c,a–d*) is also substantially more prominent than previously. In the occipital region, the supraoccipital bone has two distinct centres of ossification (one on either side of the midline) (*Plate 82,a,b*). The ossification centres within the exoccipital bones (*Plate 40b,c,d*; *Plate 40c,a,b*) have also increased in size over this period.

In the cervical region, the ossification centre located within the anterior arch of the atlas vertebra (*Plate 40c,c*), is more clearly seen than previously, as are the two components of its neural arch (*Plate 40c,c,d*). While there are minute centres of ossification present in the dens (odontoid process of C2 vertebra), and in the C7 vertebral body by 16.5–17 days p.c., no additional centres of ossification are normally seen in this region at 17.5–18 days p.c. However, in the developmentally more advanced embryos at this stage, a minute centre of ossification may be observed in the vertebral body of C6. In the sagittally sectioned embryo (*Plate 41*), the progressively increasing size of the centres of ossification within the vertebral bodies in the thoracic, lumbar and proximal coccygeal regions is clearly illustrated. The increasing size of the centres of ossification within the thoracic and lumbar vertebrae is also clearly seen in the transverse sections through these regions (e.g. *Plate 40g,a*; *Plate 40h,a*; *Plate 40i,b*; *Plate 40j,b*; *Plate 40k,d*) and in alizarin stained embryos (*Plate 82,a,b*). The centres of ossification within the basiocciput, basisphenoid, vomer and palatal shelves of the maxillae are also particularly well seen in the sagittally sectioned embryo (*Plate 41*). In the pelvic region, an ossification centre is invariably seen in the pubic bone during the early part of this stage of development (*Plate 82,b*).

In the extremities of the forelimb and hindlimb, centres of ossification are now clearly seen in the metacarpals and metatarsals, and minute centres of ossification may be seen in the phalanges towards the end of this stage of development (*Plate 82,c,d*).

Between the vertebrae, the primordia of the intervertebral discs are clearly seen, particularly in sagittally sectioned embryos (*Plate 41*). Close to the median plane, it is just possible to see the remnant of the notochordal sheath. In the central part of the intervertebral discs, the nucleus pulposus is particularly prominent.

By 18.5 days p.c., the two centres of ossification in the supraoccipital bone have fused to form a single ossified mass. Minute centres of ossification are now seen in all of the cervical vertebrae that previously showed no evidence of ossification, and more of the centra of the coccygeal vertebrae, up to the level of about the 10th coccygeal vertebra, show evidence of ossification (*Plate 82,e,f*). In the distal part of the limbs, all of the phalanges now have centres of ossification, and both the talus and calcaneus (but none of the other carpal or tarsal bones) have ossification centres (*Plate 82,g,h*). The sequential pattern of ossification in the peripheral parts of the forelimb and hindlimb during the period between 17.5 and 18.5 days p.c. is quite complex, and is described in detail in the text associated with *Plates 81–84*.

C. SCANNING ELECTRON MICROGRAPHS, AND SPECIAL SYSTEMS

1. External appearance of intact embryos isolated between 7.5 and 9.5 days p.c.

Text associated with Plate 42

Scanning electron micrographs of late presomite stage embryos isolated at about 7.5 days p.c. (*Plate 42,a–c*), reveal that almost the entire surface of the embryonic disc is covered with neural ectoderm, and that there is no distinct boundary evident between it and the prospective surface ectoderm when this method of visualizing the material is used. However, histological analysis of embryos isolated at a very similar stage of development (*Plate 7,i–n*; *Plate 8,e–h*), indicates that while the neural ectoderm has a columnar profile, the surface ectoderm at the junctional zone between these two cell types tends to have either a squamous or low cuboidal profile, while elsewhere the surface ectoderm has a squamous morphology. The junctional zone extends for a distance of several cells away from the lateral border of the neural folds (*Plate 8,f–h*).

The two neural folds (the prospective head folds) are the most prominent features at this stage, and are separated by the V-shaped neural groove (*Plate 42,a,c,d*). More caudally, the embryonic axis merges into the primitive streak region, and the midline primitive groove has a more rounded profile than has the neural groove at this stage (*Plate 8,e–j*). At its lateral boundaries, the ectoderm overlying the primitive streak, by contrast to the situation described above with regard to the relatively sharp boundary observed between the neural folds in the pre-headfold region and the adjacent surface ectoderm, gradually merges into the surface ectoderm in this region (*Plate 8,e–h*).

In the most advanced embryos recovered at this time, it is just possible to recognize a transversely running furrow in the region of the prospective hindbrain. This is termed the pre-otic sulcus (*Plate 42,c*). This feature becomes progressively deeper, so that by about 8 days p.c., in early somite-stage embryos, it becomes a useful landmark in this region. Note also that in order to view embryos at this stage by scanning electron microscopy, it is necessary to open into the amniotic cavity, and the torn edges of the amnion and visceral yolk sac are also clearly seen in these specimens (*Plate 42,a,b*). In early headfold stage embryos with 1–2 pairs of somites, the cephalic neural folds are seen to be substantially larger structures, and the pre-otic sulcus more obvious than observed previously (*Plate 42,d*). Furthermore, the neural groove is seen to extend caudally, so that it now merges into the proximal part of the primitive groove (*Plate 42,d*).

In early somite-stage "unturned" embryos (for description of the "turning" process, see Section 5 of Introductory Remarks) isolated between 8 and 8.5 days p.c., embryos at various stages of somite development may be recovered, there being a moderate degree of inter-litter variation at this stage. In embryos with about 3–6 pairs of somites, the most obvious changes seen relate to the very considerable degree of enlargement of the cephalic neural folds observed at this time. The latter is associated with a narrowing and elevation of the more caudally located neural folds, which correspond approximately to the level of the future cervical region of the neural axis (*Plate 42,e,f*). The other important feature at this stage relates to the flattened areas of neural ectoderm in the prospective forebrain region, which correspond to the location of the optic placodes (*Plate 10,d,e*; *Plate 42,g*; *Plate 49,a*).

The wide entrance (or portal) to the foregut diverticulum is also clearly seen (*Plate 42,g*), as is the exposed prospective midgut region with its endodermal lining. This occupies almost the entire convex surface of the middle region of the embryo at this stage. Analysis of a dorsal view of a similar embryo at this stage allows the relationship between the different components of the embryonic axis/primitive streak region to be seen (*Plate 42,e*). It is clear from such a view that the majority of the embryonic axis at this stage only extends caudally as far as the future cervical region of the neural axis, and that the more caudal part of the axis at this stage consists of the primitive streak region (which plays no significant part in subsequent embryonic development). The caudal part of the primitive streak is seen to merge into the allantois (*Plate 9,a–f*; *Plate 42,h–j*).

In embryos with about 5–6 pairs of somites, the neural folds in the mid-lordotic U-shaped part of the embryonic axis, which corresponds to the future mid- to lower cervical region, are seen to be closely apposed (*Plate 10,p–s*; *Plate 42,i–j*). Histological sections through this region of the embryonic axis reveal that the somites are closely associated with the neural folds in this location (*Plate 9,o,p*). In slightly more advanced embryos, more somites are seen to be present in the postlordotic rather than in the prelordotic segment of the embryonic axis (*Plate 10,o–t*).

At 8–8.5 days p.c., embryos with about 6–10 pairs of somites may be recovered. Such embryos are "unturned", and a limited degree of neural fold fusion has occurred, which is principally confined to the prelordotic segment and the concavity of the lordotic segment of the neural axis (*Plate 11,m,n*; *Plate 12,n–t*). In embryos at this stage, the optic pits are clearly distinguishable as indentations in the central flattened and anteriorly directed part of the forebrain neural folds, and the heart is at the early looping stage (*Plate 42,k–m*). A line of demarcation is seen between the neural component of the headfolds and the surface ectoderm on their lateral aspects. While it is not

(*continued on page 374*)

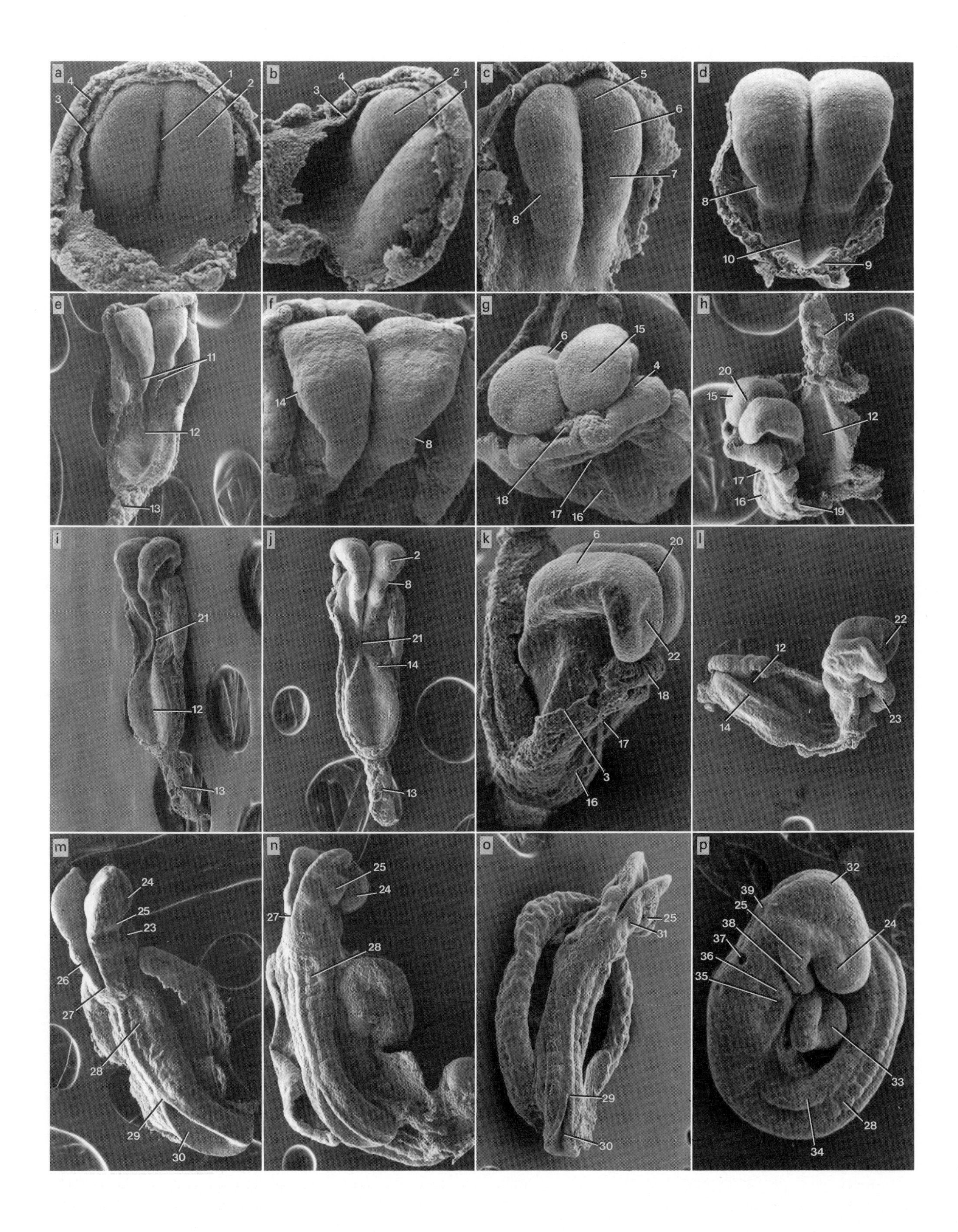
a
1
2
3
4
b
1
2
3
4
c
5
6
7
8
d
8
9
10
e
11
12
13
f
8
14
g
4
6
15
16
17
18
h
12
13
15
16
17
19
20
i
12
13
21
j
2
8
13
14
21
k
3
6
16
17
18
20
22
l
12
14
22
23
m
23
24
25
26
27
28
29
30
n
24
25
27
28
o
25
29
30
31
p
24
25
28
32
33
34
35
36
37
38
39

Plate 42

Overview of early post-implantation embryonic development (scanning electron microscopy)

a–c 7.5 days p.c.; late presomite stage embryos
d 7.5–8 days p.c.; embryo with 1–2 pairs of somites
e,f 8 days p.c.; embryos with 3–5 pairs of somites
g–j 8 days p.c.; embryos with 5–6 pairs of somites
k 8–8.25 days p.c.; embryo with 6–8 pairs of somites
l,m 8.25–8.5 days p.c.; embryos with 8–10 pairs of somites
n 8.5 days p.c.; embryo with 10–12 pairs of somites
o 8.5–9 days p.c.; embryo with 12–14 pairs of somites
p 9–9.5 days p.c.; embryo with 17–20 pairs of somites

1. neural groove in headfold region
2. cephalic neural fold (headfold)
3. torn edge of amnion
4. torn edge of visceral yolk sac
5. prospective forebrain region
6. prospective midbrain region
7. prospective hindbrain region
8. pre-otic sulcus
9. neuroepithelium of neural plate in mid-lordotic region of embryonic axis
10. neural groove in mid-lordotic region of embryonic axis
11. elevated neural folds in future lower cervical region
12. primitive streak region
13. allantois
14. junctional zone between neural and surface ectoderm
15. optic placode
16. endodermal lining of future midgut
17. entrance to foregut diverticulum
18. body wall overlying pericardial cavity
19. mid-lordotic region of embryonic axis
20. future diencephalic region
21. apposition of neural folds in lower cervical region
22. early stage in differentiation of optic pit (optic evagination)
23. primitive heart tube at "looping" stage
24. optic eminence
25. maxillary component of first branchial arch
26. neural fold in prospective hindbrain region
27. rostral extremity of neural tube formation
28. somite
29. caudal extremity of neural tube formation
30. neuroepithelium in region of caudal neuropore
31. early stage in differentiation of otic pit
32. region overlying future midbrain
33. common ventricular chamber of primitive heart tube (thoracic wall has been removed)
34. forelimb bud
35. second branchial groove
36. second branchial arch (hyoid arch)
37. otic pit
38. mandibular component of first branchial arch
39. region overlying roof of fourth ventricle

possible with any degree of certainty to assign different regions to the cephalic neural folds, the impression is formed that the prospective fore-, mid- and hindbrain regions are distinguishable at this stage though the boundaries between them are not sharply defined.

In embryos at this stage, it is also possible to recognize the maxillary component of the first branchial (pharyngeal) arch, though this feature is probably more easily distinguished in histological sections than in scanning electron micrographs of similar stage embryos (*Plate 11,d–f*; *Plate 12,c–e*; *Plate 42,l,m*). In embryos isolated at 8.5–9 days p.c., that possess about 10–15 pairs of somites, advanced "unturned" embryos, and embryos at various stages during the "turning" sequence may be observed (see Fig. 5). Posterolateral views of such embryos (*Plate 42,n,o*) are instructive in that they clearly show that a substantial length of the neural tube has closed, but that the neural folds in the cephalic and most caudal regions of the embryo are still wedge-shaped and widely separated. Just lateral to the dorsal midline, the outlines of the somites are clearly visible (due principally to shrinkage during fixation). In the cephalic region, the optic eminence is separated from the maxillary component of the first branchial (pharyngeal) arch by a deep groove (*Plate 42,n*). The two embryos illustrated at this stage (*Plate 42,n,o*) have "opened out" when they were removed from within the constraints imposed by their extra-embryonic membranes, and would probably have been at an advanced "unturned" stage, or more likely at an early stage during the "turning" sequence, at the time of their isolation.

By 9 – 9.5 days p.c., early "turned" embryos with about 15–20 pairs of somites may be recovered. The embryo illustrated (*Plate 42,p*) has about 17–20 pairs of somites. More interestingly, the neural tube has closed along the entire neural axis with the exception of the region of the caudal neuropore. In the cephalic region, the optic eminence is just seen, and the first branchial (pharyngeal) arch has yet to differentiate into a discrete maxillary and mandibular component. A second branchial arch is also present, and the two are separated by a deep groove (the first cleft). Just dorsal to this region, the deeply indented otic pit is a prominent feature, being located at the mid-hindbrain level. The thoracic wall has been removed, and the ventricular and bulbar regions of the primitive heart tube are exposed. A substantial lateral ridge runs caudally from the mid-trunk region (along the dorsal border of the somatopleure). The forelimb bud is seen to be located at the most rostral extent of this ridge, and the hindlimb bud will subsequently develop close to its caudal extremity. The forelimb buds are located at about the level of somites 8–12 (*Plate 18b,l–n*).

2. External appearance of intact embryos isolated between 9.5 and 11.5 days p.c., and differentiation of the limbs over the period from 9.5 to 17.5 days p.c.

Text associated with Plates 43–45

Most of the embryos isolated at about 9.5 days p.c. have recently "turned", and show early evidence of forelimb bud differentiation. Embryos at this stage usually possess about 20–25 pairs of somites. Embryos at the earlier end of the range may still possess otic pits (*Plate 43*, unlabelled) or incompletely formed otocysts (*Plate 19a,l*), while in more advanced embryos the otic pits will have disappeared with the formation of the otocysts (*Plate 43,b*). The overall dimensions of the cephalic region are somewhat greater than that observed previously, principally because the three primitive brain vesicles are more dilated than at earlier stages. The first evidence of the division of the forebrain is seen, with the first indication of the presence of the two telencephalic vesicles. This corresponds with the presence of a groove overlying the lamina terminalis, which is more clearly seen in histological sections through this region than on scanning electron micrographs (cf. *Plate 19a,g–l*; *Plate 19b,a,b*). While the optic eminence is not significantly more prominent than previously, the (primary) optic vesicles are fully differentiated and have reached their maximum volume (*Plate 51,e–h*), and will soon collapse (in embryos with about 25–30 pairs of somites) to form the (secondary) optic cup (*Plate 51,i,j*).

The first branchial arch is seen to be more C-shaped (*Plate 43,b*; cf. *Plate 42,p*), and its maxillary and mandibular components are easier to distinguish than previously. The enlargement of the maxillary component is largely due to the migration into this region of trigeminal (V) neural crest mesenchyme (*Plate 19a,g–l*; *Plate 19b,a,b*). The second branchial arch (the hyoid arch) is also more clearly defined than previously, as is the first branchial groove. An increase is also evident in the volume of the forelimb bud. While its rostral margin is reasonably well defined at this stage, its caudal margin is considerably more diffuse, and merges into the lateral ridge from which the limb buds differentiate (extremitätenleiste). The somites in the caudal part of the trunk and in the tail region are well defined, this being principally an artefact of shrinkage during the processing of the material for scanning electron microscopy, but the boundaries of the more rostral somites are now fairly indistinct. Indeed, the position of the first few somites at this stage can only be estimated from a knowledge of their relationship to the rostral margin of the forelimb bud. One of the most useful features for estimating the developmental stage of early forelimb bud stage embryos is the morphological appearance of the caudal neuropore. In embryos with 20–25 pairs of somites, this is reasonably widely open on the dorsal part of the tail close to its caudal extremity (*Plate 43,b*). The neural folds in the region of the caudal neuropore are particularly widely open in embryos with about 15–20 pairs of somites (*Plate 18a,a–f*), are virtually closed in embryos with about 25–30 pairs of somites, and are completely closed in embryos with about 30–35 pairs of somites (*Plate 23c,a–h*).

By 9.5–10 days p.c., embryos usually possess about 25–30 pairs of somites (*Plate 43,c–e*). In the three views of the embryos used to illustrate this stage, it is now possible to clearly see the groove overlying the lamina terminalis, and note the presence of the two telencephalic vesicles. These have fairly rounded profiles during the early part of this stage, particularly in the region overlying the future nasal (olfactory) placodes (*Plate 43,d*), though in developmentally slightly more advanced embryos with closer to 30 pairs of somites, this region (on the inferolateral aspect of the telencephalic vesicles) may be slightly flattened or even minimally indented (*Plate 43,e*, unlabelled; *Plate 47,e*). The latter area represents the location of the olfactory placode (*Plate 20b,a–e*; *Plate 21a,a*; *Plate 21b,c,d*). The location of the forelimb bud and its relationship to the lateral ridge which runs along the posterolateral body wall in close proximity to the somites is also clearly seen. The site of the future hindlimb bud can also be seen for the first time, being located at this stage at the caudal extremity of the lateral ridge, in the middle part of the tail region (*Plate 43,c,d*). As indicated above, in embryos at this stage of development, the caudal neuropore has almost completely closed.

In embryos with 30–35 pairs of somites, the caudal neuropore has completely closed. In the specimen used to illustrate this stage of development the cephalic region has been removed in order to display the degree of differentiation of the limb buds. Two views of the trunk and tail region are presented (*Plate 43,f,g*), which clearly display the relationship between the fore- and hindlimb buds and the lateral ridge from which they arise. At this stage, the hindlimb buds are reasonably well differentiated, and have a clear proximal and distal margin (*Plate 22a,a–d*; *Plate 22b,a,b*; *Plate 23c,a–l*; *Plate 43,f,g*). They are located opposite somites 23–28, but are significantly less well differentiated than the forelimb buds. This is evident by the fact that while it is just possible to recognize the apical ectodermal ridge running along the rostrocaudal margin of the forelimb bud, this feature has yet to appear in relation to the hindlimb bud.

By 10.5–11 days p.c., the limb buds become increasingly prominent, and begin their rostral "ascent". The apical ectodermal ridge is now clearly seen to be present at the peripheral margin of the hindlimb bud. A prominent marginal vein is located just subjacent to the apical ectodermal ridge both in the fore- and hindlimbs, being particularly clearly seen in the forelimb bud at this stage (*Plate 24c,l–o*; *Plate 24d,a–d*).

In embryos at about 11.5 days p.c., the limb buds are seen to be divided into proximal and distal parts. In the forelimb, the proximal part includes the region of the future pectoral girdle and arm, while the distal paddle-shaped part constitutes the hand plate. In the hindlimb bud, the proximal part includes the region of

(*continued on page 382*)

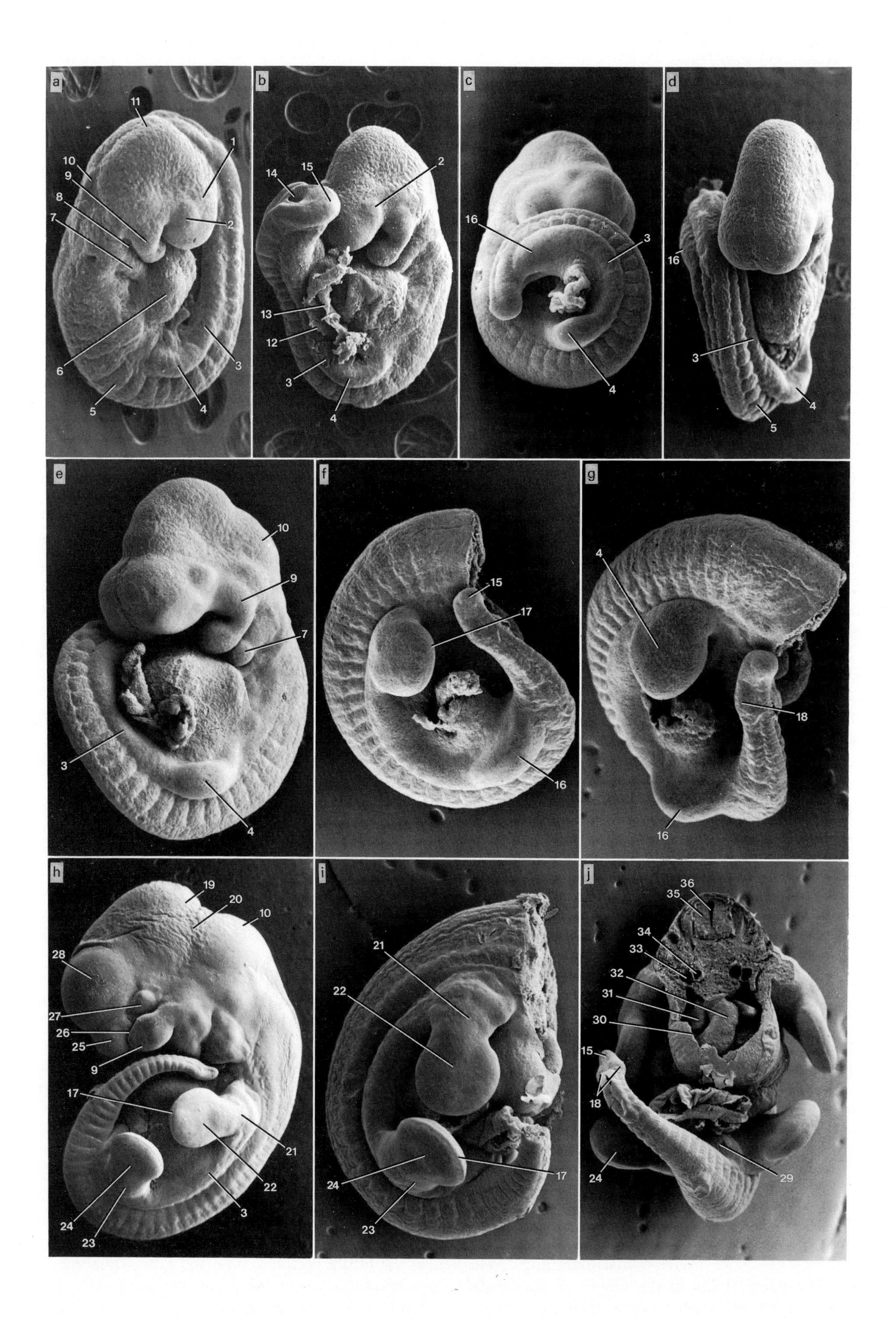
a
b
c
d
e
f
g
h
i
j
1
2
3
4
5
6
7
8
9
10
11
12
13
14
15
16
17
18
19
20
21
22
23
24
25
26
27
28
29
30
31
32
33
34
35
36

Plate 43

Early postimplantation embryonic development – differentiation of the limb buds 1 (scanning electron microscopy)

a,b 9.5 days p.c.; embryo with 20–25 pairs of somites
c–e 9.5–10 days p.c.; embryo with 25–30 pairs of somites
f,g 10 days p.c.; embryo with 30–35 pairs of somites
h–j 11.5 days p.c.; embryos with 45–50 pairs of somites

1. region overlying forebrain
2. optic eminence
3. lateral ridge along body wall from which the limb buds develop
4. forelimb bud
5. somite
6. pericardial region
7. second branchial arch (hyoid arch)
8. first branchial cleft (groove)
9. maxillary component of first branchial arch
10. region overlying roof of fourth ventricle
11. region overlying midbrain
12. cut edge of amnion
13. cut edge of visceral yolk sac
14. neuroepithclium in region of caudal neuropore
15. caudal extremity of tail
16. hindlimb bud
17. apical ectodermal ridge
18. flattened and laterally expanded region just proximal to the caudal extremity of the tail
19. region overlying cerebellar plate (of metencephalon)
20. region overlying isthmus rhombencephali
21. proximal part of forelimb bud
22. distal paddle-shaped part of forelimb bud (the hand plate)
23. proximal part of hindlimb bud
24. distal paddle-shaped part of hindlimb bud (the foot plate)
25. lateral nasal process
26. naso-lacrimal groove
27. corneal ectoderm overlying lens vesicle
28. region overlying telencephalic vesicle (future cerebral vesicle)
29. genital ridge
30. cut edge of thoracic wall
31. auricular part of right atrium
32. truncus arteriosus
33. right anterior cardinal vein
34. right dorsal aorta
35. neuroepithelium of neural tube
36. neural lumen

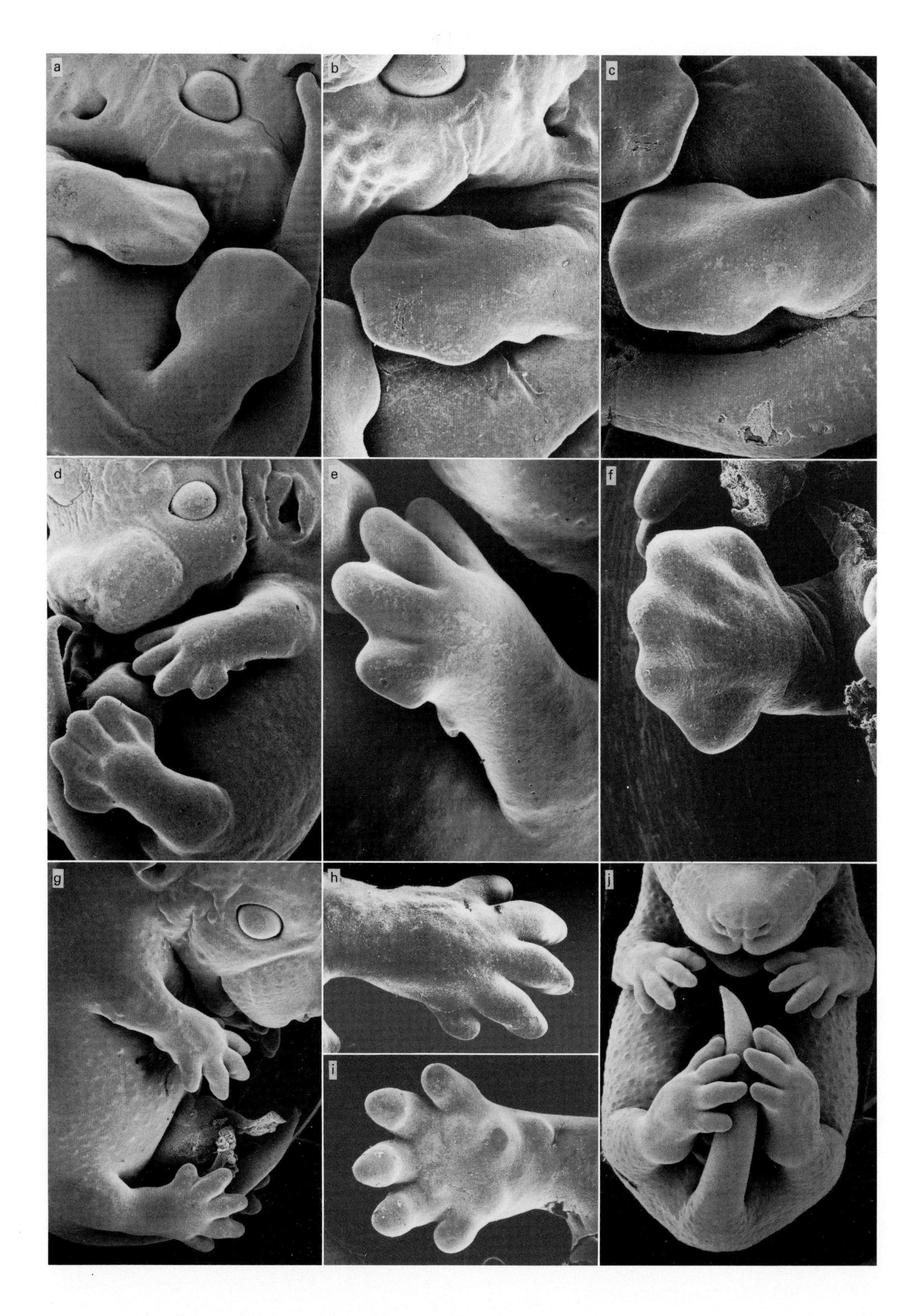
a
b
c
d
e
f
g
h
i
j

Plate 44

Differentiation of the limb buds/limbs, 2

a 12 days p.c., dorsal surface of forelimb and hindlimb
b 12.5 days p.c., dorsal surface of forelimb
c 12.5 days p.c., dorsal surface of hindlimb
d 13.5 days p.c., dorsal surface of forelimb and hindlimb
e 13.5 days p.c., dorsal surface of forelimb
f 13.5 days p.c., plantar surface of hindlimb
g 14.5 days p.c., dorsal surface of forelimb and hindlimb
h 14.5 days p.c., dorsal surface of forelimb
i 14.5 days p.c., palmar surface of forelimb
j 15.5 days p.c., dorsal surface of forelimbs and hindlimbs

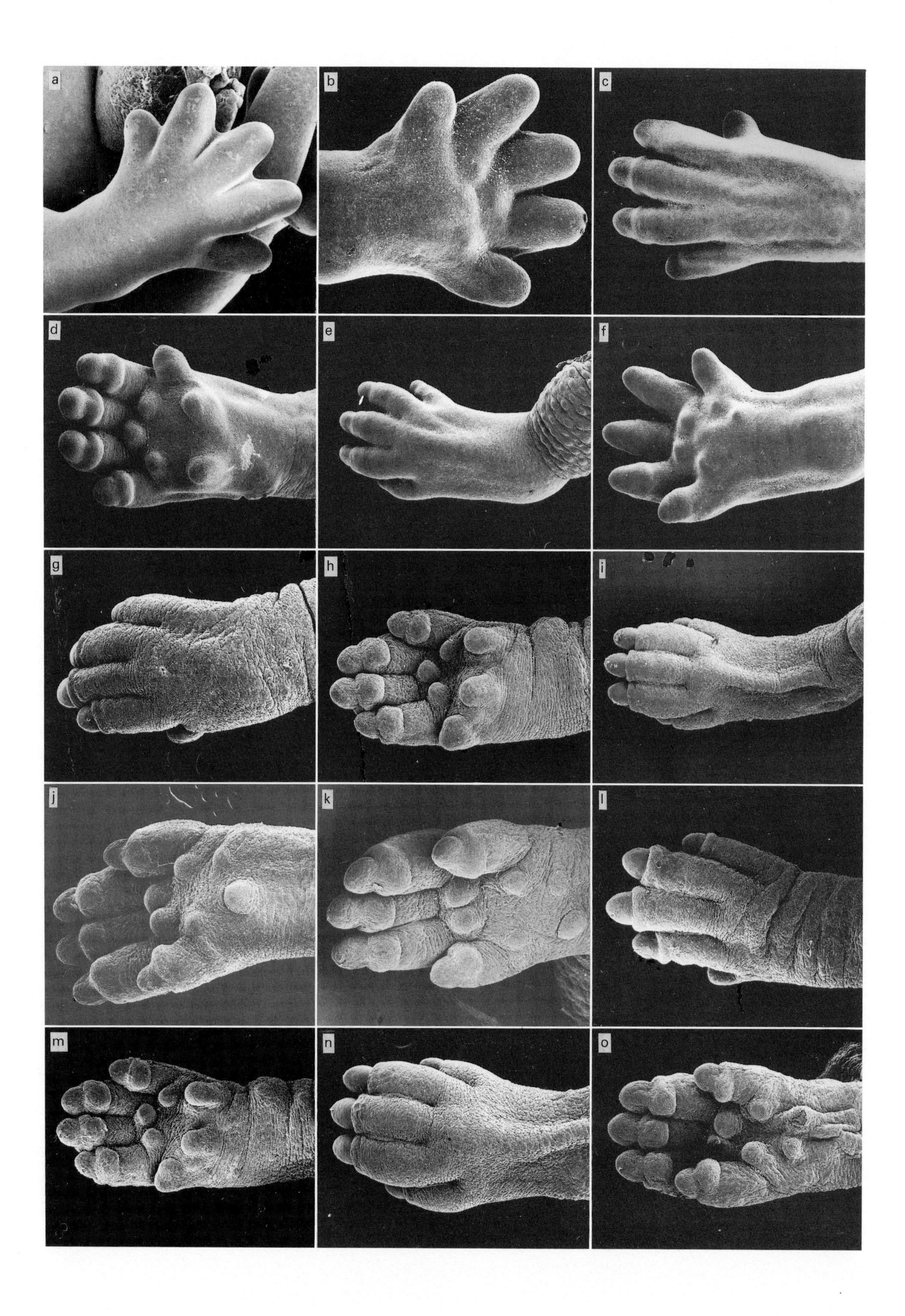
a
b
c
d
e
f
g
h
i
j
k
l
m
n
o

Plate 45

Differentiation of the limb buds/limbs, 3

a 14.5 days p.c., dorsal surface of hindlimb
b 14.5 days p.c., plantar surface of hindlimb
c 15.5 days p.c., dorsal surface of forelimb
d 15.5 days p.c., palmar surface of forelimb
e 15.5 days p.c., dorsal surface of hindlimb
f 15.5 days p.c., plantar surface of hindlimb
g 16.5 days p.c., dorsal surface of forelimb
h 16.5 days p.c., palmar surface of forelimb
i 16.5 days p.c., dorsal surface of hindlimb
j 16.5 days p.c., plantar surface of hindlimb
k 16.5 days p.c., different view of plantar surface of same hindlimb as illustrated in **j**
l 17.5 days p.c., dorsal surface of forelimb
m 17.5 days p.c., palmar surface of forelimb
n 17.5 days p.c., dorsal surface of hindlimb
o 17.5 days p.c., plantar surface of hindlimb

the future pelvic girdle and leg, while the distal paddle-shaped part constitutes the foot-plate. In all regards, the forelimb bud is at a slightly more advanced stage of limb differentiation than the hindlimb bud. In more advanced embryos in this group, isolated at about 11.5–12 days p.c., it is possible to see that the hand plate shows early evidence of developing angular contours at its peripheral margin, and these correspond to the location of the future digits. The foot plate, by contrast, is still clearly paddle-shaped. The other features of embryos at this stage are described in the appropriate section of Stage 19.

At about 12 days of gestation, the peripheral margins of both the handplates and the footplates have a polygonal shape, though it is quite clear where the digits will subsequently differentiate because of the presence of the distinct digital interzones (*Plate 44,a*). While the forelimb at this stage is developmentally slightly more advanced than the hindlimb, the difference is not particularly marked. By about 13.5 days p.c., the difference is slightly more evident, with the digital interzones of the forelimb being considerably more indented than the corresponding regions in the hindlimbs (*Plate 44,a–c*). In the interzones, the increasing degrees of indentation seen are believed to be associated with programmed (physiological) cell death in these locations. It is of interest that the homeobox-containing gene Hox–7.1 is expressed in regions where cell death occurs in the developing limbs (Hill *et al.*, 1989). By 14.5 days p.c., relatively little of the webbing that is a feature of the earlier stages of digital differentiation is seen (*Plate 44,g–i*). Up to about 14.5 days p.c., all of the digits of both the fore- and hindlimbs are fairly splayed out (*Plate 44,h,i*). While this feature is still seen to some degree in relation to the hindlimbs at about 16.5 days p.c. (*Plate 45,i*), in the forelimbs digits 2 to 5 are seen to be almost parallel to each other (*Plate 45,g*). By about 17.5 days p.c., all of the digits of the hindlimb are seen to be lying almost parallel to each other (*Plate 45,n,o*).

It is also of interest to note that while the digits of the hindlimb at all stages of development are reasonably symmetrically arranged, with the first digit of the hindlimb (the hallux) being only slightly shorter than the fifth digit (digiti quinti) (*Plate 45,f,i,n*), a similar arrangement is not seen in relation to the forelimb. Up to about 14.5 days p.c., the hand plate is reasonably symmetrical (*Plate 44,h,i*), but gradually over the next few days the first digit (the pollex) fails to increase in length to the same degree as the other four digits, and by about 16.5 days p.c., the first digit is seen to be significantly smaller than the other four digits (*Plate 45,g,h*).

The first suggestions of nail (claw) primordia are seen in association with the dorsal surfaces of the terminal phalanges of the medial four digits (i.e. digits 2–5) of the forelimb (*Plate 45,c*), and in association with all of the digits of the hindlimb (*Plate 45,e*) by about 15.5 days p.c. However, a nail (claw) primordium is not seen in association with the first digit of the forelimb until after 17.5 days p.c. (*Plate 45,l*). The latter digit in fact appears to be quite vestigial when compared to the degree of development of the other digits both in the fore- and hindlimbs, and the nail that does eventually form has the appearance of a flat plate, rather than a distinct claw as seen in association with the other digits.

The prominent pads characteristically seen on the plantar surface of the metatarsal/tarsal region of the hindlimb, and on the palmar surface of the metacarpal region of the forelimb, are first evident as small but discrete elevations by about 14.5 days p.c. (hindlimb, *Plate 45,b*; forelimb, *Plate 44,i*), but are more readily seen after about 15.5 days p.c. (hindlimb, *Plate 45,f*; forelimb, *Plate 45,d*). The latter are located approximately overlying the plantar surface of the metatarso-phalangeal joints, and overlying the palmar surface of the metacarpo-phalangeal joints. In the forelimb, there are five digital pads (*Plate 45,m*), whereas in the hindlimb there are, in addition to the five digital pads (*Plate 45,o*), four metatarsal and two tarsal pads.

3. External appearance of intact embryos isolated between 11.5 and 18.5 days p.c.
Text associated with Plate 46

Throughout the explanatory notes associated with each of the plates of the intermittent serially sectioned embryos that constitutes the main body of the text, detailed descriptive accounts are provided of the external morphological features of embryos at each of the various stages studied. In order that this material should not be duplicated to any major extent in the text associated with this plate, only an abbreviated summary is provided here of the principal changes that are observed in the external morphology of embryos isolated between 11.5 and 18.5 days p.c. Therefore, for a considerably more detailed account of the external features associated with particular stage(s) of development, the reader is recommended to refer to the appropriate stage(s) of interest for additional relevant information. It should be noted, however, that while all of the intact embryos and dissections of specific organs or regions of embryos illustrated elsewhere in the Atlas were viewed by scanning electron microscopy, following appropriate fixation, the embryos illustrated in this plate were fixed for 24 h in Bouin's solution, then transferred into 70% ethanol in which they were retained for varying periods of time before they were photographed by conventional photographic techniques.

For each of the stages studied, a frontal and left lateral view of a single representative embryo has been provided, though, as has been indicated elsewhere, at each stage of development a range of embryos may be isolated in which there may be a considerable difference in the features observed between the least and the developmentally most advanced embryos, even within a single litter. With this proviso in mind, it is of interest to briefly consider the gross changes that occur in the external features of embryos spanning the period from 11.5 to 18.5 days p.c. Embryos in the latter category, vary only minimally in appearance from embryos at full term.

The most obvious features with regard to embryos at 11.5 days p.c. (*Plate 46,a,b*) relate to the cephalic region, so that at this stage it is possible to recognize from a superficial analysis of such embryos all of the major subdivisions of the primitive brain, namely the two telencephalic vesicles, the midbrain and the hindbrain as well as the location of the cerebellar plate. The lens vesicles have recently separated from the surface ectoderm, and the latter (in the region overlying the lens vesicles) is now in the process of differentiating to form the primitive corneal epithelium, while the peripheral margins of the orbital region will soon differentiate to form the upper and lower eyelids. The entrances to the olfactory pits are reduced to relatively narrow slits, and the maxillary component of the first branchial arch is now somewhat more prominent than the mandibular component. The auditory hillocks, located on either side of the first branchial groove, are clearly seen, and will subsequently coalesce to give rise to the pinna of the ear, while the groove itself gives rise to the external acoustic meatus.

Caudal to the cephalic region, the fore- and hindlimb buds are reasonably well differentiated, the development of the forelimb bud being slightly in advance of that of the hindlimb bud. The limbs at this stage consist of a proximal part associated with the limb girdle, and a distal part consisting of a paddle-shaped handplate or foot plate, both of which still have well-defined apical ectodermal ridges. Between the hindlimb buds, a prominent transversely running genital ridge (tubercle) is clearly seen (*Plate 70,a*). The distal part of the tail is relatively long and narrow. Reference to Fig. 11 reveals that the cranial/postcranial ratio at this stage is close to 1:1.

By 12.5 days p.c. (*Plate 46,c,d*), the pigmentation of the exposed peripheral part of the outer (pigmented) layer of the retina (in non-albino strains) is first clearly seen. Due to a combination of factors, such as a slight proportionate decrease in the volume of some of the brain vesicles, an increase in the thickness of their walls and an increase in the volume of cephalic mesenchyme, the outlines of the components of the primitive brain are less clearly seen than previously. In the maxillary region, the first evidence of the primordium of the vibrissae is seen at this stage, and the primordia of the pinna of the ear is now recognized as a distinct entity. The most dramatic change, compared with the situation observed at 11.5 days p.c., however, relates to the appearance of the hand and foot plates. The hand plates, in particular, are no longer paddle-shaped, but have angular contours which correspond to the location of the future digits, and these are separated by indented "rays" which correspond to the digital interzones. The foot plates by this stage are only slightly less differentiated than the hand plates. In embryos at both 11.5 and 12.5 days p.c., the appearance of the lateral aspect of the caudal part of the hindbrain and upper cervical region of the neural tube is quite characteristic, being quite prominent and angular in shape. The latter corresponds approximately to the location of the cervical flexure. This characteristic appearance is not seen in embryos at more advanced stages of development. This is also the first stage that the physiological umbilical hernia is clearly seen. The body wall surrounding the latter is extremely thin, and loops of midgut are seen to be present within the sac. The genital tubercle is also enlarged compared to the situation observed previously.

By 13.5 days p.c. (*Plate 46,e,f*), the characteristic "hump" previously associated with the caudal hindbrain/upper cervical region has almost completely disappeared, as has the demarcation previously present between these regions. The latter corresponded approximately to the region where the caudal part of the fourth ventricle joined the rostral part of the central canal of the spinal cord. The spinal cord in the cervical and much of the thoracic region produces a prominent "bulge" in the dorsal midline, but caudal to this level the prominence overlying the spinal cord is considerably less marked. The general shape of the head, but particularly the facial region, is seen to have more adult features than previously. The pinna is now well formed, and directed anteriorly so that it partially covers the first branchial groove. The external ear appears to be more rostrally located than previously, presumably because of differential growth of the neck region. In the maxillary region, the five rows of the primordia of the vibrissae are now clearly seen (*Plate 49,i*), as are the primordia of the tactile or sinus hair follicles which are located at various characteristic sites in the periorbital and facial region.

In both the hand plates and foot plates, their distal borders are now indented and have a characteristically "webbed" appearance in the region of the digital interzones. The degree of indentation seen is slightly more marked in the hand plate than in the foot plate. The location of the wrist and elbow regions are clearly seen for the first time, though in the hindlimb, only the ankle region is distinguishable at this stage. While the hand plates are now directed inferomedially, the plantar surfaces of the footplates are directed medially, and are often in close contact with the tail. The proportionate length of the tail compared with that of the vertebral axis has diminished slightly, and its distal part ends in a fairly sharp point. Unlike the situation observed previously where the distal part of the tail normally deviates to one or other side of the midline, often in close proximity to the external ear, at this stage the tail is invariably located in the median plane, and its tip does not extend rostrally beyond the level of the forelimbs. While it is possible to distinguish each of the somites from the level of the forelimb caudally at 11.5 days p.c., by 12.5 days p.c. these are only clearly distinguishable from the lumbar region caudally, and by 13.5 days p.c., no external evidence of vertebral segmentation is seen. In the mid-abdominal region, the physiological umbilical hernia has a substantially greater volume than previously.

By 14.5 days p.c. (*Plate 46,g,h*), the cephalic region is more rounded than previously, and the primordia of the tactile or sinus hair follicles are particularly clearly seen, as is the arrangement of the primordia of the vibrissae in the maxillary region. The first evidence of primordia of hair follicles is seen at this stage, but these are largely confined to the thoracic and trunk regions, and in the limbs they do not extend distally beyond the wrist and ankle regions. The pinna extends further anteriorly than previously and now almost covers the posterior one-third of the external acoustic meatus. The neck region is also seen to have elongated slightly compared with the situation observed previously. Not infrequently the tip of the tongue extends forwards and may appear over the labial surface of the mandible and between the two components of the upper lip at this stage (*Plate 49,j,l*).

The most characteristic feature associated with this stage of development, however, relates to the appearance of the hand plate and foot plate regions of the limbs. In the hand plate, in particular, the digital interzones have retreated proximally, so that the digits are now present as quite discrete units. In the foot plates, the degree of differentiation of the digits is less marked than in the hand plates, and some residual "webbing" in the digital interzones is still present. In both the forelimb and hindlimb, the digits are characteristically quite splayed out, and in the foot plate the first indication of digital asymmetry is seen, which at later stages of development is a characteristic feature of this region. At this stage of development, the physiological umbilical hernia is at its maximum size, and from this stage onwards the loops of midgut begin to return to the peritoneal cavity. While the primordia of the mammary glands were first seen at about 12.5–13 days p.c., by 14.5–15 days p.c. they are clearly seen to extend from the thoracic to the inguinal region along the mammary ridges.

By 15.5 days p.c., the maxillary component of the facial region is narrower than previously. Similarly, the eyelid margins are more prominent than previously, and in the developmentally most advanced embryos at this stage, the first evidence of eyelid closure may be seen (*Plate 49,d,m*). The pinna of the ear now covers just about one-half of the external acoustic meatus. By 14.5–15 days p.c., the distribution of the hair follicles extends rostrally into the cephalic region, and by 15.5 days p.c. their generalized distribution is clearly seen, principally because of the absence of body creases (wrinkles) at this stage of development. However, the total number of hair follicles distributed over the entire surface of the embryo is relatively small, and on histological sections, they are not particularly prominent structures.

As in the previous stage, it is in relation to the overall appearance of the limbs that the most obvious changes are seen compared to the situation observed previously. For example, all evidence of "webbing" has disappeared from the hand plate region though some "webbing" is still seen, particularly in relation to the dorsal surface of the foot plate. The first suggestion of the nail primordia is seen in relation to the dorsum of the terminal phalanges of the hand plate at this stage of development, but particularly in relation to the medial four digits (i.e. digits 2–5) (*Plate 44,j*; *Plate 45,c*), and slightly less marked changes are observed in relation to the dorsum of the terminal phalanges in the foot plate at this stage (*Plate 45,e*). Changes are also evident in relation to the palmar surface of the hand plate, with regard to the appearance of the prominent pads characteristically seen in these locations (*Plate 45,d,f*). The former are barely seen at 14.5 days p.c. (*Plate 44,i*), and even less is seen of these structures on the plantar surface of the foot plate at this time (*Plate 45,b*).

In the mid-abdominal region, the physiological umbilical hernia is seen to have decreased in volume compared to the situation observed previously, and this is associated with the initiation of the return of the midgut loop into the peritoneal cavity that occurs at about this time. The overall shape of the back of the embryo is slightly different than previously, principally due to the gradual straightening out of the postcranial vertebral axis. This is first evident at about 14.5–15 days p.c., and is more marked at 15.5–16 days p.c., as the degree of vertebral axial flexion progressively diminishes. The latter is particularly evident in the lower thoracic and upper lumbar regions. By this stage, the cranial component of the crown–rump length is just over 40%, and this change from the situation observed previously is illustrated graphically in Fig. 11.

By 16.5 days p.c. (*Plate 46,k,l*), a number of

particularly obvious changes may be observed in relation to the external appearance of the embryo. In the cephalic region, the eyelids are invariably closed at this stage, and the closure process occurs in most embryos between 15.5 and 16 days p.c. The underlying mechanism involved is described in detail in the appropriate section of Stage 25. At an early stage following eyelid closure, an extensive ridge of cellular excrescences is seen along the outer surface of the line of fusion, and this is usually sloughed off by the time of birth or shortly afterwards. In the maxillary region of the upper lips, the vibrissae are now seen to have erupted through the epidermal evaginations overlying them, leaving a characteristic mound of cellular debris at their bases, though the tactile or sinus hair follicles have yet to erupt. The pinna of the ear now covers virtually all of the external acoustic meatus.

Obvious differences are also observed, compared with the situation observed previously, in relation to the appearance of the skin. Whereas previously this was smooth and closely followed the contours of the body, the skin, particularly in the neck, trunk and limbs distally as far as the carpal and tarsal regions, contains large numbers of wrinkles. These largely run around the trunk region, but have a less obvious pattern elsewhere, and the facial region at this stage is devoid of wrinkles. The skin is also thicker than previously, and the superficial veins are no longer seen through the skin. In the lower abdominal region, as a consequence of the return of the midgut to the peritoneal cavity, the physiological umbilical hernia has now completely disappeared. In relation to the limbs, additional changes have occurred compared with those described previously. In particular, the first digit of the forelimb (the pollex) is substantially smaller than the other four digits. In both the fore- and hindlimbs, the pads characteristically seen on their palmar and plantar surfaces, respectively, are considerably more prominent than previously, as are the nail (claw) primordia.

By 17.5 days p.c. (*Plate 46,m,n*), the principal difference observed in the external appearance of the embryo, compared with the situation observed previously, relates to the skin. Whereas previously the skin creases or wrinkles principally involved the trunk and lower neck region, at this stage the creases are now seen to extend more caudally, and are particularly prominent in the hindlimbs where they extend distally as far as the ankle region. In the facial region, in the vibrissae-bearing areas of the upper lips, the bases of the vibrissae are more prominent, and their length and diameter is somewhat greater than previously. Otherwise, the features of the rest of the facial region are similar to those observed at 16.5 days p.c.

By 18.5 days p.c. (*Plate 46,o,p*), the external features of the embryo are essentially identical to those seen in the full-term embryo, and are only minimally different from those observed at 17.5 days p.c., with the exception, that is, of the tail. The latter is proportionately shorter and wider, and its tip less pointed, than at any previous stage of development. The wrinkles now cover all but the most distal parts of the limbs, the tail and the facial region. The skin is also slightly thicker than previously, and it is more difficult to see the eyes through the closed eyelids. At this stage, the cranial component of the crown–rump length is just less than 35%, and this relationship is illustrated graphically in Fig. 11. The latter relationship is clearly a reflection of the fact that while the cranial region of the embryo gradually increases in size, it does not keep pace with the growth that occurs in the trunk region of the embryo.

The relationship between gestational age (10.5–17.5 days p.c.), crown–rump length and embryonic weight is presented in Table 5, and in graphical form in Figs 12 and 13 (see p. 388).

Table 5 Relationship between gestational age, crown–rump length and embryonic weight

Day of gestation	Total embryos analysed	Crown–rump length (mm) (mean ± s.e.m.)	Weight (mg) (mean ± s.e.m.)
10.5	17	4.57 ± 0.09	12 ± 1
11.5	17	6.87 ± 0.06	56 ± 4
12.5	9	8.69 ± 0.14	117 ± 4
13.5	9	10.17 ± 0.14	175 ± 6
14.5	9	11.97 ± 0.17	263 ± 8
15.5	8	13.59 ± 0.24	399 ± 16
16.5	10	16.60 ± 0.12	715 ± 19
17.5	8	19.31 ± 0.34	974 ± 21

(*continued on page 388*)

PLATE 46

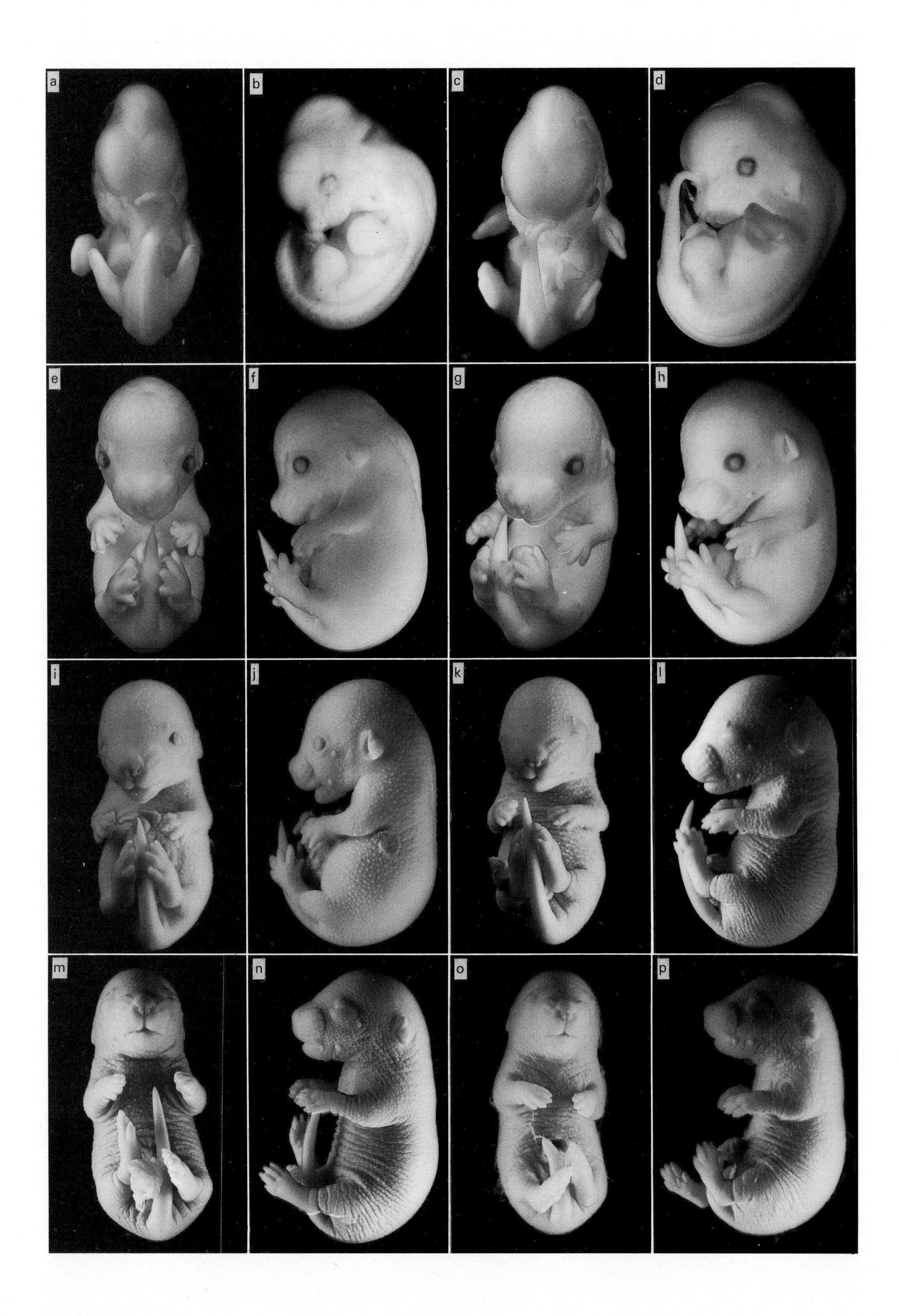

Plate 46

Frontal and left lateral views of representative mouse embryos isolated at intervals throughout the second half of gestation

All of the embryos illustrated in this plate were fixed for 24 h in Bouin's solution then transferred into 70% ethanol in which they were retained for varying periods of time. Note the gradual change that takes place in both the overall proportions of the different components of the body, but particularly in relation to the craniofacial region, as well as the differentiation of the limbs, skin, tail etc. between 11.5 and 18.5 days of gestation. The range of crown–rump lengths of fresh embryos isolated at these times as well as their approximate Theiler Stage are given in parentheses.

a,b 11.5 days p.c. (6–7.5 mm) (Stage 18/19)
c,d 12.5 days p.c. (7.5–9.5 mm) (Stage 20/21)
e,f 13.5 days p.c. (9.5–11.5 mm) (Stage 21/22)
g,h 14.5 days p.c. (11.5–12.5 mm) (Stage 22/23)
i,j 15.5 days p.c. (12.5–14.5 mm) (Stage 24)
k,l 16.5 days p.c. (14.5–17.5 mm) (Stage 25)
m,n 17.5 days p.c. (17.5–20.5 mm) (Stage 26)
o,p 18.5 days p.c. (20–23 mm) (Stage 26/27)

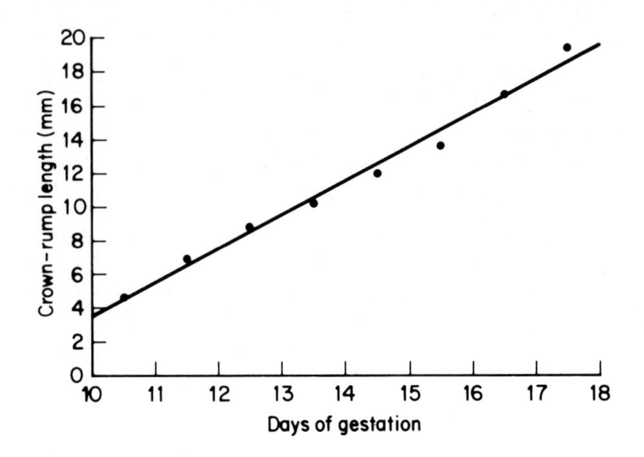

Figure 12 **The relationship between crown–rump length and day of gestation.**

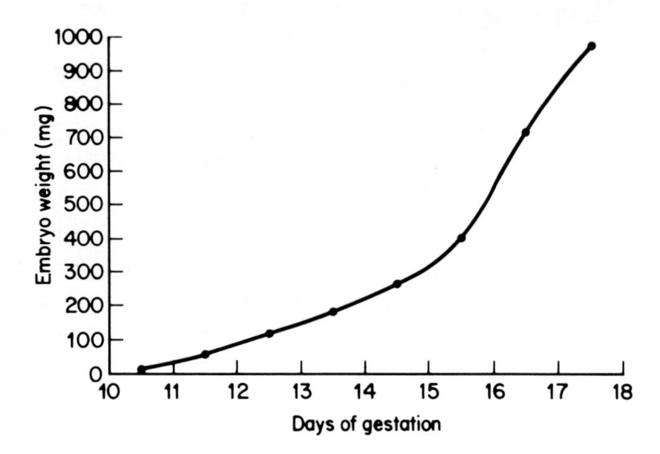

Figure 13 **The relationship between embryo weight and day of gestation.**

4. Differentiation of the cephalic region in embryos isolated between 9.5 and 11.5 days p.c. – scanning electron microscopy series

Text associated with Plate 47

The changes that occur in the external appearance of intact embryos between 7.5 and 9.5 days p.c. have been described in association with *Plate 42*, and the postcephalic morphology has been described, with particular reference to the development of the limbs in *Plates 43–45*. In *Plate 46*, views of intact embryos are presented in order to emphasize the gradual evolution that occurs in the overall configuration of embryos, on a day-to-day basis, as they progress from about 11.5 to 18.5 days p.c. Because of the important changes that are occurring between 9.5 and 11.5 days p.c. in the cephalic region, this part of the embryo justifies special consideration, particularly as an understanding of the features of this region greatly facilitates the interpretation of the histological sections through this part of the embryo at these stages of development. Over this relatively brief period, fundamental changes occur with regard to the development of the brain, the eye, the olfactory region and the pituitary gland, as well as in the pharyngeal region of the embryo, and aspects of all of these changes may be recognized in these scanning electron micrographs. For this reason, both lateral and frontal views of this region are presented here, in order to provide a more complete picture of these changes as they evolve.

In early "turned" embryos which possess about 17–20 pairs of somites (*Plate 42,p*), the cephalic part of the neural tube has only recently closed. This event coincides with the formation of the three primary brain vesicles, the walls of which are termed the prosencephalon, the mesencephalon and the rhombencephalon (i.e. the primitive forebrain, midbrain and hindbrain, respectively). At this time, the primitive forebrain surrounds a single large dilated midline vesicle, with the two optic vesicles as relatively small evaginations on either side. By about 9.5 days p.c., when the embryo possesses about 20–25 pairs of somites, the first external indication of the subdivision of the prosencephalon into the two telencephalic hemispheres is seen (*Plate 47,a,b*). The anterior midline groove lies over the region of the lamina terminalis.

On either side of the cephalic region, an optic eminence overlies the immediately subjacent optic vesicle. More caudally, the two components of the first branchial (pharyngeal) arch, namely its maxillary and mandibular components, are well differentiated. At this stage, though, the one merges into the other, there being no reasonably clear line of demarcation between the two, as occurs at about 11–11.25 days p.c. (*Plate 47,j*). The medial surfaces of the two mandibular processes meet in the midline, but show no evidence of fusion at this time. A slight indentation is already evident inferior to the optic eminence, between the posterior part of the fronto-nasal process and the anterior part of the maxillary component of the first branchial arch. This is the first indication of the naso-lacrimal groove, which will subsequently close in embryos on or about 12–12.5 days p.c. to form the naso-lacrimal duct. The deep groove between the first and second branchial arches (externally) is the first branchial cleft, and will eventually give rise to the external auditory meatus. The second branchial arch is also well differentiated at this stage, through the third arch is only poorly developed.

By 9.5–10 days p.c., the most important new feature relates to the differentiation of the olfactory placode. The latter is only barely evident as a slightly flattened region on the inferolateral surface of the swelling overlying the telencephalic vesicles (*Plate 20b,a–d*; *Plate 21a,a*; *Plate 21b,c,d*; *Plate 47,d*). The gap between the maxillary and mandibular components of the first branchial arch is wider (*Plate 47,c*), and the mandibular processes are more prominent than previously. In the roof of the oropharynx the first histological evidence of Rathke's pouch is seen as a localized region of ectodermal thickening (*Plate 19a,k*; *Plate 54,a–d*), though in developmentally slightly more advanced embryos, the wide entrance to Rathke's pouch is now clearly seen (*Plate 47,f*; *Plate 54,e–h*). The optic eminence is also slightly more obvious than previously, largely due to the substantial growth of the optic vesicle at this time (*Plate 47,e*; cf. *Plate 47,a*).

Between 10 and 10.5 days p.c., the first evidence of the olfactory pit is seen, and it is clear from the histological sections through this region that the cells lining the pit are more differentiated than those at its peripheral margins (*Plate 22a,b–d*; *Plate 23b,h–l*), though the pit is more clearly defined at 10.5–11 days p.c. (*Plate 24b,e–h*; *Plate 47,g,h*). With the formation of the olfactory (nasal) pit, the primitive nasal process becomes divided into a medial and a lateral nasal process (*Plate 47,g,h*).

In the optic region, the lens placode has been induced at about 9.5–10 days p.c., and early evidence of lens pit formation may be seen at about 10–10.5 days p.c. (*Plate 51,k,l*), though the lens pit is more clearly defined at about 10.5–11 days p.c. (*Plate 47,g*; *Plate 51,n*). The entrance to Rathke's pouch is considerably smaller than previously, and this coincides with its differentiation and rostral migration that occurs at this time (*Plate 47,h*). On the lateral side of the neck, the third and fourth pharyngeal arches, though considerably smaller than the first and second arches, are now clearly seen. The first indication of the cervical sinus is also seen at this time (*Plate 47,g,h*), being the site where the second arch overgrows the third and fourth arches. In the mouse, the third and fourth arches are not seen as external swellings after about 11 days p.c. (*Plate 47,i*).

By 11–11.25 days p.c., the olfactory pits have deepened considerably, and the dimensions of their entrance have diminished slightly (*Plate 47,j*), compared to the situation observed previously. By this stage, the two telencephalic hemispheres are particularly prominent, and each is clearly separated (inferiorly) by a deep transverse groove from the

(continued on page 392)

PLATE 47

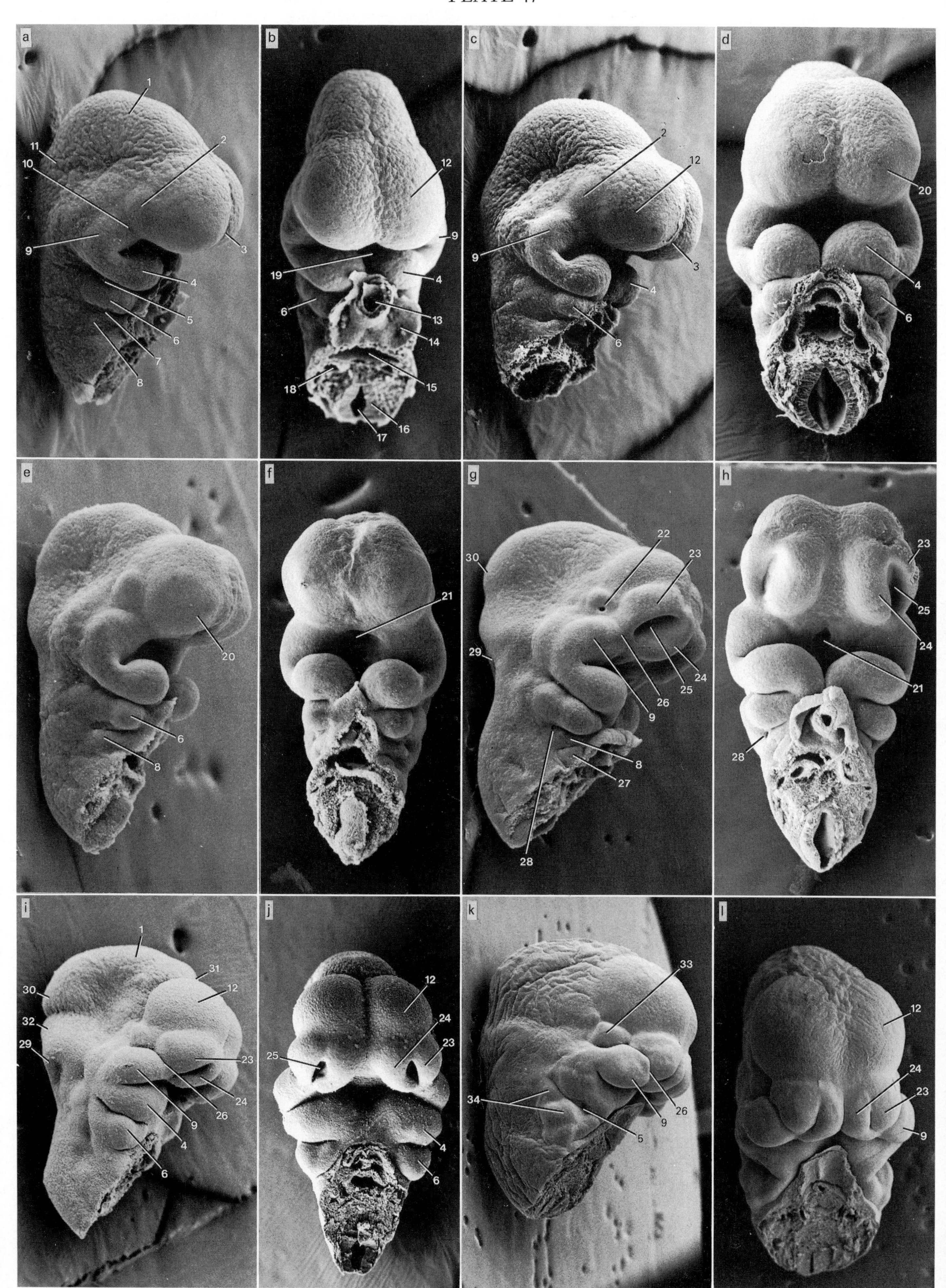

a
1
11
2
10
9
3
4
5
6
7
8
b
12
9
19
4
6
13
14
15
18
16
17
c
2
12
9
3
4
6
d
20
4
6
e
20
6
8
f
21
g
22
23
30
29
9
26
25
24
8
27
28
h
23
25
24
21
28
i
1
31
30
12
32
29
23
24
26
9
4
6
j
12
24
23
25
4
6
k
33
34
9
26
5
l
12
24
23
9

Plate 47

Development of the cephalic region (scanning electron microscopy)

Differentiation of the branchial/pharyngeal arches, and fronto-nasal derivatives. Paired lateral and frontal views of staged embryos.

a,b 9.5 days p.c.; embryo with 20–25 pairs of somites
c,d 9.5–10 days p.c.; embryo with about 25 pairs of somites
e,f 9.5–10 days p.c.; embryo with 25–30 pairs of somites
g,h 10.5–11 days p.c.; embryo with 35–40 pairs of somites
i,j 11–11.25 days p.c.
k,l 11.5 days p.c.

1. region overlying midbrain
2. optic eminence (region overlying optic vesicle)
3. groove overlying lamina terminalis of forebrain
4. mandibular component of first branchial arch
5. first branchial cleft (groove)
6. second branchial arch (hyoid arch)
7. second branchial cleft (groove)
8. third branchial arch
9. maxillary component of first branchial arch
10. groove between optic eminence and maxillary component of first branchial arch
11. region overlying hindbrain
12. region overlying telencephalic vesicle
13. lumen of truncus arteriosus/aortic sac
14. rostral extremity of pericardial cavity
15. pharyngeal region of foregut
16. neuroepithelium of neural tube in cervical region
17. neural lumen
18. dorsal aorta
19. oropharyngeal region
20. region of olfactory placode
21. entrance to Rathke's pouch
22. lens pit
23. lateral nasal process
24. medial nasal process
25. nasal (olfactory) pit
26. groove between lateral nasal process and maxillary component of first branchial arch (naso-lacrimal groove)
27. fourth branchial arch
28. first indication of cervical sinus
29. region overlying roof of hindbrain
30. region overlying caudal part of midbrain
31. region overlying diencephalon
32. region overlying cerebellar plate (of metencephalon)
33. corneal ectoderm overlying lens vesicle
34. hillocks (or tubercles) on either side of the first branchial cleft associated with the formation of the external ear

adjacent olfactory region (*Plate 47,i,j*). Similarly, since the lens pit has been transformed into the lens vesicle, the indentation in the central part of the optic eminence, which previously represented the lens pit, is no longer seen. The appearance of the maxillary and mandibular swellings are now slightly different than the situation observed previously, since the medial surfaces of the two mandibular processes have merged together across the midline, so that the sharp boundary between them, evident up to about 10.5–11 days p.c., is no longer seen (*Plate 47,i,j*; cf. *Plate 47,g,h*). The second arch is more prominent than previously, while the external prominences associated with the third and fourth arches have effectively disappeared. In the roof of the oropharynx, the entrance to Rathke's pouch is now extremely small (*Plate 54,i–l*), and is often difficult to see even on scanning electron micrographs of this region (*Plate 47,j*).

By 11.5 days p.c., the external appearance of the cephalic region is modified by the further differentiation of the fronto-nasal processes (prominences), in that the entrances to the olfactory pits have diminished in size, and are now only represented by a pair of narrow slits (*Plate 47,k,l*). The optic eminence is, however, more prominent than previously, and is covered by primitive corneal ectoderm. The latter principally overlies the lens vesicle (*Plate 47,k*; *Plate 51,o,p*). The maxillary process is considerably more prominent than previously, as is the naso-lacrimal groove, which is now located between the anterior border of the maxillary process and the posterior border of the lateral nasal process. The first branchial cleft (groove) is slightly less obvious than previously, though the first evidence of the auditory hillocks are now seen to be located on either side of it, but particularly in association with the surface of the second arch. The first groove is destined to form the external acoustic (auditory) meatus, while the auditory hillocks subsequently amalgamate to form the pinna of the ear. The auditory hillocks are particularly difficult to recognize in histological sections through this region at this time.

5. Differentiation of the eye–scanning electron microscopy and histological series

Text associated with Plates 48–53

The analysis of transverse sections through the cephalic region of a series of embryos spanning the period from about 8 to 9.5 days p.c. (*Plate 50,a–i*; *Plate 51,a–h*) provides information on the sequential changes that are associated with the early development of the brain, as well as that of the eye. This material clearly complements the scanning electron micrographs presented in *Plate 49,a–h*, in which the external appearance of the cephalic region over this relatively restricted period of time is closely followed.

At about the time that the cephalic neural folds are first clearly recognizable, in the early somite stage embryo, a slightly flattened area of neural ectoderm in the central part of the prospective forebrain region (*Plate 48,a*) represents the first evidence of the optic placode. Histological sections through this region reveal that the neural ectoderm is slightly thickened at this site (*Plate 50,a*), though its peripheral boundaries are fairly diffuse. Shortly afterwards, a slight indentation in the central part of the optic placode represents the first indication of the optic pit (*Plate 48,b*; *Plate 50,b*). Over the relatively short period between about 8 and 8.5 days p.c., the cephalic neural folds increase in volume, become elevated, and eventually apposed in the region overlying the middle of the prospective forebrain (*Plate 48,c–f*). Fusion of the neural folds at this initial site of apposition occurs at about 8.5–9 days p.c. (*Plate 48,g*). During this period, the optic pits appear as an increasingly deepening pair of indentations of the neural ectoderm (they are evaginations of the prospective third ventricle) which are connected by a pair of shallow grooves, the optic sulci (*Plate 48,c*), in the region of the future optic chiasma.

On the side (lateral aspect) of the developing cephalic region, two prominences appear which overly the optic pits, these are the optic eminences (*Plate 48,d–f*). Each of the latter is separated from the other prominence that appears slightly further posteriorly on the side of the cephalic region at this time, namely the maxillary component of the first branchial arch, by a deep groove (*Plate 48,e*). Histological sections through the cephalic region over this period clearly reveal the sequential changes that occur in the region of the developing forebrain. The increasing depth of the optic pits is seen to be closely associated with the elevation of the cephalic neural folds in the region of the prospective forebrain, and the development of the optic eminences (*Plate 50,c*; *Plate 50,d–f*). The initial site of apposition is seen to be just above the site of the developing optic pits (*Plate 50,d*), but once fusion has occurred, it extends rapidly over a relatively lengthy section, so that the entire region overlying the optic pits is almost simultaneously involved in the closure process (*Plate 48,g*; *Plate 50,g–i*).

The eventual closure of the cephalic neural folds to form the future forebrain also results in the formation of not only the optic vesicles, but also the optic stalks which connect them to the forebrain vesicle (*Plate 50,h*). At this stage of development, this is the only site where the cephalic neural folds have fused. Analysis of the neural ectoderm at an advanced stage in the differentiation of the optic pits, and at every stage in the differentiation of the optic vesicles, confirms the earlier observation that throughout these early stages of eye development, the boundary between the future optic apparatus and the neural ectoderm of the future diencephalic region of the primitive forebrain is extremely diffuse.

Between about 8.5 and 9 days p.c., the cephalic neural folds in the adjacent regions overlying the future midbrain and hindbrain have also now elevated, become apposed and eventually fused. The latter results in the formation of the primitive midbrain (mesencephalon) and hindbrain (rhombencephalon), with their associated mesencephalic vesicle and fourth ventricle, respectively (*Plate 48,h*; *Plate 51,a–d*). The actual pattern of closure in this region is quite species specific, and in even closely related species this pattern may be markedly different (see Kaufman, 1979). Observations on this process in hamster embryos (Marin-Padilla, 1970; Shenefelt, 1972; Waterman, 1976) and rat embryos (Adelmann, 1925; Christie, 1964; Keibel, 1937) indicate that the same general pattern of closure is followed in all three species. This is in marked contrast to the pattern of closure of the cephalic part of the neural tube in human embryos (Streeter, 1942; O'Rahilly, 1966) which is similar to the pattern in the pig embryo (Heuser and Streeter, 1929) and in the macaque embryo (Heuser and Streeter, 1941) where fusion of the neural folds proceeds rostrally as a continuous process from the cervical region, with the eventual closure of the rostral neuropore (see also O'Rahilly and Müller, 1987).

In the mouse, at about 8.5–9 days p.c., the most rostral site of neural tube closure is seen to be at about the level of the otic pits (*Plate 42,n,o*), and therefore is at about the level of the future upper cervical region/caudal hindbrain. Neural tube closure in this region occurs slightly before the stage when the neural folds in the region of the future forebrain first become apposed in the ventral midline and eventually fuse (see above). From the latter site, neural fold fusion progresses in a zipper-like fashion both rostrally, so that the whole of the prosencephalon is now formed, and caudally to form initially the primitive midbrain, and then, when this region of cephalic neural fold fusion eventually meets up with that in the upper cervical region, this eventually results in the formation of the primitive hindbrain.

Initially, the optic stalks are difficult to distinguish from the optic vesicles, because they have a similar luminal diameter (*Plate 51,d*), but over the period between about 9 and 9.5 days p.c., the diameter of the optic stalk progressively narrows, while that of the optic vesicle increases compared to the situation observed previously (*Plate 51,f*). It is believed that the underlying mechanism which facilitates the dilatation of the various brain vesicles as well as the optic vesicles, is the process of neural luminal occlusion. In the mouse, evidence of this process is first seen when

(*continued on page 406*)

PLATE 48

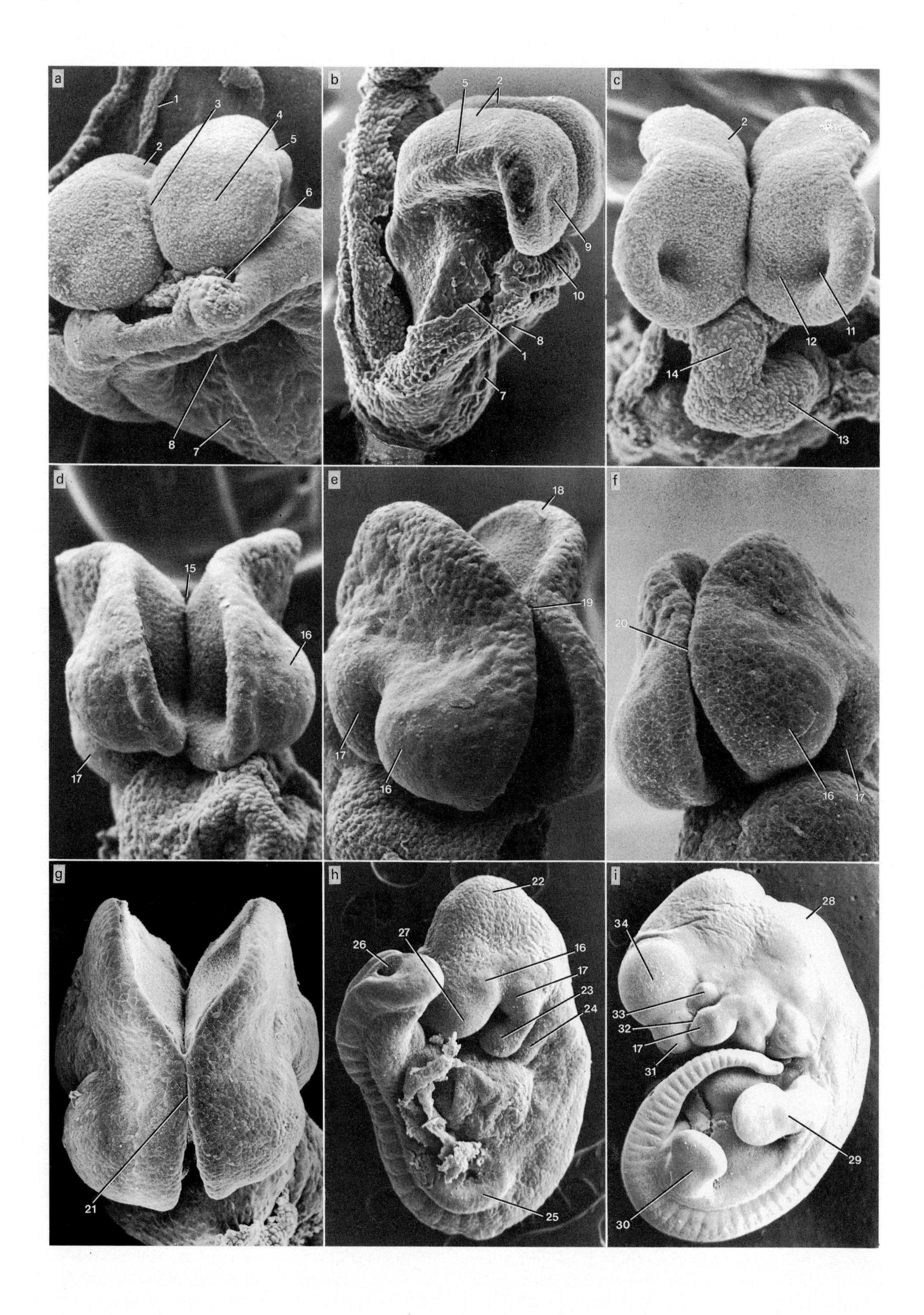
a
1
3
4
2
5
6
8
7
b
5
2
9
10
8
1
7
c
2
11
12
14
13
d
15
16
17
e
18
19
17
16
f
20
16
17
g
21
h
22
27
26
16
17
23
24
25
i
28
34
33
32
17
31
29
30

Plate 48

Development of the eye (scanning electron microscopy series)

a 8 days p.c.; embryo with 5–6 pairs of somites
b 8–8.25 days p.c.; embryo with 6–8 pairs of somites
c 8.25–8.5 days p.c.; embryo with 8–10 pairs of somites
d 8–8.5 days p.c.; embryo with 9–11 pairs of somites
e 8–8.5 days p.c.; embryo with 10–12 pairs of somites
f 8.5 days p.c.; embryo with 11–13 pairs of somites
g 8.5–9 days p.c.; embryo with 12–14 pairs of somites
h 9.5 days p.c.; embryo with 20–25 pairs of somites
i 11.5 days p.c.; embryo with 45–50 pairs of somites

1. torn edge of amnion
2. future midbrain region
3. neural groove in future diencephalic region
4. optic placode
5. junctional zone between neural and surface ectoderm
6. torn edge of visceral yolk sac
7. endodermal lining of future midgut
8. entrance to foregut diverticulum
9. early stage in differentiation of optic pit (optic evagination)
10. primitive heart tube
11. optic pit (optic evagination)
12. optic sulcus
13. common ventricular component of primitive heart tube
14. outflow tract of primitive heart tube
15. neural groove in future thalamic region
16. optic eminence
17. maxillary component of first branchial arch
18. cephalic neural fold in future midbrain region
19. first site of apposition and fusion of cephalic neural folds (at junction between future forebrain and midbrain)
20. limited site of apposition and fusion of cephalic neural folds
21. extensive site of fusion of cephalic neural folds
22. region overlying future midbrain
23. mandibular component of first branchial arch
24. second branchial arch (hyoid arch)
25. forelimb bud
26. neuroepithelium in region of caudal neuropore
27. region overlying future forebrain
28. region overlying hindbrain
29. forelimb
30. hindlimb
31. lateral nasal process
32. naso-lacrimal groove
33. corneal ectoderm overlying lens vesicle
34. region overlying telencephalic vesicle

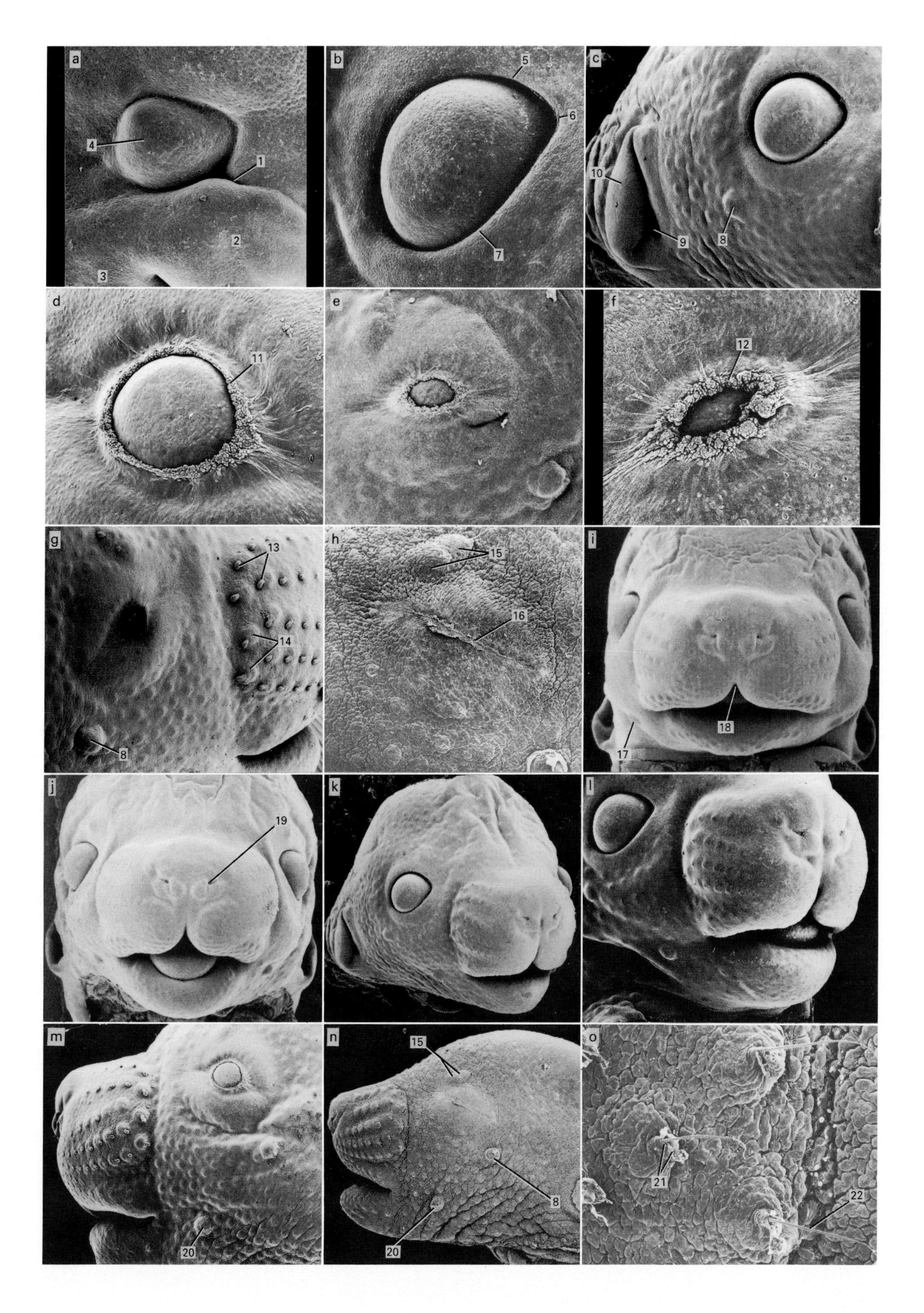
a
4
1
2
3
b
5
6
7
c
10
9
8
d
11
e
f
12
g
13
14
8
h
15
16
i
17
18
j
19
k
l
m
20
n
15
8
20
o
21
22

Plate 49

Development of the eye and face (scanning electron microscopy series)

a 11.5 days p.c.; development of the eye, formation and closure of the eyelids
b 14.5 days p.c.; development of the eye, formation and closure of the eyelids
c 15.5 days p.c.; development of the eye, formation and closure of the eyelids
d 15.5 days p.c.; development of the eye, formation and closure of the eyelids
e 15.5 days p.c.; development of the eye, formation and closure of the eyelids
f 15.5 days p.c.; development of the eye, formation and closure of the eyelids
g 15.5 days p.c.; development of the eye, formation and closure of the eyelids
h 16.5 days p.c.; development of the eye, formation and closure of the eyelids
i 13.5 days p.c.; appearance of the face, and development of the vibrissae
j 14.5 days p.c.; appearance of the face, and development of the vibrissae
k 15.5 days p.c.; appearance of the face, and development of the vibrissae (same embryo as illustrated in **c**)
l 14.5 days p.c. (same embryo as illustrated in **b**)
m 15.5 days p.c. (same embryo as illustrated in **d**)
n 16.5 days p.c. (same embryo as illustrated in **h**)
o 16.5 days p.c. (same embryo as illustrated in **h**, **n**)

1. naso-lacrimal groove
2. maxillary component of first branchial arch
3. upper part of mandibular component of first branchial arch
4. optic eminence covered by corneal ectoderm which overlies the lens vesicle
5. boundary between skin (surface ectoderm) and conjunctival epithelium in region of future upper eyelid
6. inner (nasal) canthus
7. region of future lower eyelid
8. primordium of prominent tactile or sinus hair follicle characteristically found in this (infraorbital) location, and which overlies the caudal portion of the masseter muscle
9. entrance to external auditory meatus
10. pinna of ear
11. high level of cellular activity in region of upper eyelid at early stage in fusion of the upper and lower eyelids
12. advanced stage in fusion of the upper and lower eyelids
13. vibrissae just prior to their eruption
14. follicles of vibrissae
15. primordia of two prominent tactile or sinus hair follicles characteristically found in this (supraorbital) location
16. line of fusion of upper and lower eyelids
17. early indication of primordium or tactile or sinus hair follicle characteristically found in this location (just caudal to the angle of the mouth)
18. region of philtrum
19. anterior (external) naris (nostril)
20. advanced stage in differentiation of prominent hair follicle characteristically found in this location (see 17, above)
21. site of eruption of vibrissa with characteristic mound of cellular debris
22. shaft of vibrissa

PLATE 50

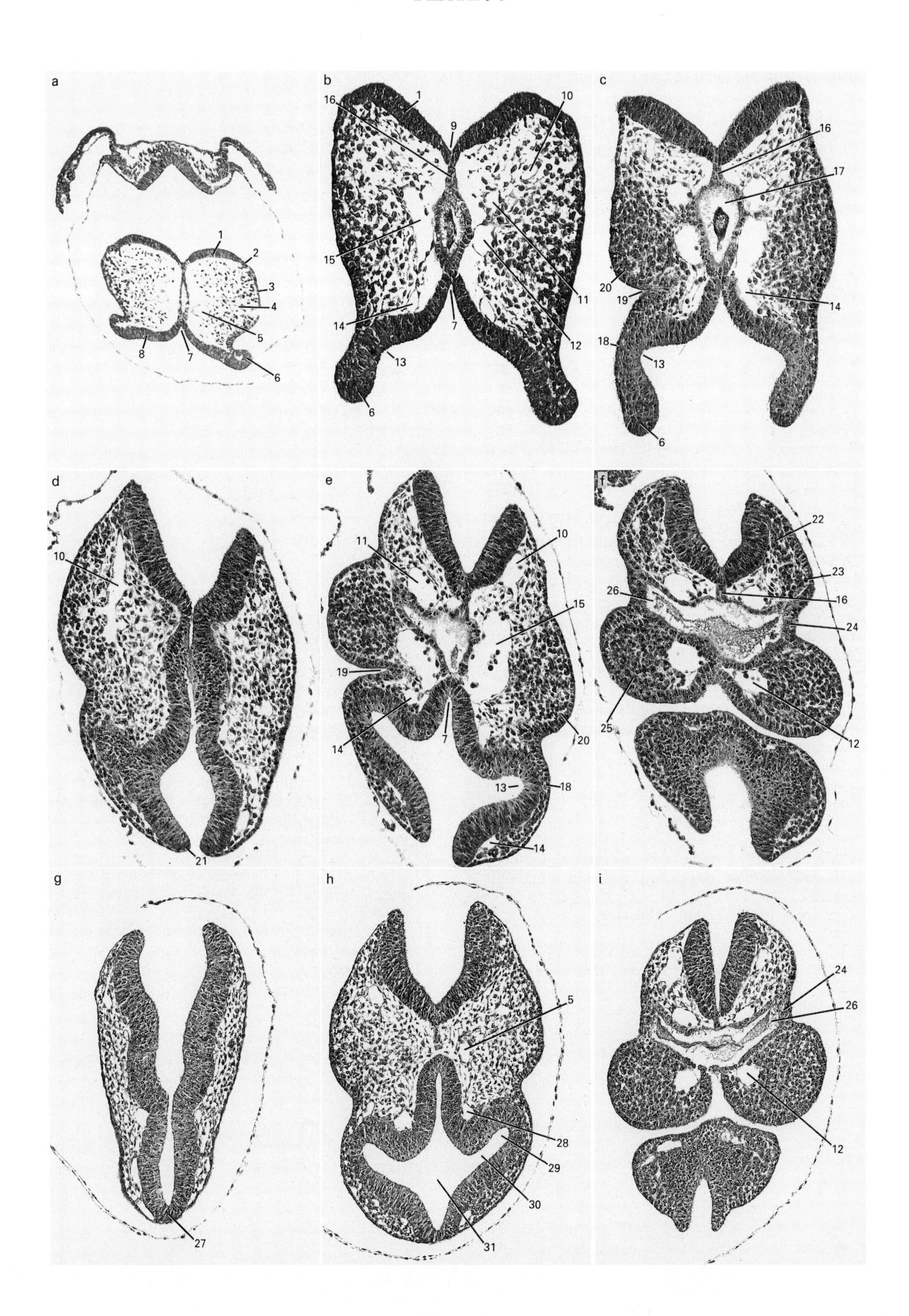
a
1
2
3
4
5
6
7
8
b
16
1
9
10
15
11
12
14
7
13
6
c
16
17
20
19
14
18
13
6
d
10
21
e
11
10
15
19
14
7
20
13
18
14
f
22
23
26
16
24
25
12
g
27
h
5
28
29
30
31
i
24
26
12

Plate 50

Development of the eye (histological series 1)

a 8 days p.c.; embryo with 5–7 pairs of somites (ref. Plate 10)
b 8 days p.c.; embryo with 6–8 pairs of somites (ref. Plate 11)
c 8–8.5 days p.c.; embryo with 8–10 pairs of somites (ref. Plate 12)
d–f 8–8.5 days p.c.; embryo with 10–12 pairs of somites (ref. Plate 14)
g–i 8.5–9 days p.c.; embryo with 12–14 pairs of somites (ref. Plate 16)
(Only the principal features of the cephalic region are labelled)

1. neuroepithelium in prospective hindbrain region
2. surface–neural ectodermal junction
3. surface ectoderm
4. cephalic mesenchyme tissue
5. internal carotid artery (rostral extension of dorsal aorta)
6. cephalic neural fold
7. neural groove in prospective diencephalic region of forebrain
8. optic placode
9. neural groove in prospective hindbrain region
10. branch of primary head vein
11. dorsal aorta
12. first branchial arch artery
13. optic pit (optic evagination)
14. arterial branch from first branchial arch artery to future perioptic vascular anastomosis
15. communication between dorsal aorta and first branchial arch artery
16. notochord
17. rostral extremity of foregut diverticulum
18. optic eminence
19. groove between optic eminence and first branchial arch
20. maxillary component of first branchial arch
21. neural fold at forebrain–midbrain junction
22. facio-acoustic (VII–VIII) neural crest tissue
23. second branchial arch
24. first branchial membrane
25. mandibular component of first branchial arch
26. first branchial pouch
27. forebrain–midbrain junction
28. perioptic vascular anastomosis
29. primary optic vesicle
30. optic stalk
31. forebrain vesicle

PLATE 51

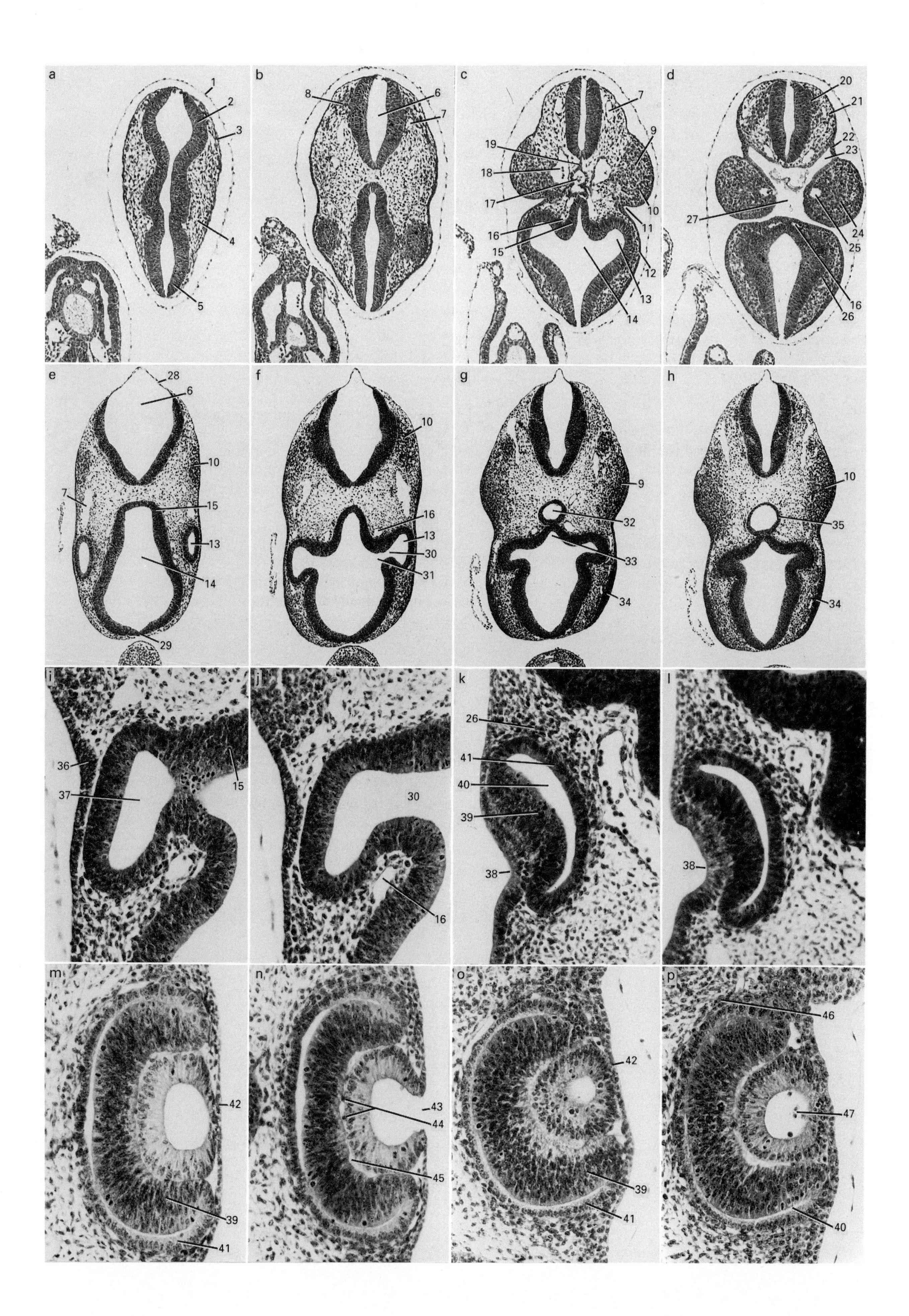

a
b
c
d
e
f
g
h
i
j
k
l
m
n
o
p
1
2
3
4
5
6
7
8
9
10
11
12
13
14
15
16
17
18
19
20
21
22
23
24
25
26
27
28
29
30
31
32
33
34
35
36
37
38
39
40
41
42
43
44
45
46
47

Plate 51

Development of the eye (histological series 2)

a–d 9 days p.c.; embryo with 15–20 pairs of somites (ref. Plate 18)
e–h 9.5 days p.c.; embryo with 20–25 pairs of somites
i,j 9.5–10 days p.c.; embryo with 25–30 pairs of somites
k,l 9.5–10 days p.c.; embryo with 25–30 pairs of somites
m,n 10.5–11 days p.c.; embryo with 35–40 pairs of somites
o,p 11.5 days p.c.; embryo with 40–45 pairs of somites
(Only the principal features of the cephalic region are labelled)
(**i–p**, ×250)

1. amnion
2. neuroepithelium in rostral part of hindbrain
3. surface ectoderm
4. cephalic mesenchyme tissue
5. neuroepithelium at forebrain–midbrain junction
6. fourth ventricle
7. primary head vein
8. trigeminal (V) neural crest tissue
9. maxillary component of first branchial arch
10. trigeminal (V) neural crest tissue within maxillary component of first branchial arch
11. groove between first branchial arch and optic eminence
12. optic eminence
13. optic vesicle
14. third ventricle
15. neuroepithelium of future diencephalon
16. perioptic vascular plexus – arterial branches from internal carotid artery, and venous branches from anterior cerebral venous plexus
17. rostral extension of foregut diverticulum
18. site of communication between dorsal aorta and first branchial arch artery
19. rostral extremity of notochord
20. facio-acoustic (VII–VIII) neural crest tissue
21. second branchial arch
22. first branchial membrane
23. first branchial pouch
24. mandibular component of first branchial arch
25. first branchial arch artery
26. perioptic neural crest tissue of trigeminal (V) crest origin
27. oropharyngeal region
28. roof of hindbrain
29. lamina terminalis
30. optic stalk
31. optic recess
32. Rathke's pouch
33. infundibular recess of diencephalon
34. olfactory placode
35. peripheral boundary of entrance to Rathke's pouch
36. lens placode differentiated from surface ectoderm induced by underlying optic vesicle
37. lumen of optic cup
38. early stage in differentiation of lens pit
39. inner (neural) layer of optic cup (future nervous layer of retina)
40. intra-retinal space
41. outer layer of optic cup (future pigment layer of retina)
42. corneal ectoderm
43. lens pit (stage of deep lens indentation)
44. primitive hyaloid vessels within hyaloid cavity
45. location of future hyaloid cavity
46. condensation of perioptic mesenchyme (ectomeninx) which subsequently differentiates in this location to form the sclera
47. the so-called "epitrichial cells" frequently observed within the cavity of the lens vesicle (possibly discarded cellular debris)

PLATE 52

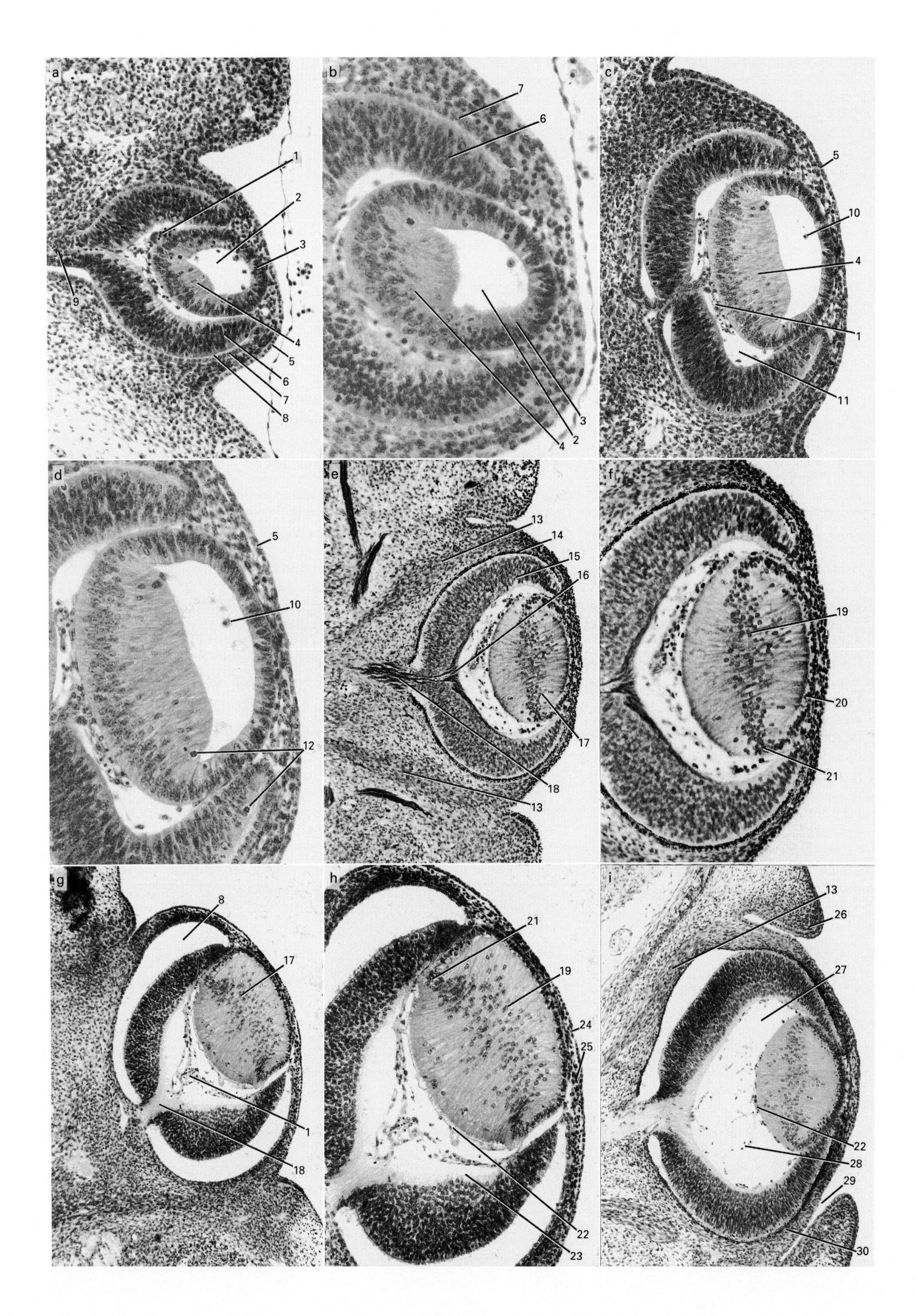

a
1
2
3
9
4
5
6
7
8
b
7
6
4
2
3
c
5
10
4
1
11
d
5
10
12
e
13
14
15
16
17
18
13
f
19
20
21
g
8
17
1
18
h
21
19
24
25
22
23
i
13
26
27
22
28
29
30

Plate 52

Development of the eye (histological series 3)

a,b 11.5–12 days p.c. (**a** ×160; **b** ×300)
c,d 12–12.5 days p.c. (ref. Plate 27) (**c** ×160; **d** ×250)
e,f 12.5–13 days p.c. (Linder's silver stain) (**e** ×100; **f** ×160)
g,h 13.5 days p.c. (**g** ×100; **h** ×160)
i 14.5 days p.c. (×100)

1. primitive hyaloid plexus of vessels within hyaloid cavity
2. cavity of lens vesicle
3. anterior wall of lens vesicle
4. elongated cells of posterior wall of lens vesicle, with nuclei mostly located towards the periphery
5. corneal ectoderm
6. inner (neural) layer of optic cup (future nervous layer of retina)
7. outer layer of optic cup (future pigment layer of retina)
8. intra-retinal space
9. optic stalk
10. the so-called "epitrichial cells" frequently observed within the cavity of the lens vesicle (possibly discarded cellular debris)
11. hyaloid cavity
12. mitotic figure
13. mesodermal condensation (future extrinsic ocular muscle)
14. pigment layer of retina
15. neural layer of retina
16. region of future optic disc, where nerve cells grow back towards the optic stalk from the neural retina; the lumen of the optic stalk is almost completely obliterated
17. lens fibres
18. heavy concentration of unmyelinated nerve fibres within optic stalk (future optic (II) nerve)
19. nuclei of lens fibres in middle region of cells
20. characteristic cuboidal epithelium comprising anterior part (or capsule) of the lens
21. equatorial region of the lens where the anterior capsular epithelial cells become gradually incorporated into the lens proper and subsequently develop into lens fibres
22. branches of hyaloid plexus in contact with lens capsule posteriorly (tunica vasculosa lentis)
23. nerve fibre layer of optic cup (only easily seen in silver-stained material)
24. surface epithelium of cornea
25. mesothelial layer of cornea (region of future substantia propria)
26. eyelid, at early stage of its differentiation (prior to fusion of upper and lower eyelids)
27. vitreous humour
28. branches of hyaloid plexus which ramify in the vitreous (vasa hyaloidea propria)
29. future conjunctival sac
30. condensation of perioptic mesenchyme (ectomeninx) which subsequently differentiates in this location to form the sclera

PLATE 53

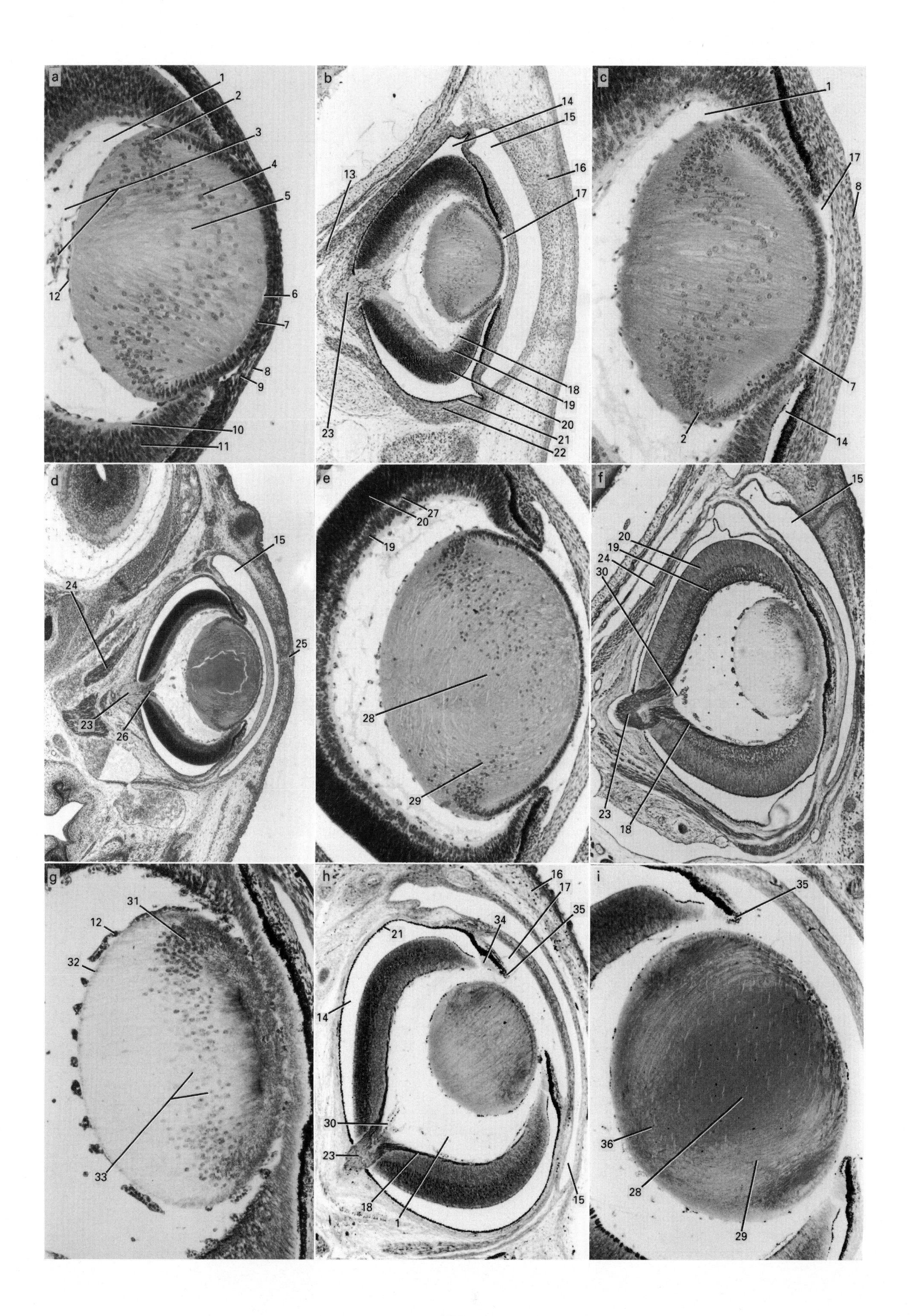
a
1
2
3
4
5
6
7
8
9
10
11
12
b
13
14
15
16
17
18
19
20
21
22
23
c
1
17
8
7
2
14
d
15
24
25
23
26
e
27
20
19
28
29
f
15
20
19
24
30
23
18
g
31
12
32
33
h
16
17
35
34
21
14
30
23
18
1
15
i
35
36
28
29

Plate 53

Development of the eye (histological series 4)

a 14.5 days p.c. (×160)
b,c 15.5–16 days p.c. (**b** ×63; **c** ×160)
d,e 16.5 days p.c. (**d** ×40; **e** ×100)
f,g 17.5 days p.c. (**f** ×63; **g** ×160)
h,i 18.5 days p.c. (**h** ×63; **i**×100)
The sections illustrated in **f–h** are stained using Linder's silver method

1. vitreous humour occupying the hyaloid cavity
2. equatorial region of the lens where the anterior capsular epithelial cells become gradually incorporated into the lens proper and subsequently develop into the lens fibres
3. branches of hyaloid plexus which ramify in the vitreous (vasa hyaloidea propria)
4. nucleus of lens fibre
5. lens fibre
6. cavity of lens vesicle almost obliterated by elongation of cells of its posterior wall
7. characteristic cuboidal epithelium comprising anterior part (or capsule) of the lens
8. surface epithelium of cornea
9. mesothelial layer of cornea (region of future substantia propria)
10. innermost (marginal) zone of inner (neural) layer of optic cup, free from nuclei
11. outermost zone of neural retina, consisting of proliferating layer of primitive neuroepithelial cells
12. branches of hyaloid plexus in contact with lens capsule posteriorly (tunica vasculosa lentis)
13. mesodermal condensation (future extrinsic ocular muscle)
14. intra-retinal space (largely an artefact of fixation at this stage)
15. conjunctival sac
16. fused eyelids
17. anterior chamber of eye
18. nerve fibre layer of optic cup
19. inner nuclear (neuroblastic) layer, which differentiates following migration of cells into the marginal layer
20. outer nuclear (neuroblastic) layer
21. pigment layer of retina
22. sclera
23. optic (II) nerve (the lumen of the optic stalk is completely obliterated)
24. extrinsic ocular muscle
25. line of fusion of upper and lower eyelids
26. optic disc
27. transient non-nucleated layer of Chievitz (deep part of marginal layer)
28. location of primary (primitive) lens fibres derived from posterior wall of lens vesicle (this gives rise to the embryonic/fetal "nucleus" of the lens)
29. secondary lens fibres derived from the cells in the equatorial region
30. hyaloid artery (central artery of retina)
31. darkly-staining and closely packed lens fibre nuclei located towards periphery of lens, associated with secondary lens fibres
32. posterior capsule of lens
33. faintly staining nuclei (in process of disintegration) of primary (primitive) lens fibres
34. pars iridis retinae
35. medial margin of primitive iris
36. location of posterior lens "suture"

the cephalic and caudal extremities of the neural tube are still widely open. After the initial contact is made between the medial (inner) surfaces of the neural tube at the level of the lower thoracic/upper lumbar region, in embryos with about 10–12 pairs of somites present, this event is followed by direct contact being established over a much more extensive area. Finally, complete luminal occlusion involving this much more extensive region occurs, and eventually extends from the caudal part of the hindbrain to the proximal part of the tail.

Complete luminal occlusion is a transient process and lasts for about 1–2 days in this species, and interestingly at no time do junctional complexes form across the midline. Immediately after luminal occlusion occurs in the postcephalic region, a build up of pressure can now occur within the lumina of the brain vesicles, presumably mediated by the increased production (or reduced absorption) of cerebrospinal fluid (or its precursor). Since the caudal neuropore has yet to close at this stage, the increased cerebrospinal fluid pressure is not dissipated elsewhere, because the lumen of the neural tube is effectively occluded. Experimental studies have demonstrated that intubation of either the neural tube (Desmond and Jacobson, 1977), the eye (Coulombre, 1956) or the hindbrain (Coulombre and Coulombre, 1958) in avian embryos invariably resulted in abnormalities of the brain and the eyes. While the lumen recanalizes along the majority of its length shortly after the closure of the caudal neuropore, it is still possible to see localized regions of neural luminal occlusion, often in the lumbar region, in embryos of 13.5 days p.c. (e.g. *Plate 30l,c,d*), even in paraffin-embedded material.

The detailed events described above in relation to neural luminal occlusion are usually only clearly seen in plastic-embedded material, as the shrinkage that occurs in paraffin-embedded material almost always results in the pulling apart of the somewhat tenuously apposed neuroepithelial surfaces (for detailed description of neural luminal occlusion in the mouse, see Kaufman, 1983b, 1986).

It is also interesting to note that at all stages before the (primary) optic vesicles have fully differentiated, cephalic mesenchyme tissue intervenes between the outer (lateral) region of the neural ectoderm and the overlying surface ectoderm. However, by about 9.5 days p.c., with the expansion of the optic vesicles, little or no mesenchyme tissue is seen to be located in this region (*Plate 51,f*), it being displaced (presumably) by the growth of the optic vesicle. The displacement of the mesenchyme tissue is of critical importance in relation to the next stage of eye differentiation, as the optic vesicle is believed to induce the differentiation of the overlying surface ectoderm to form the lens placode.

At an early stage in the differentiation of the lens placode, this is seen to be a localized thickened area of surface ectoderm which is closely apposed to the outer surface of the optic vesicle (*Plate 51,i*). During the next phase of eye development, which occurs at about 9.5–10 days p.c., the central part of the lens placode indents to form the lens pit and subsequently the lens vesicle, while the outer part of the optic vesicle collapses inwards to form the inner layer of the (secondary) optic cup.

The lens pit progressively indents, and is located in the middle part of the optic eminence (*Plate 47,g*). The duration of this stage, however, is relatively brief, and the lens pit disappears with the completion of lens vesicle formation (*Plate 47,i*). This sequence of events is most clearly seen from an analysis of the histological sections through this region during the process of lens vesicle formation (9.5–10 days p.c., *Plate 51,k,l*; 10.5–11 days p.c., *Plate 51,m,n*; 11.5 days p.c., *Plate 51,o,p*).

The majority of the cells that constitute the lens pit have a columnar morphology, and this is similar to the morphology of the cells that initially line the concavity (i.e. constitute the inner layer) of the optic cup. The morphological features of these cells are first seen to be different from those of the cells of the outer layer of the optic cup at about 10.5–11 days p.c. The latter cells have a characteristic cuboidal morphology, which changes little throughout the subsequent events that occur during the differentiation of the eye. While the inner layer of cells gives rise to the neural layer of the retina, the outer layer gives rise to the pigment layer of the retina. Even at about 11.5 days p.c., it is possible to discern that the sclera is formed by the condensation of perioptic mesenchyme in the region subjacent to the pigment layer of the retina (*Plate 51,p*). With the formation of the lens vesicle, the latter separates from and sinks beneath the (overlying) surface ectoderm. The rapidly diminishing remnant of the lens pit then completely disappears, and the surface ectoderm is reconstituted in this location (*Plate 51,o,p*).

At the same time as the lens vesicle is forming, it is possible to see the first evidence of the primitive hyaloid vessels within the hyaloid cavity (*Plate 51,n*). These are branches of the central vessels of the retina (the arteries arise from the ophthalmic artery), that enter the hyaloid cavity within the groove located on the ventral aspect of the optic stalk (this groove is termed the choroidal, or "fetal", fissure). The latter is seen most clearly in transverse sections through the optic stalk at this and at slightly later stages of development (*Plate 25c,c*; *Plate 27a,j*; *Plate 28b,a*). The choroidal fissure is still seen to be present in embryos up to about 13.5 days p.c. (*Plate 30d,b*, unlabelled), but has become narrowed by the growth of its margins around the blood vessels during 13–14 days p.c. The fusion process involves almost the whole length of the choroidal fissure, so that the blood vessels that pass to and from the optic cup eventually become surrounded by stalk tissue. The closure process is completed, and the fissure is no longer seen, by about 14.5 days p.c.

Substantial changes are next seen in relation to the lens vesicle. At about 11.5 days p.c., as indicated above, the lens pit is converted into the lens vesicle (*Plate 51,o,p*). At this stage of development, all of the cells that constitute its wall are of fairly uniform columnar morphology. However, by about 11.5–12 days p.c., the cells of the posterior wall of the lens vesicle become increasingly elongated, and their nuclei tend to be located towards the basal region of these cells. These cells progressively increase in length (*Plate 52,a–d*) until, at about 12.5–13 days p.c., the lumen of the lens vesicle is virtually completely obliterated (*Plate 52,e,f*). It is likely that the space occasionally observed in this location at slightly later stages of development is in fact an artefact of fixation (e.g. *Plate 53,a*). By about 12.5–13 days p.c., the posterior lens fibres are now maximally elongated. Their nuclei subsequently migrate away from the basal region of these cells and are now located at the mid-point of their length. It is also possible to see that new lens fibres are added from the cuboidal cells at the equatorial region of the lens. This allows the lens to increase in volume when the overall volume of the globe of the eye increases during subsequent stages of embryonic development.

The changes that occur in the histological appearance of the lens during the late embryonic period are discussed in detail in the appropriate section of Stage 26, and reference should be made to this section for this information. The most important feature observed at this time in relation to the lens, is the finding that the nuclei of the lens fibres that are located towards the centre of the lens (in the region of the embryonic/fetal nucleus) are almost translucent at this stage (*Plate 53,g*), and this is even more clearly seen in developmentally more advanced embryos (e.g. *Plate 55,i*).

During the period between 11.5 and 13 days p.c., the hyaloid cavity also increases in volume, and it is now possible to see that the hyaloid vessels form a considerable vascular network on the posterior surface of the lens capsule (termed the tunica vasculosa lentis) (e.g. *Plate 52,c,d,f*), as well as being found in association with the surface of the neural retina. The other important feature of note observed at about 12.5–13 days p.c. is the first appearance of nerve fibres which originate in the primitive ganglion cells of the neural retina, and grow towards the optic pit, and thence into the optic stalk. Fairly soon after these nerve fibres are first recognized, the lumen of the optic stalk becomes completely obliterated by them. These fibres, that now constitute the inner layer of the neural retina, are not well displayed by haematoxylin and eosin, and are best seen in silver-stained material (e.g. *Plate 52,f*). These nerve fibres are unmyelinated throughout their length, and are first directed towards the region of the future optic chiasma.

While the cells of the pigment layer of the retina show little evidence of change, the neural layer becomes increasingly differentiated, and is many cell layers thick at this stage. However, it is only at the most advanced stages of embryonic development that the stratification characteristically associated with this layer of the retina is clearly seen (e.g. at 16.5 days p.c. (*Plate 53,e*); at 17.5 days p.c. (*Plate 53,f*); at 18.5 days p.c. (*Plate 53,h*)). Silver staining enhances the difference between these layers of the neural retina. At about 16.5–17 days p.c., the neural retina is seen to consist of an outer nuclear (neuroblastic) layer, which is destined to form the horizontal cells and the nuclei of the photoreceptor cells, a transient intermediate anuclear layer (of Chievitz), and an inner nuclear layer which is destined to give rise principally to the ganglion cells. It is also relevant to note that under normal circumstances, the intra-retinal space disappears by about 11.5 days p.c. (*Plate 51,p*; *Plate 52,b*). However, if excessive shrinkage occurs, because the degree of adhesion between the neural and pigment layers of the retina is fairly slight, these two layers not infrequently separate (e.g. *Plate 52,h,i*; *Plate 53,b,d,f,h*).

The first indication of eyelid margins is seen at about 13.5–14 days p.c., but these structures are more clearly seen at about 14.5 days p.c. (*Plate 49,b*; *Plate 52,i*). A detailed description of the underlying events associated with the closure process is presented in the appropriate section of Stage 24, and reference should be made to that section for this information. Suffice it to say here that in embryos at about 15.5 days p.c., the eyelids are often widely open (*Plate 49,c*; *Plate 35b,b,c*; *Plate 35c,a,b*; *Plate 36c,b*), while in developmentally more advanced embryos isolated at about 16 days p.c., the eyelids are usually almost completely closed (*Plate 49,g*). By about 16.5–17 days p.c., the eyelids are invariably closed (*Plate 49,h*). Because of the wide variation in the degree of eyelid closure seen even within a single litter, it is believed that the total duration of this process is probably in the region of about 12–18 hours. What is probably of greatest interest is the actual pattern of eyelid closure seen in this species, which appears to be quite dissimilar to the standard description of comparable events described in the human embryo (Pearson, 1980). However, no published account is yet available in which the pattern of eyelid closure in the human embryo is described, in which the material has been studied by scanning electron microscopy. The closure of the eyelids results in the formation of the conjunctival sac (*Plate 53,d*), and this probably facilitates the differentiation of the cornea (*Plate 53,b,c*). An account of the possible significance of the timing of eyelid closure in various species is given in the appropriate section of Stage 25. In the mouse, the eyelids remain closed until about 12–14 days after birth.

As far as the differentiation of the cornea, iris and ciliary body is concerned, these are difficult to see in the histological sections illustrated. However, reference should be made to the appropriate sections of Stages 24–26, where the changes that occur in these structures during the latter part of gestation are briefly discussed. For additional information on the development of the mouse eye from the optic vesicle stage to term, see Pei and Rhodin (1970).

6. Differentiation of the pituitary
Text associated with Plates 54–57

The pituitary gland is of considerable interest because it develops from two quite different sources, namely from an ectodermal diverticulum from the roof of the stomatodaeum (the oral or buccal part of the primitive oropharynx) termed Rathke's pouch, and from a downgrowth from the floor of the diencephalic portion of the primitive forebrain. The derivatives of these two sources are as follows: Rathke's pouch gives rise to the adenohypophysis, while the diencephalic downgrowth gives rise to the neurohypophysis. The pituitary stalk (infundibular stem) is derived from the neurohypophysis and connects the neural lobe (pars nervosa or pars posterior) with the median eminence of the tuber cinereum. The pars tuberalis is formed from the adenohypophysis and surrounds most of the pituitary stalk. The major part of the adenohypophysis consists of the pars anterior (pars distalis) and the pars intermedia, and these are separated by the hypophyseal cleft, the remnant of the cavity of Rathke's pouch.

Rathke's pouch develops from the dorsal site at the junction between the ectoderm and endoderm in the roof of the primitive buccal cavity, just in front of the buccopharyngeal membrane. The latter is seen in embryos at about 8.5–9 days p.c. (*Plate 16a,f,g*; *Plate 17,a*), but is already in the process of breaking down, and by 9.5 days p.c. all traces of it have disappeared. Nine days p.c. also coincides with the first indication of Rathke's pouch (*Plate 18a,g*), though differentiation of its wall is first seen at about 9.5 days p.c. (*Plate 19a,k*), and its peripheral boundary is quite wide at this stage (*Plate 19a,l*). The latter is more clearly seen in scanning electron micrographs of this region of embryos at about 9.5–10 days p.c. (*Plate 47,f*), and on sagittal sections through this part of the cephalic region (*Plate 54,a–d*). No indication of differentiation is seen in the floor of the diencephalon at this stage, though the anterior wall of Rathke's pouch is seen to be in direct apposition with the floor of the diencephalon, and it is possible that this region may be "induced" to form the neurohypophysis by the presence of Rathke's pouch. As in other sites of induction, no intervening mesenchyme is present between these two tissues at this stage (*Plate 54,c*).

By 10 days p.c., Rathke's pouch is a much more substantial diverticulum (*Plate 54,e–h*), and its entrance from the oral (buccal) cavity is still quite wide (*Plate 54,f,g*). This is the first occasion that an infundibular recess is present (*Plate 54,f*), and its walls are seen to be in continuity with (and histologically identical to) the neural ectoderm in the region of the floor of the third ventricle. As indicated earlier, no cephalic mesenchyme appears to intervene between the outer part of the wall of Rathke's pouch and the diencephalic diverticulum (the future neurohypophysis) (*Plate 54,f*). By 10.5 days p.c., the width of the entrance to Rathke's pouch has diminished slightly compared with the situation observed previously (*Plate 24b,g*; *Plate 47,h*).

By 11–11.5 days p.c., the wall of Rathke's pouch is seen to be well differentiated (*Plate 54,i–l*), and the width of its entrance from the oral cavity is substantially smaller than previously (*Plate 54,k*). The entrance to the infundibular recess from the third ventricle is now quite large (*Plate 54,l*), and its wall is also well differentiated. It is of interest to note that the boundary between the surface ectoderm that lines the oral cavity and the wall of Rathke's pouch is now well demarcated (*Plate 54,k*), but there appears to be no obvious morphological difference between the ectoderm-derived tissue of the oral cavity and the endoderm-derived pharyngeal tissue (*Plate 54,l*). Transverse sections through this region clearly illustrate that the diencephalic diverticulum has a relatively narrow lumen (*Plate 56,a–c*), while the rostral part of Rathke's pouch is quite wide (*Plate 56,d,e*), and embraces the inferior surface of the pituitary stalk before expanding upwards on either side of it (*Plate 56,c,d*). The lateral part of Rathke's pouch is also seen to be in close association with the internal carotid arteries (*Plate 56,e*) which are rostral extensions of the paired dorsal aortae.

For the more advanced embryos in this series, representative median sections are provided, illustrated at daily intervals, covering the period between 12.5 and 17.5 days p.c. These are complemented by sets of transverse sections covering most of these days of gestation.

By 12–12.5 days p.c., the connection between Rathke's pouch and the roof of the oral cavity is lost, though for a brief period a remnant of the connecting stalk is seen in the median plane (*Plate 55,a*). The anterior wall of Rathke's pouch shows considerable evidence of proliferative cellular activity. Deep indentations from the lumen into the anterior wall are also seen at this stage, but over the next 2 days these become filled in as a result of the cellular activity that occurs in association with the anterior wall of Rathke's pouch. The region where all of this cellular activity is occurring is destined to form the pars anterior of the pituitary (*Plate 55,b,c*). The diencephalic diverticulum still has a wide lumen at this stage, and its walls also show increased evidence of differentiation. The increased cellular activity in this location results in the obliteration of its lumen, and this is first seen in some embryos at about 15.5 days p.c., but has occurred in all embryos by about 16–16.5 days p.c. (*Plate 55,f*). Transverse sections through this region (*Plate 56,g–l*), clearly illustrate that the entrance to the infundibular recess has a slightly narrower neck than previously (*Plate 56,g*; cf. *Plate 56,b*). The two rostral "wings" of Rathke's pouch origin are now seen to be closely associated with the sides of the diencephalic diverticulum, and indeed appear to cause the transverse diameter of its stalk region to be narrowed (*Plate 56,g,h*). The rostral part of Rathke's pouch is at its widest just inferior to the stalk (*Plate 56,i*), and below this level the anterior wall shows the first evidence of increased cellular proliferative activity (*Plate 56,j,k*). The internal carotid arteries are also seen to be in close association with the caudal part of Rathke's pouch, and these arteries will soon assist in the provision of a

(*continued on page 418*)

PLATE 54

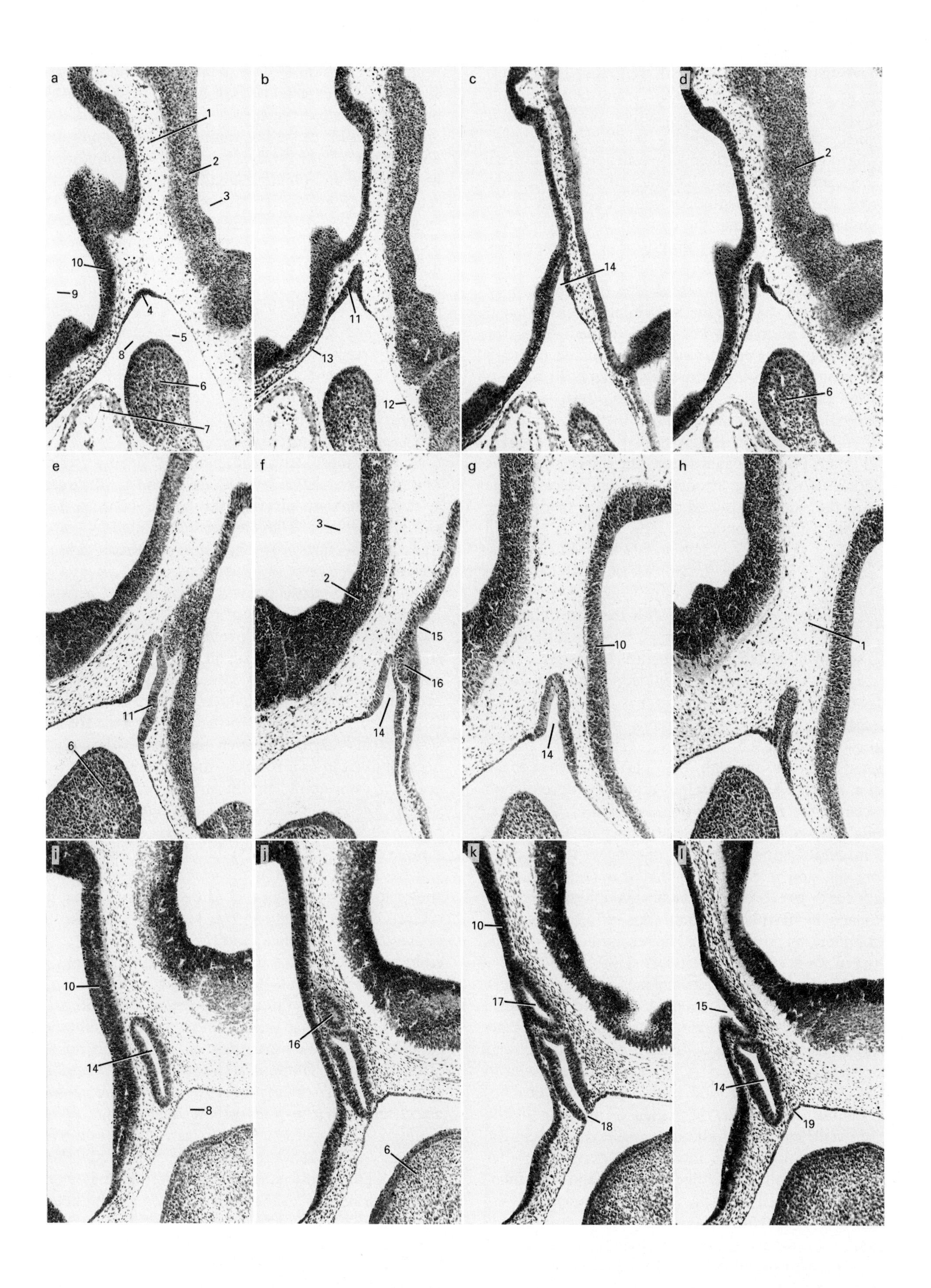
a
1
2
3
10
9
4
5
8
6
7
b
11
13
12
c
14
d
2
6
e
11
6
f
3
2
15
16
14
g
10
14
h
1
i
10
14
8
j
16
6
k
10
17
18
l
15
14
19

Plate 54

Differentiation of the pituitary 1 (sagittal sections)

a–d 10–10.25 days p.c.; embryo with 25–30 pairs of somites (ref. Plate 21)
e–h 10.25–10.5 days p.c.; embryo with 30–35 pairs of somites
i–l 11–11.5 days p.c.; embryo with 40–45 pairs of somites (ref. Plate 25)
(**a–d**, **e–h**, **i–l** ×100)

1. cephalic mesenchyme tissue
2. neuroepithelial cells forming the wall of the hindbrain
3. fourth ventricle
4. peripheral boundary of entrance to Rathke's pouch
5. rostral part of pharyngeal region of foregut
6. mandibular component of first branchial arch
7. truncus arteriosus region of outflow tract of primitive heart
8. oropharynx (buccal cavity, oral cavity)
9. third ventricle
10. neuroepithelial cells forming the wall of the diencephalon
11. ectodermal cells lining Rathke's pouch
12. endodermal lining of pharyngeal region of foregut
13. ectodermally lined roof of oral cavity (formerly, region immediately in front of the buccopharyngeal membrane)
14. lumen of Rathke's pouch
15. infundibular recess
16. lateral wall of neural component of pituitary
17. lumen of neural component of pituitary
18. narrowed entrance to Rathke's pouch
19. approximate site of transitional zone between surface ectodermal tissue and pharyngeal-derived endodermal tissue (i.e. formerly the location of the dorsal part of the buccopharyngeal membrane)

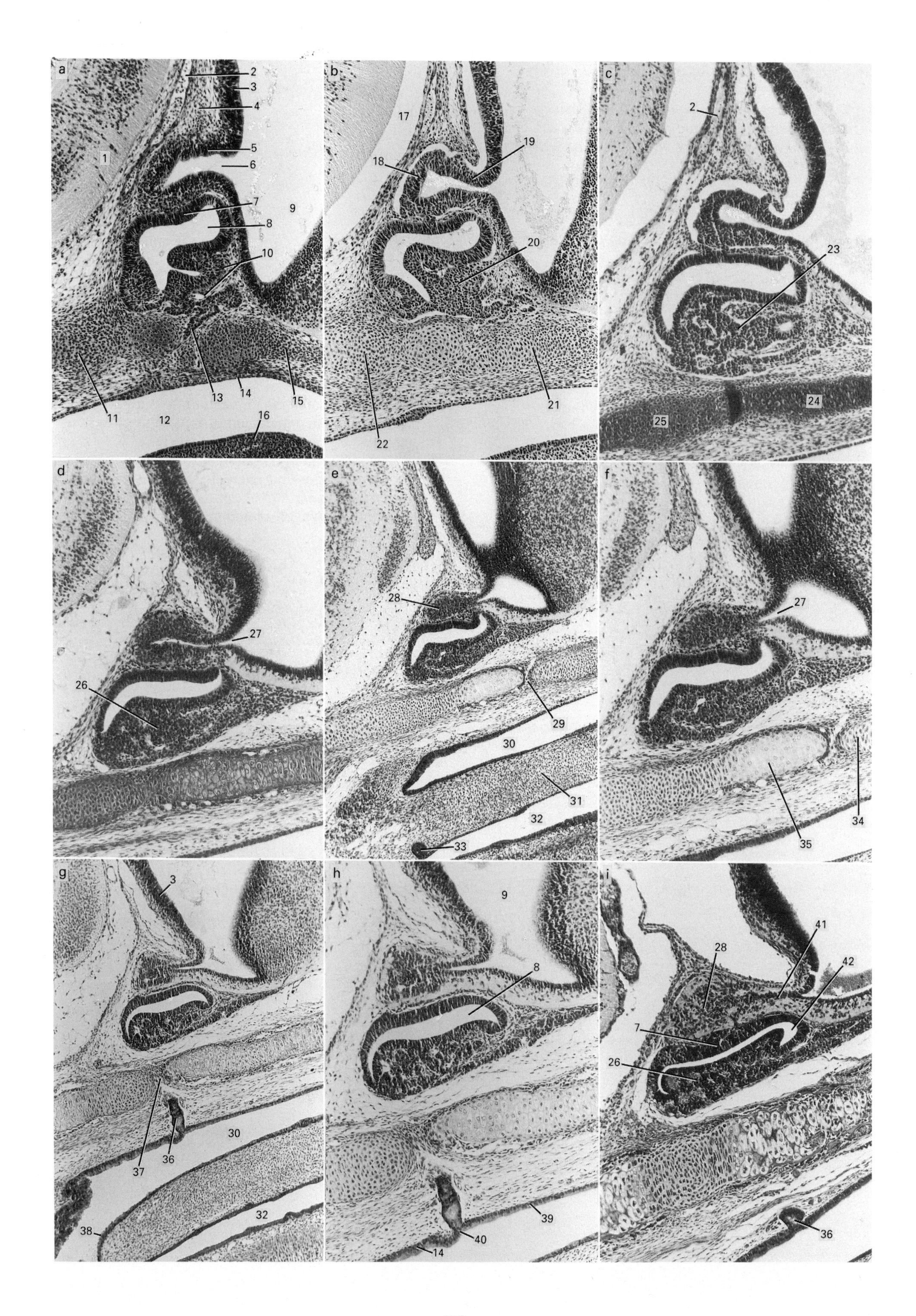
a
2
3
4
5
6
1
7
9
8
10
11
12
13
14
15
16
b
17
18
19
20
21
22
c
2
23
24
25
d
27
26
e
28
29
30
31
32
33
f
27
34
35
g
3
30
37
36
32
38
h
9
8
39
40
14
i
41
28
42
7
26
36

Plate 55

Differentiation of the pituitary 2 (sagittal sections)

a 12.5 days p.c. (×100)
b 13.5 days p.c. (×100)
c 14.5 days p.c. (×100)
d 15.5 days p.c. (early) (×100)
e,f 15.5 days p.c. (more advanced stage of differentiation) (**e** ×63; **f** ×100)
g,h 16.5 days p.c. (**g** ×63; **h** ×100)
i 17.5 days p.c. (×100)

1. medulla oblongata–pons junction
2. basilar artery
3. wall of diencephalon (hypothalamus)
4. cephalic mesenchyme tissue
5. wall of infundibular recess of third ventricle
6. entrance to infundibular recess
7. pars intermedia
8. ectodermally lined lumen of Rathke's pouch
9. third ventricle
10. early evidence of vascular differentiation which develops in close association with anterior wall of Rathke's pouch
11. mesenchymal condensation (which forms prior to cartilage model) in ventral part of future basioccipital bone (clivus)
12. oropharynx
13. remnant of connecting stalk between roof of oropharynx and Rathke's pouch
14. endodermal lining of roof of pharyngeal region of oropharynx
15. mesenchymal condensation in region of future basisphenoid bone
16. dorsal surface of tongue
17. pontine cistern
18. wall of infundibular recess in region of future pars nervosa
19. wall of infundibular recess in region of future stalk of pituitary
20. region of future pars anterior
21. precartilage condensation of basisphenoid bone
22. precartilage condensation of basioccipital bone (clivus)
23. active cellular proliferation in anterior wall of Rathke's pouch which gradually encroaches on the lumen of the pouch until it is reduced to a narrow cleft
24. cartilage primordium of basisphenoid bone
25. cartilage primordium of basioccipital bone (clivus)
26. further stage in differentiation of pars anterior, with cells arranged in intermingling columns to form a meshwork around the increasingly rich blood supply (from branches of internal carotid artery, but principally from the plexus on the surface of floor of diencephalon)
27. entrance to infundibular recess reduced to a narrow slit
28. pars nervosa – lumen now obliterated
29. location of future "sphenoidal" canal (site of transient synchondrosis between presphenoid and postsphenoid bones – normally obscured by articulation beween the vomer and the sphenoid – and very variable in size)
30. posterior part of nasopharynx
31. primordium of palatal shelf of maxilla
32. oropharynx
33. developing glandular duct opening into pharynx
34. cartilage primordium of presphenoid bone
35. cartilage primordium of postsphenoid bone
36. "pharyngeal" pituitary (neck and connecting stalk of Rathke's pouch that has failed to separate from roof of the oropharynx – when present, volume of tissue involved quite variable)
37. possible location of craniopharyngeal canal
38. posterior border of palatal shelf of maxilla
39. ectodermally lined roof of oral cavity
40. approximate location of junctional zone between oral (ectodermally-derived) and pharyngeal (endodermally-derived) epithelium
41. stalk of pituitary
42. narrow cleft representing remnant of lumen of Rathke's pouch

PLATE 56

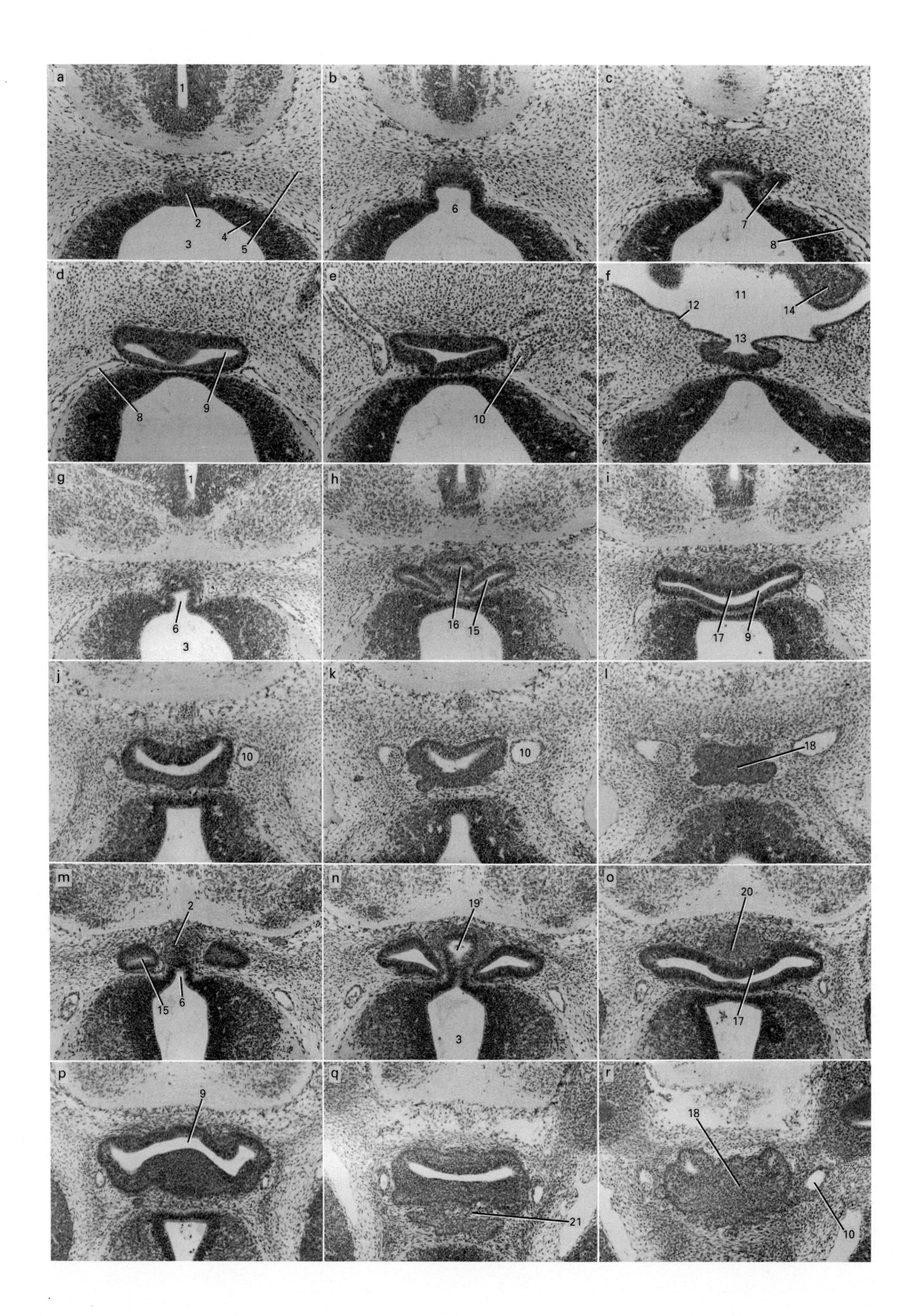
a
1
2
3
4
5
b
6
c
7
8
d
8
9
e
10
f
11
12
13
14
g
1
6
3
h
16
15
i
17
9
j
10
k
10
l
18
m
2
15
6
n
19
3
o
20
17
p
9
q
21
r
18
10

Plate 56

Differentiation of the pituitary 3 (transverse sections)

a–f 11.5 days p.c. (×100)
g–l 12.5 days p.c. (×100)
m–r 13.5 days p.c. (×100)

1. lumen of fourth ventricle
2. dorsal wall of infundibular recess
3. third ventricle
4. wall of diencephalon (hypothalamus)
5. cephalic mesenchyme tissue
6. infundibular recess
7. dorsal wall of rostral extremity of lateral part of Rathke's pouch
8. middle cerebral artery
9. ectodermally lined lumen of Rathke's pouch
10. internal carotid artery
11. oropharyngeal region (buccal cavity)
12. ectodermal lining of roof of the buccal cavity
13. narrowed entrance to Rathke's pouch
14. mandibular component of first branchial arch in region of tongue primordium
15. lumen of rostrolateral extremity of Rathke's pouch (this partly surrounds the pituitary stalk, and will form the pars tuberalis of the pituitary)
16. roof of neural component of pituitary
17. pars intermedia
18. ventrally located floor of Rathke's pouch now completely separated from roof of the buccal cavity
19. lumen of neural component of pituitary
20. ventral wall of neural component of pituitary
21. first evidence of vascular differentiation which develops in close association with anterior wall of Rathke's pouch

PLATE 57

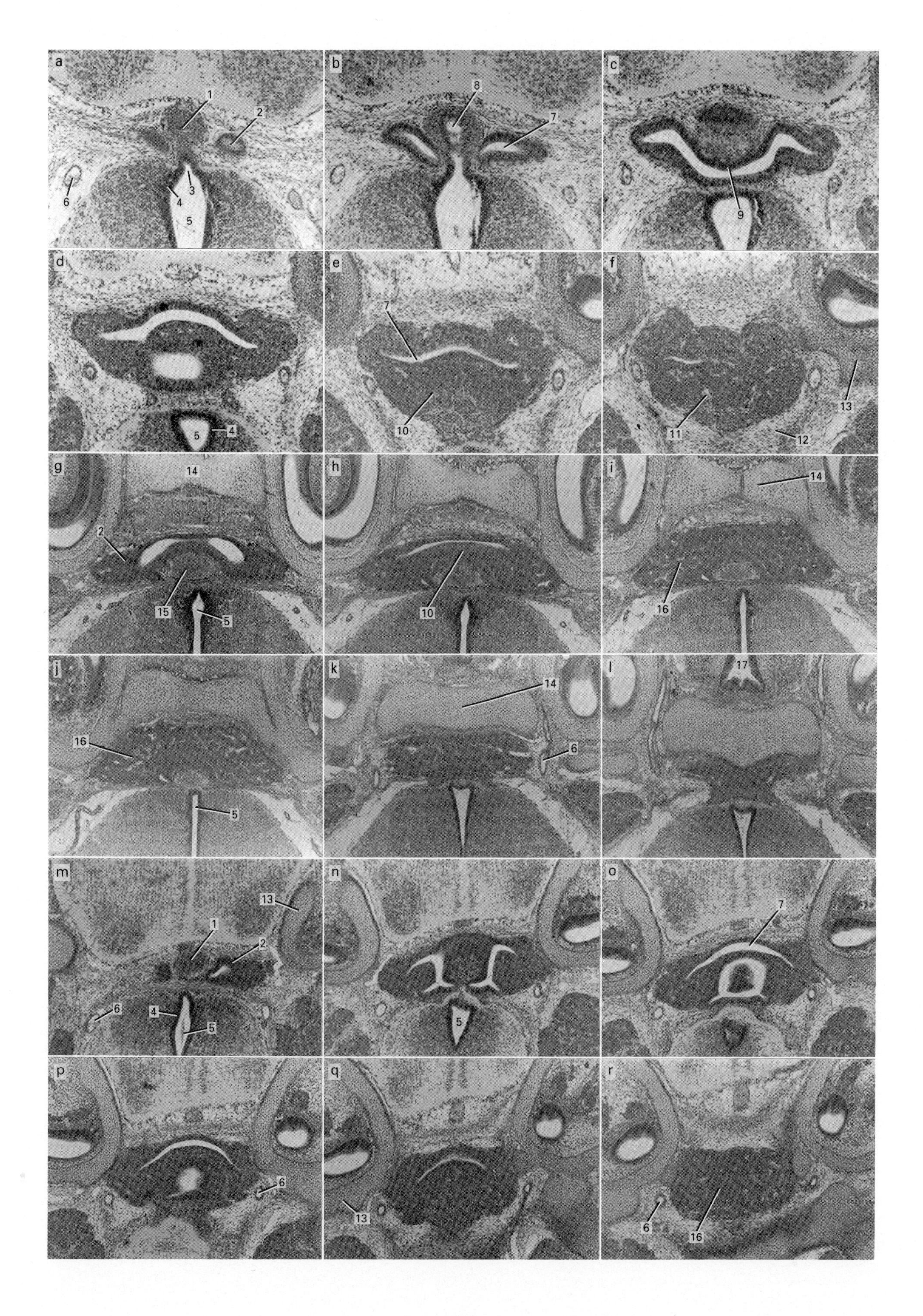

a
1
2
3
4
5
6
b
8
7
c
9
d
4
5
e
7
10
f
11
12
13
g
14
2
15
5
h
10
i
14
16
j
16
5
k
14
6
l
17
m
13
1
2
6
4
5
n
5
o
7
p
6
q
13
r
6
16

Plate 57

Differentiation of the pituitary 4 (transverse sections)

a–f 14.5 days p.c. (×100)
g–l 15.5 days p.c. (×63)
m–r 17.5 days p.c. (×63)

1. dorsal wall of neural component of pituitary
2. dorsal wall of rostral extremity of lateral part of anterior lobe of the pituitary (destined to form pars tuberalis of pituitary)
3. infundibular recess in floor of diencephalon
4. wall of diencephalon (hypothalamus)
5. third ventricle
6. internal carotid artery
7. residual lumen of Rathke's pouch
8. lumen of future stalk and pars nervosa of the pituitary
9. pars intermedia
10. site of active cellular proliferation in anterior wall of Rathke's pouch which reduces its lumen to a narrow cleft, with formation of pars anterior, and more rostrally it extends around the pituitary stalk to form the pars tuberalis
11. early stage of vascularization of the pars anterior – the blood supply is partly from the internal carotid artery and from the plexus on the surface of the floor of the diencephalon
12. cephalic mesenchyme tissue
13. cartilage primordium of petrous part of temporal bone
14. cartilage primordium of basisphenoid bone
15. pars nervosa
16. intermingling columns of cells forming a meshwork surrounding the vascular plexus which develops within the pars anterior
17. fourth ventricle

vascular plexus that proliferates within the developing pars anterior (*Plate 55,b*; *Plate 56,q,r*). This increased vascularity is first seen at about 13.5 days p.c., and is principally derived from the vascular plexus on the surface of the floor of the diencephalon.

The most obvious feature at 13.5 days p.c. is the degree of vascularity seen in the anterior wall of Rathke's pouch, and the fact that the previous connection with the roof of the oral cavity has now disappeared (*Plate 55,b*). Transverse sections through this region reveal that the two rostral "wings" of Rathke's pouch have now substantially increased in size, and their superior walls extend rostrally to the level of the dorsal wall of the pituitary stalk (*Plate 56,m,n*). These two rostral extensions of Rathke's pouch are destined to form the pars tuberalis of the pituitary, and will eventually almost completely surround the pituitary (infundibular) stalk. A limited degree of proliferation is evident for the first time in relation to the dorsal wall of Rathke's pouch, where it is closely apposed to the ventral surface of the infundibular stalk (*Plate 56,o*), and this region is destined to form the pars intermedia of the pituitary. The lumen in the region of the most caudal part of Rathke's pouch gradually becomes obliterated, principally as a result of the proliferative activity within the pars anterior (*Plate 56,q,r*).

By 14.5 days p.c., the anterior wall of Rathke's pouch is almost unrecognizable compared with the situation observed previously (*Plate 55,c*), because of the cellular proliferation and increasing vascularity of this part of the developing pituitary. The net effect of the latter is to progressively obliterate the lumen of Rathke's pouch. An impressive change is also observed at 15.5 days p.c. in the histological appearance of the pars anterior, where the cells are seen to be arranged in columns, and form a meshwork around the increasingly rich blood supply to this region. In slightly less advanced embryos analysed at this stage, the lumen of the diencephalic diverticulum is seen to be particularly narrow, principally due to the increased degree of differentiation of its walls (*Plate 55,d*). In developmentally slightly more advanced embryos, the lumen is seen to be completely obliterated, and only the relatively narrow entrance from the third ventricle remains to indicate its origin as a diverticulum from the floor of the diencephalon (*Plate 55,e,f*). The distal part of the diverticulum is destined to form the pars nervosa of the pituitary.

It is also of interest to note that by about 13.5 days p.c., the precartilage condensations in the region of the future basisphenoid bone are clearly seen, and just dorsal to this region, the precartilage primordium of the basioccipital bone is also evident (*Plate 55,b*). By 14.5 days p.c., the precartilage tissue is now seen to be replaced by the primitive cartilage primordia of these two bones, and by 15.5 days p.c., this tissue is itself replaced by well differentiated cartilage (*Plate 55,d*). These two bones represent two of the most important components of the chondrocranium (the cartilage precursor of the skull base). Little additional information is gained from an analysis of the transverse sections of embryos at this stage of development, except that they illustrate the very considerable degree of cellular and vascular activity present in the pars anterior (*Plate 57,h–j*), and the fact that the lumen of much of the lower part of Rathke's pouch is now obliterated, this being particularly clearly seen in its most caudal part, close to the pars anterior (*Plate 57,j,k*). It is evident that the degree of differentiation observed in the neural ectoderm in the wall of the third ventricle also extends to include the wall of the pars nervosa and pituitary stalk (*Plate 55,d*), though this is more clearly seen at later stages of development (*Plate 55,f,h,i*).

By 16.5 days p.c., little change is observed in the features of the pituitary gland, compared to the situation reported previously, except for the narrowing of the lumen of Rathke's pouch (*Plate 55,g,h*), and this tends to be even narrower at 17.5 days p.c. (*Plate 55,i*). The transverse sections at 17.5 days p.c. reveal that most of the lumen of Rathke's pouch is now either obliterated or substantially narrowed, due to the increased cellular activity that occurs in the walls of Rathke's pouch. This is particularly clearly seen in the region of the pars tuberalis and in the pars anterior (*Plate 57,m–r*). The degree of differentiation observed in relation to the pars intermedia is now seen to be considerably more marked than previously, and this region is now in close apposition to the inferior surface of the pars nervosa (*Plate 55,i*). At the histological level, in the pars anterior, while a few of the cells have a granular cytoplasm, the majority are seen to be uniformly darkly stained. Within the pars nervosa, moderate numbers of pituicytes are located. The pituitary stalk, by contrast, contains relatively few cells, but large numbers of nerve fibres are found in this location. These originate in some of the hypothalamic nuclei and are directed towards the pars nervosa.

It is of interest to note that in some of the median sections, it is possible to recognize aberrant "rests" of Rathke's pouch tissue still associated with the roof of the oropharynx (*Plate 55,h,i*). These may vary quite considerably in size, and are not particularly commonly encountered. In one of the sections illustrated (*Plate 55,h*), there appears to be a considerable degree of cellular activity present, and this might well be associated with functional activity, and would constitute a small "pharyngeal" pituitary. In the most extreme case of this condition, Rathke's pouch fails to "ascend" beyond this site, and only the neurohypophysis is located in the pituitary fossa. The "rest" observed in the other section illustrated (*Plate 55,i*) may well not be associated with any significant degree of functional activity. In addition, it is almost certain that the discontinuity seen in the cartilage primordium of the basisphenoid bone in the region of the future sella turcica (pituitary or hypophyseal fossa) represents a persistent craniopharyngeal canal. This is invariably present at 12.5 days p.c., in association with the remnant of the connecting stalk located between the roof of the oropharynx and Rathke's pouch, but this usually disappears with the growth of the basisphenoid primordia. A discontinuity is also seen in the cartilage primordium in one of the other median sections (*Plate 55,e*), but its location suggests that this is much more likely to be a "sphenoidal" canal. This is the site of a transient synchondrosis located between the presphenoid and postsphenoid bones, and if present in the full-term (human) skull, is normally obscured by the articulation between the vomer and the sphenoid (for a full description of the sphenoidal canal, see Sprinz and Kaufman, 1987).

7. Development of the tongue
Text associated with Plate 58

The tongue develops in the floor of the pharynx over the relatively restricted period between about 11 and 16 days p.c. There are numerous accounts in standard textbooks of human embryology which provide a useful but simplified explanation for the embryological basis of the complex nerve supply observed in the adult tongue, and analysis of scanning electron micrographs and histological sections through the developing tongue reveal that the sequential changes seen in the mouse fit the general story reasonably well. However, much additional work will need to be undertaken in this species before, for example, the specific innervation of its component parts is fully established. The terminology used in the literature for the various pharyngeal arch derivatives that amalgamate together to form the tongue is quite confusing, as identical structures have been given a variety of different names over the years, but more confusingly, different structures have also been given the same name in the extensive literature on this topic. The account presented here is more simplified than the standard textbook account, and is based on the scanning electron micrographs and the histological sections presented in these plates.

The scanning electron micrographs presented here are views of the floor of the pharynx analysed after the rest of the surrounding cephalic region has been removed to facilitate access to this region. The earliest evidence of the presence of any tongue primordium is seen at about 11–11.5 days p.c., with the appearance of a pair of discrete swellings associated with the dorsal surface of the mandibular component of the first pharyngeal arch. These are the lateral lingual swellings, and are initially separated from each other by a deep median groove (sulcus) which represents the site of fusion in the midline of the left and right mandibular processes. These swellings grow quite rapidly over this period, and their appearance precedes the appearance of the other tongue primordia (*Plate 58,a,b*). In the dorsal midline, a suggestion of a third swelling also associated with the first pharyngeal arch appears slightly later, and this is probably the anlage of the tuberculum impar (*Plate 58,b*).

While the mandibular component of the first arch, and the second pharyngeal arches are readily recognized, particularly in sagittal sections through embryos at this stage of development (*Plate 25a,a–d*; *Plate 25b,a–d*; *Plate 25c,a,b*), it is less easy to recognize the various tongue primordia shortly after their first appearance, particularly in transverse sections through this region. However, in appropriate sections, it is possible to recognize the thyroid primordium at about 10.5–11 days p.c. (*Plate 24c,g*), and this shows early evidence of canalization to form the proximal part of the thyroglossal duct (*Plate 22a,c*; *Plate 25b,a,b*) which at this stage is still connected to the floor of the pharynx. Analysis of the thyroid primordium is relevant at this stage of tongue development, since the thyroid gland forms from an endodermal downgrowth from the floor of the pharynx between the first and second pharyngeal arches (just caudal to the tuberculum impar), at the site of the future foramen caecum (*Plate 58,c*).

By about 11.5–12 days p.c., the two lateral lingual swellings have increased in size and their medial borders have fused across the midline, though the deep dorsal median sulcus still delineates their site of fusion in the region of the tip of the tongue primordium (*Plate 58,c,d*). At about this time, a pair of smaller swellings are seen in relation to the dorsal surface of the second pharyngeal arch, and these are separated by a deep V-shaped median notch. It is also now possible to recognize the tongue primordia on transverse sections through this region (*Plate 26b,d–h*; *Plate 61,a–f*), as well as the lingual vessels (*Plate 26b,e*) and the thyroid diverticulum (*Plate 26b,f*) which has recently separated from the floor of the pharynx. By 12–12.5 days p.c., the first arch derivatives continue to enlarge, and begin to amalgamate with tissues of second pharyngeal arch origin (*Plate 58,e,f*), and slightly later on, with tissues of third and fourth arch origin (*Plate 58,g*). From this stage, therefore, it becomes increasingly difficult to follow with my confidence the fate of individual arch components.

From about 11.5–12 days p.c., a median prominence is seen (*Plate 58,c*) which appears to be derived from both the third and fourth pharyngeal arches. This structure is more clearly seen at about 12.5–13 days p.c. (*Plate 58,g*), and is likely to be the hypobranchial eminence. By 13–13.5 days p.c., a dramatic increase occurs in the overall volume of the tongue, and it is now seen to protrude forwards over the dorsal surface of the anterior part of the fused mandibular processes (*Plate 58,h,i*). This is also the first stage when the intermolar eminence is seen as a median swelling located towards the caudal part of the oral surface of the tongue (*Plate 58,i*). The origin of the intermolar eminence is unclear, but it may be derived from the tuberculum impar, and consequently would be of first pharyngeal arch origin. By 13.5–14 days p.c., a smaller median swelling develops towards the most caudal part of the dorsum of the tongue, and this represents the primordium of the epiglottis (*Plate 58,j*). The latter is in close proximity to the entrance to the larynx and pharynx. On sagittal sections through this region, the relationship between the epiglottis and the entrance to the laryngeal aditus is clearly seen (*Plate 31a,a*; *Plate 36b,b*; *Plate 38c,a*).

On the surface of the tongue at this stage, it is possible to recognize numerous fungiform papillae, and these are principally dispersed over the dorsal and lateral part of the oral surface of the tongue. These become increasingly prominent, so that by 15–15.5 days p.c., their presence is clearly seen (*Plate 58,l,m*). The other important structure that is first seen at about 13–13.5 days p.c. is the single (median) circumvallate papilla (*Plate 30d,a,b*). The latter is supplied bilaterally by the glossopharyngeal (IX) nerves, and has a complex structure consisting of a central dome-like region, surrounded by a deep sulcus,

(*continued on page 424*)

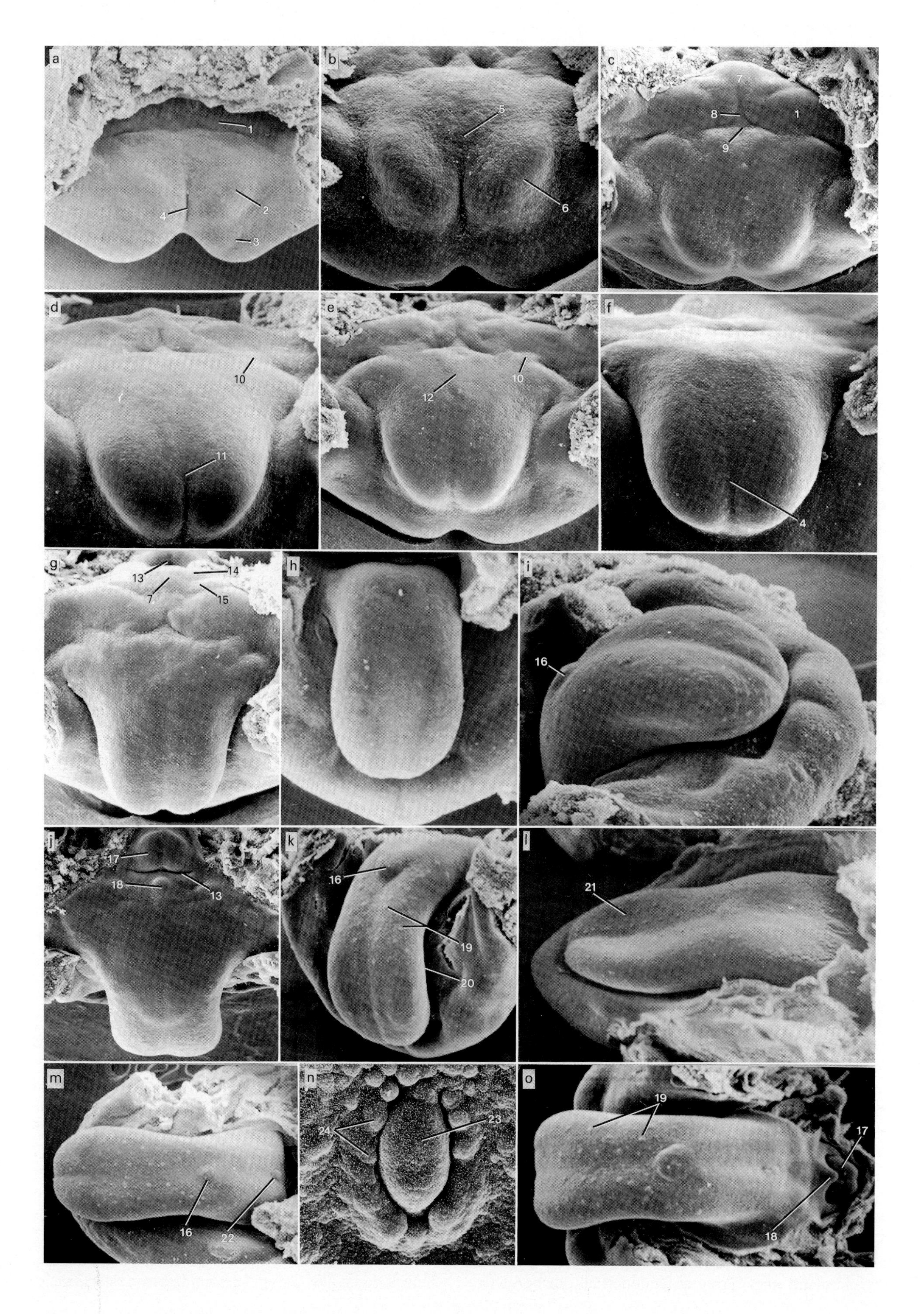

a
1
2
4
3
b
5
6
c
7
8
1
9
d
10
11
e
12
10
f
4
g
13
14
7
15
h
i
16
j
17
18
13
k
16
19
20
l
21
m
16
22
n
24
23
o
19
17
18

Plate 58

Development of the tongue

a,b 11.5 days p.c.
c–e 11.5–12 days p.c.
f 12.5 days p.c.
g 12.5–13 days p.c.
h,i 13.5 days p.c.
j,k 13.5–14 days p.c.
l 15 days p.c.
m 15.5 days p.c.
n,o 16 days p.c.

1. second pharyngeal arch
2. first indication of lateral lingual swelling
3. dorsal surface of mandibular process (first pharyngeal arch)
4. groove (median sulcus) representing site of fusion in the midline of the left and right mandibular processes
5. first indication of tuberculum impar
6. more prominent lateral lingual swelling
7. hypobranchial eminence (amalgamation of tissue derived from the third and fourth pharyngeal arches)
8. deep V-shaped notch which separates the medial ends of the left and right second pharyngeal arches
9. location of the foramen caecum (in the dorsal midline) at the junction between the first and second pharyngeal arches
10. site of amalgamation of tissue of first and second pharyngeal arch origin
11. posterior extent of dorsal median sulcus, overlying the region where the two lateral lingual swellings appear to have completely merged together
12. in this location, the tuberculum impar appears to be merging into the posterior part of the tissue derived from the lateral lingual swellings
13. entrance to pharynx
14. fourth pharyngeal arch
15. third pharyngeal arch
16. intermolar eminence (believed to develop from the tuberculum impar)
17. anlage of arytenoid swelling
18. anlage of epiglottis
19. fungiform papillae
20. lateral border of the tongue
21. anterior spatulate portion of the tongue
22. median circumvallate papilla
23. central dome-like region of circumvallate papilla
24. accessory papillae associated with the lateral wall of the furrow surrounding the circumvallate papilla

on the lateral walls of which are located a considerable number of accessory papillae (*Plate 32d,b*; *Plate 33a,a*; *Plate 58,m,n*) (for a detailed description of the innervation and development of the circumvallate papilla in the mouse, see AhPin *et al.*, 1989). The overall shape of the cross-sectional profile of the tongue changes between about 13.5 and 15 days p.c., so that whereas initially it has a rounded profile (*Plate 58,h,i*), by about 14 days p.c. its surface is more flattened (*Plate 58,k*), and by about 15 days p.c., the profile is more concave, and its anterior part is now spatulate (*Plate 58,l*).

At the histological level, all of the components that form the intrinsic musculature of the tongue are recognizable, as are the individual pairs of extrinsic muscles (*Plate 35,d,b–d*; *Plate 35e,a,b*). The critical relationship between the hyoid bone and mandible as sites of attachment of some of the extrinsic muscles is also clearly revealed in these sections.

From the above brief description of tongue development in the mouse, it appears likely that, as in the human embryo, the majority of the tongue is derived from the first pharyngeal arch. This constitutes its oral part, and in the human occupies the anterior two-thirds of the adult tongue. The latter is more correctly termed the pre-sulcal part (Warwick and Williams, 1973), as it lies anterior to the sulcus terminalis. Unfortunately, this dividing line is not recognizable in the mouse. General sensation for this region is via the lingual nerve, which is a branch of the mandibular division of the trigeminal (V) nerve (the cranial nerve that supplies the first pharyngeal arch derivatives). The second pharyngeal arch derivative (sometimes termed the copula) is largely overgrown by the derivatives of the first and third arches, but does contribute towards the anterior part of the tongue as evidenced from the nerve supply for taste (special sensation) for this region, which travels via the chorda tympani branch of the facial (VII) nerve (the cranial nerve that supplies the second pharyngeal arch derivatives), being distributed in the lingual nerve. The posterior one-third (or pharyngeal, or post-sulcal part) of the tongue is largely of third arch origin, and the glossopharyngeal (IX) nerve (the cranial nerve that supplies the third pharyngeal arch derivatives) supplies general and special sensation to this region, as well as to the circumvallate papilla. The fourth arch contributes to the most extreme caudal part of the tongue, and the region around the epiglottis. This general region is supplied by the internal laryngeal part of the superior laryngeal branch of the vagus (X) nerve (the cranial nerve that supplies the fourth, and sixth, pharyngeal arch derivatives).

Other observations on the relationship between the tongue and the developing palate may be found in the text associated with *Plates 59–62*, as well as in the appropriate sections of the various stages of development studied.

8. Closure of the palate
Text associated with Plates 59–62

The closure of the palate is of critical importance because it separates the oropharynx (the definitive mouth) and its various (principally alimentary) functions from the nasopharynx (which is exclusively involved in respiration and olfaction). The closure process is a quite complex one, and involves the elevation, apposition and eventual fusion of the palatal shelves of the maxillae (the future secondary palate) across the midline to form the roof of the oropharynx. The superior aspect of the palatal shelves in the midline becomes apposed to and eventually fuses with the inferior border of the nasal septum, and at about the same time the anteromedial borders of the palatal shelves fuse with the primary (or primitive) palate. The origin of the latter is unclear at the present time, and it is believed to be either of frontal, fronto-nasal or maxillary origin (the so-called inter-maxillary segment).

Much information is now available from both descriptive and experimental studies, in rodents and in other species, which sheds light on the underlying mechanisms involved, and this will be briefly alluded to where appropriate (for review of mechanisms of palatal shelf elevation, see Ferguson, 1981, 1988). The account that follows is principally a descriptive one based on the analysis of the scanning electron micrographs presented in *Plates 59* and *60*, supplemented by analysis of the histological sections illustrated in *Plates 61* and *62* and, where necessary, reference to other relevant material in the Atlas. It should also be noted that the development of the palate cannot be viewed in isolation without considering contemporaneous events associated with the development of the tongue. The role played by reflex opening and closing movements of the lower jaw are also believed to play a critical part in facilitating the initial elevation of the palatal shelves, and maintaining them in this position once elevation has occurred. For a useful staging system for palate formation in the mouse, see Biddle (1980).

As early as about 10.5–11 days p.c. it is possible to recognize the first indication of the folds which overly the future palatal shelves (*Plate 59,a*). These are initially directed posteromedially, but by about 11.5–12 days p.c., the medial borders of the two folds tend to lie almost parallel to each other (*Plate 59,b*). While the narrowed entrance to Rathke's pouch is just visible during the early part of this period, the surface ectoderm is reconstituted immediately after the stalk of Rathke's pouch has lost contact with the roof of the oral cavity. The latter event normally occurs at about 11–11.5 days p.c. (see text associated with (*Plates 54–57*). It is equally relevant to note the relationship that exists between the olfactory pit and the oral cavity, as initially the posterior border of the pit is in continuity with the anterior border of the maxillary process as well as with the oral cavity (*Plate 47,g,h*).

The boundaries of the entrance to the olfactory pit, provided by the medial and lateral nasal processes, get progressively smaller, and eventually become converted into the primitive anterior naris. This communicates with the roof of the pharynx via the primitive posterior naris, which in due course will form the definitive (secondary) choana, after the formation of the palate is completed (*Plate 59,b*). The anterior nares are seen to progressively "migrate" rostrally during the period between 10.5 and 13 days p.c. (*Plate 59,d*), and the two components of the upper lip form in the intervening region on either side of the midline, being separated by the primitive philtrum.

By 12.5–13 days p.c., the previous arrangement, described above, is seen to be more consolidated (*Plate 59,c–g*). During the early part of this period, the medial borders of the palatal shelves are seen to be sharper and more obvious than previously. With the differentiation of the primitive posterior naris, this facilitates the formation of the anterolateral part of the nasal septum (*Plate 59,c,f,g*). The latter is a particularly broad structure at this stage, and is in continuity anteriorly with the primordium of the primary palate (*Plate 59,c*).

While the dorsal surface of the anterior half of the tongue has a rounded profile, and is in direct contact with the inferior surface of the nasal septum, the palatal shelves in this location are only poorly differentiated. The lateral borders of the posterior half of the tongue, by contrast, are in close apposition with the medial borders of the posterior halves of the palatal shelves, the most posterior parts of which are now seen to be vertically directed. An oblique groove is present about one-third of the way along the palatal shelves, behind which the palatal shelves are seen to be vertically directed (*Plate 59,c–g*). On the inferior surface of the lateral part of the palatal shelves, rugae are first seen (*Plate 59,f*), and these become progressively more prominent with increased gestational age of embryos. At this stage, about four rugae may be distinguished. Histological sections through this region confirm the close apposition that exists between the dorsal and lateral surfaces of the posterior part of the tongue and the roof of the oropharynx, and the medial part of the palatal shelves, which are clearly seen to be vertically directed in this location (*Plate 61,a,b*). More anteriorly located sections illustrate both the change in the profile of the anterior part of the tongue, as well as the fact that the tongue in this region is less well applied to the palatal shelves, and that its dorsal surface is (in this location) an inferior relation of the primitive (and very broad) nasal septum (*Plate 61,d–f*).

By about 13.5–14 days p.c. (*Plate 59,h–k*), the medial borders of the palatal shelves are seen to be thinner than previously. While the anterior one-third or so of each palatal shelf tends to be directed medially and horizontally, the posterior one-third is clearly vertically directed, and the middle one-third is in an intermediate position. The nasal septum still appears to be closely related to the anterior part of the tongue, but is seen to be considerably narrower than previously (*Plate 59,k*). This analysis of the relationship

(*continued on page 434*)

PLATE 59

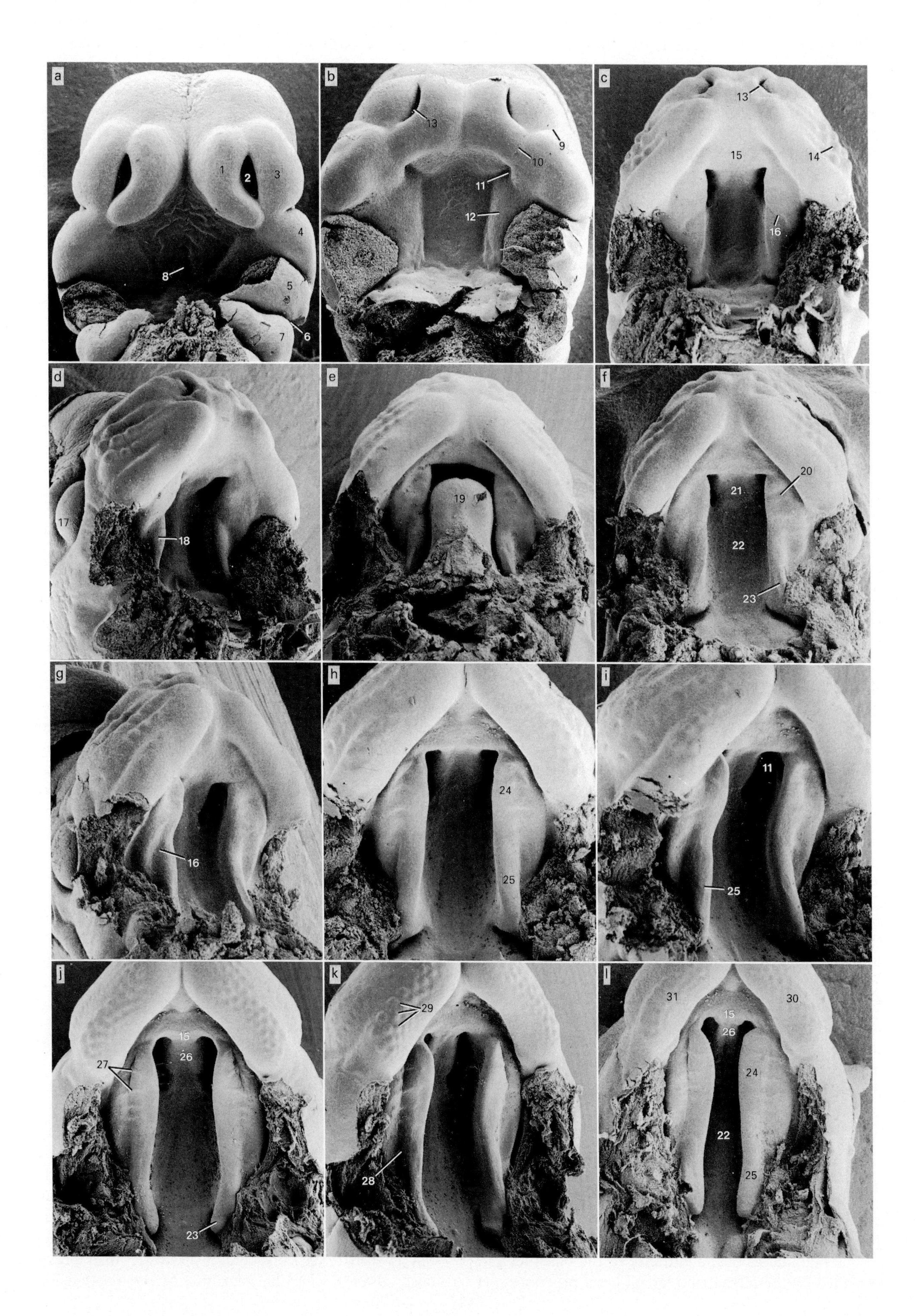

a
1
2
3
4
8
5
6
7
b
13
9
10
11
12
c
13
15
14
16
d
17
18
e
19
f
20
21
22
23
g
16
h
24
25
i
11
25
j
15
26
27
23
k
29
28
l
31
15
30
26
24
22
25

Plate 59

Closure of the palate, I

a 10.5–11 days p.c.
b 11.5–12 days p.c.
c–g 12.5–13 days p.c. (**c** and **d**, same embryo; **f** and **g**, same embryo)
h–k 13.5–14 days p.c. (**h** and **i**, same embryo; **j** and **k**, same embryo)
l 14–14.5 days p.c. (**l** and **a** (**Closure of the palate**, II), same embryo)

1. medial nasal process
2. olfactory pit (subsequently the entrance to the olfactory pit becomes the primitive anterior naris)
3. lateral nasal process
4. maxillary process (of first branchial/pharyngeal arch)
5. mandibular process (of first branchial/pharyngeal arch)
6. first branchial groove/cleft
7. second branchial arch
8. narrowed entrance to Rathke's pouch
9. groove overlying site of fusion of lateral nasal and maxillary processes
10. groove overlying site of fusion of medial nasal and maxillary processes
11. pit (primary choana, primitive posterior naris) which will subsequently form the definitive (secondary) choana after the formation of the palate
12. first evidence of palatal shelf of the maxilla
13. primitive anterior naris
14. bulge overlying primordium of vibrissa
15. location of primary palate
16. obliquely directed groove characteristically seen in this location
17. corneal epithelium
18. vertically directed medial border of palatal shelf (of secondary palate)
19. ventral surface of tongue (showing its relationship to the palatal shelves)
20. first evidence of rugae formation
21. broad ventral surface of nasal septum
22. roof of the primitive nasopharynx
23. posterior extremity of palatal shelf
24. horizontally directed anterior half of the palatal shelf
25. vertically directed posterior half of the palatal shelf
26. short, narrowing, nasal septum
27. prominent rugae
28. deep groove at the boundary between the body of the maxilla and its palatal shelf
29. bulges overlying prominent primordia of vibrissae (whiskers, tactile or sinus hairs) which are now clearly seen to be arranged in rows on each side of the upper lip
30. left side of upper lip
31. right side of upper lip

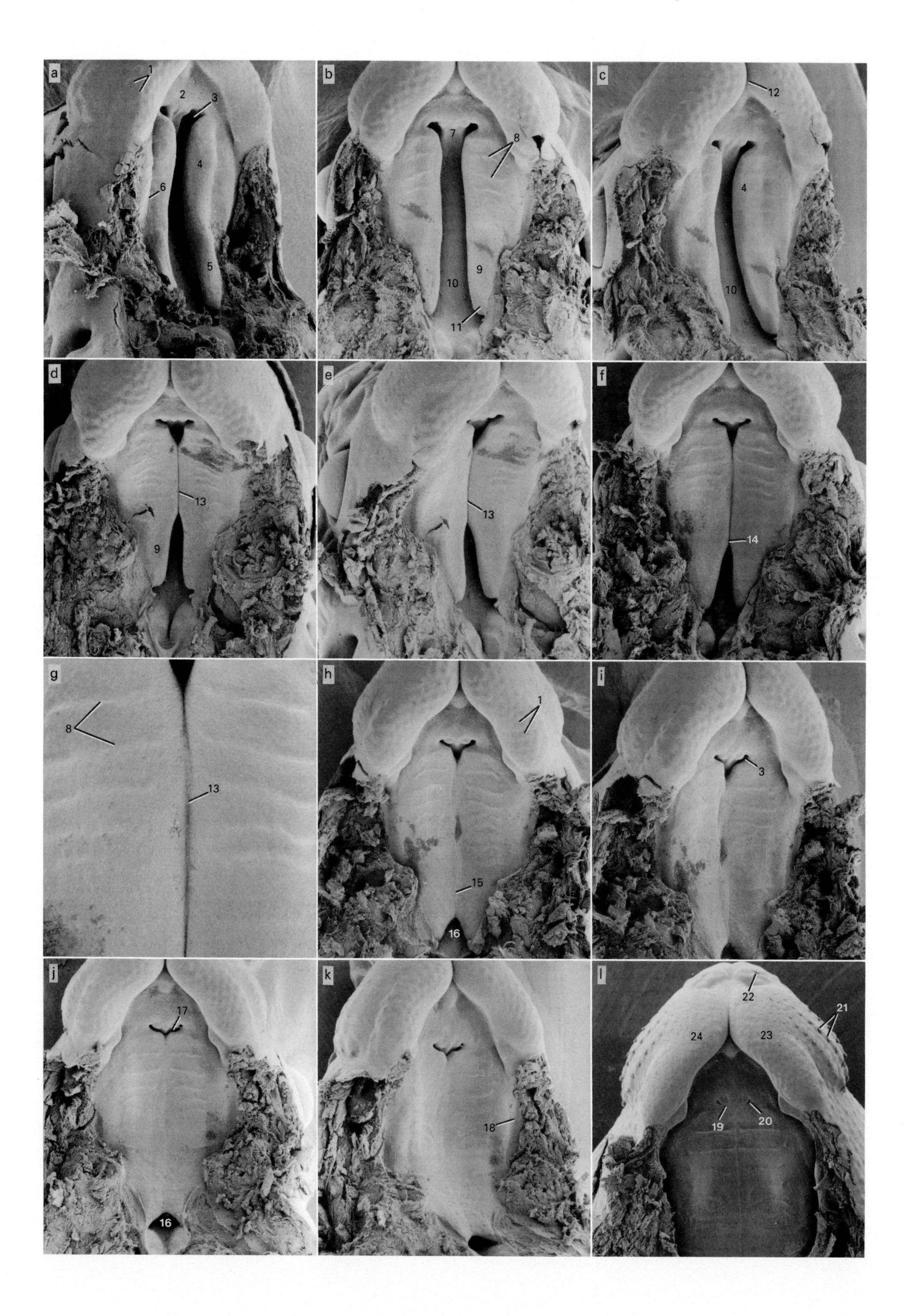
a
1
2
3
4
6
5
b
7
8
9
10
11
c
12
4
10
d
13
9
e
13
f
14
g
8
13
h
1
15
16
i
3
j
17
16
k
18
l
22
21
24
23
19
20

Plate 60

Closure of the palate, II

a–c 14–14.5 days p.c. (**a** and **l** (**Closure of palate**, I), same embryo; **b** and **c**, same embryo)
d–g 15 days p.c. (**d** and **e**, same embryo; **f** and **g**, same embryo)
h–k 15.5 days p.c. (**h** and **i**, same embryo; **j** and **k**, same embryo)
l 16 days p.c.

1. bulges overlying prominent primordia of vibrissae (whiskers, tactile or sinus hairs) arranged in rows on each side of the upper lip
2. location of primary palate
3. entrance to primary choana (primitive posterior naris) which will subsequently form the definitive (secondary) choana after the formation of the palate
4. horizontally directed anterior half of the palatal shelf
5. vertically directed posterior half of the palatal shelf
6. deep groove at the boundary between the body of the maxilla and its palatal shelf
7. anterior extremity of narrowing nasal septum
8. prominent rugae
9. posterior half of the palatal shelf, now horizontally directed
10. roof of the primitive nasopharynx
11. posterior extremity of palatal shelf
12. median groove (philtrum) between the two (maxillary) components of the upper lip
13. initial site of apposition and subsequent fusion of the palatal shelves
14. posterior (dorsal) extension of apposition and subsequent fusion of the palatal shelves
15. completion of palate formation, with the fusion of the posterior halves of the two palatal shelves across the midline (in the region of the soft palate)
16. posterior entrance to the nasopharynx
17. deep groove (naso-palatine canal) which separates the primary palate (intermaxillary segment) anteriorly, from the secondary palate more posteriorly
18. prominent longitudinal bulge at lateral extremity of palate, representing thc inner gum (alveolar process) subjacent to which are located the dental laminae (tooth primordia) of the upper molar teeth
19. final site of fusion of the primary and secondary palates in region of the future incisive papilla
20. slit-like orifice of incisive canal
21. erupted tip of vibrissa
22. entrance to nasal cavity (external naris)
23. left side of upper lip
24. right side of upper lip

PLATE 61

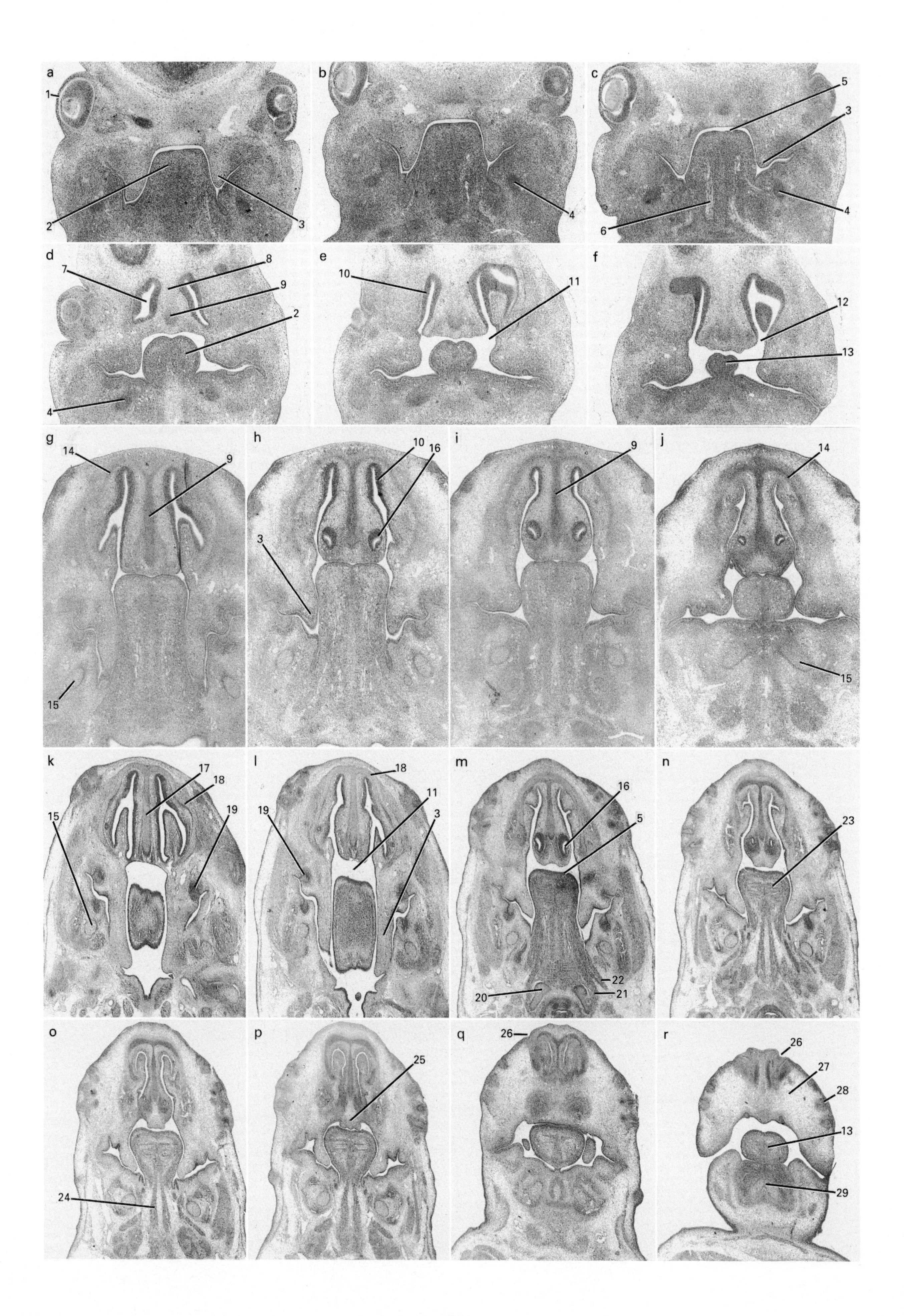

a
b
c
d
e
f
g
h
i
j
k
l
m
n
o
p
q
r
1
2
3
4
5
6
7
8
9
10
11
12
13
14
15
16
17
18
19
20
21
22
23
24
25
26
27
28
29

Plate 61

Formation of the palatal shelves and closure of the palate, I

a–f 12.5 days p.c. (coronal sections, posterior to anterior) (×40)
g–j 13.5 days p.c. (coronal sections, posterior to anterior) (×40)
k–r 14 days p.c. (transverse sections) (×25)

1. optic eminence/corneal epithelium
2. premuscle mass of intrinsic muscles of tongue
3. vertically directed palatal shelf/process of maxilla
4. precartilage primordium of Meckel's cartilage
5. dorsal surface of tongue
6. lingual nerve
7. lumen of primitive nasal (olfactory) cavity
8. nasal septum
9. precartilage primordium of nasal septum
10. olfactory epithelium
11. primitive nasopharynx
12. primary choana
13. tip of tongue
14. precartilage primordium of nasal capsule
15. Meckel's cartilage
16. vomeronasal organ (Jacobson's organ)
17. cartilage primordium of nasal septum
18. nasal capsule
19. dental lamina (tooth primordium)
20. cartilage primordium of hyoid bone
21. extrinsic muscle of tongue (hyoglossus)
22. extrinsic muscle of tongue (styloglossus)
23. intrinsic muscle of tongue (transverse component)
24. extrinsic muscle of tongue (genioglossus)
25. site of fusion between nasal septum and precursor of primary palate
26. entrance to primitive nasal cavity (anterior naris)
27. ventral extremity of upper jaw
28. primordium of vibrissa
29. rostral extremity of Meckel's cartilages (site of fusion in ventral midline)

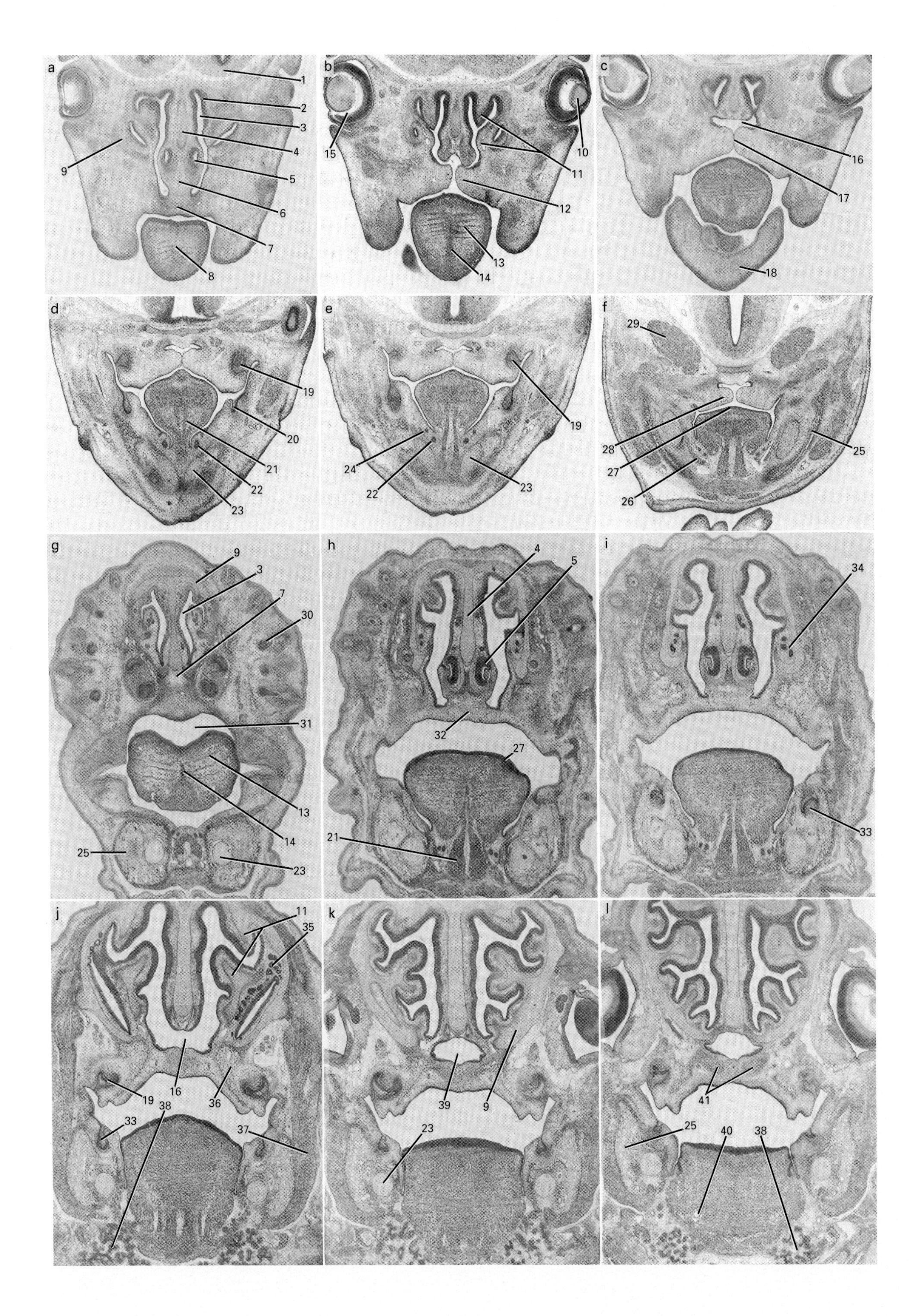
a
1
2
3
4
5
6
7
8
9
b
10
11
12
13
14
15
c
16
17
18
d
19
20
21
22
23
e
19
22
23
24
f
25
26
27
28
29
g
9
3
7
30
31
13
14
25
23
h
4
5
32
27
21
i
34
33
j
11
35
16
19
36
38
33
37
k
39
9
23
l
41
25
40
38

Plate 62

Formation of the palatal shelves and closure of the palate, II

a–f 14.5 days p.c. (coronal sections) (×25)
g–l 15.5 days p.c. (coronal sections) (×25)

1. telencephalic vesicle
2. olfactory epithelium
3. lumen of primitive nasal (olfactory) cavity
4. cartilage primordium of nasal septum
5. vomeronasal organ (Jacobson's organ)
6. nasal septum
7. site of fusion between nasal septum and primary palate
8. tip of tongue
9. cartilage primordium of nasal capsule
10. lens
11. nasal conchae
12. horizontally directed palatal shelf/process of maxilla
13. intrinsic muscle of tongue (transverse component)
14. precursor of median fibrous septum of tongue
15. hyaloid cavity
16. ventral extremity of nasopharynx
17. palatal shelves in direct apposition, but not yet fused
18. ventral extremity of lower jaw
19. tooth primordium of upper molar tooth
20. alveolar sulcus
21. extrinsic muscle of tongue (genioglossus)
22. submandibular duct
23. Meckel's cartilage
24. sublingual duct
25. ossification in body of mandible around but principally lateral to Meckel's cartilage
26. mylohyoid muscle
27. dorsal surface of tongue
28. medial surface of dorsal region of palatal shelf, yet to reach the midline
29. trigeminal (V) ganglion
30. base of follicle of vibrissa
31. oral cavity
32. region of palatal shelf fusion (secondary palate)
33. tooth primordium of lower molar tooth
34. tubules of serous glands that course rostrally in the lateral wall of the middle meatus
35. nasal glandular tissue in lateral wall of middle meatus
36. ossification within maxilla
37. masseter muscle
38. submandibular gland
39. nasopharynx
40. lingual artery
41. condensations of precartilage/cartilage in posterior part of palate

between the various parts of the tongue and the palatal shelves and the nasal septum is again confirmed by the histological sections through this region. These sections additionally demonstrate the first appearance of the primordium of the vomeronasal organ (Jacobson's organ), which is seen to be located on either side of the precartilage primordium of the nasal septum (*Plate 61,g–j*).

The relationship between the dorsal surface of the tongue and the roof of the oropharynx is also well illustrated in sagittal sections through this region (*Plate 31a,a,b*; *Plate 31b,a*). Transverse sections through this region are slightly more complex to interpret, principally because of the curvature of the tongue and that of the roof of the oropharynx. These sections do, however, provide evidence of the close relationship that exists between the lateral borders of the tongue mass and the medial borders of the palatal shelves before they display evidence of elevation to the horizontal position (*Plate 61,k–p*). The relationship between the external naris and the primitive nasopharynx is also seen (*Plate 61,o–q*). It is also clear from these sections, that the inferior borders of the palatal shelves are very closely apposed to the inferolateral surfaces of the middle and anterior parts of the tongue (*Plate 61,n–p*).

During the period between 14.5 and 15 days p.c., the first reflex opening and closing movements of the mouth occur (Humphrey, 1969), and this consequently allows the lower jaw and dorsal surface of the tongue to descend below the lower margins of the vertically directed palatal shelves for the first time. This factor, associated with the so-called "internal shelf forces" (for further details, see text associated with appropriate section of Stage 22), facilitates the elevation of the palatal shelves to the horizontal position. Due to the lowering of the dorsal surface of the tongue, the medial borders of the two palatal shelves elevate (*Plate 60,l*; *Plate 61,a–c*), and over a relatively short period of time, become closely apposed. Histological sections through this region at about 14.5 days p.c. illustrate lengthy regions of apposition (*Plate 62,c–e*), and confirm that this event initially occurs in the middle region of the shelves (opposite to the second to fourth pairs of rugae), so that both anterior and posterior to this site, the shelves have yet to become apposed (*Plate 62,b,f*). At this stage, the inferior surface of the nasal septum is only fused to the dorsal part of the primary palate (*Plate 62,a*).

Fusion of the medial surfaces of the palatal shelves is first seen at about 15 days p.c., and occurs simultaneously over a relatively lengthy region (*Plate 61,d,e*), so that probably over a period of hours, most of the apposing surfaces fuse. By 15.5 days p.c., the fusion process is virtually complete. The final part to close (at about 16 days p.c.) is close to the site where the primary and secondary palates meet in the region of the incisive canals (*Plate 60,l*). By 15.5 days p.c., the palatal shelves are seen to have fused along their entire length (*Plate 62,g–l*), and (except in a relatively localized region, where communication exists between the two sides of the nasal cavity, *Plate 62,j*) with the inferior border of the nasal septum (*Plate 62,g–i,k,l*). Fusion of the medial edge epithelia of the shelves and that of the inferior border of the nasal septum initially occurs to form a midline epithelial seam. This is followed by disruption of the seam, so that mesenchymal continuity occurs.

Over the period between about 14–14.5 days and 16 days p.c., the rugae associated with the mucous membrane on the inferior surface of the palate become more prominent. A total of nine rugae of various sizes are seen in the mouse (Sakamoto *et al.*, 1989), and may be used as landmarks to locate specific positions on the palatal shelves (see Fig. 14). The site of fusion in the midline is also marked by a small raphe which ends anteriorly in a small papilla overlying the incisive fossa. This is also the first occasion that the nasopharynx is completely separated from the oropharynx, the two only being in communication at the posterior border of the soft palate. The olfactory epithelium in the nasal cavity is now seen to cover the surface of the conchae, and groups of serous glands are now found in association with both the lateral wall of the middle meatus as well as in association with the nasal septum (*Plate 62,h–j*).

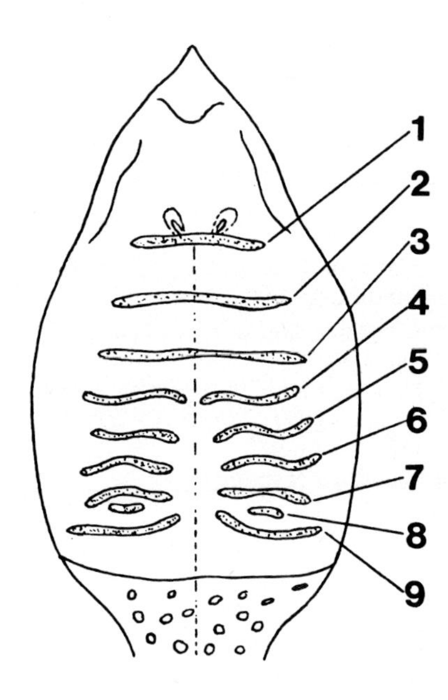

Figure 14 Diagram of numbering system of rugae, after Sakamoto *et al.* (1989).

9. Morphology of the spinal cord
Text associated with Plate 63

Analysis of representative sections through the thoracic region of the spinal cord over the period between 11.5 and 17.5 days p.c. is instructive in that these sections reflect the major changes that are occurring in the organization of the spinal cord over this period. For overviews of the changes that occur in the relationship between the grey and white matter of the cord, and the shape changes that occur both in relation to the central canal of the cord and indeed the overall cross-sectional shape of the cord, paraffin-embedded haematoxylin and eosin-stained sections are adequate for this purpose. For more detailed studies, particularly of the cellular components of the grey matter, it is essential that appropriately (i.e. silver) stained sections are analysed. For other purposes, for example, to analyse the extent of neural luminal occlusion that occurs during the earlier stages of spinal cord development, the analysis of plastic-embedded material may be preferable, since ultrathin sections through isolated regions of particular interest may then be viewed by transmission electron microscopy (Kaufman, 1983b, 1986).

During the period between 11.5 and 15.5 days p.c., only representative sections through the mid-thoracic region of the spinal cord are illustrated (*Plate 63,a–f*), as these are fairly similar to sections through all but the most rostral and caudal extremities of the cord. However, since by 16.5 days p.c. considerable differences are observed in the cross-sectional appearance of the cord at different vertebral levels, representative sections through the upper cervical region and the mid-lumbar region are illustrated, in addition to sections through the mid-thoracic region (*Plate 63,g–i*), in order to allow comparisons to be made between these different levels. Representative sections through the cord at these three levels from an embryo isolated at 17.5 days p.c. are also illustrated (*Plate 63,j–l*), as their appearance is similar to the situation observed at these levels at the time of birth. It should be noted, though, that at 15.5 days p.c., the situation observed is transitional between that seen at 14.5 days p.c., where the cross-sectional pattern is similar along the entire length of the cord, and the situation observed at 16.5 days p.c., as indicated above, where regional differences are already clearly apparent.

At 11.5–12 days p.c., in transverse sections through the mid-thoracic region of the spinal cord (*Plate 63,a*), the majority of the section consists of the ependymal and mantle (the future grey matter of the cord) layers, and there is only a thin outer rim of the marginal layer (the future white matter of the cord). Only a relatively small proportion of the mantle layer is present in the region of the alar plate (the future dorsal grey column) at this stage, whereas the basal plate (the future ventral grey horn) is seen to be quite voluminous. The ependymal layer is wide in all regions around the central canal, except in the regions of the floor and roof plates where it is relatively narrow. The overall shape of the section through the thoracic region of the spinal cord at this stage is almost square, though the ventral region is slightly wider than the dorsal region. The shape of the central canal is fairly similar throughout the entire length of the cord, with the dorsal half being considerably more voluminous than the ventral half. In the latter region, the medial (luminal) surface of the ependymal layers often meet across the midline (*Plate 26c,a–i*; *Plate 26d,a–i*; *Plate 26e,a–j*), this feature being more commonly seen in plastic-embedded than in paraffin-embedded material, and represents the residual sites of apposition left over after the two sides have separated following neural luminal occlusion. The dorsal (posterior) root ganglia are relatively large structures, and both the spinal cord and the ganglia are embedded in loose mesenchyme, which tends to disperse over the period of the next few days to allow for the formation of the spinal canal (*Plate 63,e,f*).

Relatively little difference is observed in the cross-sectional appearance of the spinal cord at 12–12.5 days p.c., compared with the situation observed previously, except that the width of the dorsal half of the ependymal layer is diminished, and the volume of the mantle layer is correspondingly increased (*Plate 63,b*). The change in the overall shape of the cord that is observed at this stage is largely accounted for by the substantial increase that occurs in the volume of the alar component of the mantle layer. The shape of the central canal is similar to that observed previously.

Over the next few days, between 12.5–13 and 15.5 days p.c., the overall shape of the cord gradually changes, so that its width increases to a greater extent than its dorsiventral (anteroposterior) diameter (*Plate 63,c–f*). The width of the ependymal layer also gradually diminishes, as does the volume of the section of the cord occupied by the central canal. The proportion of white matter also gradually increases, largely as a consequence of the differentiation of the ascending and descending pathways that pass rostrally and caudally in this region of the cord. The dorsal and ventral grey columns are also now seen to gradually assume the more familiar "butterfly" shape characteristic of this region in the adult spinal cord. Over this period, the dorsal horns (columns) gradually tend to have a greater volume (and spread rather further towards the periphery of the cord) than that of the ventral horns. It is also just possible to see the first evidence of the lateral grey columns in some of these sections.

A gradual evolution in the cross-sectional appearance of the cord occurs, as described above, so that by about 16.5–17.5 days p.c., the definitive arrangement is established, and the lumen of the central canal is seen to be almost obliterated. The spinal cord extends caudally towards the base of the tail, where the conus medullaris is located (*Plate 35l,c,d*), and more distally the filum terminale (e.g. *Plate 35j,a*) extends towards the tip of the tail. Because of the regionalization that is evident in the spinal cord, sections through the upper cervical (*Plate 63,g,j*) and mid-lumbar region (*Plate 63,i,l*) are

(*continued on page 438*)

PLATE 63

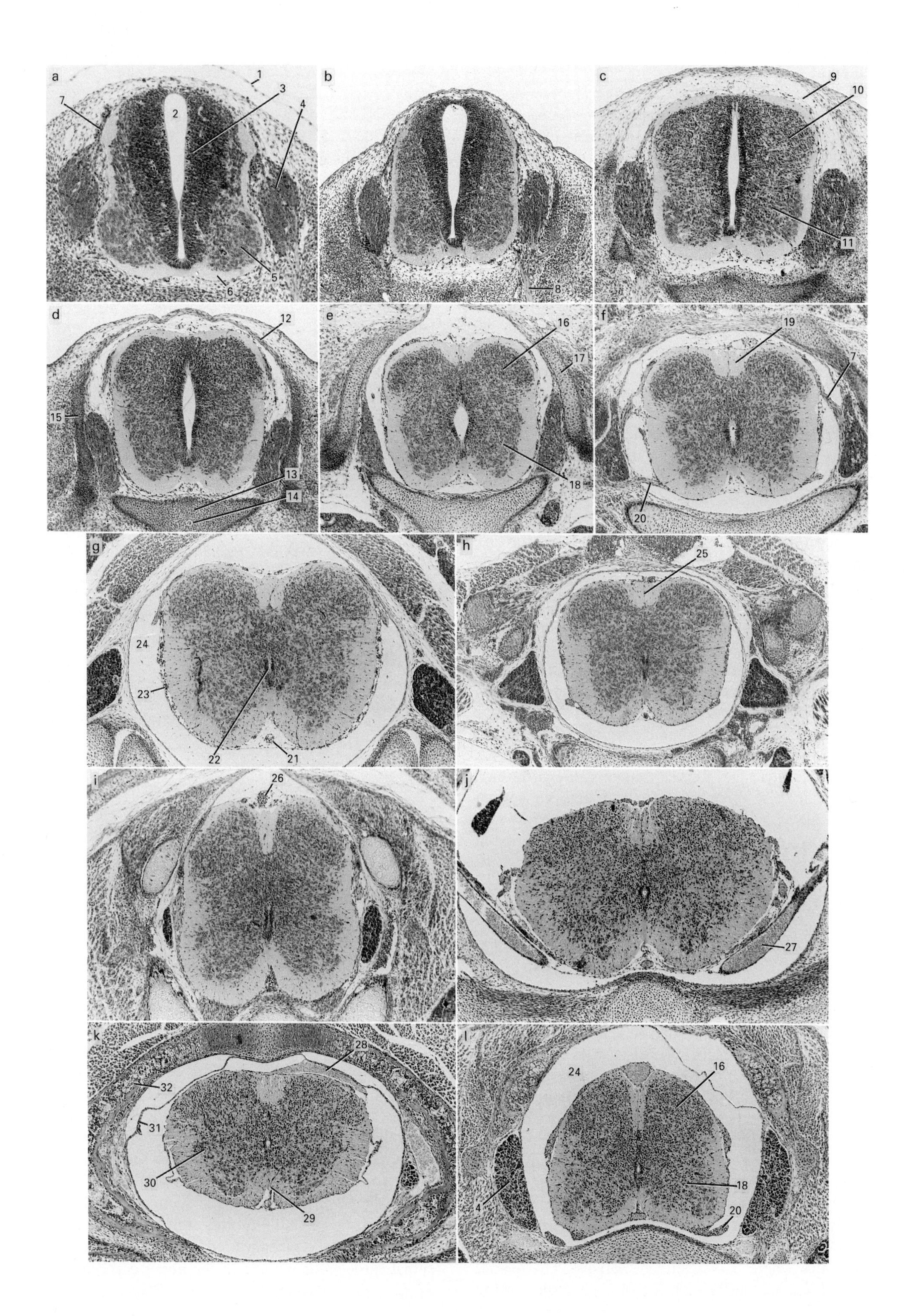

a
b
c
d
e
f
g
h
i
j
k
l
1
2
3
4
5
6
7
8
9
10
11
12
13
14
15
16
17
18
19
20
21
22
23
24
25
26
27
28
29
30
31
32

Plate 63

Morphology of the spinal cord

All sections through the mid-thoracic region unless otherwise stated.

a 11.5–12 days p.c. (×100)
b 12–12.5 days p.c. (×63)
c 12.5–13 days p.c. (×63)
d 13.5–14 days p.c. (×63)
e 14.5 days p.c. (×63)
f 15.5 days p.c. (×63)
g–i 16.5 days p.c. (×63)
j–l 17.5 days p.c. (×63)

g upper cervical region; **h** mid-thoracic region; **i** mid-lumbar region;
j upper cervical region; **k** mid-thoracic region; **l** mid-lumbar region.

1. amnion
2. central canal
3. ependymal layer
4. dorsal (posterior) root ganglion
5. mantle layer (grey matter)
6. marginal layer (white matter)
7. dorsal root (sensory)
8. mixed spinal nerve
9. mesenchyme in future spinal canal
10. mantle layer (alar plate of grey matter)
11. mantle layer (basal plate of grey matter)
12. mesenchyme precartilage primordium of dorsal part of neural arch
13. cartilage primordium of vertebral body (centrum)
14. degenerating notochord (future nucleus pulposus)
15. cartilage primordium of ventral part of neural arch
16. dorsal grey horn (column)
17. cartilage primordium of lateral part of neural arch
18. ventral grey horn (column)
19. dorsal (posterior) median fissure
20. ventral root (motor)
21. anterior (median) spinal artery
22. central canal (almost obliterated in this location)
23. vascular elements of pia mater
24. spinal canal
25. dorsal (posterior) median septum
26. principal posterior spinal vein with associated smaller spinal arteries and veins
27. left vertebral artery
28. dorsal spinal branch of lumbar artery which contributes to longitudinal anastomotic arterial channels along the spinal cord
29. ventral median fissure
30. lateral grey column
31. dura mater
32. ossification within cartilage primordium of neural arch

illustrated as well as sections through the mid-thoracic region of the spinal cord (*Plate 63,h,k*). These serve to emphasize the very considerable difference evident in the volume (i.e. cross-sectional area) of the cord at these levels, as well as in their overall appearance. The thoracic sections tend to have a considerably smaller volume than those in the upper cervical region and mid-lumbar region, due to the proximity of the latter regions to the brachial and lumbo-sacral plexuses, respectively. The caudal extent of the cord diminishes slightly with increased gestational age, so that the vertebral level of the conus medullaris progressively rises, and the length of the filum terminale gradually increases until the adult dimensions are achieved.

10. Early stages in the development of the heart
Text associated with Plates 64 and 65

It is not entirely clear at the present time where the precursors of the cardiac myoblasts originate, for example, whether they migrate into the prospective pericardial region of the intra-embryonic coelom from either side of the foregut diverticulum, before the splanchnic and somatic mesoderm separate to form the pericardial cavity, similar to the situation described in avian embryos (De Haan, 1965; Rosenquist and De Haan, 1966), or develop *in situ* by the differentiation of coelomic mesothelial cells that line the wall of the prospective pericardial component of the intra-embryonic coelom (Kaufman and Navaratnam, 1981). This issue will no doubt be resolved when labelling studies using appropriate gene probes are employed, as these will enable the migration of any possible cardiogenic precursor cells to be followed.

The analysis of histological sections through a representative series of embryos progressing from the late presomite stage (at about 7.5 days p.c., *Plate 64,a–c*) to embryos with about 5–7 pairs of somites (*Plate 64,g–i*), and scanning electron micrographs of embryos spanning the period from the early headfold stage (*Plate 64,j,k*) to embryos at about 10 days p.c., with about 30–35 pairs of somites (*Plate 65,m–o*), is particularly instructive, as these span the critical early stages of cardiogenesis. At these early stages, it is of particular value to study the appropriate histological sections in association with the scanning electron micrographs of the dissections in which "windows" have been created into the anterior aspect of the pericardial component of the intraembryonic coelom in order to visualize its contents. By this complementary approach, it is possible to observe the differentiation not only of the cardiogenic plate but also the subjacent endocardial elements. The latter subsequently amalgamate together (initially with the angiogenetic cell clusters within the embryo, and slightly later with similar elements that differentiate in the mesodermal component of the visceral yolk sac) to form the continuous endothelial lining of the intra- and extra-embryonic vascular systems. Later, when the extra-embryonic vascular system degenerates, these cells provide the endothelial elements which line the complete intra-embryonic vascular system. Initially, it is not possible to distinguish on conventional histological analysis between the cardiogenic plate cells and those cells which constitute the wall of the arch arteries, nor of other components of the vascular system, though, once again, the use of appropriate gene probes would almost certainly facilitate this exercise.

From a descriptive point of view, the first critical event associated with the process of cardiogenesis must be the splitting of the intra-embryonic mesoderm to form the intra-embryonic coelom. This event occurs at about 7–7.5 days p.c. in presomite early headfold stage embryos. Slightly later, at about 7.5–8 days p.c., it is possible to observe that the coelomic mesothelial cells that line the ventral wall of the intra-embryonic coelom show evidence of differentiation, and develop initially a cuboidal and later a columnar morphology (*Plate 64,a*). These cells constitute the cardiogenic plate, and span the ventral midline in the prospective pericardial region of the intra-embryonic coelom in advanced presomite stage embryos (*Plate 64,a*). A similar degree of cellular differentiation is also evident at this time within the arms of the U-shaped intra-embryonic coelom in the region of the prospective pericardio-peritoneal canals (*Plate 64,b,c*). It is likely that the latter cells will be involved in the differentiation of the walls of the two horns of the sinus venosus (and possibly also of components of the atria of the heart).

In early headfold stage embryos with 4–5 pairs of somites, the situation described above is consolidated, and the cardiogenic plate cells and the elements that are in continuity with them (located on either side of the foregut diverticulum) in the pericardio-peritoneal canals (*Plate 64,e*) are now seen to protrude from the ventral splanchnic wall into the prospective pericardial cavity (*Plate 64,d–f*). Analysis of scanning electron micrographs of embryos in which these events are occurring reveals that the cardiogenic plate cells are localized to one region of the wall of the intra-embryonic coelom, and are morphologically clearly distinguishable (since they have a columnar morphology) from the rest of the cells which line this region, since the latter retain their original squamous morphology (*Plate 64,j,k*). The latter is particularly clearly seen in the sagittal section through the pericardial coelom (*Plate 64,f*). This is also the first stage that endocardial elements are seen to have differentiated in the mesoderm subjacent to the cardiogenic plate (*Plate 64,d,e*). Initially, the endocardial (endothelial) elements are present as localized pockets of cells, but these gradually amalgamate together to form the continuous endothelial lining of the intra-embryonic vasculature.

In embryos with about 3–7 pairs of somites, two cardiogenic rudiments are seen which abut against each other in the midline, and are partially separated by a deep furrow. They appear to be separate entities, and each contains a medially directed blind-ending lumen which is lined by a single layer of endothelial cells (*Plate 64,g–i*; *Plate 64,l,m*) which pass caudally to line the horns of the sinus venosus (*Plate 64,i*). In the midline, the primitive heart rudiment is suspended by a relatively narrow dorsal mesocardium (*Plate 64,h*). In embryos with 5–6 pairs of somites (at about 8 days p.c.), the two vitelline veins pass on either side of the foregut pocket to join their respective horns of the sinus venosus, which then drain into the venous end of the primitive heart. The heart is now seen to be a single midline structure due to the amalgamation of the two heart primordia and the disappearance of the median furrow. At this stage of cardiogenesis, the heart and vascular system is still symmetrical, but over the next few hours this symmetrical arrangement is lost as the heart enters the so-called "looping" stage.

The changes that take place in the pericardial coelom in embryos between 8 and 8.5 days p.c. occur

(*continued on page 444*)

PLATE 64

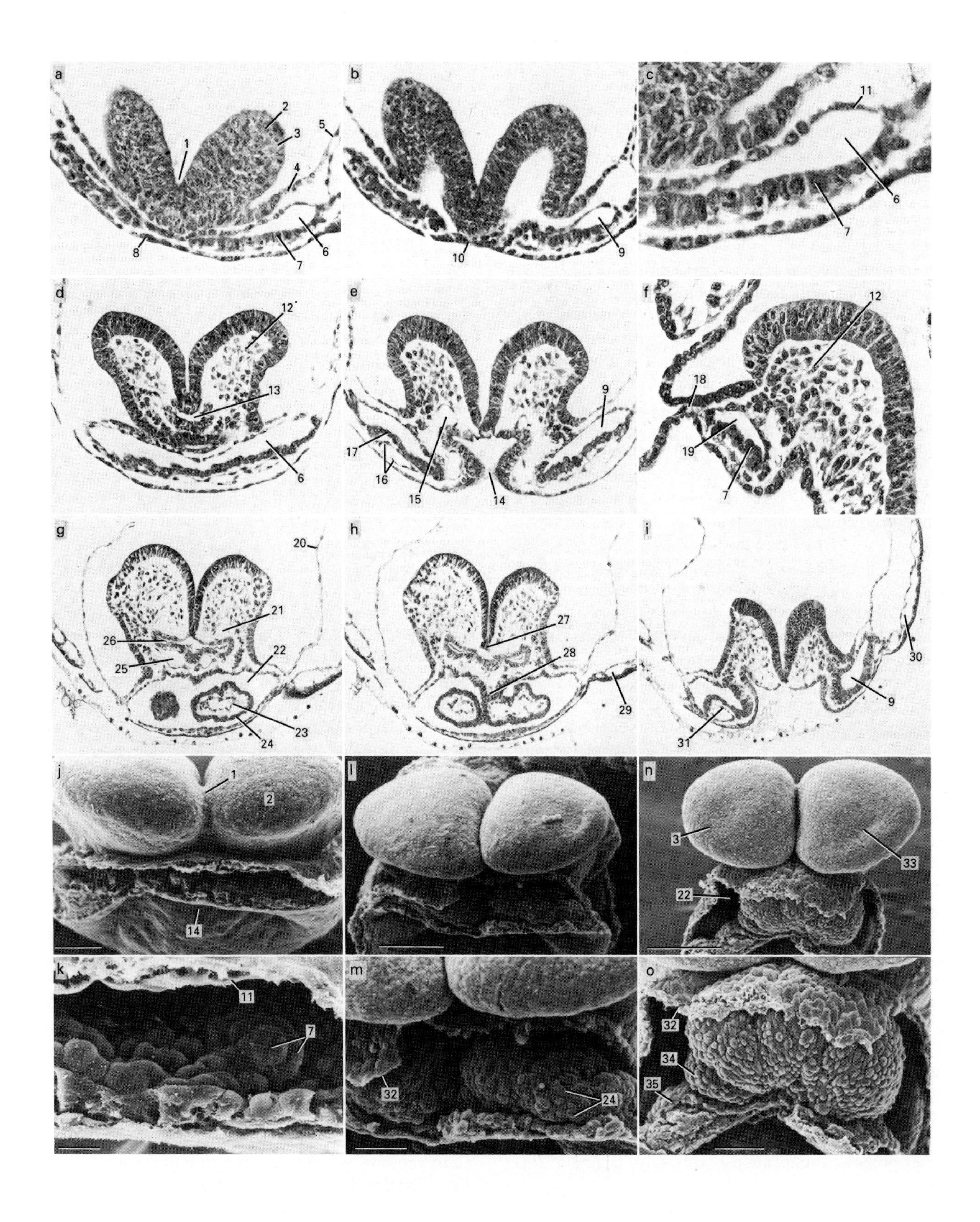

Plate 64

Differentiation of the heart. I

a–c about 7.5 days p.c.; late presomite/early somite stage embryo (**a, b** ×310; **c** ×800) transverse.
d,e 7.5–8 days p.c.; early headfold embryo with 4–5 pairs of somites (×160) transverse
f 7.5–8 days p.c.; early headfold embryo with 4–5 pairs of somites (×310) sagittal
g–i 7.5–8 days p.c.; early headfold embryo with 5–7 pairs of somites (×125) transverse
j,k 7.5–8 days p.c.; SEM of frontal view of early headfold embryo with 1–2 pairs of somites (**j**, scale bar 50 μm; **k**, scale bar 10 μm)
l,m 7.5–8 days p.c.; SEM of frontal view of embryo with 3–5 pairs of somites (**l**, scale bar 100 μm; **m**, scale bar 40 μm)
n,o 8 days p.c.; SEM of frontal view of embryo with 5–6 pairs of somites (**n**, scale bar 100 μm; **o**, scale bar 40 μm)

In the embryos illustrated in **j–o**, the layer of surface ectoderm cells and adherent subjacent presumptive parietal pericardial layer of cells have been removed to create a window on the anteroventral aspect of the pericardial cavity.

1. neural groove
2. cephalic neural fold (headfold)
3. neural ectoderm (neuroepithelium)
4. surface ectoderm
5. ectodermal component of amnion
6. lateral part of presumptive pericardial region of intra-embryonic coelomic cavity
7. presumptive myocardial cells of cardiogenic plate
8. embryonic endoderm
9. pericardio-peritoneal canal (left horn of intra-embryonic coelomic channel)
10. notochordal plate
11. mesothelial cells lining the non-cardiogenic plate region of the presumptive pericardial cavity
12. cephalic mesenchyme cells
13. rostral extension of foregut diverticulum
14. entrance to foregut diverticulum
15. rostral extension of right dorsal aorta
16. endocardial cells subjacent to cardiogenic plate
17. cardiogenic plate in region of future right horn of sinus venosus
18. lateral part of bucco-pharyngeal membrane
19. central part of pericardial region of intra-embryonic coelomic cavity
20. amnion
21. dorsal aorta
22. pericardial cavity
23. endocardial cells lining the primitive heart tube
24. myocardial cells forming the outer wall of the primitive heart tube
25. first branchial (pharyngeal) arch artery
26. first branchial (pharyngeal) pouch
27. notochord
28. dorsal mesocardium
29. visceral yolk sac
30. yolk sac blood island
31. endothelial lining of right horn of sinus venosus
32. torn edge of surface ectoderm with subjacent presumptive parietal pericardial layer of cells
33. early stage of development of optic placode
34. right atrial part of primitive heart tube
35. wall of right horn of sinus venosus at junction with right vitelline vein

PLATE 65

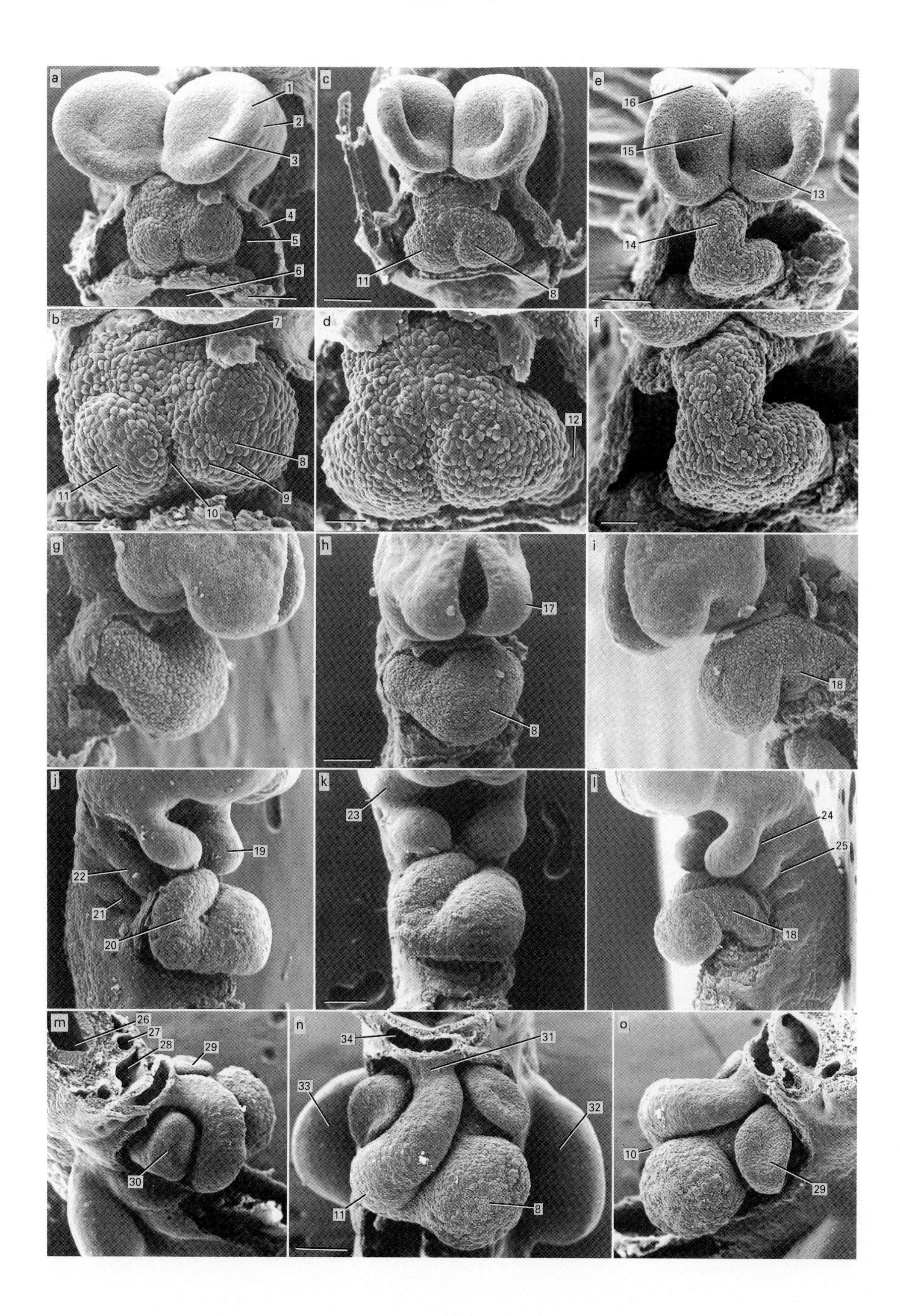
a
1
2
3
4
5
6
c
11
8
e
16
15
13
14
b
7
8
9
10
11
d
12
f
g
h
17
8
i
18
j
19
22
21
20
k
23
l
24
25
18
m
26
27
28
29
30
n
34
31
33
32
11
8
o
10
29

Plate 65

Differentiation of the heart. II

a,b 8–8.5 days p.c.; embryo with 5–7 pairs of somites
c,d 8–8.5 days p.c.; embryo with 7–9 pairs of somites
e, f 8.25–8.5 days p.c.; embryo with 8–10 pairs of somites
g–i 8.5 days p.c.; embryo with 11–13 pairs of somites
j–l 9.5–10 days p.c.; embryo with 25–30 pairs of somites
m–o 10–10.5 days p.c.; embryo with 30–35 pairs of somites

SEMs of embryos in which the thoracic wall has been removed to expose the pericardial cavity and primitive heart.
a,c,e,h,k, scale bar 100 μm; **b,d,f,** scale bar 40 μm; **n,** scale bar 200 μm

1. cephalic neural fold (headfold)
2. boundary between neural ectoderm (neuroepithelium) and surface ectoderm
3. optic pit
4. torn edge of surface ectoderm with subjacent presumptive parietal pericardial layer of cells
5. pericardial cavity
6. entrance to foregut diverticulum
7. myocardial cells forming the outer wall of the outflow tract of the primitive heart tube
8. primitive ventricle
9. myocardial cells in region of wall of primitive ventricle
10. bulbo-ventricular groove
11. bulbus cordis
12. left horn of sinus venosus
13. optic sulcus
14. outflow tract of heart at primitive "looping" stage
15. future forebrain region
16. future midbrain region
17. left optic eminence
18. left part of common atrial chamber of heart
19. mandibular component of first branchial (pharyngeal) arch
20. bulbo-truncal junction region of outflow tract of heart
21. third branchial (pharyngeal) arch
22. second branchial (pharyngeal) arch
23. maxillary component of first branchial (pharyngeal) arch
24. first branchial (pharyngeal) groove
25. second branchial (pharyngeal) groove
26. caudal part of fourth ventricle
27. dorsal aorta
28. lumen of pharynx
29. auricular part of left side of common atrial chamber of heart
30. auricular part of right side of common atrial chamber of heart
31. aortic sac
32. left forelimb bud
33. right forelimb bud
34. lumen of right first branchial (pharyngeal) arch artery

quite rapidly, and the enlargement of the primitive ventricular and bulbar regions of the primitive heart results in the appearance of two characteristic "bulges", which are separated by a shallow groove, one being located on either side of the midline (*Plate 65,a–d*). With the changes that are occurring in the configuration of the primitive heart, the venous (inflow) side now lies dorsally and caudally, while the outflow tract lies rostrally and ventrally in relation to the main mass of heart tissue. In embryos with about 8–10 pairs of somites, the primitive heart is now clearly S-shaped, with a prominent ventral curvature (*Plate 65,e,f*). The venous end of the primitive heart is in direct continuity rostrally with the primitive ventricle (the future left ventricle), which is then in continuity rostrally with the bulbus cordis region (the future right ventricle). The latter is then in continuity rostrally with the outflow tract which is itself continuous with the (ventral) aortic sac (via the truncus arteriosus, though these regions are not clearly demarcated). The latter then passes rostrally and dorsally before dividing to form the two first aortic arch arteries, which pass on either side of the pharyngeal region of the foregut diverticulum.

The series of scanning electron micrographs that illustrate these events complement the serially-sectioned embryos illustrated in *Plates 10–13*. Analysis of the pericardial cavity and the pericardio-peritoneal canals and their contents clearly demonstrates the continuity that exists between the heart and the primitive vasculature during this critical period. It should also be noted that very few primitive (nucleated) red blood cells are seen to be present within the embryonic vasculature in embryos with less than 10 pairs of somites. The simple reason for this is that the intra-embryonic and extra-embryonic vasculature have only very recently amalgamated, and it is within the mesodermal component of the visceral yolk sac that erythropoiesis initially exclusively occurs. For a more detailed account of these events the reader should refer to the appropriate section of Stage 13.

By this stage, the heart is seen to beat regularly, and the ultrastructural morphology of the precursors of the cardiac muscle cells, from the earliest stage of cardiogenesis to this stage, bear a very close functional relationship (Challice and Virágh, 1973; Virágh and Challice, 1973; Kaufman, 1981; Navaratnam *et al.*, 1986). During this period, it is possible to observe the transition of cardiac cells from the "undifferentiated" myoblast stage to the "differentiated" myocyte and myotube stages. Initially, the contractile elements within the cytoplasm are seen to be randomly orientated. They then become aligned and orientated to form recognizable sarcomeres (this constitutes the process of myofibrillogenesis). Slightly later, primitive intercalated disks are seen, and these subsequently increase in their complexity. The branching of the sarcomere bundles is a characteristic early feature and serves to distinguish these from striated (non-cardiac) muscle fibres.

Only relatively minor changes are evident in the external morphology of the heart between 8.5 days p.c. (*Plate 65,g–i*), in embryos with 11–13 pairs of somites, and 9.5–10 days p.c. (*Plate 65,j–l*), in embryos with 25–30 pairs of somites. During this period, the S-shaped loop of the primitive heart tube is accentuated, and this change in the conformation of the heart is facilitated by the breaking down of a relatively small segment of the dorsal mesocardium between the inflow and outflow regions of the heart which occurs in embryos with about 10–12 pairs of somites (*Plate 14b,a,b*; *Plate 15a,l–n*) with the formation of the transverse pericardial sinus. The venous inflow and much of the common atrial chamber is now seen to be located on the left of the midline, as is the primitive ventricle. The bulbus cordis and proximal part of the outflow tract, by contrast, is now located on the right of the midline, while the distal part of the outflow tract (the distal truncus and aortic sac) is in the median plane.

In the developmentally most advanced embryos illustrated in this series, with about 30–35 pairs of somites, isolated at about 10–10.5 days p.c., the head has been removed to facilitate observation of the heart (*Plate 65,m–o*). The right and left auricular appendages of the common atrial chamber of the heart are seen on either side of the cranial (distal) part of the outflow tract. The proximal part of the bulbar region is at an early stage of becoming absorbed into the right ventricle. The origins of the first aortic arch arteries are also seen, as is the pronounced bulbo-ventricular sulcus. Even at this stage of development, the heart is the most prominent of all of the organ systems in the embryo, and was the first organ system to differentiate and to function.

11. Differentiation of the lungs
Text associated with Plate 66

The initial event that is of critical importance in relation to the development of the lungs, is the differentiation of the laryngo-tracheal groove. This is first seen at about 9–9.5 days p.c. (*Plate 19c,a,b*), and forms an anteriorly directed diverticulum of the pharyngeal region of the primitive foregut, and subsequently forms the primitive trachea. The first evidence of the tracheal bifurcation to give rise to the two main bronchi is seen shortly after the first appearance of the laryngo-tracheal groove, and this is subsequently associated with the first indication of the paired lung buds (*Plate 19c,b*). At this stage, the latter are fairly symmetrical structures, and bulge outwards (i.e. laterally) into the pericardio-peritoneal canals, and are seen to be inferior relations of the common atrial chamber of the heart (*Plate 20c,e*; *Plate 22a,c*; *Plate 22b,b*).

By about 10.25–10.5 days p.c., the lung buds are seen to have ascended slightly, and their rostral (apical) parts are at about the level of the lower part of the common atrial chamber of the heart (*Plate 23c,g–j*). The pulmonary arteries are first seen at this time, as spurs from the paired sixth branchial arch arteries (see Fig. 10), though they are more clearly seen at about 10.5–11 days p.c., as they descend into the lung buds (*Plate 24c,l–o*). This is also the first stage that a degree of asymmetry is seen with regard to the disposition of the right and left main bronchi (*Plate 66,a*). The asymmetry observed at this time is clearly related to the fact that on the left side only a single lobe forms, whereas on the right side the main bronchus rapidly divides to form four lobar bronchi, associated with the formation, at a slightly later stage, of four distinct lobes on this side. An intermediate stage in the rostral "ascent" of the lungs is seen at 11–11.5 days p.c., when the lungs are seen to be slightly more rostrally located than previously (*Plate 25b,a*). A similar situation is observed at 11.5–12 days p.c. (*Plate 26c,g–i*; *Plate 26d,a,b*).

By 12–12.5 days p.c., the lung buds largely appear to consist of homogeneous cellular tissue which surrounds the lobar and segmental bronchi. Somewhat surprisingly, each of the latter have a larger luminal diameter than the oesophagus at this stage (*Plate 66,c*). Both of the lung buds, and the pericardio-peritoneal canals in which they are located, are covered by a single layer of epithelial (coelomic mesothelial) cells. These are the primordia of the visceral and parietal layers of pleura, respectively (*Plate 66,d*).

By 12.5–13 days p.c., the progressive branching of the bronchial tree, with the formation of numerous segmental bronchi, is clearly seen (*Plate 66,e,f*), though the parenchyma of the lungs is still fairly homogeneous at this stage. While the apical regions of the lungs are at about the same vertebral level as previously, the lungs have expanded dramatically, and now occupy a substantial volume of the peritoneal cavity posterior to the upper part of the liver. The liver and the two lung buds are separated anteriorly by the primordium of the central tendon portion, and posterolaterally by the pleuroperitoneal primordia of the diaphragm (*Plate 28d,b–i*; *Plate 28e,a,b*).

By 13.5–14 days p.c., the right lung is first clearly seen to be subdivided into its four lobes (before this stage, the lobes are only incompletely separated), namely the cranial, middle, caudal and accessory lobes (*Plate 66,g*), and progressive branching of the bronchial tree is apparent (*Plate 66,h*). The bifurcation of the trachea (the carina) is now at the level of the origins of the pulmonary trunk and ascending aorta (*Plate 30g,a,b*). The apical lobar branch of the right main bronchus is given off shortly after the bifurcation of the trachea, and this lobe of the lung is somewhat more rostrally located than the apical region of the left lung (*Plate 30g,b–d*). While the primordium of the accessory lobe of the right lung is first seen at about 12–12.5 days p.c. (*Plate 27c,f*), it is particularly well seen at this stage of development, when a large proportion of it is seen to be located either in the median plane, or to the left of the midline (*Plate 30h,c,d*; *Plate 30i,a,b*).

By 14.5–15 days p.c., the first indication of the detailed architecture of the future lung is now seen, with the formation of the terminal bronchi/bronchioles (*Plate 66,i,j*), and these are now seen to be dispersed throughout the substance of the lungs. During 15.5–16 and 16.5–17 days p.c., the peripheral components of the bronchial tree continue to branch, and the luminal diameter of these structures gets progressively smaller (e.g. *Plate 66,k,l*). By 17.5–18 days p.c., the histological features of the lung are almost identical to those seen at birth, with the exception of that of the cells that line the alveoli. These cells are initially cuboidal in shape, but by about 18.5–19 days p.c., they have acquired a squamous morphology (to facilitate gaseous exchange between the air within the alveoli and the capillary network of the pulmonary vasculature).

Towards the time of birth, the volume of the pleural cavities has expanded to a greater degree than that of the lungs, in order to accommodate the rapid expansion of the lungs that occurs very shortly after birth. As indicated in the main text, the appearance of the lungs of the 17.5–18 days p.c. embryos (illustrated in *Plate 66,o,p*; *Plate 41*) may not accurately reflect the situation *in utero*, as a limited degree of alveolar expansion is evident in these sections, presumably because these embryos must have made attempts to breathe between the time of their isolation and fixation. In addition, in the fully expanded state during the postnatal period, all of the potential space within the pleural cavities would be expected to have been filled by lung tissue (cf. *Plate 40g,e,f*; *Plate 40h,a–d*).

The wall of the trachea is seen to have a quite different histological appearance to that of the oesophagus as early as about 11.5–12 days p.c. (e.g. *Plate 26c,c*). The difference increases over the next few days, so that by about 14.5 days p.c., the mesenchyme condensations of precartilage of the primordia of the tracheal "rings" are clearly seen, particularly in sagittally sectioned material (*Plate*

(*continued on page 448*)

PLATE 66

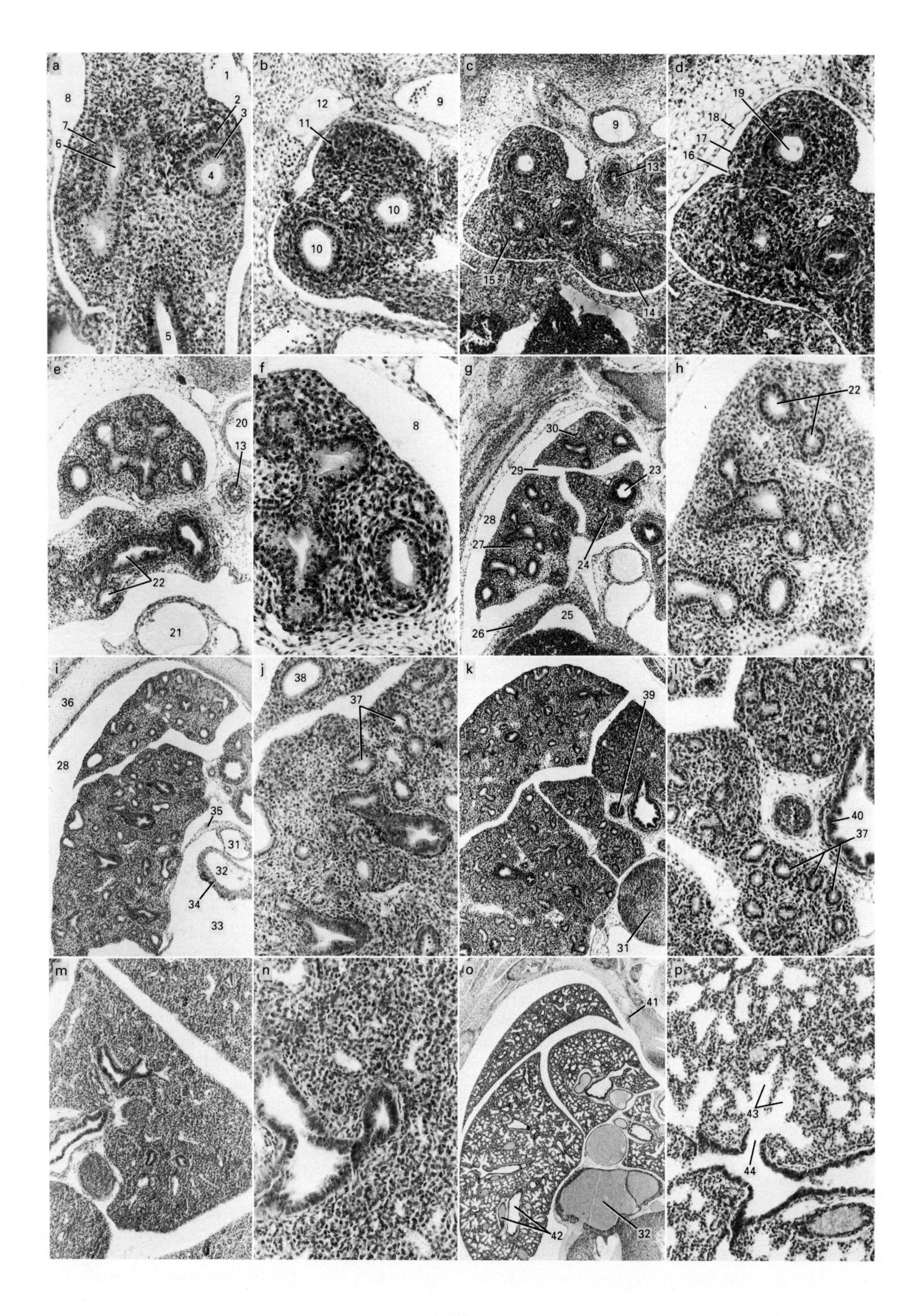
a
b
c
d
e
f
g
h
i
j
k
l
m
n
o
p
1
2
3
4
5
6
7
8
9
10
11
12
13
14
15
16
17
18
19
20
21
22
23
24
25
26
27
28
29
30
31
32
33
34
35
36
37
38
39
40
41
42
43
44

Plate 66

Differentiation of the lung

a 10.5–11 days p.c. (×160)
b 11.5–12 days p.c. (×160)
c,d 12–12.5 days p.c. (**c** ×100; **d** ×160)
e,f 12.5–13 days p.c. (**e** ×100; **f** ×250)
g,h 13.5–14 days p.c. (**g** ×63; **h** ×160)
i,j 14.5 days p.c. (**i** ×63; **j** ×160)
k,l 15.5 days p.c. (**k** ×63; **l** ×160)
m,n 16.5 days p.c. (**m** ×100; **n** ×160)
o,p 17.5 days p.c. (**o** ×40; **p** ×160)

1. left pericardio-peritoneal canal
2. left lung bud formed by endodermal outgrowth of foregut surrounded by splanchnopleuric mesenchyme
3. columnar/cuboidal epithelial lining (of endodermal origin) of left main bronchus
4. lumen of left main bronchus
5. lumen of stomach
6. lumen of right main bronchus
7. right lung bud
8. right pericardio-peritoneal canal
9. midline dorsal aorta
10. lumen of lobar bronchus
11. region of right lung bud destined to form cranial lobe of right lung
12. right posterior cardinal vein
13. oesophagus
14. region of right lung bud destined to form accessory lobe of right lung
15. region of right lung bud destined to form middle lobe of right lung
16. furrow which partially separates the cranial lobe from the middle lobe of the right lung
17. mesothelial cell layer covering surface of primitive lung bud (future visceral layer of pleura)
18. mesothelial cell layer which will differentiate into parietal layer of pleura
19. right cranial lobe bronchus
20. thoracic (descending) aorta
21. posthepatic component of inferior vena cava
22. segmental bronchi
23. right caudal lobe bronchus
24. caudal lobe of right lung
25. right hepatic recess of peritoneal cavity
26. right pleuro-peritoneal membrane (fold) (i.e. future right dome of the diaphragm)
27. middle lobe of right lung
28. right pleural cavity
29. fissure between cranial lobe and middle lobe of right lung
30. cranial lobe of right lung
31. lumen of inferior vena cava
32. lumen of right atrium
33. caudal extremity of pericardial cavity
34. wall of right atrium
35. pleuro-pericardial membrane
36. cartilage primordium of rib
37. terminal bronchi/bronchioles
38. dilated peripheral branch of bronchial tree
39. branch from right pulmonary artery to caudal lobe of right lung
40. epithelial cells lining main bronchus some of which bear cilia on their luminal border
41. right sympathetic trunk
42. peripheral branches of right pulmonary vessels
43. alveoli lined with squamous epithelial cells
44. junction between bronchiole and alveolar duct

33a,a). By 15.5 days p.c., early evidence of cartilage formation is seen in the "rings" (eg. *Plate 35e,c,d*; *Plate 36c,a*), and the latter become increasingly more differentiated over the next few days (e.g. 16.5 days p.c., *Plate 37d,b–d*; *Plate 38c,a*; 17.5 days p.c., *Plate 40e,a–c*; *Plate 41*).

12 Histochemical demonstration of primordial germ cells
Text associated with Plate 67

While various regions of the early postimplantation mouse embryo show positive evidence of intracellular alkaline phosphatase enzyme activity, most noticeably in parts of the neural tube (in cells within the mantle zone of the basal plate, which are possibly prospective motor neuron cells) (Kwong and Tam, 1984), somite mesoderm, intestinal epithelium (Rossi and Reale, 1957), in the apical ectodermal ridge etc., using a combination of such criteria as their characteristic location, cellular morphology, and histochemical staining appearance, it is also possible to unequivocally identify primordial germ cells (PGCs) during the early organogenetic stages of mouse development (Chiquoine, 1954; Ozdzeński, 1967).

Embryos were isolated and fixed in an 80% solution of ethanol in distilled water maintained at 4°C, transferred to 70% alcohol after approximately 1 hour, and routinely sectioned after embedding in paraffin wax. The slides were processed in order to demonstrate the presence of intracellular alkaline phosphatase enzyme activity using a modification of the technique originally described by Ackerman (1962) and Kaplow (1968). The appropriate reagents were supplied in kit form by the Sigma Chemical Company (Sigma Catalogue Reference No. 86–R) and the sections counterstained with 0.5% methyl green.

Using this technique, the author has demonstrated the presence of PGCs in normal diploid and diploid parthenogenetic mouse embryos (Kaufman and Schnebelen, 1986), as well as in diandric and digynic triploid (Kaufman *et al.*, 1990) and tetraploid mouse embryos (Kaufman, 1991). In all of these normal and genetically abnormal embryos, the location of the PGCs is closely related to the stage of development studied. The findings of a number of authors in this regard are summarized in Table 4. Alkaline phosphatase positive cells are first evident at the early primitive streak stage, and are presumed to be the precursors of the PGCs. At this stage, these cells are principally located in the primitive streak region. In developmentally slightly more advanced primitive streak stage embryos, the alkaline phosphatase positive cells (now assumed to be PGCs), are located in the mesodermal tissue of the visceral yolk sac, at the base of the allantois, and at the caudal end of the primitive streak. By the early somite stage (embryos with up to about 10 pairs of somites), PGCs are mostly located in association with the hindgut endoderm, though some are still seen to be present at the base of the allantois (*Plate 67,a*) and at the caudal end of the primitive streak, or occasionally they may be found in the mesodermal tissue of the visceral yolk sac (*Plate 67,a*). In embryos with 11–20 pairs of somites, the majority of the PGCs are associated within the hindgut endoderm (*Plate 67,c–g*), though a few of these cells are still seen to be located in the wall of the visceral yolk sac (*Plate 67,h*). In embryos with about 15–20 pairs of somites, a few of the PGCs are seen to have migrated into the mesentery of the hindgut. In embryos with 21–30 pairs of somites, about two-thirds of the PGCs present are found in association with the hindgut endoderm, about one-third are now present within the hindgut mesentery, and a relatively small number may be found in the medial part of the genital ridge. In embryos with about 31–36 pairs of somites, the majority of the PGCs are present in the hindgut mesentery, while a significant number have now migrated into the genital ridge. In the developmentally more advanced embryos in this group, a higher proportion of PGCs would be expected to be located within the genital ridges (*Plate 67,i–n*).

It is certainly clear from sections through embryos with about 35 pairs of somites, stained to demonstrate the presence of alkaline phosphatase enzyme activity, that the genital ridge is located on the medial aspect of the urogenital ridge, and that the mesonephros (which is a more substantial structure at this stage of development than the genital ridge) and its duct are located on the lateral aspect of the urogential ridge (*Plate 67,m*). In the embryo illustrated, additional (extra-genital ridge) sites of alkaline phosphatase-positive cells are clearly present, most obviously in association with the ventral part of the neural tube (see above), and in the apical ectodermal ridge (*Plate 67,i–k*). No clear relationship has so far been observed, however, between the genetic sex of an embryo and the location of its PGCs, nor in the total number of PGCs present during the early stages of development (Kaufman, unpublished). In the latter studies, the genetic sex of embryos was established from a cytogenetic analysis of their extra-embryonic tissues.

(*Plate 67 over page*)

PLATE 67

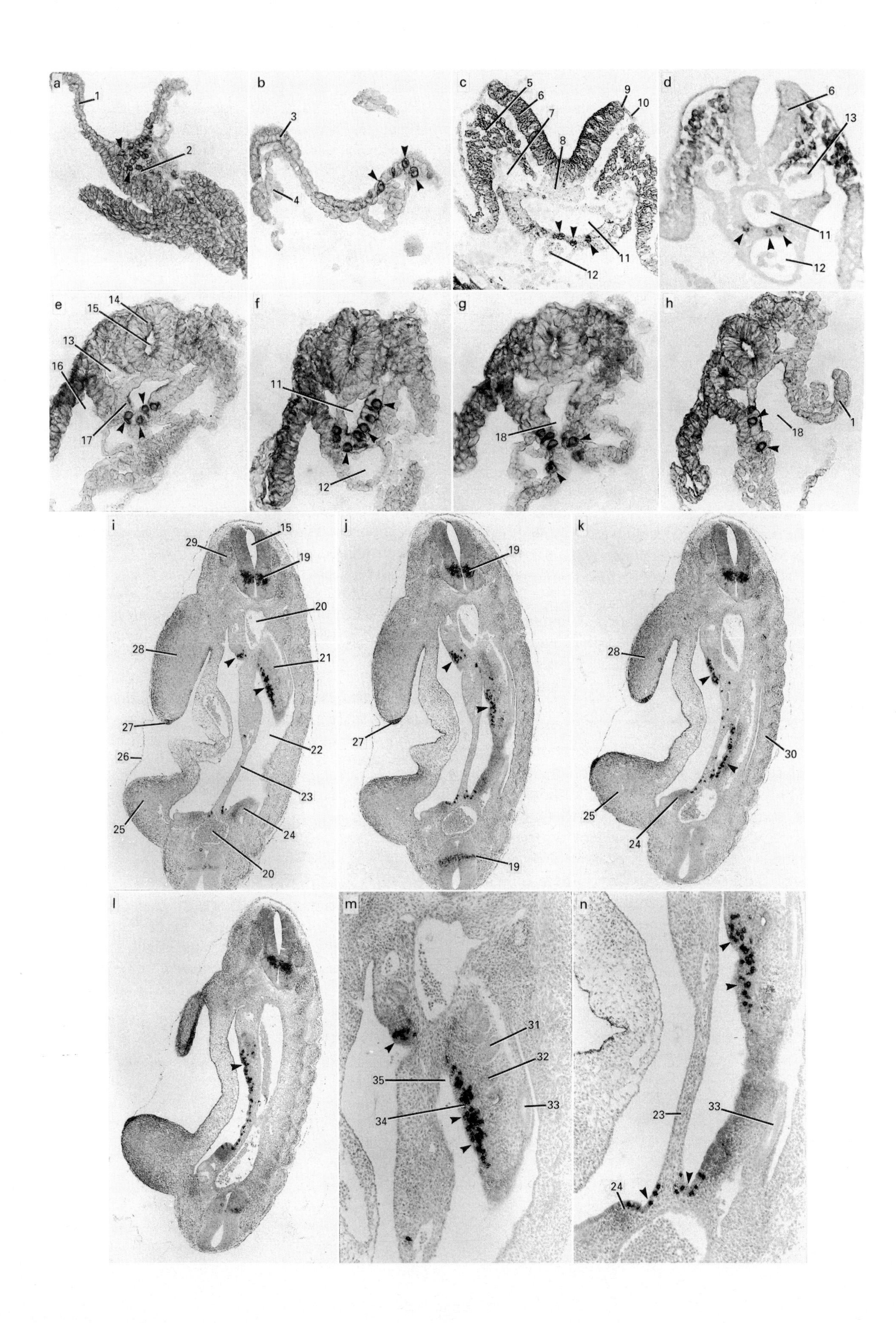

a
b
c
d
e
f
g
h
i
j
k
l
m
n
1
2
3
4
5
6
7
8
9
10
11
12
13
14
15
16
17
18
19
20
21
22
23
24
25
26
27
28
29
30
31
32
33
34
35

Plate 67

Primordial germ cells – histochemical demonstration due to their high level of intracellular alkaline phosphatase enzyme activity

a,b diandric triploid embryo with 6–8 pairs of somites (**a,b** ×160)
c diandric triploid embryo with 10–12 pairs of somites (×100)
d diandric triploid embryo with 12–14 pairs of somites (×100)
e–h diandric triploid embryo with 12–14 pairs of somites (**e–h** ×250)
i–n 10.5 days p.c.; normal diploid embryo with 35 pairs of somites (**i–l** ×40; **m,n** ×100)

Arrowheads have been used to draw attention to the location of a selection of the primordial germ cells illustrated in these sections.

1. visceral yolk sac
2. base of allantois
3. endodermal component of visceral yolk sac
4. mesodermal component of visceral yolk sac
5. intra-embryonic mesoderm
6. neuroepithelium
7. dorsal aorta
8. condensation of tissue (poorly defined) in caudal region of notochord
9. junction between surface ectoderm and neural ectoderm
10. surface ectoderm
11. hindgut diverticulum
12. vitelline (omphalomesenteric) artery
13. primitive nucleated red blood cells within dorsal aorta
14. neural tube
15. neural lumen
16. intra-embryonic coelom (future peritoneal cavity)
17. wall of hindgut diverticulum
18. entrance to hindgut diverticulum
19. high level of intracellular alkaline phosphatase enzyme activity characteristically found in cells within the mantle zone of basal plate of neural tube (possibly prospective motor neuron cells)
20. midline dorsal aorta
21. rostral part of urogenital ridge
22. intra-embryonic coelomic cavity (peritoneal cavity)
23. dorsal mesentery of hindgut
24. caudal part of urogenital ridge
25. hindlimb bud
26. amnion
27. high level of intracellular alkaline phosphatase enzyme activity characteristically found within the cells of the apical ectodermal ridge
28. forelimb bud
29. dorsal (posterior) root ganglion
30. somite
31. mesonephric tubule
32. mesonephric component of urogenital ridge
33. mesonephric duct
34. gonadal component of urogenital ridge
35. medial coelomic bay

13. Differentiation of the urogenital ridges, gonads and genital duct system
Text associated with Plates 68–71

At the earliest stage of urogenital differentiation, which occurs at about 9 days p.c., it is only possible to recognize the nephric component, which represents the first stage in the differentiation of the intermediate plate mesoderm. At this stage, it is just possible to recognize the pronephric primordium with its associated pronephric duct, though the latter has yet to canalize (*Plate 18a,m–o*; *Plate 18b,a–c*). These structures are initially located in the caudal part of the trunk region. By about 9.5 days p.c., however, the urogenital ridges are first clearly seen, and are located on either side of the dorsal mesentery of the hindgut, being particularly prominent in the mid-trunk region (*Plate 19c,a–f*; *Plate 20c,b–h*). The urogenital ridges extend caudally from this level to the mid-tail region, and protrude into the dorsolateral parts of the peritoneal cavity along almost its entire length. On close inspection, it is possible to recognize the pronephric vesicles, and note that the pronephric ducts are now canalized along much of their length (*Plate 20c,a–g*). The latter are also seen to be directed caudally towards the cloaca.

By 10 days p.c., mesonephric differentiation is now apparent, and the two mesonephroi are seen to be located on the lateral aspect of the urogenital ridges. Large numbers of mesonephric vesicles are present, several being seen at each segmental level. The mesonephric ducts are now canalized throughout their length (*Plate 23d,e,g*), but they do not yet drain into the cloaca (*Plate 23c,a*). The situation is more consolidated by 10.5 days p.c. (*Plate 68,a*), and the mesonephric ducts are now seen to make contact with the urogenital sinus (*Plate 24b,c,d*). The absence of well differentiated mesonephric glomeruli makes it unlikely that these organs function in mice. The mesonephric duct is easily recognized, and is located on the lateral aspect of the mesonephros.

By 11 days p.c., the first indication of the gonadal primordium is seen. This is located on the medial aspect of the urogenital ridge, and at this stage, the histological and morphological appearance of the gonadal primordium is such that it is not possible to distinguish between the developing ovary or testis. This is therefore referred to as the "indifferent" gonad stage.

By 11.5 days p.c., while the mesonephros is at about its maximum volume, the gonad continues to increase in size. The urogenital mesentery is also quite broad (*Plate 68,c*), but from this stage onwards progressively narrows. Within the mesonephros, the mesonephric tubules are particularly prominent structures, as is the mesonephric duct. At this stage, large numbers of primordial germ cells (PGCs) are present within the gonadal blastema, and are particularly well seen if sections are stained histochemically to demonstrate the presence of intracellular alkaline phosphatase enzyme activity (*Plate 67,i–n*). In the lower part of the trunk, in the future pelvic region, just lateral to the urogenital sinus, the distal part of the ureteric buds is seen to be dilated, and these are surrounded by the metanephric blastema (*Plate 26e,c,d*). The ureteric buds are diverticulae of the mesonephric ducts which branch just before they make contact with the urogenital sinus, and are relatively short structures at this stage. Over the next few days, however, they substantially increase in length and differentiate to form the ureters, while at the same time the metanephric blastemas differentiate to form the definitive kidneys.

By 12–12.5 days p.c., the proportionate size of the mesonephros is somewhat smaller than previously, and at the histological level it shows early evidence of regressing, while the gonad continues to increase in volume (*Plate 68,d,e*). Scanning electron microscopy of the lower part of the peritoneal cavity at this stage (*Plate 70,a*) is particularly instructive, in that it clearly reveals the very close relationship that initially exists between the gonad and the mesonephros, and that the gonad is elongated and sausage-shaped. At the histological level, the gonad has a homogeneous granular appearance (*Plate 68,d,e*), and on histological and morphological analysis it is still not possible to establish whether it is either an ovary or a testis. The disposition and appearance of the other accessory structures at this stage are equally ambiguous and unhelpful in this regard. For a descriptive account of the development of the ureteric bud and metanephros (i.e. the ureter and definitive kidney, respectively) from 12.5 to 17.5 days p.c., reference should be made to the text associated with *Plate 72*, as well as to the accounts associated with the relevant sections of these stages of development.

Between about 13 and 13.5 days p.c., critical changes occur in the cellular architecture and histological appearance of the gonads, so that it is now possible unequivocally to distinguish for the first time between an ovary and a testis, though both at this stage have a very similar size and shape. If the gonads are isolated and examined (either in the fresh or "fixed" state) by transmitted light under a dissecting microscope, the ovary is seen to have a homogeneous granular or "spotty" appearance (*Plate 68,f*), while the testis appears to contain wide regularly arranged bands of tissue (the testicular cords, which are destined to form the seminiferous tubules), and consequently has a characteristic "striped" appearance (*Plate 68,g,h*). The urogenital mesentery is still relatively wide at this stage, and is destined to form the mesovarium in the female, and the mesorchium in the male. In both sexes, the mesonephros has all but regressed, leaving only a few seemingly disorganized tubules, though both the mesonephric duct and a more laterally located paramesonephric duct are clearly seen (*Plate 68,f–h*).

The location of the gonads at this stage is similar in the two sexes, both being found in close association with the anterolateral surface of the adrenals, and the upper halves of the kidneys (*Plate 30j,c,d*; *Plate 30k,a,b*). At the histological level, the testes are seen to contain solid cords of tissue in which the germ cells are embedded (*Plate 68,j,k*). Early events associated with spermatogenesis may be seen, including the presence

(continued on page 462)

PLATE 68

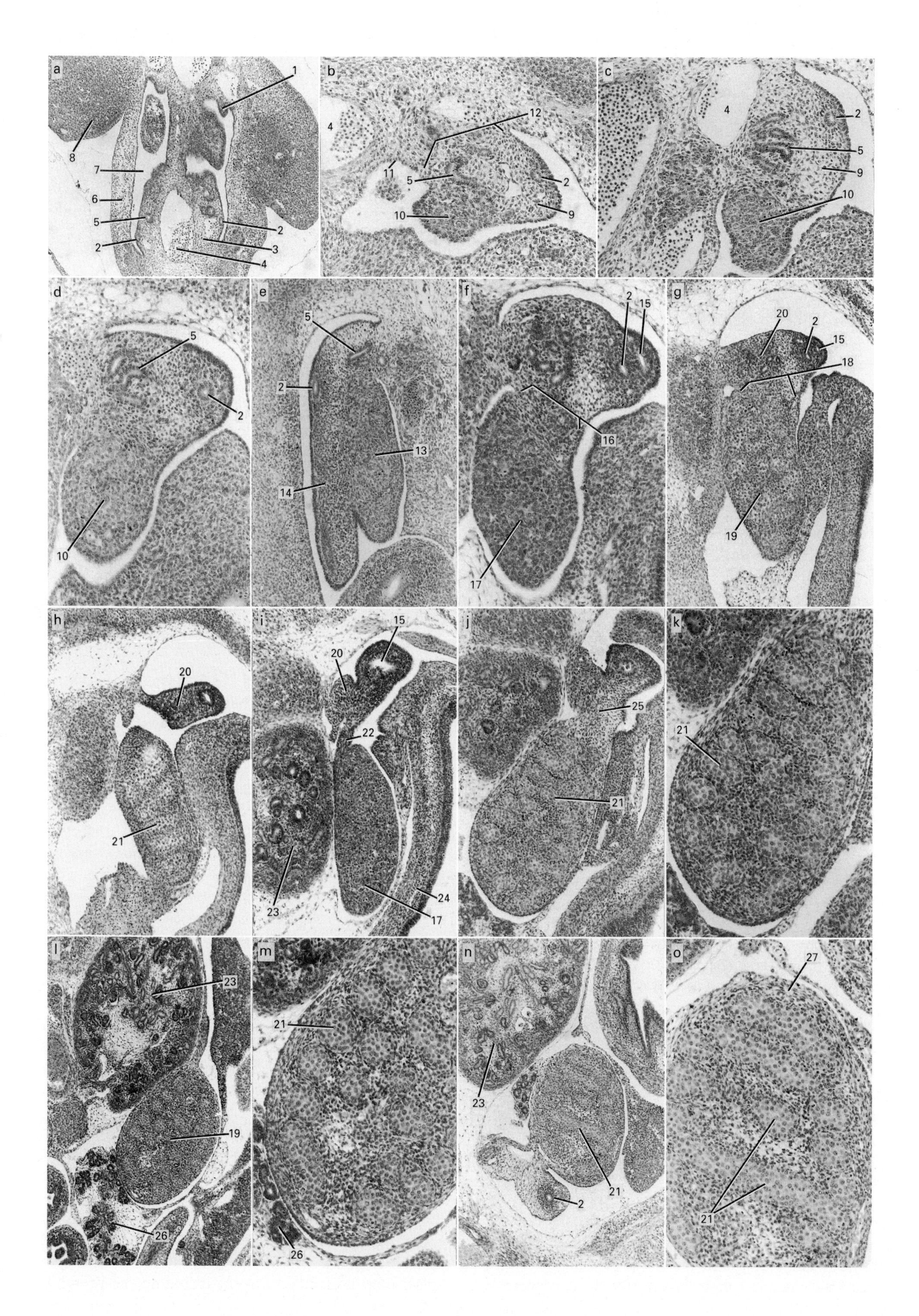
a
b
c
d
e
f
g
h
i
j
k
l
m
n
o
1
2
3
4
5
6
7
8
9
10
11
12
13
14
15
16
17
18
19
20
21
22
23
24
25
26
27

Plate 68

Differentiation of the urogenital ridge and gonads I

a 10.5 days p.c. (×63)
b 11 days p.c. (×160)
c 11.5 days p.c. (×160)
d,e 12.5 days p.c. (**d** ×160; **e** ×100)
f 13.5 days p.c. (female) (×160)
g 13 days p.c. (male) (×100)
h 13.5 days p.c. (male) (×100)
i 14.5 days p.c. (female) (×100)
j,k 14.5 days p.c. (male) (**j** ×100; **k** ×160)
l,m,n,o 15.5 days p.c. (male) (**l** ×63; **m** ×160; **n** ×63; **o** ×160)

1. rostral extremity of mesonephric ridge
2. mesonephric duct (Wolffian duct)
3. caudal extremity of mesonephric ridge
4. midline dorsal aorta
5. mesonephric tubule
6. body wall in mid-abdominal region
7. intra-embryonic coelomic cavity (peritoneal cavity)
8. forelimb bud
9. mesonephric component of urogenital ridge
10. gonadal component of urogenital ridge ("indifferent" gonad stage)
11. coelomic epithelium lining medial coelomic bay
12. wide urogenital mesentery
13. gonad (at advanced "indifferent" stage)
14. mesonephros
15. paramesonephric duct (Müllerian duct)
16. wide gonadal mesentery (mesovarium)
17. ovary (recognized by the absence of testicular cords and relatively homogeneous granular or "spotty" appearance)
18. wide gonadal mesentery (mesorchium)
19. testis (recognized by the presence of testicular cords, which give the testis a "striped" appearance)
20. mesonephros (regressing)
21. seminiferous tubule (solid at this stage) formed from testicular cord
22. mesovarium
23. metanephros
24. wall of stomach
25. mesorchium
26. pancreas
27. condensation of mesenchyme tissue to form the tunica albuginea of the testis

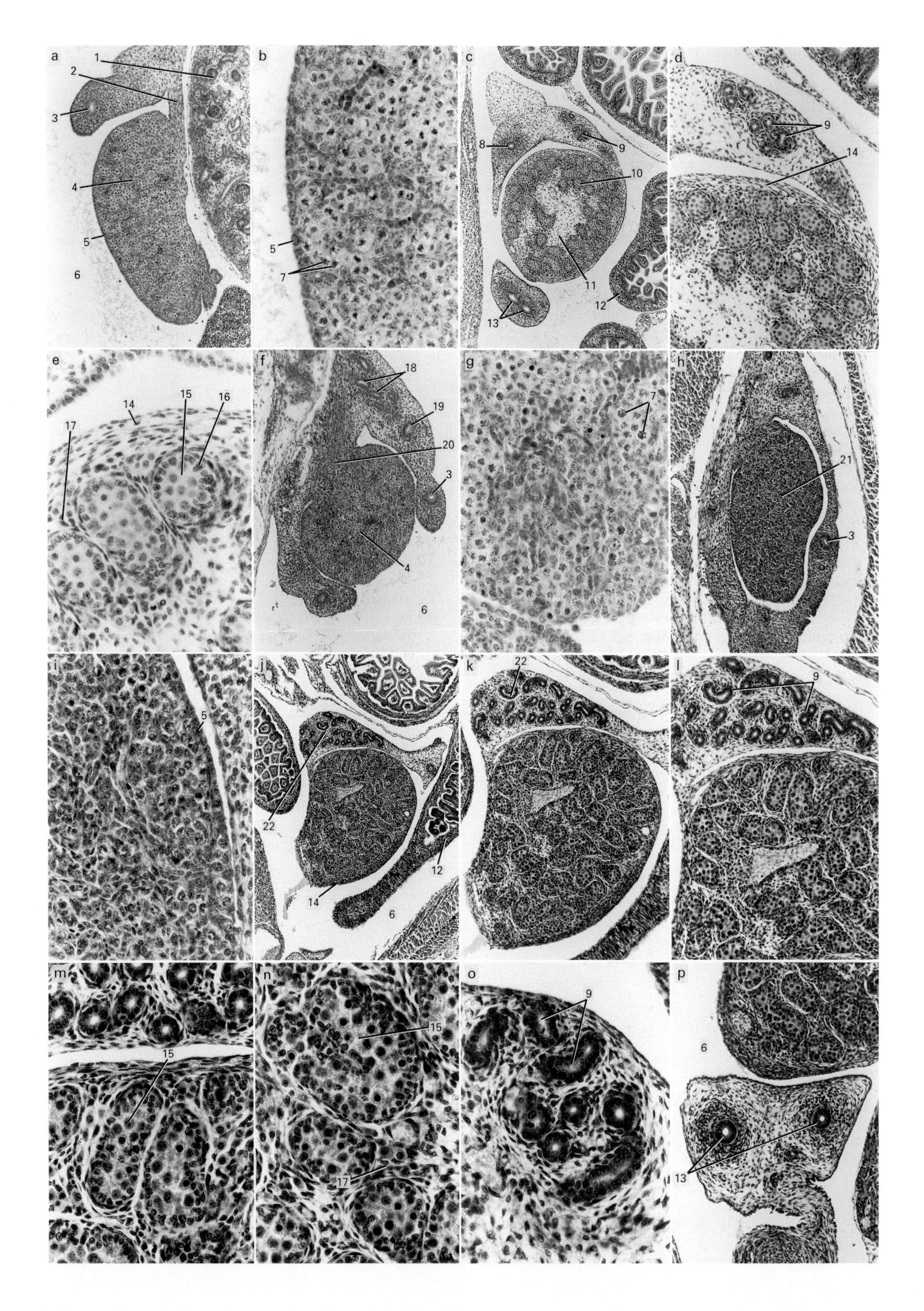
a
1
2
3
4
5
6
b
5
7
c
8
9
10
11
12
13
d
9
14
e
14
15
16
17
f
18
19
20
3
4
6
g
7
h
21
3
i
5
j
22
12
14
6
k
22
l
9
m
15
n
15
17
o
9
p
6
13

Plate 69

Differentiation of the urogenital ridge and gonads II

a,b 15.5 days p.c. (female) (**a** ×100; **b** ×400)
c,d,e 16.5 days p.c. (male) (**c** ×63; **d** ×160; **e** ×400)
f,g 16.5 days p.c. (female) (**f** ×100; **g** ×400)
h,i 17.5 days p.c. (female) (**h** ×100; **i** ×400)
j–p 17.5 days p.c. (male) (**j** ×63; **k** ×100; **l** ×160; **m,n,o** ×400; **p** ×160)

1. definitive kidney (metanephros)
2. mesovarium
3. paramesonephric duct (region of future oviduct/ uterine horn)
4. ovary (with characteristic "spotty" appearance due to presence of ovigerous cords (egg clusters) and whorl-like appearance of stromal cells, and absence of "stripes" associated with the absence of testicular cords/seminiferous tubules)
5. "germinal" epithelium (in continuity with peritoneal mesothelium) which forms the single-layered covering of the ovary
6. peritoneal cavity
7. germinal cells (oogonia) in mitosis, and primary oocytes at leptotene stage of prophase of meiosis I at this stage of ovarian differentiation
8. mesonephric duct – differentiating in this location to form proximal part of ductus (vas) deferens
9. mesonephric tubules – differentiating in this location to form a single duct which is disposed in numerous coils and termed the epididymis
10. seminiferous tubule (solid at this stage) formed from testicular cord
11. mesenchyme tissue in central region of medulla of testis
12. wall of loop of small intestine
13. mesonephric duct (middle region of ductus (vas) deferens)
14. condensation of mesenchyme tissue (covered by single layer of peritoneal mesothelium) to form the tunica albuginea of the testis
15. region of future lumen of seminiferous tubule containing primitive spermatogenic cells
16. primitive Sertoli cell attached to the basement membrane of the seminiferous tubule (elsewhere at the periphery of the tubule, the Sertoli cells are interspersed with spermatogonial cells)
17. interstitial cells (it is not possible to distinguish morphologically between mesenchyme cells and Leydig cells at this stage)
18. degenerating mesonephric tubules
19. degenerating mesonephric duct
20. region of rete ovarii in mesovarium
21. lower pole of ovary
22. region of prospective epididymis; this coiled tube makes contact with the seminiferous tubules via numerous efferent ductules in the region of the rete testis (mediastinum testis)

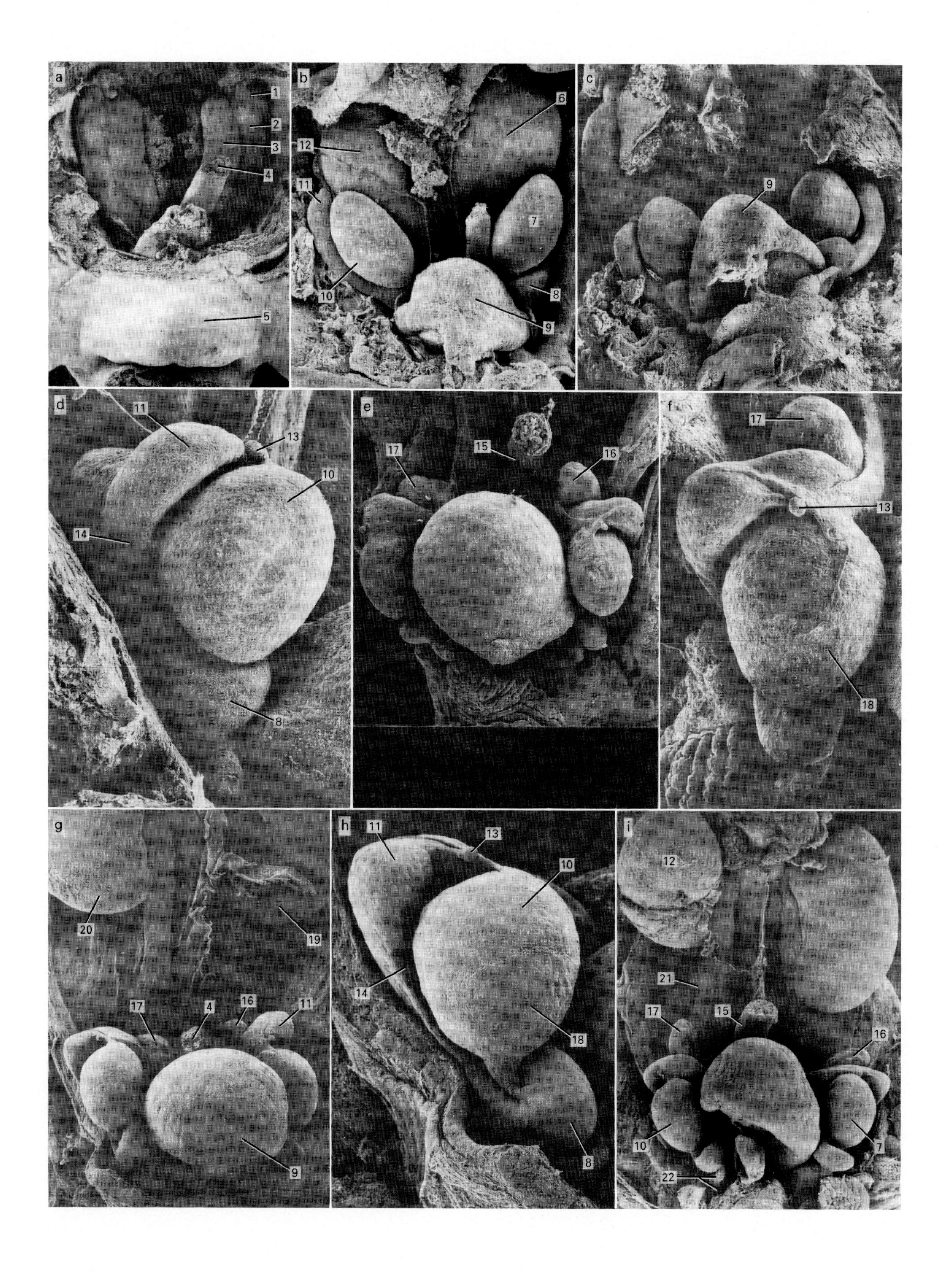
a
1
2
3
4
5
b
6
12
11
7
10
8
9
c
9
d
11
13
10
14
8
e
15
17
16
f
17
13
18
g
20
19
17
4
16
11
9
h
11
13
10
14
18
8
i
12
21
17
15
16
10
7
22

Plate 70

Differentiation of the gonads and reproductive ducts in the male

a 12.5 days p.c.; late "indifferent" gonad stage, overview of gonads and associated ducts
b 14.5 days p.c.; overview of testes, epididymes, bladder and kidneys
c 15.5 days p.c.; overview of testes, epididymes, bladder and kidneys
d 15.5 days p.c.; close-up view of testis and epididymis
e 16.5 days p.c.; overview of testes, epididymes, seminal vesicles and bladder
f 16.5 days p.c.; close-up view of testis and epididymis
g 17.5 days p.c.; overview of testes, epididymes, seminal vesicles, bladder and lower poles of kidneys
h 17.5 days p.c.; close-up view of testis and epididymis
i 18.5 days p.c.; overview of testes, epididymes, seminal vesicles, bladder and kidneys

1. upper part of urogenital mesentery
2. mesonephros
3. gonad (at "indifferent" stage)
4. cut section of hindgut (rectum, proximal part removed)
5. genital tubercle
6. left kidney
7. left testis
8. cauda (tail) epididymis
9. wall of bladder
10. right testis
11. caput (head) epididymis
12. right kidney
13. appendix testis
14. corpus (body) epididymis
15. wall of hindgut (rectum)
16. rostral part of left seminal vesicle
17. rostral part of right seminal vesicle
18. covering of tunica vaginalis testis, with subjacent tunica albuginea
19. lower pole of left kidney
20. lower pole of right kidney
21. lateral border of right psoas major muscle
22. ductus deferens and (inferiorly) the gubernaculum testis

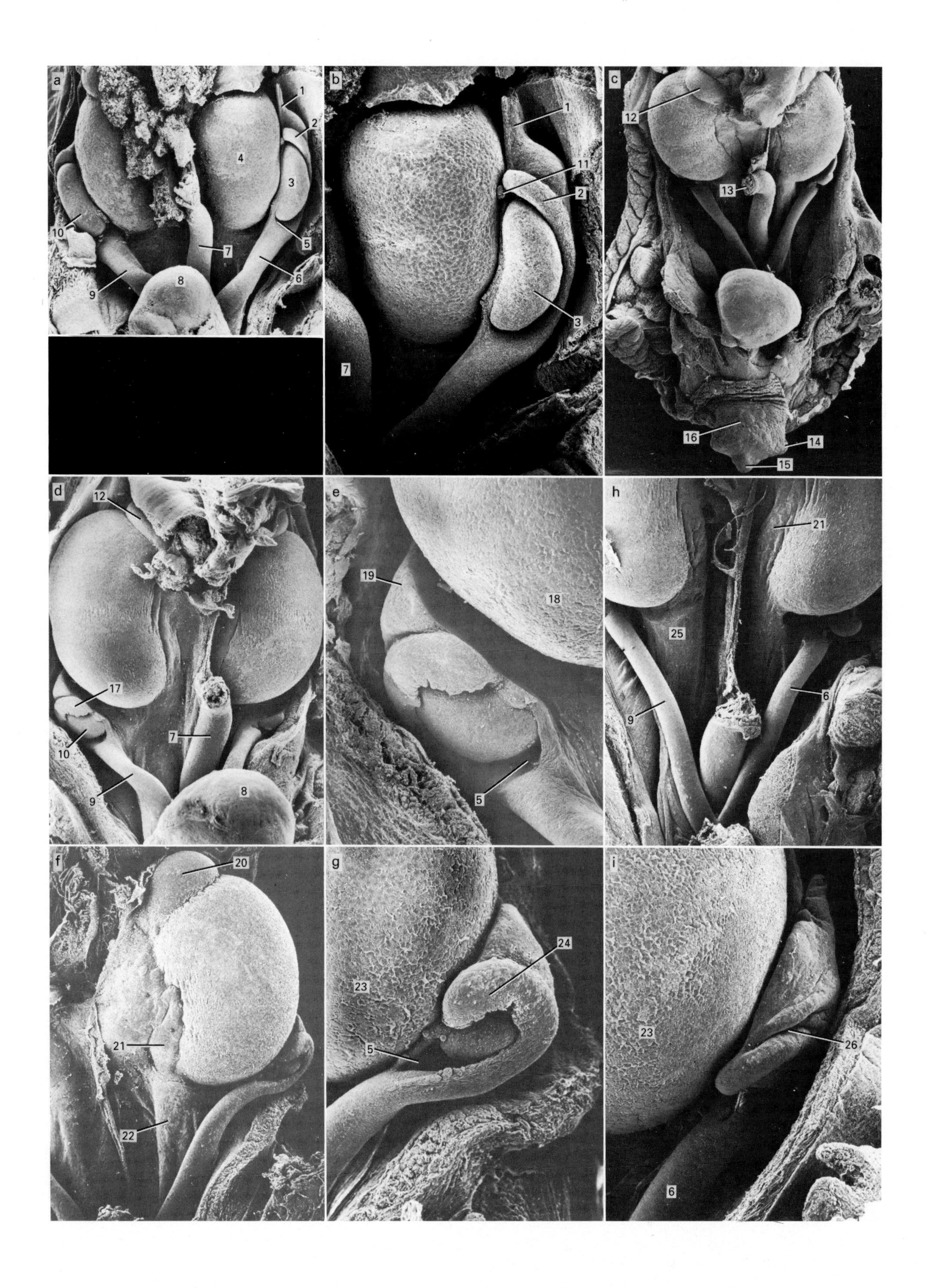
a
1
2
4
3
10
5
7
6
9
8
b
1
11
2
3
7
c
12
13
16
14
15
d
12
17
10
7
9
8
e
19
18
5
h
21
25
9
6
f
20
21
22
g
24
23
5
i
23
26
6

Plate 71

Differentiation of the ovaries and reproductive duct system in the female

a 15.5 days p.c.; overview of ovaries, reproductive ducts (paramesonephric ducts), bladder and kidneys
b 15.5 days p.c.; close-up view of left ovary and surrounding structures (same embryo as illustrated in **a**)
c 16.5 days p.c.; overview of ovaries, reproductive ducts, bladder, genital tubercle and kidneys
d 16.5 days p.c.; overview of ovaries, reproductive ducts and surrounding structures
e 16.5 days p.c.; close-up view of right ovary and surrounding structures (same embryo as illustrated in **d**)
f 17.5 days p.c.; overview of left ovary, reproductive duct and surrounding structures
g 17.5 days p.c.; close-up view of left ovary and surrounding structures (same embryo as illustrated in **f**)
h 18.5 days p.c.; overview of reproductive ducts and surrounding structures
i 18.5 days p.c.; close-up view of left ovary and surrounding structures

1. suspensory ligament of left ovary
2. upper (rostral) pole of left paramesonephric (Müllerian) duct
3. left ovary
4. left kidney
5. opening into left paramesonephric duct from peritoneal (coelomic) cavity
6. left paramesonephric duct (in this location, will give rise to the left uterine horn, and more rostrally the left oviduct and the ovarian capsule)
7. wall of hindgut (rectum)
8. fundus of the bladder
9. right paramesonephric duct
10. right ovary
11. cystic vesicular appendage (believed to be of either mesonephric duct or paramesonephric duct origin)
12. right adrenal gland
13. cut end of hindgut
14. primordium of prepuce of clitoris
15. glans clitoridis
16. genital tubercle
17. upper (rostral) pole of paramesonephric duct at early stage in formation of the ovarian capsule
18. lower pole of right kidney
19. suspensory ligament of right ovary
20. left adrenal gland
21. pelvic region of left kidney
22. peritoneum overlying left ureter
23. lower pole of left kidney
24. upper (rostral) pole of paramesonephric duct at intermediate stage in formation of the ovarian capsule
25. peritoneum overlying right ureter
26. upper (rostral) pole of paramesonephric duct at advanced stage in formation of the ovarian capsule

of numbers of type A spermatogonia in division. The cords represent the primitive seminiferous tubules, and located between the latter are relatively modest amounts of interstitial tissue. At this stage, however, it is not possible to distinguish between the various cell types present in this location. The ovaries, by contrast, are still relatively featureless, and continue to have a homogeneous cellular morphology (*Plate 68,i*).

In appropriate sections through the pelvis of female embryos, it is possible to see that both the mesonephric and paramesonephric ducts run in parallel inferomedially towards the urogenital sinus, the paramesonephric ducts being the more lateral of the two throughout the upper part of their course. Both are canalized throughout their entire length, and at this stage have a fairly similar luminal diameter caudally (*Plate 30k,c,d*), while the paramesonephric ducts tend to have a greater luminal diameter in the rostral part of their course (*Plate 68,i*). The mesovarium is seen to be considerably narrower in places than previously (*Plate 30j,d*; *Plate 30k,a,b*; *Plate 68,i*) The mesorchium, by contrast, tends to be slightly wider than the mesovarium and, particularly in the region of the rete testis, tubular structures are seen which drain into the mesonephric duct (*Plate 68,j*). These tubules are destined to form the efferent ductules, and will subsequently connect the seminiferous tubules with the upper part (caput) of the epididymis.

By 14–14.5 days p.c., it is now possible to distinguish between the gonads of the two sexes, principally because of the different relationship that now exists between the latter and their associated genital ducts, and because the ovaries and testes no longer have a similar size and shape. The testes tend to be fairly ovoid structures, whereas the ovaries tend to be more elongated than ovoid in shape, with their craniocaudal axis being almost three times as long as their width. Furthermore, the testes tend to have an increasingly greater volume than that of the ovaries (*Plate 70,b*). At this stage, the kidneys have only initiated their "ascent" to their definitive location, and are still to be found in the lower to mid-abdominal region. The testes are now seen to be located in close proximity to the anterolateral surface of the lower half of the kidneys, whereas the ovaries tend to be at a slightly lower level, and are more closely associated with the inferolateral surface of the kidneys.

It is unclear whether this change in the relationship between the kidneys and the gonads, compared with the situation observed previously, results from the "descent" of the gonads, or is a consequence of the "ascent" of the kidneys. In addition, it is possible that the gonads may be relatively "fixed" in their position by either the ovarian ligament or by the gubernaculum testis (its homologue in the male), and any changes observed may result from the disproportionate growth of the caudal compared to the rostral half of the embryo. Most likely, all of the above factors are involved to some extent in accounting for the changing relationship that is observed between the gonads and the kidneys over this period. The testis is also seen to contain mesenchymatous elements located just subjacent to the outer coelomic epithelial covering of the gland (termed the tunica vaginalis), and these elements are destined to form its fibrous capsule (termed the tunica albuginea).

By 15.5–16 days p.c., the testes are seen to be located in a more pelvic position than previously, and are now situated on either side of the upper half of the bladder (*Plate 70,c*). The ovaries, by contrast, show little further "descent", and are located at a similar level to that described previously (*Plate 71,a,b*). It should be noted, though, that the relationship of the ovaries to the kidneys can be deceptive, since the kidneys are quite large retroperitoneal organs, and at this stage not only do they protrude into the dorsal part of the lower half of the peritoneal cavity (*Plate 35k,a–d*; *Plate 70,c*; *Plate 71,b*), but their lower poles still protrude well into the pelvis (*Plate 35c,a*).

Whereas previously the testes were ovoid in shape, they now have an almost rounded profile (*Plate 70,c,d*). The ovaries, by contrast, are more ovoid and less elongated than previously, and the length of their long axis is now only about twice that of their short axis (*Plate 71,b*). The accessory structures in the male are now extremely prominent, and this is particularly the case with regard to the epididymis, which is clearly divided into a caput (head), corpus (body), and a caudal (tail) region (*Plate 70,d*). A small vesicular appendage is now clearly seen to be located between the upper pole of the testis and the caput of the epididymis (*Plate 70,d*), while a gubernaculum testis-like structure is seen to be associated with the wall of the lower part of the tail of the epididymis and the ductus deferens (*Plate 70,d*, unlabelled). From its size and location, it appears most likely that this cystic appendage is the appendix testis (which is of paramesonephric duct origin), though it is not possible to exclude the possibility that this structure is in fact the appendix epididymis (which is derived from the blind cranial end of the mesonephric duct).

In the case of the ovaries, these are at this stage partially enveloped on their posterior aspect by the paramesonephric duct. As in the male, it is uncertain whether the minute cystic appendage seen in close proximity to the cranial pole of the paramesonephric duct is the mesonephric or paramesonephric appendix, and it could therefore be derived either from the mesonephric duct or from the paramesonephric duct, respectively (for discussion, see Keibel and Mall, 1910, 1912; Hamilton and Mossman, 1972). The suspensory ligament (of the gonad) is a substantial structure at this stage in the female, and is attached to the rostral part of the paramesonephric duct (*Plate 71,a,b*). In the male, it is attached to the upper pole of the testis, close to the caput epididymis (*Plate 70,f*). This structure is present in both sexes, and passes from the rostral pole of the gonad to the atrophying mesonephros, and more rostrally to the diaphragm (Hamilton and Mossman, 1972). By the time of birth, it has virtually completely regressed.

The histological appearance of the testis is similar to that described previously, though the volume occupied by mesenchyme tissue, which is located between the seminiferous tubules, varies quite considerably in different embryos at this stage of development (*Plate 68,l–o*). At least some of the mesenchyme tissue represents the precursor elements of the interstitial tissue which, even at this stage, plays a critical role in hormone production, and in influencing the differentiation of the genital duct system and the secondary sexual features. In the female, the ovary has a "spotty" appearance, is swollen by the presence of ovigerous cords (egg clusters), and already has whorl-like stromal cells which are a characteristic feature of the adult gland (*Plate 69,b*). Most of the oogonia are in mitosis, and the primary oocytes that are present have entered the leptotene stage of the first meiotic division. The ovary is covered by a modification of the coelomic (peritoneal) mesothelium, termed the "germinal" epithelium (*Plate 69,a,b*), there being no equivalent in the female to the capsule of the testis.

By 16–16.5 days p.c., it is possible to see that the testes have now descended well below the brim of the pelvis, being guided in their descent, it is believed, by the gubernaculum testis (*Plate 70,e,f*). The latter passes via the internal inguinal ring and through the inguinal canal to the inner aspect of the scrotum. The structures that are noted for the first time at this stage are the seminal vesicles, whose upper parts are directed rostrally, and are now easily seen on scanning electron micrographs of this region (*Plate 70,e,f*). The seminal vesicles are diverticulae of the caudal part of the mesonephric ducts (ductus deferens at this stage) which branch just before they pass through the wall of the prostatic region of the urethra, as the ejaculatory ducts.

The ovaries at this stage have become increasingly enveloped by the paramesonephric duct tissue that is destined to form the capsule of the ovary (*Plate 71,d,e*). The suspensory ligament (of the gonad) is still a particularly prominent structure at this stage of development in the female (*Plate 71,e*). The kidneys are now close to their definitive level, and the ovaries are largely hidden by their inferolateral border. The two paramesonephric ducts (the future uterine horns) descend towards the midline and meet just below the neck of the bladder, though their respective lumina are separated by their two intervening walls. The paramesonephric ducts pass in front of the two ureters as they descend into the pelvis. At the histological level, the contents of the seminiferous tubules now show increased evidence of differentiation. Primitive Sertoli cells are attached to their basement membrane, and are interspersed with numerous spermatogonial cells (*Plate 69,e*). In the spaces between the seminiferous tubules, it is still not possible to unequivocally distinguish between mesenchyme cells and the precursors of the interstitial cells (of Leydig). The histological appearance of the ovary is similar to that described previously.

At 17–17.5 days p.c., the distance between the lower poles of the kidneys and the testes and their closely allied structures (including the bladder) is quite considerable (*Plate 70,g*), but their overall appearance and relationship to each other is very similar to that described previously (*Plate 70,h*). In the case of the ovaries, while their relationship to the kidneys is similar to that described previously (*Plate 71,f*), they tend to be enveloped to an even greater degree by the paramesonephric duct tissue that is destined to form the ovarian capsule (*Plate 71,g*). At the histological level, the most significant change seen relates to the appearance of the seminiferous tubules and their contents. These tubules are now seen to be canalized, and filled with large numbers of darkly staining spermatogenic cells (*Plate 69,m,n*). The volume of the spaces between the seminiferous tubules is also reduced (*Plate 69,j–n*), but it is still not possible to distinguish between the different cell types present in this location. The differentiation of the mesonephric duct to form the efferent ducts and the epididymis is now progressing apace (*Plate 69,o*), and more distally the thick wall of the ductus deferens allows this structure to be readily recognized (*Plate 69,p*). The histological appearance of the ovary is not significantly different from that described previously (*Plate 69,h,i*). However, for a more detailed description of the gross and microscopic features of the gonads and their closely allied structures in both the male and female, the reader should refer to the text associated with the appropriate section of Stage 26.

By 18–18.5 days p.c., the overall disposition of both the testes and their closely allied structures (*Plate 70,i*) and the ovaries (*Plate 71,h,i*) are very similar to the situation observed at birth. While the gross morphological and histological appearance of these structures in the male is similar to the situation observed previously, the situation in the female is slightly different. In particular, it is now apparent that the ovary is virtually completely surrounded by the primitive ovarian capsule (*Plate 71,i*). Similarly, at the histological level, it is possible to see that some of the primary oocytes are at the dictyate stage of the first meiotic division, and are surrounded by a single layer of follicle cells, these complexes being termed the primary follicles. The latter increase in number over the next few days, so that the ovary tends to contain large numbers of primary follicles during the early postnatal period.

14. Differentiation of the kidney
Text associated with Plate 72

It is unclear at the present time whether the mesonephros functions as an excretory organ in the mouse. Certainly, at no stage during its differentiation are recognizable glomeruli seen, though by 11.5–12 days p.c., the mesonephric duct and tubules are quite prominent structures (*Plate 26e,a–f*; *Plate 68,c*). Shortly before this time, from a site just proximal to the caudal end of the mesonephric duct, a diverticulum is given off, termed the ureteric bud, which is directed rostrally, and its dilated end makes contact with the metanephric blastema. The latter is the most caudally located of the structures associated with the intermediate plate mesoderm (nephrogenic cord). Two other structures differentiate from this tissue prior to the formation of the metanephric blastema, namely the pronephros, and slightly later on, the mesonephros. These structures develop in a craniocaudal sequence, and after the first system differentiates and subsequently regresses, the next system takes over whatever roles it may have had, and takes on additional new functions. In addition, the mesonephros is believed to take over and utilize the pronephric duct, while the metanephros utilizes a diverticulum of the mesonephric duct, the ureteric bud, as indicated above, and this subsequently gives rise to the drainage system of the definitive kidney. In the male, the mesonephric duct is largely taken over by the testis, and becomes modified to form much of its "drainage" system, while in the female, this system virtually completely regresses. In both sexes, the mesonephros itself completely regresses.

By 12–12.5 days p.c., the distal part of the ureteric bud shows evidence of branching, and in doing so initiates the differentiation of the metanephros by the process of induction. As a result of the successive branching of the distal part of the ureteric bud, the metanephric tissue becomes broken up into smaller and smaller segments, each associated with a discrete component of the ureteric bud. Repeated branching of the latter occurs, until eventually each functional metanephric unit (or nephron) is associated with the differentiation of a single glomerulus. It also gives rise to the proximal and distal convoluted tubules, and the loop of Henle. By contrast, the ureteric bud gives rise to all of the components of the collecting duct system. At this stage of development, the mesonephros appears to be rapidly regressing (*Plate 68,d,e*), and is already substantially smaller than the gonad. The mesonephros is located on the lateral aspect of the urogenital ridge, while the gonad is located on its medial aspect (*Plate 68,d,e*). The part of the lumen of the ureteric bud which is almost completely surrounded by metanephric tissue expands dramatically (possibly by the incorporation of some of the early generations of its branches), and forms the primitive pelvis of the kidney. Several major branches of the renal pelvis are often present at this stage, and these may represent the primordia of the major and minor renal calyces (*Plate 27d,e,f*). Shortly afterwards, by about 12.5–13 days p.c., the metanephros is already seen to contain an outer primitive cortical region, which largely consists of undifferentiated metanephric cap tissue, and an inner medullary region in which tubules at various stages of differentiation may be found (*Plate 72,a,b*). Initially, the branches of the ureteric buds form ampullae, and these subsequently divide to form the primitive intrarenal collecting system. However, despite the presence of the latter, much of the medullary tissue at this stage consists of undifferentiated mesenchyme (*Plate 72,b*).

Within both the outer cortical region as well as in the inner medullary region of the metanephros, metanephric (nephrogenic) vesicles may be seen, and these represent an early stage in the differentiation of the glomeruli (*Plate 72,b*). The mesonephros by this stage has all but regressed in the female, and only a few degenerating tubules are seen (*Plate 68,d–f*), though whether these give rise to the rete ovarii (occasionally found in association with the medulla of the ovary) has yet to be convincingly established. Another feature of this stage is the first appearance of the paramesonephric ducts. These are rostrally more laterally located than the mesonephric ducts, but more caudally are medial relations of the mesonephric ducts, and are believed to be induced by the presence of the latter (*Plate 68,f*) (for a discussion of the fate of the mesonephric and paramesonephric ducts, see text associated with *Plates 68–71*, and appropriate sections of the various stages studied). The metanephros continues to enlarge and differentiate throughout this stage, and gradually "ascends" from its original location in the pelvis, so that by about 13 days p.c., its upper pole is slightly below the level of the 13th rib (*Plate 29b,a,b*). The ureters correspondingly increase in length and make contact with, but do not yet open into, the urogenital sinus.

By about 13.5–14 days p.c., the kidney is seen to be considerably larger, and at the histological level it is more consolidated than previously (*Plate 72,c,d*). In particular, the cortical region is slightly wider, and an increased number of both tubules and vesicles are now seen, though unlike the situation observed at later stages of development, these tend to be dispersed throughout the inner part of the cortical region and in the outer part of the medullary region of the kidney at this stage. By 14.5 days p.c., primitive glomeruli are seen for the first time, and these are also dispersed throughout much of the kidney at this stage (*Plate 72,e*). An increase in the number of collecting tubules is also seen, most of which have a relatively narrow luminal diameter (*Plate 72,f*). While a limited degree of nephric differentiation is seen throughout the kidney, it is of interest to note that in the most peripheral part of the cortex, just subjacent to the capsule of the kidney, poorly differentiated metanephric cap tissue is still in abundance, and the medullary region still principally contains undifferentiated mesenchyme tissue (*Plate 72,f*).

By 15.5 days p.c., the kidney has intermediate features between those seen at 14.5–15 and 16.5–17 days p.c., this being particularly the case with regard

(*continued on page 468*)

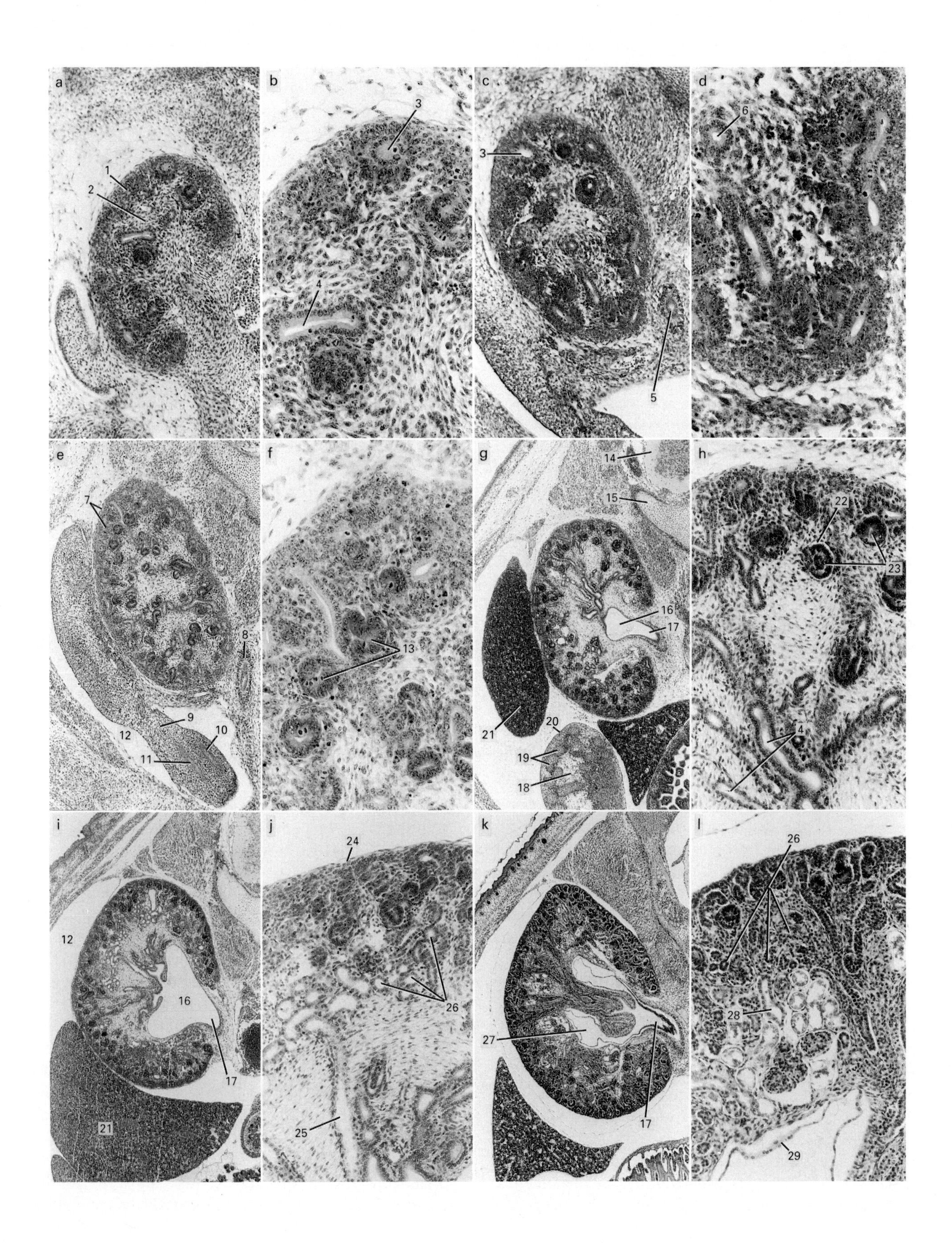
a
1
2
b
3
4
c
3
5
d
6
e
7
8
9
10
11
12
f
13
g
14
15
16
17
18
19
20
21
h
22
23
4
i
12
16
17
21
j
24
26
25
k
27
17
l
26
28
29

Plate 72

Differentiation of the kidney

a,b 12.5–13 days p.c. (**a** ×100; **b** ×250)
c,d 13.5 days p.c. (**c** ×100; **d** ×250)
e,f 14.5 days p.c. (**e** ×63; **f** ×250)
g,h 15.5 days p.c. (**g** ×40; **h** ×160)
i,j 16.5 days p.c. (**i** ×40; **j** ×160)
k,l 17.5 days p.c. (**k** ×40; **l** ×160)

1. cortical region of metanephros containing metanephric cap tissue
2. medullary region of metanephros containing mesenchyme tissue
3. metanephric vesicle
4. metanephric (collecting) tubule (collecting system derived from ureteric bud)
5. lumen in region of proximal part of ureter
6. metanephric tubule (on transverse section)
7. capsule of kidney with subjacent zone of metanephric cap tissue
8. wall of ureter
9. urogenital mesentery (future mesometrium/broad ligament)
10. paramesonephric duct (future uterine horn)
11. mesonephric duct (degenerating)
12. peritoneal cavity
13. primitive glomeruli
14. marginal zone of spinal cord
15. cartilage primordium of lumbar transverse process
16. renal pelvis
17. lumen of proximal part of right ureter
18. mesenchyme (future interstitial tissue) in medullary region of right testis
19. seminiferous tubules
20. capsule of right testis
21. caudal part of right lobe of liver
22. Bowman's capsule
23. glomeruli
24. fibrous capsule of kidney with undifferentiated metanephric cap tissue subjacent to it
25. large collecting tubule (duct)
26. proximal and distal convoluted tubules
27. major calyx
28. loop of Henle
29. endothelial lining of renal calyx

to its subdivision into well-defined cortical and medullary regions. At earlier stages, it was noted that the primitive glomeruli were dispersed throughout the cortical as well as the medullary region. By this stage, however, the glomeruli are tending to be concentrated in the outer one-third of the kidney, the region destined to form the cortex (*Plate 72,g,h*). The collecting ducts now tend to have a thinner wall, are more radially arranged than previously, and drain into a fairly dilated renal pelvis (*Plate 72,g*). By 16.5 days p.c., the kidney is seen to be even more clearly divided into an outer cortical region in which the vast majority of the glomeruli are concentrated (*Plate 72,i,j*), and an inner medullary region which principally contains the collecting ducts, which drain into a substantially enlarged renal pelvis (*Plate 72,i*). Despite all of this increased evidence of nephric differentiation, undifferentiated metanephric cap tissue may still be seen in the region subjacent to the capsule of the kidney (*Plate 72,j*).

By 17.5 days p.c., in addition to the presence of large numbers of glomeruli, the cortical region is seen to contain substantial numbers of proximal and distal convoluted tubules (*Plate 72,l*). These are only uncommonly encountered at earlier stages of development. Similarly, the medullary region now contains relatively little undifferentiated mesenchyme, and is principally occupied by the ascending and descending components of the loops of Henle, which are themselves interspersed between the more conspicuous elements of the collecting duct system. The hilar region of the kidney is now seen to be occupied by the dilated renal pelvis into which the major calyces drain (*Plate 72,k*), and more peripherally the minor calyces are seen to receive the larger elements of the collecting duct system.

Various aspects of kidney development have only very briefly been alluded to here, one of which is the factors that are believed to be involved in their "ascent" from their initial pelvic to their definitive upper lumbar location. This aspect is considered briefly in relation to the changes observed in the relative positions of the gonads and kidneys during the development of the gonads, and reference should be made to the text associated with *Plates 68–71* for additional observations on this topic. In the mouse, as in the rat, the right kidney is almost always more rostrally located than the left (by contrast to the situation in man, where the upper pole of the right kidney is almost always lower than the upper pole of the left kidney) at all stages of its development apart from the very earliest stage, shortly after the first appearance of the metanephroi, when both kidneys are at a similar vertebral level. However, the situation observed in the mouse may be more variable than that observed in man. Thus analysis of the various embryos used to illustrate this Atlas, while generally confirming that the right kidney is normally more rostrally located than the left, would seem to indicate that in this species this is not invariably the case. For example, the left kidney is slightly more rostrally located than the right in the embryo used to illustrate Stage 23 (see *Plate 32j,a*). Since this is an almost perfectly symmetrically sectioned embryo, the unusual arrangement of the kidneys in this embryo is not accounted for by any degree of obliquity of the sections through the trunk region. An analysis of the levels of the two kidneys in the embryos used to illustrate the various stages of development in the Atlas is given in the appropriate section of the text associated with Stage 23.

15. a. Discontinuities occasionally found in the visceral yolk sac

Text associated with Plate 73 and Plate 74,a–f

b. Differentiation of the placenta

Text associated with Plate 74,g–i and Plate 75

A. DISCONTINUITIES OCCASIONALLY FOUND IN THE VISCERAL YOLK SAC

A proportion of primitive streak stage embryos, at a similar stage to those illustrated in *Plates 5* and 6, possess a canal which may exceed 20 μm in diameter which traverses the outer cell layer of the visceral yolk sac. The canal appears to allow continuity between the yolk sac cavity and either the subjacent mesodermal tissue, or, in other instances, the amniotic cavity or proamniotic canal and/or the exocoelomic cavity. The features of such canals suggest that they represent small but genuine discontinuities in the visceral yolk sac, in that they have been demonstrated in semithin sections of plastic-embedded material, and in each case their features have been confirmed in ultrathin sections viewed by transmission electron microscopy. The latter has invariably revealed that the yolk sac endoderm cells that border these canals have intact microvillous borders (*Plate 73,a*; *Plate 74,b*). The presence of a continuous microvillous border on the endoderm cells that border on the canal suggests that this is probably not an artefactual phenomenon, and this seems to be confirmed by the fact that in at least several of the examples studied, an intact Reichert's membrane was still present around the conceptus. Moreover, in each of the six canals studied, cellular debris was present within the lumen of the canal, and some of the dead cells were clearly in continuity with the viable endoderm cells that bordered the canal. The latter feature in particular would seem to indicate that the canal may have been formed in these embryos as a result of programmed cell death (Kaufman, 1984).

Four examples of such canals are illustrated with low magnification views of the entire transverse or sagittal section through the embryo which indicate the general location of the canal, and these sections are complemented by higher magnification views of the canal region itself, all of which display the constant features indicated above. In the case of one of the embryos illustrated (*Plate 74,a,b*), additional views of this embryo are displayed elsewhere (*Plate 77,c–e*), as this was the smaller of a pair of asynchronous twin conceptuses that shared a single implantation site. In four of these embryos, the canal was located close to the anterior amniotic fold, at a site commonly associated with a deep indentation (Theiler, 1989), and was almost immediately opposite the caudal end of the primitive streak, or, in the case of the less advanced embryos, opposite the posterior amniotic fold.

Since possibly thousands of embryos at this stage have been studied (see, for example, Bonneville, 1950; Reinius, 1965; Poelmann, 1980, 1981a, 1981b; Beddington, 1983; Theiler, 1989, and many others) without this feature being reported in the literature, it is unclear whether this is peculiar to the strain of mice studied (CFLP), or was a pathological feature in embryos that were destined to die shortly afterwards. In the absence of confirmation in other strains, it is difficult to be certain whether this is a genuine feature of normal development, a pathological feature, or an artefact. If the former, then it shares several features with, and may be analogous to, the notochordal or archenteric canal, which is observed in other mammals at about the same stage of embryonic development. Similarly, if it is genuine, then what possible function might it serve? It might be a mechanism that allows the equilibration of fluid pressure within the conceptus and the yolk sac cavity. Such a canal may allow potentially toxic metabolites to be released from the embryonic compartment into the yolk sac cavity to be removed by the process of dilution, before toxic levels might build up, or conversely, it might allow the rapid ingress into the embryo of nutrients from the yolk sac cavity without the need for them to pass through the visceral yolk sac. These views are clearly speculative, and further studies will need to be undertaken to investigate this topic in more detail. Brief details of the sections through four of the embryos that display such a canal follow.

Transverse sections through an embryo indicate that the discontinuity in the yolk sac endoderm cells in this embryo is located just to one side of the primitive streak region (*Plate 73,b–d*). This embryo is one of two in this series in which the location of the canal is close to the primitive streak region. The sections illustrated in *Plate 73,b,c*, are at a similar level to that displayed in *Plate 6,i*, while the section illustrated in *Plate 73,d* is at a similar level to that displayed in *Plate 6,e*. A higher magnification view of a semithin section of the canal region illustrated in *Plate 73,c* is shown in *Plate 73,d* while an ultrathin section through a small region similar to that shown in *Plate 73,d* is displayed in *Plate 73,a*. The intact microvillous border on the cells on either side of the canal, and the presence of cellular debris within the lumen of the canal are clearly seen. The endoderm cells that border on the canal and some of the subjacent mesoderm cells have pathological features, but the cells more distant from the canal appear to be normal.

A sagittal section through an embryo in which the discontinuity allows communication between the yolk sac cavity and the exocoelomic cavity is displayed in *Plate 74,a*. Analysis of a low magnification ultrathin section through the canal region (*Plate 74,b*) reveals that all of the features described above are present. The cells on both sides of the canal appear to have a high level of phagocytic activity.

A sagittal section through an embryo in which the discontinuity appears to allow communication between the yolk sac cavity and both the amniotic and exocoelomic cavities is shown in *Plate 74,c*. A higher magnification view of a section through this region (*Plate 74,d*), allows the presence of the continuous microvillous border to be seen. In this embryo, Reichert's membrane was intact around the conceptus.

Analysis of a sagittal section through an embryo with a proamniotic canal (*Plate 74,e*) reveals that in

(*continued on page 476*)

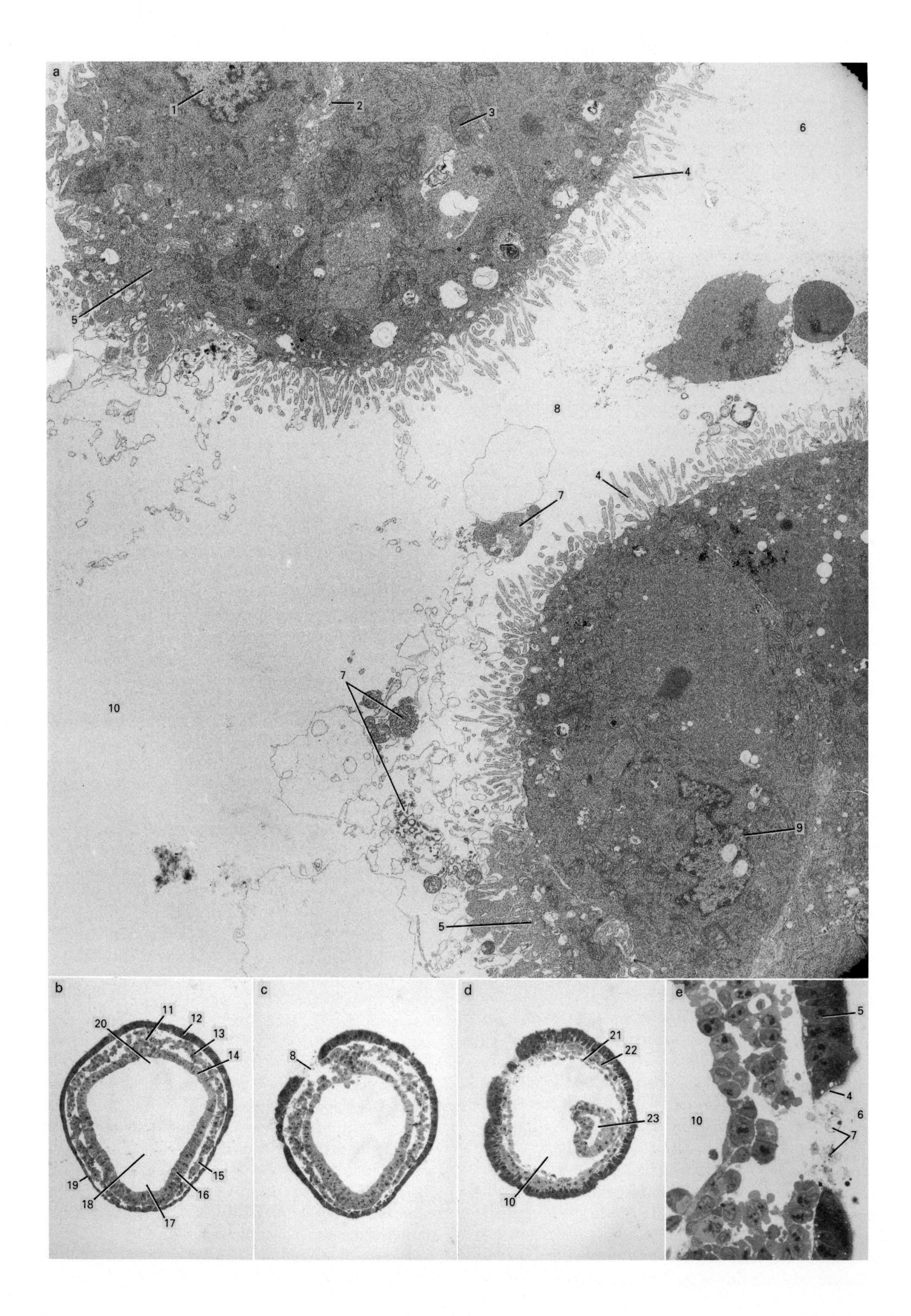
a
1
2
3
4
5
6
7
8
9
10
b
11
12
13
14
15
16
17
18
19
20
c
8
d
21
22
23
10
e
5
4
6
7
10

Plate 73

Discontinuities in the visceral yolk sac, associated with a yolk sac canal, I

- **a** transmission electron micrograph of localized region of visceral yolk sac of primitive streak stage embryo showing ultrastructural morphology of cells on either side of the yolk sac canal
- **b–d** representative semithin transverse histological sections through the embryo illustrated in **a**. Section **b** is closest to the embryonic pole, while section **d** is closer to the ectoplacental cone region, and section **c** is from an intermediate location (×160)
- **c** close-up view of canal region illustrated in section **c** (×640)

1. lobulated nucleus of visceral yolk sac mesoderm cell
2. cell boundary between visceral yolk sac endoderm and mesoderm cells
3. mitochondrion
4. intact microvillous border of visceral yolk sac endoderm cell
5. visceral yolk sac mesoderm cell
6. yolk sac cavity
7. cellular debris
8. yolk sac canal
9. lobulated nucleus of yolk sac endoderm cell
10. exocoelomic cavity
11. condensation of notochordal plate tissue (poorly defined) in caudal region of primitive streak
12. intact layer of extra-embryonic endoderm
13. intra-embryonic mesoderm subjacent to neuroepithelium of primitive streak region
14. neuroepithelium in lateral part of primitive streak region
15. intra-embryonic mesoderm subjacent to neuroepithelium in future headfold region
16. neuroepithelium in future headfold region
17. neural groove
18. amniotic cavity
19. embryonic endoderm
20. primitive groove
21. mesothelial lining (extraembryonic mesodermal cells) of extra-embryonic coelomic (exocoelomic) cavity
22. extra-embryonic mesoderm
23. ventral extension of ectoplacental cavity

PLATE 74

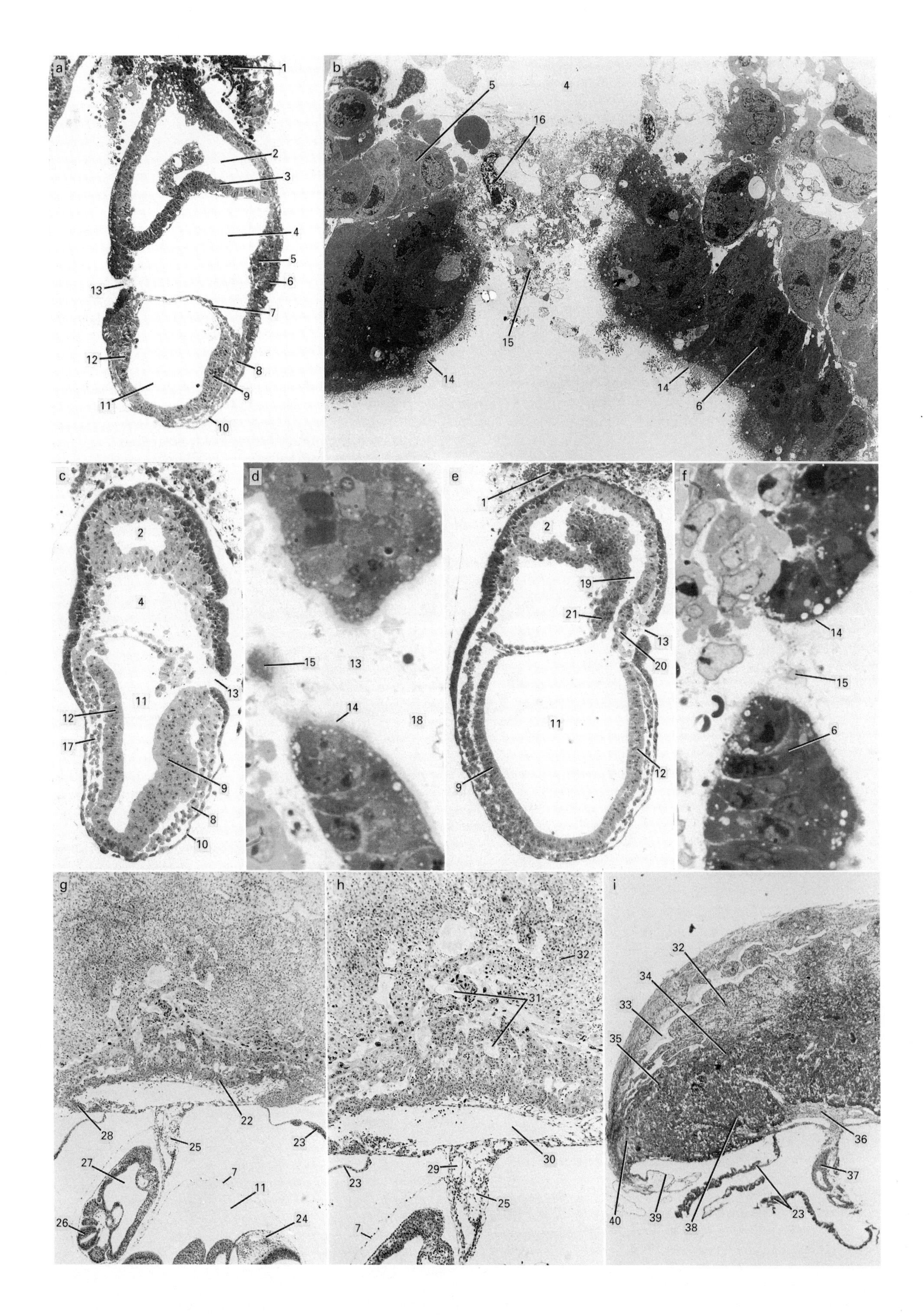
a
1
2
3
4
5
6
13
7
12
8
9
11
10
b
5
4
16
15
14
14
6
c
2
4
11
13
12
17
9
8
10
d
15
13
14
18
e
1
2
19
21
13
20
11
9
12
f
14
15
6
g
22
28
25
23
27
7
11
26
24
h
32
31
30
29
23
25
7
i
32
34
33
35
36
37
40
39
38
23

Plate 74

Discontinuities in the visceral yolk sac, associated with a yolk sac canal, II (a–f). Differentiation of the placenta, I (g–i)

a representative semithin sagittal section through primitive streak stage embryo with yolk sac canal (×100)
b transmission electron micrograph of localized region of visceral yolk sac, showing ultrastructural morphology of cells on either side of the yolk sac canal (same embryo as illustrated in **a**)
c representative semithin sagittal section through primitive streak stage embryo with yolk sac canal (×160)
d close-up view of canal region illustrated in section **c** (×1250)
e representative semithin sagittal section through primitive streak stage embryo with continuity between yolk sac canal and proamniotic canal (×100)
f close-up view of yolk sac canal region illustrated in section **e** (×1250)
g implantation site, embryo with about 20–25 pairs of somites (×40)
h close-up view of part of section illustrated in **g** (×63)
i implantation site at 11.5 days p.c., radial section through lateral half of placenta (×25)

1. ectoplacental cone
2. ectoplacental cavity
3. chorion
4. exocoelomic cavity
5. visceral yolk sac mesoderm cell
6. visceral yolk sac endoderm cell
7. amnion
8. intra-embryonic mesoderm subjacent to neuroepithelium of primitive streak
9. neuroepithelium of primitive streak
10. embryonic endoderm
11. amniotic cavity
12. neuroepithelium of future headfold region
13. yolk sac canal
14. microvillous border of visceral yolk sac endoderm cell
15. cellular debris
16. nucleus of pycnotic cell in continuity with visceral yolk sac endoderm cell bordering yolk sac canal
17. intra-embryonic mesoderm in future headfold region
18. yolk sac cavity
19. proamniotic canal
20. anterior amniotic fold
21. posterior amniotic fold
22. ectoplacental plate
23. visceral yolk sac
24. cephalic region of embryo
25. allantois
26. neural tube in caudal part of trunk region
27. peritoneal cavity (intra-embryonic coelomic cavity)
28. inverted lateral margin (lamina) of ectoplacental plate
29. dilated allantoic blood vessel
30. allantoic cavity
31. dilated materal blood vessels within future labyrinthine part of placenta
32. decidual cells within maternal part of future placenta
33. large maternal blood vessels at periphery of placental site
34. spongy or basal region (spongiotrophoblast) of primitive placenta
35. single layer of trophoblast giant cells at junction between basal layer and maternal (decidual) component of the placenta
36. chorionic plate with dilated embryonic blood vessel containing large numbers of primitive nucleated red blood cells
37. wall of umbilical cord
38. labyrinthine part of primitive placenta
39. Reichert's membrane with associated layer of parietal endoderm cells
40. trophoblast giant cells at peripheral margin of placenta

PLATE 75

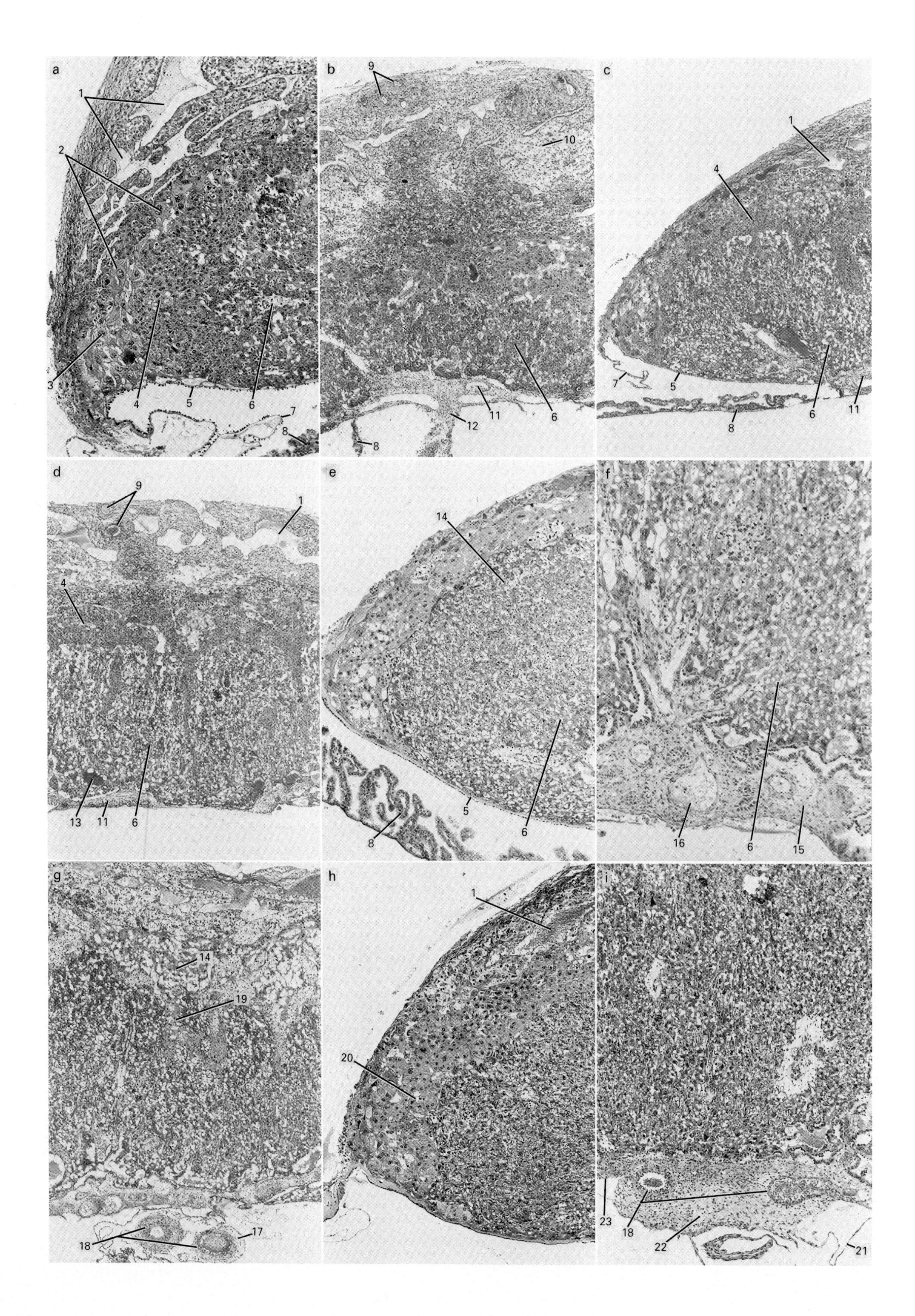

a
1
2
3
4
5
6
7
8
b
9
10
11
12
6
8
c
1
4
7
5
6
11
8
d
9
1
4
13
11
6
e
14
5
6
8
f
16
6
15
g
14
19
18
17
h
1
20
i
23
18
22
21

Plate 75

Differentiation of the placenta, II

Representative radial sections through the placental site at different stages of gestation.

a close-up view of section through lateral part of placental site, same placental site as that illustrated in *Plate 74*,**i**, 11.5 days p.c. (×63)
b section through middle part of placental site, close to site of insertion of the umbilical cord, same placenta as that illustrated in **a** (×40)
c,d peripheral (**c**) and central (**d**) parts of placental site at 13.5 days p.c. (×40)
e,f peripheral part of placental site (**e**), and central part of chorionic plate and subjacent region of placenta (**f**) at 14.5 days p.c. (**e** ×63; **f** ×100)
g central part of chorionic plate close to insertion of the umbilical cord, and subjacent region of placenta at 15.5 days p.c. (×40)
h,i peripheral part of placental site (**h**), and central part of chorionic plate close to insertion of the umbilical cord and subjacent region of placenta (**i**) at 17.5 days p.c. (×63)

1. dilated maternal blood vessels (venous sinusoids) at peripheral part of placental site
2. single layer of trophoblast giant cells at junction between basal layer and maternal (decidual) component of the placenta
3. trophoblast giant cells at peripheral margin of placenta
4. spongy or basal region (spongiotrophoblast) of primitive placenta
5. Reichert's membrane covering embryonic surface of placenta with associated layer of parietal endoderm cells
6. labyrinthine part of primitive placenta
7. Reichert's membrane and associated layer of parietal endoderm cells at lateral margin of yolk sac cavity
8. visceral yolk sac
9. maternal "feeder" vessels from uterine artery
10. maternal decidual cells
11. large embryonic blood vessels within chorionic plate
12. extension of chorionic plate into wall of umbilical cord
13. dilated maternal blood vessel subjacent to chorionic plate
14. spongy or basal layer of placenta which has a more "spongy" appearance than previously
15. chorionic plate
16. large umbilical vessel which now contains relatively few primitive nucleated red blood cells
17. Wharton's jelly forming cellular component of wall of umbilical cord
18. umbilical vessels
19. "peg" of spongiotrophoblast cells into labyrinthine part of placenta
20. spongiotrophoblast layer at periphery of placenta
21. amnion
22. site of insertion of umbilical cord into chorionic plate of placenta
23. Reichert's membrane with discontinuous layer of parietal endoderm cells associated with its surface

this embryo the discontinuity appears to allow communication between the yolk sac cavity and the cavity of the anterior amniotic fold. Analysis of a higher magnification view of a section through this region (*Plate 74,f*) reveals that all of the features indicated above are present in relation to this canal.

B. DIFFERENTIATION OF THE PLACENTA

By about 7–7.5 days p.c., when the embryo is at the primitive streak stage, the conceptus is divided into three distinct compartments, namely the amniotic cavity, the exocoelomic cavity and the ectoplacental cavity. The expansion of the amniotic and exocoelomic cavities then takes place, but at the expense of the ectoplacental cavity, which then progressively collapses. Its two walls fuse together, and eventually, by about 8–8.5 days p.c., this results in the complete obliteration of its lumen, though at its periphery the two layers (laminae) are still recognizable. Over this same period, the allantois grows, and expands across the exocoelomic cavity towards and eventually fuses with the surface cells at the periphery of the ectoplacental (or chorionic) plate. This normally occurs in embryos with about 7–10 pairs of somites. At this stage, the allantois has a very loose structure with cavities lined by endothelial cells which are seen to contain primitive nucleated red blood cells.

After the completion of the "turning" sequence, the embryo is surrounded by its extra-embryonic membranes, namely the amnion and the yolk sac. In embryos with about 10 pairs of somites, the yolk sac is seen to have a substantial vascular plexus associated with its mesodermal (inner) layer. At the primitive streak/early headfold stage, the yolk sac vasculature consists of scattered and isolated blood islands, but these rapidly amalgamate together to form the yolk sac circulation. The latter eventually fuses with the embryonic vasculature (which develops within the embryo as a completely independent system), so that the yolk sac-derived primitive red blood cells can then gain access to the embryonic circulation.

The visceral (proximal) yolk sac at this stage consists of a vascularized inner mesodermal component, and this is subjacent to an outer layer of principally columnar endodermal cells. The microvillous border to these cells faces into the yolk sac cavity (or yolk cavity), and acts as a primitive placenta by absorbing nutrients from the yolk sac cavity and transferring them into the embryonic compartment. The parietal (distal) layer of the yolk sac, by contrast to the visceral layer, consists of a single layer of endoderm cells with a squamous morphology which produce (on their outer aspect), and are adherent to, the acellular Reichert's membrane. On the outer aspect of the latter, an intermittent layer of trophoblast giant cells is found. These are of ectoplacental cone (i.e. embryonic) origin, and migrate away from this region, and eventually completely surround the implantation site with a loose network of giant cells. As these cells migrate away from the (possibly inhibitory) influence of the ectoplacental cone, they tend to substantially increase in volume. It should be noted that the ectoplacental cone is formed from the proliferation of the extra-embryonic ectoderm, and is directed away from the embryonic pole at the early egg cylinder stage. The cells of the ectoplacental cone grow rapidly and invade into the maternal tissues, rupturing into maternal blood vessels as they do so. The ectoplacental cone characteristically contains a large number of glycogen-rich cells, and these will shortly give rise to the cells at the junctional zone of the placenta.

By about 9–9.5 days p.c. the number of trophoblast giant cells that surround the embryonic pole of the conceptus has increased considerably, as has both their cellular and nuclear volume, as indicated above. The ectoplacental plate at this stage has a reasonably clear boundary zone. On the embryonic side of the latter, the primitive allantoic vasculature spreads out, whereas on the outer (decidual) aspect, numerous clefts appear which become filled with maternal blood, to form the labyrinth of the placenta (*Plate 74,g,h*). It is therefore at this site that the embryonic and maternal circulations subsequently come into their closest contact. A single or occasionally several layers of trophoblast giant cells also tend to be associated with the outer part of this layer, and these are in continuity with the trophoblast giant cells that surround the conceptus. The original lateral margins (laminae) of the ectoplacental plate invert, and at this site make contact with the interface between the visceral and parietal layers of the yolk sac (*Plate 74,g*).

By about 10 days p.c., the allantoic vessels are now seen to pierce the chorionic plate, and the ectoplacental plate is now transformed into the labyrinthine part of the placenta. The placenta at this stage is now divisible into several reasonably well-defined layers. The region closest to the conceptus, where the umbilical vessels make contact with the central region of the placenta, is termed the chorionic plate. Outside this zone is the labyrinthine layer, and outside this is the spongy or basal layer. At the junction between the basal layer and the maternal (decidual) component of the placenta, are located generally a single or occasionally several layers of trophoblast giant cells. These cells (or possibly sub-populations of them) are believed to have several functions including the phagocytosing of maternal red blood cells (as a mechanism for transferring iron to the conceptus), and the production of trophic (steroid) hormones, and it is possible that they may serve an additional critical role in facilitating normal implantation. Within the maternal component of the placenta at this stage are large sinusoidal blood vessels with feeder arteries. The various regions of the placenta are particularly clearly seen at 11.5 (*Plate 74,i*; *Plate 75,a,b*), and at 13.5 (*Plate 75,c,d*) days p.c.

In general terms, as the placenta matures, the labyrinthine zone increases in volume, and this occurs from about 12 to 17.5 days p.c. (*Plate 75,c–i*). After this stage, the volume of this zone remains unchanged until the time of birth. By about 16 days p.c., (*Plate 75,g,h*), the labyrinthine zone occupies slightly more than one-half of the volume of the entire placenta. The trophoblastic giant cell layer is readily recognized during the early period of placental differentiation, but by about 14 days p.c. these cells have largely disappeared from the boundary zone between the embryonic and maternal components of the placenta, although a few giant cells are usually found at the periphery of the placenta (*Plate 75,e,h*), even up to the time of birth.

Within the labyrinthine zone of the placenta, components of the embryonic and maternal circulations may be recognized, though this is not technically feasible in the low magnification sections illustrated here. The embryonic vessels may be identified by the presence within their lumina of nucleated primitive red blood cells, while the maternal blood sinuses contain mature non-nucleated red blood cells. Before 11.5 days p.c., all of the embryonic red blood cells are nucleated, but between 12 and 15 days p.c., the proportion of nucleated red blood cells decreases dramatically, and by 16.5 days p.c. all of the red blood cells in the embryonic circulation are non-nucleated (see Fig. 7).

The placental barrier in the labyrinthine zone by 12 days p.c. consists of three layers of trophoblast cells (termed layers 1, 2 and 3), together with an endothelial layer which lines the embryonic vessels (for further details of placental fine structure, see Enders, 1965; Kirby and Bradbury, 1965). It is believed that the trophoblast cells of layer 3 are the likely source of trophic hormone production. Reichert's membrane usually ruptures shortly before birth, but it continues to cover a large part of the placental surface, as previously. Generally, in the mature mouse placenta, a single large centrally located artery (being a branch from the uterine artery) is present which supplies maternal blood directly to the region of the chorionic plate.

The cytoplasm of the endodermal cells of the visceral yolk sac becomes increasingly swollen from about 14.5 days p.c. (*Plate 75,e*) and this is particularly apparent after about 15.5 days p.c. By about 17.5 days p.c., much of the yolk sac close to the placenta has a very florid appearance, and many (in some areas most) of the endoderm cells appear to lack nuclei, and indeed seem to be desquamating from the surface of the yolk sac.

16. Disposition of the extra-embryonic membranes associated with various types of twinning
Text associated with Plates 76–79

The types of twinning illustrated here are:

(a) Synchronously developing monochorionic monoamniotic monozygotic twinning (*Plate 76*).
(b) Synchronously developing dichorionic diamniotic twinning (*Plate 77,a*).
(c) Asynchronously developing dichorionic diamniotic twinning (*Plate 77,b,c–e*; *Plate 78*). The twins of type (b) and (c) illustrated in *Plates 77* and *78* could be either monozygotic or dizygotic in origin.
(d) Asynchronously developing dichorionic diamniotic monozygotic twinning (*Plate 79*).

The presence of two embryos associated within a single implantation site (decidual swelling) is occasionally encountered in mice, but no detailed information appears to be available to indicate the incidence of twinning in this species. It is likely that this varies quite considerably between strains, and it would be interesting to establish the spontaneous occurrence of this phenomenon. Twinning may also be induced experimentally in mice, for example by the exposure of pregnant females at an appropriate time during the early postimplantation period to certain agents, such as to a single injection of vincristine sulphate on either 6.5 or 7.5 days p.c. (Kaufman and O'Shea, 1978), or colcemid at similar stages of gestation (Kaufman, unpublished). In the former study, five pairs of monozygotic twins (with one conjoined pair) were observed out of a total of 189 living embryos isolated on either 9.5 or 11.5 days p.c. Apart from the conjoined twins which were obviously monoamniotic, at least three of the remaining four pairs were also monoamniotic, and all eight embryos from the four pairs with two singleton embryos each appeared to be morphologically normal.

Hsu and Gonda (1980) reported that approximately 1% of all 4-day expanded blastocysts which they explanted *in vitro* gave rise to monozygotic twins. The twinning stimulus appeared to be related to the orientation of the blastocyst at the time of attachment. Typically, in those instances when twinning occurred, two independent egg cylinders formed following the division of the inner cell mass region shortly after blastocyst attachment, and all the twins in this study were of the dichorionic diamniotic variety. In another study, Wan *et al.* (1982) reported the occasional spontaneous occurrence of dizygotic (presumed) twinning in which pairs of embryos (in one case, three embryos) shared a single ectoplacental cone. In a number of other mouse studies, initially Tarkowski (1959; see also Tarkowski and Wróblewska, 1967; and for an earlier study in the rat, see Nicholas and Hall, 1942); subsequently various others, have demonstrated that it is technically possible to produce monozygotic twins, triplets etc. in mice by bisecting early cleavage stage embryos, or by dividing precompacting morulae into three or more groups of blastomeres, with in some instances the successful development to term of most or occasionally all of these individuals (see Kelly, 1975). Since these approaches result in the production of a series of normal singleton embryos (albeit monozygotic in origin) each within their own implantation sites, these are not relevant to the present discussion, and will not be considered further.

The twinning situation in the mouse is not equivalent to that in man, since the mouse is a polytocous species in which all of the litter normally result from a single ovulation event (which usually occurs over a period of an hour or more), and would therefore be the equivalent of dizygotic twinning in man. Similarly, mating is usually to a single male, though more exotic events may occasionally occur in nature, or may be induced experimentally, such as superfetation or superfecundation, multiple ovulation from polyovular follicles, asynchronous implantation (see *Plate 80*), etc. Furthermore, the fact that many researchers tend to work with inbred strains of mice, in which all individuals are essentially genetically identical, makes the distinction between monozygotic and dizygotic (or polyzygotic) twinning a purely academic one. However, despite these reservations, the phenomenon of particularly monozygotic twinning, and the presence of more than one embryo within a single implantation site, is of considerable interest, and justifies some, albeit brief, discussion (for review of experimental means of inducing monozygotic twinning, see Kaufman, 1985).

Since much information is available on the incidence of the different types of monozygotic (identical) twinning in man, and the possible timing of the twinning event, this information will be very briefly presented, as it may be relevant to the situation in the mouse, particularly in relation to the possible timing of the twinning event. In man, approximately 30% of all identical twins are of the dichorionic diamniotic variety, and the twinning event is believed to occur at some stage during cleavage. As a result, two blastocysts form, and each embryo subsequently has its own placenta and chorionic sac. This mechanism gives rise to the dichorionic, diamniotic variety of monozygotic twin placentation, and can be technically extremely difficult to distinguish from the placentation in dizygotic twins, particularly when two separate placentas are present, as occurs in a significant proportion of these cases. In the other cases, an apparently single placenta is present in which the two placentas are in fact fused together. The latter situation is also commonly encountered in dizygotic twinning.

In 66–70% of human monozygotic twins, placentation is of the monochorionic diamniotic variety (Gedda, 1961; Strong and Corney, 1967; Boyd and Hamilton, 1970). In this group, splitting of the zygote is believed to occur at the early blastocyst stage, so that the inner cell mass splits into two separate groups of cells which are located within the single blastocoelic cavity. The two embryos that form share a common placenta and have a common chorionic cavity, but

(*continued on page 488*)

PLATE 76

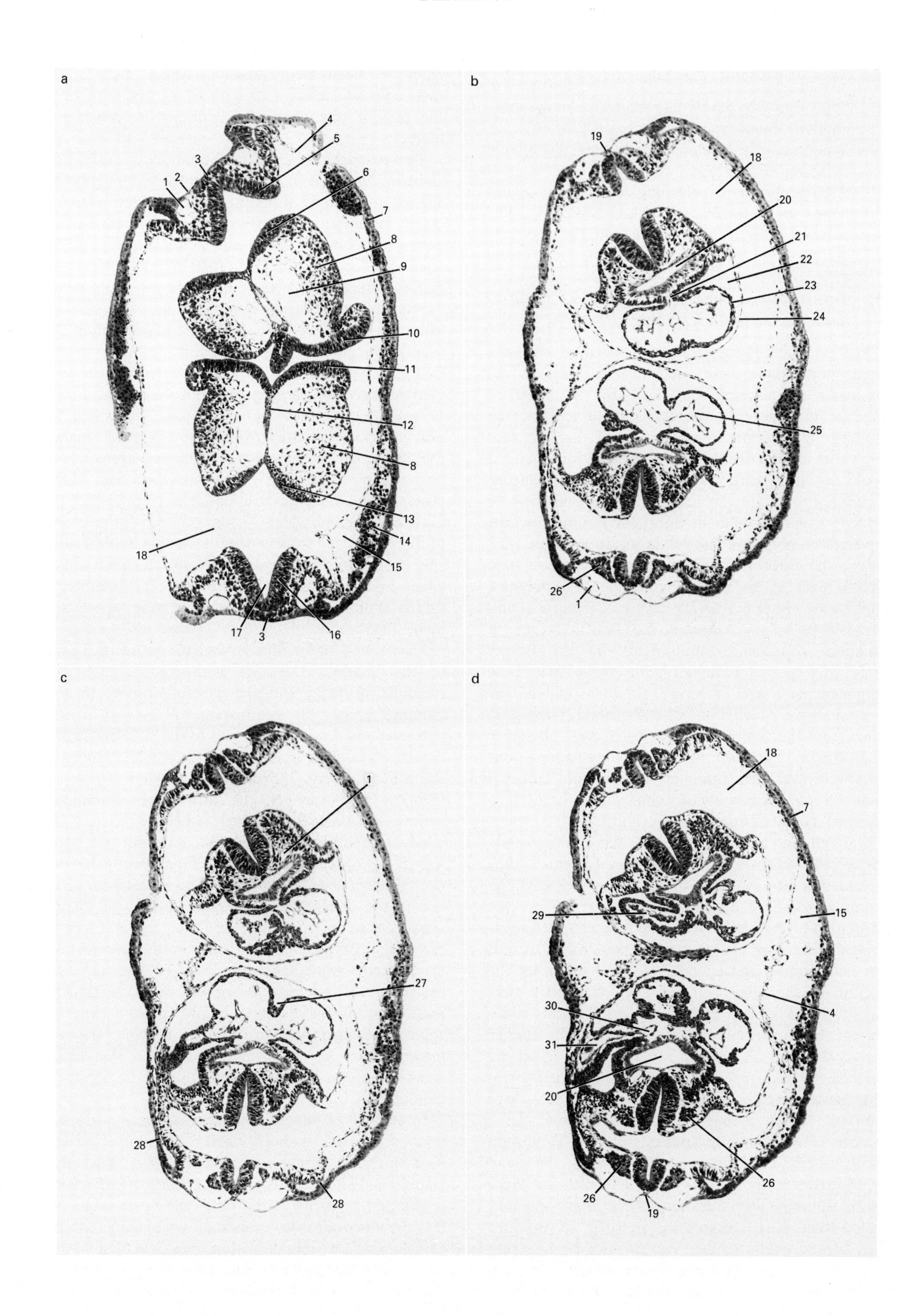
a
1
2
3
4
5
6
7
8
9
10
11
12
8
13
14
15
18
17
3
16
b
19
18
20
21
22
23
24
25
26
1
c
1
27
28
28
d
18
7
29
15
4
30
31
20
26
26
19

Plate 76

Example of synchronously developing monochorionic monoamniotic monozygotic twins

(**a–d** ×63)

1. dorsal aorta
2. endodermal lining of hindgut pocket (diverticulum)
3. notochordal plate
4. amnion
5. neuroepithelium of primitive streak region (of embryo 1)
6. neuroepithelium of prospective hindbrain (of embryo 1)
7. visceral yolk sac
8. cephalic mesenchyme tissue
9. internal carotid artery (rostral extension of dorsal aorta)
10. neuroepithelium of prospective forebrain (of embryo 1)
11. neuroepithelium of prospective forebrain (of embryo 2)
12. rostral extremity of notochord
13. neuroepithelium of prospective hindbrain (of embryo 2)
14. blood island within mesodermal component of visceral yolk sac containing large numbers of primitive nucleated red blood cells (erythroblasts)
15. extra-embryonic coelomic cavity
16. neuroepithelium of primitive streak region (of embryo 2)
17. neural groove
18. amniotic cavity
19. notochord
20. pharyngeal region of foregut
21. mesentery of heart (dorsal mesocardium)
22. pericardial cavity
23. myocardial wall of primitive heart
24. body wall overlying pericardial cavity (thoracic wall)
25. endothelial lining of primitive heart (endocardium)
26. somite
27. ventral median furrow
28. site of communication between intra- and extra-embryonic coelomic cavities
29. right horn of sinus venosus
30. left horn of sinus venosus
31. left vitelline vein

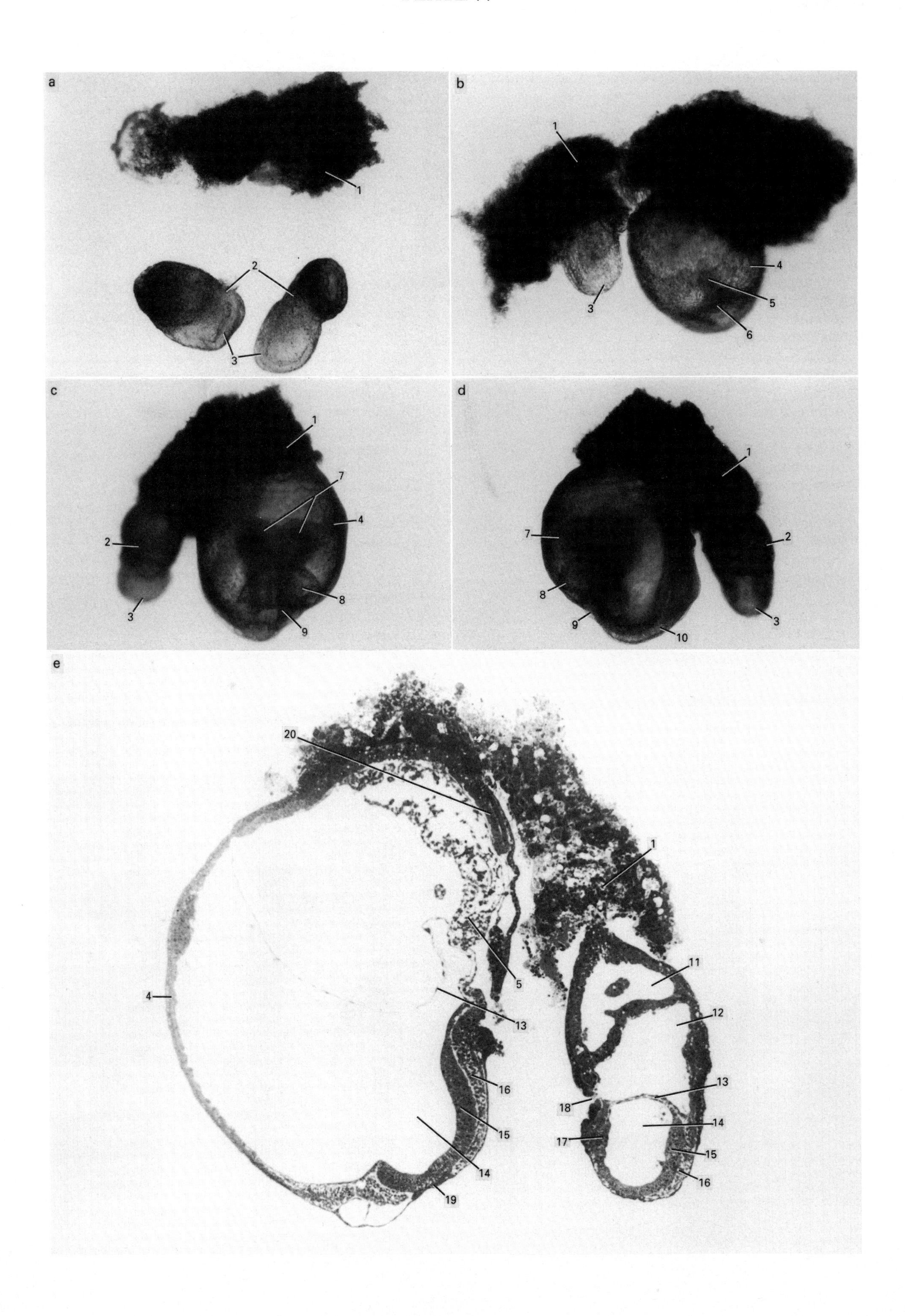
a
1
2
3
b
1
3
4
5
6
c
1
7
4
2
3
8
9
d
1
7
2
8
9
3
10
e
20
1
5
11
4
13
12
16
13
18
14
15
17
15
14
16
19

Plate 77

Examples of implantation sites containing two embryos which may develop either synchronously or asynchronously.

These may either be dizygotic twins or dichorionic diamniotic monozygotic twins.

a an example where the two embryos, both of which are advanced egg cylinders, are at a similar stage of development
b asynchronously developing "twins", where one is an egg cylinder and the other an advanced primitive streak stage embryo
c,d asynchronously developing "twins", where one is an advanced egg cylinder/early primitive streak stage embryo, and the other a headfold stage embryo with 6–8 pairs of somites present
e section through the "twins" illustrated in **c,d** showing that the two embryos appear to share a single ectoplacental cone (×100)

1. ectoplacental cone
2. advanced egg cylinder/early primitive streak stage embryo
3. embryonic pole
4. visceral yolk sac
5. allantois
6. primitive streak region
7. cephalic neural folds (headfolds)
8. primitive heart tube
9. entrance to foregut diverticulum
10. somite
11. ectoplacental cavity
12. extra-embryonic coelomic cavity
13. amnion
14. amniotic cavity
15. neuroepithelium in primitive streak region
16. mesenchyme tissue subjacent to primitive streak region
17. neuroepithelium of future headfold region
18. yolk sac "canal"
19. notochordal plate
20. chorion

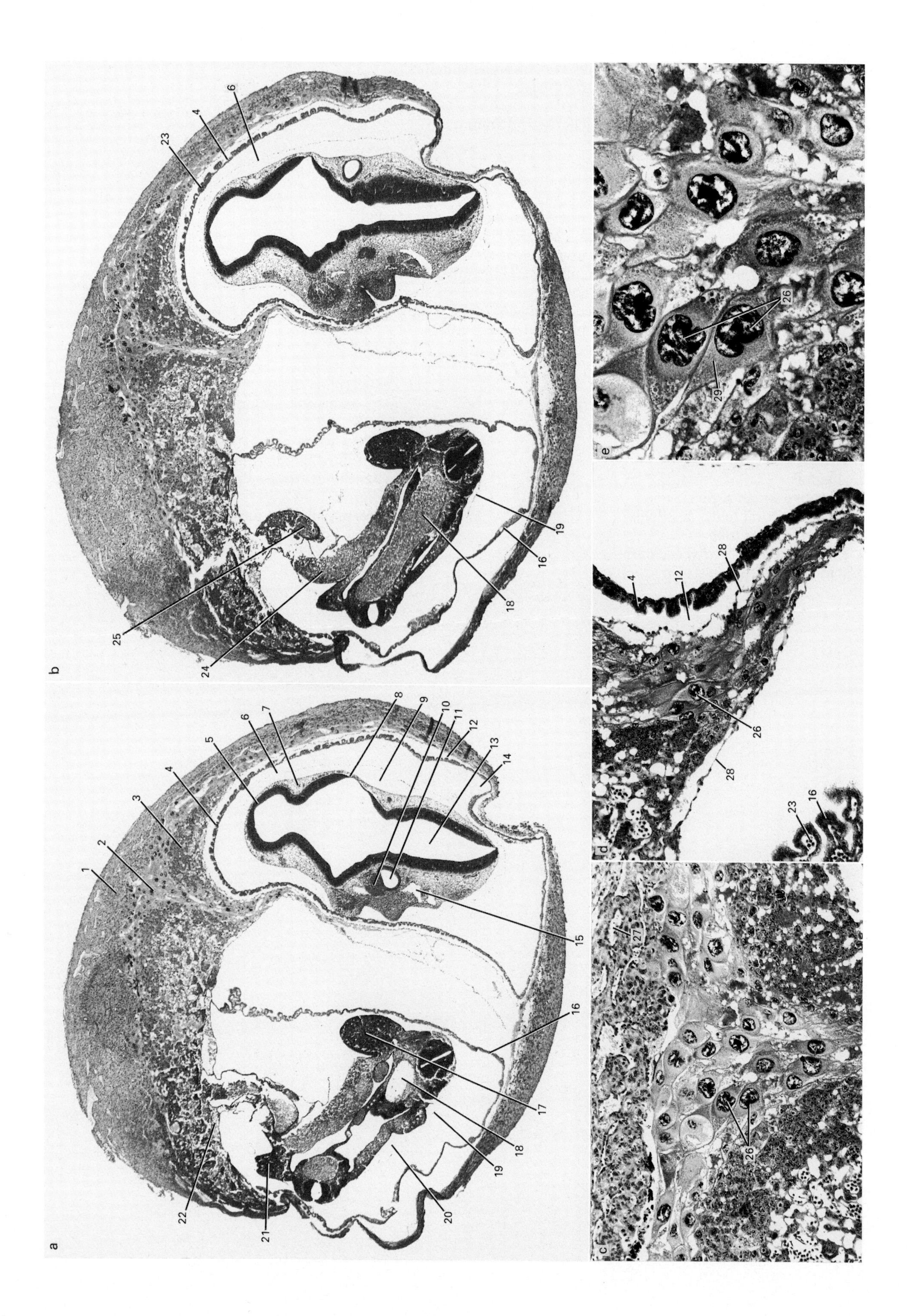
a
b
c
d
e
1
2
3
4
5
6
7
8
9
10
11
12
13
14
15
16
17
18
19
20
21
22
23
24
25
26
27
28
29

Plate 78

An example of an implantation site containing two asynchronously developing embryos.

These may either be dizygotic twins or dichorionic diamniotic monozygotic twins. Twin 1 has about 25–30 pairs of somites, whereas twin 2 has about 30–35 pairs of somites present (**c,d** ×100; **e** ×250)

1. maternal component of decidual reaction
2. group of trophoblast giant cells
3. embryonic component of decidual reaction
4. visceral yolk sac (of embryo 2)
5. neuroepithelium at forebrain/midbrain junction
6. amnion (of embryo 2)
7. cephalic region (of embryo 2)
8. rhombic lip
9. amniotic cavity (of embryo 2)
10. facio-acoustic (VII–VIII) ganglion complex
11. otocyst (otic vesicle)
12. yolk sac cavity
13. fourth ventricle
14. extraembryonic coelomic cavity
15. primary head vein
16. visceral yolk sac (of embryo 1)
17. forelimb bud
18. midline dorsal aorta
19. amnion (of embryo 1)
20. amniotic cavity (of embryo 1)
21. hindlimb
22. chorionic plate
23. blood island/blood vessel within mesodermal component of visceral yolk sac
24. right umbilical vein
25. umbilical artery
26. nuclei of trophoblast giant cells
27. maternal sinusoid
28. Reichert's membrane with attached parietal endoderm cells on embryonic surface
29. cytoplasm of trophoblast giant cell

PLATE 79

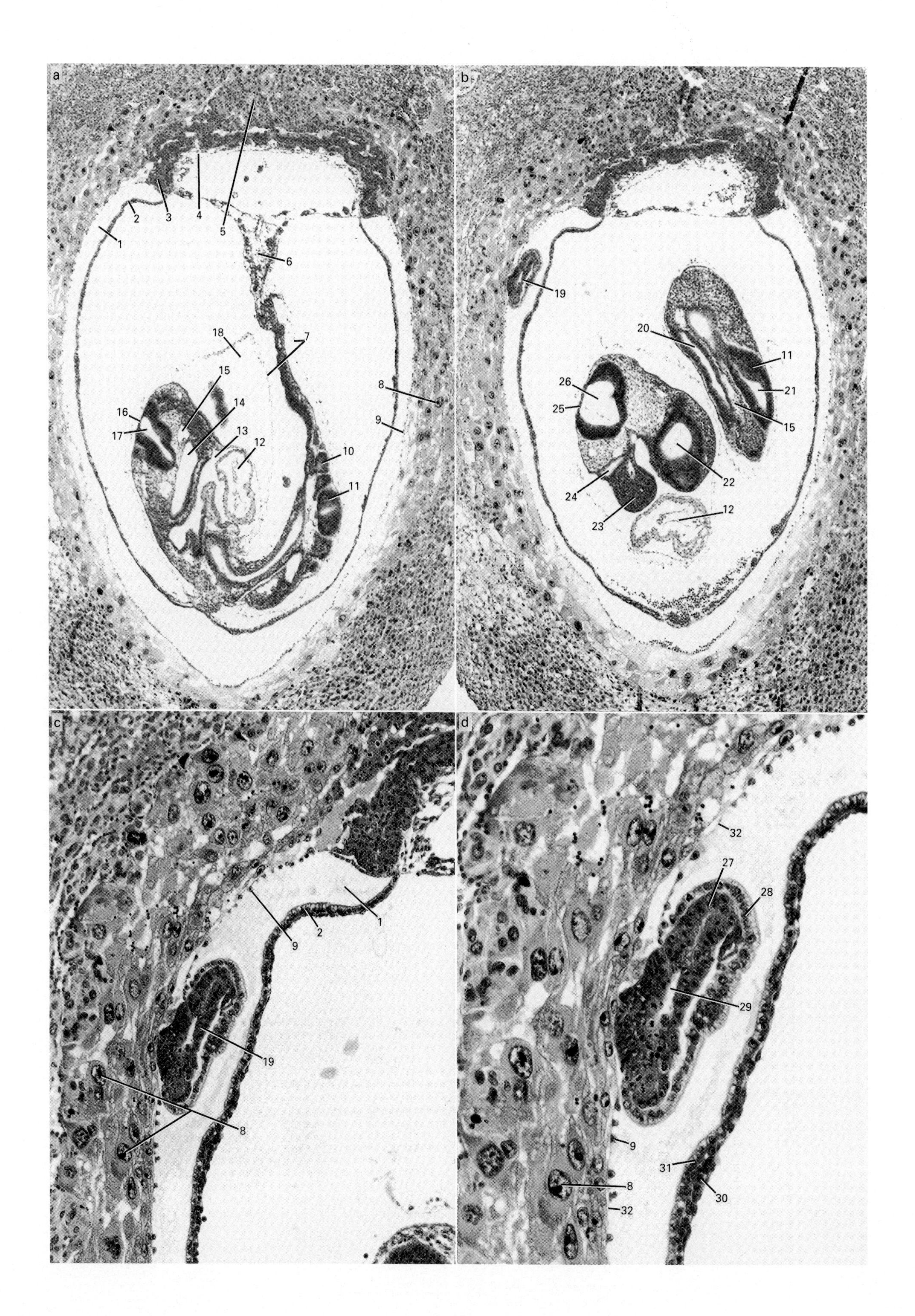

a
1
2
3
4
5
6
7
8
9
10
11
12
13
14
15
16
17
18
b
19
20
11
21
15
22
12
23
24
25
26
c
1
2
9
19
8
d
32
27
28
29
9
31
30
8
32

Plate 79

Example of asynchronously developing dichorionic diamniotic twins

These are diandric triploids produced by the technique of nuclear micromanipulation – a single male pronucleus was inserted into the perivitelline space of a normal diploid fertilized 1-cell stage embryo in the presence of inactivated Sendai virus.
The early "turned" embryo possesses 15–20 pairs of somites, while its morphologically slightly abnormal monozygotic twin is an advanced egg cylinder. (**a**,**b** ×40; **c** ×100; **d**×160)

1. yolk sac cavity
2. visceral yolk sac
3. chorionic plate
4. spongiotrophoblast
5. maternal component of placenta
6. allantois
7. amnion
8. trophoblast giant cell
9. parietal endoderm cell associated with inner (i.e. embryonic) aspect of Reichert's membrane
10. somite
11. neural tube
12. primitive heart tube
13. pericardial cavity
14. pharyngeal region of foregut
15. dorsal aorta
16. roof plate of caudal region of hindbrain
17. neural lumen (caudal region of fourth ventricle)
18. amniotic cavity
19. egg cylinder stage embryo
20. intra-embryonic coelomic cavity
21. neural (spinal) lumen
22. lumen of prospective forebrain
23. mandibular component of first branchial arch
24. first branchial pouch
25. roof of hindbrain
26. fourth ventricle
27. extra-embryonic ectoderm cells
28. extra-embryonic endoderm cells
29. proamniotic canal
30. mesodermal component of visceral yolk sac
31. endodermal component of visceral yolk sac
32. Reichert's membrane

each possesses its own separate amniotic cavity. In probably only 1–3% of all cases of monozygotic twinning in man, the twinning event occurs at the stage of the bilaminar disc, just before the appearance of the primitive streak, and results in the situation in which the two embryos not only share a single placenta, but have a common chorionic and amniotic sac. This mechanism therefore gives rise to the monochorionic monoamniotic variety (for further observations on the possible timing of the twinning event, see Bulmer, 1970; Fox, 1978).

SYNCHRONOUSLY DEVELOPING MONOCHORIONIC MONOAMNIOTIC MONOZYGOTIC TWINNING

The example illustrated (*Plate 76*) was observed in a study in which diploid parthenogenetic embryos were induced following the exposure of unfertilized eggs to culture medium lacking in calcium and magnesium salts (Surani and Kaufman, 1977; Kaufman, 1983a). The embryos were retained in culture to the blastocyst stage, then transferred to the uteri of pseudopregnant recipients. The recipients were ovariectomized at the time of transfer and given a subcutaneous injection of 1 mg Depo-Provera (Upjohn). After an interval of 3 days, during which time the blastocysts entered a "delayed" state, the recipients received daily a single combined injection of oestrogen and progesterone (Kaufman *et al.*, 1977). The recipients were autopsied between 9.5 and 10.5 days p.c., by which time the most advanced embryos had usually achieved the forelimb bud stage, and possessed about 20–25 pairs of somites.

Out of about 35 implantation sites analysed in this study, most of which contained embryos with between 10 and 25 pairs of somites, two sites each contained a pair of monochorionic monoamniotic monozygotic twins. The first pair consisted of an "unturned" early headfold stage embryo with about 3 pairs of somites which shared a common amniotic cavity with a partially "turned" embryo with 10–12 pairs of somites, and it was observed that the more advanced embryo had a beating heart. Histological examination confirmed the disposition of the extra-embryonic membranes, and the fact that both embryos appeared to be morphologically normal and healthy.

The second pair of embryos was contained within a considerably smaller sac, and removal of the single ectoplacental cone revealed the presence of two headfold stage embryos which were facing each other. The sac was fixed and embedded in resin, and then serially-sectioned at a thickness of about 0.5–0.75 μm. Both embryos were at an almost identical stage of development, and each possessed about 4–6 pairs of somites. A selection of four histological sections through the sac is presented in this plate. These clearly reveal that both embryos share a single amniotic cavity and yolk sac. Wax reconstruction of the two hearts revealed that they were at slightly different stages of cardiogenesis, but were both rotated normally. While the thoracic regions of the two embryos were closely apposed, they were not fused together, and the two embryos were morphologically normal (for further details regarding these embryos and associated discussion, see Kaufman, 1982).

SYNCHRONOUSLY DEVELOPING DICHORIONIC DIAMNIOTIC TWINNING

The example illustrated (*Plate 77,a*) shows two advanced egg cylinder/early primitive streak stage embryos that shared a single ectoplacental cone. While the embryos at this stage appear to be at an approximately similar stage of development, and this is relatively commonly seen to be the case at this stage, synchronous development at more advanced stages is only very rarely encountered. As with the asychronously developing embryos (see below), it is unclear whether these twins were monozygotic or dizygotic in origin.

ASYNCHRONOUSLY DEVELOPING DICHORIONIC DIAMNIOTIC TWINNING

Various examples of this form of configuration of the extra-embryonic membranes are presented here. In the first example illustrated (*Plate 77,b*), of the two embryos that share a single ectoplacental cone, one is at the egg cylinder stage, while the other is an advanced primitive streak stage embryo. The presence of a small allantois is seen through the wall of the visceral yolk sac of the larger embryo. In a second example, two views of the intact pair of twins with their shared ectoplacental cone are illustrated (*Plate 77,c,d*), as well as a representative histological section through both conceptuses to show their relationship to each other. These embryos were embedded in plastic, and both appeared to be morphologically healthy at the time of their isolation. One is an advanced egg cylinder/early primitive streak stage embryo, while the other is a headfold stage embryo which possesses about 6–8 pairs of somites. The smaller embryo contains a yolk sac canal (for further details, see text associated with *Plates 73–75*).

In a third example, two representative sections are illustrated through a pair of asynchronously developing twins in which one twin possesses about 25–30 pairs of somites, while the second twin possesses about 30–35 pairs of somites. A single placental site is seen to be present, and the disposition of the extra-embryonic membranes around each of these embryos is displayed (*Plate 78,a,b*). Higher magnification views of the intervening region between the two embryos, and of trophoblast giant cells associated with the placental site are also illustrated (*Plate 78,c–e*).

In a fourth example, representative sections through a pair of asynchronously developing dichorionic diamniotic twins are illustrated (*Plate 79,a–d*). The example illustrated here was observed in a study in which diandric triploid mouse embryos were produced by standard micromanipulatory techniques in which a single male pronucleus with a small volume of cytoplasm was isolated microsurgically from a donor fertilized egg and transferred in the presence of inactivated Sendai virus into the perivitelline space of a recipient 1-cell stage fertilized embryo, using the technique described by McGrath and Solter (1983). The tripronuclear eggs were transferred to pseudopregnant recipients that were subsequently autopsied on the 10th day of gestation. In most of these studies, the embryos were analysed histologically, while the extra-embryonic membranes were analysed cytogenetically to confirm their diandric triploid genetic constitution. The latter was possible, because in most studies F1 hybrid females were mated to homozygous Rb(1.3)1Bnr males (each haploid set of which contains a large metacentric "marker" chromosome, being a Robertsonian translocation involving chromosomes 1 and 3), so that the presence of two such "marker" chromosomes in a triploid chromosome spread therefore provides an unequivocal means of confirming its diandric origin. In a few cases, intact decidual swellings and their contents were analysed histologically, in order to examine the relationship between these conceptuses and their extra-embryonic tissues. It was from within one of these implantation sites, that the twins illustrated (*Plate 79,b–d*) were observed. One embryo has just completed the "turning" sequence, and possesses about 15–20 pairs of somites, and appears to be morphologically normal, while the other embryo is at the advanced egg cylinder stage, and, though obliquely sectioned, appears to be morphologically abnormal (for further details of the methodology involved, see Kaufman *et al.*, 1989).

17. Asynchronous implantation
Text associated with Plate 80

In the longitudinal histological section through the uterine horn illustrated (*Plate 80*), two morphologically normal embryos at different stages of gestation are seen to be developing simultaneously within the reproductive tract of a female. Such a situation could result from superfetation which, according to Deanesly (1966), is "the evolution in the same animal of two pregnancies at differing stages of development". This scenario implies that the two sets of conceptuses implanted at different times, either because: (a) an additional ovulation and fertilization event took place after the beginning of an already established pregnancy, or (b) a difference occurs in the timing of implantation between a group of embryos released at the same time. In fact, a variation of this phenomenon termed "asynchronous implantation" was induced experimentally (see below). These conditions all occur spontaneously in nature, but are believed to be extremely rare events.

Examples of "accelerated development", however, where one group of conceptuses develops normally and is eventually delivered, while a second group of conceptuses derived from a pre-partum oestrus and mating is retained in the reproductive tract in a "delayed" state, and only implants after the delivery of the first set of conceptuses, represent a normal reproductive strategy in some species. This is quite different from the situation where two sets of conceptuses derived from two separate ovulations actually implant at different times, and subsequently develop simultaneously within the uterus, the one group derived from the first ovulation being at a more advanced stage throughout gestation than the other, and this scenario is, as indicated above, only very rarely encountered in nature. In other situations, which may easily be confused with superfetation, the progressive retardation in growth of one or more members of a litter may occur compared with the growth of the normal members of the litter. This situation may arise in human pregnancy, for example in the twin transfusion syndrome (Gedda, 1961), and may lead to the death of the smaller twin (and occasionally give rise to a fetus papyraceus), or occasionally of both twins.

In the experimental study from which the section illustrated here was obtained, this clearly demonstrates that it is possible to induce the simultaneous development of two distinct sets of conceptuses derived from a single ovulation. If this experimental approach is employed, the first group of embryos to implant could be up to 7–10 days more advanced than the second set of conceptuses. In the section illustrated, one embryo is at the forelimb bud stage and possesses about 20–25 pairs of somites, and is equivalent to about 9.5 days p.c., while the adjacent embryo is at the egg cylinder stage, and is equivalent to about 6.5 days p.c.

In this study (Kaufman unpublished), 129/E strain female mice were mated to males of the same strain. The females were ovariectomized at about midday on the fourth day of pregnancy, and while anaesthetized were given a subcutaneous injection of Depo-Provera (Upjohn). When females were autopsied between 2 and 10 days after the progesterone injection, three possible situations were observed, namely either all or none of the embryos implanted (in the second group free floating zona-free blastocysts could be flushed from the uterine horns), or a proportion of the embryos implanted while others were free-floating zona-free blastocysts.

In a subsequent study, the free-floating population of blastocysts present in the third group could be induced to implant and continue developing, despite the presence of implanted embryos that were (in developmental terms) already several days more advanced than they were. This group of females was given combined daily subcutaneous injections of oestradiol and progesterone, the first dose being given not less than 48 h after the ovariectomy. In all, a total of 32 females were subjected to this experimental procedure, and in 20 cases two groups of normal embryos differing by at least 3 days were found to be present and developing simultaneously in their uterine horns. In four additional females only large embryos were present, and in an additional eight females only small embryos were observed. The two simulaneously developing groups of embryos in these 20 females were consistent with the timing of the predicted first or second implantation event induced by the exogenous hormonal regimes. While this technique for inducing asynchronous implantation has not previously been described, the general phenomenon is known to occur in rodents, and the hormonal basis for this condition, as well as other techniques for inducing this condition are discussed in detail by Psychoyos (1973).

(*Plate 80 over page*)

PLATE 80

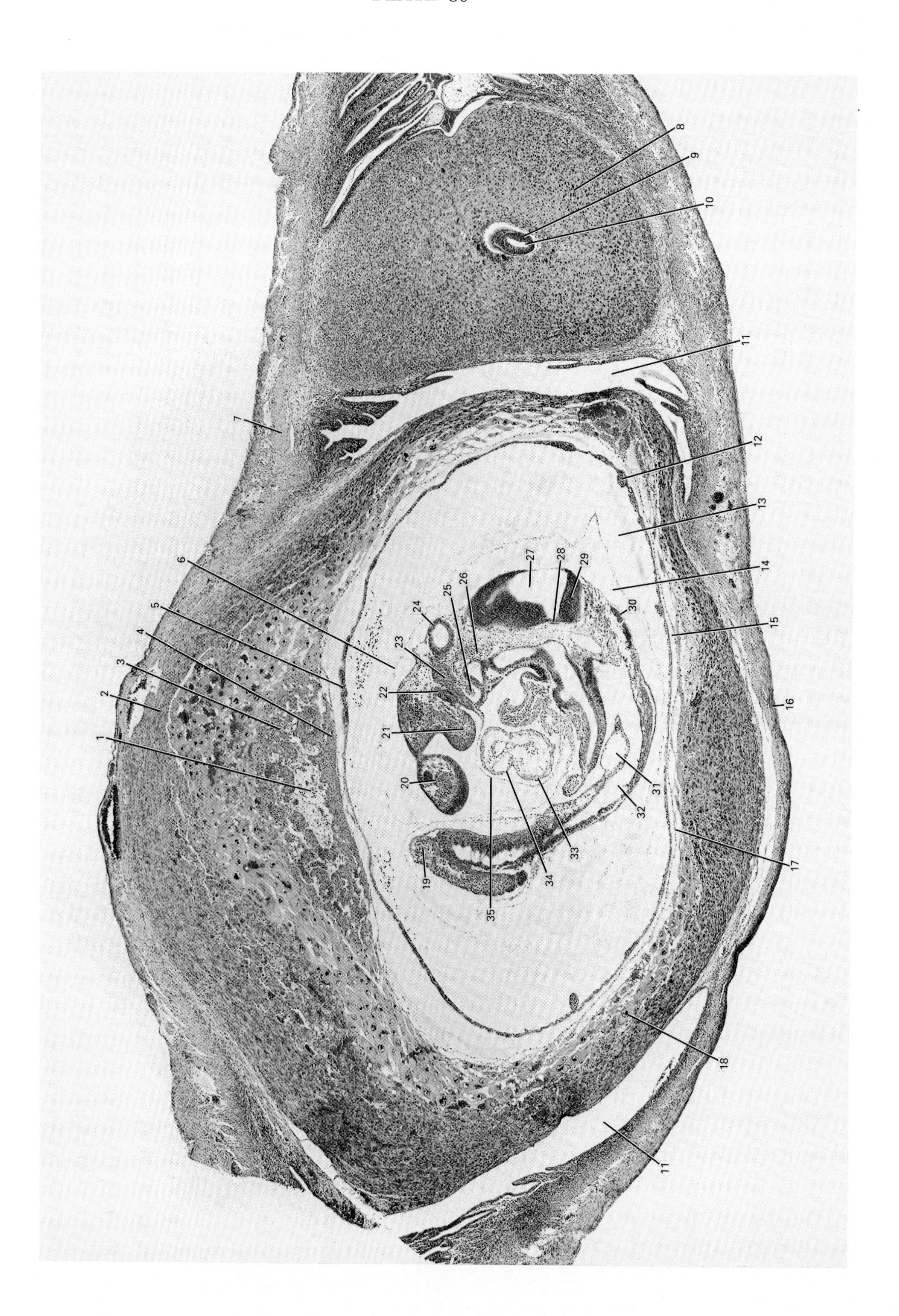

1
2
3
4
5
6
7
8
9
10
11
12
13
14
15
16
17
18
11
19
20
21
22
23
24
25
26
27
28
29
30
31
32
33
34
35

Plate 80

Example of asynchronous implantation

1. embryonic allantoic blood lake
2. maternal component of decidual reaction
3. spongiotrophoblast
4. chorionic plate
5. visceral yolk sac
6. amnion
7. myometrium
8. decidual reaction
9. egg cylinder stage embryo
10. proamniotic canal
11. remnant of uterine lumen between implantation sites
12. blood vessel of yolk sac circulation containing nucleated red blood cells (erythroblasts) formed within mesodermal component of visceral yolk sac
13. extra-embryonic coelomic cavity
14. amniotic cavity
15. yolk sac cavity
16. muscular wall of uterus (outer layer of longitudinal muscle)
17. Reichert's membrane with associated parietal endoderm cells attached to its inner (i.e. embryonic) aspect
18. trophoblast giant cell
19. caudal extremity of tail region
20. region overlying telencephalic vesicle
21. mandibular component of first branchial arch
22. facial (VII) component of facio-acoustic (VII–VIII) neural crest complex
23. acoustic (VIII) component of facio-acoustic (VII–VIII) neural crest complex
24. wall of otocyst (otic vesicle)
25. second branchial arch artery
26. dorsal aorta
27. neural lumen
28. notochord
29. neuroepithelium of neural tube
30. somite
31. umbilical vein
32. intra-embryonic coelomic cavity (peritoneal cavity)
33. trabeculated wall of common ventricular chamber of the heart
34. wall of bulbus cordis
35. body wall overlying pericardial cavity

18. Differentiation of the skeletal system
Text associated with Plates 81–84

GENERAL OBSERVATIONS

It is often extremely difficult to recognize the earliest stages of ossification in serially-sectioned paraffin-embedded material. This equally applies to the recognition of the earliest stages of both endochondral and intramembranous ossification. The former process is principally confined to the long bones and the latter process to the bones of the facial skeleton and cranial vault. However, early evidence of cartilage models is seen in certain bones in the mouse embryo which (at least in the human) are said to form principally or exclusively by intramembranous ossification. This occurs with regard to the clavicle where, in man, the pattern of ossification is clearly quite complex (Warwick and Williams, 1973), and in parts of some of the bones of the cranial vault. The two processes are closely related, however, and in the mandible, for example, the body and much of the ramus forms in membrane whereas the condylar region develops within a cartilaginous model (Hamilton and Mossman, 1972).

By far the most efficient method of demonstrating the presence of ossification centres (as well as areas of calcification in postnatal material) involves "bulk" staining using alizarin, and the subsequent "clearing" of intact embryos. This technique is capable of successfully demonstrating both very small and very early centres of ossification and, more particularly, allows such sites to be accurately localized within the embryo. Examples of the final products obtained using this technique are illustrated in *Plates 81* and *82*, and bear a close resemblance to radiographic images. Such images can be extremely instructive particularly for the analysis of skeletal abnormalities which may have been induced following the exposure of pregnant females to a teratogenic insult.

The alizarin staining technique employed here (for details, see appendix to this section) has the distinct advantage over other published methods in that the duration of the staining step is short (i.e. hours rather than days, or even longer). Other complementary techniques involving the "bulk" staining of embryos with alcian blue or methyl green may also be employed, in similarly "cleared" specimens, in order to allow the visualization of the cartilaginous skeleton. Combined methods may also be used in order to visualize both ossification centres and areas of cartilage, and localize them within the skeleton. It is in fact only from an analysis of this "double-stained" material that very accurate localization of specific centres of ossification can be determined. Examples of the final products obtained using this technique are illustrated in *Plates 83* and *84*.

THE SEQUENTIAL PATTERN OF OSSIFICATION IN SPECIFIC COMPONENTS OF THE SKELETON

The earliest centres of ossification to appear in the base of the skull are seen at about 14–14.5 days p.c., and are associated with the basioccipital and exoccipital bones, while ossification appears somewhat later in the other components of the chondrocranium such as the sphenoid and basisphenoid bones (at about 15.5–16 and 16.5–17 days p.c., respectively). A very small centre is seen in the cartilaginous outer component of the temporal bone at about 14.5–15 days p.c., and this subsequently becomes a more extensive centre (at about 16.5–17 days p.c.) with the first appearance of intramembranous ossification within the lower region of the future squamous part of the temporal bone (*Plate 81,g*). The other component of the future occipital bone, namely the supraoccipital bone (tectum posterius), is first seen in alizarin-stained material at about 17.5 days p.c. as two distinct centres of ossification, one located on either side of the dorsal midline (*Plate 82,a,b*). Only by about 18.5 days p.c. have these two areas of ossification amalgamated to form the definitive supraoccipital bone (*Plate 82,e,f*).

In the facial region, ossification is seen in various elements of the viscerocranium at an early stage of its differentiation. Ossification centres are particularly clearly seen in the maxillary region where centres for the premaxilla and maxilla are present by about 14.5 days p.c. (*Plate 81,a*), and much more extensive areas of ossification are seen in these sites by about 14.5–15 days p.c. (*Plate 81,c*). At the histological level, additional centres are seen in the maxilla, such as in the lateral margins of the palatal shelves (*Plate 32d,a*), in the periorbital region both lateral and infero-lateral to the nasal capsule (*Plate 32c,a,b*), and on either side just deep to the first upper molar tooth primordium (*Plate 32d,c,d*). Amalgamation of the various maxillary centres indicated here occurs by about 16.5–17 days p.c., and these are in close proximity to an ossification centre which develops within the zygomatic bone (which is first clearly seen at about 14.5–15 days p.c.), and centres which appear within the frontal and nasal bones (*Plate 81,g*). Analysis of double-stained material reveals that the upper/medial part of the nasal bones appears to ossify within a cartilage rather than within a membranous model (*Plate 83,d*), as might have been expected.

The intramembranous nature of the ossification seen in the majority of these regions is also clearly apparent (*Plate 81,a,c,e,g*). By 17.5–18.5 days p.c., however, it is no longer possible to distinguish between the individual components of the viscerocranium in alizarin stained material (*Plate 82,a,e*). Much of the nasal capsule, however, is cartilaginous at 18.5 days p.c., as is the region around the external nares (*Plate 83,g*).

The elements of the calvarium also show impressive early evidence of ossification, this feature being particularly marked in relation to the frontal and parietal bones. Centres are seen in both of these bones from as early as 14.5 days p.c. (*Plate 81,a*), and are particularly clearly seen by about 14.5–15 days p.c. (*Plate 81,c*), when their characteristic pattern of intramembranous ossification is especially impressive. Two centres of ossification are first seen within the interparietal bone at about 15.5–16 days p.c. (*Plate 81,f*), and these expand considerably and subsequently fuse across the dorsal midline by about 17.5 days p.c. (*Plate 82,a,b*). The interparietal bone subsequently fuses with the supraoccipital bone to form the upper (interparietal) part of the squamous part of the occipital bone, and this is located above the highest nuchal line.

It is of interest to note in relation to ossification within the bones of the calvarium, that analysis of double-stained material reveals that the orbital part of the frontal bone appears to develop within a cartilage rather than a membranous model (*Plate 83,d*), though it is difficult to distinguish this ossification centre from that of the orbito-sphenoid bone which might normally be expected to develop within a cartilage model. The lower part of the parietal bone also appears to develop in a similar fashion. Possibly more surprisingly, the whole of the interparietal bone appears to develop within a cartilage primordium. The latter is clearly seen to be the case from the analysis of double-stained embryos isolated between 15.5–16.5 days p.c. (*Plate 83,d,e*).

In histological sections, centres of ossification are also seen in the zygomatic part of the temporal bone at about 15.5 days p.c. (*Plate 35b,c*). Slightly later, at about 16–16.5 days p.c., the first evidence of ossification is seen in the tympanic ring (os tympanicum) (*Plate 81,g*). The overall appearance of the latter changes very little by the time of birth (*Plate 82,e*), principally because the ossification observed in this part of the future temporal bone does not extend medially into the cartilage primordium of the petrous part of the temporal bone (*Plate 83,g*) before birth.

As indicated earlier, the mandible is one of the earliest sites to show ossification. Much of the body and inferior part of the ramus is ossified within a membranous primordium by 14.5–15 days p.c., though at this stage the condylar region and the angle of the mandible have yet to ossify. Even at 15.5 days p.c., it is possible to recognize the relationship between the ossification in these sites and the right and left Meckel's cartilages (*Plate 83,d*). The latter extend anteriorly and fuse across the midline in the region close to the site of the future symphysis menti, and extend posteromedially towards the tubo-tympanic recess of the middle ear (*Plate 83,d*), where their relationship to the cartilage primordia of the malleus and incus is particularly clearly seen at about 17–17.5 days p.c., in double-stained "cleared" specimens. By about 18–18.5 days p.c., virtually all of the condylar region is ossified.

Ossification centres are seen in the middle ear ossicles shortly before birth (*Plate 82,e*; *Plate 83,f,g*). The cartilaginous primordia of the malleus and the incus are associated with the dorsal extremity of Meckel's cartilage (see above), while that of the stapes is formed from the dorsal part of the second arch (hyoid) cartilage. In each of these bones, a single primary ossification centre is present, though the anterior process of the malleus and the lentiform process of the incus may have a separate ossification centre. All three middle ear ossicles first display a limited degree of ossification by about 17–17.5 days p.c., and the ossification present is seen to involve a greater proportion of the cartilage primordia of these bones by about 18.5 days p.c., so that the features of the ossicles are more easily seen in "cleared" specimens at this time (*Plate 83,g*). The cartilaginous "skeleton" of the external ear is also readily seen at about 16–16.5 days p.c.

In the neck, an ossification centre is first observed in the cartilage primordium of the body of the hyoid bone at about 17.5 days p.c. (*Plate 83,f*), but no ossification centres are observed before birth in any of the laryngeal cartilages (*Plate 83,d*). In the pectoral region, however, the clavicle is one of the first of the skeletal elements to display an extensive area of ossification, and this is seen at 14–14.5 days p.c.. This is initially largely confined to the periosteal region of its middle one-third (*Plate 81,b,d*), and ossification spreads relatively rapidly laterally and only slowly medially from this extensive primary centre. Even at about 18.5 days p.c., there is a considerable deficiency in ossification at the medial end of the clavicle close to the region of the future sterno-clavicular joint.

The scapula also has an early primary centre of ossification within the middle one-third of the cartilage primordium of this bone which is clearly seen to involve both the blade and the spine, and even extends laterally to delineate the proximal part of the acromion from a very early stage (*Plate 81,b*, unlabelled; *Plate 81,d,f,h*; *Plate 84,a,c*). The characteristic complex three-dimensional shape of the adult scapula is readily recognized from about 16.5–17 days p.c. (*Plate 81,g,h*). Between about 16.5 and 18.5 days p.c., the scapula substantially increases in size, and its adult features become more prominent, so that by the time of birth it is one of the largest of the individual bones of the skeleton (*Plate 82,e,f*). However, it is of interest to note that even at 18.5 days p.c., approximately the medial one-fifth of the cartilage primordium (i.e. the vertebral border) of the blade of the scapula, the distal part of the acromion and the glenoid region have yet to ossify (*Plate 84,g*).

The other component of the pectoral region, namely the manubrium sterni (as well as the sternebrae) first develops a centre of ossification at about 17–17.5 days p.c. (*Plate 82,a*; *Plate 84,j*). While the lower part of the manubrium and upper three sternebrae usually each develop from a single centre of ossification (*Plate 84,j*), the fourth sternebra, and the xiphoid process (xiphisternum) appear to develop from two centres one on either side of the ventral midline (*Plate 40g,b–f*; *Plate 40h,a–d*; *Plate 84,k*). It is of interest that even at 18.5 days p.c. the upper part of the cartilage primordium of the manubrium sterni shows no evidence of ossification, the latter process being confined to the region between the insertion of the first and second costal cartilages. At 17.5 days p.c., the ossification centre in the lower part of the manubrium is about twice as long as it is wide, while the ossification centre in the upper two sternebrae tend to be square-shaped. The ossification centre in the third sternebra is as wide as the upper sternebrae, but about half of their length (*Plate 84,j*). No ossification centre is present between the fifth and sixth costo-sternal articulations at this time. The latter centre is first observed at about 18–18.5 days p.c., and is often in two parts, one on either side of the midline, and these rapidly fuse together across the ventral midline (*Plate 84,k*). At 18.5 days p.c., the upper part of the manubrium sterni, the regions between the sternebrae, and the lower half of the xiphisternum (xiphoid cartilage) show no evidence of ossification (*Plate 84,k*). Often a large "punched out" hole is observed in the central part of the xiphoid cartilage.

The ribs, by contrast to the elements of the sternum, show early evidence of endochondral ossification, the first signs of which are observed at about 14–14.5 days p.c. (*Plate 81,a,b*). The first site of ossification in the ribs is close to their angles, though even by about 14.5 days p.c., this has spread to involve about one-third to two-thirds of these bones (*Plate 81,a,b*). By about 14.5–15 days p.c., ossification is seen in all 13 pairs of ribs (*Plate 81,c,d*), and by about 16–16.5 days p.c., ossification has extended further posteriorly towards, and in the rostral ribs actually extends into the region of the head of the rib, almost to the site of the future costo-vertebral articulation (*Plate 82,b*), though this occurs somewhat later in the more caudal of the ribs (*Plate 84,i*). The site of articulation between the anterior end of the ribs and the costal cartilages is clearly seen in these preparations (*Plate 81,g*; *Plate 82,a,e*; *Plate 84,j,k*). The upper seven costal cartilages articulate with the sternum anteriorly (at the site of the future sterno-costal joints), and posteriorly with small concavities at the anterior end of the ribs, while the next six costal cartilages articulate with the costal cartilage of the preceding rib, and the last three costal cartilages have no direct attachment to the costal arch.

In the pelvic region, three primary centres of ossification are involved in the formation of the innominate (hip) bone. The first of these centres to appear is seen at about 14.5–15 days p.c., and is located in the middle of the cartilage primordium of the iliac bone (*Plate 81,c,d*; *Plate 84,b*). This centre is of quite substantial size, and the next ossification centre to appear is that of the ischium. Slightly later, the centre in the pubic bone appears (*Plate 81,h*; *Plate 84,d*). These three centres subsequently unite (after birth) in the acetabular region, and will in due course form the majority of the innominate bone. However, even at full term, three extensive regions of the innominate bone are still cartilaginous, namely in the region of the iliac crest, in the floor of the acetabular fossa, and along the inferior margin of the hip bone in the region of the ischio-pubic ramus (*Plate 84,h*). The latter region includes the ischial tuberosity and extends towards the lower region of the pubic symphysis. Additional small secondary centres of ossification may also appear (after birth) in the pubic tubercle and crest, and in the symphyseal surface of the pubic bone.

A total of four sacral vertebrae are present, and constitute the posterior component of the pelvic girdle. These vertebrae do not fuse to form a single unit in mice (in man, by contrast, five sacral vertebrae are normally present, and these fuse to form a single triangular-shaped bone termed the sacrum).

In the limbs, the ossification centres for the long bones appear in a proximal to distal sequence. Since the forelimb is, in developmental terms, always slightly in advance of the hindlimb (see text associated with *Plates 43–45*), it should not be surprising that a primary centre appears within the cartilage primordium of the humerus slightly before the comparable centre appears in the femur. The ossification centre within the ulna appears either at the same time or slightly before that in the radius, and all of these centres are present by about 14–14.5 days p.c. (*Plate 81,b*). The next centre to appear is that in the tibia, and slightly later that in the fibula, these additional centres being present by about 14.5–15 days p.c. (*Plate 81,a*; *Plate 81,c,d*). In all of these bones, the primary ossification centre is located in the mid-shaft region, and is of the endochondral variety. The cartilage primordium of the patella (a sesamoid bone that develops in the quadriceps femoris tendon) is first seen at about 16.5 days p.c. (*Plate 84,d*), and those of the fabellae (a pair of considerably smaller sesamoid bones that develop in the medial and lateral heads of gastrocnemius) are also seen at about this time (*Plate 84,d*).

In general terms, the ossification centres that develop within the cartilage primordia of the metacarpal bones appear very shortly before those of the metatarsal bones, and most of both sets are clearly seen by about 18.5 days p.c. (*Plate 82*). At about 17.5 days p.c., it is just possible to recognize the first evidence of ossification centres in the phalanges of both the fore- and hindlimbs (*Plate 82,c,d*, unlabelled; *Plate 84,e,f*). However, by about 18–18.5 days p.c., ossification centres are found in the majority of the cartilage primordia of the phalanges (*Plate 82,g,h*; *Plate 84,g,h*). Of the numerous carpal and tarsal cartilage primordia, the only two in which primary ossification centres are present before birth are those of the talus and calcaneus (*Plate 82,h*; *Plate 84,h*).

If the pattern of ossification in the distal parts of the fore- and hindlimbs is studied in double-stained material covering the period between 16.5 and 18.5 days p.c., a detailed pattern emerges of the normal sequence of events which is not altogether predictable. In relation to the forelimb, ossification centres are first seen in the middle region of the cartilage primordia of the metacarpals of digits 3 and 4 at about 16.5 days p.c. (*Plate 84,c*), though very early evidence of ossification may also be observed in comparable regions of digits 2 and 5. In the phalanges at this time, the first evidence of ossification is confined to the middle part of the cartilage primordia of the proximal and intermediate phalanges, but not of the thumb (pollex). In the hindlimb at this time, small but condensed regions of ossification are seen in the middle parts of the cartilage primordia of the metatarsals of digits 3 and 4 (*Plate 84,d*). The cartilage primordia of the other metatarsals and those of all of the phalanges show no evidence of ossification at this time.

(*continued on page 506*)

PLATE 81

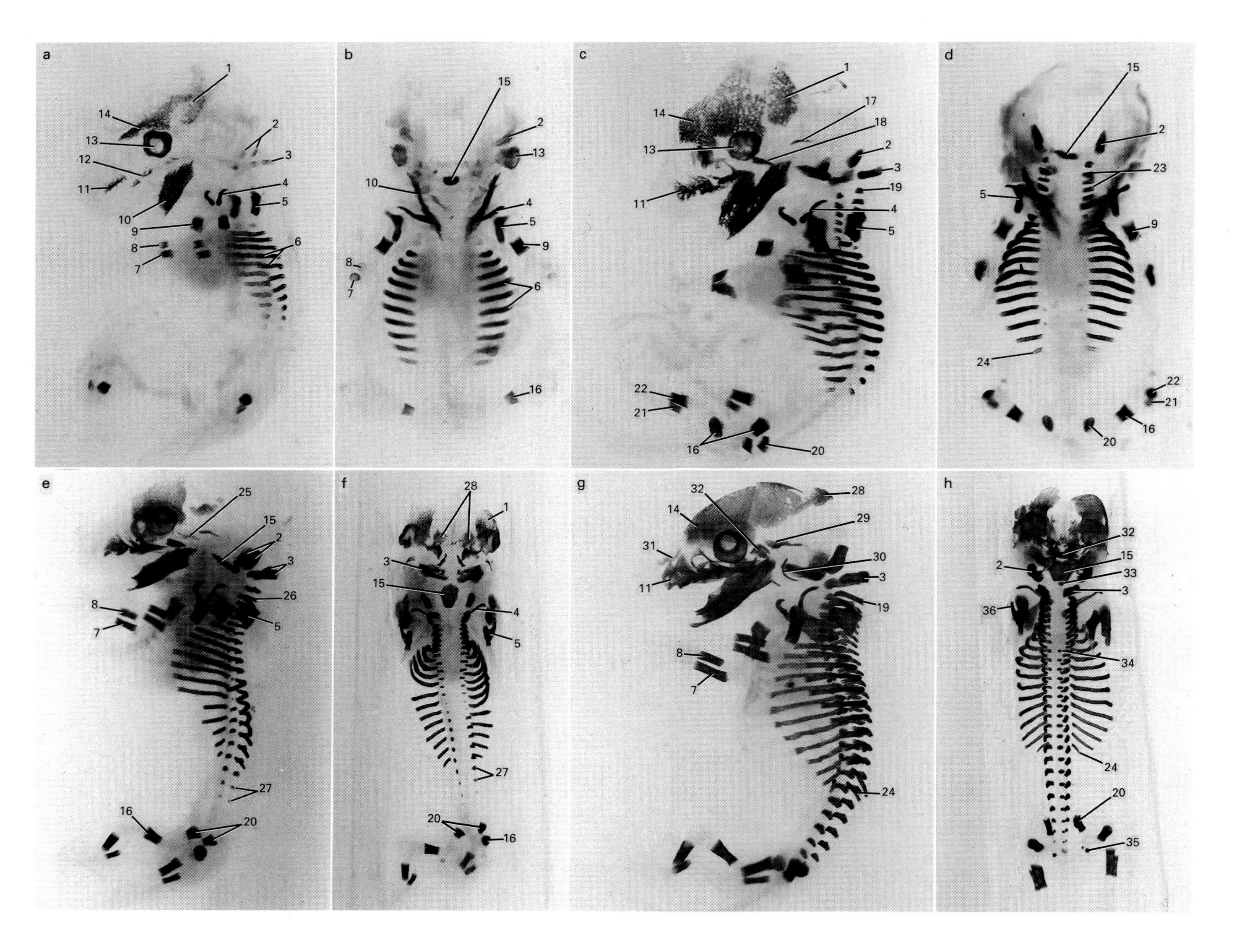

a
1
14
13
2
3
12
11
4
5
10
9
6
8
7
b
15
2
13
10
4
5
9
8
7
6
16
c
1
17
14
18
13
2
3
11
19
4
5
22
21
16
20
d
15
2
23
5
9
24
22
21
16
20
e
25
15
2
3
26
8
5
7
27
16
20
f
28
1
3
15
4
5
27
20
16
g
32
28
14
29
31
30
11
3
19
8
7
24
h
32
15
2
33
3
36
34
24
20
35

Plate 81

Alizarin-stained "cleared" embryos which display centres of ossification. I, 14.5–17 days p.c.

a,b approximately 14.5 days p.c.
c,d 14.5–15 days p.c.
e,f 15.5–16 days p.c.
g,h 16.5–17 days p.c.

a,c,e,g lateral view; **b,d,f,h** dorsal view

1. parietal bone
2. exoccipital bone
3. dorsal (neural or vertebral) arch (also termed "lateral mass") of C1 vertebra (atlas)
4. clavicle
5. scapula
6. ribs
7. ulna
8. radius
9. humerus
10. mandible
11. premaxilla
12. maxilla
13. eye
14. frontal bone
15. basioccipital bone
16. femur
17. temporal bone
18. zygomatic bone
19. dorsal arch of C2 vertebra (axis)
20. ilium (iliac bone)
21. fibula
22. tibia
23. proximal part of dorsal arch of cervical vertebrae (at junction of pedicle and lamina)
24. thirteenth rib
25. sphenoid bone
26. dorsal arch of C3 vertebra
27. proximal part of dorsal arch of upper lumbar vertebrae
28. interparietal bone
29. squamous part of temporal bone
30. tympanic ring (os tympanicum)
31. nasal bone
32. basisphenoid bone
33. anterior arch of C1 vertebra (atlas)
34. vertebral body (centrum) of T3 vertebra
35. ischium
36. acromion of scapula

PLATE 82

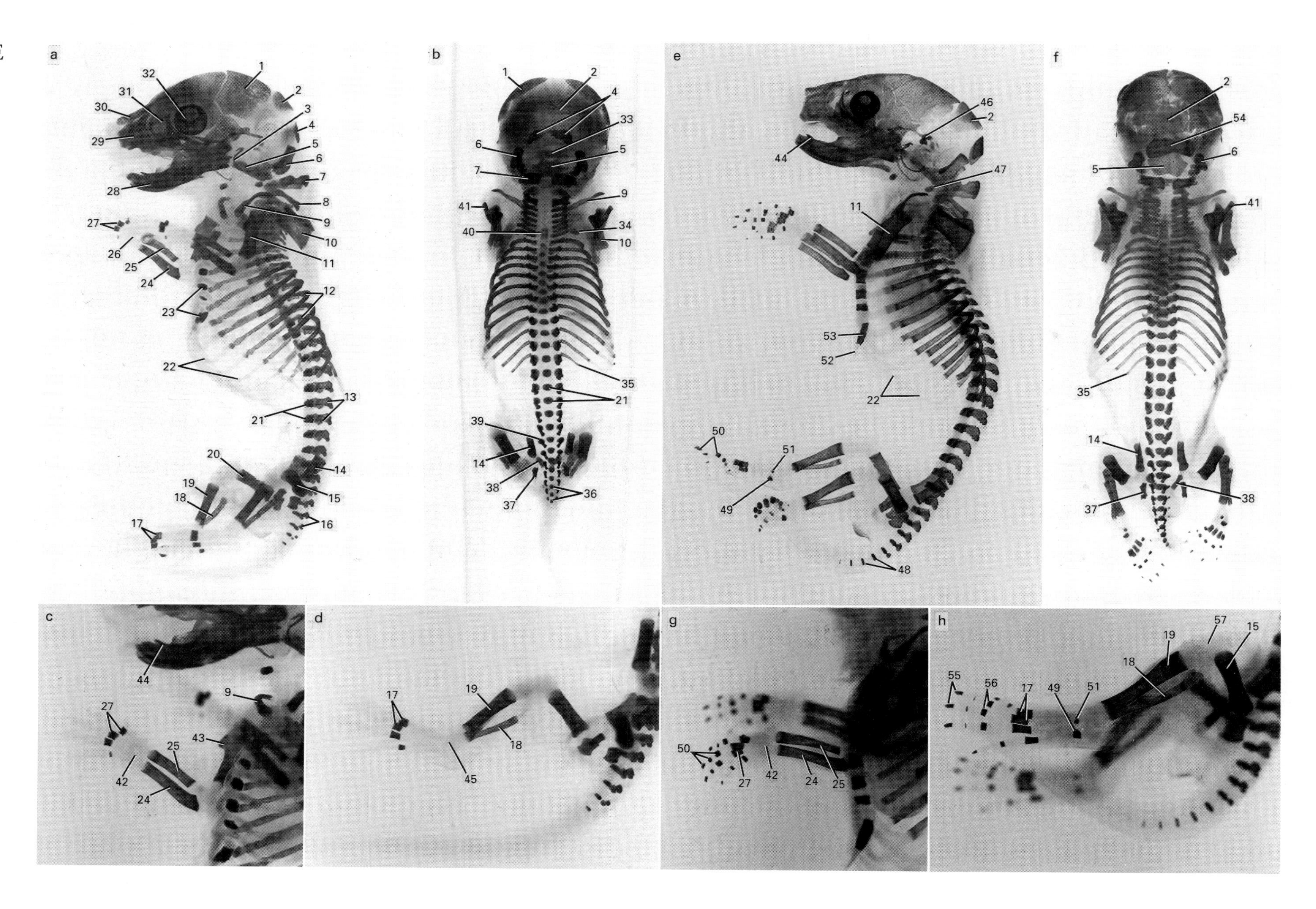

a
1
2
3
4
5
6
7
8
9
10
11
12
13
14
15
16
17
18
19
20
21
22
23
24
25
26
27
28
29
30
31
32
b
33
34
35
36
37
38
39
40
41
c
42
43
44
d
45
e
46
47
48
49
50
51
52
53
f
54
g
h
55
56
57

Plate 82

Alizarin-stained "cleared" embryos which display centres of ossification. II, 17.5–18.5 days p.c.

a–d 17.5 days p.c.
e–h 18.5 days p.c.
a,e lateral view; **b,f** dorsal view
c,g close-up view of forelimb; **d,h** close-up view of hindlimb

1. parietal bone
2. interparietal bone
3. tympanic ring (os tympanicum)
4. supraoccipital bone (with two centres of ossification at this stage, one located on either side of the midline)
5. basioccipital bone
6. exoccipital bone
7. dorsal arch of C1 vertebra (atlas)
8. dorsal arch of C2 vertebra (axis)
9. clavicle
10. blade of scapula
11. humerus
12. ribs
13. proximal part of dorsal arch of lumbar vertebrae (at junction of pedicle and lamina)
14. ilium (iliac bone)
15. left femur
16. proximal part of dorsal arch of coccygeal vertebrae
17. metatarsal bones
18. fibula
19. tibia
20. right femur
21. vertebral bodies (centra) of lumbar vertebrae
22. costal cartilages
23. sternebrae
24. ulna
25. radius
26. carpal region
27. metacarpal bones
28. mandible
29. premaxilla
30. nasal bone
31. maxilla
32. eye
33. basisphenoid bone
34. first rib
35. thirteenth rib
36. vertebral bodies (centra) of coccygeal vertebrae
37. ischium
38. pubic bone
39. proximal part of dorsal arch of first sacral vertebra
40. vertebral bodies (centra) of C7 and C8
41. acromion of scapula
42. carpal region (carpus, wrist)
43. deltoid tuberosity of humerus
44. lower incisor tooth
45. tarsal region (tarsus, ankle)
46. middle ear ossicles
47. vertebral body (centrum) of C1 vertebra (atlas)
48. vertebral bodies (centra) of coccygeal vertebrae in mid-tail region
49. calcaneus
50. phalanges
51. talus
52. cartilaginous part (xiphoid cartilage) of lower region of xiphoid process (xiphisternum)
53. ossification centre in upper region of xiphoid process
54. supraoccipital bone (in one part at this stage, since the two areas of ossification previously present have recently fused across the midline)
55. distal phalanges
56. proximal phalanges
57. knee region

PLATE 83

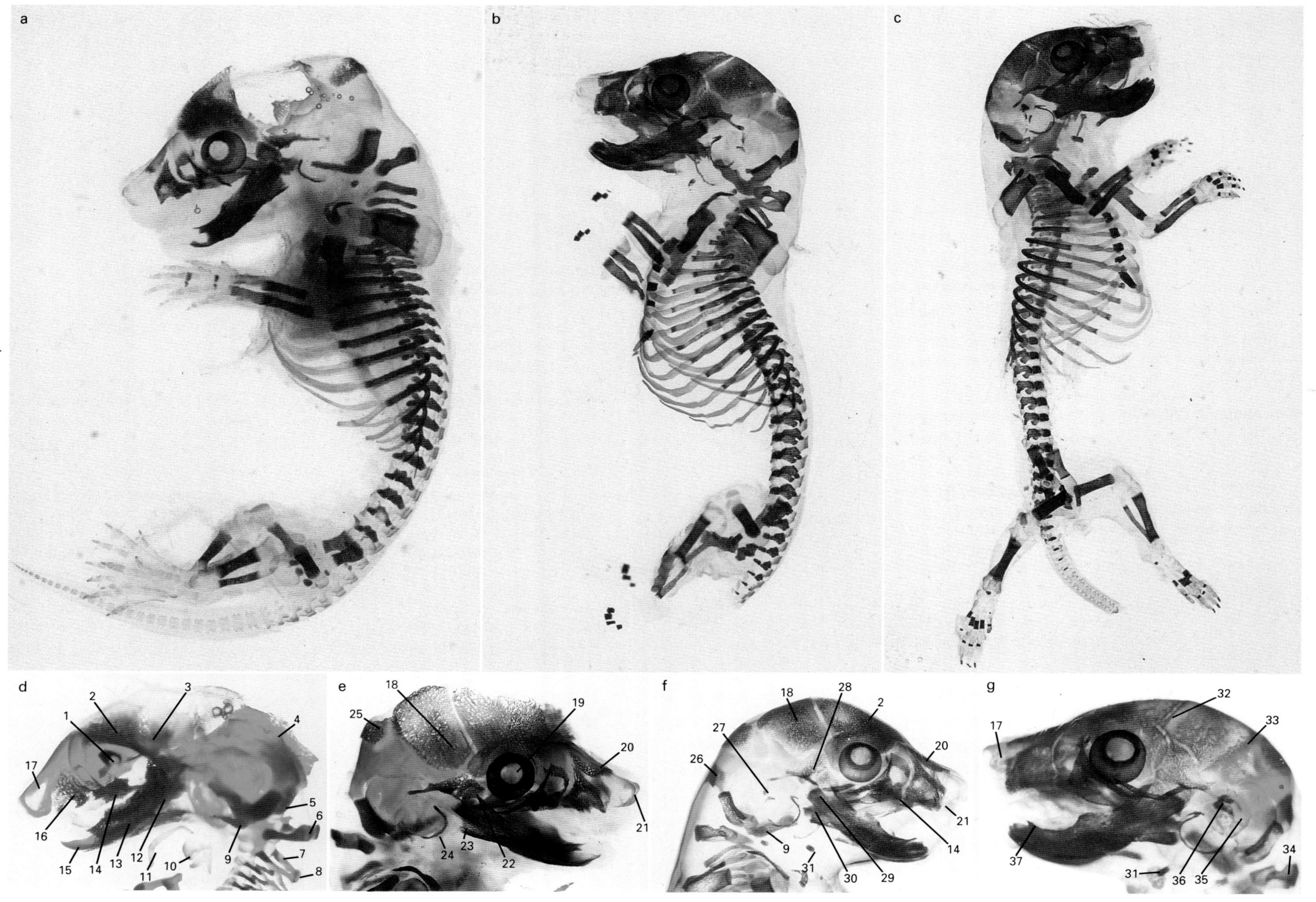

a
b
c
d
1
2
3
4
5
6
7
8
9
10
11
12
13
14
15
16
17
e
18
19
20
21
22
23
24
25
f
26
27
28
29
30
31
g
32
33
34
35
36
37

Plate 83

Alizarin and alcian blue stained "cleared" embryos which display centres of ossification and the cartilaginous skeleton. I, intact embryos and the cephalic region

a intact embryo, 16.5 days p.c.
b intact embryo, 17.5 days p.c.
c intact embryo, 18.5 days p.c.
d–g close-up view of cephalic region
d 15.5 days p.c.; **e** 16.5 days p.c.; **f** 17.5 days p.c.; **g** 18.5 days p.c.

for detailed labelling of forelimb (including scapula) and hindlimb (including innominate bone), thoracic region and lumbar vertebrae, see Plate 84.

1. eye
2. membranous ossification in frontal bone
3. membranous ossification in inferior part of parietal bone
4. cartilaginous precursor of interparietal bone
5. ossification in cartilage primordium of exoccipital bone
6. ossification in proximal part of neural arch of C1 vertebra (atlas)
7. ossification in proximal part of neural arch of C2 vertebra (axis)
8. cartilage primordium of dorsal part of neural arch of C2 vertebra
9. ossification in cartilage primordium of basioccipital bone
10. thyroid cartilage
11. cartilage primordium of hyoid bone (displaced)
12. dorsal part of Meckel's cartilage
13. membranous ossification in body of mandible
14. membranous ossification in maxilla
15. anterior extremity of Meckel's cartilage
16. membranous ossification in premaxilla
17. cartilaginous nasal capsule
18. membranous ossification in parietal bone
19. lens of eye
20. membranous ossification in nasal bone
21. external naris
22. temporo-mandibular joint
23. angle of mandible
24. tympanic ring (os tympanicum)
25. membranous ossification in interparietal bone
26. ossification in cartilage primordium of supraoccipital bone
27. ossification centres of malleus and incus
28. membranous ossification in squamous part of temporal bone
29. membranous ossification in ramus of mandible
30. ossification centre in cartilage primordium of body of basisphenoid bone
31. ossification centre in cartilage primordium of body of hyoid bone
32. coronal suture
33. lambdoidal suture
34. ossification in dorsal part of neural arch of C1 vertebra
35. cartilage primordium of petrous part of temporal bone
36. ossification centres of malleus, incus and stapes
37. lower incisor tooth

PLATE 84

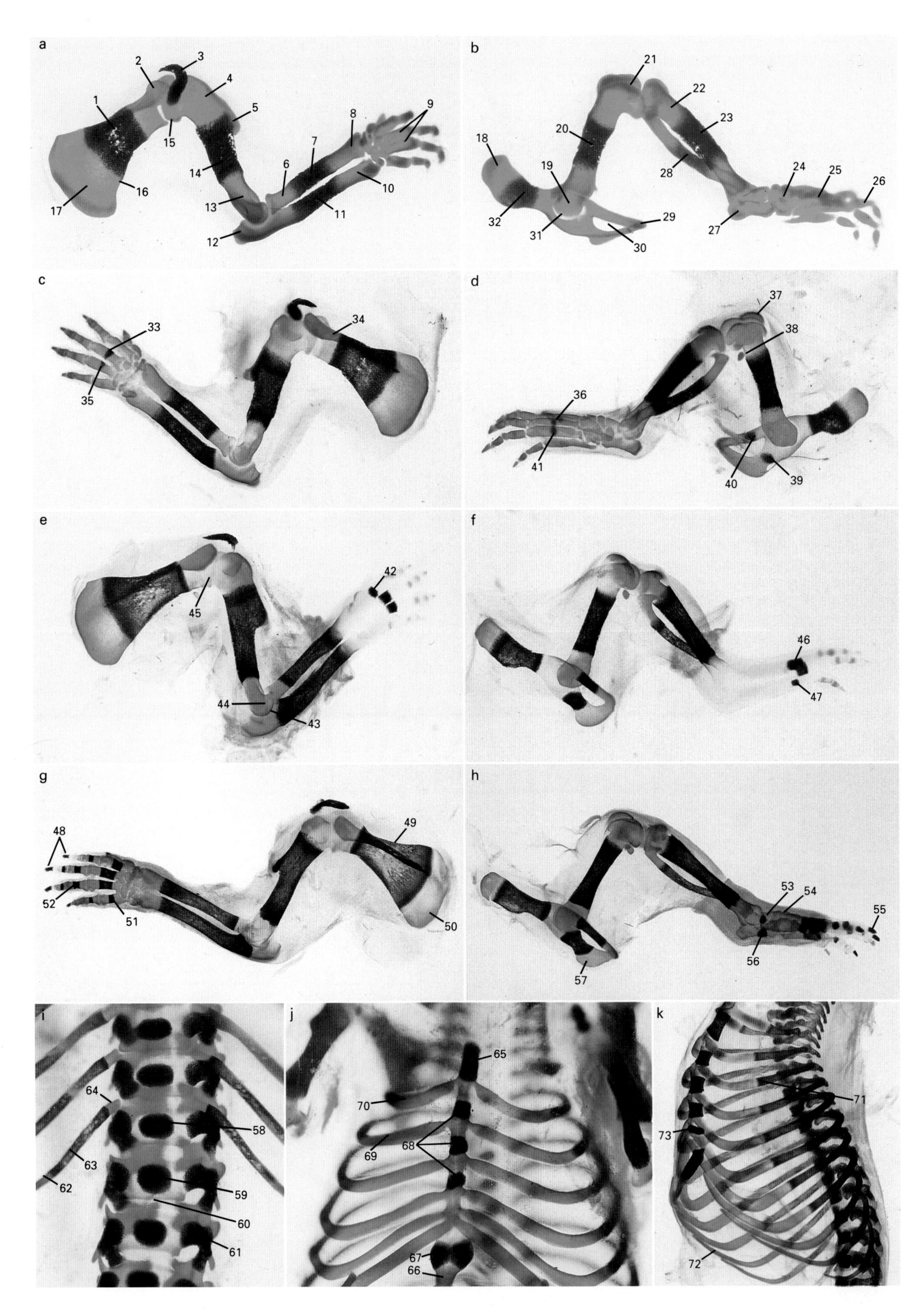

a
1
2
3
4
5
6
7
8
9
10
11
12
13
14
15
16
17
b
18
19
20
21
22
23
24
25
26
27
28
29
30
31
32
c
33
34
35
d
36
37
38
39
40
41
e
42
43
44
45
f
46
47
g
48
49
50
51
52
h
53
54
55
56
57
i
58
59
60
61
62
63
64
j
65
66
67
68
69
70
k
71
72
73

Plate 84

Alizarin and alcian blue stained "cleared" embryos which display centres of ossification and the cartilaginous skeleton. II, isolated limbs and the thoracic region

a,b 15.5 days p.c.
c,d 16.5 days p.c.
e,f,i,j 17.5 days p.c.
g,h,k 18.5 days p.c.
a,c,e,g close-up view of forelimb; **b,d,f,h** close-up of hindlimb

1. ossification centre in middle region of cartilage primordium of blade of scapula
2. acromion
3. clavicle
4. cartilage primordium of proximal part (head and neck region) of humerus
5. deltoid tuberosity
6. cartilage primordium of proximal part (head and neck region) of radius
7. ossification centre in mid-shaft region of radius
8. cartilage primordium of distal part of shaft of radius
9. cartilage primordia of metacarpal bones
10. cartilage primordium of distal part of shaft of ulna
11. ossification centre in mid-shaft region of ulna
12. cartilage primordium of olecranon process of ulna
13. cartilage primordium of distal part of shaft of humerus
14. ossification centre in mid-shaft region of humerus
15. shoulder joint
16. inferior border of scapula
17. cartilage primordium of medial (i.e. vertebral) part of scapula
18. cartilage primordium of iliac crest
19. cartilage primordium of head of femur
20. ossification centre in mid-shaft region of femur
21. cartilage primordium of distal part (condylar region) of femur
22. cartilage primordium of proximal part of shaft of tibia
23. ossification centre in mid-shaft region of tibia
24. cartilage primordia of tarsal bones
25. cartilage primordia of metatarsal bones
26. cartilage primordia of phalangeal bones
27. cartilage primordium of calcaneus
28. ossification centre in mid-shaft region of fibula
29. location of pubic symphysis
30. obturator foramen
31. hip joint
32. ossification centre in iliac bone
33. ossification centre in mid-shaft region of third metacarpal bone
34. cartilage primordium of distal part of spine of scapula where it forms the proximal part of the acromion
35. small ossification centre in mid-shaft region of fourth metacarpal bone
36. small ossification centre in mid-shaft region of third metatarsal bone
37. cartilage primordium of patella
38. cartilage primordium of fabella
39. ossification centre in ischial bone
40. ossification centre in pubic bone
41. small ossification centre in mid-shaft region of fourth metatarsal bone
42. ossification centre in mid-shaft region of second metacarpal bone
43. elbow joint
44. cartilage primordium of distal part (condylar region) of humerus
45. cartilage primordium of glenoid region of scapula
46. ossification centre in mid-shaft region of second metatarsal bone
47. ossification centre in mid-shaft region of fifth metatarsal bone
48. ossification centre in distal phalangeal bones of forelimb
49. superior border of scapula
50. medial (vertebral) border of scapula
51. ossification centre in mid-shaft region of fifth metacarpal bone
52. ossification centre in mid-shaft region of proximal phalangeal bone of fourth digit
53. small ossification centre in talus
54. cartilage primordia of distal row of tarsal bones
55. ossification centre in distal phalangeal bones of hindlimb
56. ossification centre in calcaneus
57. cartilage primordium of ischial tuberosity
58. ossification in centrum of thirteenth thoracic vertebral body
59. ossification in centrum of first lumbar vertebral body
60. region of intervertebral disc
61. ossification in proximal part of neural arch, initially at root of transverse process (at junction of pedicle and lamina)
62. thirteenth costal cartilage
63. ossification in shaft region of thirteenth rib
64. cartilage primordium of proximal part (head and neck region) of thirteenth rib
65. ossification centre in manubrium sterni
66. xiphoid cartilage at lower region of xiphoid process (xiphisternum)
67. two ossification centres in upper region of xiphoid process
68. ossification centres in first, second and third sternebrae
69. third costal cartilage
70. second costal cartilage
71. costo-chondral junction
72. costal margin
73. ossification centre in fourth sternebra

By 17.5 days p.c., an additional pronounced centre of ossification has appeared in the central part of the metacarpal cartilage primordium of digit 2, and a very early centre is seen in the comparable part of the metacarpal primordium of digit 5 (*Plate 84,e*). Very early centres of ossification are also seen in the phalanges, with those of the proximal phalanges of digits 3 and 4 being more pronounced than the others. The ossification centres appear in the middle part of the phalangeal primordia of those in the proximal and intermediate rows, and in the distal part of the cartilage primordium of those in the distal row. In the hindlimb, condensed ossification centres are seen at this time in the middle region of the cartilage primordia of metatarsals 2–5 (*Plate 84,f*).

By 18.5 days p.c., with regard to the forelimb, substantial centres of ossification are seen in the primordia of metacarpals 2–5, and centres are also seen in the proximal row of phalanges of digits 2–5, in the intermediate phalanges of digits 3 and 4 (but *not* of digits 2 and 5), and in all of the distal phalanges (namely, of digits 1–5) (*Plate 84,g*). No evidence of ossification is, however, seen in any of the primordia of the carpal bones. In the hindlimb at this time, a small ossification centre is seen in the talus, and a more substantial centre in the calcaneus, as indicated previously. In relation to the metatarsals, substantial ossification centres are seen in those associated with digits 2–5, and a small centre is seen in relation to that of the first digit. Substantial centres are now present in the proximal and distal rows of phalanges, while a lesser degree of ossification is present in the intermediate phalanges (*Plate 84,h*). In the hallux, the proximal and distal phalanges have substantial centres of ossification at this time.

All of the vertebrae form from three primary centres of ossification, namely one in each half of the dorsal arch, and one in the vertebral body (centrum) (*Plate 84,i*). The sequence for the appearance of centres in the dorsal arches is roughly cranio-caudal, with that associated with C1 vertebra (atlas) being the first to appear, and evident at about 14–14.5 days p.c. This is by far the most prominent of all of the dorsal arches, and is a useful landmark in the upper cervical region. By about 14.5–15 days p.c., all of the centres associated with the dorsal arches of the cervical vertebrae are present, and possibly also those of some of the upper thoracic vertebrae (*Plate 81,c,d*). By about 15.5–16 days p.c., all of the centres associated with the thoracic and most of the lumbar vertebrae are present (*Plate 81,e,f*), while those associated with the sacral vertebrae are present by about 16.5–17 days p.c. (*Plate 81,g,h*), and those of the proximal 3–5 coccygeal vertebrae are present by about 18–18.5 days p.c. (*Plate 82,e*; *Plate 83,b,c*).

The ossification centres for the vertebral bodies, however, appear in a curious sequence, so that the first centres to appear, which are seen at about 16.5–17 days p.c., are associated with the C1 vertebra (atlas) and the cartilage primordia of the bodies of T3 to S4 vertebrae (*Plate 81,h*). By about 17–17.5 days p.c., centres are also seen in the centra of C7 and C8 as well as in the upper thoracic and proximal coccygeal vertebrae (*Plate 82,b*). By 18–18.5 days p.c., a reasonable-sized centre is now present in C6, and minute centres are also present in the centra of C2–C5 vertebrae (*Plate 82,f*, unlabelled). In the tail region, centres of ossification are now seen in the proximal centra, and these extend distally to about the level of the tenth coccygeal vertebra (*Plate 82,h*; *Plate 83,c*).

For an interesting comparative analysis of the time of appearance of ossification centres in the pre- and postnatal mouse, rat and human, see Johnson (1933). The data on the rat in the latter report was obtained from the earlier studies by Strong (1925) and Spark and Dawson (1928), and the data on the time of appearance of the primary ossification centres in man were obtained principally from Mall (1906). Johnson's (1933) study is of alizarin-stained material, and it should be noted that the time of appearance of individual ossification centres is in all instances between 0.5–1.5 days later than that reported here.

APPENDIX DIFFERENTIAL STAINING OF BONE IN INTACT EMBRYOS BY ALIZARIN RED S* (METHOD OF C. ARNOTT)

Process	Solution		
Fixation	80% ethanol		minimum of 24 h before removing skin and viscera – particularly liver, kidneys and gut
Dehydration	96% ethanol acetone		12–24 h 1–3 days (depends on size of embryo)
Staining	0.1% alizarin red S in 95% ethanol 1% acid-alcohol	1 ml 99 ml	2–6 h (depends on size of embryo) at 37°C (allow at least 20 ml per embryo.
Clearing sequence	96% ethanol 1% aqueous KOH		1 h 12–48 h Proceed to next step as soon as skeleton clearly visible
	20% glycerine in 1% aqueous KOH 50% glycerine in 1% aqueous KOH 80% glycerine in 1% aqueous KOH		The exact timing of these steps is a matter of trial and error, and clearing through the graded glycerine/KOH series can take 2–4 weeks
Storage	100% glycerine		

* Alizarin sulphonate sodium (G.T. Gurr Ltd)

If double staining is required to demonstrate the presence of ossification and cartilage, then the staining solution is modified as follows, though the duration of staining is as indicated above:

0.1%	alizarin red S in 95% ethanol	1 ml
0.3%	alcian blue in 95% ethanol	1 ml
1%	acid-alcohol	98 ml

Embryo code numbers for serially-sectioned embryos illustrated in Plates 1c–41

Plate	Embryo code
1c	**a–c**, D5.9.15.AM.A; **d–f**, D5.9.15.AM.C; **g–i**, D5.9.15.AM.F
2	**a,b**, D6.5.PM.C; **c,d**, D6.5.PM.B
3	**a–e**, D8.2.15.4A; **f–j**, D8.2.15.4D; **k–o**, I6
4	**a–h**, D8.4.30.3C; **i–p**, D8.4.30.3A
5	**a–e**, Sag II.3; **f–j**, Sag II.4
6	D8.4.00.2B
7	D8.8.00.1C
8	NHF.5
9	NUT.7
10	NUT.6
11	NUT.3
12	NUT.4
13a,b	NUT.9
14a,b	PT.H
15a,b	N7.8.T4
16a,b	N7.8.T2
17	D10 (early) Sag
18a,b	NT.1
19a–c	D10.3.30.C2
20a–c	D10.3.30.C3
21a,b	D10.Sag.20284
22a–c	D11.Sag.A
23a–d	SUSA.H
24a–d	SUSA.J
25a–c	D12.Sag
26a–e	D12.MD.1
27a–d	8659A
28a–e	87148
29a,b	90111
30a–l	89192A
31a,b	90163
32a–k	89195
33a,b	90123
34a,b	8699A
35a–l	89198A
36a–e	90162
37a–l	89199
38a–c	90164
39a–d	91009
40a–l	9017A.trans
41	9017A.sag

References

Ackerman, G.A. (1962). Substituted naphthol AS phosphate derivatives for the localization of leukocyte alkaline phosphatase activity. *Lab Invest.* **11**, 563–567.

Adelmann, H.B. (1925). The development of the neural folds and cranial ganglia of the rat. *J. Comp. Neur.* **39**, 19–171.

AhPin, P., Ellis, S., Arnott, C. and Kaufman, M.H. (1989). Prenatal development and innervation of the circumvallate papilla in the mouse. *J. Anat.* **162**, 33–42.

Amoroso, E.C. (1952). Placentation. In: *Marshall's Physiology of Reproduction.* (A.S. Parkes, Editor), Vol. 2, 3rd edn. Longmans Green, London, pp. 127–311.

Bancroft, J.D. and Stevens, A. (1982). *Theory and Practice of Histological Techniques*, 2nd edn. Churchill Livingstone, Edinburgh.

Beddington, R.S.P. (1983). The origin of the foetal tissues during gastrulation in the rodent. In: *Development in Mammals*, (M.H. Johnson, Editor), Elsevier, Oxford, pp. 1–32.

Biddle, F.G. (1980). Palate development in the mouse: a quantitative method that permits the estimation of time and rate of palate closure. *Teratology* **22**, 239–246.

Bonneville, K. (1950). New facts on mesoderm formation and proamnion derivatives in the normal mouse embryo. *J. Morph.* **86**, 495–545.

Boyd, J.D. and Hamilton, W.J. (1970). *The Human Placenta.* W. Heffer & Sons Ltd, Cambridge.

Brown, N.A. (1990). Routine assessment of morphology and growth: scoring systems and measurements of size. In: *Postimplantation Mammalian Embryos. A Practical Approach.* (A.J. Copp and D.L. Cockroft, Editors). IRL Press, Oxford, pp. 93–108.

Brown, N.A. and Fabro, S. (1981). Quantitation of rat embryonic development in vitro: a morphological scoring system. *Teratology* **24**, 65–78.

Bulmer, M.G. (1970). *The Biology of Twinning in Man.* Clarendon Press, Oxford.

Butler, H. and Juurlink, B.H.J. (1987). *An Atlas for Staging Mammalian and Chick Embryos.* CRC Press, Boca Raton, Florida.

Challice, C.E. and Virágh, S. (1973). The embryonic development of the mammalian heart. In: *Ultrastructure of the Mammalian Heart.* (C.E. Challice and S. Virágh, Editors). Academic Press, New York, pp. 91–126.

Chiquoine, A.D. (1954). The identification, origin, and migration of the primordial germ cells in the mouse embryo. *Anat. Rec.* **118**: 135–146.

Christie, G.A. (1964). Developmental stages in somite and post-somite rat embryos, based on external appearance, and including some features of the macroscopic development of the oral cavity. *J. Morph.* **114**, 263–286.

Cole, R.J. (1967). Cinemicrographic observations on the trophoblast and zona pellucida of the mouse blastocyst. *J. Embryol. Exp. Morph.* **17**, 481–490.

Copp, A.J. and Cockroft, D.L. (Editors) (1990). *Postimplantation Mammalian Embryos. A Practical Approach.* IRL Press, Oxford.

Coulombre, A.J. (1956). The role of intraocular pressure in the development of the chick eye. I. Control of eye size. *J. Exp. Zool.* **133**, 211–225.

Coulombre, A.J. and Coulombre, J.L. (1958). The role of mechanical factors in brain morphogenesis. *Anat. Rec.* **130**, 289–290 (Abstract).

Culling, C.F.A., Allison, R.T. and Barr, W.T. (1985). *Cellular Pathology Technique*, 4th edn. Butterworths, London.

Deanesly, R. (1966). The endocrinology of pregnancy and foetal life. In: *Marshall's Physiology of Reproduction*, Vol. 3. (A.S. Parkes, Editor), Longmans Green, London, pp. 891–1063.

De Haan, R.L. (1965). Morphogenesis of the vertebrate heart. In: *Organogenesis.* (R.L. De Haan and H. Ursprung, Editors). Holt, Rinehart & Winston, New York, pp. 377–419.

Desmond, M.E. and Jacobson, A.G. (1977). Embryonic brain enlargement requires cerebrospinal fluid pressure. *Devl. Biol.* **57**, 188–198.

Enders, A.C. (1965). A comparative study of the fine structure of the trophoblast in several hemochorial placentas. *Am. J. Anat.* **116**, 29–68.

Fananapazir, K. and Kaufman, M.H. (1988). Observations on the development of the aortico-pulmonary spiral septum in the mouse. *J. Anat.* **158**, 157–172.

Ferguson, M.W.J. (1981). Developmental mechanisms in normal and abnormal palate formation with particular reference to the aetiology, pathogenesis and prevention of cleft palate. *Br. J. Orthodontics* **8**, 115–137.

Ferguson, M.W.J. (1987). Palate development: mechanisms and malformations. *Irish J. Med. Sci.* **156**, 309–315.

Ferguson, M.W.J. (1988). Palate development. *Development, Suppl.* **103**, 41–60.

Fox, H. (1978). *Pathology of the Placenta*. Saunders, London.

Fujinaga, M., Brown, N.A. and Baden, J.M. (1992). Comparison of staging systems for the gastrulation and early neurulation period in rodents: a proposed new system. *Teratology* **46**, 183–190.

Gasser, R.F. (1975). *Atlas of Human Embryos*. Harper & Row, Hagerstown, Maryland.

Gedda, L. (1961). *Twins in History and Science*. C.C. Thomas, Springfield, Illinois.

Ginsburg, M., Snow, M.H.L. and McLaren, A. (1990). Primordial germ cells in the mouse embryo during gastrulation. *Development* **110**, 521–528.

Hamburger, V. and Hamilton, H.L. (1951). A series of normal stages in the development of the chick embryo. *J. Morph.* **88**, 49–92.

Hamilton, W.J. and Mossman, H.W. (1972). *Hamilton, Boyd and Mossman's Human Embryology. Prenatal Development of Form and Function*. Fourth Edition. W. Heffer & Sons Ltd, Cambridge.

Harris, M.J. and McLeod, M.J. (1982). Eyelid growth and fusion in fetal mice. A scanning electron microscope study. *Anat. Embryol.* **164**, 207–220.

Hebel, R. and Stromberg, M.W. (1986). *Anatomy and Embryology of the Laboratory Rat*. BioMed Verlag, Worthsee.

Henery, C. and Kaufman, M.H. (1992). Relationship between cell size and nuclear volume in nucleated red blood cells of developmentally matched diploid and tetraploid mouse embryos. *J. Exp. Zool.* **261**, 472–478.

Heuser, C.H. and Streeter, G.L. (1929). Early stages in the development of pig embryos, from the period of initial cleavage to the time of the appearance of limb-buds. *Contr. Embryol.* **20**, 1–29.

Heuser, C.H. and Streeter, G.L. (1941). Development of the macaque embryo. *Contr. Embryol.* **29**, 15–55.

Hill, R.E., Jones, P.F., Rees, A.R., Sime, C.M., Justice, M.J., Copeland, N.G., Jenkins, N.A., Graham, E. and Davidson, D.R. (1989), A new family of mouse homeobox-containing genes: molecular structure, chromosomal location and the developmental expression of Hox-7.1. *Genes and Development* **3**, 26–37.

Hogan, B., Constantini, F. and Lacy, E. (1986). *Manipulating the Mouse Embryo. A Laboratory Manual.* Cold Spring Harbor Laboratory, New York.

Hsu, Y.C. and Gonda, M.A. (1980). Monozygotic twin formation in mouse embryos in vitro. *Science* **209**, 605–606.

Humphrey, T. (1969). The relation between human fetal mouth opening reflexes and closure of the palate. *Am. J. Anat.* **125**, 317–344.

Hunt, P., Whiting. J., Muchamore, I., Marshall, H. and Krumlauf, R. (1991). Homeobox genes and models for patterning the hindbrain and branchial arches. *Development Suppl. 1. Molecular and Cellular Basis of Pattern Formation*, 187–196.

Johnson, M.L. (1933). The time and order of appearance of ossification centers in the albino mouse. *Am. J. Anat.* 52, 241–271.

Johnston, M.C. (1966). A radioautographic study of the migration and fate of cranial neural crest cells in the chick embryo. *Anat. Rec.* **156**, 153–156.

Kaplow, L.S. (1968). Leukocyte alkaline phosphatase cytochemistry: applications and methods. *Ann. N.Y. Acad. Sci.* **155**, 911–947.

Kaufman, M.H. (1979). Cephalic neurulation and optic vesicle formation in the early mouse embryo. *Am. J. Anat.* **155**, 425–444.

Kaufman, M.H. (1981). The role of embryology in teratological research, with particular reference to the development of the neural tube and heart. *J. Reprod. Fert.* **62**, 607–623.

Kaufman, M.H. (1982). Two examples of monoamniotic monozygotic twinning in diploid parthenogenetic mouse embryos. *J. Exp. Zool.* **224**, 277–282.

Kaufman, M.H. (1983a). *Early Mammalian Development: Parthenogenetic Studies*. Cambridge University Press, Cambridge.

Kaufman, M.H. (1983b). Occlusion of the neural lumen in early mouse embryos analysed by light and electron microscopy. *J. Embryol. Exp. Morph.* **78**, 211–228.

Kaufman, M.H. (1984). A re-evaluation of the morphology and function of the yolk sac in the primitive streak stage mouse embryo. *J. Anat.* **139**, 730–731 (Abstract).

Kaufman, M.H. (1985). Experimental aspects of monozygotic twinning. In: *Implantation of the Human Embryo*, (R.G. Edwards, J.M. Purdy and P.C. Steptoe, Editors). Academic Press, London, pp. 371–391.

Kaufman, M.H. (1986). Occlusion of the lumen of the neural tube, and its role in the early morphogenesis of the brain. In: *Spina Bifida – Neural Tube Defects* (D. Voth and P. Glees, Editors), Walter de Gruyter, Berlin, pp. 29–46.

Kaufman, M.H. (1990). Morphological stages of post-implantation embryonic development. In: *Postimplantation Mammalian Embryos: A Practical Approach* (A.J. Copp and D.L. Cockroft, Editors), IRL Press, Oxford, pp. 81–91.

Kaufman, M.H. (1991). Histochemical identification of primordial germ cells and differentiation of the gonads in homozygous tetraploid mouse embryos. *J. Anat.* **179** 169–181.

Kaufman, M.H. and Navaratnam, V. (1981). Early differentiation of the heart in mouse embryos. *J. Anat.* **133**, 235–246.

Kaufman, M.H. and O'Shea, K.S. (1978). Induction of monozygotic twinning in the mouse. *Nature* **276**, 707–708.

Kaufman, M.H. and Schnebelen, M.T. (1986). The histochemical identification of primordial germ cells in diploid parthenogenetic mouse embryos. *J. Exp. Zool.* **238**, 103–111.

Kaufman, M.H., Barton, S.C. and Surani, M.A.H. (1977). Normal postimplantation development of mouse parthenogenetic embryos to the forelimb bud stage. *Nature* **265**, 53–55.

Kaufman, M.H., Speirs, S. and Lee, K.K.H. (1989). The sex-chromosome constitution and early postimplantation development of diandric triploid mouse embryos. *Cytogenetics Cell Genetics* **50**, 98–101.

Kaufman, M.H., Lee, K.K.H. and Speirs, S. (1990). Histochemical identification of primordial germ cells in diandric and digynic triploid mouse embryos. *Mol. Reprod. Devel.* **25**, 364–368.
Keibel, F. (1937). *Normentafl zur Entwicklungsgeschichte der Wanderratte (Rattus norvegicus Erxleben)*. Fischer, Jena.
Keibel, F. and Mall, F.P. (1910). *Manual of Human Embryology*, Vol. 1, J.P. Lippincott, Philadelphia.
Keibel, F. and Mall, F.P. (1912). *Manual of Human Embryology*, Vol. 2. J.P. Lippincott, Philadelphia.
Kelly, S.J. (1975). Studies of the potency of the early cleavage blastomeres of the mouse. In: *The Early Development of Mammals* (M. Balls and A.E. Wild, Editors). Cambridge University Press, Cambridge, pp. 97–105.
Kirby, D.R.S. and Bradbury, S. (1965). The hemochorial mouse placenta. *Anat. Rec.* **152**, 279–282.
Kwong, W.H. and Tam, P.P.L. (1984). The pattern of alkaline phosphatase activity in the developing mouse spinal cord. *J. Embryol. Exp. Morph.* **82**, 241–251.
Lovell-Badge, R. (1992). The role of **Sry** in mammalian sex determination. In: *Postimplantation Development in the Mouse*. CIBA Symposium No. 165, 162–182.
McGrath, J. and Solter, D. (1983). Nuclear transplantation in the mouse embryo by microsurgery and cell fusion. *Science* **220**, 1300–1303.
Mackay, S., Strachan, L. and McDonald, S.W. (1993). Increased vascularity is the first sign of testicular differentiation in the mouse. *J. Anat.* **183**, 171 (Abstract).
Mall, F.P. (1906). On ossification centers in human embryos less than one hundred days old. *Am. J. Anat.* **5**, 433–458.
Marin-Padilla, M. (1970). The closure of the neural tube in the golden hamster. *Teratology* **3**, 39–46.
Moffat, D.B. (1959). Developmental changes in the aortic arch system of the rat. *Am. J. Anat.* **105**, 1–35.
Navaratnam, V., Kaufman, M.H., Skepper, J.N., Barton, S. and Guttridge, K.M. (1986). Differentiation of the myocardial rudiment of mouse embryos: an ultrastructural study including freeze-fracture replication. *J. Anat.* **146**, 65–85.
Nieto, M.A., Bradley, L.C., Hunt, P., Das Gupta, R., Krumlauf, R. and Wilkinson, D.G. (1992). Molecular mechanisms of pattern formation in the vertebrate hindbrain. In: *Postimplantation Development in the Mouse*. CIBA Symposium No. 165, 92–107.
Nicholas, J.S. and Hall, B.V. (1942). Experiments on developing rats. II. The development of isolated blastomeres and fused eggs. *J. Exp. Zool.* **90**, 441–461.
O'Rahilly, R. (1966). The early development of the eye in staged human embryos. *Contr. Embryol.* **38**, 1–42.
O'Rahilly, R. and Müller, F. (1987). *Developmental Stages in Human Embryos*. Carnegie Instit. of Washington, Publication No. 637. Carnegie Institute, Washington.
Otis, E.M. and Brent, R. (1954). Equivalent ages in mouse and human embryos. *Anat. Rec.* **120**, 33–64.
Ozdzeński, W. (1967). Observations on the origin of primordial germ cells in the mouse. *Zool. Polon.* **17**, 367–379.
Paxinos, G., Tork, I., Tecott, L.H. and Valentino, K.L. (1991). *Atlas of the Developing Rat Brain*. Academic Press, San Diego.
Pearson, A.A. (1980). The development of the eyelids. Part I. External features. *J. Anat.* **130**, 33–42.
Pei, Y.F. and Rhodin, J.A.G. (1970). The prenatal development of the mouse eye. *Anat. Rec.* **168**, 105–126.
Poelmann, R.E. (1980). Differential mitosis and degeneration patterns in relation to the alterations in the shape of the embryonic ectoderm of early post-implantation mouse embryos. *J. Embryol. Exp. Morph.* **55**, 33–51.
Poelmann, R.E. (1981a). The formation of the embryonic mesoderm in the early post-implantation mouse embryo. *Anat. Embryol.* **162**, 29–40.
Poelmann, R.E. (1981b). The head-process and the formation of the definitive endoderm in the mouse embryo. *Anat. Embryol.* **162**, 41–49.
Psychoyos, A. (1973). Endocrine control of egg implantation. In: *American Physiology Society's Handbook of Physiology-Endocrinology II*, Part 2 (R.O. Greep and E.B. Astwood, Editors), American Physiological Society, Washington, pp. 187–215.
Reinius, S. (1965). Morphology of the mouse embryo from the time of implantation to mesoderm formation. *Z. Zellforsch. Mikrosk. Anat.* **68**, 711–723.
Rosenquist, G.C. and De Haan, R.L. (1966). Migration of precardiac cells in the embryonic chick heart. *Contrib. Embryol.* **38**, 113–121.
Rossi, F. and Reale, E. (1957). The somite stage of human development studied with the histochemical reaction for the demonstration of alkaline glycerophosphatase. *Acta Anat.* **30**, 656–681.
Rugh, R. (1968). *The Mouse. Its Reproduction and Development*. Burgess Publishing Company, Minneapolis (reprinted 1990 by Oxford University Press, Oxford).
Sakamoto, M.K., Nakamura, K., Handa, J., Kihara, T. and Tanimura, T. (1989). Morphogenesis of the secondary palate in mouse embryos with special reference to the development of rugae. *Anat. Rec.* **223**, 299–310.
Shenefelt, R.E. (1972). Morphogenesis of malformations in hamsters caused by retinoic acid: relation to dose and stage at treatment. *Teratology* 5, 103–118.
Sidman, R.L. (1970). Cell proliferation, migration and interaction in the developing mammalian central nervous system. In: *The Neurosciences, A Second Study Program* (F.O. Schmitt, G.C. Quarton, T. Melnechuk and G. Adelman, Editors). Rockefeller University Press, New York, pp. 100–107.
Snell, G.D. and Stevens, L.C. (1966). Early embryology. In: *Biology of the Laboratory Mouse*, 2nd edn. (E.L. Green, Editor). McGraw-Hill, New York, pp. 205–245.
Snow, M.H.L. and Monk, M. (1983). Emergence and migration of mouse primordial germ cells. In: *Current Problems in Germ Cell Differentiation* (A. McLaren and C.C. Wylie, Editors). *Br. Soc. Devel. Biol. Sym.* **7**, 115–135.
Spark, C and Dawson, A.B. (1928). The order and time of appearance of the centers of ossification in the fore and hind limbs of the albino rat, with special reference to the possible influence of the sex factor. *Am. J. Anat.* **41**, 411–445.
Sprinz, R. and Kaufman, M.H. (1987). The sphenoidal canal. *J. Anat.* **153**, 47–54.
Steven, D.H. (1975). *Comparative Placentation*. Academic Press, London.
Streeter, G.L. (1942). Developmental horizons in human embryos. Description of age group XI, 13–20 somites, and age group XII, 21–29 somites. *Contr. Embryol.* **30**, 211–245.
Strong, R.M. (1925). The order, time, and rate of ossification of the albino rat. *Am. J. Anat.* **36**, 313–355.
Strong, S.J. and Corney, G. (1967). *The Placenta in Twin Pregnancy*. Pergamon Press, Oxford.

Surani, M.A.H. and Kaufman, M.H. (1977). Influence of extracellular Ca^{2+} and Mg^{2+} ions on the second meiotic division of mouse oocytes: relevance to obtaining haploid and diploid parthenogenetic embryos. *Devl. Biol.* **59**, 86–90.

Tam, P.P.L. and Beddington, R.S.P. (1992). Establishment and organization of germ layers in the gastrulating mouse embryo. In: *Postimplantation Development in the Mouse*. CIBA Symposium No. 165, 27–49.

Tam, P.P.L. and Snow, M.H.L. (1981). Proliferation and migration of primordial germ cells during compensatory growth in mouse embryos. *J. Embryol. Exp. Morph.* **64**, 133–147.

Tarkowski, A.K. (1959). Experiments on the development of isolated blastomeres of mouse eggs. *Nature* **184**, 1286–1287.

Tarkowski, A.K. and Wróblewska, J. (1967). Development of blastomeres of mouse eggs isolated at the 4- and 8-cell stage. *J. Embryol. Exp. Morph.* **18**, 155–180.

Theiler, K. (1989). *The House Mouse: Atlas of Embryonic Development*. Springer-Verlag, New York.

Theiler, K. (1972). *The House Mouse: Development and Normal Stages from Fertilization to 4 weeks of Age*. Springer-Verlag, Berlin, (reprinted 1989, with minor changes).

Virágh, S. and Challice, C.E. (1973). Origin and differentiation of cardiac muscle cells in the mouse. *J. Ultrastruct. Res.* **41**, 1–24.

Wan, Y.J., Wu, T.C. and Damjanov, I. (1982). Twinning and conjoined placentation in mice. *J. Exp. Zool.* **221**, 81–86.

Warwick, R. and Williams, P.L. (1973). *Gray's Anatomy*, 35th edn. Longman, Edinburgh.

Waterman, R.E. (1976). Topographical changes along the neural fold associated with neurulation in the hamster and mouse. *Am. J. Anat.* **146**, 151–172.

Witschi, E. (1962). Development: rat. In: *Growth Including Reproduction and Morphological Development*, (P.L. Altman and D.S. Dittmer, Editors). Biological Handbooks of the Federation of American Societies for Experimental Biology, Washington, pp. 304–314.

The Index: guide to facilitate usage

In order to increase the user-friendliness of the index, many tissues have been grouped together according to the organ system or region to which they are most closely associated, e.g. the brain, ear, eye, gut, heart, mouth region, muscle, renal/urinary, reproductive, respiratory, skeleton, vascular system, etc.

To emphasise the chronology of development, this index is given in terms of the timing of the first appearance of specific tissues, organs, etc., as indicated by Theiler Stage. The stage given refers to when a tissue or organ is first noted in the legends to the histological sections and/or scanning electron micrographs, or in the text which complements these Plates. Because tissues and structures in all developing systems gradually differentiate from precursor elements this stage inevitably represents only a *general guide* to their first appearance: in some instances, a more detailed analysis may reveal that a particular tissue or item may be recognised at one or more stages earlier than indicated here. Occasionally, for example where a particular structure is of relatively small dimensions, this may also result in it being assigned to a slightly more advanced stage of development than may be found on close inspection of the observer's own histological sections. It must also be appreciated that as development proceeds, a certain structure may differentiate, serve a particular function, then regress and finally disappear, its role(s) having been taken over by its successor from the same or another cell lineage. Thus, for example, the metanephros (the definitive kidney) succeeds the mesonephros, which in turn succeeds the pronephros, all three being the nephrogenic cord derivatives of intermediate-plate mesoderm.

One of the problems that is particularly difficult to address in an index such as this, is the terminology that should be employed when, for example, an item gradually evolves into a more differentiated structure which is then (by convention) given a new, and possibly more appropriate, name related to its newly acquired functional status. When, for example, does the *neural tube* become the *spinal cord*, or the *embryo* become a *fetus*? In a few cases, general guidelines have been drawn up over the years, but in most instances, the terminology used has to a greater or lesser extent reflected the whims of individuals. Similarly, and this poses even greater problems, the terminology used to describe specific structures occasionally varies considerably in the literature, and in some instances (e.g. the development of the tongue) different investigators have given the same name to a variety of structures. It is for this reason that Committees of "experts" have been established over the years to standardise the nomenclature employed in all fields of science.

The terminology used in embryological studies is still evolving and that employed in this Atlas largely represents an anglicised vernacular version of the Nomina Anatomica and Nomina Embryologica, with the terms selected being those believed by the author to be in most common usage in the appropriate (principally mouse) literature. This terminology is complicated by the fact that a name for a specific item may be entirely suitable in one context, while another name for the same item may be more suitable in another context. For example, the terms pharyngeal arch, aortic arch, branchial arch and visceral arch are usually considered to be synonymous, and therefore interchangeable, but in some contexts one term may be more suitable than another. While, to a generalist, this may not pose problems, the apparently inappropriate use of a particular term may distress a specialist in the field. Since most individuals would prefer to think of themselves as specialists, rather than generalists, this only serves to increase the problems encountered by the indexer. The latter are compounded by the difficulties encountered in attempting to delineate anatomical boundaries where they do not naturally exist, as evidenced by the recent results of gene-probe analysis which delineates functional rather than anatomical boundaries. This exercise, at best, can therefore only be a compromise and, while the names used are not immutable, they do provide a linguistic basis on which the field can advance.

Since the ultimate aim of the indexer is to provide a series of universally-accepted terms which facilitate rather than hinder the analysis of, in this instance, mouse embryonic development, the choice of names used is often a particularly difficult (and sometimes idiosyncratic) one. Clearly, if some of the names used do not find favour with researchers in this field, then they will ultimately be replaced by those names that are more acceptable, or more relevant in a particular context.

Because mouse embryonic development is, in general terms, not significantly different from that of other mammalian species, it is a relatively simple matter to determine by analogy with the findings reported in the Atlas, at what stage during early rat or human development, for example, a particular event or series of events occurs. As far as the rat is concerned, the Witschi Scale may be used, and covers the complete gestation period in this species. In man, the Carnegie Stages may be employed, but are limited by the fact that they, so far, only cover development up until the end of the so-called "embryonic" period. Thus a mouse embryo of about 13.5 days p.c. is, in developmental terms, very similar to a human embryo of about 50–51 to 56–57 days post-ovulation (O'Rahilly & Müller, 1987). At earlier Theiler Stages the relationship is closer and the translation of events between the two species can be more accurately undertaken. Clearly, while the species-specific differences would have to be taken into account, the close similarities in the timing of differentiation of the tissues and organ systems of even unrelated mammalian species would allow such an exercise to be undertaken with some degree of confidence. Since no universally accepted staging system exists to-date for the human embryo that extends to the so-called "fetal" period, it is probably inappropriate to extend the comparison further for these two species, as the human fetus at term is, in developmental terms, far more advanced than its mouse equivalent.

Finally, it should be stressed that no attempt has been made to provide a fully comprehensive list of, for example, all of the individual muscles or skeletal elements present in the pre-term mouse embryo. Only those that have been labelled in the Plates are listed in the index. A similar approach has been taken with regard to the inclusion in the index of only those nerves and blood vessels that have been labelled in the Plates or referred to in the text, though the majority of the most important and substantive nerves and blood vessels are indexed.

Abbreviations used

PCC pre-cartilage condensation (usually differentiates into cartilage primordium one or two stages later)

PMM pre-muscle mass (mesenchyme condensation, usually evolves after one or two stages into differentiated muscle)

mes indicates that ossification occurs in mesenchyme/membrane rather than in cartilage precursor

Index of Stages 1–26